本书英文版于2021年由世界卫生组织（World Health Organization）出版，书名为：*WHO Study Group on Tobacco Product Regulation: Report on the Scientific Basis of Tobacco Product Regulation: Eighth Report of a WHO Study Group (WHO Technical Report Series, No. 1029)*
© World Health Organization 2021

世界卫生组织（World Health Organization）授权中国科技出版传媒股份有限公司（科学出版社）翻译出版本书中文版。中文版的翻译质量和对原文的忠实性完全由科学出版社负责。当出现中文版与英文版不一致的情况时，应将英文版视作可靠和有约束力的版本。

中文版《烟草制品管制科学基础报告：WHO研究组第八份报告》
©中国科技出版传媒股份有限公司（科学出版社）2024

WHO技术报告系列 1029

WHO烟草制品管制研究小组

烟草制品管制科学基础报告

——WHO研究组第八份报告

胡清源 侯宏卫 主译

科学出版社

北京

内 容 简 介

本报告介绍了 WHO 烟草制品管制研究小组第十次会议的结论和建议。该次会议讨论了新型烟草制品，如电子烟碱传输系统（ENDS）、电子非烟碱传输系统（ENNDS）和加热型烟草制品（HTP）管制科学基础方面的优先事项。研究小组审查了专门为会议编写的九份背景文件和两份补充文件，涉及以下专题：① HTP 中的有害物质及其暴露、健康影响和风险降低声明；② HTP 的吸引力和致瘾性：对认知和使用的影响及关联效应；③ HTP 之间的差异：考虑因素和影响；④ HTP 的使用：产品转换以及双重或多重使用；⑤ HTP、ENDS 和 ENNDS 相关法规以及国家措施和监管障碍；⑥ 电子烟碱传输系统和传统卷烟的烟碱暴露评估；⑦ ENDS、ENNDS 和 HTP 相关个体风险的量化方法探索：对人体健康和监管的影响；⑧ 新型烟草制品的风味；⑨ 新型烟草制品的全球营销和推广及其影响；⑩ 烟草中的烟碱形态、化学修饰及其对电子烟碱传输系统的影响；⑪ 电子烟碱传输系统或雾化产品使用相关肺损伤。研究组关于每个主题的建议在相关章节末尾列出，最后一章为总体建议。

本书会引起吸烟与健康、烟草化学和公共卫生学诸多应用领域科学家的兴趣，为客观评价烟草制品的管制和披露措施提供必要的参考。

图书在版编目（CIP）数据

烟草制品管制科学基础报告．WHO 研究组第八份报告 / WHO 烟草制品管制研究小组著；胡清源，侯宏卫主译. -- 北京：科学出版社，2024.9. -- (WHO 技术报告系列) . -- ISBN 978-7-03-079497-0

Ⅰ．TS45

中国国家版本馆 CIP 数据核字第 202486KY97 号

责任编辑：刘　冉 / 责任校对：杜子昂
责任印制：徐晓晨 / 封面设计：北京图阅盛世

科学出版社 出版
北京东黄城根北街 16 号
邮政编码：100717
http://www.sciencep.com

北京厚诚则铭印刷科技有限公司印刷
科学出版社发行　各地新华书店经销
*
2024 年 9 月第 一 版　开本：720×1000　1/16
2024 年 9 月第一次印刷　印张：39 1/4
字数：790 000
定价：198.00 元
（如有印装质量问题，我社负责调换）

WHO Technical Report Series 1029

WHO Study Group on Tobacco Product Regulation

Report on the Scientific Basis of Tobacco Product Regulation: Eighth Report of a WHO Study Group

译者名单

主　译：胡清源　侯宏卫

副主译：陈　欢　李　晓　崔利利　周　静

译　者：胡清源　侯宏卫　陈　欢　李　晓
　　　　崔利利　周　静　韩书磊　付亚宁
　　　　王红娟　田雨闪　张　远　任培培
　　　　苗瑞娟　王永秀

译 者 序

2003年5月，第56届世界卫生大会*通过了《烟草控制框架公约》(FCTC)，迄今已有包括我国在内的180个缔约方。根据FCTC第9条和第10条的规定，授权世界卫生组织（WHO）烟草制品管制研究小组（TobReg）对可能造成重要公众健康问题的烟草制品管制措施进行鉴别，提供科学合理的、有根据的建议，用于指导成员国进行烟草制品管制。

自2007年起，WHO陆续出版了八份烟草制品管制科学基础报告，分别是945、951、955、967、989、1001、1015和1029。WHO烟草制品管制科学基础系列报告阐述了降低烟草制品的吸引力、致瘾性和毒性等烟草制品管制相关主题的科学依据，内容涉及烟草化学、代谢组学、毒理学、吸烟与健康等烟草制品管制的多学科交叉领域，是一系列以科学研究为依据、对烟草管制发展和决策有重大影响意义的技术报告。将其引进并翻译出版，可以为相关烟草科学研究的科技工作者提供科学性参考。希望引起吸烟与健康、烟草化学和公共卫生学等诸多应用领域科学家的兴趣，为客观评价烟草制品的管制和披露措施提供必要的参考。

第一份报告（945）由胡清源、侯宏卫、韩书磊、陈欢、刘彤、付亚宁翻译，全书由韩书磊负责统稿；

第二份报告（951）由胡清源、侯宏卫、刘彤、付亚宁、陈欢、韩书磊翻译，全书由刘彤负责统稿；

第三份报告（955）由胡清源、侯宏卫、付亚宁、陈欢、韩书磊、刘彤翻译，全书由付亚宁负责统稿；

第四份报告（967）由胡清源、侯宏卫、陈欢、刘彤、韩书磊、付亚宁翻译，全书由陈欢负责统稿；

第五份报告（989）由胡清源、侯宏卫、陈欢、刘彤、韩书磊、付亚宁翻译，全书由陈欢负责统稿；

第六份报告（1001）由胡清源、侯宏卫、韩书磊、陈欢、刘彤、付亚宁、王红娟翻译，全书由韩书磊负责统稿；

---

* 世界卫生大会 (World Health Assembly，WHA) 是世界卫生组织的最高决策机构，每年召开一次。

第七份报告（1015）由胡清源、侯宏卫、陈欢、张小涛、田永峰、付亚宁、刘彤、韩书磊、李国宇、王红娟、田雨闪翻译，全书由刘彤负责统稿；

第八份报告（1029）由胡清源、侯宏卫、陈欢、李晓、崔利利、周静、韩书磊、付亚宁、王红娟、田雨闪、张远、任培培、苗瑞娟、王永秀翻译。

由于译者学识水平有限，本中文版难免有错漏和不当之处，敬请读者批评指正。

译　者

2024年4月

WHO 烟草制品管制研究小组第十次会议

网络会议，2020年9月28日至10月2日

参加者

D. L. Ashley博士，美国公共卫生部退役海军少将；美国佐治亚州立大学亚特兰大分校人口健康科学系研究教授

O. A. Ayo-Yusuf教授，Sefako Makgatho卫生科学大学（南非比勒陀利亚）副校长

A. R. Boobis教授，英国伦敦帝国理工学院国家心肺研究所名誉教授

Mike Daube教授，科廷大学（澳大利亚西澳大利亚州珀斯）卫生科学系名誉教授

P. Gupta博士，Healis Sekhsaria公共卫生研究所（印度孟买）所长，环境健康科学教授

S. Katharine Hammond教授，美国加利福尼亚大学伯克利分校公共卫生学院工业卫生计划主任，环境卫生学教授

D. K. Hatsukami博士，美国明尼苏达大学（明尼阿波利斯）精神病学教授

A. Opperhuizen博士，荷兰乌得勒支食品和消费品安全局风险评估与研究办公室主任

G. Zaatari博士，WHO烟草制品管制研究小组主席；贝鲁特美国大学（黎巴嫩贝鲁特）病理学与实验医学系教授兼主任

WHO烟草实验室网络（TobLabNet）主席

Nuan Ping Cheah博士，新加坡卫生科学局应用科学司制药部药物、化妆品和卷烟检测实验室主任

WHO地区办事处

Nivo Ramanandraibe博士，WHO非洲地区办事处（刚果布拉柴维尔）烟草和降低其他非传染性疾病风险因素小组代理组长

Francisco Armada博士，WHO美洲地区办事处（美国华盛顿）烟草管制顾问

Rosa Sandoval Correa de Kroeger女士，WHO美洲地区办事处（美国华盛顿）烟草管制顾问

Jagdish Kaur博士，WHO东南亚地区办事处（印度新德里）健康促进小组健康风险因素区域顾问

Kristina Mauer Stender女士，WHO欧洲地区办事处（丹麦哥本哈根马尔莫维吉联合国城）项目经理

Angela Ciobanu博士，WHO欧洲地区办事处（丹麦哥本哈根马尔莫维联合国城）技术干事

Fatimah El-Awa博士，WHO东地中海地区办事处（埃及开罗）无烟草行动组区域顾问

Joung-eun Lee女士，WHO西太平洋地区办事处（菲律宾马尼拉）无烟草行动组

发言人

Ranti Fayokun博士，世界卫生组织（瑞士日内瓦）全民健康覆盖和健康人口司健康促进部无烟草行动组

Maciej Goniewicz博士，美国Roswell Park综合癌症中心（纽约布法罗）癌症预防和人口研究部健康行为系烟碱和烟草制品评估资源组肿瘤学副教授

Anne Havermans博士，荷兰国家公共卫生与环境研究所（比尔托芬）卫生保护中心

Ryan David Kennedy博士，美国约翰·霍普金斯大学Bloomberg公共卫生学院（巴尔的摩）健康行为与社会系全球烟草管制研究所副教授

Farrah Kheradmand博士，美国贝勒医学院（休斯敦）生物医学研究生院临床科学家培训项目联合主任

Suchitra Krishnan Sarin教授，美国耶鲁医学院（纽黑文）精神病学系人类调查委员会副主席

Richard O'Connor教授，美国Roswell Park综合癌症中心（纽约布法罗）癌症预防和人口研究部健康行为系肿瘤学教授

Vinayak Prasad博士，世界卫生组织（瑞士日内瓦）全民健康覆盖和健康人口司健康促进部无烟草行动组负责人

Najat A. Saliba教授，贝鲁特美国大学（黎巴嫩贝鲁特）文理学院化学系教授

Yvonne Staal博士，荷兰国家公共卫生与环境研究所（比尔托芬）卫生保护中心

Irina Stepanov博士，美国明尼苏达大学（明尼阿波利斯）环境健康科学与共济会肿瘤中心副教授

Reinskje Talhout博士，荷兰国家公共卫生与环境研究所（比尔托芬）卫生保护中心高级科学家

合作者

Nuan Ping Cheah博士，新加坡卫生科学局应用科学司制药部药物、化妆品和卷烟检测实验室主任

Caleb Clawson先生，美国约翰·霍普金斯大学Bloomberg公共卫生学院（巴尔的摩）健康行为与社会系全球烟草管制研究所研究助理

Laura E. Crotty Alexander博士，美国加利福尼亚大学圣地亚哥分校医学院副教授，肺重症监护与睡眠科主任；美国圣地亚哥退伍军人管理局圣地亚哥医疗系统科长

Danielle R. Davis博士，美国耶鲁医学院（纽黑文）精神病学系博士后研究员

Enrico Davoli博士，意大利Mario Negri药理学研究所（米兰）环境健康科学部质谱实验室主任

Thomas Eisenberg博士，美国弗吉尼亚联邦大学（里士满）烟草制品研究中心联合主任，心理学教授

Ahmad El Hellani博士，贝鲁特美国大学（黎巴嫩贝鲁特）文理学院化学系

Ranti Fayokun博士，世界卫生组织（瑞士日内瓦）全民健康覆盖和健康人口司健康促进部无烟草行动组

Frank Henkler Stephani博士，德国联邦风险评估研究所（柏林）化学品和产品安全部

Federica Mattioli女士，意大利Mario Negri药理学研究所（米兰）环境健康科学部质谱实验室

Marine Perraudin女士，世界卫生组织（瑞士日内瓦）全民健康覆盖和健康人口司健康促进部无烟草行动组技术干事

Armando Peruga博士，西班牙Bellvitge生物医学研究所（巴塞罗那）烟草管制组研究员；智利Desarrollo大学（圣地亚哥）医学院流行病学和卫生政策中心研究员

Meagan Robichaud女士，美国约翰·霍普金斯大学公共卫生学院（巴尔的摩）健康行为与社会系全球烟草管制研究所博士生

Moira Sy女士，世界卫生组织（瑞士日内瓦）健康促进部无烟草行动组顾问

Andrew Barnes博士，美国弗吉尼亚联邦大学（里士满）

欧盟及其他国家专家

Matus Ferech博士，欧盟委员会（比利时布鲁塞尔）跨境医疗和烟草管制部门卫生和食品安全总局烟草管制小组政策官员

Andre Luiz Oliveira da Silva博士，巴西卫生管理局（里约热内卢）烟草或非烟草制品注册和检验总干事，烟草或非烟草管制过程协调健康监管和监督专家

Denis Chonière先生，加拿大卫生部（安大略渥太华）管制物质和烟草局烟草制品监管办公室主任

Peyman Altan博士，土耳其卫生部（安卡拉）烟草管制司

WHO FCTC秘书处（瑞士日内瓦）

Adriana Blanco博士，WHO FCTC秘书处主任

秘书处（世界卫生组织健康促进部无烟草行动组，瑞士日内瓦）

Jennifer Brown博士，顾问

Ranti Fayokun博士，科学家

Ruediger Krech博士，主任

Vinayak Prasad博士，无烟草行动组负责人

Anne Sikanda女士，部门主管

Simone St Claire女士，顾问

Moira Sy女士，顾问

Participants in the 10th meeting of the WHO study group on tobacco product regulation
Virtual meeting, 28 September–2 October 2020

Members

Dr D.L. Ashley, Rear-Admiral (retired), Public Health Service; Research Professor, Department of Population Health Sciences, Georgia State University, Atlanta (GA), United States of America (USA)

Professor O.A. Ayo-Yusuf, Acting Vice-Chancellor, Postgraduate Studies and Innovation, Sefako Makgatho Health Sciences University, Pretoria, South Africa

Professor A.R. Boobis, Emeritus Professor, National Heart and Lung Institute, Imperial College London, London, United Kingdom of Great Britain and Northern Ireland

Professor Mike Daube AO, Emeritus Professor, Faculty of Health Sciences, Curtin University, Perth, Western Australia, Australia

Dr P. Gupta, Director, Healis Sekhsaria Institute for Public Health, Mumbai, India

Professor S. Katharine Hammond, Professor, Environmental Health Sciences; Director, Industrial Hygiene Program; Associate Dean, Academic Affairs, School of Public Health, University of California, Berkeley (CA), USA

Dr D.K. Hatsukami, Professor of Psychiatry, University of Minnesota, Minneapolis (MN), USA

Dr A. Opperhuizen, Director, Office for Risk Assessment and Research, Food and Consumer Product Safety Authority, Utrecht, Netherlands

Dr G. Zaatari (Chair), Professor and Chairman, Department of Pathology and Laboratory Medicine, American University of Beirut, Beirut, Lebanon

Chair of the WHO Tobacco Laboratory Network

Dr Nuan Ping Cheah, Director, Pharmaceutical, Cosmetics and Cigarette Testing Laboratory, Pharmaceutical Division, Applied Sciences Group, Health Sciences Authority, Singapore

WHO regional offices[1]

Dr Nivo Ramanandraibe, Acting Team Lead, Tobacco and Reduction of Other NCD Risk Factors, WHO Regional Office for Africa, Brazzaville, Congo

Dr Francisco Armada, Adviser, Tobacco Control, WHO Regional Office for the Americas, Washington (DC), USA

---

1 Apologies: Ms Kristina Mauer-Stender, Programme Manager, WHO Regional Office for Europe, UN City, Marmorvej, Copenhagen, Denmark

Ms Rosa Sandoval Correa de Kroeger, Adviser, Tobacco Control, WHO Regional Office for the Americas, Washington (DC), USA

Dr Jagdish Kaur, Regional Adviser, Health Risk Factors, Health Promotion, WHO Regional Office for South-East Asia, New Delhi, India

Dr Angela Ciobanu, Technical Officer, WHO Regional Office for Europe, UN City, Marmorvej, Copenhagen, Denmark

Dr Fatimah El-Awa, Regional Adviser, Tobacco Free Initiative, WHO Regional Office for the Eastern Mediterranean, Cairo, Egypt

Ms Joung-eun Lee, Tobacco Free Initiative, WHO Regional Office for the Western Pacific, Manila, Philippines

Presenters

Dr Ranti Fayokun, No Tobacco Unit (Tobacco Free Initiative), Department of Health Promotion, Division of Universal Health Coverage and Healthier Populations, World Health Organization, Geneva, Switzerland

Dr Maciej Goniewicz, Associate Professor of Oncology, Nicotine and Tobacco Product Assessment Resource, Department of Health Behavior, Division of Cancer Prevention and Population Studies, Roswell Park Comprehensive Cancer Center, Buffalo (NY), USA

Dr Anne Havermans, Centre for Health Protection, National Institute for Public Health and the Environment, Bilthoven, Netherlands

Dr Ryan David Kennedy, Associate Professor, Institute for Global Tobacco Control, Department of Health, Behavior and Society, Johns Hopkins Bloomberg School of Public Health, Baltimore (MD), USA

Dr Farrah Kheradmand, Co-Director, Clinical Scientist Training Program, Graduate School of Biomedical Sciences, Baylor College of Medicine, Houston (TX), USA

Professor Suchitra Krishnan-Sarin, Department of Psychiatry, Vice-Chair, Human Investigations Committee, Yale School of Medicine, New Haven (CT), USA

Professor Richard O'Connor, Professor of Oncology, Department of Health Behavior, Division of Cancer Prevention and Population Studies, Roswell Park Comprehensive Cancer Center, Buffalo (NY), USA

Dr Vinayak Prasad, Unit Head, No Tobacco (Tobacco Free Initiative), Department of Health Promotion, Division of Universal Health Coverage and Healthier Populations, World Health Organization, Geneva, Switzerland

Professor Najat A. Saliba, Professor of Chemistry, Faculty of Arts and Sciences, American University of Beirut, Beirut, Lebanon

Dr Yvonne Staal, Centre for Health Protection, National Institute for Public Health and the Environment, Bilthoven, Netherlands

Dr Irina Stepanov, Associate Professor, Division of Environmental Health Sciences and the Masonic Cancer Center, University of Minnesota, Minneapolis (MN), USA

Dr Reinskje Talhout, Senior Scientist, Centre for Health Protection, National Institute for Public Health and the Environment, Bilthoven, Netherlands

Co-authors[1]

Dr Nuan Ping Cheah, Director, Pharmaceutical, Cosmetics and Cigarette Testing Laboratory, Pharmaceutical Division, Applied Sciences Group, Health Sciences Authority, Singapore

Mr Caleb Clawson, Research Assistant, Department of Health Behavior and Society, Institute for Global Tobacco Control, Johns Hopkins Bloomberg School of Public Health, Baltimore (MD), USA

Dr Laura E. Crotty Alexander, Associate Professor of Medicine, Division of Pulmonary Critical Care and Sleep, University of California at San Diego; Section Chief, Veterans Administration San Diego Healthcare System, San Diego (CA), USA

Dr Danielle R. Davis, Postdoctoral Fellow, Department of Psychiatry, Yale School of Medicine, New Haven (CT), USA

Dr Enrico Davoli, Head, Laboratory of Mass Spectrometry, Environmental Health Sciences Department, Mario Negri Institute for Pharmacological Research, Milan, Italy

Dr Thomas Eissenberg, Professor of Psychology and Co-director, Center for the Study of Tobacco Products, Virginia Commonwealth University, Richmond (VA), USA

Dr Ahmad El-Hellani, American University of Beirut, Department of Chemistry, Faculty of Arts and Sciences, Beirut, Lebanon

Dr Ranti Fayokun, No Tobacco Unit (Tobacco Free Initiative), Department of Health Promotion, Division of Universal Health Coverage and Healthier Populations, World Health Organization, Geneva, Switzerland

Dr Frank Henkler-Stephani, Department of Chemical and Product Safety, Federal Institute for Risk Assessment, Berlin, Germany

Ms Federica Mattioli, Mass Spectrometry Laboratory, Environmental Health Sciences Department, Istituto di Ricerche Farmacologiche Mario Negri, Milan, Italy

Ms Marine Perraudin, Technical Officer, No Tobacco Unit (Tobacco Free Initiative), Department of Health Promotion, Division of Universal Health Coverage and Healthier Populations, World Health Organization, Geneva, Switzerland

Dr Armando Peruga, Researcher, Tobacco Control Group, Bellvitge Biomedical Research Institute, Barcelona, Spain; Centre for Epidemiology and Health Policy, School of Medicine Clínica Alemana, Universidad del Desarrollo, Santiago, Chile

---

1   Apologies: Dr Andrew Barnes, Virginia Commonwealth University, Richmond (VA), USA

Ms Meagan Robichaud, PhD student, Department of Health Behavior and Society, Institute for Global Tobacco Control, Johns Hopkins Bloomberg School of Public Health, Baltimore (MD), USA

Ms Moira Sy, Consultant, No Tobacco Unit (Tobacco Free Initiative), Department of Health Promotion, World Health Organization, Geneva, Switzerland

European Union and selected countries

Dr Matus Ferech, Policy Officer, Tobacco Control Team, Directorate-General for Health and Food Safety, Unit B2 - Cross-border Healthcare and Tobacco Control, European Commission, Brussels, Belgium

Dr Andre Luiz Oliveira da Silva, Health Regulation and Surveillance Specialist, Coordination of Tobacco Derived or Non-tobacco Control Processes, General Manager of Registration and Inspection of Tobacco Derived or Non-tobacco Products, Health Regulatory Agency, Rio de Janeiro, Brazil

Mr Denis Chonière, Director, Tobacco Products Regulatory Office, Controlled Substances and Tobacco Directorate, Health Canada, Ottawa, Ontario, Canada

Dr Peyman Altan, Tobacco Control Department, Ministry of Health, Ankara, Turkey

Secretariat of the WHO Framework Convention on Tobacco Control, Geneva, Switzerland

Dr Adriana Blanco, Head, Convention Secretariat

Secretariat of the No Tobacco Unit, Tobacco Free Initiative, Department of Health Promotion, WHO, Geneva, Switzerland

Dr Jennifer Brown, Consultant

Dr Ranti Fayokun, Scientist

Dr Ruediger Krech, Director

Dr Vinayak Prasad, Unit Head, No Tobacco, Tobacco Free Initiative

Ms Anne Sikanda, Assistant to Unit Head

Ms Simone St Claire, Consultant

Ms Moira Sy, Consultant

致　　谢

世界卫生组织烟草制品管制研究小组（TobReg）对提供本报告基础背景文件的作者表示感谢。本报告在无烟草行动组负责人Vinayak Prasad博士和WHO健康促进部主任Ruediger Krech博士的监督和支持下，由Ranti Fayokun博士多方协调得以完成。

感谢以下世界卫生组织人员提供的行政支持：Miriamjoy Aryee-Quansah女士、Anne Sikanda女士、Gareth Burns先生、Moira Sy女士、Simone St Claire女士和Jennifer Brown博士。

感谢德国政府基金资助TobReg第十次会议背景文件的编写，并感谢所有背景文件的作者，他们以其专业知识对本报告的编写作出了贡献。

感谢Elisabeth Heseltine女士执笔编写本报告，Teresa Lander女士对报告进行校对，Ana Sabino女士对报告进行排版。

还要感谢世界卫生组织《烟草控制框架公约》秘书处协助起草缔约方大会的要求，作为背景文件的基础。

Acknowledgements

The WHO study group on tobacco product regulation (TobReg) expresses its gratitude to the authors of the background papers used as the basis for this report. Production of the report was coordinated by Dr Ranti Fayokun, with the supervision and support of Dr Vinayak Prasad, Unit Head, No Tobacco Unit, Department of Health Promotion, and Dr Ruediger Krech, Director of the WHO Department of Health Promotion.

Administrative support was provided by the following WHO personnel: Ms Miriamjoy Aryee-Quansah, Ms Anne Sikanda, Mr Gareth Burns, Ms Moira Sy, Ms Simone St Claire and Dr Jennifer Brown.

TobReg expresses its gratitude to Germany for providing the funds to support the preparation of the background papers for the 10th meeting of TobReg and acknowledges all authors of the background papers, as listed, for their expertise and contribution to the development of the report.

Our thanks go to Ms Elisabeth Heseltine for editing the report, Ms Teresa Lander for proofreading and Ms Ana Sabino for typesetting the report.

The Study Group also thanks the Secretariat of the WHO Framework Convention on Tobacco Control (WHO FCTC) for facilitating drafting of the requests of the Conference of the Parties to the WHO FCTC, which served as a basis for some of the background papers.

缩 略 语

BAT	英美烟草公司
CC	传统卷烟
CC16	club棒状细胞16 kDa蛋白
CI	置信区间
CO	一氧化碳
COP	缔约方大会
CORESTA	烟草科学研究合作中心
COVID-19	新型冠状病毒感染
ENDS	电子烟碱传输系统
ENNDS	电子非烟碱传输系统
EVALI	电子烟碱传输系统或雾化产品使用相关肺损伤
FDA	美国食品药品监督管理局
HCI	加拿大深度抽吸模式
HTP	加热型烟草制品
ISO	国际标准化组织
JTI	日本烟草国际公司
KT&G	韩国烟草与人参公司
MRTP	风险降低的烟草制品
NAT	N-亚硝基新烟碱
NNK	4-(甲基亚硝基氨基)-1-(3-吡啶基)-1-丁酮
NNN	N-亚硝基降烟碱
PAH	多环芳烃
PATH	烟草与健康人口评估
PM	颗粒物
$PM_{2.5}$	空气动力学直径小于2.5 μm的细颗粒物
PMI	菲利普·莫里斯国际公司
SD	标准偏差
THC	Δ^9-四氢大麻酚
THP	烟草加热产品
THS	烟草加热系统
TSNA	烟草特有亚硝胺
TTC	毒理学关注阈值
USA	美国
US CDC	美国疾病控制和预防中心
WHO FCTC	世界卫生组织《烟草控制框架公约》

Abbreviations and acronyms

BAT	British American Tobacco
CC	conventional cigarette
CC16	club cell 16-kDa protein
CI	confidence interval
CO	carbon monoxide
COP	Conference of the Parties
CORESTA	Cooperation Centre for Scientific Research Relative to Tobacco
COVID-19	coronavirus disease
ENDS	electronic nicotine delivery system
ENNDS	electronic non-nicotine delivery system
EVALI	e-cigarette- or vaping product use-associated lung injury
FDA	United States Food and Drug Administration
HCI	Health Canada method for the testing of tobacco products "Intense puffing regime"
HTP	heated tobacco product
ISO	International Organization for Standardization
JTI	Japan Tobacco International
KT&G	Korea Tobacco and Ginseng Corporation
MRTP	modified risk tobacco product
NAT	N′-nitrosoanatabine
NNK	4-(methylnitrosamino)-1-(3-pyridyl)-1-butanone
NNN	N′-nitrosonornicotine
PAH	polycyclic aromatic hydrocarbons
PATH	Population Assessment of Tobacco and Health
PM	particulate matter
PM2.5	particulate matter < 2.5 μm
PMI	Philip Morris International
SD	standard deviation
THC	Δ^9-tetrahydrocannabinol
THP	tobacco heating product
THS	tobacco heating system
TSNA	tobacco-specific nitrosamines
TTC	threshold of toxicological concern
USA	United States of America
US CDC	United States Centers for Disease Control and Prevention
WHO FCTC	WHO Framework Convention on Tobacco Control

目　录

译者序	i
WHO 烟草制品管制研究小组第十次会议	iii
致谢	xi
缩略语	xv
1. 引言	1
参考文献	3
2. 加热型烟草制品中的有害物质及其暴露、健康影响和风险降低声明	5
摘要	5
2.1　背景	5
2.2　加热型烟草制品释放物中的有害物质	7
2.2.1　测量有害物质的实验室方法	7
2.2.2　烟碱	8
2.2.3　其他有害物质	9
2.3　加热型烟草制品的体外暴露和实验动物暴露及影响	15
2.3.1　体外研究	15
2.3.2　实验动物研究	17
2.4　加热型烟草制品中有害物质的人体暴露及健康影响	18
2.4.1　产品用途和使用模式	18
2.4.2　暴露和效应生物标志物	19
2.4.3　被动暴露	22
2.4.4　健康影响	23
2.5　加热型烟草制品风险或危害降低的证据审查	25
2.5.1　烟草制品危害降低	25
2.5.2　风险降低声明	25
2.6　小结和公共卫生影响	28

2.6.1　数据摘要 ·· 28
　　2.6.2　公共卫生影响 ··· 29
2.7　研究差距和优先事项 ··· 30
2.8　政策建议 ··· 31
2.9　参考文献 ··· 31

3. 加热型烟草制品的吸引力和致瘾性：对认知和使用的影响及关联效应 ······· 40
摘要 ·· 40
3.1　背景 ··· 41
3.2　HTP 的吸引力 ·· 42
　　3.2.1　WHO FCTC 第 9 条和第 10 条中吸引力的定义 ···················· 42
　　3.2.2　HTP 的吸引力特性 ·· 42
　　3.2.3　了解 ENDS 和 ENNDS 并应用到 HTP ································· 45
3.3　HTP 的致瘾性 ·· 46
　　3.3.1　致瘾性 ··· 46
　　3.3.2　了解 ENDS 和 ENNDS 并应用到 HTP ································· 49
　　3.3.3　HTP 的总体滥用倾向 ··· 49
3.4　ENDS、ENNDS 和 HTP 的吸引力和致瘾性对风险、危害和使用认知的影响 ·· 50
　　3.4.1　吸引力和致瘾性对开始、转换、补偿和戒除传统卷烟的贡献 ···· 50
　　3.4.2　了解 ENDS 和 ENNDS 并应用到 HTP ································· 51
3.5　讨论 ··· 52
　　3.5.1　不同群体不同使用方式的行为影响 ······································ 52
　　3.5.2　公共卫生影响 ··· 53
　　3.5.3　研究差距、优先事项和问题 ··· 53
　　3.5.4　政策建议 ··· 54
3.6　结论 ··· 54
3.7　参考文献 ··· 55
附录　IQOS、卷烟和 JUUL 产品中的薄荷醇浓度 ································· 62

4. 加热型烟草制品之间的差异：考虑因素及影响 ······································ 63
摘要 ·· 63
4.1　背景 ··· 63

	4.1.1 概述	63
	4.1.2 缔约方大会 FCTC/COP8(22) 号决议	64
	4.1.3 范围与目标	64
4.2	市场上产品的差异	64
	4.2.1 产品类别和类型的概述	64
	4.2.2 加热型烟草制品之间的差异	66
	4.2.3 产品类型和类别的市场分布	66
4.3	产品特征和设计特点	67
	4.3.1 产品的温度分布情况和操作性能	67
	4.3.2 电池性能	68
	4.3.3 烟草插片、烟棒和胶囊的特性	68
4.4	产品的成分、释放物及总体设计	69
	4.4.1 成分和释放物	69
	4.4.2 烟碱递送	70
	4.4.3 风险概况	71
	4.4.4 加热型烟草制品成分的监管意义	72
	4.4.5 释放物的监管意义	73
4.5	产品、制造商和销售点的差异	73
	4.5.1 制造商和销售点	73
	4.5.2 对客户吸引力的影响	74
4.6	讨论	75
4.7	结论	75
4.8	研究差距、优先事项和问题	76
4.9	政策建议	77
4.10	参考文献	77

5. 加热型烟草制品的使用：产品转换以及双重或多重使用 ... 82
    摘要 ... 82
    5.1 概述 ... 82
    5.2 关于 HTP 在人群层面使用的信息 ... 83
    5.3 从传统卷烟转向 HTP 的动态情况：双重或多重产品联用是过渡态还是永久态？ ... 85

- 5.4 HTP 作为传统卷烟替代品的潜在作用 ... 87
- 5.5 双重 / 多重使用者的烟碱暴露及潜在健康风险 ... 88
- 5.6 动物药代动力学 ... 91
- 5.7 人体药代动力学 ... 91
- 5.8 使用 HTP 和传统卷烟的主观效应 ... 95
- 5.9 讨论和影响 ... 96
- 5.10 研究差距 ... 97
- 5.11 政策建议 ... 97
  - 5.11.1 戒烟政策 ... 97
  - 5.11.2 监督政策 ... 97
  - 5.11.3 研究政策 ... 97
  - 5.11.4 合作政策 ... 98
- 5.12 参考文献 ... 98

6. 加热型烟草制品和电子烟碱 / 非烟碱传输系统相关法规以及国家措施和监管障碍 ... 106
   - 摘要 ... 106
   - 6.1 背景 ... 106
     - 6.1.1 缔约方大会的介绍和要求（FCTC/COP8(22)） ... 106
     - 6.1.2 范围和目标 ... 108
     - 6.1.3 来源 ... 108
   - 6.2 新型烟草制品的监管情况 ... 108
     - 6.2.1 HTP 的可用性 ... 108
     - 6.2.2 产品分类 ... 109
     - 6.2.3 减少烟草需求的监管框架和措施 ... 113
   - 6.3 监管、实施和执行政策的考虑事项和障碍 ... 118
     - 6.3.1 实施政策时的监管考虑 ... 118
     - 6.3.2 实施和执行政策的障碍 ... 119
     - 6.3.3 其他考虑和意外后果 ... 119
   - 6.4 讨论 ... 120
   - 6.5 结论 ... 121
   - 6.6 研究差距 ... 121

6.7	政策建议	121
6.8	参考文献	122

7. 电子烟碱传输系统和传统卷烟的烟碱暴露评估 ·········· 127

摘要 ·········· 127

7.1	背景	128
7.2	电子烟碱传输系统的烟碱暴露	129
	7.2.1　电子烟碱传输系统的烟碱释放	129
	7.2.2　电子烟碱传输系统电功率对烟碱释放的影响	129
	7.2.3　电子烟碱传输系统烟油中烟碱和其他化合物浓度对烟碱释放的贡献	130
7.3	伴生物质的暴露概述	131
7.4	电子烟碱传输系统的烟碱递送	131
7.5	与使用有关的暴露行为模式	132
	7.5.1　使用者群体和使用者模式的定义	132
	7.5.2　影响行为模式的因素	133
7.6	被动暴露于烟碱和其他有害成分	135
7.7	烟碱通量	135
7.8	讨论	137
7.9	结论	138
7.10	研究差距、优先事项和问题	138
7.11	政策建议	139
7.12	参考文献	139

8. 电子烟碱/非烟碱传输系统和加热型烟草制品相关个体风险的量化方法探索：对人体健康和监管的影响 ·········· 148

摘要 ·········· 148

8.1	背景	148
	8.1.1　量化风险的挑战	150
8.2	使用 ENDS 和 ENNDS 相关的风险评估与量化	151
	8.2.1　与特定成分或无意添加的物质相关的健康风险	152
	8.2.2　ENDS 和 ENNDS 的潜在健康影响	154
	8.2.3　非使用者的风险	156

8.3 量化风险的方法 ... 156
  8.3.1 毒理学关注的阈值 ... 158
  8.3.2 基于单个化合物的风险评估 ... 158
  8.3.3 相对风险方法 ... 159
  8.3.4 暴露边界值法 ... 159
  8.3.5 非致癌作用 ... 159
  8.3.6 评估框架 ... 160
8.4 加热型烟草制品 ... 162
8.5 对监管的影响 ... 162
8.6 讨论 ... 164
8.7 建议 ... 165
8.8 结论 ... 166
8.9 参考文献 ... 166

9. 新型烟草制品的风味 ... 175
摘要 ... 175
9.1 背景和引言 ... 175
9.2 风味电子烟的流行病学（使用频率、模式和原因，按社会人口统计变量）... 177
  9.2.1 电子烟碱/非烟碱传输系统 ... 178
  9.2.2 传统卷烟和无烟烟草制品 ... 179
  9.2.3 加热型烟草制品 ... 180
9.3 调味物质对吸引力、尝试使用、使用和持续使用的影响 ... 180
  9.3.1 青少年和年轻人 ... 180
  9.3.2 成年人 ... 182
9.4 常见调味物质特性、健康效应及其对公众健康的影响 ... 183
  9.4.1 电子烟和烟草制品中常见的调味物质 ... 183
  9.4.2 风味电子烟中常见调味物质的化学和物理性质 ... 183
  9.4.3 调味物质的毒性 ... 184
9.5 风味产品的监管 ... 185
  9.5.1 全球风味电子烟监管 ... 185
  9.5.2 全球风味烟草制品监管 ... 186

	9.5.3	常用方法的优缺点	188
	9.5.4	对烟碱产品特定风味监管产生的影响	188
	9.5.5	调味物质的未来监管	189
9.6	讨论		191
9.7	研究差距、优先事项和问题		191
9.8	政策建议		192
9.9	参考文献		192

10. 新型烟草制品的全球营销和推广及其影响 206

摘要 206

- 10.1 背景 206
- 10.2 电子烟碱传输系统和电子非烟碱传输系统 208
  - 10.2.1 引言 208
  - 10.2.2 电子烟市场、产品和营销策略 209
  - 10.2.3 电子烟的全球使用情况和使用流行程度 212
  - 10.2.4 电子烟的广告、促销和赞助趋势 216
  - 10.2.5 控制电子烟的广告、促销和赞助的措施 217
  - 10.2.6 建议 218
  - 10.2.7 电子烟的研究差距 220
- 10.3 加热型烟草制品 221
  - 10.3.1 引言 221
  - 10.3.2 市场参与者、产品和策略 221
  - 10.3.3 加热型烟草制品的全球使用和流行情况 224
  - 10.3.4 加热型烟草制品的广告、促销和赞助趋势 225
  - 10.3.5 加热型烟草制品广告、促销和赞助的控制措施 227
  - 10.3.6 对监测加热型烟草制品的广告、促销和赞助趋势的建议 227
  - 10.3.7 对监管机构的建议 228
  - 10.3.8 加热型烟草制品研究的研究差距 229
- 10.4 总结 229
- 10.5 参考文献 230

11. 烟草中的烟碱形态、化学修饰及其对电子烟碱传输系统的影响 241

摘要 241

11.1	背景	241
11.2	烟碱的化学修饰及其对烟碱传输的影响	243
	11.2.1 烘烤对烟碱的影响	243
	11.2.2 碱改性	244
	11.2.3 酸改性	244
11.3	对电子烟碱传输系统产品和多样性的影响	245
	11.3.1 电子烟碱传输系统中游离态烟碱与烟碱盐	245
	11.3.2 有效浓度和滥用倾向	246
	11.3.3 对产品刺激性可能的掩饰	247
	11.3.4 健康影响和监管考虑	247
11.4	讨论	248
11.5	研究差距、优先事项和问题	248
11.6	建议	249
11.7	考虑因素	249
11.8	参考文献	249

12. 电子烟碱传输系统或雾化产品使用相关肺损伤 258

摘要 258

12.1	背景	258
	12.1.1 与电子烟和雾化产品相关的呼吸影响	258
	12.1.2 电子烟碱传输系统或雾化产品使用相关肺损伤（EVALI）	259
	12.1.3 与 EVALI 有关的产品和化学制品	259
12.2	EVALI	260
	12.2.1 详细说明和历史记录	260
	12.2.2 症状	260
	12.2.3 临床表现	261
	12.2.4 报告病例	261
12.3	EVALI 的识别	261
12.4	对 EVALI 的监测	262
	12.4.1 国家监测机制	262
	12.4.2 地区监测机制	262
	12.4.3 国际监测机制和验证	262

12.5　讨论 262
　　12.6　考虑事项 263
　　12.7　建议 263
　　　　12.7.1　主要建议 263
　　　　12.7.2　其他建议 264
　　12.8　参考文献 264
13. 总体建议 266
　　13.1　主要建议 267
　　13.2　对公共卫生政策的意义 268
　　13.3　对 WHO 规划的影响 269
　　13.4　参考文献 269

Contents

Participants in the 10th meeting of the WHO study group on tobacco product regulation vii
Acknowledgements xiii
Abbreviations and acronyms xvii
1. Introduction 273
    References 277
2. Toxicants in heated tobacco products, exposure, health effects and claims of reduced risk 278
    Abstract 278
    2.1 Background 279
    2.2 Toxicants in heated tobacco product emissions 281
        2.2.1 Laboratory methods for measuring toxicants 281
        2.2.2 Nicotine 282
        2.2.3 Other toxicants 284
    2.3 Exposure and effects of HTPs in vitro and in laboratory animals 291
        2.3.1 In-vitro studies 291
        2.3.2 Studies in laboratory animals 293
    2.4 Exposure of humans to toxicants in HTPs and implications for health 295
        2.4.1 Product use and topography 295
        2.4.2 Biomarkers of exposure and effect 296
        2.4.3 Passive exposure 300
        2.4.4 Impact on health outcomes 302
    2.5 Review of the evidence for reduced risk or harm with use of HTPs 304
        2.5.1 Harm reduction in the context of tobacco products 304
        2.5.2 Claims of reduced risk 304
    2.6 Summary and implications for public health 308
        2.6.1 Summary of data 308

    2.6.2 Implications for public health ··310
  2.7 Research gaps and priorities ···311
  2.8 Policy recommendations ···312
  2.9 References ···312
3. The attractiveness and addictive potential of heated tobacco products: effects on perception and use and associated effects ··320
  Abstract ···320
  3.1 Background ··321
  3.2 Attractiveness of HTPs ···323
    3.2.1 Definition of attractiveness in the context of Articles 9 and 10 of the WHO FCTC ···323
    3.2.2 Attractive features of HTPs ···323
    3.2.3 What we can learn from studies on ENDS and ENNDS and relevance to HTPs ··327
  3.3 Addictiveness of HTPs ··328
    3.3.1 Addictiveness ···328
    3.3.2 What we can learn from ENDS and ENNDS and relevance to HTPs ··332
    3.3.3 Overall abuse liability of HTPs ···332
  3.4 Effects of the attractiveness and addictiveness of ENDS, ENNDS and HTPs on perceptions of risk and harm and use ··333
    3.4.1 Contributions of attractiveness and addictiveness to initiation, switching, complementing and quitting conventional tobacco products ···333
    3.4.2 Learning from ENDS and ENNDS and application to HTPs ···········334
  3.5 Discussion ···335
    3.5.1 Behavioural implications of different patterns of use in different groups ···335
    3.5.2 Implications for public health ··336
    3.5.3 Research gaps, priorities and questions ···337
    3.5.4 Policy recommendations ··338
  3.6 Conclusions ··339
  3.7 References ···339
  Annex 3.1. Menthol concentrations in IQOS, cigarettes and JUUL products ····345
4. Variations among heated tobacco products: considerations and implications ····346

Abstract ··346
4.1 Background ···347
    4.1.1 Overview ···347
    4.1.2 Decision FCTC/COP8(22) of the Conference of the Parties ···············347
    4.1.3 Scope and objectives ···347
4.2 Variations among products on the market ···348
    4.2.1 Overview of product categories and types ···································348
    4.2.2 Variations among heated tobacco products ·································350
    4.2.3 Market distribution of product types and categories ····················350
4.3 Characteristics and design features of products ··································352
    4.3.1 Temperature profiles of products and operational capabilities ········352
    4.3.2 Battery characteristics ··352
    4.3.3 Properties of tobacco inserts, sticks and capsules ························352
4.4 Content, emissions and general design of products ·····························353
    4.4.1 Content and emissions ··353
    4.4.2 Nicotine delivery ··356
    4.4.3 Risk profiles ··356
    4.4.4 Regulatory implications of the contents of heated tobacco products
    ··357
    4.4.5 Regulatory implications of emissions ···358
4.5 Variations among products, manufacturers and selling points ·············359
    4.5.1 Manufacturers and selling points ···359
    4.5.2 Implications for customer pulling power ···································360
4.6 Discussion ···361
4.7 Conclusions ···362
4.8 Research gaps, priorities and questions ···363
4.9 Policy recommendations ··363
4.10 References ···365

5. Use of heated tobacco products: product switching and dual or poly product use ··369

Abstract ··369
5.1 Introduction ··370
5.2 Information on HTP use at population level ·······································370
5.3 Dynamics of switching from conventional cigarettes to HTPs: Is dual or poly use a transitional or permanent state? ··373

- 5.4 Potential role of HTPs as a substitute for conventional cigarettes ... 376
- 5.5 Exposure to nicotine and potential health risks among poly users ... 377
- 5.6 Pharmacokinetics in animals ... 380
- 5.7 Pharmacokinetics in people ... 380
- 5.8 Subjective effects of use of HTPs and conventional cigarettes ... 385
- 5.9 Discussion and implications ... 386
- 5.10 Research gaps ... 387
- 5.11 Policy recommendations ... 387
  - 5.11.1 Cessation policy ... 387
  - 5.11.2 Surveillance policy ... 387
  - 5.11.3 Research policy ... 388
  - 5.11.4 Cooperation and partnership policy ... 388
- 5.12 References ... 388

6. **Regulations on heated tobacco products, electronic nicotine delivery systems and electronic non-nicotine delivery systems, with country approaches, barriers to regulation and regulatory considerations** ... 395
   - Abstract ... 395
   - 6.1 Background ... 396
     - 6.1.1 Introduction and the request of the Conference of the Parties (FCTC/COP8(22)) ... 396
     - 6.1.2 Scope and objectives ... 397
     - 6.1.3 Sources ... 398
   - 6.2 Regulatory mapping of novel and emerging nicotine and tobacco products ... 398
     - 6.2.1 Availability of HTPs ... 398
     - 6.2.2 Product classification ... 399
     - 6.2.3 Regulatory frameworks and measures to reduce tobacco demand ... 404
   - 6.3 Considerations and barriers to regulation, implementation and enforcement of policies ... 411
     - 6.3.1 Regulatory considerations in implementing policies ... 411
     - 6.3.2 Barriers to implementing and enforcing policies ... 413
     - 6.3.3 Other considerations and unintended consequences ... 413
   - 6.4 Discussion ... 414
   - 6.5 Conclusions ... 415
   - 6.6 Research gaps ... 416

6.7 Policy recommendations ··416
  6.8 References ···417

7. **Estimation of exposure to nicotine from use of electronic nicotine delivery systems and from conventional cigarettes** ··421
   Abstract ···421
   7.1 Background ···422
   7.2 Exposure to nicotine from ENDS ···424
       7.2.1 ENDS nicotine emission ···424
       7.2.2 Influence of ENDS electrical power on nicotine emission ·············424
       7.2.3 Contribution of the concentrations of nicotine and other
             compounds in ENDS liquids to nicotine emissions ······················425
   7.3 Overview of exposure to accompanying substances ··································426
   7.4 Nicotine delivery from ENDS ···427
   7.5 Behavioural patterns of exposure according to use ··································428
       7.5.1 Definition of user groups and user patterns ···································428
       7.5.2 Factors that influence behavioural patterns ···································428
   7.6 Passive exposure to nicotine and other toxicants ······································430
   7.7 Nicotine flux ··431
   7.8 Discussion ··433
   7.9 Conclusions ···434
   7.10 Research gaps, priorities and questions ··435
   7.11 Policy recommendations ···436
   7.12 References ··436

8. **Exploration of methods for quantifying individual risks associated with electronic nicotine and non-nicotine delivery systems and heated tobacco products: impact on population health and implications for regulation** ···········444
   Abstract ···444
   8.1 Background ···445
       8.1.1 Challenges to quantifying risk ···447
   8.2 Risk assessment and quantification of risks associated with use of ENDS
       and ENNDS ···449
       8.2.1 Health risks associated with specific ingredients or unintentionally
             added substances ··450
       8.2.2 Potential health effects of ENDS and ENNDS ·······························453
       8.2.3 Risks for bystanders ··455

8.3 Methods for quantifying risk 455
    8.3.1 Threshold of toxicological concern 458
    8.3.2 Risk assessment based on individual compounds 458
    8.3.3 Relative risk approaches 459
    8.3.4 Margin-of-exposure approach 459
    8.3.5 Non-carcinogenic effects 460
    8.3.6 Evaluation frameworks 460
8.4 Heated tobacco products 463
8.5 Implications for regulation 463
8.6 Discussion 466
8.7 Recommendations 467
8.8 Conclusions 468
8.9 References 469

9. Flavours in novel and emerging nicotine and tobacco products 476
Abstract 476
9.1 Background and introduction 477
9.2 Epidemiology of flavoured products (frequency, patterns and reasons for use by sociodemographic variables) 479
    9.2.1 Electronic nicotine and non-nicotine delivery systems 480
    9.2.2 Traditional smoked and smokeless tobacco products 481
    9.2.3 Heated tobacco products 482
9.3 Effects of flavours on appeal, experimentation, uptake and sustained use ...483
    9.3.1 Adolescents and young adults 483
    9.3.2 Adults 484
9.4 Common flavours, properties, health effects and implications for public health 486
    9.4.1 Common flavours in electronic nicotine delivery systems and tobacco products 486
    9.4.2 Chemical and physical properties of common flavours in flavoured products 486
    9.4.3 Toxicity of flavours 487
9.5 Regulation of flavoured products 488
    9.5.1 Global regulation of flavoured ENDS 488
    9.5.2 Global regulation of flavoured tobacco products 490
    9.5.3 Pros and cons of common approaches 492

9.5.4 Impact of regulation of specific flavoured nicotine products ············ 492
9.5.5 Future regulation of flavours ·· 493
9.6 Discussion ·· 495
9.7 Research gaps, priorities and questions ····································· 496
9.8 Policy recommendations ·· 497
9.9 References ·· 497

10. Global marketing and promotion of novel and emerging nicotine and tobacco products and their impacts ·· 509

Abstract ·· 509
10.1 Background ·· 510
10.2 Electronic nicotine delivery systems and electronic non-nicotine delivery systems ·· 512
  10.2.1 Introduction ··· 512
  10.2.2 Markets, products and strategies used in marketing ENDS and ENNDS ··· 513
  10.2.3 Global use of ENDS and prevalence of use ························ 518
  10.2.4 Trends in advertising, promotion and sponsorship of ENDS products ··· 521
  10.2.5 Measures to control advertising, promotion and sponsorship of ENDS products ·· 523
  10.2.6 Recommendations ·· 524
  10.2.7 Research gaps for ENDS and ENNDS ······························· 527
10.3 Heated tobacco products ··· 528
  10.3.1 Introduction ··· 528
  10.3.2 Market players, products and strategies ·························· 529
  10.3.3 Global use and prevalence of use of HTPs ························ 532
  10.3.4 Trends in advertising, promotion and sponsorship of HTPs ········· 533
  10.3.5 Measures to control advertising, promotion and sponsorship of HTPs ··· 535
  10.3.6 Recommendations for monitoring trends in marketing, advertising, promoting and sponsorship of HTPs ···························· 536
  10.3.7 Recommendations for regulators ···································· 536
  10.3.8 Gaps in research on HTPs ·· 538
10.4 Summary ··· 538
10.5 References ··· 539

Supplementary sections 548
11. Forms of nicotine in tobacco plants, chemical modifications and implications for electronic nicotine delivery systems products 549
   Abstract 549
   11.1 Background 550
   11.2 Chemical modification of nicotine and influence on nicotine delivery 553
      11.2.1 Brief summary of the effect of curing on nicotine 553
      11.2.2 Modification with alkali 553
      11.2.3 Modification with acid 554
   11.3 Implications for ENDS products and diversity 554
      11.3.1 Free-base nicotine vs nicotine salt in ENDS 554
      11.3.2 Feasible concentrations and abuse liability 556
      11.3.3 Potential masking of the harshness of products 557
      11.3.4 Health implications and potential regulations 557
   11.4 Discussion 558
   11.5 Research gaps, priorities and questions to members regarding further work or a full paper 558
   11.6 Recommendation 559
   11.7 Considerations 559
   11.8 References 560
12. EVALI: e-cigarette or vaping product use-associated lung injury 567
   Abstract 567
   12.1 Background 568
      12.1.1 Respiratory effects associated with e-cigarettes and vaping 568
      12.1.2 E-cigarette or vaping product use-associated lung injury (EVALI) 568
      12.1.3 Products and chemicals implicated in EVALI 569
   12.2 EVALI 569
      12.2.1 Detailed description and history 569
      12.2.2 Symptoms 570
      12.2.3 Clinical presentation 570
      12.2.4 Reported cases 571
   12.3 Identification of EVALI 571
   12.4 Surveillance for EVALI 571
      12.4.1 National surveillance mechanisms 571

      12.4.2 Regional surveillance mechanisms 572
      12.4.3 International surveillance mechanisms and validation 572
  12.5 Discussion 572
  12.6 Considerations 573
  12.7 Recommendations 574
      12.7.1 Key recommendations 574
      12.7.2 Other recommendations 574
  12.8 References 574

13. Overall recommendations 576
  13.1 Main recommendations 577
  13.2 Significance for public health policies 579
  13.3 Implications for the Organization's programmes 580
  13.4 References 581

1. 引　　言

有效的烟草制品管制是综合烟草控制计划的重要组成部分，包括通过授权测试并披露测试结果来管制烟草制品成分和释放物，设定适当限值，公布产品信息，以及制定产品包装和标签标准。世界卫生组织《烟草控制框架公约》(WHO FCTC)[1]第9、10和11条以及实施WHO FCTC第9条和第10条的部分指南[2]涵盖了烟草制品管制的内容。《烟草制品管制：基本手册》[3]和《烟草制品管制：实验室检测能力建设》[4]以及2018年出版的在线模块课程[5]等世界卫生组织的其他资料也在这些方面为成员国提供了支持。

世界卫生组织总干事于2003年正式组建了世界卫生组织烟草制品管制研究小组（TobReg），以填补烟草制品管制空白。其任务是向WHO总干事提供有关烟草制品管制的政策建议。TobReg由产品监管、烟草依赖治疗、毒理学以及烟草制品成分和释放物实验室分析等领域的国际科学专家组成。这些专家来自WHO六大地区的成员国[6]。作为WHO的正式实体，TobReg通过总干事向WHO执行委员会提交技术报告，为烟草制品管制提供科学依据，以提请成员国注意WHO在该领域的工作。WHO技术报告系列包括以前未发表的背景文件，这些文件综合了已发表并由TobReg进行讨论、评估和审查的科学文献。根据WHO FCTC第9条和第10条、WHO FCTC缔约方会议（COP）的相关决定以及提交给COP的相关WHO报告，TobReg报告确定了根据循证方法来监管所有形式的烟草制品，包括新型产品，如电子烟碱传输系统（ENDS）、电子非烟碱传输系统（ENNDS）、加热型烟草制品（HTP）和烟碱袋。这些报告是对世界卫生大会WHA54.18（2001年）、WHA53.17（2000年）和WHA53.8（2000年）决议的响应，也已被视为世界卫生组织全球公共卫生产品。"全球公共卫生产品"是由世界卫生组织制定及实施，对全球许多国家或地区有益的举措[7]。这对TobReg来说是一个独特的机会，可以直接与成员国对话并影响国家、地区和全球政策。

TobReg第十次会议由世界卫生组织日内瓦总部协调，于2020年9月28日至10月2日举行，包括TobReg成员在内的50多名参与者讨论了关于新型烟草制品（特别是加热型烟草制品）的致瘾性、有害性、吸引力、差异性、营销、健康影响和监管的科学文献。加热型烟草制品有较长的历史，但在过去十年中吸引了越来越多的国际关注。由于监管能力不足，面临各种挑战：循证干预措施分散，产品类别混淆，未经证实的健康主张，相关反对和干扰有效监管政策及烟草相关行业针对儿童和青少年的营销活动。该会议为讨论九份背景文件提供了平台：

- 加热型烟草制品中的有害物质及其暴露、健康影响和风险降低声明（第2章）；
- 加热型烟草制品的吸引力和致瘾性：对认知和使用的影响及关联效应（第3章）；
- 加热型烟草制品之间的差异：考虑因素及影响（第4章）；
- 加热型烟草制品的使用：产品转换以及双重或多重使用（第5章）；
- 加热型烟草制品和电子烟碱/非烟碱传输系统相关法规以及国家措施和监管障碍（第6章）；
- 电子烟碱传输系统和传统卷烟的烟碱暴露评估（第7章）；
- 电子烟碱/非烟碱传输系统和加热型烟草制品相关个体风险的量化方法探索：对人群健康和监管的影响（第8章）；
- 新型烟草制品的风味（第9章）；
- 新型烟草制品的全球营销和推广及其影响（第10章）。

TobReg还讨论了两篇补充文件：

- 烟草中的烟碱形态、化学修饰及其对电子烟碱传输系统的影响（第11章）；
- 电子烟碱传输系统或雾化产品使用相关肺损伤（第12章）。

背景文件和补充文件是由主题专家根据WHO组织秘书处为每篇论文起草大纲，并经由专家审评员和研究组成员审查和修订的文件。补充文件是关于产品监管中出现的新问题的简短文件，供成员国参考以决定是否有必要根据这些主题开展进一步的工作，例如为研究组未来的会议准备一份完整的背景文件。在缔约方第八次会议（COP8）上，关于HTP和新型烟草制品的文件涉及FCTC/COP8(22)[8]决议第2a段的要求：

要求WHO FCTC秘书处邀请世界卫生组织，并酌情邀请世界卫生组织烟草实验室网络与烟草行业外的专家以及各国家相关部门共同编写一份综合报告，并提交给缔约方第九次会议。该报告主要针对新型烟草制品，特别是加热型烟草制品的健康影响，包括对非使用者的健康影响、致瘾性、认知和使用、吸引力、在吸烟和戒烟过程中的潜在作用、营销（包括促销策略和影响）、风险降低声明、产品的差异性、缔约方的监管经验和监测、对烟草管制工作的影响和监管空白，并随后提出可行的政策以实现决议第5段的目标和措施。

世界卫生组织审查了通过WHO FCTC秘书处向WHO提出的关于新型烟草制品的请求，并将其与烟草制品管制方面的其他新问题（包括成员国向WHO提出的技术援助请求）一起作为决定文件主题的基础。秘书处与研究组协商，邀请的专家不仅为讨论作出了贡献，而且还在背景文件中提供了有关烟碱和烟草制品的最新科学证据和相关法规。每篇文献都标明了检索时间，大多数检索时间为2020年第二季度。会议前后，独立技术专家、WHO秘书处及其他相关部门人员、地区干事和研究组成员对文献进行多轮审查后，将其编入技术报告。TobReg关于烟草制品管制的科学基础第八份报告旨在指导成员国寻找最有效的、基于证据的手段来填补监管空白及应对烟草管制方面的挑战，从而促进制定烟碱和烟草制品的协调监管框架。包括研究组成员在内的所有会议专家和其他参与者，都必须完成由WHO评估以确保独立于烟草行业之外的利益声明。

本报告第2~6章和第7~8章分别是关于HTP和ENDS的内容；第9章和第10章分别是关于新型烟草制品的风味以及全球营销和推广的一般性评论；第11章和12章概述烟草中的烟碱形态和电子烟碱传输系统或雾化产品相关肺损伤（EVALI）；第13章为总结建议。这些建议综合了复杂的研究和证据，促进了监管的国际协调和产品监管最佳实践的采用，加强了世界卫生组织所有地区的产品监管能力建设，为成员国提供了可靠的科学依据，并支持缔约方实施世界卫生组织《烟草控制框架公约》。

本报告涉及ENDS、ENNDS和HTP，但并未涵盖这些产品的所有方面，因为许多文献是为满足FCTC/COP8(22)决议的要求（即审查对新型烟草制品的理解）而编写的，因此需要对这些产品进行持续研究。研究组将根据各国的监管要求并针对烟草制品监管的相关问题，在下一份报告中涵盖其他感兴趣的产品（包括传统产品，如水烟、卷烟和无烟烟草）。这将确保向所有国家提供持续、及时的技术支持并涉及所有产品，同时认识到其可用性取决于司法管辖区。

总之，TobReg的审议结果及其建议将提高成员国对ENDS、ENNDS和HTP证据的理解，有助于建立产品监管知识体系，为WHO的工作提供信息，特别是向成员国提供技术支持，并通过各种平台使成员国、监管机构、民间社团、研究机构和其他利益相关方了解最新的产品监管信息。WHO FCTC缔约方将根据FCTC/COP8(22)决议，通过WHO FCTC秘书处向COP9提交一份关于新型烟草制品的更新综合报告，包括本报告的信息和建议。因此，研究组的工作将有助于实现可持续发展目标的具体目标，即加强世界卫生组织《烟草控制框架公约》的实施[9]。

参 考 文 献

[1] WHO Framework Convention on Tobacco Control. Geneva: World Health Organization; 2003 (http://www.who.int/fctc/en/, accessed 10 January 2021).

[2] Partial guidelines on implementation of Articles 9 and 10. Geneva: WorldHealthOrganization;2012(https://www.who.int/fctc/guidelines/Guideliness_Articles_9_10_rev_240613.pdf, accessed January 2019).

[3] Tobacco product regulation: basic handbook. Geneva: World Health Organization;2018(https://www.who.int/tobacco/publications/prod_regulation/basic-handbook/en/, accessed 10 January 2021).

[4] Tobacco product regulation: building laboratory testing capacity. Geneva: WorldHealthOrganization; 2018(https://www.who.int/tobacco/publications/prod_regulation/building-laboratory-testing-capacity/en/, accessed 14 January 2019).

[5] Tobacco product regulation courses. Geneva: World Health Organization (https://openwho.org/courses/TPRS-building-laboratory-testing-capacity/items/3S11LKUFGyoTksZ5RSZblD; https://openwho.org/courses/TPRS-tobacco-product-regulation-handbook/items/7zq7S1jxAtp-bUWiZdfH98l).

[6] TobReg members. In: World Health Organization [website]. Geneva: World Health Organization(https://www.who.int/groups/who-study-group-on-tobacco-product-regulation/about, accessed 10 January 2021).

[7] Feacham RGA, Sachs JD. Global public goods for health. The report of working group 2 of the Commission on Macroeconomics and Health. Geneva:WorldHealthOrganization;2002(https://apps.who.int/iris/bitstream/handle/10665/42518/9241590106.pdf?sequence=1, accessed 10 January 2021).

[8] Decision FCTC/COP8(22). Novel and emerging tobacco products. Decision of the Conference of the Parties to the WHO Framework Conventionon Tobacco Control, Eighth session. Geneva: World Health Organization; 2018 (https://www.who.int/fctc/cop/sessions/cop8/FCTC__COP8(22).pdf, accessed 7 November 2020).

[9] Sustainable development goals. Geneva: World Health Organization (https://www.who.int/health-topics/sustainable-development-goals#tab=tab_2, accessed 10 January 2021).

2. 加热型烟草制品中的有害物质及其暴露、健康影响和风险降低声明

Irina Stepanov, Professor, Division of Environmental Health Sciences and Masonic Cancer Center, University of Minnesota, Minneapolis (MN), USA

Federica Mattioli, Mass Spectrometry Laboratory, Environmental Health Sciences Department, Istituto di Ricerche Farmacologiche Mario Negri, Milan, Italy

Enrico Davoli, Head, Mass Spectrometry Laboratory, Environmental Health Sciences Department, Istituto di Ricerche Farmacologiche Mario Negri, Milan, Italy

摘　　要

新型加热型烟草制品（HTP）的市场营销和日益普及要求紧急评估其对公众健康的潜在影响。本报告评估了关于HTP化学成分、产品暴露及其对模型毒理学系统（体外和实验动物）和人体影响的已发表文献。此类评估是表征HTP的致瘾性、有害性和致癌性潜力并将其与市场上其他烟草制品进行比较的关键初始步骤。数据表明，HTP气溶胶的化学特征与传统卷烟或电子烟碱传输系统的化学特征有很大不同。然而，在缺乏标准分析方法的情况下，应谨慎解释这些发现。研究还表明，某些HTP的烟碱摄入量可能与传统卷烟相似或更高。体外和实验动物及人体暴露的研究通常证实减少了对某些燃烧产生的有害物质的暴露；然而，一些研究引起了对使用者潜在的心肺毒性和肝毒性以及非使用者的二手暴露危害的担忧。该报告强调了企业外研究的缺乏，并建议了研究和监管优先事项。

2.1　背　　景

加热型烟草制品（HTP）也被制造商和一些监管机构称为"加热不燃烧"或烟草加热型产品，通过在电池供电的加热系统中以低于传统卷烟燃烧烟草的温度（通常<600℃）加热烟草来产生气溶胶。吸烟者对早期HTP（如Eclipse和Accord卷烟）的接受度很低，这些产品没有获得大的市场份额；然而，过去十年推出的

HTP已开始在世界某些地区流行起来。由于传统卷烟的燃烧会产生许多有害释放物，包括许多有害物质和致癌物，加热型烟草制品的目标是为吸烟者提供一种比传统卷烟"危害小的替代品"。因此，加热型烟草制品以"降低风险产品"、"传统卷烟的清洁替代品"和"无烟替代品"的宣传进行营销。当前销售的加热型烟草制品的品牌有IQOS（菲利普·莫里斯国际公司，PMI）、glo和iFuse（英美烟草公司，BAT）和PloomTECH（日本烟草公司）。尽管它们都是加热型烟草制品，但它们的结构和成分以及加热烟草和产生气溶胶的机制不同。例如，IQOS和glo通过在240~350℃下加热卷烟状的烟草棒来产生含有烟碱的气溶胶，而Ploom-TECH和iFuse通过加热甘油和丙二醇的混合物产生气溶胶，然后将其通过含有烟草材料的胶囊。

评估使用者和非使用者对释放的化学成分（其中许多是重要的有害物质和致癌物）的暴露以及与当前销售的加热型烟草制品相关的影响存在几个主要挑战。第一，IQOS等产品是近几年才推出的，最初在2015年仅在有限的市场上销售。第二，是加热型烟草制品的多样性，正如前文所述，这些设备产生的气溶胶的化学成分可能不同。虽然加热型烟草制品气溶胶中许多与燃烧相关的化学物质的浓度可能低于传统卷烟，但有些物质的浓度更高，而且加热型烟草制品因具有独有的特性和使用方式可能会释放独特的有害化学物质。第三，一种设备生产的烟棒可以与其他设备一起使用，并且可以重复使用；产品的互换性和误用使得很难确定使用者接触到的有害释放物。第四，目前上市的加热型烟草制品的科学数据大部分是由烟草企业或其资助的附属机构生成和发布的，严重缺乏关于加热型烟草制品的独立学术文献。鉴于某些烟草制品制造商误导营销和曲解研究数据的历史，他们的出版物必须在严格审查原始数据和方法后谨慎解读。制造商向监管机构报告的未公开数据应以同样严格的方式进行审查。

本报告是应世界卫生组织《烟草控制框架公约》缔约方大会第八次会议（2018年10月1~6日，瑞士日内瓦）向公约秘书处提出的要求，邀请世界卫生组织编写关于新型烟草制品，特别是加热型烟草制品的研究和证据的综合报告，并提出政策选择以实现相关决议（FCTC/COP8(22)）中概述的目标和措施。本报告部分解决了这一要求，涵盖了加热型烟草制品对健康影响的方方面面，包括对非使用者的健康影响、致癌性、认知和使用、吸引力、在吸烟和戒烟中的潜在作用、营销（促销策略和影响）、降低危害的声明、产品之间的差异、缔约方的监管经验和监测、对烟草管制的影响和研究差距。

本章我们回顾了当前关于加热型烟草制品释放物有害性和暴露以及相关健康影响的文献，并评估了加热型烟草制品降低风险的说法。特别包括：

■ 加热型烟草制品释放物和其他烟草制品中的有害物质；
■ 加热型烟草制品释放物及其对模型系统（体外和实验动物）暴露和人体

暴露的影响以及健康效应；
■ 评估加热型烟草制品降低风险或危害的声明的依据；
■ 对公众健康的影响；
■ 研究差距和优先事项；
■ 相关政策建议。

文献检索主要基于PubMed数据库和SciFinder检索工具，SciFinder从Medline和CAplus数据库中检索数据。检索获得的出版物中引用的相关文章也包括在内。此外，还访问了制造商的网站以及美国疾病控制和预防中心、美国食品药品监督管理局（FDA）以及其他提供加热型烟草制品化学和健康影响信息的相关网站。

2.2 加热型烟草制品释放物中的有害物质

2.2.1 测量有害物质的实验室方法

如上所述，市场上的HTP设备在设计上各不相同，其中的主要区别在于烟草的加热方式和设备达到的温度。温度对烟草制品（包括传统卷烟和电子烟碱传输系统）释放物中有害成分形成的影响已得到充分证明。第三方生产的加热型烟草制品设备也可用，主要是在线的，通常没有产品规格（使用"IQOS sticks compatible"作为关键词进行简单的谷歌或亚马逊网络检索，便可以找到许多品牌，如Uwoo、Luckten、Kacig、Hotcig、Uwell、Vaptio、G-taste和SmokNord）。为了从加热型烟草制品有害释放水平的评估中得出可靠的推论，并将其与传统卷烟和电子烟碱传输系统等其他产品的释放物进行比较，必须采用可靠、经过分析验证的测试方法。

迄今为止，企业和独立研究小组使用的加热型烟草制品释放物的分析方法大多源自于传统卷烟的分析方法，主要用于确定常见的释放物和其他物质（例如烟碱和其他一些烟草衍生成分）[1,2]。为了分析加热型烟草制品特有的化合物，一些实验室使用了复杂的分析方法，例如多维气相色谱与质谱联用或液相色谱-串联质谱[1,3]。2019年，国际标准化组织（ISO）烟草和烟草制品技术委员会TC126成立了加热型烟草制品工作组，但尚未发布加热型烟草制品的特定检测方法。

虽然原则上用于分析传统卷烟的方法可用于分析加热型烟草制品，但两种产品之间存在关键差异。首先，加热型烟草制品中的烟草材料可能含有比传统卷烟更多的保润剂和水[4]，这可能会对释放物分析产生影响。至少，应该研究确定通常的玻璃纤维剑桥滤片在捕集卷烟烟气及主要来自甘油、丙二醇和添加剂热降解的加热型烟草制品特定成分中发现的有害物质方面的效率。其次，由于缺乏有关人体使用加热型烟草制品的抽吸模式信息，因此不确定在加热型烟草制品分析中应使用何种抽吸方案（标准化的抽吸容量和频率）。因加热型烟草制品设备差异，加热型烟草制品的抽吸参数受固件限制；制造商仅提供"正确"产品使用和操作的规范。因此，使用具有不同建议抽吸方案的不同设备[5]可能会导致不同研究中解吸温度和有害物质释放量的巨大差异。此外，由于热解产物生成的动力学不同于烟草材料中成分的解吸动力学，因此更高强度地抽吸加热型烟草制品可能会改变释放曲线[6]。例如，增加抽吸强度为使烟气中能产生更多的烟碱，从而提高了烟草材料的加热温度，可能会显著增加有害热解产物的释放。最后，加热型烟草制品设备或烟草填料的参比物尚不可用，因此无法进行充分的分析质量控制。

由于缺乏标准参比物、分析方法和抽吸参数，尚无法对HTP和其他烟草制品进行准确比较。由于这些原因，对这些产品使用者相对风险的一般性陈述仍然是初步的，应该谨慎使用，并且只有在认识到这种背景的情况下才可以引用。

2.2.2 烟碱

由于传统卷烟和HTP都含有烟草，因此在使用过程中都会释放烟碱。烟碱是烟草制品中已知的主要致瘾成分[7]；因此，它在HTP释放中的水平非常重要。表2.1列出了两种HTP（IQOS和glo）报告中的烟碱含量和释放量。总体而言，企业支持和独立学术研究的分析结果是一致的。表2.1还显示了参比卷烟中相应的烟碱含量，它们代表了流行的"全味"（3R4F和CM6）和通风过滤式（旧称"淡味"）传统卷烟（1R5F）。报告中的普通和薄荷醇IQOS烟支的烟草材料（含量）水平与传统卷烟相当[8,9]；然而，由于HTP烟棒比传统卷烟更短更细，因此含有较少的烟草材料，因此每支烟棒的总烟碱含量低于传统卷烟。例如，Liu等[10]报告了3种不同HTP中每支烟棒的烟碱含量，其含量范围为1.9~4.6 mg。

表2.1 IQOS和glo HTP中的烟碱含量和释放量含量报告

产品的品牌和种类	成分（烟草材料）	平均烟碱水平范围		参考文献
		释放物		
		ISO方案[a]	HCI方案[a]	
IQOS				
Regular	15.2~15.7 mg/g	0.4~0.77 mg/支 0.3 mg/14口[b]	1.1~1.5 mg/支 1.1~1.4 mg/12口	[1,6,8,9,11-13]
Menthol	15.6~17.1 mg/g	0.43 mg/支	1.2 mg/支 1.38 mg/12口	[6,8,9]
Mint	NR	0.32 mg/支	1.2 mg/支	[6]
Essence	NR	NR	1.14 mg/支	[4]
Glo				
Regular	NR	0.07 mg/支	0.27 mg/支	[6]
Bright tobacco	NR	0.09~0.15 mg/支	0.31~0.57 mg/支	[1,4,6]

续表

产品的品牌和种类	成分（烟草材料）	平均烟碱水平范围		参考文献
		释放物		
		ISO方案[a]	HCI方案[a]	
Fresh mix	NR	0.14 mg/支	0.51 mg/支	[6]
Intensely fresh	NR	0.13~0.15 mg/支	0.36~0.51 mg/支	[1,4,6]
Coolar green	NR	0.068 mg/支	0.17 mg/支	[6]
Coolar purple	NR	0.06 mg/支	0.25 mg/支	[6]
参比传统卷烟				
3R4F	15.9~19.7 mg/支	0.73~0.76 mg/支	1.4~2.1 mg/支	[1,6,12,14-16]
1R5F	15.9~17.2 mg/支	0.12 mg/支	1.0~1.1 mg/支	[6,8,15,16]
CM6	18.7 mg/支	(1.2 ± 0.13) mg/支	2.6~2.73 mg/支	[6,15,16]

注：NR，没有报道

a. 用于抽吸HTP进行释放分析的吸烟机方案：ISO，国际标准化组织（ISO）方案；HCI，加拿大卫生部深度抽吸方案

b. 基于使用相应的机器抽吸模式，消费一支HTP烟草棒所需的抽吸口数

HTP释放数据也显示不同品牌之间存在显著差异。例如，在ISO或深度抽吸条件下抽吸的IQOS气溶胶中的烟碱水平与传统卷烟中的烟碱水平相当（表2.1）。这表明IQOS的烟碱解吸或从烟草材料转移到气溶胶或烟气的效率比传统卷烟高得多。相比之下，据报道，glo气溶胶中的烟碱含量约为IQOS的40%和参比卷烟的23%[17]。其他研究表明两种产品之间存在类似差异（表2.1）。这些结果揭示系统监测和报告HTP中烟碱含量和释放的重要性。应研究这些差异对产品滥用倾向和有害性的影响。

2.2.3 其他有害物质

HTP释放物中有害化学物质的浓度通常低于传统卷烟，因为其工作温度较低，这也是传统卷烟烟气中许多有害物质的主要来源。在HTP气溶胶中测量的大多数有害和潜在有害成分的水平低于参比卷烟烟气中的水平[1,2,18-22]，但缩水甘油含量较高[23]。然而，独立研究和制造商资助的研究表明，即使HTP达到的温度不足以燃烧，但足以形成有害物质。Davis等[24]报告表明IQOS烟草填料在没有点火的情况下会出现焦化，并且每次使用后不清洁IQOS时，焦化会增加。在另一项研究[3]中也发现了一些燃烧迹象。Auer等[11]发现IQOS释放的气溶胶含有来自热解和热降解的元素，这些元素与传统卷烟烟气的有害成分相同。

英美烟草公司研究人员的一项研究表明，由于烟草混合物的蒸发转移或初始热分解，烟碱和一些卷烟烟气化合物在100~200℃之间释放。需要注意的是，加热到200℃的烟草比燃烧的卷烟产生释放物的时间长得多。随着温度的升

高，一些分析物的含量逐渐增加：在180~200℃，一氧化碳（CO）、巴豆醛和甲基乙基酮的含量为2倍，甲醛的含量为2倍；120~200℃，乙醛含量增加15倍；在160~200℃，丙酮为2倍，丙醛为3倍；而在140~200℃，丁醛的浓度增加了1倍。这些化学物质可以通过碳水化合物和烟草结构聚合物的热解形成。烟草特有亚硝胺（TSNA）是可量化的，但在不同温度下变化不同[25]。

保润剂主要对HTP产生的气溶胶的总颗粒物有贡献。在HTP气溶胶中发现的保润剂（如甘油）含量比传统卷烟烟气中高得多[4]。独立研究发现，与传统卷烟的18 μg/支相比，HCI抽吸模式中HTP产生的甘油含量约为IQOS 360 μg/支，glo 520 μg/支，PloomTECH 5900 μg/支。除水、丙二醇、甘油和丙酮醇外，HTP产生的化合物比传统卷烟少[5]。原型HTP THS2.2的总颗粒物主要由甘油（56.3%）和丙二醇组成。

尽管甘油和丙二醇对人体可能是安全的，但在IQOS和电子烟碱传输系统释放的研究中发现它们在加热时会产生有害物质，包括丙烯醛（一种强烈的呼吸道刺激物）和缩水甘油（一种致癌物）。据报道，这些羰基化合物是电子烟碱传输系统中丙二醇和甘油热分解的副产物。由于HTP含有大量甘油，其降解副产物存在于IQOS释放物中[26]。

一氧化碳

在独立研究和烟草企业研究中，发现CO是HTP释放物中不完全燃烧的产物。

企业研究

Forster等[1]报告称，当温度<180℃时，CO释放量低于报告限值，比传统卷烟释放量减少了99%以上。高于此温度（至200℃），CO含量随温度升高而增加。HTP的CO释放量低于检测限，而传统卷烟的CO释放量为31.2 mg/支[2]。调味物质的存在与CO含量之间没有相关性：调味和未调味的HTP都产生CO≤0.22 mg/支，而传统卷烟产生CO 32.8 mg/支[23]。

学术研究

在IQOS气溶胶[11]中发现了CO，尽管温度仅为330℃，但其含量低于传统卷烟的主流烟气，而传统卷烟烟气的温度为684℃。IQOS释放的CO浓度用官方的WHO TobLabNet方法测量，大约是传统卷烟释放浓度的百分之一[8]。在ISO和HCI抽吸模式中发现了类似的结果，THS2.2释放的CO比传统卷烟少90%[3]。

烟草特有亚硝胺

HTP的制造商声称HTP中烟草特有亚硝胺（TSNA）的含量较低。几项研究报告称,电子烟烟液和HTP中的TSNA水平明显低于传统卷烟[27-29]。与制造商资助的研究相比,独立研究报告的焦油含量更低,TSNA含量更高[17]。

企业研究

PMI和BAT报告称,与传统卷烟的主流烟气相比,HTP气溶胶中TSNA的水平平均降低了90%。降低的主要原因是较低的蒸发迁移和工作温度,减少了热合成和热释放[27]。据报道,原型HTP,THP1.0的TSNA释放量比传统卷烟烟气中的释放量减少了80%~98%[1]。

学术研究

非企业的学术研究人员表明,HTP气溶胶中的TSNA水平低于传统卷烟烟气中的TSNA[30,31],与HTP气溶胶相比,卷烟烟气每次抽吸中的单口TSNA减少8~22倍。HTP HeatStick气溶胶减少7~17倍。IQOS气溶胶中每口TSNA的释放量比传统卷烟烟气中的低一个数量级,但比电子烟碱传输系统气溶胶中的高一个数量级[32,33]。其他独立研究证实,IQOS烟草材料和主流烟气中TSNA的水平显著低于传统卷烟,尽管N'-亚硝基降烟碱(NNN)、N'-亚硝基新烟碱(NAT)和4-(甲基亚硝胺基)-1-(3-吡啶基)-1-丁酮(NNK)在IQOS中的迁移率略高于传统卷烟[8,13]。Ratajczak等[28]报告称,TSNA的浓度是传统卷烟的五分之一。Li等[3]发现,在ISO和HCI两种抽吸模式下,NNN、NNK和NAT的释放量比传统卷烟少92%,N'-亚硝基假木贼碱比3R4F释放量少72%。

羰基化合物

HTP会释放出由热解产生的有毒羰基化合物,其含量随着温度(从160℃或180℃升高到200℃)升高而逐渐增加[8,25]。电子烟碱传输系统中的甘油和丙二醇混合物在加热过程中形成甲醛、乙醛、丙烯醛和其他醛类[34,35]。由于这些保润剂也存在于HTP中,同样会形成醛类化合物[36]。然而,根据制造商资助和独立的研究,在大多数情况下,该物质含量水平低于传统卷烟烟气。

企业研究

醛类占THP1.0释放物中成分总估计浓度的41%[37]。Crooks等[23]测量了调味和未调味的HTP以及传统卷烟烟气中的气溶胶中某些醛的含量。发现调味HTP气溶胶中的甲醛含量为1.52 μg/支,未调味HTP气溶胶中的甲醛含量为1.79 μg/支,传统卷烟烟气中的甲醛含量为66.67 μg/支;调味HTP气溶胶中乙醛含量为35.48 μg/支,未调味HTP的气溶胶中乙醛含量为35.54 μg/支,传统卷烟的烟气中乙醛含量为2164.73 μg/支。据报道,与传统卷烟气相比,HTP气溶胶中的甲醛减少90%[2,18]。

甲醛（16.3 μg/m³）和乙醛（12.4 μg/m³）均在住宅和公共环境的平均浓度范围内[19]。

学术研究

非企业研究人员的研究证实，IQOS的羰基释放量低于传统卷烟，但高于电子烟碱传输系统[26]。HTP气溶胶中的甲醛含量比传统卷烟烟气中的甲醛含量低91.6%，乙醛降低84.9%，丙烯醛降低90.6%，丙醛降低89%，巴豆醛降低95.3%。在更深度的抽吸方案中，除了甲醛水平比HCI抽吸方案增加了3~4倍，常规IQOS从6.4 μg/支增加到17.1 μg/支之外，IQOS和传统产品之间观察到的羰基释放量差异很小。HTP气溶胶中的羰基含量高于电子烟碱传输系统气溶胶[38]。Ruprecht等[39]也得到了类似的结果，观察到使用IQOS形成的丙烯醛水平比传统卷烟的丙烯醛高2%，乙醛高6%，甲醛高7%；电子烟碱传输系统产生的醛量仅为传统卷烟的1%。其他作者观察到羰基化合物的释放量比传统卷烟的释放量低80%~96%[12,13,30,40-43]。

Uchiyama等[5]比较了HCI抽吸模式下不同HTP的乙醛释放量，发现IQOS产生的量为210 μg/支，glo™产生的量为250 μg/支，PloomTECH产生的量为0.45 μg/支，而传统卷烟产生的量为1300 μg/支。Salman等[41]预估了使用IQOS代替传统卷烟后，每日甲醛和乙醛的摄入量分别减少了70%和65%。Li等[3]在ISO模式下，观察到甲醛和乙醛分别减少了55.80%和77.34%。

氰化物甲醛可由HTP中聚合物滤嘴的热解形成[42]。这种薄塑料片在IQOS使用过程中会熔化，释放出甲醛氰醇[24]。

苯并[a]芘和其他多环芳烃

苯并[a]芘和其他多环芳烃（PAH）是不完全燃烧的典型产物。因为是致癌物，故其测定很重要[13]。

企业研究

制造商资助的研究报告称，HTP中包括苯并[a]芘的多环芳烃含量低于传统卷烟。HTP气溶胶中的多环芳烃、芳香胺、酚类和醛类的形成减少75%以上[22]，芳香族、脂环族和单环芳烃占比小于4%，而传统卷烟主流烟气中的这一比例为64%[37]。与传统卷烟主流烟气相比，所有HTP产品的气溶胶中苯并[a]芘的含量都非常低[18]。

Takahashi等[2]发现HTP气溶胶中苯并[a]芘含量小于0.531 ng，而传统卷烟烟气中其含量小于12.9 ng。Crooks等[23]发现在调味HTP的气溶胶中苯并[a]芘含量为0.44 ng/支，未调味HTP的含量为0.41 ng/支，传统卷烟烟气为12.76 ng/支，表明这些苯并[a]芘含量与调味物质的存在无关。

热解薄荷醇可能是薄荷卷烟产品烟气中苯并[a]芘的前体，尽管在THS2.2中没有观察到薄荷醇对苯并[a]芘的释放量有显著贡献[20]。

学术研究

Auer等[11]发现HTP释放物中苯并[a]芘含量为0.8 ng/支，传统卷烟的烟气中其含量为20 ng/支；这些含量高于制造商资助的研究所发现的水平。HTP比传统卷烟释放出更高水平的芘。St Helen等[12]发现的水平与烟草企业研究中的水平相似，HTP气溶胶中苯并[a]芘含量为0.736 ng/支，传统卷烟烟气中含量为13.3 ng/支，这表明苯并[a]芘减少了94%。

其他有害物质

HTP气溶胶和传统卷烟烟气粒相的大部分化学成分是含氧化合物，分别占卷烟和HTP颗粒相中分析物总估计浓度的39%和70%。HTP颗粒相中含氧化合物的含量较高，可能是因为含有大量甘油和其他保润剂[43]。

企业研究

含氮化合物。这些化合物在HTP气溶胶中的含量比传统卷烟烟气中的含量低12%，分别占卷烟和HTP颗粒相中分析物总估计浓度的58%和29%[43]。传统卷烟烟气中的氮氧化物水平随时间增加，但在HTP气溶胶中保持不变。HTP气溶胶的含量水平是传统卷烟烟气的5.5%~7.3%[21]。

制造商资助的研究报告称，HTP中氨和其他一些有害物质的含量降低了25%~50%[20]。HTP释放物中氨的含量比传统卷烟烟气中低88%[1]。Crooks等[23]报告称，调味HTP气溶胶中氨、氮氧化物和邻甲酚的含量高于未调味HTP的Neostik。

金属。在一项研究中，HTP释放物中的汞含量比卷烟主流烟气中的汞含量低69%，而其他研究报告称减少了25%~50%[1,20]。铬、镍和硒的含量低于参比卷烟烟气中的检测限[2]。

挥发性有机化合物。在一些研究中，HTP释放物中某些挥发性碳氢化合物的含量比传统卷烟烟气中的含量低97%~99%[13,40]。企业研究人员报告称，HTP原型产品THP1.0的气溶胶中霍夫曼清单挥发性化合物的浓度显著低于传统卷烟烟气中的浓度，甲苯含量降低99%，丙酮含量降低91%[37]。HTP气溶胶中苯含量为0.93 $\mu g/m^3$，电子烟气溶胶中苯含量更低。另外，HTP气溶胶中未检出甲苯，而常规卷烟的甲苯含量为151.1 $\mu g/m^3$[36]。

其他有害物质。除甲醛、丙酮和氨外，许多HTP释放成分的水平低于定量或检测水平[2]。醛、酮和杂环化合物分别占分析物总浓度的41%、32%和10%。传统卷烟主流烟气中的丙酮（152 μg/支）高于HTP气溶胶（13.3 μg/支），而HTP气溶胶中吡啶和二甲基三硫化物的水平略高[37]。Forster等[1]计算出酚类（间苯二酚、对甲酚和咖啡酸除外）的含量减少了96%~99%，环氧乙烷和环氧丙烷的含量减少了99%；然而，与传统卷烟烟气相比，丙酮和丙酮醛在HTP释放物中的含量更高。

颗粒物。制造商资助的研究发现，HTP释放物中PM$_{2.5}$的颗粒物水平比传统卷烟烟气中的水平低28倍[19]。HTP的总颗粒物释放量大约是传统卷烟的两倍。HTP中总颗粒物中的水和保润剂含量高于传统卷烟：HTP为90%（质量分数），传统卷烟为37%[2]。PMI研究表明，HTP气溶胶中颗粒的可吸入部分比传统卷烟烟气低90%[44]，相当于空气中的浓度；然而，该限值低于方法的检测范围下限[45]。

学术研究

活性氧类。独立研究报告称，HTP释放物中的总活性氧水平比传统卷烟低85%[41]。在使用IQOS期间释放活性氧可能是有害的[46]。

金属。独立研究发现HTP释放物中的金属含量低于卷烟烟气中的金属含量[13,42]。Ruprecht等[39]观察到，与不含薄荷醇的IQOS相比，含薄荷醇IQOS的金属浓度更高。在IQOS中发现了传统卷烟中不存在的铝、钛、锶、钼、锡和锑等金属，以及IQOS中检测到传统卷烟中没有的镍、铜、锌、镧和铅等金属。一篇文章报道了一例使用HTP的罕见汞中毒案例，它被添加到烟草棒中导致受害者吸入雾化的汞[47]。

挥发性有机化合物。在IQOS的主流释放物中检测到70多种挥发性有机化合物，包括异戊二烯、丙烯腈、甲酚、苯、苯酚、萘、乙醛、丙醛、丙烯醛、甲醛、2-丁酮、丙酮、巴豆醛和喹啉[26]。所有这些化合物都被FDA认为是潜在有害的。然而，一些研究发现，HTP气溶胶中某些挥发性碳氢化合物的含量比传统卷烟烟气中的含量低97%~99%[13,40]。在IQOS侧流气溶胶中未发现多环芳烃[42]。加热的烟草材料比传统卷烟释放物含有更多类型的挥发性有机化合物[48]。

黑碳。Ruprecht等[39]发现传统卷烟中黑碳浓度最高，有机化合物为78 μg/m³，元素碳为2.3 μg/m³。在IQOS的抽吸过程中检测不到元素碳，而有机化合物的含量是传统卷烟的2.81%~3.89%。

其他有害物质。在一项独立研究中获得了与制造商资助的研究相似的结果，除了羰基化合物、氨和N'-亚硝基假木贼碱外，大多数有害物质的水平降低了90%。特别是HCI抽吸模式的氨水平比传统卷烟烟气低63.4%[3]。

颗粒物。IQOS主流气溶胶中的颗粒浓度低于电子烟碱传输系统和传统卷烟的释放物[49]，使用HTP导致细颗粒物含量最低[36]。PM>0.1和PM>0.3的浓度在传统卷烟和IQOS的气溶胶中显著（对PM>0.3），但与传统卷烟相比，电子烟碱传输系统的气溶胶中的浓度甚微[39]。其他研究发现，电子烟碱传输系统和HTP释放物中的颗粒物水平约为卷烟烟气中的25%。气溶胶小于1000 nm，比更低的质量更安全。在glo和传统卷烟烟气中，颗粒的可吸入部分更高[50]。

由于这些结果表明IQOS加热卷烟的成分与传统卷烟的成分不同，包括调味物质和其他添加剂，因此IQOS气溶胶可能含有烟草烟气中不存在的其他化学成分，这些化学成分尚未在非靶向分析中进行研究[12]。

2.3 加热型烟草制品的体外暴露和实验动物暴露及影响

2.3.1 体外研究

企业研究

PMI和BAT的研究小组已经发表了几篇关于体外毒理学研究结果的报告，这些研究使用的原型HTP在装置和烟草特性上不同。大多数报告表明HTP气溶胶的细胞毒性明显低于参比卷烟烟气。

PMI研究。PMI发表了一系列关于在ISO抽吸条件和另一种抽吸模式下的原型电加热吸烟系统的细胞毒性（在中性红摄取试验中测量）和致突变性（沙门氏菌回复突变试验）[51-53]的论文。报告表明，与1R4F参比卷烟的烟气相比，这些装置产生的气溶胶的细胞毒性降低了40%，致突变性降低了90%（基于总颗粒物释放量的比较）。在2012年发表的一项研究中，PMI报告了对同一原型系统的毒理学评估，其中包含25种反映人类吸烟行为的其他吸烟方案[54]。虽然测试的HTP的整体生物活性低于参比卷烟，但吸烟强度的增加导致细胞毒性显著增加，细菌致突变性比在低强度ISO条件下产生的HTP气溶胶高出近36倍。得出的结论是，增加可能是由于有害成分释放的变化，这表明基于ISO的测试结果在对人类的潜在影响方面没有提供信息。

PMI还报告了另一个HTP原型THS2.2气溶胶的生物效应。在一项研究中，研究了THS2.2气溶胶抑制单胺氧化酶活性的能力，以评估该产品的潜在滥用倾向[55]。表明3R4F参比卷烟烟气显著抑制活性，但THS2.2没有。在另一项研究中，对暴露于THS2.2气溶胶或3R4F参比卷烟烟气（烟碱浓度相似）后不同时间的人体器官支气管上皮培养物的组织学、细胞毒性、分泌的细胞因子和趋化因子、纤毛搏动和全基因组mRNA/miRNA谱进行了评估[56]。暴露于THS2.2气溶胶4小时后，细胞命运、细胞增殖、细胞应激和炎症网络模型仅为暴露于3R4F烟气后观察到的模型的7.6%。即使暴露于THS2.2气溶胶中的烟碱浓度是3R4F烟气中的3倍，但细胞形态学未发生变化。在源自口腔和牙龈上皮[57]、小气道和鼻上皮细胞[58]以及内皮细胞[59]的人体器官培养物中，对另一种HTP原型CHTP1.2进行了类似的研究。细胞急性（28分钟）或反复暴露（28分钟/天，持续3天）于CHTP1.2气溶胶或3R4F烟气。报告表明，与暴露于3R4F烟气后相比，暴露于CHTP1.2后细胞毒性小，病理生理学改变少，毒理学和炎症生物标志物变化也更少。然而，在暴露于CHTP1.2气溶胶的小气道和鼻上皮培养物中检测到mRNA表达的改变。

BAT研究。BAT研究人员评估了IQOS和glo气溶胶的H292人支气管上皮细胞的烟碱传递和细胞毒性，并与3R4F卷烟的烟气进行比较；所有产品均在HCI[60]模式下制备。报告称，与3R4F卷烟相比，HTP的烟碱传递更多，但细胞毒性更低。还表明HTP之间没有区别；然而，报告中的数据表明IQOS比glo产品更具细胞毒性。在另一项研究中，RNA测序方法比较了3D呼吸道组织急性暴露于相同产品气溶胶后的转录组扰动[61]。与空气对照相比，IQOS中有115个RNA的差异表达，glo产品中有两个RNA的差异表达，而3R4F中有2809个RNA的差异表达。对报告中的数据和图表的检查表明，结果取决于数据分析设置的阈值（即P值和表达的倍数变化），并且炎症和外源代谢基因表达可能会受到测试的HTP的影响。随后进行了一项研究以补充这两项评估，并确定调味对Neostik体外反应的影响[62]。结果表明，添加调味物质不会改变未调味Neostik的体外基线反应。

学术研究

Leigh等[63]使用气液界面与人支气管上皮细胞（H292）的体外模型研究了吸入IQOS、电子烟碱传输系统和传统卷烟释放物的毒性作用。调整每种产品的抽吸口数以实现对细胞的类似烟碱递送，并在中性红摄取和台盼蓝试验中测量细胞毒性。IQOS的中性红试验结果表明，其细胞毒性高于空气对照，低于传统卷烟烟气。在另一项研究中，用8种不同细胞类型的3种检测方法比较了IQOS气溶胶与万宝路红和3R4F参比卷烟的烟气[64]的细胞毒性。转化小鼠NIH/3T3成纤维细胞的结果与PMI先前报道的结果相似[51-53]；但是在其他类型的细胞和更高浓度的气溶胶中进行的评估表明，IOQS和卷烟烟气溶液对线粒体和溶酶体活性的抑制相当。然而，本研究没有对不同产品中的烟碱含量进行调整，并且所使用的抽吸模式没有详细的描述。

在另一项研究中，研究了暴露于IQOS、电子烟碱传输系统气溶胶和传统卷烟烟气的培养基对3T3-L1前脂肪细胞活力和向米色脂肪细胞分化的影响[65]。作者得出结论，与空气对照相比，暴露于IQOS的影响有限或没有影响。对本报告中的数据的检查表明，在分化结束（10天）经IQOS处理的细胞中，脂肪生成标志物Ppar-γ和抵抗素的表达在统计学上显著降低，这与卷烟烟气处理的细胞中所观察到的效果相似。值得注意的是，本报告的作者有与企业相关研究资助的历史。

2.3.2 实验动物研究

企业研究

大多数关于HTP对实验动物影响的报告已由烟草企业研究小组发表，即PMI和日本烟草，它们报告的HTP气溶胶的有害性和致癌性通常比参比卷烟的烟气低得多。

JTI研究。已经发表了两项关于原型"加热"卷烟对小鼠模型皮肤致瘤性和吸入毒性影响的研究。在第一项研究中，研究了局部应用7,12-二甲基苯并[a]蒽作为肿瘤引发剂后，HTP对雌性SENCAR小鼠皮肤肿瘤促进的影响[66]。在改进的HCI吸烟方案下产生的HTP气溶胶或卷烟烟气（3R4F卷烟）的冷凝物以5个不同剂量（最高30 mg）焦油重复应用30周。在≤15 mg焦油时，与卷烟烟气处理的动物相比，HTP气溶胶处理的动物发生肿瘤的潜伏期更长，发生率和多样性更低，炎症和鳞状上皮增生的发生率也更低。然而，在处理结束时，这些影响比空白对照动物更为普遍。在剂量≥22.5 mg焦油时，HTP气溶胶和3R4F冷凝物之间的差异不太清楚。

在第二项工作中，对上述HTP在5周和13周的口鼻吸入实验进行了研究[67]。在HTP处理的动物的呼吸道中发现了比卷烟烟气处理动物更少的组织病理学变化（鼻腔中的呼吸道上皮增生和肺泡色素巨噬细胞积累）和肺部炎症（通过支气管肺泡灌洗液中的中性粒细胞百分比和γ-谷氨酰转肽酶、碱性磷酸酶和乳酸脱氢酶的活性来衡量）。然而，与空气对照相比，HTP处理对组织病理学结果有显著影响，因为HTP处理的动物在两种处理方案中的所有剂量下都显示出100%的喉部和会厌增生及角化过度发生率，鼻上皮、腹侧袋和肺组织的变化发生率显著增加（≤100%）。

PMI研究。在一项为期90天的研究中，大鼠通过口鼻吸入相同或两倍于吸入区浓度的烟碱，与3R4F卷烟烟气相比，CHTP1.2气溶胶导致的有害成分暴露显著减少，呼吸道刺激以及全身病理也减少[68]。该研究的毒理学部分包括转录组学、蛋白质组学和脂质组学分析[69]，表明CHTP1.2气溶胶在呼吸道鼻上皮和肺中的炎症和细胞应激反应比3R4F烟气弱得多。许多这些效应显示出剂量-反应关系。CHTP1.2气溶胶还可以降低暴露动物血清中的脂质浓度。

PMI研究人员还在ApoE-/-小鼠模型中进行了为期6个月的暴露研究，以研究THS2.2和CHTP1.2对心肺系统的影响。一项结合生理学、组织学和分子测量的系统毒理学方法的报告，证明3R4F对心肺系统的影响较低，肺部炎症或肺气肿变化很少甚至没有，动脉粥样硬化斑块也很少形成[70]。在另一项研究中，通过超声心动图、组织病理学、免疫组织化学和转录组学分析研究了心血管效应。表明，持续暴露于HTP气溶胶下不会影响动脉粥样硬化的进展、心脏功能、左心室结构或心血管转录组[71]。然而，对这些出版物中的数据进行审查后发现，与对照组相比，HTP处理的动物的许多结果指标均增加，尽管这些增加没有达到统计显著性。

学术研究

已知传统卷烟烟气会损害血管内皮功能[72]，在一项暴露于IQOS气溶胶是否会影响血管内皮功能的研究中，大鼠通过口鼻暴露于单个HeatStick产生的IQOS气溶胶、万宝路红卷烟的主流烟气或清洁空气中。由于动脉血流介导的扩张是血管内皮功能的衡量标准，因此在暴露前后对大鼠进行了动脉血流介导扩张的测量。报道称，暴露于IQOS气溶胶和传统卷烟烟气中，血流介导的扩张在相当程度上受到损害，而在清洁空气中没有观察到影响。暴露于IQOS的大鼠的血清烟碱和可替宁水平比暴露于卷烟的大鼠高约4.5倍，尽管IQOS气溶胶中的烟碱含量低于卷烟烟气。短时间暴露于IQOS和卷烟组的血清烟碱水平相似，导致血流介导的舒张功能受损。作者得出结论，使用IQOS并不一定能避免吸烟对心血管的不利影响。

2.4 加热型烟草制品中有害物质的人体暴露及健康影响

2.4.1 产品用途和使用模式

Davis等[24]使用两种不同的抽吸装置评估了IQOS在五种条件下的性能。在使用前后检查加热棒以确定烟草材料热解（炭化）和聚合物膜滤嘴熔化的迹象，并评估清洁对炭化的影响。烟草材料在使用后便发生炭化，气相色谱-质谱顶空分析结果也表明聚合物膜滤嘴在90℃（比正常IQOS使用期间的温度低）释放剧毒的甲醛氰醇。在连续使用HeatSticks期间未清洁设备时，观察到炭化和甲醛氰醇释放量增加。作者得出结论，该设备的局限性（即短烟草棒和在吸完一定数量的烟后关闭设备）可能会导致抽吸间隔时间缩短，这可能增加使用者对烟碱和其他有害化学物质的摄入。

一些企业研究承认使用方式和设备特性对使用者暴露的潜在影响。例如，BAT的研究人员调查了抽吸参数对释放量的影响[73]，发现抽吸参数的选择会影响释放量，HTP类型之间存在显著差异。他们建议应该研究HTP详细抽吸方式，以确定最适合实验室测试的抽吸参数。在另一项研究中，PMI研究人员提出了一种名为"烟碱桥接"的建模方法，以估计人类暴露于HTP释放量[74]。该方法涉及确定多种机器吸烟方案的有害成分和体外毒性参数对烟碱的回归；在一项临床研究中，烟碱吸收的分布是根据24小时烟碱代谢物的排泄量确定的。该方法用原型HTP[54]的数据进行了说明，该数据表明HTP使用者的暴露量低于传统吸烟者，分布曲线之间几乎没有重叠。作者提出，该方法可用于将暴露分布外推到没有特定生物标志物的烟气成分。由于这种方法依赖于不代表人类暴露的机器吸烟协议，因此推断可能不合理，需要其他的独立研究。

2.4.2 暴露和效应生物标志物

企业研究

有关细胞培养和动物模型的研究，很多关于生物标志物的报告都是由企业提供的，尤其是PMI和BAT。

PMI研究。PMI研究人员报告了吸烟者转用原型HTP的暴露和效应生物标志物，其中一些已经在体外和实验动物中进行了测试，如上所述。原型HTP在IQOS发布之前由PMI开发并测试了十多年，是当代HTP的前身[75]。2012年的一系列出版物报告了关于使用K型原型HTP（EHCSS-K3和EHCSS-K6）的随机、对照、开放标签、平行组、单中心研究。在英国的一项研究中，对160名由万宝路卷烟转用EHCSS-K3或EHCSS-K原型HTP的男性和女性吸烟者进行为期8天的9种暴露卷烟烟气成分和尿致突变性生物标志物的测量[76]。在基线和第8天之间，所有生物标志物测量值均出现统计学意义上的平均降低（$P \leqslant 0.05$），同时两种HTP的吸烟者的尿致突变性均降低。在韩国的一项类似研究中，对72名由在基线时吸食低量LarkOne卷烟改为EHCSS-K3原型HTP的男性和女性受试者进行为期8天的12种选定暴露成分和尿致突变性生物标志物的测量[77]。对于EHCSS-K3组，在基线和第8天之间发现尿致突变性和10个组成生物标志物测量值在统计学意义上显著降低。在日本的一项研究中，对128名在基线时吸食万宝路卷烟并转用EHCSS-K3或EHCSS-K6的男性和女性参与者进行了为期8天的暴露卷烟烟气成分和尿致突变性生物标志物的测量[78]。对于转用HTP的参与者，基线和第8天之间所有成分和尿致突变性测量值均出现统计学意义上的平均下降（$P \leqslant 0.05$）。另一份关于日本吸烟者的报告提供了102名抽吸原型HTP EHCSS-K6(M)男性和女性受试者为其6天的12种成分暴露生物标志物、尿致突变性和血清club棒状细胞16 kDa蛋白（CC16）数据[79]。与之前的研究一样，12种卷烟烟气成分中的10种暴露和尿致突变性在统计上显著降低；然而，肺上皮损伤的指标（血清CC16）在所有组中都没有变化，包括不使用产品的组。尽管转用HTP后暴露措施有所减少，但这些研究中的生物标志物水平仍显著高于指定戒烟和不使用任何产品的组（表2.2）。另一项针对波兰吸烟者的包括与心血管风险相关的广泛生物标志物，以及暴露的生物标志物测量的PMI研究中[80]，316名男性和女性吸烟者被随机分配继续抽吸传统卷烟或改吸EHCSS-K6原型HTP 1个月。尽管暴露的生物标志物减少，但大多数心血管生物标志物在使用HTP设备后没有改变，这与之前的研究结果相一致。使用EHCSS-K6

一个月后，高密度脂蛋白胆固醇平均显著增加，从基线中位值36.7 ng/mL（95% CI 2.1，3410）上升到59.0 ng/mL（95% CI 2.1，4535）；然而，在EHCSS-K6组中红细胞计数、血红蛋白和血细胞比容均减少。

表2.2 转用HTP的吸烟者和指定持续吸烟或戒除所有烟草制品的吸烟者的暴露生物标志物水平报告

暴露的生物标志物[来源]	原型HTP（使用时间）	转用HTP后的水平（平均值±SD）	转用其他抽吸方案的受试者的生物标志物水平比较（↑升高，↓降低）		参考文献
			HTP vs 持续抽烟	HTP vs 戒烟	
SPMA, μg/24 h [苯]	EHCSS-K3 (8天)	1.26 ± 1.26	↓80%	↑473%	[66]
	EHCSS-K3 (8天)	1.63 ± 0.55	↓66%	↑19%	[77]
	EHCSS-K3 (8天)	0.48 ± 0.28	↓78%	↑41%	[68]
	EHCSS-K6 (8天)	0.86 ± 0.81	↓86%	↑291%	[66]
	EHCSS-K6 (8天)	0.57 ± 0.57	↓75%	↑68%	[68]
	EHCSS-K6^M (6天)	0.35 ± 0.13	↓85%	↓34%	[69]
MHBMA, μg/24 h [1,3-丁二烯]	EHCSS-K3 (8天)	2.63 ± 2.78	↓49%	↑874%	[66]
	EHCSS-K3 (8天)	0.59 ± 0.58	↓75%	↑111%	[67]
	EHCSS-K3 (8天)	0.66 ± 0.69	↓57%	↑50%	[68]
	EHCSS-K6 (8天)	1.54 ± 1.52	↓70%	↑470%	[66]
	EHCSS-K6 (8天)	0.74 ± 0.56	↓51%	↑68%	[68]
	EHCSS-K6^M (6天)	0.61 ± 1.06	↓58%	↑69%	[69]
3-HPMA, mg/24 h [丙烯醛]	EHCSS-K3 (8天)	1.15 ± 0.74	↓37%	↑140%	[66]
	EHCSS-K3 (8天)	2.41 ± 0.67	↓18%	↑44%	[67]
	EHCSS-K3 (8天)	1.01 ± 0.36	↓26%	↑110%	[68]
	EHCSS-K6 (8天)	1.28 ± 0.87	↓30%	↑167%	[66]
	EHCSS-K6 (8天)	1.09 ± 0.51	↓20%	↓127%	[68]
	EHCSS-K6^M (6天)	0.83 ± 0.32	↓30%	↑110%	[69]
3-HMPMA, mg/24 h [巴豆醛]	EHCSS-K3 (8天)	2.59 ± 1.90	↓50%	↑49%	[66]
	EHCSS-K3 (8天)	1.99 ± 1.07	↓18%	↑20%	[67]
	EHCSS-K3 (8天)	0.61 ± 0.17	↓52%	↑30%	[68]
	EHCSS-K6 (8天)	2.62 ± 1.29	↓49%	↑51%	[66]
	EHCSS-K6 (8天)	0.63 ± 0.15	↓51%	↑34%	[68]
	EHCSS-K6^M (6天)	0.19 ± 0.08	↓85%	↓60%	[69]
总NNAL, ng/24 h [NNK]	EHCSS-K3 (8天)	104.3 ± 55.3	↓65%	↑76%	[66]
	EHCSS-K3 (8天)	80.4 ± 59.3	↓59%	↑78%	[67]
	EHCSS-K3 (8天)	102 ± 48	↓53%	↑23%	[68]
	EHCSS-K6 (8天)	100.6 ± 68.8	↓66%	↑70%	[66]
	EHCSS-K6 (8天)	100 ± 65	↓54%	↑20%	[68]
	EHCSS-K6^M (6天)	95 ± 53	↓49%	↑19%	[69]
1-HOP, ng/24 h [芘,多环芳烃]	EHCSS-K3 (8天)	73.1 ± 30.4	↓60%	↓3%	[66]
	EHCSS-K3 (8天)	143.62 ± 76.08	↓40%	↑13%	[67]
	EHCSS-K3 (8天)	59.0 ± 35.3	↓56%	↑40%	[68]
	EHCSS-K6 (8天)	71.9 ± 38.8	↓60%	↓4%	[66]
	EHCSS-K6 (8天)	56.0 ± 28.4	↓59%	↑33%	[68]
	EHCSS-K6^M (6天)	38.4 ± 22.7	↓64%	↑2%	[69]

注：1-HOP, 1-羟基芘；3-HPMA, 3-羟丙基硫醇酸；MHBMA, 单羟基丁基硫醇酸；NNAL, 4-(甲基亚硝胺基)-1-(3-吡啶基)-1-丁醇；SPMA, S-苯巯基尿酸

PMI发表了一系列关于波兰吸烟者转用原型HTP的研究报告。在一项研究中，对112名男性和女性成年吸烟者进行了有害暴露评估，他们转用碳加热型烟草制品MD2-E7原型，持续5天吸烟或戒烟[81]。还研究了HTP使用期间的抽吸模式。转用HTP后，吸烟强度和烟碱摄入量增加。然而，尽管采用更深度的抽吸模式，但转用原型HTP会暴露和泌尿诱变的生物标志物测量值下降，这与之前的报告一致。在一项针对转用THS2.1原型为期5天的吸烟者的研究中，吸烟的持续时间和频率比传统卷烟的基线模式增加，在研究期间[82]，使用HTP增加了27%。然而，在研究结束时，总吸入量恢复到基线，与基线的暴露烟碱的平均量相当。在THS2.2原型[83]的更多研究中也观察到了产品消耗和总抽吸容量的增加。平均抽吸持续时间长约32%，总抽吸持续时间长约37%，第4天抽吸频率比传统卷烟在基线上吸烟的频率高32%左右。虽然作者声称烟碱摄入量没有变化，但对数据的审查显示，在基线与第4天之间，烟碱生物标志物（总烟碱当量）增加了约23%。虽然在这些研究中一致观察到有害成分的暴露减少，但更高强度的HTP抽吸模式表明，为了准确评估转用HTP的吸烟者接触有害物质的减少，还需要进行更多、更长的研究。

据报道，对美国984名成年吸烟者进行了一项长期的研究[84]。参与者被随机抽调转用THS2.2装置或继续吸烟6个月，并评估了暴露和效应生物标志物小组的变化。据报道，与继续吸烟者相比，转用THS2.2装置吸烟者的暴露生物标志物和4种效应生物标志物（高密度脂蛋白胆固醇、白细胞计数、1秒的强迫呼气量占预测百分比和碳氧血红蛋白）有所下降。在转用HTP的吸烟者中，有30%是传统卷烟和HTP的双重使用者。在主要使用HTP（平均每天使用16.5根烟棒和两支传统卷烟）的人群中，暴露生物标志物的减少幅度为基线吸烟水平的16%~49%之间，这低于之前短期时间的研究结果。虽然研究过程中的一些差异可能对转用HTP吸烟者的影响较小，但HTP抽吸模式的纵向适应可能会发挥作用。报告没有提供有关产品消费量或测试HTP使用强度的信息。

BAT研究。BAT研究人员报告了在日本进行的为期5天[85]的随机对照研究的结果，该研究针对glo和IQOS两种品牌的商用HTP。180名日本吸烟者随机选择继续抽吸传统卷烟、转用glo（薄荷醇或非薄荷醇）、转用非薄荷醇IQOS、戒烟这4种方式。与PMI研究一样，与基线吸烟相比，转用HTP可以显著降低呼出一氧化碳和尿液暴露生物标志物。两个HTP的减少量相似。在研究期间，HTP消费量有所增加；然而，卷烟消费也出现了类似的增长，作者将其归因于封闭研究中常见的典型的产品使用增加现象。

自2018年3月以来，BAT一直在进行另一项长期随机、多中心、对照临床转换研究[86]。随机选择商业上可用的HTP或继续吸烟的多达280名吸烟者展开为期一年的实验。此外，多达190名登记在戒烟机构且希望戒烟的参与者和40名从未吸烟的人作为对照组。评估暴露有害物质的生物标志物和对有害物质的影响，并将其变化与戒烟者和从不吸烟人群进行比较。

日本烟草研究。2014年，日本烟草国际公司研究人员报告了一项对照、半随机、开放标签、住宅式临床研究，监测转用原型HTP的健康日本男性吸烟者暴露生物标志物水平的变化[87]。共有70名吸烟者被登记并随机加入HTP或连续4周持续吸烟。与PMI和BAT研究人员的研究一样，HTP组尿液暴露生物标志物和尿液突变性的测量显著降低。

学术研究

关于HTP使用者暴露的独立研究很少。在比利时进行了使用IQOS的随机交叉实验室行为试验[88]，其中30名参与者随机抽吸5分钟的常规卷烟、电子烟碱传输系统或IQOS。抽吸IQOS呼出的一氧化碳（0.3 ppm）导致少量但可重复的增加，尽管该水平低于吸烟后观察到的浓度（4.7 ppm）。具有烟草企业相关资助的学术研究人员在意大利进行了类似的随机交叉研究[30]。共招募了12名健康吸烟者抽吸IQOS（10次抽吸，每次间隔5分钟），并在第一次抽吸后提供呼气的CO样本。结果表明，在使用IQOS后所呼出一氧化碳含量没有增加。在美国的一项研究中，招募了18名没有IQOS或JUUL经验的吸烟者，进行IQOS、JUUL电子烟碱传输系统和常规卷烟[89]的受控（10口抽吸，30秒间隙）和随意使用（90分钟）实验室研究，测量血浆烟碱和呼出的一氧化碳。使用IQOS后，呼出的一氧化碳量没有增加，而平均血浆烟碱显著增加，10次抽吸后从2.1（0.2）ng/mL增加到12.7（6.2）ng/mL，随意使用后增加到11.3（8.0）ng/mL，血浆烟碱的增幅约为传统卷烟的一半。

2.4.3 被动暴露

非企业研究人员对被动暴露于HTP释放物质的潜在影响进行了研究。在粒度分布中，研究了传统卷烟和IQOS中亚微米颗粒（5.6~560 nm）的暴露情况，以评估它们在人类呼吸系统中的潜在沉积[90]。IQOS气溶胶所含的此类颗粒量比传统卷烟烟气中约低4倍，并且IQOS气溶胶中的颗粒消散迅速；然而，大约一半的此类颗粒小到足以在吸入时到达肺泡区域。在后续的建模研究中，同一小组在多路径粒子剂量学模型中预估了不同年龄人群的呼吸道中IQOS、电子烟碱传输系统和传统卷烟的超细颗粒沉积情况[91]。IQOS释放的二手颗粒剂量明显低于传统卷烟，但其释放的剂量比电子烟碱传输系统高50%~110%，这表明非吸烟者可能接触有效的二手暴露。在同一项研究中，沉积在3个月大婴儿头部区域的60%~80%的颗粒测量值小于100 nm，这表明它们可以通过嗅球转移到大脑。在另一项研究中，在使用来源不明的HTP期间，在模型室中分析了气溶胶颗粒、羰基化合物和烟碱的浓度。与背景测量值相比，使用HTP导致几种分析物（包括烟碱、乙醛和$PM_{2.5}$）的数量和颗粒数量在统计学上显著增加。与之前的研究一样，作者报告说，

HTP颗粒会在几秒钟内消散或蒸发，并且在相同条件下，颗粒和单个成分的含量显著低于传统卷烟烟气中的含量。作者最后得出结论，在通风有限的密闭空间中大量使用HTP会显著增加非吸烟者接触二手气溶胶释放的风险。

在德国的一项研究中，还测量了颗粒大小和浓度，用以比较在汽车中使用IQOS、电子烟碱传输系统和传统卷烟而导致的潜在被动暴露[92]。结果表明，使用IQOS对细颗粒物的平均浓度（>300 nm）或车内$PM_{2.5}$浓度几乎没有影响，但较小颗粒物的浓度（25~300 nm）在所有车辆中均有所增加。从IQOS和电子烟碱传输系统中获得的烟碱浓度相当，并且都低于传统吸烟的烟碱浓度。在二手暴露的情况下对IQOS释放进行更全面的化学分析，使用HCI吸烟方案在环境室中产生IQOS气溶胶，并分析了主流和非主流的释放物中包括醛和含氮芳香物质在内的33种挥发性有机化合物。与上述研究一样，IQOS的颗粒释放量远远低于传统卷烟，有时高于电子烟碱传输系统的颗粒释放量。浓度大于$0.35 \mu g/m^3$的丙烯醛被确定为潜在关注的暴露物质。

日本基于人群的研究也关注HTP造成的被动暴露。在一项研究中，根据国际烟草管制日本调查（2018年2~3月）[93]中的全国代表性数据预估了室内公共空间使用HTP的加权流行率。研究发现，日本目前有15.6%的烟草使用者报告在室内公共场所使用HTP。在一项关于因接触二手HTP气溶胶而感知到的症状研究中，对2015~2017年间的纵向互联网调查中的8240名15~69岁的人进行了随访[94]。在报告曾接触过二手HTP气溶胶的受访者中，37%的人报道至少出现一种症状。最常见的症状通常是感觉身体不适、眼睛不适和喉咙痛。在接触过二手HTP气溶胶的从不吸烟者中，近一半的人报告了至少一种症状。

在一项针对模拟"办公室"和"酒店"环境中不同基线室内空气质量的原型HTP的二手释放物的企业研究中，ISO条件下使用HTP产生的许多有害成分明显低于抽万宝路卷烟：29种烟气成分中的24种含量降低了90%以上，并且有五种烟气成分的浓度平均减少了80%~90%。HTP的烟碱释放量比使用万宝路卷烟的烟碱释放量平均降低97%，可吸入悬浮颗粒的总数减少了90%。这些结果与非企业研究的结果基本一致，后者表明HTP是比传统卷烟更弱的室内污染源。然而，它们对室内空气质量和被动暴露的影响不容忽视，且没有被很好地了解。

2.4.4 健康影响

关于HTP在毒理学模型或人类参与者中接触和影响的研究很少，部分原因是这些产品的广泛营销和使用是相对较新的。因此，对使用HTP对健康影响的评估是有限的，应谨慎解释现有数据。

在评估HTP对健康的影响方面，吸烟者对多种传统烟草制品释放物的暴露明显减少，这是一个重要但不充足的因素。从HTP中高效摄入烟碱是一个新兴问题，应特别注意对脆弱人群的潜在影响，例如有先天性疾病的个体、孕妇和青少年。

烟碱摄入量。烟碱是一种强力致瘾的化学物质，是使用者暴露于烟草制品中有害和致癌成分的主要驱动因素。从体外、实验动物和人类研究中新出现的迹象表明，从HTP中吸收的烟碱可能多于从卷烟烟气中吸收的烟碱，这引起了人们对这些产品潜在的高滥用风险的担忧。此外，烟碱是一种重要的生殖和神经行为毒剂，可导致心血管疾病的死亡率。这种影响预期在HTP使用者中看到。

心血管疾病。在转用HTP[80,95]的吸烟者中，许多心血管生物标志物并没有减少，这表明HTP可能与吸烟一样，存在心血管疾病的风险。虽然目前尚不清楚原因，但HTP中某些化学成分水平升高可能是关键因素。这些结果表明，转用HTP不太可能降低与烟草使用相关的心血管发病率和死亡率风险。

慢性呼吸系统疾病。虽然一些研究表明，在患有慢性阻塞性肺病的吸烟者中，转用HTP[96]的呼吸道症状有所缓解，但其他研究和专家组对电子烟碱传输系统的使用与呼吸系统疾病[97-99]之间的关系表示担忧。与电子烟碱传输系统相比，HTP含有更高的挥发性呼吸有害物质水平（表2.2）。此外，实验动物研究的数据表明，HTP暴露对呼吸器官[67]和肺损伤反应[61]相关的RNA表达有影响。一项企业试验表明，在转用HTP[78]的吸烟者中，肺上皮损伤（CC16）的生物标志物水平没有降低。因此，致瘾的成年吸烟者转用HTP可能无法降低与吸烟相关的慢性呼吸道疾病的风险，而不吸烟者使用这些产品可能会增加患肺部疾病的风险，特别是如果有其他健康状况的话。

孕妇和儿童健康影响。鉴于烟碱对子宫发育和分娩结果的不利影响，如婴儿的心肺功能受损，以及青少年期间接触烟碱的负面认知和神经行为-神经影响，怀孕妇女和儿童使用HTP尤其令人关切。

基于上述报告中对HTP中存在的特定成分影响的了解，以及HTP的体外、体内和生物标志物研究的间接证据的考虑。几乎没有研究直接考察HTP与健康结果之间的关联。下面总结了一些相关研究。

青少年健康影响。韩国对60040名中学生进行了一项大型调查，以评估卷烟、HTP和电子烟碱传输系统[100]对哮喘和过敏性鼻炎流行的影响。在调整后的模型中，一直使用HTP与哮喘和过敏性鼻炎有显著关联，尽管这种关联性比目前使用传统卷烟的关联性要弱。在HTP或电子烟碱传输系统与传统卷烟的双重使用者中，哮喘的发病率特别高。在2018年韩国青少年风险行为调查的研究[101]中，根据58336名（12~18岁）学生的横断面数据，评估了烟草制品使用与过敏性疾病风险之间的关系。使用传统卷烟、HTP和电子烟碱传输系统都显著增加了哮喘、过敏性鼻炎和特异性皮炎的风险。

成年人健康影响。哈萨克斯坦正在进行一项研究，以评估40~59岁使用IQOS与传统卷烟成年人吸烟者[102]相比的健康结果。在这份为期5年的单中心组群观察研究中,分析了呼吸道症状加重、不耐受体育锻炼、肺功能异常和其他参数以及合并状况的数据。这项研究还包括基线和每年的全面临床评估、慢性阻塞性肺病的持续监测以及急性呼吸道疾病恶化的登记。临床评估包括肺活量计、胸部计算机断层扫描、心电图、体格检查、炎症和代谢综合征生物标志物的实验室测试、人体测量和6分钟步行测试。招募于2017年12月开始,截至2020年7月,结果尚未公布。在销售IQOS和其他HTP的其他国家,也应进行类似的研究,以便尽早确定与这些产品相关的潜在健康风险。

2.5 加热型烟草制品风险或危害降低的证据审查

2.5.1 烟草制品危害降低

吸烟对健康的影响是有据可查的,也是毁灭性的。烟碱是一种强力的成瘾物质,然而戒烟是非常困难的。正如烟草管制界一些人所描述的,减少烟草危害的概念是基于这样一种观点,即不愿意或不能戒烟的吸烟者应该有一种比传统燃烧型卷烟危害较小的替代品[103,104]。在过去的二十年里,不断推出声称减少暴露和风险的新产品,包括无烟烟草制品、电子烟碱传输系统和现在的HTP。由于这些新型产品发展迅速,且市场范围较小,因此很难就它们减少危害的潜力进行独立于烟草企业的及时研究。对于HTP来说更是如此,并且对这些产品的大多数现有研究都由其制造商进行的。对数据和根据这些数据提出的声明进行分析将对研究需求和监管考虑至关重要。

2.5.2 风险降低声明

企业声明

企业进行的所有研究都基于类似的策略,生成证据支持与HTP相关的"降低风险"声明。该模型基于标准体外测定、毒理学建模、实验动物研究以及人类生物标志物的暴露和潜在危害的测量。重要的是,所有已发表的报告的结论都基于HTP与传统卷烟的比较。

为了支持与HTP相关的"降低风险"的说法,PMI构建IQOS对人口健康影响的模型[105]。作者使用PMI评估方法进行了各种模拟,以评估在日本引入"低风险烟草制品"对人口健康的影响,并对1990年以来的20年间的不同情况进行建模。他们估计,如果完全禁烟,与烟草相关的肺癌、缺血性心脏病、中风和慢性阻塞性肺疾病所导致的死亡人数将减少269916;如果全部使用低风险烟草制品,死亡人数将减少167041~232519;如果引入该产品,估计死亡人数减少65126~86885。

声明分析

虽然企业赞助的出版物充分公开了实验数据，但它们的解释和结论往往忽略了重要的观察结果或者对研究结果的介绍不够充分。因此，企业声明和数据的仔细分析显得至关重要。

分析具体声明及其准确性问题。几项独立研究审查了企业发布的研究报告的主张和结论。例如，St Helen等[12]审查了向美国食品药品监督管理局（FDA）提交的PMI申请中关于气溶胶化学和人体暴露评估的章节，以授权"风险降低的烟草制品"（MRTP）评估降低暴露和风险声明的有效性。分析的作者指出，PMI报告的IQOS气溶胶中的含量仅为FDA名单上93种有害和潜在有害成分中的40种。他们还指出，虽然PMI清单上所有58种成分在IQOS气溶胶中的水平低于3R4F卷烟烟气，但其他56种成分（不在PMI或FDA清单上）的IQOS释放物水平更高。其中一些成分的含量显著增加：在IQOS气溶胶中，22种成分的含量比在3R4F烟气中的含量高出200%以上，7种成分的浓度比在3R4F烟气中的浓度高出1000%以上（图2.1）。FDA网站提供了所报告成分的完整列表（https://www.fda.gov/media/110668/download）。应独立评估这些成分在IQOS释放中发现的水平对该产品的整体危害的影响，因为它们可能会因使用IQOS而增加产生疾病的总体风险。

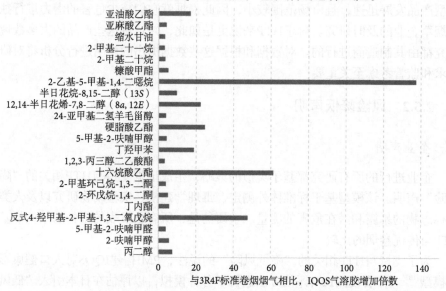

图2.1　PMI报告中比3R4F卷烟的烟气高200%的IQOS气溶胶中的成分含量
PMI向FDA提交的申请中列出的数据和化合物名称（https://www.fda.gov/media/110668/download，已于2021年1月10日生效）

PMI在其MRTP应用程序中对包含的IQOS体外和体内毒性数据也进行了

评估。Moazed等[106]回顾了IQOS在实验室动物和人类研究中的肺和免疫毒性。MRTP应用中的数据提供了使用IQOS后肺炎症和免疫调节的证据，但没有证明转向IQOS的吸烟者的肺炎症或肺功能有所改善。另一组评估了IQOS在PMI临床前研究和PMI的MRTP应用中可能的肝毒性[107,108]。他们发现，在实验动物中，暴露于HTP会增加肝脏重量、血清丙氨酸转氨酶活性和肝细胞空泡化，而暴露于传统卷烟后，并未出现这些情况。临床数据显示，转用IQOS的吸烟者的丙氨酸转氨酶活性和血浆胆红素升高，表明肝细胞损伤。但并未考察非使用者被动接触IQOS气溶胶的潜在影响。

减少暴露与降低风险。对最近发表的关于HTP使用后生物标志物的评估研究论文倾向于支持HTP减少了许多有害成分的暴露。例如，Drovandi等[109]对2010~2019年间发布的试验进行了分析，并对比使用常规燃烧型卷烟和各种HTP装置的暴露生物标志物水平。作者确定了10项非盲、随机对照实验，该实验共涉及1766名受试者，均由企业研究人员进行。分析表明，与抽吸传统卷烟的参与者相比，使用HTP人群的12种有害暴露生物标志物的含量显著降低。与戒烟人群相比，使用HTP人群的8种生物标志物的含量接近，4种生物标志物的水平在转用HTP的人群中显著提高。对转用IQOS或原型HTP[95]的受试者的PMI研究中发现，IQOS使用者与传统卷烟抽吸者在24种潜在危害的生物标志物中没有显著性差异，这表明减少烟草成分的暴露并不一定导致疾病风险的成比例降低。

分析企业用来支持其降低风险主张的策略。还分析了该企业评估HTP对公众健康影响的总体方法，特别是PMI对IQOS的建模。Max等[110]审查了向FDA申请的关于PMI将IQOS作为MRTP进行营销时所构建的"人口健康影响模型"，并将其与FDA的MRTP应用指南进行比较；还考虑了评估低风险烟草制品的一般标准。他们发现，该模型解决了IQOS对4种烟草引起的疾病（而不是发病率）死亡率的影响，它低估了死亡率，不适用于卷烟以外的烟草制品，不包括FDA建议的对非使用者的影响，并低估了对其他人群的影响。因此，企业模型低估了IQOS对健康的影响。尽管改进的模型也必须依靠企业数据和若干假设，但与HTP使用相关的健康结果和/或替代健康影响措施必须由独立研究人员评估，以便向监管机构提供关于HTP对公众健康影响的信息。

据报告，在公共信息交流和IQOS作为MRTP的申请方面，审查了以前的PMI内部秘密文件，内容涉及与IQOS类似的前体产品Accord[111]。与IQOS一样，Accord被推销为减少使用者暴露于有害烟草成分的产品；然而，PMI一直强调，这种削减并没有使Accord更安全。审查的结论是，鉴于这两种产品的相似性以及与Accord相比，IQOS气溶胶的有害物质释放量没有持续减少，因此不支持使用IQOS减少健康风险的说法。

然而，即使没有降低风险的声明，降低风险的主张也可能表明目前销售的

HTP的安全性。对PMI的MRTP申请中提交给FDA的降低风险的声明进行了定性和定量研究[112]，发现成年消费者将减少暴露的说法视为风险降低的声明。

2.6 小结和公共卫生影响

虽然HTP并不是一个新概念，但全球新一代产品使用者对这些产品的积极推广和应用是近期才开始的。尽管独立研究小组最近已经发布关于HTP成分和释放物以及HTP使用者暴露和影响的信息，但大多数研究仍然是由企业内生成和发布的。目前的文献表明，出版物之间存在着一些不同意见。企业发布的报告倾向于对转用HTP的好处作出有利的结论，即使它们在表格和数字中包含不完全支持这些结论的明确数据。

2.6.1 数据摘要

本章综述了有关HTP有害物质释放、暴露以及对模式系统和人体健康影响的出版物，主要结论概述如下。

有害物质释放和其他烟草制品的比较：用于测试HTP释放物的机器抽吸主要基于传统卷烟的抽吸方案（如ISO和HCI）。然而，使用者使用HTP和抽吸传统卷烟的方式可能存在重大差异，这增加了解释吸烟机生成数据的局限性和复杂性。然而，在没有其他数据的情况下，可以对HTP以及传统卷烟和电子烟碱传输系统的释放量进行大致比较。大多数出版物，包括非企业研究，都表明HTP和传统卷烟（按烟支计算）中的烟碱含量是可比的。据报道，从燃烧过程中产生的许多有害成分在HTP气溶胶中的含量一直比传统卷烟烟气低得多。这些包括CO、PAH，一些碳基化合物（甲醛、乙醛）和其他挥发性有害物质，以及黑碳、氮氧化物和氨的成分。HTP气溶胶中的TSNA水平也低于卷烟烟气中的TSNA水平。然而，一些报告表明，其他成分，如吡啶、二甲基三硫化物、3-羟基丁酮和甲基乙二醛，其含量可能与传统卷烟中的含量相当或更高，而诸如TSNA、CO、苯并[a]芘和羰基化合物等有害物质的释放量高于电子烟。

HTP在体外的影响：细胞培养研究可以为HTP使用相关的急性或慢性有害影响提供重要的机制性见解。企业发表的研究通常声称，HTP气溶胶比传统卷烟烟气的体外暴露的细胞毒性和致突变性明显降低，毒理学和炎症生物标志物水平范围较低。然而，吸烟强度的增加导致这些影响显著增加。此外，与暴露于参比卷烟的烟气相比，暴露于HTP的细胞被传送更多的烟碱。企业和独立出版物都表明，暴露于HTP气溶胶中的某些RNA的细胞毒性、致突变性和表达性高于空气暴露对照。不同的HTP测量结果也有所差异，例如IQOS和glo™之间。

HTP对实验动物的影响：企业研究表明，对啮齿动物皮肤致癌性、急性和慢性吸入毒性，与传统卷烟烟气处理的动物相比，使用HTP气溶胶处理的动物的肿瘤发病率和多发性较低，炎症和细胞应激反应更少，组织变化更少。然而，对这些出版物的分析揭示了许多这些效应的剂量-反应关系，并且使用HTP处理的动物的反应始终比在空气对照组中的反应更大。有限的数据表明，暴露于HTP比传统卷烟烟气产生更多的烟碱，这与体外结果一致。

转用HTP的吸烟者的暴露和影响以及与使用电子烟或戒烟的比较：企业出版物报告表明转用HTP的吸烟者的某些暴露生物标志物减少以及尿致突变性减少。然而，对出版物的审查显示，与指定戒烟和不使用任何产品的群组相比，暴露生物标志物水平明显更高（表2.2）。此外，许多心血管和其他疾病的生物标志物水平没有降低，在某些情况下，在改用超过基线水平的HTP后，有些生物标志物（CC16、丙氨酸氨基转移酶活性、血浆胆红素）水平增加。这表明HTP与传统卷烟有相似或更大的心血管毒性。最后，在转用研究中，HTP的消耗量和烟碱的摄入量明显随着时间的推移而增加，这表明暴露于其他气溶胶成分的程度增加。

HTP和其他烟草制品或清洁空气的被动暴露比较：HTP气溶胶的被动暴露研究有限。迄今的结果表明，使用HTP可能会使非吸烟者暴露于某些成分，其水平低于传统卷烟烟气被动暴露的水平，但高于清洁空气或电子烟气溶胶的被动暴露水平。

2.6.2 公共卫生影响

目前HTP的真实环境暴露及其对使用者的影响还没有被很好地描述，无法准确评估完全转用HTP或将其与传统卷烟、电子烟或其他含烟草或烟碱产品结合使用的吸烟者的长期健康结果。通过对企业和学术研究的数据进行仔细分析，提醒人们注意HTP使用产生的以下潜在公共卫生后果。

致瘾性。与其他烟草制品（包括传统卷烟）相比，HTP在向使用者提供烟碱方面可能更有效率。因此，在缺乏对HTP使用的健康后果、成瘾可能性以及不同人群长期使用这些产品（包括年轻人和多重使用的成年人）了解的情况下，HTP是一个公共卫生问题。

使用者接触有害物质的情况有显著差异。增加HTP的抽吸强度可显著增加有害物质的释放量，并以剂量-反应的方式增加HTP释放的细胞毒性和致突变效应。因此，根据产品类型和使用模式，使用相同HTP的个体的有害物质暴露和随后的风险可能存在显著差异。

转用加热型烟草制品的吸烟者的慢性疾病负担没有降低。一些吸烟者可能会选择转用HTP，以减少有害成分暴露，而不戒烟。如上文所述，数据表明，在改用研究的参与者中，一些肺和心血管指标没有改善，双重使用（与卷烟一起使用）的流行率很高。因此，使用HTP可能不会显著降低与吸烟相关慢性病的患病率。

开始使用HTP的非吸烟者患慢性疾病的风险增加。迄今为止进行的研究一致表明，与未接触HTP相比（如实验研究中的对照和人体试验中的戒烟），接触HTP的风险更高，影响更大。因此，非烟草制品使用者摄入HTP将增加患呼吸道、心血管和潜在其他疾病等不良后果的风险。

HTP未知或独特的毒性作用。对HTP气溶胶的研究大多局限于关键燃烧成分和烟草特定成分的分析，以及与传统卷烟烟气的比较。HTP气溶胶可能含有独特的有害成分，这些成分还没有被识别或很好地描述，正如在实验和临床试验中所表明的，HTP具有导致肝细胞损伤的迹象，而传统卷烟没有。

非吸烟者的二手暴露。接触颗粒物、烟碱和HTP气溶胶的其他成分可能会对非吸烟者构成风险。

使用者对安全的认知。烟草制品使用者和非使用者可能将烟草制品接触减少的声明视为风险降低的声明，这可能导致原本戒烟或不抽烟的个体开始抽烟。

最后，HTP品牌之间也有差异，表明HTP所使用的技术将继续发展。这可能包括HTP产生的气溶胶的化学成分的变化，以及随后对吸烟者的暴露和影响。因此，目前销售的HTP的有限数据（其中大部分是企业产生的）可能不会直接适用于未来。

2.7 研究差距和优先事项

本章综述了HTP中的有害物质，并对体外和体内毒性、暴露水平和对人体的影响进行了报道，指出了以下研究领域：

- 对HTP释放物进行更合理、严格、创新的实验室评估，例如进行非靶向分析，以确定可能导致转用HTP的吸烟者潜在危害生物标志物缺乏改善的有害物质；
- 更好地描述人体抽吸模式，以了解HTP在现实生活中的使用方式(而不是临床限制性试验)，包括随着时间的推移可能增加产品消费，以及它们如何影响成瘾、有害物质暴露和健康结果；
- 更好地描述HTP使用者对烟碱的摄入和对滥用倾向的潜在影响；
- 使用毒理学模型（如细胞或实验动物）了解细胞和组织中的烟碱摄入情况，并预测HTP的独特或未知毒性；
- 识别特定暴露和效应生物标志物促进HTP使用、相关暴露和潜在健康影响的研究；
- HTP与其他烟草制品（尤其是传统卷烟）双重或多重使用的影响；
- 二手烟暴露对非吸烟者，特别是对弱势群体（患者、儿童、孕妇）的影响；
- 监测不同国家/地区的暴露于HTP的使用者和非使用者的纵向人群水平健

2.8　政策建议

以下建议供决策者、研究人员和公共卫生界酌情考虑：
- 彻底检查HTP的企业研究数据以确保正确地解释。
- 优先支持HTP对公共卫生影响的独立研究。
- 制定HTP监测和研究的标准，如产品相关术语，HTP释放物的标准测试程序，包括抽吸方案以及可用于质量控制的HTP特定参比产品。
- 按国家对现有HTP的使用流行情况及其与健康结果的潜在联系进行监测。
- 在获得关于HTP对公共卫生影响的独立证据之前，禁止所有制造商和相关团体声称与其他产品相比危害减少，包括广告和健康警示的修改，以及将HTP描述为适用于停止使用任何烟草制品；并禁止在公共场所使用HTP，因为它们是烟草制品。

2.9　参考文献

[1] Forster M, Fiebelkorn S, Yurteri C, Mariner D, Liu C, Wright C et al. Assessment of novel tobacco heating product THP1.0. Part 3: Comprehensive chemical characterisation of harmful and potentially harmful aerosol emissions. Regul Toxicol Pharmacol. 2018;93:14–33.

[2] Takahashi Y, Kanemaru Y, Fukushima T, Eguchi K, Yoshida S, Miller-Holt J et al. Chemical analysis and in vitro toxicological evaluation of aerosol from a novel tobacco vapor product: A comparison with cigarette smoke. Regul Toxicol Pharmacol. 2018;92:94–103.

[3] Li X, Luo Y, Jiang X, Zhang H, Zhu F, Hu S et al. Chemical analysis and simulated pyrolysis of tobacco heating system 2.2 compared to conventional cigarettes. Nicotine Tob Res. 2019;21(1):111–8.

[4] Gasparyan H, Mariner D, Wright C, Nicol J, Murphy J, Liu C et al. Accurate measurement of main aerosol constituents from heated tobacco products (HTPs): Implications for a fundamentally different aerosol. Regul Toxicol Pharmacol. 2018;99:131–41.

[5] Uchiyama S, Noguchi M, Takagi N, Hayashida H, Inaba Y, Ogura H et al. Simple determination of gaseous and particulate compounds generated from heated tobacco products. Chem Res Toxicol. 2018;31(7):585–93.

[6] Uchiyama S, Noguchi M, Sato A, Ishitsuka M, Inaba Y, Kunugita N. Determination of thermal decomposition products generated from e-cigarettes. Chem Res Toxicol. 2020;33(2):576–83.

[7] Hukkanen J, Jacob P III, Benowitz NL. Metabolism and disposition kinetics of nicotine. Pharmacol Rev. 2005;57(1):79–115.

[8] Bekki K, Inaba Y, Uchiyama S, Kunugita N. Comparison of chemicals in mainstream smoke in heat-not-burn tobacco and combustion cigarettes. J UOEH. 2017;39(3):201–7.

[9] Farsalinos KE, Yannovits N, Sarri T, Voudris V, Poulas K. Nicotine delivery to the aerosol of a

heat-not-burn tobacco product: Comparison with a tobacco cigarette and e-cigarettes. Nicotine Tob Res. 2018;20(8):1004–9.

[10] Liu S, Li J, Zhu L, Liu W. Quantitative determination and puff-by-puff analysis on the release pattern of nicotine and menthol in heat-not-burn tobacco material and emissions. Paris: Cooperation Centre for Scientific Research Relative to Tobacco; 2020.

[11] Auer R, Concha-Lozano N, Jacot-Sadowski I, Cornuz J, Berthet A. Heat-not-burn tobacco cigarettes: Smoke by any other name. JAMA Intern Med. 2017;177(7):1050–2.

[12] St Helen G, Jacob P III, Nardone N, Benowitz NL. IQOS: examination of Philip Morris International's claim of reduced exposure. Tob Control. 2018;27(Suppl 1):s30–6.

[13] Mallock N, Pieper E, Hutzler C, Henkler-Stephani F, Luch A. Heated tobacco products: a review of current knowledge and initial assessments. Front Public Health. 2019;7:287.

[14] Counts ME, Morton MJ, Laffoon SW, Cox RH, Lipowicz PJ. Smoke composition and predicting relationships for international commercial cigarettes smoked with three machine-smoking conditions. Regul Toxicol Pharmacol. 2005;41(3):185–227.

[15] Standard operating procedure for determination of nicotine and carbon monoxide in mainstream cigarette smoke under intense smoking conditions, WHO TobLabNet SOP10. Geneva: World Health Organization; 2016.

[16] Standard operating procedure for determination of nicotine in cigarette tobacco filler, TobLabNet SOP 4. Geneva: World Health Organization; 2014.

[17] Simonavicius E, McNeill A, Shahab L, Brose LS. Heat-not-burn tobacco products: a systematic literature review. Tob Control. 2019;28(5):582–94.

[18] Jaccard G, Tafin Djoko D, Moennikes O, Jeannet C, Kondylis A, Belushkin M. Comparative assessment of HPHC yields in the tobacco heating system THS2.2 and commercial cigarettes. Regul Toxicol Pharmacol. 2017;90:1–8.

[19] Meišutovič-Akhtarieva M, Prasauskas T, Čiužas D, Krugly E, Keratayté K, Martuzevičius D et al. Impacts of exhaled aerosol from the usage of the tobacco heating system to indoor air quality: A chamber study. Chemosphere. 2019;223:474–82.

[20] Schaller JP, Keller D, Poget L, Pratte P, Kaelin E, McHugh D et al. Evaluation of the tobacco heating system 2.2. Part 2: Chemical composition, genotoxicity, cytotoxicity, and physical properties of the aerosol. Regul Toxicol Pharmacol. 2016;81(Suppl 2):S27–47.

[21] Shein M, Jeschke G. Comparison of free radical levels in the aerosol from conventional cigarettes, electronic cigarettes, and heat-not-burn tobacco products. Chem Res Toxicol. 2019;32(6):1289–98.

[22] Smith MR, Clark B, Lüdicke F, Schaller JP, Vanscheeuwijck P, Hoeing J et al. Evaluation of the tobacco heating system 2.2. Part 1: Description of the system and the scientific assessment program. Regul Toxicol Pharmacol. 2016;81(Suppl 2):S17–26.

[23] Crooks I, Neilson L, Scott K, Reynolds L, Oke T, Meredith C et al. Evaluation of flavourings potentially used in a heated tobacco product: Chemical analysis, in vitro mutagenicity, genotoxicity, cytotoxicity and in vitro tumour promoting activity. Food Chem Toxicol. 2018;118:940–52.

[24] Davis B, Williams M, Talbot P. iQOS: evidence of pyrolysis and release of a toxicant from plastic. Tob Control. 2019;28(1):34–41.

[25] Forster M, Liu C, Duke MG, McAdam KG, Proctor CJ. An experimental method to study emissions from heated tobacco between 100–200℃ . Chem Cent J. 2015;9:20.

[26] Cancelada L, Sleiman M, Tang X, Russell ML, Montesinos VN, Litter MI et al. Heated tobacco products: volatile emissions and their predicted impact on indoor air quality. Environ Sci Technol. 2019;53(13):7866–76.

[27] Jaccard G, Kondylis A, Gunduz I, Pijnenburg J, Belushkin M. Investigation and comparison of the transfer of TSNA from tobacco to cigarette mainstream smoke and to the aerosol of a heated tobacco product, THS2.2. Regul Toxicol Pharmacol. 2018;97:103–9.

[28] Ratajczak A, Jankowski P, Strus P, Feleszko W. Heat not burn tobacco product – a new global trend: Impact of heat-not-burn tobacco products on public health, a systematic review. Int J Environ Res Public Health. 2020;17(2):409.

[29] Konstantinou E, Fotopoulou F, Drosos A, Dimakopoulou N, Zagoriti Z, Niarchos A et al. Tobacco-specific nitrosamines: A literature review. Food Chem Toxicol. 2018;118:198–203.

[30] 30Caponnetto P, Maglia M, Prosperini G, Busà B, Polosa R. Carbon monoxide levels after inhalation from new generation heated tobacco products. Respir Res. 2018;19(1):164.

[31] Ishizaki A, Kataoka H. A sensitive method for the determination of tobacco-specific nitrosamines in mainstream and sidestream smokes of combustion cigarettes and heated tobacco products by online in-tube solid-phase microextraction coupled with liquid chromatography-tandem mass spectrometry. Anal Chim Acta. 2019;1075:98–105.

[32] Glantz SA. Heated tobacco products: the example of IQOS. Tob Control. 2018;27(Suppl 1):s1–6.

[33] Leigh NJ, Palumbo MN, Marino AM, O'Connor RJ, Goniewicz ML. Tobacco-specific nitrosamines (TSNA) in heated tobacco product IQOS. Tob Control. 2018;27(Suppl 1):s37–8.

[34] Ogunwale MA, Li M, Ramakrishnam Raju MV, Chen Y, Nantz MN, Conklin DJ et al. Aldehyde detection in electronic cigarette aerosols. ACS Omega. 2017;2(3):1207–14.

[35] Conklin DJ, Ogunwale MA, Chen Y, Theis WS, Nantz MH, Fu XA et al. Electronic cigarette-generated aldehydes: The contribution of e-liquid components to their formation and the use of urinary aldehyde metabolites as biomarkers of exposure. Aerosol Sci Technol. 2018;52(11):1219–32.

[36] Kaunelienė V, Meišutovič-Akhtarieva M, Martuzevičius D. A review of the impacts of tobacco heating system on indoor air quality versus conventional pollution sources. Chemosphere. 2018;206:568–78.

[37] Savareear B, Escobar-Arnanz J, Brokl M, Saxton MJ, Wright C, Liu C et al. Comprehensive comparative compositional study of the vapour phase of cigarette mainstream tobacco smoke and tobacco heating product aerosol. J Chromatogr A. 2018;1581–2.

[38] Farsalinos KE, Yannovits N, Sarri T, Voudris V, Poulas K, Leischow SJ. Carbonyl emissions from a novel heated tobacco product (IQOS): comparison with an e-cigarette and a tobacco cigarette. Addiction. 2018;113(11):2099–106.

[39] Ruprecht AA, Marco CD, Saffari A, Pozzi P, Mazza R, Veronese C et al.Environmental pollution and emission factors of electronic cigarettes, heat-not-burn tobacco products, and conventional cigarettes. Aerosol Sci Technol. 2017;51(6):674–84.

[40] Mallock N, Böss L, Burk R, Danziger M, Welsch T, Hahn H et al. Levels of selected analytes in the emissions of "heat not burn" tobacco products that are relevant to assess human health risks.

Arch Toxicol. 2018;92(6):2145-9.

[41] Salman R, Talih S, El-Hage R, Haddad C, Karaoghlanian N, El-Hellani A et al. Free-base and total nicotine, reactive oxygen species, and carbonyl emissions From IQOS, a heated tobacco product. Nicotine Tob Res. 2019;21(9):1285-8.

[42] De Marco C, Borgini A, Ruprecht AA, Veronese C, Mazza R, Bertoldi M et al.La formaldeide nelle sigarette elettronichee nei riscaldatoridi tabacco (HnB): facciamoil punto.[Formaldehyde in electronic cigarettes and in heat-not-burn products (HnB): let's make the point]. Epidemiol Prev. 2018;42(5-6):351-5.

[43] Savareear B, Escobar-Arnanz J, Brokl M, Saxton MJ, Wright C, Liu C et al. Non-targeted analysis of the particulate phase of heated tobacco product aerosol and cigarette mainstream tobacco smoke by thermal desorption comprehensive two-dimensional gas chromatography with dual flame ionisation and mass spectrometric detection. J Chromatogr A. 2019;1603:327-37.

[44] Tricker AR, Schorp MK, Urban HJ, Leyden D, Hagedorn HW, Engl J et al.Comparison of environmental tobacco smoke (ETS) concentrations generated by an electrically heated cigarette smoking system and a conventional cigarette. Inhal Toxicol. 2009;21(1):62-77.

[45] Mitova MI, Campelos PB, Goujon-Ginglinger CG, Maeder S, Mottier N, Rouget EJR et al. Comparison of the impact of the tobacco heating system 2.2 and a cigarette on indoor air quality. Regul Toxicol Pharmacol. 2016;80:91-101.

[46] Jankowski M, Brożek GM, Lawson J, Skoczyński S, Majek P, Zejda JE. New ideas, old problems? Heated tobacco products – a systematic review. Int J Occup Med Environ Health. 2019;32(5):595-634.

[47] Hitosugi M, Tojo M, Kane M, Shiomi N, Shimizu T, Nomiyama T. Criminal mercury vapor poisoning using heated tobacco product. Int J Legal Med. 2019;133(2):479-81.

[48] A study on production and delivery of volatile aroma substances from heat-not-burn products' tobacco material to emissions. Paris: Cooperation Centre for Scientific Research Relative to Tobacco; 2020.

[49] Pacitto A, Stabile L, Scungio M, Rizza V, Buonanno G. Characterization of airborne particles emitted by an electrically heated tobacco smoking system. Environ Pollut. 2018;240:248-54.

[50] Górski P.e-Cigarettes or heat-not-burn tobacco products – advantages or disadvantages for the lungs of smokers. Adv Resp Med. 2019;87(2):123-34.

[51] Roemer E, Stabbert R, Veltel D, Müller BP, Meisgen TJ, Schramke H et al. Reduced toxicological activity of cigarette smoke by the addition of ammonium magnesium phosphate to the paper of an electrically heated cigarette: smoke chemistry and in vitro cytotoxicity and genotoxicity. Toxicol In Vitro. 2008;22(3):671-81.

[52] Tewes FJ, Meisgen TJ, Veltel DJ, Roemer E, Patskan G. Toxicological evaluation of an electrically heated cigarette. Part 3: Genotoxicity and cytotoxicity of mainstream smoke. J Appl Toxicol. 2003;23(5):341-8.

[53] Werley MS, Freelin SA, Wrenn SE, Gerstenberg B, Roemer E, Schramke H et al. Smoke chemistry, in vitro and in vivo toxicology evaluations of the electrically heated cigarette smoking system series K. Regul Toxicol Pharmacol. 2008;52(2):122-39.

[54] Zenzen V, Diekmann J, Gerstenberg B, Weber S, Wittke S, Schorp MK. Reduced exposure evaluation of an electrically heated cigarette smoking system. Part 2: Smoke chemistry and in vitro

toxicological evaluation using smoking regimens reflecting human puffing behavior. Regul Toxicol Pharmacol. 2012;64(2 Suppl):S11-34.

[55] van der Toorn M, Koshibu K, Schlage WK, Majeed S, Pospisil P, Hoeng J et al. Comparison of monoamine oxidase inhibition by cigarettes and modified risk tobacco products. Toxicol Rep. 2019;6:1206-15.

[56] Iskandar AR, Mathis C, Schlage WK, Frentzel S, Leroy P, Xiang Y et al. A systems toxicology approach for comparative assessment: Biological impact of an aerosol from a candidate modified risk tobacco product and cigarette smoke on human organotypic bronchial epithelial cultures. Toxicol In Vitro. 2017;39:29-51.

[57] Zanetti F, Sewer A, Scotti E, Titz B, Schlage WK, Leroy P et al. Assessment of the impact of aerosol from a potential modified risk tobacco product compared with cigarette smoke on human organotypic oral epithelial cultures under different exposure regimens. Food Chem Toxicol. 2018;115:148-69.

[58] Iskandar AR, Martin F, Leroy P, Schlage WK, Mathis C, Titz B et al. Comparative biological impacts of an aerosol from carbon-heated tobacco and smoke from cigarettes on human respiratory epithelial cultures: A systems toxicology assessment. Food Chem Toxicol. 2018;115:109-26.

[59] Poussin C, Laurent A, Kondylis A, Marescotti D, van der Toorn M, Guedj E et al. In vitro systems toxicology-based assessment of the potential modified risk tobacco product CHTP 1.2 for vascular inflammation- and cytotoxicity-associated mechanisms promoting adhesion of monocyticcells to human coronary arterial endothelial cells. Food Chem Toxicol. 2018;120:390-406.

[60] Jaunky T, Adamson J, Santopietro S, Terry A, Thorne D, Breheny D et al. Assessment of tobacco heating product THP1.0. Part 5: In vitro dosimetric and cytotoxic assessment. Regul Toxicol Pharmacol. 2018;93:52-61.

[61] Haswell LE, Corke S, Verrastro I, Baxter A, Banerjee A, Adamson J et al. In vitro RNA-seq-based toxicogenomics assessment shows reduced biological effect of tobacco heating products when compared to cigarette smoke. Sci Rep. 2018;8(1):1145.

[62] Godec TL, Crooks I, Scott K, Meredith C. In vitro mutagenicity of gas-vapour phase extracts from flavoured and unflavoured heated tobacco products. Toxicol Rep. 2019;6:1155-63.

[63] Leigh NJ, Tran PL, O'Connor RJ, Goniewicz ML. Cytotoxic effects of heated tobacco products on human bronchial epithelial cells. Tob Control. 2018;27(Suppl 1):s26-9.

[64] Davis B, To V, Talbot P. Comparison of cytotoxicity of IQOS aerosols to smoke from Marlboro Red and 3R4F reference cigarettes. Toxicol In Vitro. 2019;61:104652.

[65] Zagoriti Z, El Mubarak MA, Farsalinos K, Topouzis S. Effects of exposure to tobacco cigarette, electronic cigarette and heated tobacco product on adipocyte survival and differentiation in vitro. Toxics. 2020;8(1):9.

[66] Tsuji H, Okubo C, Fujimoto H, Fukuda I, Nishino T, Lee KM et al. Comparison of dermal tumor promotion activity of cigarette smoke condensate from prototype (heated) cigarette and reference (combusted) cigarette in SENCAR mice. Food Chem Toxicol. 2014;72:187-94.

[67] Fujimoto H, Tsuji H, Okubo C, Fukuda I, Nishino T, Lee KM et al. Biological responses in rats exposed to mainstream smoke from a heated cigarette compared to a conventional reference

cigarette. Inhal Toxicol. 2015;27(4):224–36.

[68] Phillips BW, Schlage WK, Titz B, Kogel U, Sciuscio D, Martin F et al. A 90-day OECD TG 413 rat inhalation study with systems toxicology endpoints demonstrates reduced exposure effects of the aerosol from the carbon heated tobacco product version 1.2 (CHTP1.2) compared with cigarette smoke. I. Inhalation exposure, clinical pathology and histopathology. Food Chem Toxicol. 2018;116(Pt B):388–413.

[69] Titz B, Kogel U, Martin F, Kogel U, Sciuscio D, Martin F et al. A 90-day OECD TG 413 rat inhalation study with systems toxicology endpoints demonstrates reduced exposure effects of the aerosol from the carbon heated tobacco product version 1.2 (CHTP 1.2) compared with cigarette smoke. II. Systems toxicology assessment. Food Chem Toxicol. 2018;115:284–301.

[70] Phillips B, Szostak J, Titz B, Schlage WK, Guedj E, Leroy P et al. A six-month systems toxicology inhalation/cessation study in ApoE-/- mice to investigate cardiovascular and respiratory exposure effects of modified risk tobacco products, CHTP 1.2 and THS 2.2, compared with conventional cigarettes. Food Chem Toxicol. 2019;126:113–41.

[71] Szostak J, Titz B, Schlage WK, Guedj E, Sewer A, Phillips B et al. Structural, functional, and molecular impact on the cardiovascular system in ApoE-/- mice exposed to aerosol from candidate modified risk tobacco products, carbon heated tobacco product 1.2 and tobacco heating system 2.2, compared with cigarette smoke. Chem Biol Interact. 2020;315:108887.

[72] Nabavizadeh P, Liu J, Havel CM, Ibrahim S, Derakhshandeh R, Jacob P III et al. Vascular endothelial function is impaired by aerosol from a single IQOS HeatStick to the same extent as by cigarette smoke. Tob Control. 2018;27(Suppl 1):s13–9.

[73] McAdam K, Davis P, Ashmore L, Eaton D, Jakaj B, Eldridge A et al. Influence of machine-based puffing parameters on aerosol and smoke emissions from next generation nicotine inhalation products. Regul Toxicol Pharmacol. 2019;101:156–65.

[74] Urban HJ, Tricker AR, Leyden DE, Forte N, Zenzen V, Fueursenger A et al. Reduced exposure evaluation of an electrically heated cigarette smoking system. Part 8: Nicotine bridging – estimating smoke constituent exposure by their relationships to both nicotine levels in mainstream cigarette smoke and in smokers. Regul Toxicol Pharmacol. 2012;64(2 Suppl):S85–97.

[75] Patskan G, Reininghaus W. Toxicological evaluation of an electrically heated cigarette. Part 1: Overview of technical concepts and summary of findings. J Appl Toxicol. 2003;23(5):323–8.

[76] Tricker AR, Stewart AJ, Martin Leroy C, Lindner D, Schorp MK, Dempsey R. Reduced exposure evaluation of an electrically heated cigarette smoking system. Part 3: Eight-day randomized clinical trial in the UK. Regul Toxicol Pharmacol. 2012;64(2 Suppl):S35–44.

[77] Tricker AR, Jang IJ, Martin Leroy C, Lindner D, Dempsey R. Reduced exposure evaluation of an electrically heated cigarette smoking system. Part 4: Eight-day randomized clinical trial in Korea. Regul Toxicol Pharmacol. 2012;64(2 Suppl):S45–53.

[78] Tricker AR, Kanada S, Takada K, Martin Leroy C, Lindner D, Schorp MK et al. Reduced exposure evaluation of an electrically heated cigarette smoking system. Part 5: 8-Day randomized clinical trial in Japan. Regul Toxicol Pharmacol. 2012;64(2 Suppl):S54–63.

[79] Tricker AR, Kanada S, Takada K, Martin Leroy C, Lindner D, Schorp MK et al. Reduced exposure evaluation of an electrically heated cigarette smoking system. Part 6: 6-Day

randomized clinical trial of a menthol cigarette in Japan. Regul Toxicol Pharmacol. 2012;64(2 Suppl):S64–73.

[80] Martin Leroy C, Jarus-Dziedzic K, Ancerewicz J, Lindner D, Kulesza A, Magnette J. Reduced exposure evaluation of an electrically heated cigarette smoking system. Part 7: A one-month, randomized, ambulatory, controlled clinical study in Poland. Regul Toxicol Pharmacol. 2012;64(2 Suppl):S74–84.

[81] Lüdicke F, Haziza C, Weitkunat R, Magnette J. Evaluation of biomarkers of exposure in smokers switching to a carbon-heated tobacco product: A controlled, randomized, open-label 5-day exposure study. Nicotine Tob Res. 2016;18(7):1606–13.

[82] Lüdicke F, Baker G, Magnette J, Picavet P, Weitkunat R. Reduced exposure to harmful and potentially harmful smoke constituents with the tobacco heating system 2.1. Nicotine Tob Res. 2017;19(2):168–75.

[83] Haziza C, de La Bourdonnaye G, Skiada D, Ancerewicz J, Baker G, Picavet P et al. Evaluation of the tobacco heating system 2.2. Part 8: 5-Day randomized reduced exposure clinical study in Poland. Regul Toxicol Pharmacol. 2016;81 Suppl 2:S139–50.

[84] Lüdicke F, Ansari SM, Lama N, Blanc N, Bosilskova M, Donelli A et al.Effects of switching to a heat-not-burn tobacco product on biologically relevant biomarkers to assess a candidate modified risk tobacco product: a randomized trial. Cancer Epidemiol Biomarkers Prev. 2019;28(11):1934–43.

[85] Gale N, McEwan M, Eldridge AC, Fearon IM, Sherwood N, Bowen E et al.Changes in biomarkers of exposure on switching from a conventional cigarette to tobacco heating products: a randomized, controlled study in healthy Japanese subjects. Nicotine Tob Res. 2019;21(9):1220–7.

[86] Newland N, Lowe FJ, Camacho OM, McEwan M, Gale N, Ebajemito J et al. Evaluating the effectsof switching from cigarette smoking to using a heated tobacco product on health effect indicators in healthy subjects: study protocol for a randomized controlled trial. Intern Emerg Med. 2019;14(6):885–98.

[87] Sakaguchi C, Kakehi A, Minami N, Kikuchi A, Futamura Y. Exposure evaluation of adult male Japanese smokers switched to a heated cigarette in a controlled clinical setting. Regul Toxicol Pharmacol. 2014;69(3):338–47.

[88] Adriaens K, Gucht DV, Baeyens F. IQOSTM vs. e-cigarette vs. tobacco cigarette: A direct comparison of short-term effects after overnight-abstinence. Int J Environ Res Public Health. 2018;15(12):2902.

[89] Maloney S, Eversole A, Crabtree M, Soule E, Eissenberg T, Breland A. Acute effects of JUUL and IQOS in cigarette smokers. Tob Control. 2020: doi: 10.1136/tobaccocontrol-2019-055475.

[90] Protano C, Manigrasso M, Avino P, Sernia S, Vitali M. Second-hand smoke exposure generated by new electronic devices (IQOS® and e-cigs) and traditional cigarettes: submicron particle behaviour in human respiratory system. Ann Ig. 2016;28(2):109–12.

[91] Protano C, Manigrasso M, Avino P, Vitali M. Second-hand smoke generated by combustion and electronic smoking devices used in real scenarios: Ultrafine particle pollution and age-related dose assessment. Environ Int. 2017;107:190–5.

[92] Schober W, Fembacher L, Frenzen A, Fromme H. Passive exposure to pollutants from conventional cigarettes and new electronic smoking devices (IQOS, e-cigarette) in passenger

cars. Int J Hyg Environ Health. 2019;222(3):486–93.

[93] Sutanto E, Smith DM, Miller C, O'Connor RJ, Hyland A, Tabuchi T et al. Use of heated tobacco products within indoor spaces: Findings from the 2018 ITC Japan survey. Int J Environ Res Public Health. 2019;16(23):4862.

[94] Tabuchi T, Gallus S, Shinozaki T, Nakaya T, Kunugita N, Colwell B. Heat-not-burn tobacco product use in Japan: its prevalence, predictors and perceived symptoms from exposure to secondhand heat-not-burn tobacco aerosol. Tob Control. 2018;27(e1):e25–33.

[95] Glantz SA. PMI's own in vivo clinical data on biomarkers of potential harm in Americans show that IQOS is not detectably different from conventional cigarettes. Tob Control. 2018; 27 (Suppl 1): s9–12.

[96] Polosa R, Morjaria JB, Caponnetto P, Prosperini U, Russo C, Pennisi A et al. Evidence for harm reduction in COPD smokers who switch to electronic cigarettes. Respir Res. 2016;17(1):166.

[97] Bhatta DN, Glantz SA. Association of e-cigarette use with respiratory disease among adults: A longitudinal analysis. Am J Prev Med. 2020;58(2):182–90.

[98] Osei AD, Mirbolouk M, Orimoloye OA, Dzaye O, Iftekhar Uddin SM, Benjamin EJ et al. Association between e-cigarette use and chronic obstructive pulmonary disease by smoking status: behavioral risk factor surveillance system 2016 and 2017. Am J Prev Med. 2019;58(3):336–42.

[99] Wills TA, Pagano I, Williams RJ, Tam EK. e-Cigarette use and respiratory disorder in an adult sample. Drug Alcohol Depend. 2019;194:363–70.

[100] Chung SJ, Kim BK, Oh JH, Shim JS, Chang YS, Cho SH et al. Novel tobacco products including electronic cigarette and heated tobacco products increase risk of allergic rhinitis and asthma in adolescents: Analysis of Korean youth survey. Allergy. 2020;75(7):1640–8.

[101] Lee A, Lee SY, Lee KS. The use of heated tobacco products is associated with asthma, allergic rhinitis, and atopic dermatitis in Korean adolescents. Sci Rep. 2019;9(1):17699.

[102] Sharman A, Zhussupov B, Sharman D, Kim I, Yerenchina E. Lung function in users of a smoke-free electronic device with HeatSticks (iQOS) versus smokers of conventional cigarettes: Protocol for a longitudinal cohort observational study. JMIR Res Protoc. 2018;7(11):e10006.

[103] Hatsukami DK, Slade J, Benowitz NL, Giovino GA, Gritz ER, Leischow S et al. Reducing tobacco harm: research challenges and issues. Nicotine Tob Res. 2002;4 Suppl 2:S89–101.

[104] Warner KE. Tobacco harm reduction: promise and perils. Nicotine Tob Res. 2002;4(Suppl 2): S61–71.

[105] Lee PN, Djurdjevic S, Weitkunat R, Baker G. Estimating the population health impact of introducing a reduced-risk tobacco product into Japan. The effect of differing assumptions, and some comparisons with the US. Regul Toxicol Pharmacol. 2018;100:92–104.

[106] Moazed F, Chun L, Matthay MA, Calfee CS, Gotts J. Assessment of industry data on pulmonary and immunosuppressive effects of IQOS. Tob Control. 2018;27(Suppl 1):s20–5.

[107] Chun L, Moazed F, Matthay M, Calfee C, Gotts J. Possible hepatotoxicity of IQOS. Tob Control. 2018;27(Suppl 1):s39–40.

[108] Wong ET, Kogel U, Veljkovic E, Martin F, Xiang Y, Boue S et al. Evaluation of the tobacco heating system 2.2. Part 4: 90-day OECD 413 rat inhalation study with systems toxicology endpoints demonstrates reduced exposure effects compared with cigarette smoke. Regul Toxicol Pharmacol. 2016;81(Suppl 2):S59–81.

[109] Drovandi A, Salem S, Barker D, Booth D, Kairuz T. Human biomarker exposure from cigarettes versus novel heat-not-burn devices: A systematic review and meta-analysis. Nicotine Tob Res. 2020;22(7):1077-85.

[110] Max WB, Sung HY, Lightwood J, Wang Y, Yao T. Modelling the impact of a new tobacco product: review of Philip Morris International's population health impact model as applied to the IQOS heated tobacco product. Tob Control. 2018;27(Suppl 1):s82-6.

[111] Elias J, Dutra LM, St Helen G, Ling PM. Revolution or redux? Assessing IQOS through a precursor product. Tob Control. 2018;27(Suppl 1):s102-10.

[112] Popova L, Lempert LK, Glantz SA. Light and mild redux: heated tobacco products' reduced exposure claims are likely to be misunderstood as reduced risk claims. Tob Control. 2018;27(Suppl 1):s87-95.

3. 加热型烟草制品的吸引力和致瘾性：对认知和使用的影响及关联效应

Reinskje Talhout, National Institute for Public Health and the Environment, Centre for Health Protection, Bilthoven, Netherlands

Richard J. O'Connor, Department of Health Behavior, Roswell Park Comprehensive Cancer Center, Buffalo (NY), USA

摘 要

FCTC/COP8(22)决议要求关于新型烟草制品，特别是HTP的若干方面提出报告。本章论述了致瘾性、认知和使用、吸引力、在吸烟起始和戒烟方面的潜在作用、营销（包括促销策略和影响）以及"减少危害"的主张。回顾了HTP的吸引力和致瘾性特性，以及对消费者的认知和使用的影响。现有的关于HTP的文献得到了来自于更广泛的电子烟知识体系的信息补充。

在数据库PubMed中检索了截至2020年1月的文献，排除了关于使用者毒性（即释放物中的有害物质、体外研究、暴露生物标志物）的研究和环境烟气的研究以及非英语语言的研究。还包括菲利普·莫里斯国际（PMI）提交美国食品药品监督管理局（FDA）的风险降低的烟草制品（MRTP）IQOS的研究。

关于增加吸引力的特征，发现了感官属性、易用性、成本、声誉和形象以及假定的风险和收益方面的信息。但对于这些不同的功能如何影响消费者的认知和使用，知之甚少；最近的一项研究报告指出，六个重要因素是健康、成本、享受和满意度、易用性、使用实践和社会因素。关于致瘾性，目前销售的HTP在气溶胶中提供大量烟碱，其药代动力学、生理和主观效应与目前的ENDS产品相似，表明存在类似的滥用倾向。

HTP在一些市场很受欢迎，可能是因为作为一种"干净、现代、优雅、降低危害"的产品进行营销等因素。它们的感官特性和易用性通常低于传统卷烟，但与它们的吸引力和风险认知直接相关，因此决定了它们的摄入量。电子烟的历史表明，任何新的烟草及相关产品上市可以迅速成为流行产品。对电子烟的了解表明，HTP所关注的因素，比如烟碱水平和"击喉感"、口味种类、烟具设计、营销和减少危害的认知，这些也将是潜在的监管目标。

电子烟的共同监管原则包括最大限度地降低产品吸引力，从而减少年轻人的潜在消费，提高产品安全性以尽量减少对健康影响的误导性理念。HTP也可采用相似的策略。建议决策者监测HTP市场，向公众宣传风险，限制营销，考虑调节口味，鼓励研究，尤其是关于认知和使用。

3.1 背 景

自20世纪80年代以来，烟草公司一直试图在市场上推广HTP，称其比传统卷烟"更健康"。以前它们都失败了[1]，但是HTP现在越来越多地作为燃烧型烟草制品（主要是卷烟）的替代品进行营销，尽管围绕其营销和使用的公共卫生背景存在争议[2]。自2014年以来，已推出包括菲利普·莫里斯国际（PMI）的IQOS在内的各种新产品。企业分析师预测，到2025年，HTP将占据美国普通卷烟市场的30%[3]，如日本烟草国际公司的PloomTECH、英美烟草公司的glo和PAX实验室的PAX[4]。预计HTP的销量将快速增长[5]。

与电子烟一样，HTP是电池供电为产品加热的可充电设备；不同的是，消费的产品是烟草[6,7]。根据世界卫生组织所述[4]：

HTP加热时或激活含有烟草的装置时，会产生含有烟碱和有害化学物质的气溶胶。这些气溶胶在吸吮或设备抽吸的过程中被使用者吸入。它们含有高度致瘾的烟碱物质以及非烟草添加剂，而且通常是调味的。烟草可能以特别设计的卷烟（如"加热棒"）或烟弹的形式出现。

在FCTC/COP8(22)关于新型烟草制品的决议[8]中，世界卫生组织《烟草控制框架公约》缔约方大会在其第八次会议上指出了HTP的演变，其作为"危害降低"产品的营销以及由此产生的监管挑战，并"对HTP产品的分类和监管向缔约方提供了有限的指导"。因此，他们要求向第九次会议提交一份报告——关于新型烟草制品，特别是HTP，涉及其对健康的影响，包括对非使用者的影响、致瘾性、认知和使用、吸引力、在吸烟起始和戒烟方面的潜在作用、营销（包括促销策略和影响）、降低危害的声明、产品的差异性、缔约方的监管经验和监测、对烟草管制工作的影响和研究差距，并随后提出潜在的政策选择。

本章的目的是通过审查以下方面的现有文献，解决上述要求：

- 根据世界卫生组织对"吸引力"（第3.2节）的定义，包括期望利用电子烟研究的大量信息填补现有文献的空白；
- HTP的致瘾性特征（第3.3节），包括烟碱递送，还参考电子烟的数据，评估了HTP的整体滥用潜力；

■ 吸引力和致瘾性对消费者认知和使用的影响（第3.4节），包括以下相关内容：认识、态度、知识、意图、使用理由和风险认知。消费者使用包括流行程度、使用行为（如频率、强度、持续时间和使用地点）、使用者概况、使用起始、转换、补偿和戒除传统卷烟。同样，从电子烟中吸取的教训被用来假设可能发挥作用的因素。

讨论涵盖不同群体之间不同使用模式的行为影响以及对公共卫生的影响。最后，提出了研究和政策建议。

检索了截至2020年1月的文献数据库PubMed，包括以下关键词：加热型烟草制品、加热不燃烧的烟草制品、HTP、加热棒和加热薄片、烟草棒和IQOS。排除了关于有害性（释放物中的有害物质、体外数据、暴露生物标志物）和环境烟气的研究，以及非英语语言的研究。还使用菲利普·莫里斯国际（PMI）在2016年提交美国食品药品监督管理局（FDA）关于申请授权在美国销售其IQOS HTP系统和调味"烟弹"作为风险降低的烟草制品（MRTP）的数据，以及FDA评估结果文件[9-12]。

3.2 HTP 的吸引力

3.2.1 WHO FCTC第9条和第10条中吸引力的定义

世界卫生组织将"吸引力"定义为[13]味道、气味和其他感官属性、易用性、剂量系统的灵活性、成本、声誉或形象、假定的风险和收益，以及旨在促进使用产品的其他特征。

下面将审查所有这些方面的数据，还将推广到促销策略以及降低危害的宣称。

3.2.2 HTP的吸引力特性

味道、气味和其他感官属性

关于HTP的味道、气味和其他感官属性的研究较少。三项PMI研究[14-16]报告指出，抽吸IQOS抑制了吸烟的冲动，其程度与吸烟相同，但感觉和心理满意度低于卷烟[17]。一项类似的独立研究表明，与自主品牌卷烟相比，HTP PAX心理满意度低、口感好、镇静，但PAX对吸烟欲望没有显著影响[17,18]。日本和瑞士的参与者也表示与燃烧型卷烟相比，对IQOS的满意度较低，有奇怪或令人不愉快的味道和气味，味道较温和，感官线索较少，但咽喉不适较少[19]。

关于HTP中口味的可获得性和可变性的信息也很重要，因为众所周知，各种可用的口味在烟草和电子烟等相关产品方面起着重要作用[20]。例如在韩国，IQOS

烟弹（HEETS）有烟草、薄荷醇、泡泡糖和酸橙的口味，glo Dunhill Neosticks有烟草、薄荷和柠檬姜、樱桃和葡萄口味[7]。KT&G "lil" 棒（Fiit）含有新型口味胶囊（薄荷醇、薄荷、苹果薄荷、泡泡糖和杏仁口味）[7,21]。像胶囊卷烟一样，胶囊棒含有薄荷醇和其他味道，可以掩盖烟草的刺激性，会吸引女性和年轻的非吸烟者[7]。也许是出于对这种新奇事物的反应，英美烟草公司推出了含有胶囊的Dunhill Neosticks（强烈的薄荷醇和烟草/薄荷醇），PMI则推出了Sienna Caps胶囊（"Sienna精选薄荷醇胶囊"）。这些口味的HTP仅在美国以外的国家销售，因为在美国HEETS被认为是一种卷烟产品，不能出售薄荷醇和烟草[11]以外的口味产品[1][22]。

易用性

日本和瑞士参与者报告称，IQOS可替代燃烧型卷烟在室内使用，"因为它不会产生烟灰或气味"[19]。两国的许多使用者都表示，这款产品使用起来很不熟悉，也很复杂，使用IQOS也很麻烦，因为充电器和加热棒可能很笨重，而且IQOS必须充电和清洁。在新一代IQOS中，支架与充电器集成，产品在充电前可以使用10次[23]。使用更加容易方便，从而增加其吸引力。

剂量系统的灵活性

烟碱剂量将在致瘾性章节中的滥用倾向标题下进行描述。这里不作过多的描述。

成本

与燃烧型卷烟（可直接从包装中取出使用）不同，HTP通常需要购买外部设备。这种设备的价格远远超过消耗品的价格，例如，在韩国大约是一包HEETS价格的25倍[3]。尽管HTP的消费税通常低于燃烧型卷烟的消费税，但在不到一半的研究国家[21,24]，HTP的使用成本低于燃烧型卷烟。2018年，以色列的HEETS平均售价比卷烟[25]高出9.5%。在日本，年轻非吸烟参与者评论说，价格可能是一个潜在的障碍，但总的来说，价格使该种产品具有奢侈性和声誉[19]。

声誉或形象：销售点营销、包装和设备

HTP的广告和促销并非总是在其上市国家被禁止[6]。对新型烟草及相关产品的系统互联网检索表明，营销或推广HTP中使用的常用术语包括"降低风险"、"替代"、"清洁"、"无烟"和"创新"[26]。专家访谈和IQOS在日本和瑞士的包装和营销分析也表明，该产品作为一种清洁、别致、纯净的产品进行推销[19]。在以色列，

---

1 卷烟或其任何组成部分（包括烟草、滤嘴或纸张）不得含有人造或天然香料（烟草或薄荷醇除外）或香草或香料作为成分（包括烟气成分）或添加剂，包括草莓、葡萄、橙子、丁香、肉桂、菠萝、香草、椰子、甘草、可可、巧克力、樱桃或咖啡，使烟草制品或烟草烟气保留其特征性味道。

PMI将IQOS作为其"无烟以色列愿景"的一部分，侧重于"降低危害"，并强调该产品清洁，气味少，无烟灰[27]。零售商也描述IQOS产品危害小，不产生烟气，是一种戒烟装置[25]。

IQOS商店作为营销的核心组成部分，在选定的城市中处于突出和战略位置。当HTP在意大利、日本和瑞士的未来IQOS旗舰店发布时，人们对这些产品的认识和使用显著增加[28]。在意大利的"IQOS大使馆"和"IQOS精品店"高档概念商店中，IQOS被宣传为地位的象征，人们可以免费试用[6]。同样，在加拿大，IQOS在许多烟草零售店（安大略省为1029家）销售[29]。在IQOS精品店，促销活动包括用一包卷烟或打火机换一个IQOS设备、举办派对、"见面问候"午餐和下班后活动。商店外的促销元素是IQOS标志，宣传板上写着"建设无烟未来"的标语，销售代表经常使用IQOS。然而，在安大略，为了遵守国家和省级法律[30]，必须拆除IQOS标牌。在韩国，IQOS旗舰店位于首尔的黄金地段，店面设计和产品展示为IQOS[3]提供了干净、精致的外观和感觉。在美国，IQOS在佐治亚州亚特兰大市推出，IQOS商店位于富裕地区的购物中心[31]。

IQOS的名称、设备、包装和商店与吸引儿童和青少年的热门手机较为相似；结合购买过程，这将IQOS定位为技术熟练使用者的高需求、高档产品，与普通卷烟大相径庭[3]。日本和瑞士的参与者发现产品包装很吸引人，甚至非使用者也很感兴趣，这表明该产品的时尚外观与高科技设备相比表现良好[19]。

对于烟弹包装，在以色列，HEETS包装的展示明显地贴近消费者，在大多数情况下用年轻人喜欢的商品外观和包装颜色来表明烟草味道和强度[25]。虽然许多国家的卷烟包装都要求必须带有显示吸烟各种负面后果的图形警示标识，并带有鲜明的烟弹彩色图片，但据我们所知，这在任何国家的HTP中都不需要。例如，在韩国，HEETS包装上只有关于烟碱成瘾的黑白警示标识[3]。烟草制品的视觉设计可以通过暗示产品特征来影响消费者。对三种IQOS包装的测试表明，HEETS包装的吸引力、独特性和品牌净值明显低于万宝路包装[32]；然而，万宝路包装的认知安全性低于其他两种包装。

承担的风险和收益

HTP是烟草公司开发和销售产品的长期传统的一部分，从20世纪60年代的所谓"更安全的卷烟"开始声称这些产品比传统卷烟危害性小[33]。HTP的营销和媒体账号明确或含蓄地声称它们比卷烟更安全，烟草企业也声称一些HTP可以帮助吸烟者戒烟[5,33]。之所以声称HTP的危害性比卷烟小，是因为其有害物质[33]的暴露水平较低。尽管IQOS可能含有较低水平的有害物质，但PMI向FDA申请的关于IQOS作为MRTP的数据并不支持降低风险[34]这一主张。然而，数据确实表明，美国的成年消费者将减少暴露视为降低风险[35]，且此结果在一项针对美国成年人和青少年的独立研究中得到了证实[36]。对德国年龄≥14岁人口调查的前九波分析结果也表示，61名HTP使用者中的大多数人认为HTP危害较小（41.0%）或低很多（14.8%），37.7%认为它们与卷烟一样有害[37]。在意大利，对60名高中生进行的问卷调查的初步结果显示，40名学生认为HTP对健康有害，而24名学生表示，如果朋友推荐这些产品，他们也会尝试[38]。

日本和瑞士的参与者已经确定了使用IQOS的几项可感知的好处，包括降低喉咙不适、包装吸引人、清洁、无烟灰和无烟以及更大的社会可接受性。但只有少数报告表明使用IQOS比燃烧型烟草制品有健康益处[19]。用于戒烟可能是另一个可感知的好处。在意大利，60名高中生中有19名表示，他们会向希望戒烟的人推荐HTP[38]。

还有两个已经发表的有趣的研究，但都是企业内的相关报道。PMI正在德国、意大利和英国等国家开展问卷调查，以估计IQOS和其他含烟草或烟碱产品的流行情况和使用模式[39]。问卷还包括潜在益处（自述牙齿颜色、呼吸气味、运动能力和皮肤外观的改善）、使用经验、风险认知和强化效应体验，如满意度、心理奖励、厌恶、呼吸道感觉的享受和抽烟欲望。另一项方案[40]是一项前瞻性研究，比较随机选HTP或电子烟的吸烟者的卷烟消费量和使用率的变化，还对产品的可接受性、耐受性及其降低危害的潜力进行比较。然而，由于该研究得到了Philip Morris Products SA研究人员发起的研究奖的支持，因此存在潜在的利益冲突，尽管作者表示PMI"在研究方案的设计中没有任何作用，并且在其执行、分析、数据解释或撰写手稿期间也不存在任何作用"。此外，一些作者直接通过烟草企业或其前沿团体的资助，存在与烟草公司相关的未公开利益冲突。

3.2.3 了解ENDS和ENNDS并应用到HTP

电子烟的历史使我们对任何新型烟草或相关产品的上市持谨慎态度，因为电子烟已经吸引了年轻人，但他们可能会继续使用卷烟[33]。由于在HTP中也发现了许多电子烟的功能，因此了解电子烟的吸引力对于HTP是非常有用的。例如，一项对电子烟使用者的调查研究表明，电子烟有吸引力的特点是各种电子烟烟液口味（69%）、电子烟设计（44%）、调整电子烟烟液烟碱水平的能力（31%）、调整设备设置的能力（25%）、各种电子烟设计（21%）、做"大烟"的能力（16%）和价格（13%）[41]。较少的双重使用者、吸烟者和非吸烟者发现这些产品特征具有吸引力，但顺序相同。同样，自我报告数据的分析表明，风味（39%）、价格（39%）、烟碱量（27%）、电子烟种类（22%）、健康声明（12%）、设计（10%）、品牌（9.4%）和包装（3.7%）是使用者在选择电子烟时重点考虑的因素[20]。

在感官特性方面，使用者报告称，与卷烟相比，他们对电子烟的某些方面感到不满意，因为许多电子烟不能像卷烟那样迅速将烟碱释放到血液中，也没有卷烟那样的"击喉感"[1]。尽管这一观察结果是在含有烟碱盐的电子烟烟液上市之前做出的，但这可能是吸烟者尝试HTP而不是电子烟的原因之一。对上述HTP的感官研究表明，吸烟者对味道、气味和击喉感的评价低于卷烟。口味的多样性被认为是电子烟最吸引人的特征[20,41]。因此，禁止或限制可用的HTP口味可能有助于降低它们在非使用者中受欢迎的程度。

虽然电子烟的使用方便性和HTP似乎相似，但电子烟的使用成本要低得多。有人提到，成本是使用电子烟的一个原因[20,41]，而且可能成为一些群体，特别是儿童和青少年使用HTP的障碍。

此外，像电子烟使用者一样，HTP消费者可能不明白，他们必须完全戒烟才能实现HTP声称的健康益处，而且可能还错误地认为，烟草企业未经证实的降低风险的说法意味着HTP是无风险的[33]。

3.3 HTP的致瘾性

致瘾性是一种药物及其传递系统的滥用倾向或滥用潜力的概括指标。在一个模型[42]中，烟草制品滥用的可能性被认为包括重复使用的可能性（概括为药物代谢动力学、药物效应和强化）和使用的后果（对功能的影响、身体依赖、不良影响）。

3.3.1 致瘾性

制造商公布数据最多的三类HTP机器抽吸方案的烟碱释放量可总结如下（以mg/支为单位）：ISO条件下的Eclipse，0.18[43]；ISO条件下的电加热吸烟系统，0.313[44]；HCI条件下的烟草加热系统THS2.2（IQOS），1.32[45]；HCI条件下烟草加热产品THP1.0，0.462[46]。一般来说，HTP烟碱水平低于参比卷烟1R6F的水平，且1R6F具有关注释放物的确定值[47]，如在ISO条件下烟碱含量为0.721 mg/支，在HCI条件下烟碱含量为1.90 mg/支。在业内[48]和业外[49,50]的研究中，不同HTP之间的水平基本相同。特别要注意的是，Eclipse和电加热吸烟系统的烟碱含量较低。在2017年烟碱和烟草研究协会的会议上，英美烟草公司在其HTP（THP1.0）上发布了两张海报[46,51]，并于2017年底在一份期刊的增刊上发表了8项研究[52-60]。这些结果均表明，在标准机器条件下HTP的烟碱释放量比卷烟低。

Bekki等[50]的研究表明，在IQOS中烟碱从烟草填充物向气溶胶的转移率与传统卷烟相当或略高。Salman及其同事[61,62]表明，在ISO条件下，IQOS和传统卷烟在气溶胶中释放的总烟碱量相似（0.77 mg/支和0.80 mg/支），而在HCI条件下略低（1.5 mg/支和1.8 mg/支）。他们还表示，在ISO条件下，IQOS气溶胶中的游离烟碱含量为13%，在HCI条件下为5.7%。Meehan-Atrash及其同事[63]使用核磁共振，发现游离烟碱含量为0.53 mg。Uchiyama等[64]在对IQOS、glo和PloomTECH释放的气溶胶的广泛调查(表3.1)中发现,迄今为止IQOS在气溶胶中递送的烟碱最多,IQOS中的烟碱含量也是glo的3倍多(5.2 mg vs 1.7 mg),尽管转移率类似(IQOS为23%,glo为30%)。

表3.1 各种口味IQOS、glo和PloomTECH的烟碱释放量[64]

HTP品牌和种类	ISO方案(μg/支)	HCI方案(μg/支)
IQOS tobacco	400	1200
IQOS menthol	430	1200
IQOS mint	320	1200
glo bright tobacco	150	570
glo fresh mix	140	510
glo intensely fresh	150	440
PloomTECH regular	70	270
PloomTECH green	68	170
PloomTECH purple	60	250

FDA从PMI提交的数据中得出结论,"转用IQOS的吸烟者的烟碱药代动力学(PK)与继续抽吸传统卷烟的人相似","IQOS具有致瘾性,与燃烧型卷烟类似,具有烟碱递送、致瘾潜力和滥用的可能性",然而,并没有与电子烟等其他替代品进行比较。FDA得出结论,IQOS"提供了足够高的烟碱水平,以满足目前吸烟者的戒断和渴望症状",但没有与ENDS进行比较。图3.1显示了PMI进行的4项研究中,IQOS和卷烟对烟碱释放的药代动力学。美国吸烟者IQOS的平均C_{max}明显低于其他国家吸烟者,卷烟的C_{max}在各个国家之间基本一致。没有发现产品之间的差异可以解释这一发现。Maloney等[65]在一项实验室研究中发现,IQOS显著提高了平均血浆烟碱含量,10次抽吸后从2.1 ng/mL增加到12.7 ng/mL,随意抽吸后增加到11.3 ng/mL,与JUUL相当,但略低于自有品牌卷烟。

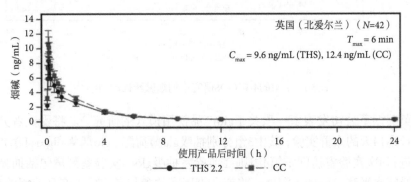

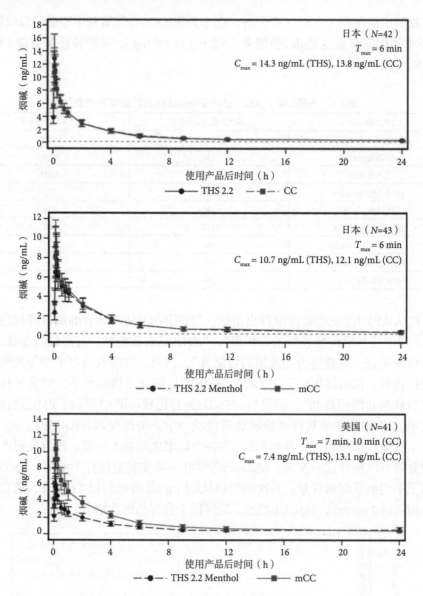

图3.1　PMI的4项IQOS研究中的烟碱药代动力学[11]

英美烟草公司发表了一项关于glo产品使用模式的研究[58]。将受试者分成4组进行长达14天的人群实验，其中3组分别抽吸glo薄荷醇、glo烟草和glo+IQOS家用，最多进行4次实验室访问；第4组则在实验室抽吸glo。实验室测量包括抽吸模式、口腔暴露水平和口腔插入深度。将产品带回家的参与者完成了产品使用的日常日记。总的来说，使用glo产品的抽吸容量约为60 mL，平均每次10~12口，持续时间为1.8~2.0秒,平均间隔时间为8.8秒。每口烟的抽吸容量和总抽吸容量明显高于对照卷烟,但与IQOS产品相当。受试者报告基线时每天抽12~15支烟,在家的4天内每天使用8~12单位的glo/IQOS产品。使用的每一种产品的口腔暴露量低于卷烟,特别是烟碱。

在一项更长时间(90天)的HTP使用研究中[16],使用烟草加热系统的参与者和抽吸卷烟的参与者尿液中的总烟碱当量类似(7 mg/g肌酐 vs 6 mg/g肌酐)。

目前市场上销售的HTP缺乏关于烟碱暴露、抽吸模式和其他衡量滥用倾向指标的独立研究成果;然而,这是一个活跃的研究领域,新的研究结果经常发表,因此应该持续监测同行评审的文献。越来越多的证据表明,IQOS具有容易被滥用的产品特征。目前还没有其他的HTP详细信息。可以得出两个结论:

- IQOS的主流气溶胶提供卷烟烟气中约70%的烟碱。与参比卷烟相比,IQOS的烟碱释放量在57%~103%之间,中位值为64.7%。由烟草企业资助的研究的中位值与独立研究的中位值在统计学上没有显著差异。
- 其他用于分析的HTP在主流烟气中提供一定比例烟碱方面似乎不如IQOS有效。

3.3.2 了解ENDS和ENNDS并应用到HTP

已经发表了许多关于电子烟滥用倾向的研究,这些研究广泛表明,在评估电子烟的滥用潜力时,应该考虑影响烟碱递送的设计特征[66-69]。由于至少有一种HTP能像ENDS一样提供烟碱并抑制渴望[65],因此可以合理地假设,在ENDS中观察到的使用模式可以推广到HTP。然而,HTP和ENDS都有很宽泛的产品类别,一种产品的研究结果可能不适用于其他产品。直接比较领先的HTP和领先的ENDS使用的研究可以解决这个问题。

3.3.3 HTP的总体滥用倾向

如前所述,滥用倾向包括重复使用的可能性(概括为药代动力学)、药物效应、强化(支持未来使用的奖励效应)和使用后果(对功能的影响、身体依赖、不良影响)。HTP似乎能像卷烟一样释放烟碱,并抑制戒断症状和对烟碱的渴望。HTP使用者也表现出烟碱依赖的迹象[16,58,65]。因此,现有的一些数据表示HTP的滥用倾向可能与传统卷烟的滥用倾向相似。根据烟碱递送、感官特性和易用性等因素,不同品牌和类型的HTP滥用倾向可能不同。

3.4 ENDS、ENNDS和HTP的吸引力和致瘾性对风险、危害和使用认知的影响

3.4.1 吸引力和致瘾性对开始、转换、补偿和戒除传统卷烟的贡献

与传统卷烟吸引力方面有关的认知、使用、使用者特征、开始、转换、补偿性或戒烟的研究很少。其中一些解决了使用HTP的原因。一项研究表明，营销和IQOS的开始使用存在直接联系。谷歌检索查询数据显示，在日本，IQOS最大的互联网检索量是在一个受欢迎的全国性电视节目推出IQOS后的那一周[70]。此外，IQOS的使用率从2015年1~2月的0.3%上升至2016年1~2月的0.6%，2017年1~2月上升至3.6%，而2017年其他HTP的预估使用率仍然很低。2016年看过该电视节目的受访者比没有看过的受访者更有可能使用IQOS（10.3% *vs* 2.7%）。

一项针对228名韩国年轻人的在线调查显示，使用IQOS的原因是相信其危害更小或对戒烟有用[71]。一项PMI上市前观察性研究表明，使用该产品的主要预测因素是喜欢气味、味道、回味和易用性[72]。同时使用常规和薄荷醇THS的参与者的接受率高于只使用一种类型的参与者。在日本进行的第一波国际烟草管制调查中，有4684名成年人参与，薄荷醇是最常见的味道（41.5%）[73]。据报道，不同市场的IQOS中薄荷醇的含量有显著差异。

在英国伦敦进行的一项探索性研究[74]中，通过对22名IQOS使用者和8名前IQOS使用者的调查分析，得出影响使用起始和使用IQOS的6个主要因素是健康（想要减少抽烟或戒烟以及降低危害的认知）、成本（启动成本高，但持续成本比吸烟更便宜）、享受和满足感（如自主性、清洁、气味少、触觉质量与燃烧型卷烟相当）、易用性（可接近性、维护或操作的缺陷限制了连续使用，但在无烟场所增加了使用）、使用实践（类似于吸烟，但开发了充电和清洁的新实践、新技术）和社交方面（使用IQOS而不是吸烟改善了社交互动，但一些人分享的社交经验较少）。

PMI发表的关于IQOS前体的研究[75,76]包括对该产品的主观反应信息，这对于理解为什么使用该产品以及它作为替代品的有效性通常很重要。卷烟评价问卷共分为5类12个项目[77]：吸烟满意度、厌恶感、渴望减少、呼吸道感觉享受和心理奖励。关于吸烟欲望的问卷由10个题目组成，单分累计的7分制[78]。

在日本一项关于THS的实验室研究中[14]，研究期间THS的平均满意度得分比传统卷烟下降得多。在波兰进行的一项类似研究[15]中，观察到THS和卷烟在评价

量表上的差异很大,在统计学上具有显著意义。总的来说,卷烟在满足、减少渴望、感觉和奖励方面的得分更高,而在厌恶方面的得分较低。早前对THS2.1的一项研究显示了类似的结果模式,THS在第5天的满意度得分比传统卷烟平均低1.4分($P<0.001$)[79]。在奖励、感觉和渴望因子尺度上也存在显著差异,在所有情况下,THS得分都低于卷烟。Adriaens及其同事[80]对IQOS和ENDS与自有品牌卷烟的主观效应进行了对比研究。他们发现,IQOS和ENDS在减少对卷烟的渴望方面相当,但两者在抑制对卷烟的渴望方面都不如自用品牌的卷烟。一项在日本长期使用行为的研究[16]表明,随着持续使用90天以上,卷烟和HTP在满意度方面的差异逐渐消失。这些研究表明,随着时间的推移,关于吸烟冲动的问卷得分会增加,戒烟的得分也会增加(以Minnesota烟碱戒断量表[81]衡量),这可能表明人们对该产品作为吸烟的长期替代品有些不满。

3.4.2 了解ENDS和ENNDS并应用到HTP

如上文所述,关于吸引力和致瘾性对使用HTP的看法和理由以及开始、转换、补偿和戒除传统卷烟的影响的数据很少。下面我们总结了这些因素与ENDS的吸引力和致瘾性有关的证据。

一项对成年人和青少年电子烟使用者、卷烟抽吸者、双重使用者和非使用者报告的电子烟使用原因的综述表明,成年人对电子烟使用的看法和理由通常与戒烟有关,而青少年喜欢电子烟的新奇性[82]。不吸烟的青少年认为电子烟是一种很酷、很时尚的产品,可以模仿吸烟习惯,而且使用起来很安全。一般来说,认为的好处包括避免吸烟限制、产品很时尚、对健康有好处、比卷烟成本低、积极体验(模仿吸烟习惯、愉快的味道、刺激喉咙、控制体重、提高注意力)、使用安全、戒烟或减少吸烟、社会接受度和无二手烟危害。另一项针对青少年[83]的研究表明,他们使用电子烟的原因是更多样的,而不仅仅是戒烟。与吸烟状态无关,好奇是最初使用电子烟的最常见原因。持续使用电子烟可能是由于吸烟的习惯或不同的烟碱摄入方式。在欧洲[84],使用电子烟最常提到的原因(61%)是停止或减少烟草消费。其他原因包括认为电子烟危害更小(31%)和成本更低(25%)。大于40岁的参与者(76%~78%)比15~24岁的参与者(59%)更常提到减少烟草消费和减少有害物质。

电子烟的口味吸引年轻人和成年人[85]。味道降低了对危害的感知,增加了尝试和开始使用电子烟的意愿。在成年人中,电子烟的味道增加了产品的吸引力,也是使用该产品的主要原因。"烟弹模块"设备由于其设计、用户友好性、不那么令人厌恶的吸烟体验、令人满意的口味和在禁止吸烟地方的自主性而变得很受欢迎,尤其是在青少年中[86]。目前市场上销售的HTP具有这些特征中的几个,这表明了一些ENDS的推广经验。

3.5 讨 论

3.5.1 不同群体不同使用方式的行为影响

关于烟草和相关产品的使用者、前使用者、卷烟抽吸者、双重使用者和从未使用过烟草和相关产品的人对使用烟草和相关产品的看法和理由的研究很少（第3.1节）。虽然烟草制品吸引人或使人成瘾的作用尚不清楚，但关于传统卷烟的认识、使用、使用者概况、开始使用、转换、补偿和戒烟方面的研究是可看到的。由于吸烟者的风险状况与从不吸烟者的风险状况不同，有关几个群体的吸烟模式和流行程度的信息对监管机构来说很重要。

一个重要的问题是HTP主要是吸烟者使用还是非吸烟者也在使用。目前看来，大多数HTP使用者都是吸烟者。例如，在日本，几乎所有使用HTP的人都是当前（67.8%）或既往（25.0%）吸烟者，只有1.0%的人从不吸烟[73]。根据FDA的上市前审查[5]：

尽管关于从不吸烟者、既往吸烟者和年轻人对IQOS的使用的数据有限，但来自IQOS营销国家（意大利和日本）的一些数据显示，年轻人和目前不吸烟者对IQOS的接受度很低。在这些国家，既往吸烟者的接受度略高，但仍然很低。

虽然大多数HTP使用者可能是吸烟者，但大多数人同时使用HTP和传统卷烟，因此都会受到这两种产品的暴露释放物的影响。PMI研究表明，大多数使用IQOS的人同时也吸烟[33]。韩国的研究表明，目前96%的HTP使用者（研究人口的2%）是双重使用者，同时使用HTP与戒烟意图无关[87]。对12~18岁青少年的调查显示，曾经使用过HTP的人群中，75.5%（2.8%）是当前吸烟的人群，45.6%是当前使用电子烟的人群，40.3%是同时使用卷烟和电子烟的人群[88]。在尝试戒烟时是否使用过HTP方面没有发现差异。

PMI和独立数据显示，IQOS会吸引青少年和年轻成年人非使用者开始吸烟[3]。在意大利，营销导致IQOS使用量增加，目的是在非吸烟者和长期前吸烟者中使用IQOS[6,89]。据FDA[5]称：

当然，青少年迅速接受一种新型烟草制品的潜力是存在的。自电子烟进入美国市场以来的十年中，青少年对电子烟的使用迅速增加，但有限的口味选择可能会降低IQOS对青少年的吸引力。在口味选择和IQOS设备的价格方面的有限选择可能会降低对青少年的吸引力。

来自日本的证据表明，较年轻（<30岁）、较富裕的人使用IQOS[90]，而加拿大、英国（英格兰）和美国的新证据表明，HTP至少对包括非吸烟者在内的年轻人有一定的吸引力[91,92]。

3.5.2 公共卫生影响

HTP在一些市场上很可能受到欢迎，因为它们的营销理念是"清洁"、现代、优雅、降低危害的产品。增强其吸引力的因素似乎是感官属性、易用性、成本、声誉和形象以及可感知的风险和收益。虽然HTP能抑制吸烟欲望，但人们普遍认为它不如卷烟令人满意；然而，可以得到不同的口味是众所周知的一个烟草和相关产品吸引人的特点。使用者发现它们产生的烟灰更少，气味更吸引人，因此可能会在室内使用。其他吸引人的特点包括其作为一种独特、现代、危害较小、适合戒烟的产品进行营销。然而，这种产品通常被认为不如卷烟容易使用，而且成本可能是一个障碍。

关于致瘾性，目前市场上销售的HTP可在气溶胶中提供大量的烟碱，它们的药代动力学、药物和主观效应与目前的ENDS产品相似，表明有类似的滥用倾向。关于这些特征如何影响消费者的认知和使用，包括开始使用、更换、补偿和戒除传统卷烟的信息很少。企业研究表明，采用的主要预测因素是对气味、味道、回味和易用性的喜爱。唯一探索吸烟者和戒烟者使用、继续和停止IQOS的原因的独立研究发现，6个重要因素是：健康、成本、享受和满意度、易用性、使用模式和社会因素。

如上所述，绝大多数当前IQOS使用者是当前吸烟者或既往吸烟者，而且大多数是HTP和传统卷烟的双重使用者，无意戒烟。因此，预期减少与吸烟有关的有害物质的暴露将比完全改变暴露方式要小得多，而且风险的降低（如果有的话）可能也会低得多。对于双重使用者来说，任何风险的降低都比完全改抽的人要小，即使是完全改抽的人，风险的降低也没有得到证实。尽管很少HTP使用者从来没有使用过卷烟或曾经使用过卷烟，但他们对HTP的潜在兴趣在人群层面上仍然是一个风险，特别是对从来没有使用过卷烟的人来说。电子烟的历史表明，任何一种新型烟草或相关产品进入市场都会很快受到欢迎。对电子烟的了解表明，有关HTP的担忧以及潜在的监管因素是烟碱水平和"击喉感"、味道种类、营销和降低伤害的认知。此外，就像电子烟[85]一样，HTP可能是抽吸卷烟的一个门户。

3.5.3 研究差距、优先事项和问题

- 几乎没有关于吸烟者、既往吸烟者、电子烟使用者、双重使用者以及从未使用烟草和相关产品的人对HTP的认知和使用理由的信息。只有一项独立研究涉及吸烟者和既往吸烟者使用、继续和停止IQOS的原因。
- 关于HTP的吸引力如何影响消费者的认知和使用，包括启动、切换、补偿和戒除传统卷烟的信息也很少。
- 没有关于HTP是否可以成为使用燃烧型烟草门户的信息。

大多数关于HTP的独立研究都侧重于释放物和有害性，而不是吸引力和致瘾性。研究人员不应该试图复制PMI释放数据和评估降低危害的声明，而应该研究吸烟者、从不吸烟者和戒烟者开始使用HTP的实际轨迹，以及吸引人的产品特性的作用。

对于本章所描述的增强HTP吸引力和致瘾性的所有特征，还应进行其他的独立研究。这些研究应包括：

- 感官研究，包括口味以及其他增强吸引力的成分和添加剂（如糖和保润剂）在使用HTP中的作用；
- 关于知识、态度和风险认知（包括健康信息）的定性和定量行为研究；
- 研究与吸烟者群体、双重使用者、HTP使用者和从不使用有关的所有增强吸引力和致瘾性的所有特征，最好是与吸烟者群体、双重使用者、HTP使用者和从不使用者一起进行，在纵向定量研究中，包括使用其他类型的烟草和相关产品感知不同类型烟草制品的影响，并研究从当前吸烟或从未吸烟到使用HTP甚至燃烧型烟草制品的转变；
- 开展IQOS以外的HTP的药代动力学研究，以建立在现实条件下使用和传输烟碱模式。

3.5.4 政策建议

决策者可考虑采用在许多司法管辖区已成功应用于烟草和相关产品的监管原则，以最大限度地减少产品的吸引力和年轻人的接受程度，提高产品安全性，并最大限度地减少有关使用HTP对健康影响的错误观念。因此，他们应考虑以下措施，重点是保护年轻人和非使用者。

- 禁止向未成年人销售、价格促销、吸引年轻人的口味和风味胶囊；限制在销售点和其他地方的营销；引入素包装，以尽量减少HTP的吸引力和青少年的使用。
- 确保公众不会被它们的吸引力和制造商的声明误导，而是充分了解HTP的风险，包括与卷烟同时使用和怀孕期间使用的风险；纠正错误的看法，反驳错误信息，并澄清减少暴露并不一定意味着减少危害。
- 监测流行率和使用者概况；建立或扩展对产品和使用者的监测，包括人口统计学、其他烟草和相关产品、设备、品牌、类型和口味的使用。

3.6 结　　论

HTP在某些市场很受欢迎，可能是由于多种因素的结合，包括作为"降低风险"、"干净"、"现代"和"优雅"的产品营销。它们的感官特性和易用性通常低于传统卷烟，但作为吸引力的重要维度，决定了使用量。对这些功能如何影响消费者的认知和使用知之甚少。包装、标签、风险宣传、价格和无烟政策等因素似乎会影响使用起始和持续。

目前销售的HTP在气溶胶中提供大量烟碱，其药代动力学、药物和主观效应与当前ENDS产品相似，表明滥用倾向相当。电子烟的历史表明，任何上市的新型烟草或相关产品都可能很快就会流行起来。关于电子烟的知识表明，与HTP相关的关注因素（因此也是潜在的监管目标）是烟碱水平和"击喉感"、口味的多样性、设备设计、营销和降低危害的看法。

3.7 参 考 文 献

[1] Caputi TL. Industry watch: heat-not-burn tobacco products are about to reach their boiling point. Tob Control. 2016;26(5):609–10.

[2] Ratajczak A, Jankowski P, Strus P, Feleszko W. Heat not burn tobacco product – A new global trend: impact of heat-not-burn tobacco products on public health, a systematic review. Int J Environ Res Public Health. 2020;17(2):409.

[3] Kim M. Philip Morris International introduces new heat-not-burn product, IQOS, in South Korea. Tob Control. 2018;27(e1):e76–8.

[4] Heated tobacco products information sheet. Second edition. Geneva: WorldHealthOrganization;2020(https://apps.who.int/iris/bitstream/handle/10665/331297/WHO-HEP-HPR-2020.2-eng.pdf?sequence=1&isAllowed=y, accessed 10 January 2021).

[5] Bialous SA, Glantz SA. Heated tobacco products: another tobacco industry global strategy to slow progress in tobacco control. Tob Control. 2018;27(Suppl 1):s111–7.

[6] Liu X, Lugo A, Spizzichino L, Tabuchi T, Gorini G, Gallus S. Heat-not-burn tobacco products are getting hot in Italy. J Epidemiol. 2018;28(5):274–5.

[7] Cho YJ, Thrasher JF. Flavour capsule heat-sticks for heated tobacco products. Tob Control. 2019;28(e2):e158–9.

[8] Decision FCTC/COP8(22). Novel and emerging tobacco products. Geneva: WorldHealthOrganization;2018(https://www.who.int/fctc/cop/sessions/cop8/FCTC__COP8(22).pdf?ua=1, accessed 10 January 2021).

[9] Philip Morris Products S.A. modified risk tobacco product (MRTP) applications. Silver Spring (MD):FoodandDrugAdministration;2020(https://www.fda.gov/tobacco-products/advertising-and-promotion/philip-morris-products-sa-modified-risk-tobacco-product-mrtp-applications, accessed 10 January 2021).

[10] Marketing order. FDA submission tracking numbers (STNs): PM0000424-PM0000426, PM0000479. Silver Spring (MD): Food and Drug Administration; 2019 (https://www.fda.gov/media/124248/download, accessed 10 January 2021).

[11] PMTA.SilverSpring(MD):Food and Drug Administration;2019 (https://www.fda.gov/media/124247/download, accessed 10 January 2021).

[12] The public health rationale for recommended restrictions on new tobacco product labeling,

[12] advertising, marketing, and promotion. Silver Spring(MD): Food and Drug Administration; 2019 (https://www.fda.gov/media/124174/download, accessed 10 January 2021).

[13] Partial guidelines for implementation of Articles 9 and 10 of the WHO Framework Convention on Tobacco Control. Regulation of the contents of tobacco products and of tobacco product disclosures (FCTC/COP4(10)). Geneva:World Health Organization; 2012 (https://www.who.int/fctc/guidelines/Guideliness_Articles_9_10_rev_240613.pdf?ua=1, accessed 10 January 2021).

[14] Haziza C, de la Bourdonnaye G, Merlet S, Benzimra M, Ancerewicz J, Donelli A et al. Assessment of the reduction in levels of exposure to harmful and potentially harmful constituents in Japanese subjects using a novel tobacco heating system compared with conventional cigarettes and smoking abstinence: A randomized controlled study in confinement. Regul Toxicol Pharmacol. 2016;81:489–99.

[15] Haziza C, de la Bourdonnaye G, Skiada D, Ancerewicz J, Baker G, Picavet P et al. Evaluation of the tobacco heating system 2.2. Part 8: 5-day randomized reduced exposure clinical study in Poland. Regul Toxicol Pharmacol. 2016;81(Suppl 2):S139–50.

[16] Lüdicke F, Picavet P, Baker G, Haziza C, Poux V, Lama N et al. Effects of switching to the tobacco heating system 2.2 menthol, smoking abstinence, or continued cigarette smoking on biomarkers of exposure: a randomized, controlled, open-label, multicenter study in sequential confinement and ambulatory settings (Part 1). Nicotine Tob Res. 2018;20(2):161–72.

[17] Simonavicius E, McNeill A, Shahab L, Brose LS. Heat-not-burn tobacco products: a systematic literature review. Tob Control. 2019;28(5):582–94.

[18] Lopez AA, Hiler M, Maloney S, Eissenberg T, Breland A. Expanding clinical laboratory tobacco product evaluation methods to loose-leaf tobacco vaporizers. Drug Alcohol Depend. 2016;169:33–40.

[19] Hair EC, Bennett M, Sheen E, Cantrell J, Briggs J, Fenn Z et al. Examining perceptions about IQOS heated tobacco product: consumer studies in Japan and Switzerland. Tob Control. 2018;27(Suppl 1):s70–3.

[20] Laverty AA, Vardavas CI, Filippidis FT. Design and marketing features influencing choice of e-cigarettes and tobacco in the EU. Eur J Public Health. 2016;26(5):838–41.

[21] Lee J, Lee S. Korean-made heated tobacco product, "lil". Tob Control. 2019;28(e2):e156–7.

[22] Reversing the Youth Tobacco Epidemic Act of 2019. 116th Congress, 1st session.H.R.2339. Washington(DC):United States Government;2019 (https://www.congress.gov/116/bills/hr2339/BILLS-116hr2339ih.xml, accessed 10 January 2021).

[23] Our tobacco heating system. IQOS. Tobacco meets technology. Lausanne: PhilipMorrisInternational;2020(https://www.pmi.com/smoke-free-products/iqos-our-tobacco-heating-system, accessed 10 January 2021).

[24] Liber AC. Heated tobacco products and combusted cigarettes: comparing global prices andtaxes. Tob Control. 2019;28(6):689–91.

[25] Bar-Zeev Y, Levine H, Rubinstein G, Khateb I, Berg CJ. IQOS point-of-sale marketing strategies in Israel: a pilot study. Isr J Health Policy Res. 2019;8(1):11.

[26] Staal YC, van de Nobelen S, Havermans A, Talhout R. New tobacco and tobacco-related products: early detection of product development, marketing strategies, and consumer interest. JMIR Public Health Surveill. 2018;4(2):e55.

[27] Rosen LJ, Kislev S. IQOS campaign in Israel. Tob Control. 2018;27(Suppl 1):s78–81.

[28] Tabuchi T, Kiyohara K, Hoshino T, Bekki K, Inaba Y, Kunugita N. Awareness and use of electronic cigarettes and heat-not-burn tobacco products in Japan. Addiction. 2016;111(4):706–13.

[29] Mathers A, Schwartz R, O'Connor S, Fung M, Diemert L. Marketing IQOS in a dark market. Tob Control. 2019;28(2):237–8.

[30] Yuen J. Health Canada orders IQOS tobacco storefront to remove its signs.TorontoSun, 1November2018(https://torontosun.com/news/local-news/health-canada-orders-iqos-tobaccostorefront-to-remove-its-signs, accessed 10 January 2021).

[31] Churchill V, Weaver SR, Spears CA, Huang J, Massey ZB, Fairman RT et al. IQOS debut in the USA: Philip Morris International's heated tobacco device introduced in Atlanta, Georgia. Tob Control. 2020:doi: 10.1136/tobaccocontrol-2019-055488.

[32] Lee JGL, Blanchflower TM, O'Brien KF, Averett PE, Cofie LE, Gregory KR. Evolving IQOS packaging designs change perceptions of product appeal, uniqueness, quality and safety: a randomised experiment, 2018, USA. Tob Control. 2019;28(e1):e52–5.

[33] Glantz SA. Heated tobacco products: the example of IQOS. Tob Control. 2018;27(Suppl 1):s1–6.

[34] Lempert LK, Glantz SA. Heated tobacco product regulation under US law and the FCTC. Tob Control. 2018;27(Suppl 1):s118–25.

[35] Popova L, Lempert LK, Glantz SA. Light and mild redux: heated tobacco products' reduced exposure claims are likely to be misunderstood as reduced risk claims. Tob Control. 2018;27(Suppl 1):s87–95.

[36] El-Toukhy S, Baig SA, Jeong M, Byron MJ, Ribisi KM, Brewer NT. Impact of modified risk tobacco product claims on beliefs of US adults and adolescents. Tob Control. 2018;27(Suppl 1):s62–9.

[37] KotzD,KastaunS.E-Zigaretten und Tabakerhitzer:repräsentative Daten zu Konsumverhalten und assoziierten Faktoren in der deutschen Bevölkerung (die DEBRA-Studie) [E-cigarettes and heat-not-burn products: representative data on consumer behaviour and associated factors in the German population (theDEBRA study)]. Bundesgesundheitsblatt Gesundheitsforschung Gesundheitsschutz. 2018;61(11):1407–14.

[38] La Torre G, Dorelli B, Ricciardi M, Grassi M, Mannocci A. Smoking E-CigaRette and HEat-noT-burn products: validation of the SECRHET questionnaire. Clin Ter. 2019;170(4):e247–51.

[39] Sponsiello-Wang Z, Langer P, Prieto L, Dobrynina M, Skiada D, Camille N et al. Household surveys in the general population and web-based surveys in IQOS users registered at the Philip Morris International IQOS user database: protocols on the use of tobacco- and nicotine-containing products in Germany, Italy, and the United Kingdom (Greater London), 2018–2020. JMIR Res Protoc. 2019;8(5):e12461.

[40] Caponnetto P, Caruso M, Maglia M, Emma R, Saitta D, Busà B et al. Non-inferiority trial comparing cigarette consumption, adoption rates, acceptability, tolerability, and tobacco harm reduction potential in smokers switching to heated tobacco products or electronic cigarettes: Study protocol for a randomized controlled trial. Contemp Clin Trials Commun. 2020;17:100518.

[41] Romijnders KA, Krüsemann EJ, Boesveldt S, Graaf K, de Vries H, Talhout R. e-Liquid flavor

preferences and individual factors related to vaping: A survey among Dutch never-users, smokers, dual users, and exclusive vapers. Int J Environ Res Public Health. 2019;16:4661–76.

[42] Carter LP, Stitzer ML, Henningfield JE, O'Connor RJ, Cummings KM, Hatsukami DK. Abuse liability assessment of tobacco products including potential reduced exposure products. CancerEpidemiol Biomarkers Prev. 2009;18(12):3241–62.

[43] Slade J, Connolly GN, Lymperis D. Eclipse: does it live up to its health claims? Tob Control. 2002;11(Suppl 2):ii64–70.

[44] Werley MS, Freelin SA, Wrenn SE, Gerstenberg B, Roemer E, Schramke H et al. Smoke chemistry, in vitro and in vivo toxicology evaluations of the electrically heated cigarette smoking system series K. Regul Toxicol Pharmacol. 2008;52(2):122–39.

[45] Schaller JP, Keller D, Poget L, Pratte P, Kaelin E, McHugh D et al. Evaluation of the tobacco heating system 2.2. Part 2: Chemical composition, genotoxicity, cytotoxicity, and physical properties of the aerosol. Regul Toxicol Pharmacol. 2016;81(Suppl 2):S27–47.

[46] Jakaj B, Eaton D, Forster M, Nicol T, Liu C, McAdam K et al. Characterizing key thermophysical processes in a novel tobacco heating product THP1.0(T). Poster. Society for Research on Nicotine and Tobacco, 7–11 March 2017, Florence, Italy; 2017.

[47] Certificate of Analysis. 1R6F certified reference cigarette. Lexington(KY):University of Kentucky; 2016 (https://www.ecigstats.org/docs/research/CoA_1R6F.pdf, accessed 10 January 2021).

[48] Li X, Luo Y, Jiang X, Zhang H, Zhu F, Hu S et al. Chemical analysis and simulated pyrolysis of tobacco heating system 2.2 compared to conventional cigarettes. Nicotine Tob Res. 2018;21(1):111–8.

[49] Farsalinos KE, Yannovits N, Sarri T, Voudris V, Poulas K. Nicotine delivery to the aerosol of a heat-not-burn tobacco product: comparison with a tobacco cigarette and e-cigarettes. Nicotine Tob Res. 2018;20(8):1004–9.

[50] Bekki K, Inaba Y, Uchuyama S, Kunugita N. Comparison of chemicals in mainstream smoke in heat-not-burn tobacco and combustion cigarettes. J UOEH. 2017;39(3):201–7.

[51] Scott JK, Poynton S, Margham J, Forster M, Eaton D, Davis P et al. Controlled aerosol release to heat tobacco: product operation and aerosol chemistry assessment. Poster. Society for Research on Nicotine and Tobacco.2–5Marc2016,Chicago,Illinois (http://www.researchgate.net/publication/298793405_Controlled_aerosol_release_to_heat_tobacco_product_operation_and_aerosol_chemistry_assessment, accessed 10 January 2021).

[52] Eaton D, Jakaj B, Forster M, Nicol J, Mavropoulou E, Scott K et al. Assessment of tobacco heating product THP1.0. Part 2: Product design, operation and thermophysical characterisation. Regul Toxicol Pharmacol. 2018;93:4–13.

[53] Forster M, Fiebelkorn S, Yurteri C, Mariner D, Liu C, Wright C et al. Assessment of novel tobacco heating product THP1.0. Part 3: Comprehensive chemical characterisation of harmful and potentially harmful aerosol emissions. Regul Toxicol Pharmacol. 2018;93:14–33.

[54] Murphy J, Liu C, McAdam K, Gaça M, Prasad K, Camacho O et al. Assessment of tobacco heating product THP1.0. Part 9: The placement of a range of next-generation products on an emissions continuum relative to cigarettes via pre-clinical assessment studies. Regul Toxicol Pharmacol. 2018;93:92–104.

[55] Taylor M, Thorne D, Carr T, Breheny D, Walker P, Proctor C et al. Assessment of novel tobacco heating product THP1.0. Part 6: A comparative in vitro study using contemporary screening approaches. Regul Toxicol Pharmacol. 2018;93:62–70.

[56] Thorne D, Breheny D, Proctor C, Gaça M. Assessment of novel tobacco heating product THP1.0. Part 7: Comparative in vitro toxicological evaluation. Regul Toxicol Pharmacol. 2018;93:71–83.

[57] Forster M, McAughey J, Prasad K, Mavropoulou E, Proctor C. Assessment of tobacco heating product THP1.0. Part 4: Characterisation of indoor air quality and odour. Regul Toxicol Pharmacol. 2018;93:34–51.

[58] Gee J, Prasad K, Slayford S, Gray A, Nother K, Cunningham A et al. Assessment of tobacco heating product THP1.0. Part 8: Study to determine puffing topography, mouth level exposure and consumption among Japanese users. Regul Toxicol Pharmacol. 2018;93:84–91.

[59] Jaunky T, Adamson J, Santopietro S, Terry A, Thorne D, Breheny D et al. Assessment of tobaccoheating product THP1.0. Part 5: In vitro dosimetric and cytotoxic assessment. Regul Toxicol Pharmacol. 2018;93:52–61.

[60] Proctor C. Assessment of tobacco heating product THP1.0. Part 1: Series introduction. Regulatory toxicology and pharmacology : RTP 2017 doi: 10.1016/j.yrtph.2017.09.010 [published Online First: 2017/10/11].

[61] Salman R, Talih S, El-Hage R, Haddad C, Karaoghlanian N, El-Hellani A et al. Free-base and total nicotine, reactive oxygen species, and carbonyl emissions from IQOS, a heated tobacco product. Nicotine Tob Res. 2019;21(9):1285–8.

[62] Albert RE, Peterson HT Jr, Bohning DE, Lippmann M. Short-term effects of cigarette smoking on bronchial clearance in humans. Arch Environ Health. 1975;30(7):361–7.

[63] Meehan-Atrash J, Duell AK, McWhirter KJ, Luo W, Peyton DH, Strongin RM. Free-base nicotine is nearly absent in aerosol from IQOS heat-not-burn devices, as determined by (1)H NMR spectroscopy. Chem Res Toxicol. 2019;32(6):974–6.

[64] Uchiyama S, Noguchi M, Takagi N, Hayashida H, Inaba Y, Ogura H et al. Simple determination of gaseous and particulate compounds generated from heated tobacco products. Chem Res Toxicol. 2018;31(7):585–93.

[65] Maloney S, Eversole A, Crabtree M, Soule E, Eissenberg T, Breland A. Acute effects of JUUL and IQOS in cigarette smokers. Tob Control. 2020:doi: 10.1136/tobaccocontrol-2019-055475.

[66] Wagener TL, Floyd EL, Stepanov I, Driskill LM, Frank SG, Meier E et al. Have combustible cigarettes met their match? The nicotine delivery profiles and harmful constituent exposures of second-generation and third-generation electronic cigarette users. Tob Control. 2017;26(e1):e23–8.

[67] FearonIM, Eldridge AC, Gale N, McEwan M, Stiles MF, Round EK. Nicotine pharmacokinetics of electronic cigarettes: A review of the literature. Regul Toxicol Pharmacol. 2018;100:25–34.

[68] Voos N, Goniewicz ML, Eissenberg T. What is the nicotine delivery profile of electronic cigarettes? Expert Opin Drug Deliv. 2019;16(11):1193–203.

[69] Yingst JM, Foulds J, Veldheer S, Hrabovsky S, Trushin N, Eissenberg TT et al. Nicotine absorption during electronic cigarette use among regular users. PLoS One/ 2019;14(7):e0220300.

[70] Tabuchi T, Gallus S, Shinozaki T, Nakaya T, Kunugita N, Colwell B. Heat-not-burn tobacco

[71] Kim J, Yu H, Lee S, Paek YJ. Awareness, experience and prevalence of heated tobacco product, IQOS, among young Korean adults. Tob Control. 2018;27(Suppl 1):s74-7.

[72] Roulet S, Chrea C, Kanitscheider C, Kallischnigg G, Magnani P, Weitkunat R. Potential predictors of adoption of the tobacco heating system by US adult smokers: An actual use study. F1000Res. 2019;8:214.

[73] Sutanto E, Miller C, Smith DM, O'Connor RJ, Quah ACK, Cummings KM et al. Prevalence, use behaviors, and preferences among users of heated tobacco products: Findings from the 2018 ITC Japan survey. Int J Environ Res Public Health. 2019;16(23):4630.

[74] Tompkins CNE, Burnley A, McNeill A, Hitchman SC. Factors that influence smokers' and ex-smokers' use of IQOS: a qualitative study of IQOS users and ex-users in the UK. Tob Control. 2020:doi: 10.1136/tobaccocontrol-2019-055306.

[75] Hanson K, O'Connor R, Hatsukami D. Measures for assessing subjective effects of potential reduced-exposure products. Cancer Epidemiol Biomarkers Prev. 2009;18(12):3209-24.

[76] Rees VW, Kreslake JM, Cummings KM, O'Connor RJ, Hatsukami DK, Parascandola M et al. Assessing consumer responses to potential reduced-exposure tobacco products: a review of tobacco industry and independent research methods. Cancer Epidemiol Biomarkers Prev. 2009;18(12):3225-40.

[77] Cappelleri JC, Bushmakin AG, Baker CL, Merikle E, Olufade AO, Gilbert DG.Confirmatory factor analyses and reliability of the modified cigarette evaluation questionnaire. Addict Behav.2007;32(5):912-23.

[78] Cox LS, Tiffany ST, Christen AG. Evaluation of the brief questionnaire of smoking urges (QSUbrief) in laboratory and clinical settings. Nicotine Tob Res. 2001;3(1):7-16.

[79] Lüdicke F, Baker G, Magnette J, Picavet P, Weitkunat R. Reduced exposure to harmful and potentially harmful smoke constituents with the tobacco heating system 2.1. Nicotine Tob Res. 2017;19(2):168-75.

[80] Adriaens K, Gucht DV, Baeyens F. IQOS(TM) vs. e-cigarette vs. tobacco cigarette: A direct comparison of short-term effects after overnight-abstinence. Int J Environ Res Public Health. 2018;15(12):2902.

[81] Hughes JR, Hatsukami D. Signs and symptoms of tobacco withdrawal. Arch Gen Psychiatry. 1986;43(3):289-94.

[82] Romijnders K, van Osch L, de Vries H, Talhout R. Perceptions and reasons regarding e-cigarette use among users and non-users: A narrative literature review. Int J Environ Res Public Health. 2018;15(6):1190.

[83] Kinouani S, Leflot C, Vanderkam P, Auriacombe M, Langlois E, Tzourio C. Motivations for using electronic cigarettes in young adults: A systematic review. Subst Abus. 2020;41(3):315-22.

[84] Attitudes of Europeans towards tobacco and electronic cigarettes (Special Eurobarometer 458, wave EB87.1). Brussels: European Commission; 2017.

[85] Fadus MC, Smith TT, Squeglia LM. The rise of e-cigarettes, pod mod devices, and JUUL among youth: Factors influencing use, health implications, and downstream effects. Drug Alcohol Depend. 2019;201:85-93.

[86] Meernik C, Baker HM, Kowitt SD, Ranney LM, Goldstein AO. Impact of non-menthol flavours in e-cigarettes on perceptions and use: an updated systematic review. BMJ Open. 2019;9(10):e031598.

[87] Hwang JH, Ryu DH, Park SW. Heated tobacco products: Cigarette complements, not substitutes. Drug Alcohol Depend. 2019;204:107576.

[88] Kang H, Cho SI. Heated tobacco product use among Korean adolescents. Tob Control. 2020;29(4):466–8.

[89] Liu X, Lugo A, Spizzichino L, Tabuchi T, Pacifici R, Gallus S. Heat-not-burn tobacco products: concerns from the Italian experience. Tob Control. 2019;28(1):113–4.

[90] Igarashi A, Aida J, Kusama T, Osaka K. Heated tobacco products have reached younger or more affluent people in Japan. J Epidemiol. 2020:doi: 10.2188/jea.JE20190260.

[91] Czoli CD, White CM, Reid JL, O'Connor RJ, Hammond D. Awareness and interest in IQOS heated tobacco products among youth in Canada, England and the USA. Tob Control. 2020;29(1):89–95.

[92] Dunbar MS, Seelam R, Tucker JS, Rodriguez A, Shih RA, D'Amico EJ. Correlates of awareness and use of heated tobacco products in a sample of US young adults in 2018–2019. Nicotine Tob Res. 2020:doi: 10.1093/ntr/ntaa007.

附录 IQOS、卷烟和JUUL产品中的薄荷醇浓度

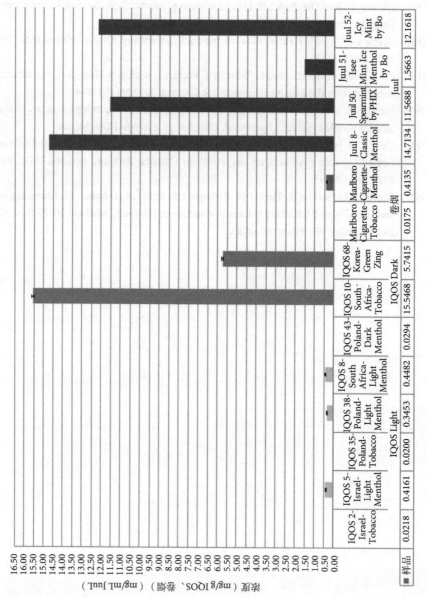

资料来源：Goniewicz ML，未发表数据，2019年

4. 加热型烟草制品之间的差异：考虑因素及影响

Maciej L. Goniewicz, Associate Professor of Oncology, Nicotine and Tobacco Product Assessment Resource, Department of Health Behavior, Division of Cancer Prevention and Population Studies, Roswell Park Comprehensive Cancer Center, Buffalo (NY), USA

摘　　要

加热型烟草制品（HTP）在过去的几年中引起了人们的关注，现已在约50个国家（地区）市场上销售。它的兴起有以下几种因素：制造商的积极营销；使用者认为HTP是其他卷烟产品"更安全的替代品"；HTP兼具向使用者提供烟碱的功能和口味的多样性，例如包含烟草和薄荷醇；技术的进步、产品和产品特性的多样性可供使用者选择。这些产品、装置及其特性之间的巨大差异导致了烟碱和有害成分的释放水平不同，这些使其具有重要的监管意义。正因为如此，必须了解HTP与传统卷烟的差异，以及这些差异如何影响烟碱和其他有害成分的释放，以制定有效的管制措施和政策。因此，本章的目的是描述市场上HTP在其特征和设计特点方面的差异，以及这些特性如何影响产品的有害性、吸引力和对监管的影响。

4.1　背　　景

4.1.1　概述

HTP是一种新型"潜在暴露降低的产品"，由制造商以"降低风险"、"更清洁的替代品"和"无烟产品"进行推广。这些产品的概念来源于一个原则：与吸烟有关的大多数危害源于燃烧过程。在传统卷烟制品中，燃烧锥的温度可以达到900℃，整个烟棒的温度中位值为600℃[1]。这导致包括燃烧、热裂解和热合成在内的大量的热化学反应，从而在烟草烟气中确定了超过7000种化合物[1]。由于燃烧烟草对烟碱的雾化是非必要的（虽然燃烧有助于烟碱的雾化），因此已经开始从可吸入形式的烟碱出发探索烟碱释放的替代方法。这些新产品可能会改善与传统卷烟产品有关的毒理学风险,并影响消费者的接受度,而老一代产品不能做到以上方面。

4.1.2 缔约方大会FCTC/COP8(22)号决议

本文件受世界卫生组织委托,旨在回应世界卫生组织《烟草控制框架公约》(WHO FCTC)缔约方大会第八次会议FCTC/COP8(22)号决议第2段中向公约秘书处提出的关于新型烟草制品的请求的一部分,以解决与这些产品相关的关键议题,并提交一份全面报告。

4.1.3 范围与目标

这份报告包含了对新型HTP产品差异性研究的回顾,以及在要求的相关方面研究结果的审查,尽可能包括对不同类型HTP的特性及其设计特点的描述,它们的成分和释放物、产品多样性、市场分布和制造商以及对使用者、非使用者和监管机构影响的证据的讨论。确定了研究的空白和重点、一些关键问题以及对决策者的一些建议。

4.2 市场上产品的差异

4.2.1 产品类别和类型的概述

HTP具有不同于电子烟碱传输系统(ENDS)和电子非烟碱传输系统(ENNDS)的操作系统(表4.1)。HTP含有热源,可加热烟草并将烟草成分蒸发成可吸入的含有烟碱及其他有害成分的气溶胶。尽管HTP含有烟草材料(与ENDS中的烟液不同),但它们不同于传统卷烟,传统卷烟必须燃烧以产生并向使用者输送烟气。HTP在燃烧过程中无法实现燃烧过程中的高温,因此在比传统卷烟(800℃)更低的温度下(当前HTP在<350℃下工作)使烟草中的烟碱雾化[2,3]。与ENDS一样,据称它们是比传统卷烟烟气中释放出更少有害成分的产品。

表4.1 加热型烟草制品(HTP)、卷烟和电子烟碱传输系统(ENDS)的产品性能特征和主要成分

特征	卷烟	HTP	ENDS
烟碱	有	有	有*
烟草	有	有	无(烟碱主要来自于烟草)
燃烧	是	否(不完全燃烧的潜在风险)	否(烟碱溶液中成分的热裂解风险)
加热	是(抽吸时温度极高)	是(通常低于传统卷烟;可能会过热)	是(通常低于传统卷烟;可能会过热)
电子系统	无	有	有

续表

特征	卷烟	HTP	ENDS
产品示例			

a. 电子非烟碱传输系统（ENNDS）不含烟碱

HTP系统具有三个常见组件：包含加工烟草的插入物（例如棒、胶囊或烟弹）、加热烟草的装置（电池、碳或气溶胶）以及用于电加热设备的充电器。制造商对HTP使用了4种基本设计方法，这些方法可以作为产品分类的依据。它们的不同之处在于烟草材料的加热方式以及是否与加热元件分离（表4.2）。

表4.2 加热型烟草制品按烟草加热机理的分类

主要特性	产品样例
集成加热元件	Premier, Eclipse, PMI "Platform 2" (TEEPS)
带有特殊"卷烟"的外部加热元件	Accord, Heatbar, IQOS, glo
混合型设备："气溶胶"混合烟草；间接加热	iFuse, PloomTECH
松散烟草材料加热室	PAX

第一种，也可以说是最古老的加热型烟草制品类型，是一种带有可用于雾化烟碱的嵌入式热源的类似于卷烟的装置（HTP类型1）。第二种类型是使用外部热源将专门设计的卷烟中的烟碱进行雾化的装置（HTP类型2）。这是菲利普·莫里斯国际公司（PMI）的IQOS（及其前身Accord和Heatbar）和英美烟草公司（BAT）的glo的基本设计。出于监管目的，它们代表了两种类型，一类基于加热装置，另一类基于含烟草的烟棒。烟棒一般符合世界海关组织对卷烟的分类（纸卷烟草；协调制度 2402.20）。在美国，IQOS设备由FDA作为附件类监管（类似于ENDS、ENNDS和水烟），而加热棒符合卷烟的法定定义[4]。第三种类型是使用ENDS技术从少量烟草中提取烟碱和烟草风味物质（HTP类型3）。英美烟草公司的iFuse产品类似于一种混合ENDS烟草制品，其气溶胶在流经烟草后到达使用者。日本烟草国际公司（JTI）的PloomTECH也是类似的操作方式，不同之处在于该公司产生气溶胶的溶液不含烟碱。第四种类型是利用加热密封腔直接从松散烟草中雾化烟碱（HTP类型4）。此类以干草本汽化器为代表，例如PAX，其主要应用于大麻，但也可以雾化烟草中的烟碱。在美国，至少从20世纪70年代以来，烟草公司一直关注于大麻以及其作为潜在产品和竞争产品的合法化。HTP等加热装置可与含有大麻的插入物一起使用，并与大麻一起销售。

4.2.2 加热型烟草制品之间的差异

非燃烧的加热型烟草的概念出现于20世纪80年代，分别来自美国的烟草公司Philip Morris和RJ Reynolds，产品名称分别是Accord和Premier。与这些产品和概念上类似的产品一直在不断发展，现在可能会占据重要的市场份额。ENDS的引入、积极的营销和日益普及可能为这些产品的成功创造了条件，另一方面，这些产品也改变了社会规范以及人们对吸烟和提供烟碱的设备的看法。类似于推广ENDS和ENNDS的策略已被用于HTP的积极推广和营销，这些产品现已在大约50个国家/地区销售，并计划扩展到其他市场。

4.2.3 产品类型和类别的市场分布

由于卷烟销量下降，人们对吸烟健康风险的认识提高，世界卫生组织《烟草控制框架公约》的贯彻实施越来越好，以及ENDS和ENNDS最近在商业上的成功，烟草公司已在全球市场上重新引入HTP。尽管它们在公共卫生的用途尚不清楚，但PMI、BAT和JTI的营销策略均基于产品降低危害的声明。自2014年以来，这些公司都在多个国家推出了新一代HTP品牌[5]。2018年国际HTP市场估值为63亿美元[6]，预计未来几年市场将大幅增长[7]。HTP已在烟草市场中占据了相当大的份额，尤其是在日本。市场分析师的报告表明，日本拥有世界上最成熟的HTP市场，占2018年HTP销售额的85%[6]，而PMI主要HTP品牌（IQOS）的插入式烟草棒占2019年7~9月所有烟草销售额的17%[8]。表4.3显示了日本三个主要HTP产品及其定价信息。在韩国，IQOS和韩国本土HTP（KT&G，lil®）于2017年同时上市[9]。HTP在其他地方也越来越受欢迎。BAT报告称，其HTP品牌glo于2019年6月在波兰、罗马尼亚和塞尔维亚的全国烟草市场中占有至少5%的份额[10]。IQOS自2016年12月起在英国（英格兰）和加拿大的特定城市进行网上零售，而BAT于2017年5月在加拿大温哥华推出了glo。早期数据表明，在这两个国家，HTP上市3~6个月内，消费者对HTP的认知有限且消费量可以忽略不计，BAT于2019年9月停止了在加拿大的销售[11]。尽管如此，自2018~2019年以来，PMI报告称，其所谓的"降低风险"产品线（包括IQOS）在英国（英格兰）和加拿大的收入增长了92.5%，在加拿大增长了44.2%，这表明对HTP的需求正在增长。与此形成鲜明对比的是，澳大利亚[12]有效地禁止销售HTP，包括印度、沙特阿拉伯和新加坡在内的其他一些国家同样禁止销售HTP，直到最近，美国还禁止销售IQOS、glo和其他HTP。然而，2019年4月，IQOS和三个品牌的"加热棒"被授权作为烟草制品销售[13]，预计其他HTP品牌也将紧随其后。2020年7月7日，FDA批准IQOS为"风险降低的烟草制品"，因为科学研究评估表明，完全从传统卷烟转向IQOS系统显著减少了对有害及潜在有害成分的暴露[14]。

表4.3　日本3种最受欢迎的加热型烟草制品的类型和价格信息[15]

	IQOS	glo	PloomTECH
设备图片			
推出时间	2014年11月	2016年12月	2016年3月
制造商	Philip Morris International	British American Tobacco	Japan Tobacco International
烟草插入类型	棒	棒	胶囊
设备代系	第一代：IQOS 第二代：IQOS2.4 第三代：IQOS3和IQOS3 Multi	第一代：glo 第二代：glo Series2和glo Series2 Mini	第一代：PloomTECH 第二代：PloomTECH+ 和 PloomS[a]
插入物品牌	Marlboro Heatsticks	Kent Neostick	Mevius for PloomTECH
插入物口味	Balanced Regular, Menthol, Mint, Purple Menthol, Smooth Regular	Bright Tobacco, Citrus Fresh, Dark Fresh, Fresh Mix, Intense Fresh, Refreshing Menthol, Regular, Rich Tobacco, Smooth Fresh, Spark Fresh, Strong Menthol	Brown Aroma, Cooler Green, Cooler Purple, Red Cooler, Regular
价格[b]	IQOS 2.4：¥ 7980 (~US$ 76) IQOS 3 Multi：¥ 8980 (~US$ 85) Marlboro Heatsticks：¥ 500 (~US$ 4.73)	glo：¥ 2980 (~US$ 28) glo Series 2：¥ 2980 (~US$ 28) glo Series 2 Mini：¥ 3980 (~US$ 38) Kent Neostick：¥ 420 (~US$ 3.97)	PloomTECH：¥ 2980 (~US$ 28) PloomECH+：¥ 4980 (~US$ 47) PloomS：¥ 7980 (~US$ 76) Mevius for PloomTECH：¥ 490 (~US$ 4.64)

a. 使用的是烟棒而非胶囊
b. 相比之下，一包传统卷烟的价格约为500日元（约合4.73美元）

4.3　产品特征和设计特点

4.3.1　产品的温度分布情况和操作性能

在PMI的IQOS中，烟草棒（"加热棒"）由插入末端的加热片加热，以便抽吸时热量通过烟草散发[2]。随后气溶胶通过一个中空的醋酸纤维管和一个聚合物膜滤嘴到达口腔内。该产品的设计温度不超过350℃，当达到350℃时供应给加热片的能量将被切断。BAT将其glo产品描述为由两个独立控制室组成的加热管，使用者可以通过设备上的按钮激活，并在30~40秒内达到240℃的工作温度[3]。BAT的iFuse产品似乎是一种混合HTP，它有烟液成分，但也含有烟草；它将液体通过烟草产生的气溶胶后到达使用者。一项研究表明，当气溶胶通过烟草腔室时（从

35℃到32℃），其热量损失很小，这意味着一些烟草会被加热[16]。据报道，在使用无烟草室产品的条件下吸烟，有害成分的释放与来自ENDS的几乎相同，这意味着烟草的贡献很小。JTI的PloomTECH以类似的方式运行，不同之处在于用于制造气溶胶的溶液不含烟碱。

4.3.2　电池性能

不同的HTP设备使用不同的热源，包括来自电池的电能和用火柴或打火机加热的碳端。大多数HTP具有锂离子电池，可充电并用于许多产品，例如笔记本电脑、手机和电动汽车。所有锂离子电池都以相同的方式工作：溶剂中的离子在两个带相反电荷的电极之间流动，两极之间由可渗透的薄片隔开，流动的方向取决于电池是充电还是放电。一般来说，锂离子电池被认为是安全的，但是，如果极间的隔板被破坏，则极间短路，导致温度升高，进而导致高度易燃的电解质溶剂燃烧，从而发生爆炸。

4.3.3　烟草插片、烟棒和胶囊的特性

PMI的IQOS（长45 mm，直径7 mm）的烟棒包含大约320 mg的烟草材料，而传统卷烟的长度为84~100 mm，直径为7.5~8.0 mm，包含大约700 mg的烟草材料[17]。IQOS中的烟草似乎不是典型的烟草切丝填料，而是一种增强型烟叶网状物（一种再造烟草），含有5%~30%（质量分数）的气溶胶形成成分，如多元醇、乙二醇酯和脂肪酸[17]。该混合物可与加热系统共同使用的气溶胶产生基质。BAT glo产品的烟草插入物是一根长82 mm、直径5 mm的烟棒，可插入加热腔。烟棒由烟杆、管状冷却段、过滤嘴和烟嘴组成。它包含大约260 mg再造烟叶和14.5%甘油作为雾化剂[3]。BAT的iFuse产品是一种混合产品，其烟弹含有浓度为1.86 mg/mL的烟碱溶液，和烟草共同的机器输送量为20~40 μg/口[16]。然而，很难估计iFuse设备中烟草对使用者吸入的雾化烟碱的贡献（如果有的话）。JTI的PloomTECH以类似的方式运行，不同之处在于类似烟液ENDS的组成成分不含烟碱（图4.1）。

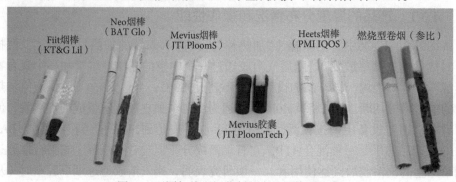

图4.1　不同加热型烟草制品中的烟棒和胶囊

4.4 产品的成分、释放物及总体设计

4.4.1 成分和释放物

目前对HTP产品的单个成分和释放物知之甚少，应对烟草中的许多雾化成分和产品的其他成分（包括保润剂和添加剂）对健康的影响进行调查，以确定它们对消费者和非消费者健康的影响。HTP烟草插入物通常含有并释放有害成分，包括致癌化学物质、呼吸道刺激物和心血管毒素，如烟草特有亚硝胺、金属、挥发性有机化合物、酚类化合物、多环芳烃、少量烟草生物碱和有机溶剂[18-21]。这些化学物质中许多被归类为致癌物，吸入时有害或潜在有害。这些有害成分以不同的量存在于HTP中，但其含量通常低于传统卷烟中的含量（图4.2）。

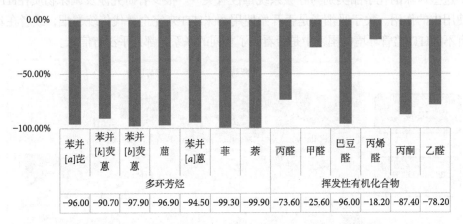

图4.2 与传统卷烟的烟气相比，在IQOS释放物中选定的有害化学物质的下降百分比[18]

FDA的毒理学审查发现IQOS气溶胶中某些有害和潜在有害成分的含量低于3R4F标准参比卷烟和市售传统卷烟的烟气；然而，FDA还发现Heat烟棒气溶胶中的80种化学物质，其中包括IQOS独有的，或者比3R4F烟气中的含量更高的4种可能致癌的化学物质。此外，IQOS气溶胶还含有15种可能具有遗传毒性的其他化学物质，以及20多种通常被认为可以安全摄入的化合物，但这些化合物对健康有潜在的不利影响[22,23]。FDA技术项目牵头审查得出的结论是："尽管有些化学物质具有遗传毒性或细胞毒性，但这些化学物质的含量非常低，相较于传统卷烟中发现的有害和潜在有害化学物质的数量和含量大幅减少，因而潜在影响被抵消，并且这些化学物质的存在不会引起公众健康方面的重大关注。"[24]

关于吸入HTP中使用的大剂量保润剂的相关风险，如丙二醇和植物甘油虽然已被批准用于其他用途，但尚未得到很好的表征。例如，丙二醇通常用作食品和

化妆品中的添加剂、药物中的溶剂、防冻剂以及气雾的成分。对暴露于此类气雾中的剧院工作人员的健康影响的研究得出结论,大量、长时间的暴露会导致呼吸道刺激[25,26]。植物甘油虽然作为一种无毒添加剂广泛用于食品和化学工业,但可能会带来与HTP一样的风险,因为它们在高温下会产生有毒醛类化合物(包括甲醛、乙醛、丙烯醛和丙酮),其中一些被列为致癌物。

一些HTP含有调味物质,包括薄荷醇和水果调味剂(图4.3,另见第6章)。虽然大多数调味物质也常用于食品和室内香水,但人们对其在吸入后对健康的影响知之甚少。在HTP中使用调味物质很普遍,并且经常作为消费者选择某种产品的主要原因[15]。据报道,调味物质在塑造消费者对新型烟草制品的看法方面也发挥着重要作用,因为它们的使用关系着产品的尝试与开始使用[27]。市场上有多种口味可供选择,包括烟草、薄荷、果味和咖啡。某些化学品(如香兰素、柠檬烯、异戊醇)和用于提供这些风味的化学品类别与呼吸系统毒性有关[28]。很少有研究涉及调味物质在HTP使用中的作用。关于吸烟者是否开始使用调味HTP来完全替代传统卷烟,或者在具有不同HTP监管环境的国家中是否看到了相同的联系,都几乎没有信息。

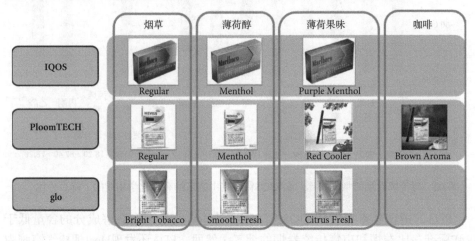

图4.3 用于加热型烟草制品的各种加香卷烟插入物、烟弹和胶囊

作为一个产品类别,HTP非常多样,在热源、加热元件和加热温度方面存在差异。这些特征中的每一个都会影响烟碱和非烟碱有害成分的释放及其向使用者的递送。加热温度尤其会影响保润剂热裂解产生的呼吸系统有害物质(包括甲醛、乙醛和丙烯醛等羰基化合物)的释放。一些HTP可能会产生高浓度的有害化学物质。

4.4.2 烟碱递送

大多数HTP中的烟碱是通过加热烟草达到使烟碱雾化但烟草无法燃烧的温度从烟草中释放出来的。在一些混合HTP中,烟碱存在于制造可吸入气溶胶的溶液中。原则上，非燃烧产生的挥发烟碱会产生一种不太复杂的气溶胶，与传统卷烟相比，其有害成分更少。烟碱不能以气体形式有效传递，要将烟碱输送到使用者的肺部，必须添加气溶胶形成剂以将烟碱悬浮在气溶胶颗粒上。

烟碱以两种形态存在于HTP气溶胶中：非质子化游离态烟碱和质子化烟碱盐。烟碱是一种弱碱，可以通过改变溶液的pH来增加或减少任何一种形式的烟碱含量。游离态烟碱是挥发性的，比单质子化形式更容易吸收，在人体中产生增强的电生理和主观效应，因此可能更容易上瘾[29]。然而，当吸入时，游离态烟碱比质子化烟碱盐对喉咙更具刺激性[29]。实验室研究表明，IQOS气溶胶中的烟碱主要以质子化盐的形式存在[20,30]。

烟碱是一种具有广泛健康影响的具有药理活性的化合物。虽然它可供成年人在受控条件下安全使用，但它与发育中的胎儿的各种不良健康效应有关，包括胎儿生长受限、早产和死胎的风险，并可能影响青少年的大脑发育[31-33]。此外，有证据表明，摄入高剂量的烟碱会造成急性中毒风险[34,35]。

4.4.3 风险概况

由于HTP近年来才出现在烟草市场上，因此其有害性和可能的减害潜力（包括生化和行为）的科学证据仍在积累。企业资助的信息来源报告称，在传统卷烟完全转用HTP后，血清和尿液中的有害成分浓度降低[36-43]，而独立研究报告称，与传统卷烟相比，使用HTP后细胞毒性[44]和烟草特有亚硝胺水平均降低，虽然其水平仍高于ENDS[21]。与ENDS不同，HTP含有烟草，因此，即便没有燃烧，预计使用者仍会接触烟草材料中存在的多种化学物质。此外，在燃烧过程中会完全燃烧和分解的一些烟草成分可能存在于HTP的释放物中。因此，HTP可能提供具有独特毒理学特征的混合物，并且减少接触特定有害物质的潜在好处可能会被新的健康风险所取代。侧流释放物和二手暴露也是HTP的一个问题。一些烟草制造商声称某些HTP会导致最小的侧流暴露，而其他研究则显示出更高的暴露水平[45]。这可能部分取决于每个产品的设计。

在目前可用的人体暴露研究（均由制造商进行）中，IQOS似乎比卷烟释放的有害物质更少，并且可以作为卷烟的有效短期替代品，这是通过烟碱释放和主观效应来评估的。尽管最近发表了一项随机试验的研究方案，表明BAT在这一领域的工作，但关于BAT HTP的已发表数据有限。目前还没有关于JTI Ploom产品的已发表研究。然而，最近的研究表明，HTP与传统卷烟和/或ENDS同时使用的患病率很高[46-48]。关于同时使用ENDS和传统卷烟的研究通常没有显示出对烟草特定有害物质的暴露有任何显著减少[49-51]。因此，HTP和传统卷烟的双重使用者对有害物质的暴露不太可能显著减少。

4.4.4 加热型烟草制品成分的监管意义

将HTP引入当前的烟碱和烟草制品格局，是一个监管挑战（另见第3章），因为目前无法确定这些产品是否可以在降低吸烟者风险或危害方面发挥作用。各国对新出现的烟碱和烟草制品管制的差异可能会影响烟碱的含量和释放、设备的使用、非吸烟者的吸收以及吸烟者对传统卷烟的替代。监管环境的差异也可能通过影响这些产品的可用性来影响整体使用和消费者行为。因此，重要的是要了解新型烟草制品市场的性质，包括HTP的多样性及其使用方式，以充分了解监管对环境的影响、不同类型设备对使用的潜在影响及其潜在危害。

HTP通常不受强制性制造标准的约束，因此缺少防止成分标签不准确或有害成分的监督。虽然香料配方的商业敏感性可能会限制成分的披露，特别是对于天然原料成分，但是政府和卫生当局应要求向有关政府机构报告HTP成分。一些地区制定了政策来规范加热卷烟（包括HTP）的制造和营销，以减少其对公众的吸引力和健康影响；其中包括欧盟、加拿大、埃塞俄比亚和摩尔多瓦共和国[52]。一些国家将HTP视为烟草制品，适用于其他烟草制品的规定也适用于HTP。监管机构可能会对HTP施加各种各样的产品标准，例如最大烟碱含量和释放量以及烟草相关有害成分、重金属、杀虫剂、残留溶剂、霉菌、酵母、霉菌毒素的阈值，以及其他化学和生物杂质。HTP的产品标准还可能指定禁用的化学品，包括某些调味剂、着色剂和甜味剂。

监管机构还应警惕产品设计和成分的变化。在向纽约消费者分析师小组[53]的报告中，PMI声称IQOS是由2014~2017年间日本的原始设计改进而来的，包括美观性、刀片自清洁技术、改进的用户界面、更快的充电、蓝牙连接、随附的移动应用程序和颜色的使用，从而增加设备的吸引力。因此，一种产品在推出后可能会发生变化，正如烟草行业对其卷烟产品随着时间的推移和市场的不同进行细微调整的做法，例如2017年万宝路不一定与2010年万宝路相同，并且法国销售的万宝路不一定与美国销售的相同。此外，对产品的研究可能不是针对消费者当前可用的产品，而是针对原型（甚至一系列不同的原型）。虽然这种做法本身并无恶意，但应确定所研究产品和上市产品之间在设计、功能或外观方面的任何差异，以及这些差异是否影响以及如何影响消费者使用。在美国，FDA遵循营销授权途径，在该途径中，必须报告产品的变更并可以对其进行监管，对影响公众健康的变更（例如有害成分递送的变更，实质性设计变更）进行更严格的审查和要求。欧盟成员国在烟草制品指令中也有类似的规定。其他国家/地区的监管机构应考虑类似的关于产品变更通知和理由的规定。

4.4.5 释放物的监管意义

鉴于已发现HTP含有多种有害成分,产品测试和成分分析可以显示出消费者可能接触到与健康相关的化学品。新型烟草制品可以在获得分析测试许可的独立实验室进行测试。目前尚不清楚HTP可能释放的有害成分是否因市场而异。例如,无烟立法是否涵盖这些产品取决于具体的措辞[53,54]。预防原则则支持将它们纳入此类法规。

与HTP释放有关的另一个问题是实验室间报告的结果存在差异,这可能是分析方法、测试产品和所用雾化方案的不同造成的。例如,气溶胶采样的抽吸持续时间可能因实验室而异,这将影响释放物的有害成分,因为增加抽吸持续时间会增加吸入的气溶胶质量并令使用者暴露于更高水平的有害成分。包括抽吸口数、流速和抽吸间隔在内的抽吸方案可能影响释放物的其他方面。使用为传统卷烟设计的标准抽吸方案,例如国际标准化组织或加拿大卫生部(深度抽吸)的方案,可以减少释放结果的差异。然而,标准化的雾化方案可能并不总是适用的,因为某些产品可能受限于其设计。例如,IQOS旨在确保与卷烟相同的抽吸持续时间和抽吸口数,即最多14次抽吸或使用6分钟。因此,抽吸方案可能由所测试的设备决定。HTP的这些特点对设计监管提出了更多的挑战。

主要的ENDS和HTP公司在全球范围内销售他们的产品的同时,有一些较小的公司销售整个设备和设备部件。例如,一些非PMI制造的烟草插入物可用于IQOS设备[55,56]。英国Imperial Brands Plc为其Pulze HTP系统制造的烟草棒也适用于IQOS设备。由于电源等特性的变化,这种组合可能会产生与原始产品不同的有害物质。因此,在设计规范时应考虑产品的组合。

4.5 产品、制造商和销售点的差异

4.5.1 制造商和销售点

目前大多数HTP是由大型跨国烟草公司制造和销售的。虽然HTP的分销具有独特性,但HTP的插入物通常在传统零售店出售,而这些设备在专卖店和网上销售。因此,专卖店依靠单一产品,销售代表解释设备、免费清洁客户的设备并提供免费试用服务(图4.4)。HTP专卖店通常拥有干净、时尚、现代的设计(如Apple专卖店的美感)。

图4.4　日本和韩国的HTP商店（2019年）

4.5.2　对客户吸引力的影响

即使在许多国家吸烟流行率持续下降，吸烟仍然是世界范围内可预防性发病率和死亡率的主要原因之一。在使用过程中吸入的燃烧副产品是吸烟造成健康影响的主要因素。因此，如果吸烟者转而只使用改良了化学和物理特性的新的烟碱和烟草制品，则可能降低吸烟者的健康风险。关于HTP减害的潜在用途已有研究。企业数据表明HTP可以作为高度受控环境中的长期替代品[36-41]。然而，正如本报告其他地方所指出的，这些问题应该通过独立研究来解决。还有人担心"现实世界"同时使用卷烟和HTP可能会延长吸烟行为，即吸烟者根据情况使用HTP，而不是完全戒烟[57]。此外，基于几个亚洲国家的人口研究发现，其他烟草制品和HTP的一起大量使用引起了人们对这些产品是否可以替代传统卷烟或作为补充产品使用的担忧[15,46-48,58]。大型烟草公司使用各种营销策略向不同的社会人群推广HTP，与卷烟营销方式的差异导致其对青少年和年轻人、女性、少数民族和关注健康的吸烟者等人群具有不成比例的吸引力（另见第10章）。

人们对HTP使用者的隐私和安全以及HTP设备和烟草公司如何收集和处理使用者的个人信息提出了一些担忧。HTP是第一种可以收集使用者吸烟习惯个人数据的烟草制品。PMI已经为在IQOS公司注册的客户建立数据库[59]。一些HTP设备，包括PMI的IQOS，配备了微控制器芯片，可以存储使用信息并可能将信息传输给生产商。数据可能包括详细信息，例如抽吸口数以及每天吸烟的次数。获得的数据可能会被烟草行业用于营销。根据PMI代表的声明，该公司在调查故障时从设备中提取数据[59]。

4.6 讨 论

HTP是电池供电的设备，可将烟草中雾化的烟碱以及其他有害成分递送给使用者。它们是一种新型的"潜在暴露降低产品"，由制造商宣传为"降低风险"、"更清洁的替代品"和"无烟"产品。HTP已在世界各地销售，宣称它们比传统卷烟危害更小，因为它们让使用者接触到较低水平的某些有害成分。然而，关于各种类型和品牌的HTP产生的气溶胶的化学性质、毒理学、对临床措施的影响、对产品及其包装和行为因素的认知，几乎没有来自独立研究的证据。

HTP引发了许多监管挑战。烟碱和非烟碱有害成分的含量取决于产品特性，例如烟草插入物的成分、加热元件可以达到的温度以及设备设计和特性。了解这些特征如何影响重要的产品特性，如温度、烟碱和非烟碱有害成分的释放，对于设计有效的法规和限制这些产品的有害性至关重要。目前市场上的设备差异很大，使用者可以控制许多设备功能从而影响释放，包括许多有害化学物质的释放，如醛、金属、挥发性有机物和活性氧。由于HTP可能会释放传统卷烟中不存在的化学物质，因此对HTP释放的化学评估应超出卷烟烟气中发现的范围。此外，HTP技术正在迅速发展，新的、更先进的设备不断进入市场。这些设备可能具有增加释放物中有害成分水平的新功能。跟踪市场上的HTP及其可能具有的任何新功能，将有助于评估新功能如何影响气溶胶有害物质的释放和对使用者健康的影响。

一个重要的问题是HTP可能会增加其他烟草制品的伴随使用。虽然企业数据表明HTP可以在高度可控的环境中用作长期替代品，但有人担心"现实世界"同时使用卷烟和HTP可能会延长吸烟行为，即吸烟者根据情况使用HTP而不是完全戒烟。当前的吸烟者可能不明白"完全转换"意味着他们必须戒掉传统卷烟才能实现HTP声称的健康益处。然而，根据世界卫生组织的说法："戒烟"是指在不使用戒烟辅助工具的情况下完全戒烟至少6个月。由于HTP是烟草制品，从使用传统卷烟转换为HTP不是戒烟[60]。目前，尚无关于HTP如何影响吸烟者戒烟意愿的信息。

4.7 结 论

- HTP是一类新型的"潜在暴露降低产品"，由制造商宣传为"降低风险"、"更清洁的替代品"和/或"无烟"产品。
- 作为一个产品类别，HTP是特异的，在材料、配置、烟草插入物的含量和加热元件可以达到的温度方面都不同。这些特征中的每一个都会影响烟碱和非烟碱有害物质的释放。
- HTP含有并释放烟碱。
- HTP会释放大量有害化学物质，包括烟草特有亚硝胺、醛类和金属，但是这些产品的单一使用者似乎比吸烟者接触到的有害成分的水平更低。
- 与ENDS不同的是，HTP含有烟草，因此预计使用者会接触到烟草材料中存在的多种化学物质。
- HTP可能会释放出传统卷烟中不存在的化学物质。
- 虽然HTP可能使使用者接触到的某些有害成分含量低于卷烟，但它们也可能使使用者接触到更高含量的其他有害物质。
- 由于这些产品最近被引入烟草市场，关于其有害性和长期健康影响的科学证据仍在积累。
- 虽然这些新型烟草制品的公共卫生效益尚不清楚，但烟草公司正在广泛使用基于潜在危害降低的营销策略。
- 尽管企业数据表明，在高度可控的环境中，HTP可以作为长期替代品，但基于人群的独立研究引起了人们的担忧，即"现实世界"同时使用卷烟和HTP可能会延长吸烟行为。

4.8 研究差距、优先事项和问题

独立于企业进行的研究需要向产品使用者、公共卫生专业人员和监管机构通报HTP的潜在公共卫生影响。随着烟草市场在各种监管环境中不断发展，评估HTP认知和使用趋势至关重要。

必须彻底调查所有新型烟草制品（包括HTP）的化学特征和有害性。重要的是要了解这些产品相对于传统卷烟、ENDS和其他烟草制品在风险序列中的位置。

对传统卷烟使用和替代流行程度的研究是有限的。其中许多产品在一个国家或小地理区域进行了试销。因此，很难预测新产品的使用情况，尤其是在年轻人中。在评估新型烟草制品的潜在公共卫生影响时，应考虑新型烟草制品当前和潜在使用者的特征。危害序列的概念通常基于产品与卷烟烟气相比的有害性特征，较少关注使用者的特征和其他重要因素。

重要的是要了解与传统卷烟烟气相比，HTP是否可以通过减少某些有害化学物质的暴露来降低吸烟者的风险。它们可能会降低吸烟者的风险，也可能对健康造成严重风险。如何看待这种平衡取决于二者之间的关系。

4.9 政策建议

主要建议

- 应要求烟草公司调查所有HTP的化学特征和有害性。HTP制造商应公开产品测试的结果,并向相关监管机构提供所用测试方法的详细说明。应调查产品中因素组合的影响。
- 决策者应持续监测市场,以识别新型HTP和类似产品以及产品和释放物特征的变化。必须向监管机构报告现有产品的变化,以便对其进行监管,并对影响公众健康的变化(例如有害成分的变化或实质性设计变化)进行更严格的审查和要求。
- 新产品或改良产品在上市前应接受监管机构的上市前审查。所有HTP成分都应像其他烟草制品一样受到严格监管,包括对标签、广告、向未成年人的销售、价格和税收政策以及无烟措施的限制。

其他建议

- 应进行独立于烟草业的研究,以告知产品使用者、公共卫生专业人员和监管机构HTP对公共卫生的潜在影响,并评估人们对HTP的认识和使用趋势。
- 应确定所有新型烟草制品(包括HTP)的化学特征和有害性。
- 应在多个市场进行卷烟使用和替代流行程度的研究,同时考虑新型烟草制品的当前和潜在用户的特征,并评估产品对公众健康的潜在影响。
- 应进行精心设计的独立研究,了解HTP释放的化学物质的有害性特征,以评估HTP使用者相对于非吸烟者和传统卷烟吸烟者的暴露风险。此类评估应解释其他烟草制品在流行率、使用者行为和人群风险方面的差异。
- 应保护HTP使用者的隐私和安全。应对烟草公司在HTP设备上收集和处理个人信息进行规范。

4.10 参考文献

[1] Stedman RL. Chemical composition of tobacco and tobacco smoke. Chem Rev. 1968;68(2):153-207.

[2] Smith MR, Clark B, Lüdicke F, Schaller JP, Vanscheeuwijck P, Hoeing J et al. Evaluation of the tobacco heating system 2.2. Part 1: Description of the system and the scientific assessment program. Regul Toxicol Pharmacol. 2016;81(Suppl 2):S17-26.

[3] Eaton D, Jakaj B, Forster M, Nicol J, Mavropoulou E, Scott K et al. Assessment of tobacco heating product THP1.0. Part 2: Product design, operation and thermophysical characterisation. Regul Toxicol Pharmacol. 2018;93:4–13.

[4] US Code: Federal Food, Drug, and Cosmetic Act, 21 U.S.C. §§ 301–392 (Suppl 3 1934). Washington (DC): US Congress; 1934 (https://www.loc.gov/item/uscode1934-005021009). 5.

[5] Bialous SA, Glantz SA. Heated tobacco products: another tobacco industry global strategy to slow progress in tobacco control. Tob Control. 2018;27(Suppl 1):s111–7.

[6] Uranaka T, Ando R. Philip Morris aims to revive Japan sales with cheaper heat-not-burn tobacco. Reuters, 23 October 2018 (https://www.reuters.com/article/us-pmi-japan/philip-morris-aims-to-revive-japan-sales-with-cheaper-heat-not-burn-tobacco-idUSKCN1MX06E, accessed 29 August 2019).

[7] Heat-not-burn tobacco products market by product and geography – Forecast and analysis 2020–2024. Toronto: Technavio; 2020.

[8] Third-quarter results 2019. Lausanne: Philip Morris International; 2019 (https://www.pmi.com/investor-relations/reports-filings, accessed 10 January 2021).

[9] Lee MH. KT&G's heat-not-burn cigar overcomes downsides of competitors. The Korea Times, 22 December 2017 (http://www.koreatimes.co.kr/www/tech/2017/12/133_241123.html, accessed 20 February 2019).

[10] Half-year report for the six months to 30 June 2019. London: British American Tobacco;2019(https://www.bat.com/group/sites/UK__9D9KCY.nsf/vwPagesWebLive/DOBELLYE, accessed 10 January 2021).

[11] glo is being discontinued. British American Tobacco (glo.ca).

[12] Greenhalgh EC. Heated tobacco ("heat-not-burn") products. In: Tobacco in Australia: Facts and issues. Melbourne: Cancer Council Victoria; 2018.

[13] FDA permits sale of IQOS Tobacco Heating System through premarket tobacco product application pathway. Silver Spring (MD): Food and Drug Administration; 2019.

[14] FDA authorizes marketing of IQOS Tobacco Heating System with "reduced exposure" information. Press release; Silver Spring (MD): Food and Drug Administration; 2020 (https://www.fda.gov/news-events/press-announcements/fda-authorizesmarketing-IQOS-tobacco-heating-system-reduced-exposure-information, accessed 10 January 2021).

[15] Sutanto E, Miller C, Smith DM, O'Connor RJ, Quah ACK, Cummings KM et al. Prevalence, use behaviors, and preferences among users of heated tobacco products: Findings from the 2018 ITC Japan survey. Int J Environ Res Public Health. 2019;16(23):4630.

[16] Poynton S, Margham J, Forster M et al. Controlled aerosol release to heat tobacco: product operation and aerosol chemistry assessment. In: Society for Research on Nicotine and Tobacco, 2–5 March 2016, Chicago (IL) (https://cdn.ymaws.com/www.srnt.org/resource/resmgr/Conferences/2016_Annual_Meeting/Program/FINAL_SRNT_Abstract_WEB02171.pdf).

[17] Batista RNM. Reinforced web of reconstituted tobacco. Neuchatel: Philip Morris Products SA; 2017.

[18] Auer R, Concha-Lozano N, Jacot-Sadowski I, Cornuz J, Berthet A. Heat-not-burn tobacco cigarettes: Smoke by any other name. JAMA Intern Med. 2017;177(7):1050–2.

[19] Davis B, Williams M, Talbot P. IQOS: evidence of pyrolysis and release of a toxicant from plastic.

Tob Control 2018;28(1).

[20] Salman R, Talih S, El-Hage R, Karaoghlanian N, El-Hellani A, Saliba NA et al. Free-base and total nicotine, reactive oxygen species, and carbonyl emissions from IQOS, a heated tobacco product. Nicotine Tob Res. 2019;21(9):1285–8.

[21] Leigh NJ, Palumbo MN, Marino AM, O'Connor RJ, Goniewicz ML. Tobacco-specific nitrosamines (TSNA) in heated tobacco product IQOS. Tob Control. 2018;27(Suppl 1):s37–8.

[22] Lempert LK, Glantz S. Analysis of FDA's IQOS marketing authorisation and its policy impacts. Tob Control. 2020. doi: 10.1136/tobaccocontrol-2019-055585.

[23] St Helen G, Jacob III P, Nardone N, Benowitz NL. IQOS: examination of Philip Morris International's claim of reduced exposure. Tob Control. 2018;27:s30–6.

[24] PMTA coversheet: Technical Project Lead Review (TPL), 29 April 2019. Silver Spring(MD):Food and Drug Administration; 2019 (https://www.fda.gov/media/124247/download, accessed 10 March 2020).

[25] Varughese S, Teschke K, Brauer M, Chow Y, van Netten C, Kennedy SM. Effects of theatrical smokes and fogs on respiratory health in the entertainment industry. Am J Ind Med. 2005;47:411–8.

[26] Teschke K, Chow Y, van Netten C, Varughese S, Kennedy SM, Brauer M. Exposures to atmospheric effects in the entertainment industry. J Occup Environ Hyg. 2005;2(5):277–84.

[27] Meernik C, Baker HM, Kowitt SD, Ranney LM, Goldstein AO. Impact of non-menthol flavours in e-cigarettes on perceptions and use: an updated systematic review. BMJ Open. 2019;9(10):e031598.

[28] Leigh NJ, Lawton RI, Hershberger PA, Goniewicz ML. Flavourings significantly affect inhalation toxicity of aerosol generated from electronic nicotine delivery systems (ENDS). Tob Control. 2016;25(Suppl 2):ii81–7.

[29] Voos N, Goniewicz ML, Eissenberg T. What is the nicotine delivery profile of electronic cigarettes? Exp Opinion Drug Deliv. 2019:1–11.

[30] Meehan-Atrash J, Duell AK, McWhirter KJ, Luo W, Peyton DH, Strongin RM. Free-base nicotine is nearly absent in aerosol from IQOS heat-not-burn devices, as determined by 1H NMR spectroscopy. Chem Res Toxicol. 2019;32(6):974–6.

[31] Benowitz NL. Pharmacology of nicotine: addiction, smoking-induced disease, and therapeutics. Annu Rev Pharmacol Toxicol. 2009;49:57–71.

[32] Dempsey DA, Benowitz NL. Risks and benefits of nicotine to aid smoking cessation in pregnancy. Drug Saf. 2001;24(4):277–322.

[33] Benowitz NL. Toxicity of nicotine: implications with regard to nicotine replacement therapy. Prog Clin Biol Res. 1988;261:187–217.

[34] Appleton S. Frequency and outcomes of accidental ingestion of tobacco products in young children. Regul Toxicol Pharmacol. 2011;61(2):210–4.

[35] Solarino B, Rosenbaum F, Riesselmann B, Buschmann CT, Tsokos M. Death due to ingestion of nicotine-containing solution: case report and review of the literature. Forensic Sci Int. 2010;195(1–3):e19–22.

[36] Martin Leroy C, Jarus-Dziedzic K, Ancerewicz J, Lindner D, Kulesza A, Magnette J. Reduced exposure evaluation of an electrically heated cigarette smoking system. Part 7: A one-month,

randomized, ambulatory, controlled clinical study in Poland. Regul Toxicol Pharmacol. 2012;64(2 Suppl):S74-84.

[37] Tricker AR, Kanada S, Takada K, Martin Leroy C, Lindner D, Schorp MK et al. Reduced exposure evaluation of an electrically heated cigarette smoking system. Part 6: 6-Day randomized clinical trial of a menthol cigarette in Japan. Regul Toxicol Pharmacol. 2012;64(2 Suppl):S64-73.

[38] Tricker AR, Jang IJ, Martin Leroy C, Lindner D, Dempsey R. Reduced exposure evaluation of an electrically heated cigarette smoking system. Part 4: Eight-day randomized clinical trial in Korea. Regul Toxicol Pharmacol. 2012;64(2 Suppl):S45-53.

[39] Urban HJ, Tricker AR, Leyden DE, Forte N, Zenzen V, Feuersenger A et al. Reduced exposure evaluation of an electrically heated cigarette smoking system. Part 8: Nicotine bridging – estimating smoke constituent exposure by their relationships to both nicotine levels in mainstream cigarette smoke and in smokers. Regul Toxicol Pharmacol. 2012;64(2 Suppl):S85-97. Erratum in: Regul Toxicol Pharmacol. 2015;71(2):185.

[40] Tricker AR, Kanada S, Takada K, Leroy CM, Lindner D, Schorp MK et al. Reduced exposure evaluation of an electrically heated cigarette smoking system. Part 5: 8-Day randomized clinical trial in Japan. Regul Toxicol Pharmacol. 2012;64(2 Suppl):S54-63.

[41] Tricker AR, Stewart AJ, Leroy CM, Lindner D, Schorp MK, Dempsey R. Reduced exposure evaluation of an electrically heated cigarette smoking system. Part 3: Eight-day randomized clinical trial in the UK. Regul Toxicol Pharmacol. 2012;64(2 Suppl):S35-44.

[42] Schorp MK, Tricker AR, Dempsey R. Reduced exposure evaluation of an electrically heated cigarette smoking system. Part 1: Non-clinical and clinical insights. Regul Toxicol Pharmacol. 2012;64(2 Suppl):S1-10.

[43] Gale N, McEwan M, Eldridge AC, Sherwood N, Bowen E, McDermott S et al. A randomised, controlled, two-centre open-label study in healthy Japanese subjects to evaluate the effect on biomarkers of exposure of switching from a conventional cigarette to a tobacco heating product. BMC Public Health. 2017;17(1):67.

[44] Leigh NJ, Tran PL, O'Connor RJ, Goniewicz ML. Cytotoxic effects of heated tobacco products (HTP) on human bronchial epithelial cells. Tob Control. 2018;27(Suppl 1):s26-9.

[45] Cancelada L, Sleiman M, Tang X, Russell ML, Montesinos VN, Litter MI et al. Heated tobacco products: volatile emissions and their predicted impact on indoor air quality. Environ Sci Technol. 2019;53(13):7866-76.

[46] Sutanto E, Miller C, Smith DM, Borland R, Hyland A, Cummings KM et al. Concurrent daily and non-daily use of heated tobacco products with combustible cigarettes: Findings from the 2018 ITC Japan Survey. Int J Environ Res Public Health. 2020;17(6):2098.

[47] Hwang JH, Ryu DH, Park SW. Heated tobacco products: Cigarette complements, not substitutes. Drug Alcohol Depend. 2019;204:107576.

[48] Kim J, Yu H, Lee S, Paek YJ. Awareness, experience and prevalence of heated tobacco product, IQOS, among young Korean adults. Tob Control. 2018;27(Suppl 1):s74-7.

[49] Shahab L, Goniewicz ML, Blount BC, Brown J, McNeill A, Udeni Alwis K et al. Nicotine, carcinogen, and toxin exposure in long-term e-cigarette and nicotine replacement therapy users: A cross-sectional study. Ann Intern Med. 2017;166(6):390-400.

[50] Goniewicz ML, Smith DM, Edwards KC, Blount BC, Caldwell KL, Feng J et al. Comparison of nicotine and toxicant exposure in users of electronic cigarettes and combustible cigarettes. JAMA Netw Open. 2018;1(8):e185937.

[51] Czoli CD, Fong GT, Goniewicz ML, Hammond D. Biomarkers of exposure among "dual users" of tobacco cigarettes and electronic cigarettes in Canada. Nicotine Tob Res. 2019;21(9):1259–66.

[52] How other countries regulate flavored tobacco products. Saint Paul (MN): Tobacco ControlLegalConsortium;2020(https://www.publichealthlawcenter.org/sites/default/files/resources/ tclc-fs-global-flavored-regs-2015.pdf, accessed 12 August 2020).

[53] Event details. Philip Morris International Inc. presents at the Consumer Analyst Group of New York (CAGNY) conference. Philip Morris International, 19 February 2020 (http://www.pmi.com/ 2020cagny).

[54] Sutanto E, Smith DM, Miller C, O'Connor RJ, Hyland A, Tabuchi T et al. Use of heated tobacco products within indoor spaces: Findings from the 2018 ITC Japan Survey. Int J Environ Res Public Health. 2019;16(23):4862.

[55] Gretler C. Philip Morris has a Nespresso problem. Bloomberg, 21 June 2019 (https://www.bloomberg.com/news/articles/2019-06-20/philip-morris-arms-itself-to-battle-emerging- iqos-knockoffs, accessed 12 August 2020).

[56] Asun. IQOS sets to launch "cartridge recognition" technology to counter ImperialTobacco. VapeBiz,2July2019(https://vapebiz.net/iqos-sets-to-launch-cartridge-recognition-technology-to-counter-imperial-tobacco/, accessed 12 August 2020).

[57] Miller CR, Sutanto E, Smith DM, Hitchman SC, Gravely S, Yong HH et al. Awareness, trial, and use of heated tobacco products among adult cigarette smokers and ecigarettes users: Findings from the 2018 ITC Four Country Smoking & Vaping Survey. Tob Control. 2020. [In press.]

[58] Kim SH, Kang SY, Cho HJ. Beliefs about the harmfulness of heated tobacco products compared with combustible cigarettes and their effectiveness for smoking cessation among Korean adults. Int J Environ Res Public Health. 2020;17(15):E5591.

[59] Lasseter T, Wilson D, Wilson T, Bansal P. Every puff you take. Part 5. Philip Morris device knows a lot about your smoking habit. Reuters, 15 May 2018 (https://www.reuters.com/investigates/special-report/tobacco-iqos-device, accessed 12 August 2020).

[60] WHO report on the global tobacco epidemic, 2019. Geneva: World Health Organization, 2019(https://www.who.int/teams/health-promotion/tobacco-control/who-report-on-the-global-tobacco-epidemic-2019#:~:text=The%20%22WHO%20report%20on%20the,bans%20to%20no%20smoking%20areas, accessed 10 January 2021).

5. 加热型烟草制品的使用：产品转换以及双重或多重使用

Richard O'Connor, Department of Health Behavior, Roswell Park Comprehensive Cancer Center, Buffalo (NY), USA

Armando Peruga, Tobacco Control Group, Bellvitge Biomedical Research Institute, Barcelona, Spain; Centre for Epidemiology and Health Policy, School of Medicine Clínica Alemana, Universidad del Desarrollo, Santiago, Chile

摘 要

制造商声称新一代加热型烟草制品（HTP）可帮助传统卷烟的吸烟者戒烟并完全转用HTP，将其作为烟碱来源"更安全的替代"。同时使用两种或多种烟碱或卷烟产品（多重使用）包括各种类型的行为，产品使用频率不同，健康风险不同，受使用模式的影响。新一代HTP完全替代传统卷烟的能力可能取决于产品特性和吸烟者的特征，包括经验、转换和长期使用HTP的准备情况以及烟草监管环境。我们回顾了在实验室和现实世界中使用HTP的文献。当前的HTP似乎以与卷烟或ENDS类似的方式及具有药理学意义的剂量输送烟碱。新发布的独立研究表明，卷烟和HTP的混合使用比最初由企业赞助的研究所暗示的更为普遍。关于从吸烟过渡到HTP的信息很少，但有证据表明使用HTP的吸烟者更依赖烟碱。需要对HTP的认识和引入HTP的国家的使用行为进行研究，包括在提供产品的国家进行国家层面的调查。在允许出售HTP的地区这种区分具有法律意义，就无烟立法、税收、营销和购买的目的，HTP应被视为卷烟而不是ENDS。

5.1 概 述

制造商声称新一代加热型烟草制品（HTP）可帮助传统卷烟的吸烟者戒烟，并转用HTP作为烟碱来源"更安全的替代"。本章受WHO委托，旨在探讨HTP在从传统卷烟（CC）和其他烟草制品过渡中的潜在作用。与使用ENDS和ENNDS一样，我们特别讨论了HTP的营销是否导致HTP和CC或其他烟草制品的同时使用，以及HTP与卷烟产品的双重或多重联用是否有助于或妨碍传统卷烟草制品的吸烟者完全切换到HTP。由于企业迫使监管机构对HTP实施比其他烟草制品更宽松的监管，因此必须仔细评估这些产品"降低风险"或可以帮助"吸烟者"转用其他产品的说法。

回答这些问题的困难在于缺乏关于人群层面使用HTP的信息，无论是双重还是多重使用的信息，抑或这种使用对健康的影响的信息。单独或多重使用的暴露数据是有限的。另外，我们没有发现关于HTP是否可以使吸烟者完全从吸烟到戒烟或烟碱使用到烟碱戒断转变的实证研究，因此，无法在单一HTP使用者或多重使用者中探讨戒烟与HTP之间的关系。最后，没有关于停止使用HTP的研究报告。

5.2 关于HTP在人群层面使用的信息

表5.1总结了关于HTP使用流行率的13项研究[1-13]，其中一半描述了日本使用这些产品的经验。2015~2019年间仅有6个国家的研究可用。所有数据均来自同时进行的研究，除了一项在英国进行的研究和一项在日本进行的研究。因此，除英国外，很难确定普通人群中HTP使用的任何趋势。根据最近的研究，在2018年和2019年，日本约有3%的成年人目前使用HTP，而英国（英格兰）和波兰的这一数字要低得多。这些研究没有提供任何关于HTP和卷烟或其他卷烟产品（雪茄、比迪烟、水烟）双重使用频率的实际意义。

表5.1 HTP使用流行率的研究：按国家和数据收集年份排列（2015~2019年）

年份	德国	意大利	日本	波兰	韩国	英国
2015			15~69岁成年人ªIQOS当前使用率0.3%，Ploom当前使用率0.3% 15~69岁成人ᵇHTP既往使用率0.5%，HTP非吸烟人群使用率0.1%，HTP既往吸烟者使用率1.0%，HTP当前吸烟者使用率1.8% 40~69岁成年人NCD患者ᶜHTP既往使用率，男性1.7%，女性0.6%；HTP当前使用率，男性0.8%，女性0%			
2016			15~69岁成年人ªIQOS当前使用率0.6%，Ploom当前使用率0.3%			

续表

年份	德国	意大利	日本	波兰	韩国	英国
2017	年龄在14岁[d]以上的当前吸烟者和最近曾吸烟者当前HTP使用率0.3%	年龄在15岁以上的年轻人[e]IQOS既往使用率1.4%	15~69岁成年人[a]IQOS当前使用率3.6%，Ploom当前使用率1.2%，glo当前使用率0.8%成年人[f]IQOS当前使用率1.8%		19~24岁年轻人[g]IQOS当前使用率3.5%	成年人[h]HTP既往使用率1.7%，成年人[i]HTP当前使用率，Quarter 1 0.1%，Quarter 2 0.1%，Quarter 3 0.1%，Quarter 4 0.1%
2018			成年人[f]IQOS当前使用率3.2%，15岁以上的年轻人[j]HTP当前使用率2.7%，HTP日常使用率1.7%		中学生[k]HTP既往使用率，男生4.4%，女生1.2%	英国成年人[i]HTP当前使用率，Quarter 1 0.1%，Quarter 2 0.1%，Quarter 3 0.1%，Quarter 4 0.1%
2019				15岁以上的年轻人[l]HTP当前使用率0.4%		英国成年人[i]HTP当前使用率，Quarter 1 0%，Quarter 2 0.1%，Quarter 3 0.1%，Quarter 4 0.2%

注：HTP，加热型烟草制品；NCD：非传染性疾病

a. 当前HTP使用率（即在过去30天内使用）是根据对8240名日本人（2015年为15~69岁）的全国代表性样本进行的纵向互联网调查计算得出的，该调查随访至2017年[1,2]

b. 2015年1月31日至2月17日期间，Rakuten Research对7338名18~69岁的受访者进行了互联网调查[3]

c. 2015年互联网调查的4432名日本40~69岁慢性病患者[4]

d. 2016年6月至2017年11月期间参与家庭调查的代表性样本中，18415名14岁以上的德国人；0.3%（95% CI 0.09~0.64）的当前吸烟者和新的前吸烟者（<12个月戒烟）当前使用HTP。HTP的使用随着教育和收入的增加而增加[5]

e. 通过多阶段抽样选出的3086人样本，该样本代表了15岁及以上的意大利人口，并进行了面对面访谈[6]

f. 2017年的4878名日本成年人中，1.8%使用了IQOS。2018年上半年的2394名日本成年人中，3.2%使用了IQOS。在"平台1"注册用户中，2017年和2018年分别有1.3%和1.6%从不吸烟，2016年和2017年分别有98%和98.6%的双重烟草使用者[7]

g. 2017年对228名19~24岁年轻人的在线调查[8]

h. 2017年2~3月，英国市场研究公司YouGov Plc对12696名17岁以上成年人的全国代表性样本进行了采访[9]

i. 来自Smoking Toolkit Study的月度家庭调查，新的代表性样本为每月15岁以上的约1800名英国受访者。实地考察由英国市场研究局进行。同意再次联系的吸烟者和最近的前吸烟者（前一年吸烟者）在3个月和6个月后通过邮寄问卷进行跟踪。自2017年1月以来，HTP的数据涵盖了63499名成年人[10]

j. 2018年2~3月进行的具有全国代表性的互联网调查的4684名日本成年人参与者中，2.7%每月至少使用一次HTP，1.7%每天使用一次。在当前吸烟者中，1.8%的人每月至少使用一次HTP，每天约50%的人使用HTP。在从不吸烟的人中，0.02%的人每月至少使用一次HTP，并且每天都使用[11]

k. 截至2018年4月，在韩国60040名就读中学的青少年中，4.4%的男生和1.2%的女生曾经使用过HTP。大约6%是当前吸烟者。其中，32.4%的人曾经使用过HTP。在86%从不吸烟者中，只有0.3%曾经使用过HTP[12]

l. 2019年9月在波兰15岁及以上的1011名全国代表性样本中，0.4%的人使用HTP，他们都是当前吸烟者，占当前吸烟者的1.9%[13]

5.3 从传统卷烟转向 HTP 的动态情况：双重或多重产品联用是过渡态还是永久态？

同时使用两种或多种烟碱或烟草制品包括多种行为，产品使用频率和健康风险，受使用者特征影响。在美国，多重产品联用的吸烟者往往是男性，使用其他药物并且更依赖烟碱[14-23]。随着时间的推移，几种烟草制品的使用往往变得不稳定[14-27]。表5.2概述了一些关于多重联用的常见定义，这些定义来自于已发表的ENDS和ENNDS相关文献。

表5.2　烟草和烟碱研究中描述的多重产品联用类型

使用类型	使用报告	优点	缺点
终身多重联用	曾使用过两种或更多产品	最广泛的测量方式；获得一些潜在使用方式	可能获取的是几年前发生的使用或者可能对当前行为或疾病风险几乎没有影响的少量尝试
近期多重联用	在过去30天内使用过两种或更多产品	获得当前使用数据	不考虑使用的数量或频率；一次使用两种产品被认为相当于每天使用每种产品
主导性多重使用	在过去30天内使用过两种或更多产品，每天或几乎每天使用一种产品多于另一种产品	更全面的使用模式评估	需要更多的讨论。可能会受到召回偏见的影响，特别是对于不太常用的产品
均衡联用	在过去30天内每天或几乎每天等量使用两种或更多产品	更全面的使用模式评估	需要更多的讨论。可能会受到召回偏见的影响，特别是对于不太常用的产品
间歇联用	在过去30天内至少在某些日子使用过两种或更多产品，但没有一致的使用任何产品的模式	更全面的使用模式评估	需要更多的讨论。可能会受到召回偏见的影响，特别是对于不太常用的产品

虽然很少有关于HTP混合使用的研究，但对ENDS的研究可能具有指导意义。Borland及其同事[28]分析了来自澳大利亚、加拿大、英国（英格兰）和美国的调查数据，并描述了同时使用卷烟和ENDS的四个亚组，这些亚组在烟碱依赖、戒烟行为和认知方面存在差异[28]：①每日双重使用者；②主要抽吸卷烟者（每天抽吸卷烟且ENDS不是每天使用）；③主要使用电子烟者（每天使用ENDS且卷烟不是每天使用）；④双重使用的非日常使用者（非每日抽吸卷烟和使用电子烟）。虽然许多同时使用卷烟和ENDS的人报告显示试图减少吸烟[29,30]，但这种说法往往没有反映在暴露生物标志物中[31,32]，并且每天减少卷烟可能不会有效地降低吸烟导致的死亡风险[33-36]。Baig和Giovenco[37]关于ENDS和卷烟双重使用的研究表明了不同双重联用行为的一些可能的过渡途径。从广义上讲，受过高等教育或收入较高的双重使用者更有可能在两年内完全转用电子烟或戒烟。

目前，没有足够的证据可得出HTP比CC危害更小的结论。事实上，有人担心，虽然HTP可能令使用者接触比CC更低水平的某些有害成分，但它们可能令使用者接触更高水平的其他有害成分[38-40]。研究表明，日本高达65%的HTP使用者和韩国几乎所有（96.2%）的HTP使用者同时也抽吸传统卷烟[2,11,41-43]。Sutanto及其同事[44]分析了日本多重产品联用者的子群，发现总体分布如表5.3所示。

表5.3 2018年日本HTP四个子群多重联用者比例（%）[44]

	加权百分比（95%置信区间）	
	每日HTP使用者(n=594)	非每日HTP使用者(n=265)
	51.5 (46.7~56.3)[a]	48.5 (43.7~53.3)[a]
每日吸烟者(n=3686)	每日双重使用者	主用卷烟者
94.4 (91.9~96.2)[b]	51.0 (46.2~55.7)	43.4 (38.6~48.4)
非每日吸烟者(n=213)	主用HTP者	非每日双重使用者
5.6 (3.8~8.1)[b]	0.5 (0.2~1.3)	5.1 (3.4~7.6)

a. 整列总和
b. 整行总和

2018年，日本大多数HTP使用者同时抽吸卷烟，并且大多数人每天同时使用这两种产品[44]。虽然每日单纯吸烟者和每日双重使用者之间在每日烟支数量上没有差异，但主要抽吸卷烟者报告的每天吸烟量比单纯吸烟者多，而主要抽吸卷烟者每天使用的含烟草插入物少于每日双重使用者；单纯HTP使用者每天使用的含烟草插入物比每日双重使用者更多。除了使用频率更高之外，这表明HTP可能无法有效替代卷烟，这与韩国的数据一致[41]。卷烟-HTP双重使用者比单独吸烟者年轻，而美国的一项实际使用研究发现中年吸烟者对该产品更感兴趣[45]。新型烟草制品通常出于各种原因吸引年轻用户，包括较低风险的说法、营销信息以及图像和产品外观[29,46-49]。只有大约10%的同时使用卷烟和HTP的人计划在未来6个月内戒烟，这一发现与Borland等[28]的发现相反。50%的同时使用ENDS和卷烟的人计划戒烟，与Baig等[37]的研究结果一致。

英美烟草公司（BAT）报告了一项关于使用其glo产品的研究[50]。三组受试者将glo薄荷醇、glo烟草、glo和IQOS带回家共计14天（共计4次实验室访问）。据报道，受试者在基线时每天使用12~15支卷烟，每天使用8~12单位的glo和IQOS产品。

菲利普·莫里斯国际公司（PMI）向美国FDA申请将IQOS注册为"风险降低的烟草制品"，其中包括PMI在德国、意大利、日本、韩国、瑞士和美国进行的关于产品转换的一系列观察性研究[51]。美国的这项研究包括1106名当前每日吸烟者，他们在1周基线后4周内免费获得IQOS。使用的产品（卷烟和加热棒）的数量记录在日记中。在这项研究中，"转换"到IQOS被定义为使用加热棒超过总消费量的70%。到研究结束时，大约15%的参与者符合这一定义，而22%是双重使用者（消费量的30%~70%为加热棒），63%是主要吸烟者。在第二项研究中，该产品及其相关营销提供给德国、日本、波兰、韩国和美国的2089名每日吸烟者。4周试用后完全转用烟草加热系统（THS）的流行率范围由德国的10%到韩国的37%不等，而双重使用的范围从日本的32%到韩国的39%不等。2013~2014年在美国进行了一项为期90天的IQOS使用研究，160人中有88人完成了研究。值得注意的是，这项研究中关于禁止使用其他烟草制品的依从性远低于日本的一项类似设计的研究，这表明在一种情况下的经验不能推广到其他情况。

IQOS的PMI申请中提供了有限的上市后数据。可用的数据主要来自日本，利用PMI的IQOS购买者登记册[51]。2016年1~7月，IQOS单独使用比例（>95%的总消费）从52%增加到65%。日本两个IQOS购买者队列（2015年9月和2016年5月）的过渡马尔可夫模型表明，那些转而使用IQOS的人不太可能重新使用卷烟。

美国FDA烟草制品科学咨询委员会对IQOS的一些证据表示担忧，特别是"完全转换"的定义以及消费者对风险降低声明的理解研究的局限性。委员会仅对暴露降低的声明提供有保留的支持，并对措辞无法有效传达风险表示担忧[52]。

2017年12月，英国食品、消费品和环境中化学产品的毒性、致癌性和致突变性委员会评估了市场上的两种HTP并得出结论[53]：

虽然转用"加热不燃烧"烟草制品，吸烟者的风险可能会降低，但是会有残留风险，完全戒烟对吸烟者更有益。这应该成为所有以尽量减少烟草使用带来的风险的长期战略的一部分。

由于很少有研究专门针对当前HTP与其他产品联用的暴露和健康影响，对已有HTP的研究进行了评估。一项针对早期HTP（Accord）的研究[54]分析了受试者同时使用自有品牌卷烟的情况。使用6周后，Accord似乎减少了吸烟和一氧化碳的暴露剂量，即更多地使用Accord与吸烟量减少有关，当参与者减少每日烟支数量时，他们的抽吸强度似乎没有增加。Fagerstrom及其同事[55]报道了对早期HTP Eclipse的研究。在最初为期4周的随机研究后，参与者自行选择再使用Eclipse（n=10）、烟碱吸入器（n=13）或卷烟（n=13）8周。在基线内，选择Eclipse的人平均每天吸烟量（18.0）少于其他组（吸入器20.4，卷烟21.3）。在8周内，报告中30%~60%的参与者完全不抽烟，平均每天吸2.6支卷烟，随着时间的推移几乎没有变化。总体而言，较早的文献表明HTP不完全替代卷烟通常与显著降低的健康风险无关[56-58]，不是完全切换为HTP可促进减少危害的预测。

5.4　HTP作为传统卷烟替代品的潜在作用

PMI在2018年1月表示，"仅在两年内，就有超过370万美国以外的吸烟者完全转用IQOS。同时，不吸烟者和既往吸烟者对该产品几乎没有兴趣。"[59]然而，我们无法找到任何实证研究来证实这一说法或任何其他关于使用新一代HTP来戒烟的说法。在英国（英格兰）进行了一些监测，在4155名吸烟并试图戒烟或者在调查前12个月内已戒烟的成年人中，有0~1.4%提到HTP是2016年之后的替代方法，以季度累计，平均季度流行率中位值为0.4%[10]。

新一代HTP替代CC的能力可能取决于产品特性、吸烟者的特征，包括他们的经验以及对长期转用HTP的准备，也许还取决于烟草管制环境的特征。在缺乏HTP帮助转换潜力和有效性的直接经验证据的情况下，我们使用有关产品特性的信息，例如它们的可获得性以及它们是否提供足够剂量的烟碱以减少对CC的渴望或戒断症状。以下描述了已发表的烟碱药代动力学研究和HTP对吸烟者吸引力的证据，假设烟碱释放量越大，吸引力越大，可能会使卷烟的替代量越大。

5.5 双重/多重使用者的烟碱暴露及潜在健康风险

本节我们比较了当前吸烟者中HTP和CC的使用情况，包括主流释放物中总烟碱及血浆和尿液中的关键药代动力学。

表5.4描述了截至2020年1月[60-71]我们发现的11项研究中的12篇论文，其中5篇是由烟草行业进行或资助的。该表更新并扩展了Simonavicius等[72]的表。HTP气溶胶和CC主流烟气中烟碱递送的比较因产品的多样性和使用的方法而变得复杂。研究的HTP包括IQOS（PMI）、glo（BAT）、iFuse（BAT）和烟草雾化器。不同研究中的参比产品不同，它们包括为研究开发的3R4F、1R5F和1R6F（3R4F最为常用）和市售卷烟，其中烟碱的释放量因品牌、国家和年份而异。在大多数研究中，主流释放物是在HCI方案下获得的，但在其他研究中采用ISO方案。值得注意的是，没有任何一种机器吸烟方案可对应于人类吸烟和暴露，并且它们与HTP使用的相关性尚未得到验证。有了这些警告，可以从研究中得出两个结论：

■ IQOS可释放CC烟气中约70%的烟碱。当对照商用卷烟时，IQOS的相对烟碱释放率为40.7%~102.8%（中位值为76.9%），而当对照为研究而开发的参比卷烟时，则为57%~103%（中位值为64.7%）。烟草行业资助的研究中的中位值与独立研究中的中位值在统计学上没有显著差异。

■ 与CC相比，其他HTP研究在提供烟碱方面的效率似乎低于IQOS（<50%）。

表5.4 主流HTP气溶胶和传统卷烟主流烟气中的烟碱释放

参考文献	[69]	[66]	[65]	[67]	[63]	[62]	[60]	[61]	[68]	[64]	[70]	[71]
出版年份	2019	2018	2018	2018	2018	2017	2017	2017	2017	2017	2016	2016
烟草行业进行资助	否	否	否	否	是, BAT	否	否	否	是, BAT	是, PMI	是, PMI	是, PMI
吸烟机	自行设计	LM4E	5M450 linear	NR	LM	自行设计	自行设计	NR	NR	LM set to 12 puffs (1/305)	LM M405X8 5M405RH LM20X	NR
模式	ISO 6口 HCI 12口	HCI 12口	ISO HCI	NR	HCI	HCI 12口	ISO 14口	HCI 9~11口	HCI (CC CORESTA HTP (持续3 s))	HCI	HCI, ISO和5种非干观察的HTP使用模式	NR
产品 型号	Marlboro Red	48 commercial PM+1R4F 50, MDPH and HC smoking regimes[a]	3R4F	Marlborg Red,1R6F	3R4F,1R6F	Malboro Regular	Lucky Strike Blue Lights	1R5F 3R4F	3R4F	3R4F	3R4F	3R4F
传统卷烟 烟碱(mg/支) 合计	ISO 0.8 HCI 1.8	Min 1.07 maz 2.7	ISO 0.71 HCI 1.9	MR 1.07 1R6F 0.65	1.86	1.99 mg/支	0361 mg/支	10~1.7mg/支	1.8 mg/支	1.86 mg/支	1.89 mg/支	1.88 mg/支
游离态	0.12~0.10	—	—	—	—	—	—	—	—	—	—	—

续表

参考文献	[69]	[66]	[65]	[67]	[63]	[62]	[60]	[61]	[68]	[64]	[70]	[71]
出版年份	2019	2018	2018	2018	2018	2017	2017	2017	2017	2017	2016	2016
烟草企业隶属或资助	否	否	否	否	是, BAT	否	否	否	是, BAT	是, PMI	是, PMI	是, PMI
加热型烟草制品 型号	IQOS	IQOS	IQOS	IQOS	THS (IQOS) THP1.0 (glo)	IQOS	IQOS	IQOS	Disposable ncopod Fus	IQOS THS2.2 regular	IQOS THS2.2 regular and menthol	IQOS THS2.2 with 43 different tobacco blends
烟碱 单位[b]	mg/支	mg/支	mg/支	mg/支		mg/12口	mg/支	mg/支			mg/支	平均mg/支
烟碱 合计	0.77~15	1.1	0.50~1.35	0.67	IQOS 1.16 glo 0.462	1.4	0.30	1.1(regular) 1.2 (menthol)	首个100口平均 2.56 mg/口	1.14	1.3 (regular) 1.2 (menthol)	Reference blend, 1.38 Range of 43 blends, 1.6-6
烟碱 游离态	0.10~0.09											
对照传统卷烟	83.3%~96.3% 81.9%~90.3%	40.7%~102.8%[c]	70.4%~71.1%	MR 63% 1R6F 103%	IQOS 57% glo 23%	70.4%	83.4%	64.7%	14.5%[d]	61%	70% (regular)[d] 64% (menthol)	Reference blend 73% Range of 43 blends: 87-33%

注: BAT, 英美烟草公司; CC, 传统卷烟; CORESTA, 烟草科学研究国际合作中心; HCI, 加拿大卫生部深度抽吸模式; ISO, 国际标准化组织抽吸模式; PMI, 菲利普·莫里斯国际公司; THS, 烟草加热系统

a. 2005年一项独立研究的评估[73]。

b. 与参比卷烟烟碱释放量最大值和最小值的比较。作者报告了烟碱递送的不一致: 烟碱水平最初低于吸烟过程中发现的烟碱水平的50%, 因此IQOS占总烟碱释放量的10%~12%。他们认为烟碱递送的不一致可能会影响消费者满意度, 烟碱血液水平并不会改变吸烟行为

c. 14.5%是作者的计算; 但是, 如果14口的浓度与其他研究相同 (2.56×0.14=0.358), 浓度为19%

d. 作者报告了模拟地中海、热带和沙漠气候条件下的百分比较低, 但没有报告确切数字

5.6 动物药代动力学

我们只发现了一项实验动物研究。在这项独立研究中，Nabavizadeh等[67]将3组每组各8只大鼠暴露于来自单个加热棒的IQOS产生的气溶胶、来自单个万宝路红卷烟的主流烟气或清洁空气。在一系列连续30秒的循环中暴露1.5~5分钟，每个循环由5秒或15秒的暴露组成。暴露后，暴露于IQOS的大鼠的血清烟碱浓度比暴露于卷烟的大鼠高约4.5倍，即使IQOS气溶胶的烟碱量约为烟气中测得的烟碱量的63%。当暴露于IQOS释放物的时间较短时，暴露于IQOS和卷烟释放物的大鼠的血清烟碱浓度相似。

5.7 人体药代动力学

在大多数研究中报告了在使用HTP、ENDS和CC后生物标志物的数值。仅报告使用HTP和CC后烟碱的药代动力学。这些方法总结在表5.5中。

独立研究

发现了两项独立研究，一项使用IQOS进行，另一项使用PAX的松散烟草雾化器。Biondi-Zoccai等[85]进行了一项随机、交叉试验，比较使用一根IQOS、ENDS和一根CC对吸烟者的影响。通过在使用前、从一种烟草或烟碱产品中戒除1周后以及产品使用后立即测量血清可替宁来评估烟碱暴露情况。使用CC和IQOS显著增加了血浆可替宁水平：对于CC，从使用CC之前的34.4（SD ± 19.3）ng/mL升到65.5（SD ± 10.2）ng/mL，其次，对于IQOS，从30.4 ng/mL ± 12.0 ng/mL升至61.0 ng/mL ± 16.7 ng/mL。在每种情况下，差异在$P<0.001$处具有统计学意义，但产品之间没有发现显著差异。

Lopez等[86]比较了当前吸烟者在使用来自PAX（Ploom）的松散烟草雾化器HTP、ENDS和参与者自选品牌的CC前后的血浆烟碱浓度。在使用两轮预定产品后平均血浆烟碱浓度立即显著增加，第一轮中，使用CC后从基线（所有$P<0.025$）增加到24.4（SD ± 12.6）ng/mL，使用HTP后增加到14.3（SD ± 8.1）ng/mL，在第二轮使用CC后增加到23.7（SD ± 14.5）ng/mL，使用HTP后增加到16.4（SD ± 11.3）ng/mL。CC组在每轮后立即获得的血浆烟碱水平比HTP高；然而，只有在第一轮实验之后CC和HTP中烟碱水平之间的差异具有统计学意义（所有$P<0.017$）。从实验开始到结束，使用CC后的平均血浆烟碱浓度显著高于使用HTP后的血浆烟碱浓度。

表5.5 用于研究来自HTP和CC的烟碱人体药代动力学方法

参考文献	烟草企业	国家	n	参与者	设计	产品 HTP	产品 CC	暴露	暴露生物标志物
[74]	PMI	波兰	160	年龄21~65岁，有3年吸烟史且当前吸烟的健康白人吸烟者	对照，三臂平行，单中心禁闭研究，参与者随机分为单独使用HTP、CC或戒断	IQOS THS2.2	自有品牌	每天随意使用指定产品5天	尿NEQ
[75]	PMI	日本	160	年龄23~65岁，有3年吸烟史、过去4周吸烟≥10例	对照，三臂平行，单中心研究，参与者随机分为单独使用HTP、CC或戒断	IQOS THS2.2	自有品牌	每天随意使用指定产品5天	尿NEQ
[76]	PMI	日本	62	年龄23~65岁，有3年吸烟史、过去4周吸烟≥10例	开放标签，交叉研究，参与者随机分为单独使用HTP、CC或NRT口香糖	IQOS THS2.2 常规和薄荷醇	自有品牌，常规和薄荷醇	1天洗脱后，按顺序每天1单位指定产品（CC:1支卷烟，HTP:1支抽14吸或6分钟，口香糖:1次咀嚼35分钟±5分钟），1天任意使用一种产品，1天戒断，1天任意使用一种产品；1: THS-CC；2: CC-THS；3: THS-Gum；4: Gum-THS	血浆烟碱
[77]	PMI	英国	28	年龄23~65岁，有3年吸烟史、过去4周内非薄荷醇卷烟≥10例的健康白人吸烟者，如果使用其他烟草制品、ENDS或ENNDS，则排除	开放标签，随机，两期、两序列交叉研究，参与者按1:1随机分配到单独使用HTP或CC	IQOS THS2.1	自有品牌	HTP:每天1支，随意使用，1天任意使用；天单支，1天戒断，与其他产品重复序列	血浆烟碱
[78, 79]	PMI	日本	160	健康吸烟者，年龄23~65岁，BMI 18.5~32 kg/m²，吸烟≥10支/天、在过去4周内发现薄荷醇卷烟，并且报告吸烟薄荷醇卷烟≥3年	对照，三臂平行，单中心研究，参与者随机分为单独使用HTP、CC或戒断	IQOS THS2.2 薄荷醇	自有品牌薄荷醇	每天随意使用指定产品5天	尿NEQ
[80]	PMI	波兰	112	健康的白人吸烟者，BMI 18.5~27.5 kg/m²，年龄23~55岁，每天吸烟10~30支，至少连续吸烟5年	对照，开放标签，单中心对照研究，三臂平行组，参与者以2:1:1的比例随机分为HTP、CC或戒烟组	CHTP 原型 MD2-E7	自有品牌	随意使用指定产品5天	尿NEQ，可替宁
[81]	PMI	波兰	40	健康吸烟者，报告在过去连续3年每天吸10支节售非薄荷醇CC至少4周	对照，开放标签，单中心对照研究，双臂平行组，参与者以1:1随机分为HTP或CC	IQOS THS 2.1	自有品牌	随意使用指定产品5天	血浆烟碱，可替宁和尿NEQ

续表

参考文献	烟草企业	国家	n	参与者	设计	产品 HTP	产品 CC	暴露	暴露生物标志物
[82]	PMI	美国	160	至少22岁的健康吸烟者，BMI 18.5~35 kg/m³，在过去连续3年报告吸烟，并在研究开始前至少4周每天吸烟10支市售非薄荷醇CC（根据尿可替宁测试验证）。受试者不打算在接下来的6个月内戒烟	受试者以2:1:1的比例随机分配（第0天）至薄荷醇THS、薄荷醇CC和戒烟组。随机分组按性别和每日CC消费配额进行分层。5天的治疗期之后是86天的动态期和额外的28天的安全随访期	IQOS THS 2.2	自有品牌	随意使用指定产品5天	血浆烟碱，可替宁和尿NEQ
[83]	BAT	日本	180	健康，每天吸烟10~30支，有≥3年吸烟史；排除前2周内经常使用其他烟碱和烟草制品或计划在未来12个月内戒烟的人	随机，对照，平行组开放标签，临床5天禁闭研究	IQOS, glo, THP1.0 非薄荷醇和薄荷醇	自有品牌	有限随意使用指定产品5天	尿NEQ
[84]	JTI	日本		年龄在21~65岁的健康吸烟者，在筛查时平均每天吸烟≥11支人造烟，并且在试验前吸烟≥12个月	开放标签，双序列，双周期，随机交叉，禁闭研究	烟草蒸气产品原型	商业卷烟	HTP: 10次抽吸3分钟，间隔约20秒 CC: 10次抽吸3分钟，间隔约20秒 每种产品单独使用1天，另一种产品在第二天使用	血浆烟碱
[85]		意大利	20	健康吸烟者	交叉随机试验	IQOS THS2.2	万宝路金	在六个分为期一天的周期中使用一种指定的产品，每个周期旋转指定的产品一周的周期间冲洗。每个循环中使用的产品为：CC: 1支平均每次烟碱含量为0.60 mg的卷烟ENDS: 从一种平均烟液中抽取16 mg的电子烟烟液中抽9口，因此在9口中产生0.58 mg的烟碱含量HTP: 1根，每根平均烟碱含量为0.50 mg	血浆烟碱
[86]	否	美国	15	18~25岁的健康吸烟者，每天吸烟≥10支，在过去30天内未使用大麻，一生中未使用ENDS≤20次，LLTV≤4次		LLTV HTP (PAX) 预充1 g LLT	Own brand	三次，每天两次，每次10次，每次同间隔30秒，每次间隔60分钟，每次同间隔≥48小时，每次前禁欲≥2小时	血浆烟碱

注：BAT，英美烟草公司；BMI，身体质量指数；CC，传统卷烟；CHTP，碳燃烧型烟草制品；ENDS，电子烟烟碱传输系统，电子非烟烟碱传输系统；HTP，加热型烟草制品；JTI，日本烟草国际公司；LLTV，松散烟草雾化器；NEQ，烟碱等效物；NRT，烟碱替代疗法；PMI，菲利普·莫里斯国际公司；THS，烟草加热系统

烟草行业研究

发现了烟草行业进行的10项研究:来自PMI的8项研究的9篇论文[74-82],1项来自BAT[83],1项来自JTI[84](表5.5)。所有PMI研究都是随机试验,主要分别使用IQOS(1项研究中使用了碳头HTP)、参与者的常规CC品牌,或者在5项研究中使用戒断或烟碱替代疗法。暴露通常是约5天内随意使用指定产品。BAT和JTI的研究是随机试验,参与者被分配使用HTP(在BAT的情况下为IQOS或glo,在JTI的情况下为烟草雾化器)或商业CC。参与者在5天内的BAT试验中随意使用指定产品,在JTI试验中使用一根烟棒2天。

血浆生物标志物水平

3项研究[76,77,84]比较了使用产品后的药代动力学(表5.6),从开始到结束可量化的血浆浓度比时间曲线下面积($AUC_{0\text{-last}}$)是假设所有参与者的药物清除率相等的情况下烟碱总暴露量的指标;观察到的最大血浆浓度(C_{max})是烟碱吸收的指标;达到C_{max}的时间(t_{max})是达到C_{max}的速度的指标;半衰期($t_{1/2}$)是显著药理作用持续时间的指标。IQOS THS2.2的值与CC的值相似,t_{max}为6分钟。在烟碱的吸收和提供相同的最大浓度方面,IQOS THS2.1似乎不如CC和IQOS THS2.2有效。然而,它呈现出与CC非常相似的t_{max}(8分钟)和比CC稍长的半衰期。Yuki等[84]测试的烟草雾化器与CC同时达到t_{max}(3.8分钟),但C_{max}达到CC的一半以下,并产生了<70%的从CC吸收的烟碱。3项研究中,将IQOS、THS2.1和THS2.2与CC进行比较,报告了在随意使用HTP 5天后,血浆中烟碱和可替宁的平均水平,使用CC后达到的水平为THS2.1的约85%,THS2.2的100%。

尿液生物标志物水平

6项研究提出了24小时尿液中烟碱当量的值。THS2.2和碳头HTP的总和≥使用CC后的100%。对于THS2.1,总和是CC的87%。使用glo导致尿液中烟碱当量分别为CC的57%和74%,具体取决于它们是否含有薄荷醇。

表5.6 单次使用IQOS和传统卷烟后血浆烟碱的药代动力学

参考文献	[76]	[76]	[77]	[84]
发表年份	2017	2017	2016	2017
企业进行或资助	是	是	是	是
参比卷烟	自有非薄荷醇品牌	自有薄荷醇品牌	自有品牌	CC1
HTP	THS2.2 IQOS	THS2.2 menthol IQOS	THS2.1	PNTV

续表

参考文献	[76]	[76]	[77]	[84]
HTP（单独使用）达到C_{max}的时间（min）	6	6	8	3.8
AUC_{0-last} (ng·h/mL) 几何LS均值比	96.3% 85.1%~109.7%	98.1% 80.6%~119.5%	77.4% 60.5%~85.0^a	68.3% 54.3%~85.9%
C_{max}(ng/mL) 几何最小二乘均值比	103.5% 84.9%~126.1%	88.5% 68.6%~114%	60.3% 60.0%~82.2%	45.7% 34.1%~61.4%
tC_{max}(min)中位值差异	0.04 −1.0~1.05	1.0 0.0~2.5	0.1 −1.0~2.0	−0.5 −1.1~0.03
$t_{1/2}$(h) 几何最小二乘均值比	93.1% 84.6%~102.4%	102.3% 85.3%~122.7%	110.9% 101.7%~120.9%	89.1% 78.2%~102%

PNTV: 新型烟草蒸发器原型
a. 90%的置信区间

5.8 使用HTP和传统卷烟的主观效应

我们确定了10项研究，其中比较了HTP和CC的主观效应[74-77,79,81,82,86-88]。IQOS是测试所用的全部HTP，但一项除外。最常用于评估HTP和CC主观效应的心理测量工具是关于吸烟冲动的简短问卷（QSU-brief）[89]和改进的卷烟评估问卷（mCEQ）[90]。QSU简报是一份10个问题的问卷，通常在使用指定产品之前提交，然后在使用结束时再次测量渴望程度。分数可以作为它的两个组成部分的总量表的报告，期望吸烟带来的愉悦感和缓解烟碱戒断或负面影响带来的迫切和压倒性的吸烟欲望。mCEQ评估产品使用的强化效应包括"吸烟满意度"、"心理奖励"和"厌恶"三个多维度领域，以及"享受呼吸道感觉"和"减少渴望"两个单维度领域。

QSU-brief用于8项研究，其中2项独立研究[86,88]，6项与企业相关[74-77,79,82]。这些研究都发现，正如预期的那样，在干预开始之前，所有组的渴望得分都很高。使用IQOS或CC后得分立即显著下降；然而，唯一关于IQOS的独立研究报告称，与使用IQOS后相比，吸烟导致的渴望评分更低（所有$P<0.001$）[88]，但6项企业研究没有得出该结论。IQOS和CC组在企业研究中QSU总分的最小二乘平均差异，涵盖从产品使用开始到结束的所有时间，通常很小，没有统计学意义。两项独立研究[86,88]将分数报告为QSU简要量表的双因素组成，随着时间的推移发现了类似的结果。报告一次性使用和随意使用的QSU简要量表的研究[77]发现平均总分对比一次性使用没有差异（最小二乘均差，1.4（95% CI：−1.0，3.7）vs 0.2（95% CI：−2.9, 5.3））。这项研究将松散烟草雾化器与CC进行了比较，发现使用CC后的渴望比

使用蒸发器后显著降低。

mCEQ用于七项研究[74,75,77,79,81,82,88]。在所有研究中，问卷在暴露结束后进行，有时立即进行。唯一的独立研究[88]发现，使用IQOS和CC这两种产品对所有分量表都有强化作用；然而，在满意度、心理奖励、呼吸道感觉的享受和减少渴望方面，CC比IQOS具有更实质性的主观效应。企业研究表明，除了厌恶之外，IQOS的使用对所有或部分mCEQ分量表分数的增强作用明显低于CC使用。两项研究[75,79]在暴露期开始时和一项研究在暴露期结束时报告了所有分量表的显著差异，满意度和渴望减少的差异最大。另一项研究[81]发现，与基线相比，在暴露的最后一天，CC抽吸者比IQOS使用者的吸烟满意度更高。在另一项研究[82]中，在调整基线、吸烟满意度、渴望减少、享受呼吸道感觉和心理奖励后，转用IQOS的参与者在MCEQ上的平均结果明显低于继续吸烟的参与者。一项研究[77]表明两种产品在随意使用后，吸烟满意度、心理奖励、渴望减少和呼吸道享受方面的差异感觉更显著。在一项独立研究中，Biondi-Zoccai等[85]既不使用QSU-brief也不使用mCEQ，而是使用七个问题的产品满意度问卷[91]，在每次产品使用后进行测量，CC的满意度得分高于HTP。

这项有限的研究表明，与CC相比，HTP总体上以更低的剂量和更慢的速度递送烟碱。在分析的HTP中，只有IQOS THS2.2达到了CC的烟碱递送。IQOS可以减少对吸烟的渴望，可能比CC的程度要小。与企业相关的研究显示几乎没有差异，一项独立研究显示，与CC相比，渴望的减少明显更少。对其他HTP的唯一研究表明，松散烟草雾化器比CC更难抑制吸烟的渴望；然而，IQOS和雾化器被认为不如CC令人满意。

5.9 讨论和影响

关于在人群层面使用HTP，或多重使用，或停止使用CC或烟碱，与之相关的信息很少。虽然这方面的研究正在增加，并且一些调查（例如国际烟草管制、日本协会和新烟草互联网调查）正在评估HTP的使用，但发表需要几个月到几年的时间[44,92-107]。独立研究确实表明，卷烟和HTP的双重使用比最初由企业赞助的研究所暗示的更为普遍[44,98,100]。然而，鉴于社会人口统计学混杂因素，需要更多关于使用模式的信息[95,99,106]。对于新一代HTP总体上有助于吸烟者从卷烟过渡的建议几乎没有经验支持，并且没有发表关于烟碱戒断的研究。大多数可用的研究都与企业相关，并且研究最多的是IQOS。实验室研究表明，分析的HTP中只有一种可以释放与卷烟相当剂量的烟碱[76,77,84]；然而，其他因素，包括吸引力和致瘾性，在替代行为中往往很重要。HTP似乎主观上减少了对吸烟的渴望，尽管可能不如CC显著。然而，很明显的是HTP不像CC那样让吸烟者满意。

5.10 研究差距

- 关于从未使用过该类产品的人群层面使用模式的独立数据。关键指标包括与其他产品的使用、使用量、日常或非日常使用和口味偏好（如适用），以确定降低风险声明的有效性。
- 对lil（KT&G）的研究，这是一种在韩国售卖的HTP，现在由PMI在其他地方销售，没有找到公开数据。
- 对IQOS以外的HTP中烟碱的药代动力学进行独立研究，最好在受试者之间进行，以便进行直接比较。
- 研究HTP、领先的ENDS产品和/或烟碱替代疗法的烟碱递送药代动力学，以比较HTP与用于戒烟的产品的潜在滥用倾向。
- 对专门采用HTP用于常规卷烟的戒断和使用行为独立研究。
- 使用HTP的戒烟研究。

5.11 政策建议

决策者应考虑以下有关HTP在戒断传统卷烟中的潜在作用的政策建议，特别是在使用多种烟草制品情况下。

5.11.1 戒烟政策

没有足够的证据表明HTP有助于戒烟。因此，不应就此提出声明，即使未来的证据支持HTP作为有效的转换辅助手段（即用一种烟草制品替代另一种烟草制品），它们也不应被视为戒烟的治疗方法，包括戒掉烟碱的使用。

5.11.2 监督政策

在国家层面很少有长期对各种社会人群中HTP的流行率和使用模式进行的监测，因此应立即实施。了解弱势群体（例如青少年、少数民族、孕妇）的使用模式尤为重要。调查的变量应包括使用频率（日常或非日常使用）、使用量、同时使用其他烟草和烟碱产品（多重使用）和使用调味产品（如适用）。监测系统可能还包括强制在医疗记录中记录烟草使用情况，包括HTP。

5.11.3 研究政策

应该对消费者使用HTP完全替代传统卷烟进行研究。鼓励决策者优先考虑如何增加循证吸烟治疗的覆盖面、需求、质量、传播、实施和可持续性的资助研究。

5.11.4 合作政策

鉴于在缺乏必要的科学证据证明HTP作为转换传统吸烟的辅助手段的有效性的情况下，HTP的使用在迅速传播，因此应敦促决策者分享国家经验并合作制定适当的HTP监管框架。

5.12 参 考 文 献

[1] Tabuchi T, Gallus S, Shinozaki T, Nakaya T, Kunugita N, Colwell B. Heat-not-burn tobacco product use in Japan: its prevalence, predictors and perceived symptoms from exposure to secondhand heat-not-burn tobacco aerosol. Tob Control. 2017;27(e1):e25–33.

[2] Tabuchi T, Kiyohara K, Hoshino T, Bekki K, Inaba Y, Kunugita N. Awareness and use of electronic cigarettes and heat-not-burn tobacco products in Japan. Addiction. 2016;111(4):706–13.

[3] Miyazaki Y, Tabuchi T. Educational gradients in the use of electronic cigarettes and heat-not-burn tobacco products in Japan. PloS One. 2018;13(1):e0191008.

[4] Kioi Y, Tabuchi T. Electronic, heat-not-burn, and combustible cigarette use among chronic disease patients in Japan: A cross-sectional study. Tob Induc Dis. 2018;16:41.

[5] Kotz D, Kastaun S. E-Zigaretten und Tabakerhitzer: repräsentative Daten zu Konsumverhalten und assoziierten Faktoren in der deutschen Bevölkerung (die DEBRA-Studie) [E-cigarettes and heat-not-burn products: representative data on consumer behaviour and associated factors in the German population (the DEBRA study)]. Bundesgesundheitsblatt Gesundheitsforschung Gesundheitsschutz. 2018;61(11):1407–14.

[6] Liu X, Lugo A, Spizzichino L, Tabuchi T, Pacifici R, Gallus S. Heat-not-burn tobacco products: concerns from the Italian experience. Tob Control. 2019;28(1):113–4.

[7] Langer P, Prieto L, Rousseau C. Tobacco product use after the launch of a heat-not-burn alternative in Japan: results of two cross-sectional surveys. Poster. Global Forum on Nicotine, 2019 (https://www.pmiscience.com/resources/docs/default-source/posters2019/langer-2019-tobacco-product-use-after-the-launch-of-a-heat-not-burn-alternative-in-japan.pdf?sfvrsn=460ed806_4, accessed 10 January 2021).

[8] Kim J, Yu H, Lee S, Paek YJ. Awareness, experience and prevalence of heated tobacco product, IQOS, among young Korean adults. Tob Control. 2018;27(Suppl 1):s74–7.

[9] Brose LS, Simonavicius E, Cheeseman H. Awareness and use of "Heat-not-burn" tobacco products in Great Britain. Tob Reg Sci. 2018;4(2):44–50.

[10] West R, Beard E, Brown J. Trends in electronic cigarette use in England. Smoking Toolkit Study, 2020 (http://www.smokinginengland.info/sts-documents/, accessed 10 January 2021).

[11] Sutanto E, Miller C, Smith DM, O'Connor RJ, Quah ACK, Cummings KM et al. Prevalence, use behaviors, and preferences among users of heated tobacco products: Findings from the 2018 ITC Japan survey. Int J Environ Res Public Health. 2019;16(23):4630.

[12] Lee Y, Lee KS. Association of alcohol and drug use with use of electronic cigarettes and heat-not-burn tobacco products among Korean adolescents. PloS One. 2019;14(7):e0220241.

[13] Pinkas J, Kaleta D, Zgliczynski WS, Lusawa A, Iwona Wrześniewska-Wal, Wierzba W et al. The prevalence of tobacco and e-cigarette use in Poland: A 2019 nationwide cross-sectional survey. Int J Environ Res Public Health. 2019;16(23):4820.

[14] Berg CJ, Haardorfer R, Schauer G, Getachew B, Masters M, McDonald B et al. Reasons for polytobacco use among young adults: Scale development and validation. Tob Prev Cessation. 2016;2:69.

[15] Bombard JM, Pederson LL, Koval JJ et al. How are lifetime polytobacco users different than current cigarette-only users? Results from a Canadian young adult population. Addictive behaviors 2009;34(12):1069-72. doi: 10.1016/j.addbeh.2009.06.009 [published Online First: 2009/08/04]

[16] Bombard JM, Pederson LL, Nelson DE, Malarcher AM. Are smokers only using cigarettes? Exploring current polytobacco use among an adult population. Addict Behav. 2007;32(10):2411-9.

[17] Fix BV, O'Connor RJ, Vogl L, Smith D, Bansal-Travers M, Conway KP et al. Patterns and correlates of polytobacco use in the United States over a decade: NSDUH 2002–2011. Addict Behav. 2014;39(4):768-81.

[18] Horn K, Pearson JL, Villanti AC. Polytobacco use and the "customization generation" – new perspectives for tobacco control. J Drug Educ. 2016;46(3-4):51–63.

[19] Lee YO, Hebert CJ, Nonnemaker JM, Kim AE. Multiple tobacco product use among adults in the United States: cigarettes, cigars, electronic cigarettes, hookah, smokeless tobacco, and snus. Prev Med. 2014;62:14–9.

[20] Martinasek MP, Bowersock A, Wheldon CW. Patterns, perception and behavior of electronic nicotine delivery systems use and multiple product use among young adults. Respir Care. 2018;63(7):913–9.

[21] Sung HY, Wang Y, Yao T et al. Polytobacco use of cigarettes, cigars, chewing tobacco, and snuff among US adults. Nicotine Tob Res. 2016;18(5):817-26. doi: 10.1093/ntr/ntv147 [published Online First: 2015/07/03].

[22] Sung HY, Wang Y, Yao T, Lightwood J, Max W. Polytobacco use and nicotine dependence symptoms among US adults, 2012–2014. Nicotine Tob Res. 2018;20(Suppl 1):S88-98.

[23] Wong EC, Haardorfer R, Windle M, Berg CJ. Distinct motives for use among polytobacco versus cigarette only users and among single tobacco product users. Nicotine Tob Res. 2017;20(1):117–23.

[24] Hinton A, Nagaraja HN, Cooper S, Wewers ME. Tobacco product transition patterns in rural and urban cohorts: Where do dual users go? Prev Med Rep. 2018;12:241–4.

[25] Miller CR, Smith DM, Goniewicz ML. Changes in nicotine product use among dual users of tobacco and electronic cigarettes: Findings from the Population Assessment of Tobacco and Health (PATH) study, 2013–2015. Subst Use Misuse. 2020:1–5. doi: 10.1080/10826084.2019.1710211.

[26] Niaura R, Rich I, Johnson AL, Villanti AC, Romberg AR, Hair EC et al. Young adult tobacco and e-cigarette use transitions: Examining stability using multi-state modeling. Nicotine Tob Res. 2020;22(5):647–54.

[27] Piper ME, Baker TB, Benowitz NL, Jorenby DE. Changes in use patterns over one year among smokers and dual users of combustible and electronic cigarettes. Nicotine Tob Res.

2020;22(5):672–80.

[28] Borland R, Murray K, Gravely S, Fong GT, Thompson ME, McNeill A et al. A new classificati-on system for describing concurrent use of nicotine vaping products alongside cigarettes (so-called "dual use"): findings from the ITC-4 Country Smoking and Vaping wave 1 survey. Addiction. 2019;114(Suppl 1):24–34.

[29] Bhatnagar A, Whitsel LP, Blaha MJ, Huffman MD, Krishan-Sarin S, Maa J et al. New and emerging tobacco products and the nicotine endgame: The role of robust regulation and comprehensive tobacco control and prevention: A Presidential Advisory from the American Heart Association. Circulation. 2019;139(19):e937–58.

[30] Farsalinos KE, Romagna G, Voudris V. Factors associated with dual use of tobacco and electronic cigarettes: A case control study. Int J Drug Policy. 2015;26(6):595–600.

[31] Goniewicz ML, Smith DM, Edwards KC, Blount BC, Caldwell KL, Feng J et al. Comparison of nicotine and toxicant exposure in users of electronic cigarettes and combustible cigarettes. JAMA Netw Open. 2018;1(8):e185937.

[32] Shahab L, Goniewicz ML, Blount BC, Brown J, McNeill A, Alwis KU et al. Nicotine, carcinogen, and toxin exposure in long-term e-cigarette and nicotine replacement therapy users: a cross-sectional study. Ann Intern Med. 2017;166(6):390–400.

[33] Bjartveit K, Tverdal A. Health consequences of smoking 1–4 cigarettes per day. Tob Control. 2005;14(5):315–20.

[34] Gerber Y, Myers V, Goldbourt U. Smoking reduction at midlife and lifetime mortality risk in men: a prospective cohort study. Am J Epidemiol. 2012;175(10):1006–12.

[35] Hart C, Gruer L, Bauld L. Does smoking reduction in midlife reduce mortality risk? Results of 2 long-term prospective cohort studies of men and women in Scotland. Am J Epidemiol. 2013;178(5):770–9.

[36] Tverdal A, Bjartveit K. Health consequences of reduced daily cigarette consumption. Tob Control. 2006;15(6):472–80.

[37] Baig SA, Giovenco DP. Behavioral heterogeneity among cigarette and e-cigarette dual-users and associations with future tobacco use: Findings from the Population Assessment of Tobacco and Health study. Addict Behav. 2020;104:106263.

[38] Farsalinos KE, Yannovits N, Sarri T, Voudris V, Poulas K. Nicotine delivery to the aerosol of a heat-not-burn tobacco product: comparison with a tobacco cigarette and e-cigarettes. Nicotine Tob Res. 2018;20(8):1004–9.

[39] Farsalinos KE, Yannovits N, Sarri T, Voudris V, Poulas K, Leischow SJ. Carbonyl emissions from a novel heated tobacco product (IQOS): comparison with an e-cigarette and a tobacco cigarette. Addiction. 2018;113(11):2099–106.

[40] Leigh NJ, Palumbo MN, Marino AM, O'Connor RJ, Goniewicz ML. Tobacco-specific nitrosamines (TSNA) in heated tobacco product IQOS. Tob Control. 2018;27(Suppl 1):s37–8.

[41] Hwang JH, Ryu DH, Park SW. Heated tobacco products: Cigarette complements, not substitutes. Drug Alcohol Depend. 2019;204:107576.

[42] Tabuchi T, Gallus S, Shinozaki T, Nakaya T, Kunugita N, Colwell B. Heat-not-burn tobacco product use in Japan: its prevalence, predictors and perceived symptoms from exposure to secondhand heat-not-burn tobacco aerosol. Tob Control. 2018;27(e1):e25–33.

[43] Tabuchi T, Shinozaki T, Kunugita N, Nakamura M, Tsuji I. Study profile: The Japan "Society and New Tobacco" Internet Survey (JASTIS): A longitudinal Internet cohort study of heat-not-burn tobacco products, electronic cigarettes, and conventional tobacco products in Japan. J Epidemiol. 2019;29(11):444–50.

[44] Sutanto E, Miller C, Smith DM, Borland R, Hyland A, Cummings KM et al. Concurrent daily and non-daily use of heated tobacco products with combustible cigarettes: Findings from the 2018 ITC Japan survey. Int J Environ Res Public Health. 2020;17(6):2098.

[45] Roulet S, Chrea C, Kanitscheider C, Kallischnigg G, Magnani P, Weitkunat R. Potential predictors of adoption of the tobacco heating system by US adult smokers: An actual use study. F1000Res. 2019;8:214.

[46] Hair EC, Bennett M, Sheen E, Cantrell J, Briggs J, Fenn Z et al. Examining perceptions about IQOS heated tobacco product: consumer studies in Japan and Switzerland. Tob Control. 2018;27(Suppl 1):s70–3.

[47] Mays D, Arrazola RA, Tworek C, Rolle IV, Neff LJ, Portnoy DB. Openness to using non-cigarette tobacco products among US young adults. Am J Prev Med. 2016;50(4):528–34.

[48] McKelvey K, Popova L, Kim M, Chaffee BW, Vijayaraghavan M, Ling P et al. Heated tobacco products likely appeal to adolescents and young adults. Tob Control. 2018;27(Suppl 1):s41–7.

[49] Soneji S, Sargent JD, Tanski SE, Primack BA. Associations between initial water pipe tobacco smoking and snus use and subsequent cigarette smoking: Results from a longitudinal study of US adolescents and young adults. JAMA Pediatr. 2015;169(2):129–36.

[50] Gee J, Prasad K, Slayford S, Gray A, Nother K, Cunningham A et al. Assessment of tobacco heating product THP1.0. Part 8: Study to determine puffing topography, mouth level exposure and consumption among Japanese users. Regul Toxicol Pharmacol. 2018;93:84–81.

[51] Philip Morris Products S.A. modified risk tobacco product (MRTP) applications. Silver Spring (MD): Food and Drug Administration; 2020 (https://www.fda.gov/tobacco-products/advertising-and-promotion/philip-morris-products-sa-modified-risk-tobacco-product-mrtp-applications, accessed 10 January 2021).

[52] Tobacco Products Scientific Advisory Committee, 25 January 2018. Silver Spring (MD): Food and Drug Administration; 2018 (https://www.fda.gov/downloads/AdvisoryCommittees/CommitteesMeetingMaterials/TobaccoProductsScientificAdvisoryCommittee/UCM599235.pdf, accessed 10 January 2021).

[53] Committees on Toxicity, Carcinogenicity and Mutagenicity of Chemical Products In Food, Consumer Products And The Environment (COT, COC and COM). Statement on the toxicological evaluation of novel heat-not-burn tobacco products. London: Food Standards Agency; 2017 (https://cot.food.gov.uk/sites/default/files/heat_not_burn_tobacco_statement.pdf, accessed 10 January 2021).

[54] Hughes JR, Keely JP. The effect of a novel smoking system – Accord – on ongoing smoking and toxin exposure. Nicotine Tob Res. 2004;6(6):1021–7.

[55] Fagerstrom KO, Hughes JR, Callas PW. Long-term effects of the Eclipse cigarette substitute and the nicotine inhaler in smokers not interested in quitting. Nicotine Tob Res. 2002;4(Suppl 2):S141–5.

[56] Hackshaw A, Morris JK, Boniface S, Tang JL, Milenković M. Low cigarette consumption and

risk of coronary heart disease and stroke: meta-analysis of 141 cohort studies in 55 study reports. BMJ. 2018;360:j5855.

[57] Inoue-Choi M, Christensen CH, Rostron BL, Cosgrove CM, Reyes-Guzman C, Apelberg B et al. Dose–response association of low-intensity and nondaily smoking with mortality in the United States. JAMA Netw Open. 2020;3(6):e206436.

[58] Inoue-Choi M, Hartge P, Liao LM, Caporaso N, Freedman ND. Association between long-term low-intensity cigarette smoking and incidence of smoking-related cancer in the National Institutes of Health–AARP cohort. Int J Cancer. 2018;142(2):271–80.

[59] Gilchrist M. Presentation by Moira Gilchrist, PhD, Vice President Scientific and Public Communications, Philip Morris International, before the Tobacco Products Scientific Advisory Committee (TPSAC). Lausanne: Philip Morris International; 2018 (https://tobacco.ucsf.edu/sites/g/files/tkssra4661/f/wysiwyg/PMI-24Jan-TPSAC-Presentation.pdf).

[60] Auer R, Concha-Lozano N, Jacot-Sadowski I, Cornuz J, Berthet A. Heat-not-burn tobacco cigarettes: Smoke by any other name. JAMA Intern Med. 2017;177(7):1050–2.

[61] Bekki K, Inaba Y, Uchuyama S, Kunugita N. Comparison of chemicals in mainstream smoke in heat-not-burn tobacco and combustion cigarettes. J UOEH. 2017;39(3):201–7.

[62] Farsalinos KE, Yannovits N, Sarri T, Voudris V, Poulas K. Nicotine delivery to the aerosol of a heat-not-burn tobacco product: comparison with a tobacco cigarette and e-cigarettes. Nicotine Tob Res. 2018;20(8):1004–9.

[63] Forster M, Fiebelkorn S, Yurteri C, Mariner D, Liu C, Wright C et al. Assessment of novel tobacco heating product THP1.0. Part 3: Comprehensive chemical characterisation of harmful and potentially harmful aerosol emissions. Regul Toxicol Pharmacol. 2018;93:14–33.

[64] Jaccard G, Tafin Djoko D, Moennikes O, Jeannet C, Kondylis A, Belushkin M. Comparative assessment of HPHC yields in the tobacco heating system THS2.2 and commercial cigarettes. Regul Toxicol Pharmacol. 2017;90:1–8.

[65] Li X, Luo Y, Jiang X, Zhang H, Zhu F, Hus S et al. Chemical analysis and simulated pyrolysis of tobacco heating system 2.2 compared to conventional cigarettes. Nicotine Tob Res. 2019;21(1):111–8.

[66] Mallock N, Boss L, Burk R, Dabziger M, Welsch T, Hahn H et al. Levels of selected analytes in the emissions of "heat not burn" tobacco products that are relevant to assess human health risks. Arch Toxicol. 2018;92(6):2145–9.

[67] Nabavizadeh P, Liu J, Havel CM, Ibrahim S, Derakhshandeh R, Jacon P III et al. Vascular endothelial function is impaired by aerosol from a single IQOS HeatStick to the same extent as by cigarette smoke. Tob Control. 2018;27(Suppl 1):s13–9.

[68] Poynton S, Sutton J, Goodall S, Margham J, Forster M, Scott K et al. A novel hybrid tobacco product that delivers a tobacco flavour note with vapour aerosol (Part 1): Product operation and preliminary aerosol chemistry assessment. Food Chem Toxicol. 2017;106(Pt A):522–32.

[69] Salman R, Talih S, El-Hage R, Haddad C, Karaoghlanian N, El-Hellani A et al. Free-base and total nicotine, reactive oxygen species, and carbonyl emissions from IQOS, a heated tobacco product. Nicotine Tob Res. 2019;21(9):1285–8.

[70] Schaller JP, Keller D, Poget L, Pratte P, Kaelin E, McHugh D et al. Evaluation of the tobacco heating system 2.2. Part 2: Chemical composition, genotoxicity, cytotoxicity, and physical

[71] Schaller JP, Pijnenburg JP, Ajithkumar A, Tricker AR. Evaluation of the tobacco heating system 2.2. Part 3: Influence of the tobacco blend on the formation of harmful and potentially harmful constituents of the tobacco heating system 2.2 aerosol. Regul Toxicol Pharmacol. 2016;81(Suppl 2):S48–58.

[72] Simonavicius E, McNeill A, Shahab L, Brose LS. Heat-not-burn tobacco products: a systematic literature review. Tob Control. 2019;28(5):582–94.

[73] Counts M, Morton M, Laffoon S, Cox R, Lipowicz P. Smoke composition and predicting relationships for international commercial cigarettes smoked with three machine-smoking conditions. Regul Toxicol Pharmacol. 2005;41(3):185–227.

[74] Haziza C, de la Bourdonnaye G, Skiada D, Ancerewicz J, Baker G, Picavet P et al. Evaluation of the tobacco heating system 2.2. Part 8: 5-day randomized reduced exposure clinical study in Poland. Regul Toxicol Pharmacol. 2016;81(Suppl 2):S139–50.

[75] Haziza C, de la Bourdonnaye G, Merlet S, Benzimra M, Ancerewicz J, Donelli A et al. Assessment of the reduction in levels of exposure to harmful and potentially harmful constituents in Japanese subjects using a novel tobacco heating system compared with conventional cigarettes and smoking abstinence: A randomized controlled study in confinement. Regul Toxicol Pharmacol. 2016;81:489–99.

[76] Brossard P, Weitkunat R, Poux V, Lama N, Haziza C, Picavet P et al. Nicotine pharmacokinetic profiles of the tobacco heating system 2.2, cigarettes and nicotine gum in Japanese smokers. Regul Toxicol Pharmacol. 2017;89:193–9.

[77] Picavet P, Haziza C, Lama N, Weitkunat R, Lüdicke F. Comparison of the pharmacokinetics of nicotine following single and ad libitum use of a tobacco heating system or combustible cigarettes. Nicotine Tob Res. 2016;18(5):557–63.

[78] Lüdicke F, Picavet P, Baker G, Haziza C, Poux V, Lama N et al. Effects of switching to the menthol tobacco heating system 2.2, smoking abstinence, or continued cigarette smoking on clinically relevant risk markers: a randomized, controlled, open-label, multicenter study in sequential confinement and ambulatory settings (Part 2). Nicotine Tob Res. 2018;20(2):173–82.

[79] Lüdicke F, Picavet P, Baker G, Haziza C, Poux V, Lama N et al. Effects of switching to the tobacco heating system 2.2 menthol, smoking abstinence, or continued cigarette smoking on biomarkers of exposure: a randomized, controlled, open-label, multicenter study in sequential confinement and ambulatory settings (Part 1). Nicotine Tob Res. 2018;20(2):161–72.

[80] Lüdicke F, Haziza C, Weitkunat R et al. Evaluation of biomarkers of exposure in smokers switching to a carbon-heated tobacco product: a controlled, randomized, open-label 5-day exposure study. Nicotine Tob Res. 2016;18:1606–13. doi:10.1093/ntr/ntw022.

[81] Lüdicke F, Baker G, Magnette J et al. Reduced exposure to harmful and potentially harmful smoke constituents with the tobacco heating system 2.1. Nicotine Tob Res. 2016;19:168-175. doi:10.1093/ntr/ntw164.

[82] Haziza C, de la Bourdonnaye G, Donelli A, Poux V, Skiada D, Weitkunat R et al. Reduction in exposure to selected harmful and potentially harmful constituents approaching those observed upon smoking abstinence in smokers switching to the menthol tobacco heating system 2.2 for 3 months (Part 1). Nicotine Tob Res. 2020;22(4):539–48.

[83] Gale N, McEwan M, Eldridge A, Fearon IM, Sherwood N, Bowen E et al. Changes in biomarkers of exposure on switching from a conventional cigarette to tobacco heating products: A randomized, controlled study in healthy Japanese subjects. Nicotine Tob Res. 2018;21:122–7.

[84] Yuki D, Sakaguchi C, Kikuchi A, Futamura Y. Pharmacokinetics of nicotine following the controlled use of a prototype novel tobacco vapor product. Regul Toxicol Pharmacol. 2017;87:30–5.

[85] Biondi-Zoccai G, Sciarretta S, Bullen C, Nocella C, Violo F, Loffredo L et al. Acute effects of heat-not-burn, electronic vaping, and traditional tobacco combustion cigarettes: the Sapienza University of Rome – Vascular Assessment of Proatherosclerotic Effects of Smoking (SUR-VAPES) 2 randomized trial. J Am Heart Assoc. 2019;8(6):e010455.

[86] Lopez AA, Hiler M, Maloney S, Eissenberg T, Breland A. Expanding clinical laboratory tobacco product evaluation methods to loose-leaf tobacco vaporizers. Drug Alcohol Depend. 2016;169:33–40.

[87] Newland N, Lowe FJ, Camacho OM, McEwan M, Gale N, Ebajemito J et al. Evaluating the effects of switching from cigarette smoking to using a heated tobacco product on health effect indicators in healthy subjects: study protocol for a randomized controlled trial. Intern Emergency Med. 2019;14(6):885–98.

[88] Adriaens K, Gucht DV, Baeyens F. IQOS(TM) vs. e-cigarette vs. tobacco cigarette: A direct comparison of short-term effects after overnight-abstinence. Int J Environ Res Public Health. 2018;15(12):2902.

[89] Cox LS, Tiffany ST, Christen AG. Evaluation of the brief questionnaire of smoking urges (QSUbrief) in laboratory and clinical settings. Nicotine Tob Res. 2001;3(1):7–16.

[90] Cappelleri JC, Bushmakin AG, Baker CL, Merikle E, Olufade AO, Gilbert DG. Confirmatory factor analyses and reliability of the modified cigarette evaluation questionnaire. Addict Behav. 2007;32(5):912–23.

[91] Shiffman S, Terhorst L. Intermittent and daily smokers' subjective responses to smoking. Psychopharmacology (Berl). 2017;234(19):2911–7.

[92] Cruz-Jiménez L, Barrientos-Gutiérrez I, Coutiño-Escamilla L, Gallegos-Carrillo K, Arillo-Santillán E, Thrasher JF. Adult smokers' awareness and interest in trying heated tobacco products: perspectives from Mexico, where HTPs and e-cigarettes are banned. Int J Environ Res Public Health. 2020;17(7):2173.

[93] Czoli CD, White CM, Reid JL, O'Connor RJ, Hammond D. Awareness and interest in IQOS heated tobacco products among youth in Canada, England and the USA. Tob Control. 2020;29(1):89–95.

[94] Dunbar MS, Seelam R, Tucker JS, Rodriguez A, Shih RA, D'Amico EJ. Correlates of awareness and use of heated tobacco products in a sample of US young adults in 2018–2019. Nicotine Tob Res. 2020:doi: 10.1093/ntr/ntaa007.

[95] Gravely S, Fong GT, Sutanto E, Loewen R, Ouimet J, Xu SS, et al. Perceptions of harmfulness of heated tobacco products compared to combustible cigarettes among adult smokers in Japan: Findings from the 2018 ITC Japan Survey. Int J Environ Res Public Health.2020;17(7):2394.

[96] Hori A, Tabuchi T, Kunugita N. Rapid increase in heated tobacco product (HTP) use from 2015 to 2019: from the Japan "Society and New Tobacco" Internet Survey (JASTIS). Tob Control.

2020;5(5):652.

[97] Igarashi A, Aida J, Kusama T, Tabuchi T, Tsuboya T, Sugiyama K et al. Heated tobacco products have reached younger or more affluent people in Japan. J Epidemiol. 2020:doi: 10.2188/jea.JE20190260.

[98] Kang H, Cho SI. Heated tobacco product use among Korean adolescents. Tob Control. 2020;29(4):466–8.

[99] Kim K, Kim J, Cho HJ. Gendered factors for heated tobacco product use: Focus group interviews with Korean adults. Tob Induc Dis. 2020;18:43.

[100] Kim SH, Cho HJ. Prevalence and correlates of current use of heated tobacco products among a nationally representative sample of Korean adults: Results from a cross-sectional study. Tob Induc Dis. 2020;18:66.

[101] Kim SH, Kang SY, Cho HJ. Beliefs about the harmfulness of heated tobacco products compared with combustible cigarettes and their effectiveness for smoking cessation among Korean adults. Int J Environ Res Public Health. 2020;17(15):5591.

[102] Okawa S, Tabuchi T, Miyashiro I. Who uses e-cigarettes and why? e-Cigarette use among older adolescents and young adults in Japan: JASTIS Study. J Psychoactive Drugs. 2020;52(1):37–45.

[103] Ratajczak A, Jankowski P, Strus P, Feleszko W. Heat not burn tobacco product – A new global trend: Impact of heat-not-burn tobacco products on public health, a systematic review. Int J Environ Res Public Health. 2020;17(2):0409.

[104] Siripongvutikorn Y, Tabuchi T, Okawa S. Workplace smoke-free policies that allow heated tobacco products and electronic cigarettes use are associated with use of both these products and conventional tobacco smoking: the 2018 JASTIS study. Tob Control. 2020;doi: 10.1136/tobaccocontrol-2019-055465.

[105] Sugiyama T, Tabuchi T. Use of multiple tobacco and tobacco-like products including heated tobacco and e-cigarettes in Japan: A cross-sectional assessment of the 2017 JASTIS study. Int J Environ Res Public Health. 2020;17(6):2161.

[106] Tompkins CNE, Burnley A, McNeill A, Hitchman SC. Factors that influence smokers' and ex-smokers' use of IQOS: a qualitative study of IQOS users and ex-users in the UK. Tob Control. 2020:doi: 10.1136/tobaccocontrol-2019-055306.

[107] Wu YS, Wang MP, Ho SY, Li HCW, Cheung YTD, Tabuchi T, et al. Heated tobacco products use in Chinese adults in Hong Kong: a population-based cross-sectional study. Tob Control. 2020;29(3):277–81.

6. 加热型烟草制品和电子烟碱/非烟碱传输系统相关法规以及国家措施和监管障碍

Ranti Fayokun, Moira Sy, Marine Perraudin and Vinayak Prasad
No Tobacco Unit (Tobacco Free Initiative), Department of Health Promotion, Division of Universal Health Coverage and Healthier Populations, World Health Organization, Geneva, Switzerland

摘　要

烟草行业越来越多地将加热型烟草制品（HTP）作为更新产品组合的一部分进行营销，这些产品组合声称与传统卷烟相比，对使用者和非使用者造成的风险更小。这些产品自上市以来就获得了相当大的市场份额，目前在全球约50个国家（地区）市场上均有销售。新一代HTP由于其新颖性、非传统技术和企业声称的对低于传统卷烟的健康风险被不同国家以不同的方式分类。这些分类已经影响到各国用于监管这些产品的机制，导致各国之间的监管出现不一致，包括它们在多大程度上适用于世界卫生组织《烟草控制框架公约》（WHO FCTC）的规定。本章回顾了可获得HTP的市场、产品的常见分类以及分类如何影响监管结果。此外，描述了常用的监管框架、监管障碍、监管考虑因素和不可预见的后果。还向各国提供WHO和WHO FCTC的建议，使其根据其国家法律制定HTP的监管策略，并确保强有力地保护人体健康。

6.1 背　景

6.1.1 缔约方大会的介绍和要求（FCTC/COP8(22)）

当加热烟草材料或激活含有烟草的装置时，HTP会产生含有烟碱和有害化学物质的气溶胶[1]。烟草可以是专门设计的卷烟（例如"加热棒"和"新型棒"）、烟弹或插片。在专门为此目的设计的设备中加热烟草后,使用者会吸入由此产生的气溶胶[1]。烟草行业以多种方式积极营销和推广HTP,包括与传统卷烟相比的"无烟"、"更清洁的替代品"、"更安全的替代品"和"降低风险"的产品等声明。截至2020年7月,HTP可在世界卫生组织所有六个区域的50多个市场合法销售[2],示例见表6.1。

表6.1　可获得HTP的市场示例

装置	公司	市场
IQOS	Philip Morris International	安道尔、阿尔巴尼亚、亚美尼亚、白俄罗斯、波斯尼亚和黑塞哥维那、保加利亚、加拿大、哥伦比亚、克罗地亚、塞浦路斯、捷克、丹麦、多米尼加共和国、法国(包括留尼汪岛)、德国、希腊、危地马拉、匈牙利、以色列、意大利、日本、哈萨克斯坦、拉脱维亚、立陶宛、马来西亚、墨西哥、摩纳哥、荷兰(包括库拉索)、新西兰、波兰、葡萄牙、韩国、摩尔多瓦共和国、罗马尼亚、俄罗斯联邦、塞尔维亚、斯洛伐克、斯洛文尼亚、南非、西班牙(包括加那利群岛)、瑞典、瑞士、乌克兰、阿拉伯联合酋长国、英国、美国和巴勒斯坦被占领土,包括东耶路撒冷
iFuse,glo	British American Tobacco	加拿大、意大利、日本、韩国、罗马尼亚、俄罗斯和瑞士
Lil	KT&G	日本、韩国、俄罗斯和乌克兰
Ploom	Japan Tobacco International	意大利、日本、韩国、俄罗斯、瑞士和英国

加热型烟草制品(HTP)与电子烟碱传输系统(ENDS)和电子非烟碱传输系统(ENNDS)三大类产品在多个地区,尤其是在日本和韩国变得流行。虽然自20世纪80年代以来HTP技术就已经可用,但早期尝试引入这些产品并不成功,而这些产品的新一代产品仅在过去七年才开始流行(图6.1;另见第3章和第5章)。

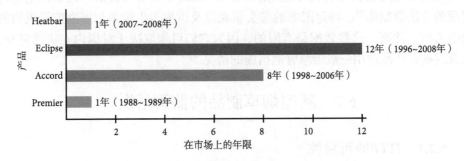

图6.1　早期加热型烟草制品上市和退市日期[3]

HTP对烟草监管提出了重大挑战,特别是因为其新颖的操作机制和健康影响的认识不足。烟草行业利用这些挑战,采用营销策略,特别是"减少危害"或"降低风险",以促进其进入市场,并认为HTP应与烟草制品,特别是传统卷烟进行不同的分类。由于认知不足,缺乏国际公认的方法来评估其使用风险,进一步使烟草管制复杂化。新颖性、错误信息和企业操纵导致了对其分类和监管的不同方法。

WHO FCTC缔约方大会（COP）第八次会议关于新型烟草制品的WHO FCTC COP8(22)号决议[4]要求公约秘书处邀请WHO准备一份关于这些产品涵盖多个研究领域的综合报告。预计WHO将向缔约方大会第九次会议报告缔约方的监管经验和监测情况、对烟草控制的影响和研究差距，并提出政策选择以实现决议第5段中概述的目标和措施。本章的目的是绘制HTP的监管图，回顾WHO成员国在HTP方面的监管经验，考虑HTP对烟草管制的影响并确定研究差距。

6.1.2 范围和目标

我们在解释了这些产品的常见分类后，描述了HTP的国家监管经验，这些分类可能决定了其特定的监管途径。例如，归入某个类别可能会导致完全禁止HTP，尽管某类别的监管含义因国家而异，并且一个国家的分类决定了烟草管制法律的适用程度。归类为"电子烟"产品的HTP可能会在一个国家被禁止而在另一个国家被监管。

本章描述了产品禁令，广告、促销和赞助，无烟场所，销售限制，包装和标签，产品设计等方面分类后的通用监管框架。由于现有数据的限制和解释，没有提供迄今为止实施政策的权威列表，但提供了有关监管机构针对HTP所采用的方法类型以及此类监管示例的信息。还描述了对这些产品使用传统烟草控制措施的监管障碍和应用的不一致性。

6.1.3 来源

立法和政策摘要来自世界卫生组织2019~2020年收集的数据、无烟儿童运动管理的立法数据库[5]、特定国家的案头研究以及世界卫生组织与国家监管机构的内部通信。然而，这些数据是有限的，因为它们可能取决于对国内立法的解释，以及这些新产品适用一般烟草管制措施的情况。

6.2 新型烟草制品的监管情况

6.2.1 HTP的可用性

HTP目前在50多个国家（地区）市场[2]上销售（表6.1）。然而，这个数字正在迅速增加，仅在过去两年中就增加了15个国家（地区）市场。最近，在一些国家（地区）这些产品的供应量有所增加，其中HTP正从传统烟草制品（主要是卷烟）中获得市场份额，尽管HTP销售收入仅占行业卷烟收入的一小部分[6]。烟草行业计划扩大市场份额，并在全球范围内增加这些产品的供应、知名度和获得渠道[7]，特别是通过保持其在市场上的供应以避免因缔约方对世界卫生组织《烟草控制框架公约》的义务而采取的严格烟草控制措施。监管豁免将推进烟草行业的长期目标，即提高各种新型烟草制品的可接受性，从而逐渐削弱WHO FCTC的管控。

HTP不仅在一些国家合法销售，而且在一些被禁止的国家以及产品不符合监管要求或没有获得上市前授权的国家也被非法交易。

6.2.2 产品分类

根据世界卫生组织的指南，HTP应归类为烟草制品[8]。世界卫生组织《烟草控制框架公约》缔约方大会第八次会议承认这些产品为烟草制品，并提醒缔约方有义务根据框架公约将其认定为烟草制品。然而，一些监管机构将HTP归入与传统卷烟或烟草制品不同的类别，因为它们具有非常规的特性。这些不同的类别主要是由于行业主张对所谓的"风险降低"产品弱化监管或不予监管，以及对产品的了解不足。迄今为止，各国已将HTP分为几类，包括：

- ■ 烟草制品；
- ■ 无烟烟草制品；
- ■ 新型或下一代烟草制品；
- ■ 电子烟[9]。

表6.2列出了使用这些类别的国家（地区）的示例。

表6.2 HTP的监管分类示例

分类	国家（地区）示例（非详尽清单）
烟草制品	韩国、阿拉伯联合酋长国 110多个国家/地区按定义将HTP归类为烟草制品
ENDS	巴西、印度、俄罗斯、沙特阿拉伯
新型烟草制品	欧盟、英国
新兴仿制烟草制品	新加坡
无烟烟草制品	新西兰

注：ENDS，电子烟碱传输系统；HTP，加热型烟草制品。

HTP也可能被归为混合或豁免类别[9]，这往往导致HTP的待遇优于传统卷烟。装置的性质也可能影响其分类：烟棒（含有烟草）可归类为烟草制品，而将烟棒插入其中的加热装置可能有不同的分类[9]。

产品的定义与其分类或归类密切相关，因为法规对"烟草制品"的定义因司法管辖区而异。一些国家将"烟草制品"定义为包含烟碱的所有烟草衍生材料，以及产品的消费方式（例如吸吮、抽吸、咀嚼），有些国家可能将定义扩展到产品的呈现方式或烟碱的来源，有些可能包括所有含烟碱的产品。烟草制品的定义方式决定了该国现有烟草管制法规和立法在多大程度上适用于HTP，除非针对

HTP或适用于HTP的具体法律有专门的例外规定。

烟草制品

许多世界卫生组织地区的国家法律和法规对烟草制品的定义足够广泛，包括HTP。如果烟草管制法没有具体规定，HTP可以通过其他适用的法规进行监管，例如消费者保护法或毒物法。例如，澳大利亚的毒物标准法将烟碱归类为附表7毒物，因此其销售和持有在很大程度上是非法的，尽管在某些州，根据治疗用品管理局的"个人进口计划"，可以凭医疗处方进口3个月的供应量。这种限制性措施毫无意外地受到来自烟草行业的挑战。菲利普·莫里斯公司向治疗用品管理局的化学品和毒物咨询委员会提交了一份监管申请，以要求毒物标准的豁免，允许合法销售为"加热"而制备和包装的烟草中的烟碱。拟议的修正案于2020年6月被治疗用品管理局驳回[10]。

电子烟

一些国家根据立法定义将HTP归类为"电子烟"或"电子吸烟设备"。这可能会导致类似于电子烟的监管，或者在其他国家禁止销售或进口整类产品。例如，在韩国，虽然HTP主要归类为烟草制品，但根据法律，使用电子设备消费烟草（例如通过加热）的烟草制品被归类为电子烟。因此，该子类别适用于HTP，因为HTP包含加热烟草的电子设备。在该国的监管环境中，这意味着大多数烟草制品法规，包括关于无烟区、税收、广告、健康警示和标签以及禁止向未成年人销售的法规，都适用于HTP；然而，只有卷烟税率的90%适用于HTP。根据国家立法，将HTP归类为电子烟或电子吸烟设备可能会导致一系列措施，从禁止（巴西）到监管。

新型烟草制品

在欧盟，根据欧盟烟草制品指令2014/40/EU[11]，HTP作为新型烟草制品受到监管。"新型烟草制品"被定义为需要遵守指令规定的烟草制品，包括禁止第13条预见的误导成分，特别是禁止暗示特定烟草制品比其他烟草制品危害更小[12]。

根据指令第19条关于新型烟草制品的通告，烟草制造商和进口商必须提供其打算投放到国内市场并属于新型烟草制品类别的所有产品的信息和支持文件。具体而言，这些产品的制造商和进口商必须在产品投放市场前6个月以电子形式向成员国的主管机构提交通告，并附上有关产品成分和释放物的信息。

制造商和进口商还必须向主管机构提供：①关于新型烟草制品的有害性、致瘾性和吸引力的现有科学研究，特别是关于其成分和释放物的研究；②现有研究、研究摘要和关于各种消费者群体（包括年轻人和当前吸烟者）偏好的市场调查；③其他可获得的相关信息，包括产品的风险/收益分析、其对戒烟的预期影响、其对开始烟草消费的预期影响以及预测的消费者认知。

此外，必须向其主管机构提交有关上述①~③中提及的研究、调查和其他信息的任何新的资料，并根据有关主管机构的要求进行补充试验或提交补充资料。除了向相关国家机构通报这些产品之外，成员国还可以在认为合适的情况下引入授权程序，并可以向制造商和进口商收取一定比例的授权费用。

包括波兰（1995年11月9日《防止吸烟危害和保护身体健康法》第110条）和西班牙（579/2017号皇家法令）在内的国家将HTP归类为新型烟草制品。法国根据《法国公共卫生法》第L3512-1条第1款将HTP规定为烟草制品，并根据第L3512-1条第2款将HTP规定为新型烟草制品。法国法律没有建立产品授权制度，但包括产品信息报告制度（通报）。HTP制造商必须向国家机构报告成分的名称、数量和相关的健康影响。法国公共卫生法典第L3512-8条禁止在公共场所使用HTP，还禁止使用健康声明、向18岁以下青少年出售HTP以及对HTP进行广告、促销和赞助，并需要警示标签。

新兴仿制烟草制品

在新加坡，HTP的监管方式与含有烟碱的电子烟和电子烟碱传输系统相同，因为两者都被认为是新兴仿制烟草制品。属于此类别的产品包括具有以下特性的任何设备或物品：

■ 类似于烟草制品或被设计成类似烟草制品；
■ 可抽吸；
■ 可以模仿吸烟行为的方式使用；
■ 包装类似于或被设计成类似通常与烟草制品相关的包装[13]。

根据2011年修订的《烟草广告和销售控制法案》[13]，禁止进口、分销、销售、购买使用或拥有此类产品。

无烟烟草制品

欧盟烟草制品指令第2条将无烟烟草制品定义为"不涉及燃烧过程的烟草制品，包括咀嚼烟草、鼻烟和口用烟草"。因此，欧盟的一些国家，例如捷克（第110/1997 Coll号法案第2(1)t条）和葡萄牙（第109/2015号法案），将HTP归类为无烟烟草制品。

在荷兰，HTP也被作为无烟烟草制品，根据《荷兰烟草法案》[14]进行监管，该法案由荷兰食品和消费品安全局执行。根据欧盟指令的通告要求，荷兰国家公共卫生与环境研究所（RIVM）分析和处理上市前通告文件[15]。对HTP的要求包括：与无烟烟草制品类似的警示标签、禁止健康声明、禁止向18岁以下青少年销售以及禁止促销和营销，只有少数例外。与其他烟气型烟草制品一样，HTP的税率为99.25欧元/千克。根据荷兰政府的说法，如果有关于HTP使用的新证据或信息，可将其归类为无烟烟草制品。RIVM目前正在为卫生部分析和研究不同IQOS HeatStick口味和其他HTP的成分和释放量。

新一代产品

在意大利，HTP作为新一代产品受到监管。由于HTP被认为可以减少危害，因此这些HTP不受烟草制品财政制度的约束。相反，它们在特定消费税结构下按"不燃烧的吸入产品"类别征税[16]。因此，这些产品享受与电子烟相同的税收减免，相当于传统卷烟的消费税的一半[17]。尽管禁止向未成年人出售HTP（第6/2016号法令第II章第24.3条），但对HTP而言，烟草管制法规的执行力度很小。健康警示只需要覆盖30%的HTP包装（而不是传统卷烟的65%），并且不需要图片[18]。禁止在所有公共场所和工作场所吸烟的全面法规不适用于HTP。此外，HTP的广告和促销没有被禁止，"IQOS旗舰店"和"IQOS精品店"（人们可以在其中免费试用产品的高档概念店）出现在意大利的几个大城市。因此，对于HTP，国家削弱了最受认可的烟草管制政策，即提价和增税、禁烟、广告禁令和健康警示。

HTP分类的国际方法

各国对HTP分类方法的差异和烟草行业提出的法律挑战引起了人们对缺乏HTP分类国际标准的担忧。世界卫生组织正在与专家和研究人员合作制定烟草制品分类树，其中将包括HTP，并将在项目完成后公布其研究结果。还将根据WHO FCTC COP8(22)号决议第3(b)段向2021年11月举行的世界卫生组织《烟草控制框架公约》下届缔约方大会提交一份关于HTP适当分类的报告，其中缔约方大会要求公约秘书处就HTP等新型烟草制品的适当分类提出建议[19]。世界海关组织正在推动修订"协调制度编码"，为HTP引入新的特定海关编码。《商品名称和编码协调制度国际公约》的附件目前指出：

- 加热型烟草制品没有具体的海关编码，属于《商品名称和编码协调制度国际公约》第24章中涉及烟草制品的"其他"(2403.99)子目；
- 用于加热型烟草制品（即HTP）的设备没有具体的海关编码，属于第85章中涉及电机的"其他机器和设备"(8543.70)的子目。

在2022年进行了强制性更改，以标准化这些品目，HTP归入新的类目24.4：

含有烟草、再造烟草、烟碱或烟草或烟碱替代品的不经燃烧吸入的产品，用于将烟碱摄入人体的其他含烟碱产品[20]。

各国将有义务在2020年修改其国内海关编码。世界海关组织将在当年晚些时候确定一次性设备的分类方式。

协调制度编码的目的不是影响HTP的国内监管,但如WHO FCTC秘书处信息说明[21]所述,实际上,这些编码影响货物在边境的进出境和消费税征收,以便在国内立法中分类。烟草行业可以利用这些措施游说政府对HTP给予更优惠的待遇。

6.2.3 减少烟草需求的监管框架和措施

如上所述,烟草制品的分类决定了适用的法规。这反过来又会影响这些产品的可获得性和使用,以及有关税收、广告、促销和赞助限制、产品在无烟场所的使用以及包装和标签要求的规定。一些国家选择通过禁止整个类别的产品(例如电子烟或新兴仿制烟草制品)来完全禁止进口、销售或使用HTP。例如,如果一个国家禁止电子烟,然后做出将HTP归类为电子烟的监管决定,该分类将确保HTP不会进入市场(例如印度)。然而,相同的产品分类(即"电子烟")可能会导致在一个国家被禁止而在另一个国家被监管限制。

世界卫生组织成员国已采用各种机制来监管HTP。虽然许多国家使用了现有的烟草管制法律,但有些国家制定了专门的条款。世界卫生组织掌握的信息表明有以下共同机制。

现行法律

HTP的定义可以与法律已涵盖的产品相同,例如在南非,HTP被视为烟草制品。如上所述(第6.2.1节),在澳大利亚,HTP受《药品和有毒物质统一目录标准》(简称《毒物标准》)的监管,含有烟碱的产品归类为附表7"危险毒物"。治疗用品管理局最近拒绝修改毒物标准以允许销售HTP[11]。

现行法律的修订

如果现行立法的定义未明确涵盖这些产品,则可以修改现有立法以包括HTP。在马来西亚,2015年对2004年《烟草制品管制条例修正案》更改了"吸烟"的定义,将使用HTP纳入其中[22]。

新的立法和其他机制

新的立法和其他机制可用于监管HTP或将其明确纳入现有立法。阿拉伯联合酋长国颁布了电子烟碱产品(传统烟草制品的等效物)标准[23],该标准对电子烟、含和不含烟碱的电子烟液和HTP进行了规定,并要求详细说明其生产、进口、零售和展示。因此,这些产品现在属于15/2009号联邦法律规定的与烟草制品相同的监管框架。在菲律宾,总统于2020年2月发布了一项行政法令,规范电子烟、HTP和其他新型烟草制品的商业化和使用[24]。该行政法令将HTP排除在烟草制品的定义之外,但将其纳入吸烟的定义中。

如前所述，HTP在WHO FCTC COP8(22)决议[19]中被确认为烟草制品。该决议提醒缔约方需履行的WHO FCTC承诺，在应对新型烟草制品（如HTP和设计用于消费此类产品的设备）的挑战时，考虑根据WHO FCTC和国家法律优先采取以下措施。这些是决议第5段中列出的：

(1) 预防开始使用新型烟草制品；

(2) 保护人们免于接触其释放物，并根据WHO FCTC第8条，将无烟立法的范围明确扩大到这些产品；

(3) 预防新型烟草制品无害健康的声明；

(4) 根据WHO FCTC第13条，针对新型烟草制品的广告、促销和赞助采取措施；

(5) 根据WHO FCTC第9条和第10条，监管和披露新型烟草制品的成分；

(6) 根据WHO FCTC第5.3条，保护烟草控制政策和行动免受与新型烟草制品相关的所有商业利益和其他既得利益，包括烟草行业的利益的影响；

(7) 管制包括限制或酌情禁止新型烟草制品的制造、进口、分销、展示、销售和使用，并根据其国家法律，考虑高度保护人体健康；

(8) 在适当的情况下，对设计用于消费此类产品的设备采取上述措施。

广告、促销和赞助（第13条）

WHO FCTC第1条提供了烟草广告、促销和赞助的全面定义。烟草广告和促销被定义为"以直接或间接促进烟草制品或烟草使用为目的、效果或可能效果的任何形式的商业传播、推荐或活动"，赞助是"对任何事件、活动或个人的任何形式的捐助，以直接或间接促进烟草制品或烟草使用为目的、效果或可能效果"[25]。

尽管大多数国家没有具体规范HTP的广告、促销和赞助，但根据WHO和COP的指导，这些产品应包含在适用于传统烟草制品的广告、促销和赞助禁令中。如果区分HTP烟棒和设备，并且烟草制品的定义仅涵盖烟棒，则可能不会禁止该设备的广告。

全面禁止烟草广告、促销和赞助不仅包括电视、广播和印刷等传统广告形式，还包括"品牌延伸"、销售点的产品展示和烟草行业赞助的企业社会责任计划等。然而，快速变化的媒体环境造成了监管漏洞，使烟草广告得以在社交媒体活动中出现，并通过有影响力的人进行宣传，通常针对年轻人。例如，烟草行业参与公共关系和企业社会责任相关活动、赞助活动并使用社交媒体和在线平台推广HTP，所有这些都促进了产品在全球的扩散。2020年初，韩国国务院通过了《国民健康促进法》修正案，禁止烟草制造商向消费者进行任何直接或间接的促销活动。卫生和福利部计划禁止在促销活动中对ENDS和HTP打折以及免费分发这些产品（包括设备）等做法[26]。

无烟空间（第8条）

HTP通常被一些人称为"加热不燃烧"产品，这是该行业创造的术语，具有积极的含义。制造商称这些产品是传统卷烟的"无灰"、"无烟"和"更清洁的替代品"，这可能会对其分类造成混淆。为减少该术语造成的混淆，尤其是在无烟法律适用法规中，世界卫生组织引入了"加热型烟草制品"一词。菲利普·莫里斯公司试图通过与罗马尼亚和乌克兰等国家[27]的数百家"IQOS友好"餐厅和酒吧建立合作伙伴关系，将IQOS与传统吸烟区分开来，这些餐厅和酒吧可能禁止卷烟但允许使用IQOS，从而破坏了室内吸烟禁令。罗马尼亚在无烟政策方面没有将HTP归类为"用于抽吸的烟草制品"，理由是这些产品不会产生烟气[28]。

包装和标签（第11条）

WHO FCTC第11条规定，监管机构应确保烟草制品的包装和标签不以任何虚假、误导、欺骗或可能对其特性、健康影响、危害或释放物产生错误印象的方式宣传烟草制品，包括直接或间接地造成某种烟草制品比其他烟草制品危害更小的错误印象的词语、描述、商标、图形或任何其他标志，这些可能包括诸如"低焦油"、"淡味"、"超淡味"或"温和"等词语。

这些禁令的目的是避免误导消费者，使其认为一种烟草制品比另一种更健康，这对于HTP而言尤其重要。然而，目前对HTP的健康警示要求往往不如传统卷烟的健康警示要求那么严格。即使在强加健康警示的地方，在某些国家（日本和荷兰），它们也仅适用于烟草插入物而不适用于设备。

欧盟烟草制品指令第9~12条涉及健康警示及其尺寸。对于被认为是"无烟烟草"的新型烟草制品，文字（但不是图片）健康警示必须覆盖两个最大表面积的30%；对于用于抽吸的新型烟草制品，组合的健康警示（图形和文字）必须覆盖两个最大表面的65%。因此，归类为无烟烟草制品对烟草行业更有利。关于产品外观的指令第13条禁止在标签或包装上使用任何会对产品特性、健康影响、风险或释放物产生错误印象的元素或特征。标签或包装不得包含有关烟草制品的烟碱、焦油或一氧化碳含量的任何信息；不得暗示特定烟草制品比其他烟草制品危害更小，可减少烟气中任何有害成分的影响，或具有振奋、提神、治疗、恢复活力、天然或有机特性或其他健康或生活方式益处；也不得提及味道、气味、任何调味剂或其他添加剂或不存在这些添加剂的情况。

美国食品药品监督管理局（FDA）要求所有HTP包装标签和广告都包含有关烟碱致瘾性的附加警告，以及卷烟所要求的其他警告。这一要求的目的是纠正使用者的一种误解，即IQOS的致瘾风险低于传统卷烟。可以向FDA提交指定为风险降低的烟草制品（MRTP）的申请，以允许产品以降低风险的声明进行销售[29]，即"任何销售或分销用于减少与商业销售的烟草制品相关的危害或烟草相关疾病风险的烟草制品"。2016年，菲利普·莫里斯公司申请销售IQOS的授权，声称该产品"可以降低烟草相关疾病的风险"、"显著减少身体接触有害或潜在有害化学物质"和"比继续吸烟危害更低"。然而，FDA的结论是，该公司没有提供足够的证据证明消费者不会被这些声明误导，因此不允许菲利普·莫里斯公司销售声称降低风险的HTP[30]。根据美国法律，FDA可能会针对风险降低的烟草制品发布两种类型的命令："风险降低令"或"暴露降低令"。菲利普·莫里斯公司同时申请了两种类型的命令。尽管FDA认定证据不支持发布第一种命令，但证据支持发布IQOS设备和烟草加热棒的暴露降低令。暴露降低令授权菲利普·莫里斯公司在产品广告和营销中就烟草的加热方式、有害和潜在有害化学物质的产生以及对这些化学物质的暴露做出声明。2020年7月7日，FDA授权"IQOS烟草加热系统"作为一种风险降低的烟草制品的销售，其中包括IQOS设备、万宝路加热棒、万宝路顺滑薄荷加热棒和万宝路新鲜薄荷加热棒[31]。FDA强调，该授权并不表明该产品是安全的或已获得FDA批准，并驳回了该公司已充分证明使用该产品比使用其他烟草制品危害更小或降低了健康风险的说法[32]。菲利普·莫里斯公司将这一授权称为公共卫生的里程碑，并称其为"政府和公共卫生组织如何监管无烟替代品，将其与卷烟区分开来以促进公众健康的一个重要例子"。

韩国主要将HTP作为烟草制品进行监管，要求HTP包装上使用图形健康警示。卫生和福利部强制要求使用有关烟草使用后果的图形图像（例如患癌症的器官）以及带有具体风险数据的更简洁的书面警示，这是2018年底实施的一系列阻止吸烟措施的一部分。由政府官员和私人专家组成的13人特别委员会进行了一年的审议，并对1500名吸烟者和非吸烟者进行了民意调查，之后加强了措施[33]。适用于烟草制品的所有法规，例如税收、无烟区、广告、包装警示和标签，也适用于HTP[1]，这符合WHO的建议，即HTP应遵守适用于所有其他烟草制品的相关政策和监管措施[34]。

在加拿大，于2019年11月9日生效的《烟草制品法规（素包装和标准化外观）》适用于烟草制品包括使用全部或部分由烟草制成的产品所必需的装置，例如HTP，因为它们根据该法案被定义为"烟草制品"[35]。以色列和新西兰也要求HTP的素包装。

销售限制

在大多数国家，对烟草制品实施的销售限制也适用于HTP。其中包括禁止某些销售方式（例如自动售货机或互联网）、限制地点、购买者的年龄限制以及对零售商的许可或要求。例如，塞浦路斯禁止从自动烟草售货机销售HTP、向未成年人销售和免费分发HTP[36]。在斯洛文尼亚，销售烟草、烟草制品和相关产品（包括HTP）的场所必须根据2017年《烟草制品使用限制法》[37]第35(1)条进行注册。该法案禁止向未成年人销售HTP，在临时和移动销售点、互联网、电信或任何其他发展中的技术或跨境远程销售或以单个单位销售（第30条）。沙特阿拉伯的法律禁止出售超过20支包装的HTP烟棒[38]，制造商的原始包装除外。

所有禁止向未成年人销售烟草制品的国家都暗中将禁令扩大到HTP；然而，一些国家有不同的年龄限制。例如，日本对20岁以下的人实施销售禁令[39]，而奥地利和比利时仅禁止向16岁以下的儿童销售HTP[40]。联邦系统，例如加拿大、瑞士和美国，可能有不同的地方限制。在瑞士，购买HTP的年龄限制为16岁或18岁，具体取决于各州，而在加拿大，年龄限制为18~19岁。2019年12月，美国颁布了一项销售禁令，禁止向21岁以下的任何人销售所有烟草制品，包括HTP。

成分和释放物（第9条和第10条）

欧盟的大多数国家都要求制造商报告成分的名称、数量和健康影响，包括调味剂。根据欧盟烟草制品指令，禁止具有特征风味的烟草制品。"特征风味"在指令第1(25)条中定义为由添加剂或添加剂组合（包括但不限于水果、香料、药草、酒精、糖果、薄荷醇或香草）在烟草制品消费前或消费过程中产生的除烟草以外的明显气味或味道。

目前禁止的特征风味仅适用于卷烟和自卷烟，不适用于HTP。然而，第7(2)条指出，委员会可以根据欧盟委员会的倡议或成员国的要求，确定特定烟草制品受此禁令的约束。因此，未来欧盟可能会禁止在HTP中加入特征性风味。

教育、交流、培训和公众意识（第12条）

烟草行业凭借其新的产品组合，主要以"降低危害"、"降低风险"和传统卷烟的替代品进行营销和促销，以此作为操纵政府向HTP开放市场的策略。然而，这些说法没有根据，因为这些产品在烟草相关风险方面并未被证明与传统卷烟不同，而且这些说法分散了人们对减少烟草使用和保护公众健康的循证烟草控制政策措施的注意力。WHO FCTC第12条作了如下规定。

缔约方应促进和加强公众对烟草控制问题的认识，酌情使用所有可用的沟通工具，并采取和实施有效的立法、实施、行政或其他措施，以促进以下方面：

(1) 广泛获得关于烟草消费烟草烟气暴露的健康风险，包括成瘾性的有效和全面的教育和公众意识计划；

(2) 公众对烟草消费和烟草烟气暴露的健康风险以及第14.2条所述的戒烟和无烟生活方式的益处的认识；

(3) 根据国家法律，公众可以获取与本公约目标相关的广泛的烟草业信息；

(4) 针对卫生工作者、社区工作者、社会工作者、媒体专业人员、教育工作者、决策者、行政管理人员和其他相关人员等的有关烟草控制的有效和适当的培训或宣传和认识计划；

(5) 不隶属于烟草行业的公共和私立机构以及非政府组织在制定和实施跨部门烟草控制战略方面的意识和参与；

(6) 烟草生产和消费对健康、经济和环境的不利后果的公众意识和信息获取。

这些都是以证据为基础的措施，用于提高公众对使用烟草制品的不良影响的认识。所有国家，不仅仅是WHO FCTC的缔约方，都应考虑优先采取这些措施以保护公众健康。

6.3 监管、实施和执行政策的考虑事项和障碍

6.3.1 实施政策时的监管考虑

由于对产品的了解不足、烟草行业游说、基于这些产品"无烟"、"无灰"或"比传统卷烟更清洁的替代品"的论点而对无烟产品进行的监管分类，以及对装置和插入物的不同监管方式，HTP可能被视为不同于传统卷烟。烟草行业积极推销这些产品，游说政府制定更宽松的法规，并对有关HTP的监管决定施加巨大压力。这导致全面烟草控制监管措施仅部分地适用于HTP，最终将破坏现有的烟草控制体系。

一项关于以色列IQOS活动的研究[41]描述了烟草行业如何试图将新产品定义为现有烟草法未涵盖的类别的一部分，在这种情况下在论证中使用了"吸烟"一词。当IQOS于2016年12月在以色列推出时，菲利普·莫里斯国际公司组织了高层会议，与以色列卫生部和财政部进行了直接沟通，向政府施加压力，要求IQOS不受现有烟草法规的限制，在三次向最高法院请愿后，这些规定被撤销。该研究的作者警告说，由于对HTP没有特定的健康警示要求，烟草行业可能会自主地在新的产品上做出警示，例如"研究表明卷烟会导致成瘾"，这可能会导致人们对有关卷烟的公认证据产生怀疑。

烟草行业以确保在适用的国家法律下获得最优惠待遇的方式对HTP进行分类。在新西兰，HTP作为"无烟产品"被禁止，但该禁令在法庭上被菲利普·莫里斯国际公司成功挑战，理由是HTP并非"无烟"。目前，新西兰将所有针对烟气产品的烟草管制法律适用于HTP，包括素包装。在罗马尼亚，企业提交的文件中基于危害降低、不燃烧和无烟的论点，对HTP作为烟气产品的监管提出了质疑。

在确定最合适的HTP监管方法时，各国应考虑以下因素：

■ 对使用者和非使用者的绝对和相对健康风险；

■ 随着有关这些产品的科学知识的获得，HTP是否可以持续受到监管；

- 烟草使用和吸烟"重新正常化"的风险；
- 非烟草制品使用者，特别是年轻人开始使用的风险；
- 已戒烟从而改善健康状况的吸烟者可能转而使用HTP，尽管这些都是烟草制品，HTP尚未被证明可以降低与烟草相关的风险；
- 与其他烟碱和烟草制品一起使用，使使用者暴露于两种或多种产品的释放物；
- 评估关于HTP相对于传统卷烟危害的企业声明的能力，并防止可能误导消费者的声明。

如前所述，根据WHO FCTC第2.1条，缔约方可以超越其规定，该条规定：为了更好地保护人体健康，鼓励缔约方执行本公约及其议定书要求以外的措施，这些文书中的任何规定均不应阻止缔约方执行符合其规定并符合国际法的更严格的要求。

6.3.2 实施和执行政策的障碍

除了游说外，企业诉讼还威胁着政策的通过、实施和执行。在新西兰，地方法院2018年的一项裁决（菲利普·莫里斯公司诉卫生部）[42]推翻了先前对HTP的分类，即"任何标明或以其他方式描述为适合咀嚼或任何其他口用（吸烟除外）的烟草制品"。根据1990年《无烟环境法》第29(2)条，这些都是被禁止的，菲利普·莫里斯公司因销售IQOS的HTP插件"Heets"而被指控违反法律[43]。他们发现，由于该法律最初旨在控制咀嚼烟草和其他口用烟草制品的销售，因此不应适用于HTP的烟草插入物。因此，地方法院裁定菲利普·莫里斯公司胜诉，根据该法案，可以在新西兰合法进口、销售、包装和分销HTP。因此，《无烟环境法》规定适用于HTP，包括禁止向未成年人销售和限制广告。该案例突出了监管这些产品的挑战以及可以适应不断变化的烟草制品格局的立法的重要性。

大部分诉讼是基于燃烧或非燃烧的声明，以决定对企业最有利的待遇。如前所述，将HTP作为烟气产品的监管可能会因不燃烧而受到企业挑战，而作为无烟产品的监管可能会因这些产品不是"无烟"而受到挑战。

6.3.3 其他考虑和意外后果

当各国将HTP作为有烟产品进行监管时，适用于其他有烟产品的健康警示要求也适用于HTP。当HTP作为无烟烟草制品进行监管时，同样的原则也适用。许多国家的印象可能是这些产品需要具体规定，而它们已被其现行烟草管制法所涵盖。泛美卫生组织已向美洲地区国家提出建议，根据现有烟草制品法规对HTP进行监管。由于某些法规中对烟草制品的定义方式可能使烟草管制法律难以适用，因此应扩大法规范围以涵盖新型烟草制品。这将限制烟草行业利用监管漏洞。世界卫生组织《2019年全球烟草流行报告》[44]为各国提供了有用的信息和建议。

- HTP含有烟草，应以与烟草制品相同的方式进行监管。
- HTP产生有害释放物，其中许多与卷烟烟气中的释放物相似。
- HTP使用者会接触到产品的有害释放物，而非使用者也可能接触到有害的二手释放物。
- 虽然HTP中几种有害物质的含量通常低于传统卷烟中的含量，但其他有害物质的含量更高。较低水平的有害成分并不意味着较低的健康风险。
- HTP含有烟碱。烟碱有高致瘾性，而且会造成伤害，尤其是对儿童、孕妇和青少年。
- HTP使用和暴露于其释放物对健康的长期影响仍然未知。目前关于相对和绝对风险的独立证据不足。应进行独立研究，以确定它们对使用者和非使用者构成的健康风险。

这些信息包括HTP的重要考虑因素，因为它们在市场上的可用性可能会对公共卫生产生意想不到的后果，在制定政策和确定HTP的监管路径时应予以考虑。

6.4 讨 论

HTP是烟草制品，WHO FCTC将其定义为"完全或部分以烟叶为原料制成的产品，供抽吸、吸吮、咀嚼或鼻吸"。HTP在一些国家获得了相当大的市场份额，目前在全球50多个国家（地区）市场上有售。它们的独特特征、密集的企业游说、对其健康风险缺乏明确性以及缺乏国际方法都对监管机构构成了挑战。

虽然可用数据有限，因为法规依赖于国家对法律的解释，不能独立评估，但不同国家显然以不同的方式监管HTP，从禁止到不监管。一些国家认为HTP与传统烟草制品属于同一类别。许多国家已经制定了关于基本烟草控制措施的国内立法和法规，包括广告、促销和赞助、无烟空间以及包装和标签。一种误解是，监管HTP将是一项新的、资源密集型的举措，而事实上，这些产品已经被现行的烟草管制法所涵盖。

然而，HTP的营销是战略性的，其监管面临挑战。由于设备和插入物单独出售，可能使设备（不含烟草）不受例如广告、促销和赞助甚至向未成年人销售的限制。烟草行业声称HTP不燃烧，许多欧盟国家将其归类为新型无烟烟草制品，因此对警示和无烟区等限制的要求可能与传统卷烟不同。本章提出了解决这些产品的监管问题的一些考虑因素。

6.5 结 论

在过去几年中，烟草行业通过新一代烟草制品（如HTP）显著扩大了其"降低风险"的产品组合。与这些产品相关的创新技术、设计、营销和健康声明削弱了一些国家的烟草控制措施,这些国家有相对强有力的法律来监管传统卷烟,并且烟草行业试图将自己重新定位为公共卫生合作伙伴。监管机构对这些新产品基本上没有准备,尤其是其"不燃烧"、"无烟"和"无灰"的声明,企业已用这些说法来游说政府以获得有利的监管待遇,特别是规避无烟法律。因此,当前关于HTP的法规因国家而异。HTP会产生含有害物质的气溶胶,许多有害物质的含量低于传统卷烟烟气中的含量,但在某些情况下更高。然而,较低水平的有害成分并不一定意味着较低的风险。由于使用和暴露于这些产品的释放物对健康的长期影响仍然未知,目前没有足够的独立证据证明相对和绝对风险,它们应完全遵守WHO FCTC的规定,包括在室内空间使用的禁令。本章的目的是提高人们对HTP监管方法不一致的认识,并在烟草行业将HTP和其他新型烟草制品引入其市场时,为监管机构做好准备。

正在审查其立法选择的国家可以从其他国家的监管成功和挑战中吸取经验教训。新的烟草管制法不仅应预见HTP,还应预见其他新兴产品,其定义应足够广泛以涵盖所有创新发展。烟草行业的干预,包括游说和误导信息,应受到监控,并受到WHO FCTC第5.3条的保护。

6.6 研究差距

- 对HTP产品和使用进行全球监测,以了解行业营销策略。
- 对限制将WHO FCTC应用于HTP并破坏烟草控制的行业机制进行系统监测。
- 对HTP立法进行全面规划,以确定行业可利用的监管漏洞和现有政策的实施水平,以改进政策并为促进最大限度保护公众健康的监管方法提供证据。
- 有效监管以更好地了解HTP的可用性、营销和使用。

6.7 政策建议

随着HTP的发展和普及,面对行业压力和科学不确定性,监管机构必须解决有关这些产品的问题。各国政府对这些产品进行分类和监管的不同方法反映了缺乏国际公认的方法。有一点很清楚:HTP是烟草制品。因此,敦促决策者考虑以下建议:

- 将HTP归类为烟草制品,除非这种分类会导致更宽松的法规、破坏现有烟草管制法规或在类似产品被禁止的情况下允许其进入市场。
- 将WHO FCTC的所有监管措施应用于HTP,尤其是第5.3、6、8、9和10、

11、12、13、14和20条中的措施。其中包括保护烟草控制活动不受所有商业利益和其他既得利益的影响，对这些产品征收消费税，要求报告和披露产品信息，要求对HTP进行组合健康警示，将HTP纳入无烟法律，以及禁止和限制烟草广告、促销和赞助。

- 根据第13.4(a)条，禁止"以任何虚假、误导或欺骗，或可能对其特性、健康影响、危害或释放物产生错误印象的手段，推销烟草制品的所有形式的广告、促销和赞助"。
- 使用现有的烟草制品法规来监管HTP，并扩大这些法规的范围以确保企业不能利用监管漏洞，即使在这些产品目前尚未（合法）上市的国家也是如此。
- 将HTP纳入监测范围，以通过现有渠道了解其使用和可用性，为这些产品的监管提供信息，并确保最大限度地保护公众健康。
- 要求制造商承担举证责任来支持有关产品的声明，并禁止关于HTP相对于其他烟草制品的相对风险或危害的未经证实的声明。
- 监管有关HTP的错误信息以及这些产品相对于其他产品的风险或危害的声明，并采取适当的监管行动来遏制此类做法。
- 要求对新型烟草制品进行上市前通报，以使政府能够评估是否批准其销售。
- 对这些产品进行定义和分类，以确保公卫共生目标得到保护并避免监管漏洞。鉴于市场上和正在开发的产品种类繁多，法律定义必须涵盖所有产品设计并适应产品创新。
- 密切监测产品及其在国内的市场，并采取有效措施强制遵守相关政策和法规。
- 明确产品和产品类别的监管区别，明确产品及其组成部分，确保监管有效。

6.8 参考文献

[1] Heated tobacco products: information sheet, 2nd edition. Geneva: World Health Organization; 2020 (https://www.who.int/publications/i/item/WHO-HEP-HPR-2020.2, accessed 10 January 2021).

[2] Our tobacco heating system. IQOS. Lausanne: Philip Morris International; 2020 (https://www.pmi.com/smoke-free-products/iqos-our-tobacco-heating-system, accessed 10 January 2021).

[3] Heated tobacco products (HTPs). Market monitoring information sheet (WHO/NMH/PND/18.7). Geneva: World Health Organization; 2018 (https://apps.who.int/iris/bitstream/handle/10665/273459/WHO-NMH-PND-18.7-eng.pdf, accessed 10 January 2021).

[4] Novel and emerging tobacco products (FCTC/COP/8/22). Geneva: World Health Organization; 2018(https://www.who.int/fctc/cop/sessions/cop8/FCTC__COP8(22).pdf?ua=1, accessed 28 December 2019).

[5] Tobacco control laws. Washington (DC): Campaign for Tobacco-Free Kids; 2020 (https://www.tobaccocontrollaws.org/, accessed 10 January 2021).

[6] Warner J. Vaping or cannabis: Where's the growth for Big Tobacco? IG, 17 February 2020(https://www.ig.com/uk/news-and-trade-ideas/vaping-or-cannabis--wheres-the-growth-forbig-tobacco--200217, accessed 10 January 2021).

[7] Bialous SA, Glantz SA. Heated tobacco products: another tobacco industry global strategy to slow progress in tobacco control. Tob Control. 2018;27:s111–7.

[8] Preamble. Novel and emerging tobacco products (FCTC/COP/8/22). Geneva: World Health Organization; 2018 (https://www.who.int/fctc/cop/sessions/cop8/FCTC__COP8(22).pdf?ua=1, accessed 28 December 2019).

[9] HTP information sheet. Geneva: World Health Organization; 2019 (https://www.who.int/tobacco/publications/prod_regulation/heated-tobacco-products/en/, accessed 10 January 2021).

[10] Notice of final decisions to amend (or not amend) the current Poisons Standard 24 August 2020. Canberra: Department of Health, Therapeutic Goods Administration; 2020 (https://www.tga.gov.au/sites/default/files/public-notice-final-decisions-acms29-accs27-joint-acmsaccs24-march-2020.pdf, accessed 10 January 2021).

[11] Directive 2014/40/EU of the European Parliament and of the Council of 3 April 2014 on the approximation of the laws, regulations and administrative provisions of the Member States concerning the manufacture, presentation and sale of tobacco and related products. Strasbourg: European Parliament; 2014 (https://eur-lex.europa.eu/legal-content/EN/TXT/?uri=OJ%3AJOL_2014_127_R_0001, accessed 10 January 2021).

[12] Parliamentary questions (P-009191/2016(ASW). Strasbourg: European Parliament; 2017(https://www.europarl.europa.eu/doceo/document/P-8-2016-009191-ASW_ES.html).

[13] Tobacco (Control of Advertisements and Sale) Act (as amended), Arts. 16(1)-(2). Singapore; 2011 (https://www.tobaccocontrollaws.org/.../Singapore%20- %20Control%20of%20Ads%20 %26%20Sale%20-%20national.pdf, accessed 10 January 2021).

[14] Act of April 26, 2016 Amending the Tobacco Act to Implement Directive 2014/40/EU on the Manufacture, Presentation and Sale of Tobacco and Related Products. The Hague: Staatsblad van het Koninkrijk der Nederlanden; 2016 (https://www.tobaccocontrollaws.org/files/live/Netherlands/Netherlands%20-%20Act of April 26%2C 2016%20-%20national.pdf, accessed 10 January 2021).

[15] Information for suppliers. Bilthoven: National Institute for Public Health and the Environment; 2017 (https://www.rivm.nl/en/tobacco/information-for-suppliers, accessed 10 January 2021).

[16] Gambaccini P. Blog. How should heated tobacco be taxed? Lausanne: Vapor Products Tax, 12 July 2017 (https://vaporproductstax.com/how-should-heated-tobacco-be-taxed/, accessed 10 January 2021).

[17] Decreto legislativo, 15 dicembre 2014, n. 188. Disposizioni in materia di tassazione dei tabacchi lavorati, dei loro succedanei, nonche' di fiammiferi, a norma dell'articolo 13 della legge 11 marzo 2014, n. 23. Rome: Government of Italy; 2014 (http://www.governo.it/sites/governo.it/files/77443-9913.pdf, accessed 10 January 2021).

[18] Legislative decree. Implementation of Directive 2014/40/EU on streamlining the legislative, regulatory and administrative provisions of the member states regarding the processing,

[19] Conference of the Parties to the WHO Framework Convention on Tobacco Control (eighth session), Decision FCTC/COP8(22). Novel and emerging tobacco products. Geneva: World Health Organization; 2018 (https://www.who.int/fctc/cop/sessions/cop8/FCTC__COP8(22).pdf?ua=1, accessed 10 January 2021).

[20] International Convention on the Harmonized Commodity Description and Coding System (Brussels, 14 June 1983). Amendments to the nomenclature as an annex to the Convention accepted pursuant to the recommendation of 28 June 2019 of the Cistoms Co-operation Council (NG0262B1). Brussels: World Customs Organization; 2019:19 (http://www.wcoomd.org/-/media/wco/public/global/pdf/topics/nomenclature/instruments-and-tools/hs-nomenclature-2022/ng0262b1.pdf?db=web, accessed 10 January 2021).

[21] Information note on classification of novel and emerging tobacco products. Geneva: World Health Organization; 2019 (https://untobaccocontrol.org/impldb/wp-content/uploads/Info-Note_Novel-Classification_EN.pdf, accessed 10 January 2021).

[22] Federal Government Gazette. 28 December 2015. Control of Tobacco Product (Amendment) (No. 2) Regulations 2015. Kuala Lumpur: Federal Government; 2015 (https://www.tobaccocontrollaws.org/legislation/country/malaysia/laws, accessed 10 January 2021).

[23] Notification detail. Draft of the UAE GCC Technical Regulation "Electronic nicotine products (equivalents of traditional tobacco products)" (G/TBTN/ARE/482). Brussels: European Commission; 2020 (https://ec.europa.eu/growth/tools-databases/tbt/en/search/?tbtaction=search.detail&num=482&Country_ID=ARE&dspLang=EN&BASDATEDEB=&basdatedeb=&basdatefin=&baspays=ARE&basnotifnum=482&basnotifnum2=482&-bastypepays=ARE&baskeywords=, accessed 10 January 2021).

[24] Executive order No. 106. Prohibiting the manufacture, distribution, marketing and sale of unregistered and/or adulterated electronic nicotine/non-nicotine delivery systems, heated tobacco products and other novel tobacco products, amending Executive Order No. 26 (S. 2017) and for other purposes. Manila: Malacañan Palace; 2020 (https://perma.cc/B7UY-BMR8, accessed 10 January 2021).

[25] Guidelines for implementation of Article 13 of the WHO Framework Convention on Tobacco Control (Tobacco advertising, promotion and sponsorship). Geneva: World Health Organization; 2008 (https://www.who.int/fctc/guidelines/article_13.pdf?ua=1, accessed 10 January 2021).

[26] Indirect promotional activities such as providing discount vouchers for electronic cigarette devices are prohibited. Sejong City: Health promotion Division; 2020 (http://www.mohw.go.kr/react/al/sal0301vw.jsp?PAR_MENU_ID=04&MENU_ID=0403&CONT_SEQ=352444, accessed 10 January 2021).

[27] Jackler RK, Ramamurthi D, Axelrod AK, Jung JK, Louis-Ferdinand NG, Reidel JE et al. Global marketing of IQOS. The Philip Morris campaign to popularize "heat not burn" tobacco. Stanford (CA): Stanford Research into the Impact of Tobacco Advertising; 2020 (http://tobacco.stanford.

edu/tobacco_main/publications/IQOS_Paper_2-21-2020F.pdf, accessed 10 January 2021).

[28] Law No. 349 of June 6, 2002. On preventing the consumption of tobacco products and combating its effects. Bucharest: Government of Romania; 2002 (https://www.tobaccocontrollaws. org/files/live/Romania/Romania%20-%20Law%20No.%20349.pdf, accessed 10 January 2021).

[29] Section 911 of the Federal Food, Drug, and Cosmetic Act – Modified Risk Tobacco Products. Silver Spring (MD): Food and Drug Administration; 2020 (https://www.fda.gov/tobacco-products/ rules-regulations-and-guidance/section-911-federal-food-drug-and-cosmetic-act-modified-risk-tobacco-products accessed 10 January 2021).

[30] McKelvey K, Popova L, Kim M, Kass Lempert L, Chaffee BW, Vijayaraghavan M et al. IQOS labelling will mislead consumers. Tob Control. 2018;27:s48–54.

[31] Modified risk tobacco orders. FDA authorizes marketing of IQOS tobacco heating system with "reduced exposure" information. Press release, 7 July 2020. Silver Spring (MD): US Food and Drug Administration; 2020 (https://www.fda.gov/news-events/press-announcements/ fda-authorizes-marketing-iqos-tobacco-heating-system-reduced-exposure-information, accessed 10 January 2021).

[32] WHO statement on heated tobacco products and the US FDA decision regarding IQOS. Geneva: World Health Organization; 2020 (https://www.who.int/news-room/detail/27-07-2020-who-statement-on-heated-tobacco-products-and-the-us-fda-decision-regarding-iqos, accessed 10 January 2021).

[33] Lee KM. Gov't to mandate graphic warnings on heated tobacco products. The Korea Times, 2020 (https://www.koreatimes.co.kr/www/nation/2018/05/119_248951.html, accessed 10 January 2021).

[34] Enforcement Decree of the National Health Promotion Act. Seoul: Ministry of Government Legislation; 2016 (https://www.tobaccocontrollaws.org/files/live/South%20Korea/South%20Korea%20-%20Enf.%20Decree%20of%20Nat'l%20Health%20Promotion%20Act.pdf, accessed 10 January 2021).

[35] Tobacco product regulations: plain and standardized appearance. Ottawa: Government of Canada; 2020 (https://www.canada.ca/en/health-canada/services/health-concerns/tobacco/legislation/federal-regulations/products-regulations-plain-standardized-appearance.html, accessed 10 January 2021).

[36] Law that provides for measures for the reduction of smoking. Nicosia: Government of Cyprus; 2002 (https://www.tobaccocontrollaws.org/files/live/Cyprus/Cyprus - Reduction of Smoking.pdf, accessed 10 January 2021).

[37] The Restriction of the Use of Tobacco Products Act (Zoutpi). Ljubljana: National Assembly; 2017 (https://www.tobaccocontrollaws.org/.../Slovenia%20- %20TC%20Act%202017.pdf, accessed 10 January 2021).

[38] Controls and requirements for electronic smoking devices. Version No. 1 (18/9/1440 AH). Riyadh: Saudi Food and Drug Authority; 2018 (https://www.tobaccocontrollaws.org/files/live/Saudi%20Arabia/Saudi%20Arabia%20-%20SFDA%20E-Cig%20Requirements.pdf, accessed 10 January 2021).

[39] Act Prohibiting Smoking by Minors Act No. 33 on March 7, 1900. Last revision: Act No. 152 on

December 12, 2001. Tokyo: Government of Japan; 2001 (https://www.tobaccocontrollaws.org/files/live/Japan/Japan%20-%20Act%20Prohibiting%20Smoking%20by%20Minors.pdf, accessed 10 January 2021).

[40] Purchasing and consuming tobacco. Vienna: European Union Agency for Fundamental Rights; 2018 (https://fra.europa.eu/en/publication/2017/mapping-minimum-age-requirements/purchase-consumption-tobacco, accessed 10 January 2021).

[41] Rosen L, Kislev S. The IQOS campaign in Israel. Tob Control 2018;27(Suppl1):s78–81.

[42] Tobacco Control Laws. Litigation by country. New Zealand MOH v. PMI. Washington (DC: Campaign for Tobacco-Free Kids; 2020 (https://www.tobaccocontrollaws.org/litigation/decisions/ nz-20180312-new-zealand-moh-v.-pmi, accessed 10 January 2021).

[43] Smokefree Environments and Regulated Products Act 1990. Auckland: Ministry of Health; 1990 (https://www.tobaccocontrollaws.org/files/live/New%20Zealand/New%20Zealand%20-%20SF%20Act%201990%20-%20national.pdf, accessed 10 January 2021).

[44] WHO report on the global tobacco epidemic, 2019. Geneva: World Health Organization; 2019.

7. 电子烟碱传输系统和传统卷烟的烟碱暴露评估

Anne Havermans, Centre for Health Protection, National Institute for Public Health and the Environment, Antonie van Leeuwenhoeklaan 9,3721 MA Bilthoven, Netherlands

Thomas Eissenberg, Center for the Study of Tobacco Products, Department of Phychology, Virginia Commonwealth University, Richmond (VA), USA

摘 要

电子烟碱传输系统是一类传输雾化烟碱的产品。电子烟碱传输系统的技术快速发展，有各种各样的类型，从第一代"类卷烟"设备到当前流行的以"烟弹"为主的设备。设备设计、烟液成分和使用者行为等因素都会影响电子烟碱传输系统气溶胶中的烟碱和非烟碱有害物质。尽管有一些证据表明电子烟碱传输系统也许有助于一些吸烟者取代传统卷烟，但是电子烟碱传输系统与燃烧型卷烟的双重使用，以及以前从未暴露过烟碱的个体中越来越普遍地开始使用电子烟碱传输系统，进一步增加了公众健康问题。我们回顾了有关电子烟碱传输系统的烟碱释放和递送的文献，并探讨了影响电子烟碱传输系统使用者暴露于烟碱和非烟碱有害物质的因素。该审查显示，电子烟碱传输系统由很多种类的产品组成，发展速度超过了监管；经过调味的电子烟碱传输系统烟液有助于以前对烟碱一无所知的人开始和持续使用；在某些情况下，有效释放烟碱的电子烟碱传输系统可能会帮助一些吸烟者戒掉燃烧型卷烟；大多数电子烟碱传输系统使用者并没有戒掉燃烧型卷烟；对电子烟碱传输系统烟碱释放物的监管很困难，因为管控释放物的投入很多；监管重点放在烟碱释放速率（例如烟碱"通量"）可能是有用的，这将涉及到只允许销售"封闭系统"的电子烟碱传输系统。在此背景下，提出了未来的研究需求和政策建议。

7.1 背 景

电子烟碱传输系统（ENDS）是旨在为使用者提供雾化烟碱的多样化产品。它们含有称为"线圈"或"雾化器"的电池供电加热元件，对包含烟碱、载体溶液（例如丙二醇、植物甘油）、调味化学品的烟液进行加热，从而产生含有一定浓度的烟碱和其他有害成分的气溶胶，被使用者吸入。电子烟碱传输系统是一类发展迅速的产品，包括各种各样的类型，从第一代"类卷烟"设备到当前流行的装有烟液的一次性"烟弹"装置[1]。产品设计特征和特性（如瓦数和线圈尺寸）、烟液组分（如载体和烟碱浓度）和使用行为（如抽吸容量和持续时间）可以以多种方式组合在一起，以此来影响使用者对气溶胶中烟碱和其他有害成分的吸入量（释放量）[2]。

在过去十年中，一些地方的电子烟碱传输系统的使用量大幅上涨[3,4]。在一些国家，特别是在欧洲、加拿大和美国，儿童和青少年的烟草使用量增加尤其迅速，以至于电子烟碱传输系统已经发展为美国青少年最常用的烟草制品[5,6]。这引起了人们的关注，因为电子烟碱传输系统释放物中含有可能对健康有害的化学物质[7]和产生依赖性的药物烟碱。烟碱是所有烟草制品（例如燃烧型卷烟、无烟烟草、加热卷烟）以及电子烟碱传输系统的主要致瘾成分。烟碱除了会导致依赖外，还会对健康产生负面影响[8]。儿童、青少年和年轻人更易受到烟碱的长期神经认知效应的影响，因为大脑的成熟持续到二十岁出头[9]。据推测，青少年通过增强多巴胺能纹状体细胞的兴奋和抑制，从而获得烟碱奖赏的增强和戒断反应的减少，这使他们比成年人更容易长期依赖烟碱[10]。此外，一些研究发现电子烟碱传输系统可能作为一个开始吸烟的途径，并会增加青少年和年轻人开始吸烟的风险[11]。

虽然有些电子烟碱传输系统可以帮助一些吸烟者通过提供相似数量和形式的烟碱（即质子化态）来取代卷烟[12]，但年轻的非吸烟者开始使用电子烟碱传输系统引起了很大的关注[11]。监管机构应该确定和监管ENDS向使用者提供烟碱的影响因素，以尽量减少其滥用倾向和健康影响，同时最大限度地减少卷烟吸烟者的风险。然而，由于烟碱输送是产品设计、烟液成分和使用行为的综合结果[2]，监管可能难以将所有这些因素综合考虑在内。由于通过电子烟碱传输系统输送烟碱受多方因素的影响（例如设备的特性和烟液成分），所以有人建议将监管重点放在电子烟释放烟碱和其他有害成分的速率上，以及能够控制释放速率的相关装置、烟液和使用者的因素。烟碱通量（即电子烟释放烟碱的速率），已被建议作为监管的目标[13]。如下文所述，对电子烟烟碱通量（以及其他有害成分潜在释放速率）的监管比其他可替代因素（例如烟液烟碱浓度）更有利于直接控制影响公众健康的因素，单独监管后者可能达不到公共卫生要求。

本背景文件对电子烟碱传输系统中烟碱释放和递送的文献（截至2020年3月）进行了叙述性综述，并探讨了影响使用者在电子烟碱传输系统释放烟碱和非烟碱有害物质的因素。我们在PubMed中使用检索词 "ENDS" or "E-cigarette" or "electronic-cigarette" or "Nicotine" or "exposure" or "emission" or "yield" or "delivery" 检索了过去五年内的相关出版物。为了找到有关使用模式的相关文献，使用检索词 "ENDS" or "E-cigarette" or "electronic cigarette" and "Topography" or "behavior" 进行了更多的检索。还对特定用户群体的信息进行了更多的搜索，检索词 "ENDS" or "E-cigarette" or "electronic cigarette" 与特定假设使用者群体相关的词语结合在一起检索，比如 "race" "ethnicity" "gender" "male" "female" "dual use"。还通过数据库检索获得出版物中引用的相关文献（即雪球方法）。由于本章的目的是提供叙述性综述，因此没有正式的选择标准适用于这些检索结果。

7.2 电子烟碱传输系统的烟碱暴露

7.2.1 电子烟碱传输系统的烟碱释放

"烟碱释放"可以被定义为离开电子烟碱传输系统设备的气溶胶中所含烟碱的量，又称烟碱释放量。可以用预先设定好参数的吸烟机来分析气溶胶中烟碱的释放量。气溶胶会被滤片捕捉，并用合适的溶剂进行萃取，最后用色谱法对萃取液进行分析[2]。使用这些方法和各种抽吸方案的研究表明，ENDS气溶胶中的烟碱含量不同，一些结果显示释放量低于通常从燃烧型卷烟中获得的烟碱量，而另一些结果则表示等于或超过燃烧型卷烟的释放量（即1.76~2.20 mg/支）[2,14,15]。重要的是，如果机器抽吸方案不能模仿人的抽吸行为，就不是人体暴露的有效措施。然而，如果任意选择的机器抽吸方式同样应用于所有研究产品时，那么就是有效的对比。

7.2.2 电子烟碱传输系统电功率对烟碱释放的影响

每一口抽吸的气溶胶中烟碱量受很多因素影响，包括该装置的功率、雾化烟油中的烟碱浓度以及使用者的抽吸行为[2]。电功率（W）是电池电压（V）和线圈电阻（Ω）的函数，$W=V^2/\Omega$。电子烟碱传输系统的功率范围从早期型号的≤10 W到目前市售型号的≥250 W[16]。通常通过将低电阻线圈（即<1 Ω）集成到设备中来实现更高的功率，俗称"低欧姆电子烟"[17]。不同电子烟碱传输系统型号之间的电池电压和默认电源设置差异很大，更高级的电子烟通常允许使用者调整电源设置。"封闭系统"的电子烟碱传输系统无法以这样的方式调整，因为它们更小，功率更低，更像燃烧型卷烟，而"开放系统"的电子烟碱传输系统更大，

因此可以包含更大的电池和更低的电阻加热元件[1,18]。一个真正"封闭系统"的电子烟碱传输系统不允许使用者更改设备或烟油中影响烟碱释放量的任何元素，例如电池电压、线圈电阻和烟液烟碱浓度，同时还会限制使用者抽吸行为（例如抽吸持续时间[19]）。

增加用于蒸发烟液的加热元件功率会增加产生的气溶胶的量，这可能导致元件过热或烟液热降解，从而产生有害物质。功率对气溶胶中烟碱释放量的影响尚未得到广泛研究，但有一项研究发现，将功率输出从3 W增加到7.5 W可使烟碱释放量增加4~5倍[2]。增加功率也会增加非烟碱有害物质的释放[20]。

7.2.3 电子烟碱传输系统烟油中烟碱和其他化合物浓度对烟碱释放的贡献

填充瓶或预填充烟弹电子烟碱传输系统的烟液中烟碱浓度范围较广，通常以mg/mL计或占总体积的百分比在标签上标示。烟碱的最大浓度可能因国家或地区法规不同而有所区别。例如，《欧盟烟草制品指令》中规定，烟液中烟碱浓度不应超过20 mg/mL[21]。根据该指令描述，电子烟在吸烟所需时间内递送的允许烟碱量与标准卷烟允许的烟碱浓度相同。然而，由于设备本身的因素（例如电功率）、烟液组成成分、使用者行为的相互作用，烟液中烟碱浓度与向电子烟碱传输系统使用者递送烟碱之间的关系并不容易了解。

截至2017年，在美国，常用烟液中的烟碱浓度通常为0.36 mg/mL[1,22-25]。然而，一些新产品烟碱含量高达67 mg/mL[26,27]，人们担心电子烟碱传输系统烟液配方的创新正在引起一场"烟碱军备竞赛"[28]。此外，电子烟碱传输系统烟液中的烟碱浓度通常与标签中的内容不匹配，其偏差高达52%[15]，并且有一些研究表明，某些标注不含烟碱的烟液中含有可测量的烟碱[24,25,29]。

一些研究表明，电子烟碱传输系统烟液中的烟碱浓度直接影响烟碱的释放量，即较高的烟液烟碱浓度会导致气溶胶中烟碱的释放增加[2,14]。电源设置也起着一定的作用，因为增加设备功率会增加烟碱的释放量[14,30]。此外，低烟碱强度的电子烟碱传输系统烟液的使用者可以通过调整其抽吸行为获得与高烟碱电子烟使用者每次抽吸相同的烟碱量[2,31]。这样的话，他们也可能暴露于更多数量的有害成分中（见7.3节）。电子烟碱传输系统烟液中的其他化学物质也会影响其气溶胶中烟碱的释放量。例如，烟液通常含有以各种比例混合的丙二醇/植物甘油，当溶剂丙二醇水平高于植物甘油时，将导致低功率设备设置的烟碱释放量升高[30]。这可能是由于丙二醇在相对较低的温度下波动性较大，导致蒸发量升高。由于植物甘油在高温下变得易挥发，推测的差异在较高功率设置下变得不那么明显[30]。

7.3 伴生物质的暴露概述

除了烟碱，电子烟碱传输系统释放物中还包含有其他有害物质。这些有害物质要么存在于烟液中，要么是烟液成分的热分解产物。烟液中的有害物质包括丙二醇、植物甘油和各种调味化学品[32,33]。此外，由于电子烟碱传输系统中的烟碱来自于烟草植物，因此烟液可能含有与烟草相关的有害成分，例如烟草特有亚硝胺[1]。电子烟碱传输系统烟液中使用的调味品尽管在添加到食物中时"通常被认为是安全的"，但是加热和吸入时的风险状况是未知的[34]。一些调味化学品，如双乙酰（黄油味）[35,36]、苯甲醛（水果味）[37,38]和肉桂醛（肉桂味）[36,38-40]在吸入时是有害的[41,42]。此外，烟草和健康人口评估研究的结果表明，水果味电子烟碱传输系统使用者体内的致癌物质丙烯腈生物标志物的浓度明显高于其他口味的使用者[43]。加热后产生羰基化合物、挥发性有机物和多环芳烃等有害成分也存在于烟草烟气中。这些有害物质受电子烟碱传输系统使用者行为以及设备类型和功率设置等因素的影响[2,44]。例如，更频繁的抽吸模式可以增加甲醛、乙醛和丙酮[14,44]等羰基化合物的产生，并且这些羰基化合物与吸烟者的肺病密切相关[45]。

电子烟碱传输系统释放物可能含有加强烟碱致瘾性的物质。例如，薄荷醇是电子烟碱传输系统和燃烧型卷烟的常见成分，即使在没有标记为含有薄荷或薄荷味的情况下，它也存在于许多电子烟碱传输系统烟液中[46]。薄荷醇能以各种方式增强烟碱的特性，例如通过促进吸入和作用于大脑中相关的烟碱乙酰胆碱受体亚型[46]。另外，电子烟碱传输系统的调味剂如香兰素和乙基香兰素，也被发现可以作为单胺氧化酶抑制剂，增强大脑对烟碱的反应[47]。青苹果调味的化学品金合欢烯可以通过刺激烟碱乙酰胆碱受体和烟碱效力来激活受体，引起奖赏相关行为[48]。酒精和烟草次要生物碱也可能增强烟碱效应并影响其代谢。尽管电子烟碱传输系统中释放的烟碱和酒精的相互作用尚未研究，但一项研究表明，电子烟碱传输系统烟液中的高酒精含量会严重影响精神运动功能[49]。此外，酒精和烟草常常在一起使用[50]，饮酒可以增加吸烟[46]。二烯烟碱是烟碱的热反应物，其释放量是卷烟释放量的2~63倍[51]。它通过抑制体内烟碱代谢，从而可能增加来自电子烟碱传输系统的烟碱递送[1,46,51]。

7.4 电子烟碱传输系统的烟碱递送

电子烟碱传输系统向使用者血液和大脑递送烟碱的能力各不相同。作为燃烧型卷烟最有效的替代品，电子烟碱传输系统同样可以有效地递送烟碱，因此评估电子烟碱传输系统的烟碱递送概况十分重要[52]。电子烟碱传输系统的烟碱递送概况同样受到设备类型和功率、烟液成分和使用者行为的组合影响[15,17,53]。例如，高功率电子烟碱传输系统模型比低瓦数的模型[16,54,55]更能有效地提供烟碱，高烟液烟碱浓度能提供更多的烟碱，特别是对于有经验的使用者来说[17,56]，丙二醇含量高于植物甘油的烟液可以增加烟碱的递送（可能是由于丙二醇蒸发阈值较低和/或更小的颗粒更有可能到达使用者肺部）[57]。

一项研究表明，与烟草口味相比，樱桃和薄荷醇调味剂可增加烟碱的递送（即血液中烟碱的最大浓度）[58]。另一项研究表明，从草莓味烟液中递送的烟碱比从烟草味烟液中递送的烟碱多，尽管吸入的烟碱数量相似，这可能与烟液pH的差异有关[59]。总体而言，不同设备和烟液烟碱递送存在明显差异，一些并没有增加血浆烟碱浓度，而另一些则以接近卷烟的水平输送烟碱（即10~30 ng/mL）[15,16,58,60-65]。

电子烟碱传输系统给使用者递送的烟碱也可能取决于烟液或气溶胶中烟碱的生物可用性。因此，在较高的pH下，烟碱更倾向于以非质子化（游离态）的形式存在，将会产生更大的刺激，从而增加烟碱的不愉快味道[28,66]。在较低的pH下，烟碱大多数是以质子化形式出现，减少了上呼吸道的吸收以及刺激和难闻的味道，令使用者在没有感到不适的情况下吸入更多，导致较多气溶胶通过增强烟碱的吸收到达肺部下部。起初，除极少例外，电子烟碱传输系统烟液大多仅含有游离态烟碱。然而，随着新的烟液进入市场，通过加入酸来增加质子化烟碱（即烟碱盐）的烟液越来越多[67]。这种烟碱浓度高、游离态烟碱比例小的烟液被认为更有可能提供烟碱有效的"类卷烟"递送[66]。根据这一概念，一项研究表明，使用含烟碱盐的"豆荚"系统电子烟的青少年尿液中可替宁（烟碱的主要代谢物）的水平明显高于经常吸烟的青少年[26]。电子烟碱传输系统烟液的pH也非常多样，不仅因为品牌和烟碱浓度不同而有所差异，而且品牌和烟碱浓度相同也可能有所差异[68]。例如，具有相同烟液烟碱浓度和相同设备特性（电功率）的电子烟碱传输系统，由于pH的不同可能具有不同的烟碱递送概况，以及吸入气溶胶时的不同感官效果。所有其他条件相同时，质子化烟碱气溶胶就变得不那么刺激。然而，这一概念尚未经过实验的检验，因为尚未有研究报告在其他所有变量保持不变时烟液pH的影响。

7.5 与使用有关的暴露行为模式

7.5.1 使用者群体和使用者模式的定义

烟碱暴露的一个重要因素是使用者的行为或抽吸模式。使用者抽吸模式包括抽吸口数、持续时间、抽吸容量和抽吸间隔等变量，并且这些模式具有高度的个性化。各种各样的因素影响电子烟碱传输系统如何被使用，如使用者的体验和烟液的组成。烟碱的暴露受到个体特征的影响，不同的使用者群体可能会根据他们使用电子烟的方式来区分。例如，有经验的电子烟碱传输系统使用者通常比没使用过电子烟的使用者抽吸得更长、更大口，从而导致烟碱的释放更高[15,56,69]。一项对"自然"抽吸模式的研究确定了三种类型的使用者：第一类几乎只有"轻的"阶段[即低抽吸容量（59.9 mL）、低流速（28.7 mL/s）和低抽吸持续时间（2 s）]，第二类主要是"重的"阶段[即高抽吸容量（290.9 mL）、高流速（71.5 mL/s）和高持续抽吸时间（4.4 s）]，第三类主要是"轻的"阶段（75%）和一些"重的"阶段（25%）[70]。

虽然有些人只使用电子烟碱传输系统，但也有很多人既使用电子烟碱传输系统也使用卷烟。在美国，目前几乎70%的成年电子烟使用者也抽吸卷烟[71]，而年轻人中双重吸烟者的比例较低，约33%[72]。一项研究表明，卷烟抽吸者在使用电子烟时比不吸卷烟的电子烟使用者有更长的抽吸持续时间和更大的抽吸容量[73]。电子烟和卷烟使用者的另两项研究显示，短期使用电子烟后血浆烟碱浓度低于标准化实验室抽吸卷烟后的血浆烟碱浓度[55,74]；然而，在同一研究中没有将这些值与只使用电子烟碱传输系统的人进行比较。其他个人特征，例如性别和种族，已被证明在抽吸卷烟导致的烟碱暴露方面的影响[75-77]，但都没有对电子烟进行过调查。

7.5.2 影响行为模式的因素

是否鼓励使用电子烟碱传输系统的不同生活方式和社会因素可能会影响烟碱的暴露（另见第3章）。例如，地方或国家的政策或法规可能禁止在某些封闭的公共场所禁止使用电子烟（例如无烟立法禁止使用电子烟），以及某些企业和机构可能禁止使用电子烟，因此使用者只有将它们限制在家庭或户外使用。无烟政策已经降低了吸烟的社会可接受性[78-80]，如果电子烟碱传输系统也包含在同样的政策下，那么其使用情况也将达到类似效果。

人们通过公共渠道看到的广告和其他信息也可能影响他们对电子烟碱传输系统[81]的看法和使用。几项研究表明，电子烟的广告可以增加人们对电子烟的购买和使用兴趣[82-84]。相反，健康警示和公共教育活动等政策措施可能会阻止人们，尤其是儿童和青少年开始使用这些产品。例如，在美国，一项名为"真实成本"的教育活动在防止年轻人开始吸烟方面取得了巨大的成功，并已推广到各个领域[85]。此类信息可能会影响非使用者和使用者对电子烟碱传输系统的使用风险和好处的认识，从而影响使用者持续使用和非使用者接受的可能性。

电子烟使用者的社交网络也在产品的接受度方面发挥了作用[86-91]。尤其是在年轻使用者中，电子烟碱传输系统的使用往往发生在同龄人之间[92-94]。

电子烟碱传输系统设备的设计和特性

影响电子烟碱传输系统使用和烟碱暴露的其他因素是电子烟碱传输系统设备的设计和特性。例如，一些较新型号的电子烟碱传输系统在外观上类似于U盘，便于隐蔽在学校和其他公共场所[95,96]。它们小巧的外表更加吸引电子烟新人使用。一些电子烟碱传输系统型号定制特征明显，允许使用者更改功率设置，使用不同烟碱浓度和口味的烟液，这会影响使用。例如，功率设置会影响抽吸行为，高功率会降低抽吸口数和持续时间[97]。设备功率的这种变化可以反映使用者试图递送烟碱和/或吸入气溶胶的感官效果。使用模式还与烟液中的烟碱浓度相关，因此较低浓度的烟碱与较长、较大口的抽吸有关[17,98]。第一次使用含烟碱的电子烟可能会增加一生中的烟碱暴露，因为一项研究表明，在高中第一年，最初使用含有烟碱电子烟的青少年往往比最初使用不含烟碱电子烟的青少年使用电子烟的天数更多[99]。

溶剂（丙二醇和植物甘油）

电子烟碱传输系统烟液中较高的丙二醇与植物甘油的比值，与较低的持续时间和容量、较多的烟碱递送密切相关[57]。在同一项研究中，丙二醇比例越高的烟液也被评吸者给出不太"令人愉快"和"不太满意"的结果。这可能是因为纯丙二醇的烟液产生很少或根本看不到气溶胶，这通常被使用者认为是积极的一面，同时丙二醇也可能是烟碱的增强剂。另一项研究表明，含有较多植物甘油的烟液比含有较多丙二醇的烟液更受欢迎，在使用和购买烟液时，"口感好"是最重要的考虑因素[100]。

调味物质（另见第6章）

ENDS烟液的味道也被证明会影响抽吸行为。例如，一项研究表明，吸烟者使用调味ENDS（香草、樱桃、薄荷、咖啡或烟草香精）的抽吸时间比卷烟的抽吸时间要长得多，使用调味ENDS比烟草味ENDS的抽吸频率更低[58]。在另一项研究中，经验丰富的ENDS使用者用草莓味烟液的抽吸时间比烟草味烟液的抽吸时间长；然而，当使用他们常用品牌的ENDS烟液时，抽吸容量更大、次数更多[101]。第三项研究发现ENDS烟液口味影响抽吸流速和抽吸容量但不影响抽吸持续时间[102]，尽管直接的影响尚不清楚。特别是对青少年和年轻人而言，口味不只影响使用行为，而且也是开始和持续使用ENDS产品的重要原因[99,103,104]。

7.6 被动暴露于烟碱和其他有害成分

ENDS使用者通过吸入设备释放的气溶胶直接暴露在烟碱和其他有害成分中（另见第8章）。非使用者可能通过"二手"暴露（又称环境暴露）或"三手"暴露（皮肤接触沉积在表面的释放物或摄入含有烟碱的烟液）于烟碱和其他有害物质[46]。越来越多的文献表明，ENDS的使用对室内空气质量有负面影响[105-109]，这支持当ENDS使用者与非使用者在同一空间时，非使用者可能会暴露在有害成分环境中的观点。几项研究报告了评估二手和三手暴露的有效方法[110]。一项研究显示，在真实的社会环境中，非吸烟者在二手环境中急性暴露到电子烟气溶胶后，会吸入不同程度的烟碱[100]。另一项研究表明，电子烟使用者呼出的气体中可能会对身边的人造成心率加快和收缩压升高等全身性影响[111]。进一步的研究证实，二手暴露于电子烟气溶胶30分钟会引起感官刺激和呼吸道症状，这与释放物中挥发性有机化合物的浓度密切相关[112]。

在孕妇中，烟碱很容易穿过胎盘[113]，并与胎儿大脑中的烟碱乙酰胆碱受体结合，这对胎儿的大脑发育有重要影响[114]。由烟碱造成这些受体在早期的激活和脱敏会影响胎儿发育，并造成长期后果[9]。虽然目前还没有研究表明使用ENDS会如何影响妊娠结束或胎儿发育，但烟碱被认为是产妇吸烟导致一系列不良影响的主要因素，以及一氧化碳被认为是出生体重较低的一个原因。在产前暴露于烟碱和烟草制品下的新生儿的出生体重相对较低，胎龄较早，患肺部和心肺疾病的风险较高，并且儿童期更易出现哮喘和过敏等症状[9,115]。另外，他们还面临较高的神经认知风险，可能导致学业成绩差和严重的行为问题，如注意力缺陷、多动障碍、攻击性行为和未来的物质滥用等[116]。虽然很难断定这些影响是由烟碱或烟草烟气的其他成分造成的，但烟碱被认为是孕产妇吸烟对胎儿造成大多数不良影响的主要原因[9,116]。研究表明，孕妇使用无烟碱产品与胎儿早产、死胎和胎儿唇腭裂密切相关[117-120]。与吸烟相比，怀孕期间使用烟碱替代法将降低烟碱暴露[121]早产儿和低体重胎儿的出生风险[122]。由于一些ENDS提供的烟碱含量与燃烧型卷烟中的烟碱含量相当，因此母亲吸烟的一些不良影响也可能发生在母亲使用ENDS后的烟碱暴露。应该指出的是，ENDS释放物含有其他可能的有害化合物，对胎儿发育的影响还没有得到全面研究。例如，一项研究表明，ENDS可换液罐中烟液的调味剂对人类胚胎干细胞具有细胞毒性[123]。

7.7 烟碱通量

ENDS烟碱"通量"是ENDS设备释放烟碱的速率，即ENDS设备每单位时间的烟碱释放量（μg/s）。长期以来，药物递送速率一直是了解药物滥用的相关指标，因为更快的药物递送会导致更大的滥用潜力[124,125]。当烟碱以不同的速率递送给吸烟者时，较快的递送速率被认为会产生更多的奖赏积极效果[126]。全世界有数百万人使用燃烧型卷烟，一般来说，燃烧型卷烟释放烟碱速率约为100 μg/s（根据Djordjevic等获得的数据统计）[127]，并快速将烟碱递送到血液和大脑[128,129]。虽然所有燃烧型卷烟品牌的烟碱通量总体上是稳定的，但ENDS没有呈现类似的稳定性，这主要是因为产品类别的差异性。当考虑到设备功率、结构、烟液和烟碱以及其他成分的所有可能组合时，ENDS通量可能是0 μg/s（即没有烟碱释放）到>100 μg/s。这种烟碱通量的巨大差异解释了ENDS中烟碱递送概况的差异，相同抽吸口数，低功率设备和低烟碱浓度的烟液只释放少量或不释放烟碱，而高功率设备和高烟碱浓度的烟液会释放与燃烧型卷烟相同或更多的烟碱[1]。重要的是，烟碱的通量与使用者的行为无关（例如，较长的抽吸持续时间不会改变通量），但通量和行为结合起来决定烟碱的释放量和暴露，以及输送到血液和大脑的药物数量。例如，100 μg/s的通量和1 s的抽吸持续时间产生100 μg烟碱，而4 s的抽吸产生400 μg烟碱。这就解释了为什么即使在通量被控制的情况下，更长的抽吸也会导致更多的烟碱释放[56]。抽吸时间越长，使用者吸入的烟碱剂量就越大。

虽然ENDS设备和烟液的异质性使得难以在所有可能的组合中测量通量，但是ENDS通量可以在基于物理的模型中进行数学预测[130]。如文献[131]所描述的，该模型考虑了线圈在电流开始流动后所需的加热时间、线圈在抽吸之间的冷却时间以及线圈传递热量的各种方式。模型的输入包括线圈的长度、直径、电阻和热容、烟液的组成和热力学性质（包括烟碱浓度）、环境温度和使用者行为（抽吸速度和持续时间、抽吸间隔）。在一项验证研究中，生成了预测模型，并在100个条件下测量了实际烟碱通量，在这些条件下，功率、设备类型、烟液成分和使用者行为都不同。在测试条件下，该模型解释了72%的烟碱通量变化，这个模型可以用来预测当今市场上任何ENDS（开放或封闭系统）以及正在设计的ENDS的烟碱通量。因此，烟碱通量的数学模型为希望监管ENDS烟碱释放的决策者提供了一个潜在的工具。

正如已经被指出的[18]，如果监管的目标是降低不吸烟的年轻人对ENDS滥用的可能性，那么减少ENDS的烟碱通量将会是一个有效的方法。由于烟碱通量是所有电子烟特性（结构、瓦数、烟液烟碱含量）的结果，因此监管机构可以专注于烟碱释放率（即烟碱通量）这一单一的产品性能指标，制造商也可以选择安全范围内的设备和烟液。通量目标不一定是一个单一的值，而是一系列允许的烟碱通量条件（烟碱释放率不低于X，不大于Y），允许设计一系列产品，以最大限度地减少滥用，最大限度地增加寻求戒烟的吸烟者的潜在益处，并最终解决烟碱依赖，如果能够证明ENDS能提供这种治疗益处的话。总之，烟碱通量的数学模型允许监管机构有效地检查一系列产品，以确定它们是否满足或落在指定的烟碱通量范围之内。不幸的是，如果使用者能够控制设备功率和烟液烟碱浓度等关键参数，这种监管方法就无法成功。因此，决策者也可能希望考虑"开放系统"设备在多大程度上可以接受有效的监管[18]。

7.8 讨 论

ENDS是一类多样化、不断发展的产品，在全球越来越受欢迎，尤其是在儿童、青少年和年轻人中。有些ENDS使用者以前是卷烟抽吸者，他们使用ENDS来戒烟，也有一些来自随机临床试验的经验证据表明，ENDS有助于戒烟，尽管此结果与一些报道结果不一致。许多ENDS使用者是双重使用者，除了ENDS，他们继续使用其他烟草制品，特别是传统卷烟。另外，没有暴露烟碱经历的ENDS使用者可能面临随后开始使用传统卷烟的风险。尽管市场上各种口味的ENDS烟液可能有助于一些吸烟者戒烟，也可能鼓励双重或多重使用，但几乎可以肯定的是，ENDS会鼓励没暴露过烟碱的年轻人开始使用电子烟。此类分析中，仅使用ENDS的使用者中可能的吸烟者和那些继续保持不吸烟者的比例是一个潜在的混淆因素。

有些ENDS可以以相同的抽吸口数提供与燃烧型卷烟一样多或更多的烟碱。有些ENDS提供比燃烧型卷烟少的烟碱。ENDS传递或不传递烟碱的程度取决于各种设备特性（例如功率、线圈尺寸）、烟液成分（例如烟碱浓度、丙二醇与植物甘油的比例）和使用者行为（例如抽吸持续时间）。这些相同的因素影响ENDS释放可能损害使用者健康的非烟碱有害物质的程度。最近对ENDS烟碱输送的影响是销售含有质子化烟碱（烟碱盐）的烟液。由质子化烟液形成的气溶胶比游离态烟碱形成的气溶胶吸起来刺激性更少，因此制造商可以增加烟液中的烟碱浓度，而不会使产生的气溶胶难以接受。

鉴于影响ENDS烟碱和非烟碱释放物的因素较多，对本产品类别的监管可能会很困难。在监管ENDS烟碱传输时，有人认为可以将注意力集中在单一因素上，例如烟液烟碱浓度；然而，这种方法可能会通过操纵不受管制的因素（如使用高功率设备和/或增加抽吸持续时间）来驱使使用者获得更多的烟碱。这种行为可能会降低监管的有效性，使烟碱的输送量仍然高于监管机构的预期，同时使用者将暴露于比使用低功率设备和/或短抽吸条件下更多有害的气溶胶中。

因此，有人建议监管机构将烟碱从ENDS中释放的速度，也就是烟碱通量作为监管目标。这一重点还要求ENDS产品不允许使用者使用影响烟碱通量的许多设备、烟液和使用者行为特征，例如内置抽吸持续时间限制的"封闭系统"ENDS。这些设备存在于一些市场中，因此显然是可行的。究竟哪些烟碱通量参数有利于促进当前吸烟者戒烟，同时限制不吸烟的年轻人的滥用倾向都还有待确定。

7.9 结　　论

审查的数据得出以下的结论：
- ENDS包括各种各样的产品类型，发展的速度超过了目前的监管工作。
- ENDS的性能特征因人而异，有些使用者暴露于较低水平的烟碱和其他有害成分中，而另一些使用者暴露于较高水平的烟碱和其他有害成分中。
- 以前没暴露过烟碱的人使用ENDS不符合公共卫生目标。
- 调味ENDS烟液有助于以前未暴露过烟碱的人开始和维持ENDS的使用。它们也可能对想戒烟的吸烟者有吸引力。
- 在某些情况下，比如在强化的行为专业咨询下，ENDS可能会帮助一些吸烟者戒烟，这对公共卫生产生积极的影响。然而，这些人中的大多数继续使用ENDS，其个人健康后果是不确定的，从而对公共卫生的影响也不确定。
- 大多数ENDS使用者不会戒掉燃烧型卷烟，而是同时使用ENDS和燃烧型卷烟，至少会保持与抽吸卷烟时相同的健康风险，并可能会增加健康风险。
- 由于对释放物的大量投入，对ENDS产生的烟碱和其他有害成分的释放物进行监管变得复杂起来。
- 可能需要对来自ENDS的烟碱和非烟碱有害物质的释放进行监管。这要求构建市场化的ENDS，以便使用者无法改变设备功率和烟液成分等重要特征。
- 目前尚不清楚最可能实现传统吸烟戒烟（理想情况下，同时也减少没暴露过烟碱的个体对烟碱的滥用倾向）的烟碱释放和递送情况。如果存在这种情况，则需要进行仔细的经验性研究，类似于对其他用于治疗的药理学化合物进行的研究，即使它们也可能以某些形式或通过某些途径被滥用（如阿片类药物）。

7.10 研究差距、优先事项和问题

这里审查的数据提出了许多研究问题，包括以下内容：
- 研究确定烟碱通量的范围，这将减少对ENDS产品的滥用倾向，并限制年轻人的使用，同时帮助吸烟者戒掉卷烟和其他ENDS类抽吸产品。
- 研究比较胎儿暴露于ENDS与传统卷烟条件下的烟碱和其他有害物质。
- 哪种方式实现给定的ENDS烟碱通量对使用者的风险最小？设备功率和烟

液烟碱浓度的许多不同组合可以达到相同的通量，其中哪些可能导致比其他组合产生更多的非烟碱有害物质？
- 为了使ENDS促进戒烟需要多大程度的调味ENDS烟液？只含烟碱、丙二醇和植物甘油制成的无调味的ENDS烟液，如果与一种能像燃烧型卷烟一样有效释放烟碱的设备搭配使用，是否能促进当前吸烟者戒烟？
- 何种程度的强化行为咨询是ENDS促进戒烟所需要的？
- 哪些吸烟者最有可能通过能有效提供烟碱的ENDS产品实现戒烟？在一些吸烟者中，使用ENDS的人比其他人更有可能戒烟吗？
- 考虑到对ENDS监管方法的多样性（即不同国家有不同的方法），哪些政策在保护公众健康方面对ENDS的使用最有效？
- 在众多零售网点提供的ENDS在多大程度上促进了没暴露过烟碱的个体开始使用？在受管制的场所（也许在提供个性化戒烟咨询的地方）销售ENDS是否更符合公共卫生目标？
- ENDS烟液中质子化烟碱含量增加对有害性（其他酸性成分）和滥用倾向（吸入高剂量烟碱刺激性较低）有何影响？

7.11 政策建议

此处审查的数据支持以下3项建议：
- 监管机构不应允许使用者操控设备功能和烟液成分（即开放系统ENDS），并且应确保允许的ENDS不会对年轻人有吸引力。
- 监管者应将烟碱释放率或通量（即效果）作为监管目标，而不是任何单一输入变量（如烟液烟碱浓度或设备功率）。
- ENDS不应比燃烧型卷烟有更高的滥用倾向，因此，ENDS烟碱释放率或通量不应该比燃烧型卷烟高。

7.12 参考文献

[1] Breland A, Soule E, Lopez A, Romôa C, El-Hellani A, Eissenberg T. Electronic cigarettes: what are they and what do they do? Ann NY Acad Sci. 2017;1394(1): 5–30.

[2] Talih S, Balhas Z, Eissenberg T, Salman R, Karaoghlanian N, El-Hellani A et al. Effects of user puff topography, device voltage, and liquid nicotine concentration on electronic cigarette nicotine yield: measurements and model predictions. Nicotine Tob Res. 2015;17(2):150–7.

[3] Filippidis FT, Laverty AA, Gerovasili V, Vardavas CI. Two-year trends and predictors of e-cigarette use in 27 European Union member states. Tob Control. 2017;26(1):98.

[4] McMillen RC, Gottlieb MA, Whitmore Schaefer RM, Winickoff JP, Klein JD. Trends in electronic cigarette use among US adults: Use is increasing in both smokers and nonsmokers. Nicotine

Tob Res. 2015;17(10):1195-202.

[5] Jamal A, Gentzke A, Hu SS, Cullen KA, Apelberg BJ, Homa DM et al. Tobacco use among middle and high school students-United States, 2011-2016. Morbid Mortal Wkly Rep. 2016;66:597-603.

[6] Cullen KA, Ambrose BK, Gentzke AS, Apelberg BJ, Jamal A, King BA. Notes from the field: Use of electronic cigarettes and any tobacco product among middle and high school students –United States, 2011-2018. Morbid Mortal Wkly Rep. 2018;67:1276-7.

[7] Hutzler C, Paschke M, Kruschinski S, Henkler F, Hahn J, Luch A. Chemical hazards present in liquids and vapors of electronic cigarettes. Arch Toxicol. 2014;88(7):1295-308.

[8] Benowitz NL, Burbank AD. Cardiovascular toxicity of nicotine: Implications for electronic ci-garette use. Trends Cardiovasc Med. 2016;26(6):515-23.

[9] England LJ, Bunnell RE, Pechacek TE, Tong VT, McAfee TA. Nicotine and the developing human: a neglected element in the electronic cigarette debate. Am J Prev Med. 2015;49(2):286-93.

[10] O'Dell LE. A psychobiological framework of the substrates that mediate nicotine use during adolescence. Neuropharmacology. 2009;56(Suppl 1):263-78.

[11] Soneji S, Barrington-Trimis JL, Wills TA, Levanthal AM, Unger JB, Gibson LA et al. Association between initial use of e-cigarettes and subsequent cigarette smoking among adolescents and young adults: a systematic review and meta-analysis. JAMA Pediatrics. 2017;171(8):788-97.

[12] Rahman MA, Hann N, Wilson A, Mnatzaganian G, Worrall-Carter L. e-Cigarettes and smoking cessation: evidence from a systematic review and meta-analysis. PLoS One. 2015;10(3):e0122544.

[13] Shihadeh A, Eissenberg T. Electronic cigarette effectiveness and abuse liability: predicting and regulating nicotine "ux. Nicotine Tob Res. 2015;17(2):158-62.

[14] El-Hellani A, Salman R, El-Hage R, Talih S, Malek N, Baalbaki R et al. Nicotine and carbonyl emissions from popular electronic cigarette products: Correlation to liquid composition and design characteristics. Nicotine Tob Res. 2018;20(2):215-23.

[15] Voos N, Goniewicz ML, Eissenberg T. What is the nicotine delivery pro#le of electronic cigarettes? Expert Opin Drug Deliv. 2019;16(11):1193-203.

[16] Wagener TL, Floyd EL, Stepanov I, Driskill LM, Frank SG, Meier E et al. Have combustible cigarettes met their match? The nicotine delivery pro#les and harmful constituent exposures of second-generation and third-generation electronic cigarette users. Tob Control. 2017;26(e1):e23-8.

[17] Hiler M, Karaoghlanian N, Talih S, Maloney S, Breland A, Shihadeh A et al. Effects of electronic cigarette heating coil resistance and liquid nicotine concentration on user nicotine delivery, heart rate, subjective effects, puff topography, and liquid consumption. Exp Clin Psychopharmacol, 2020;28(5):527-39.

[18] Eissenberg T et al. "Open-system" electronic cigarettes cannot be regulated effectively. Tob Control, 2020 (in press).

[19] Talih S, Salman R, El-Hage R, Karam E, Karaoghlanian N, El-Hellani A et al. Characteristics and toxicant emissions of JUUL electronic cigarettes. Tob Control. 2019;28(6):678.

[20] Talih S, Salman R, El-Hage R, Karam EE, Karaoghlanian N, El-Hellani A et al. Might limiting liquid nicotine concentration result in more toxic electronic cigarette aerosols? Tob Control. 2020:10.1136/tobaccocontrol-2019-055523.

[21] The European Parliament and the Council of the European Union, Directive 2014/40/EU on the approximation of the laws, regulations and administrative provisions of the Member States concerning the manufacture, presentation and sale of tobacco and related products and repealing Directive 2001/37/EC. Off J Eur Union. 2014;L127/1:127.

[22] Nicotine salts e-liquid overview. Singapore: Wingle Group Electronics Ltd; 2018 (https://vape.hk/wp-content/uploads/2019/05/NICOTINE_SALTS_E-LIQUID.pdf, accessed 10 January 2021).

[23] El-Hellani A, El-Hage R, Baalbaki R, Salman R, Talih S, Shihadeh A et al. Free-base and protonated nicotine in electronic cigarette liquids and aerosols. Chem Res Toxicol. 2015;28(8):1532–7.

[24] Goniewicz ML, Gupta R, Lee YH, Reinhardt S, Kim S, Kim B et al. Nicotine levels in electronic cigarette refill solutions: A comparative analysis of products from the US, Korea, and Poland. Int J Drug Policy. 2015;26(6):583–8.

[25] Raymond BH, Collette-Merrill K, Harrison RG, Jarvis S, Rasmussen RJ. The nicotine content of a sample of e-cigarette liquid manufactured in the United States. J Addict Med. 2018;12(2):127–31.

[26] Goniewicz ML, Boykan R, Messina CR, Eliscu A, Tolentino J. High exposure to nicotine among adolescents who use Juul and other vape pod systems ("pods"). Tob Control. 2019;28(6):676–7.

[27] Omaiye EE, McWhirter KJ, Luo W, Pankow JF, Talbot P. High-nicotine electronic cigarette products: toxicity of JUUL fluids and aerosols correlates strongly with nicotine and some flavor chemical concentrations. Chem Res Toxicol. 2019;32(6):1058–69.

[28] Jackler RK, Ramamurthi D. Nicotine arms race: JUUL and the high-nicotine product market- Tob Control, 2019;28(6):623–8.

[29] Omaiye EE, Cordova I, Davis B, Talbot P. Counterfeit electronic cigarette products with mislabeled nicotine concentrations. Tob Regul Sci. 2017;3(3):347–57.

[30] Kosmider L, Spindle TR, Gawron M, Sobczak A, Goniewicz ML. Nicotine emissions from electronic cigarettes: Individual and interactive effects of propylene glycol to vege表 glycerin composition and device power output. Food Chem Toxicol. 2018;115:302–5.

[31] Robinson RJ, Hensel EC. Behavior-based yield for electronic cigarette users of different strength eliquids based on natural environment topography. Inhal Toxicol. 2019;31(13– 14):484–91.

[32] Madison MC, Landers CT, Gu BH, Chang CY, Yung HY, You R et al., Electronic cigarettes disrupt lung lipid homeostasis and innate immunity independent of nicotine. J Clin Invest. 2020;129(10):4290–304.

[33] Chaumont M, Bernard A, Pochet S, Mélot C, El Khattabi C, Ree F et al. High-wattage e-cigarettes induce tissue hypoxia and lower airway injury: a randomized clinical trial. Am J Resp Critical Care Med. 2018;198(1):123–6.

[34] Safety assessment and regulatory authority to use flavors – Focus on electronic nicotine delivery systems and flavored tobacco products. Washington (DC): Flavor and Extract Manufacturers Association of the United States; 2018.

[35] Farsalinos KE, Kistler KA, Gillman G, Voudris V. Evaluation of electronic cigarette liquids and aerosol for the presence of selected inhalation toxins. Nicotine Tob Res. 2015;17(2):168–74.

[36] Muthumalage T, Prinz M, Ansah KO, Gerloff J, Sundar IK, Rahman I. Inflammatory and oxidative responses induced by exposure to commonly used e-cigarette flavoring chemicals and flavored

e-liquids without nicotine. Front Physiol. 2017;8:1130.

[37] Kosmider L, Sobczak A, Prokopowicz A, Kurek J, Zaciera M, Knysak J et al. Cherry-flavoured electronic cigarettes expose users to the inhalation irritant, benzaldehyde. Thorax. 2016;71(4):376-7.

[38] Hickman E, Herrera CA, Jaspers I. Common e-cigarette flavoring chemicals impair neutrophil phagocytosis and oxidative burst. Chem Res Toxicol. 2019;32(6):982-5.

[39] Behar RZ, Davis B, Wang Y, Bahl V, Lin S, Talbot P. Identification of toxicants in cinnamon-flavored electronic cigarette refill fluids. Toxicol In Vitro. 2014;28(2):198-208.

[40] Clapp PW, Lavrich KL, van Heusden CA, Lazarowski ER, Carson JL, Jaspers I. Cinnamaldehyde in flavored e-cigarette liquids temporarily suppresses bronchial epithelial cell ciliary motility by dysregulation of mitochondrial function. Am J Physiol Lung Cell Mol Physiol. 2019;316(3):L470-86.

[41] Barrington-Trimis JL, Samet JM, McConnell R. Flavorings in electronic cigarettes: An unrecognized respiratory health hazard? JAMA. 2014;312(23):2493-4.

[42] Tierney PA, Karpinski CD, Brown JE, Luo W, Pankow JF. Flavour chemicals in electronic cigarette fluids. Tob Control. 2016;25(e1):e10-5.

[43] Smith DM, Schneller LM, O'Connor RJ, Goniewicz ML. Are e-cigarette flavors associated with exposure to nicotine and toxicants? Findings from wave 2 of the Population Assessment of Tobacco and Health (PATH) study. Int J Environ Res Public Health. 2019;16(24):5055.

[44] Kosmider L, Kimber CF, Kurek J, Corcoran O, Dawkins LE. Compensatory puffing with lower nicotine concentration e-liquids increases carbonyl exposure in e-cigarette aerosols. Nicotine Tob Res. 2017;20(8):998-1003.

[45] Talhout R, Schulz T, Florek E, van Benthern J, Wester P, Opperhuizen A. Hazardous compounds in tobacco smoke. Int J Environ Res Public Health. 2011;8(2):613-28.

[46] DeVito EE, Krishnan-Sarin S. e-Cigarettes: Impact of e-liquid components and device characteristics on nicotine exposure. Curr Neuropharmacol. 2018;16(4):438-59.

[47] Truman P, Stanfill S, Heydari A, Silver E, Fowles J. Monoamine oxidase inhibitory activity of flavoured e-cigarette liquids. NeuroToxicology. 2019;75:123-8.

[48] Cooper SY, Akers AT, Henderson BJ. Green apple e-cigarette flavorant farnesene triggers reward-related behavior by promoting high-sensitivity nAChRs in the ventral tegmental area. eNeuro. 2020;7(4):0172.

[49] Valentine GW, Jatlow PI, Coffman M, Nadim H, Gueorguieva R, Sofuoglu M. The effects of alcohol-containing e-cigarettes on young adult smokers. Drug Alcohol Depend; 2016;159:272-6.

[50] Van Skike CE, Maggio SE, Reynolds AR, Casey EM, Bardo MT, Dwoskin LP et al. Critical needs in drug discovery for cessation of alcohol and nicotine polysubstance abuse. Prog Neuropsychopharmacol Biol Psychiatry. 2016;65:269-87.

[51] Son Y, Wackowski O, Weisel C, Schwander S, Mainelis G, Delnevo et al. Evaluation of e-vapor nicotine and nicotyrine concentrations under various e-liquid compositions, device settings, and vaping topographies. Chem Res Toxicol. 2018;31(9):861-8.

[52] Vansickel AR, Cobb CO, Weaver MF, Eissenberg TE. A clinical laboratory model for evaluating the acute effects of electronic "cigarettes": nicotine delivery profile and cardiovascular and subjective effects. Cancer Epidemiol Biomarkers Prev. 201019(8):1945-53.

[53] Blank MD, Pearson J, Cobb CO, Felicione NJ, Hiler MM, Spindle TR et al. What factors reliably predict electronic cigarette nicotine delivery? Tob Control. 2019;29(6):644–51.

[54] Yingst JM, Foulds J, Veldheer S, Hrabovsky S, Trushin N, Eissenberg T et al. Nicotine absorption during electronic cigarette use among regular users. PLoS One; 2019;14(7):e0220300.

[55] Hajek P, Przulj D, Phillips A, Anderson R, McRobbie H. Nicotine delivery to users from cigarettes and from different types of e-cigarettes. Psychopharmacology. 2017;234(5):773–9.

[56] Hiler M, Breland A, Spindle T, Maloney S, Lipato T, Karaoghlanian N et al. Electronic cigarette user plasma nicotine concentration, puff topography, heart rate, and subjective effects: In"uence of liquid nicotine concentration and user experience. Exp Clin Psychopharmacol. 2017;25(5):380–92.

[57] Spindle TR, Talih S, Hiler MM, Karaoghlanian N, Halquist MS, Breland AB et al. Effects of electronic cigarette liquid solvents propylene glycol and vege表 glycerin on user nicotine delivery, heart rate, subjective effects, and puff topography. Drug Alcohol Depend. 2018;188:193–9.

[58] Voos N, Smith D, Kaiser L, Mahoney MC, Bradizza CM, Kozlowski LT et al. Esect of e-cigarette "avors on nicotine delivery and puffing topography: results from a randomized clinical trial of daily smokers. Psychopharmacology (Berl). 2020;237(2):491–502.

[59] St Helen G, Dempsey DA, Havel CM, Jacob P III, Benowitz NL. Impact of e-liquid "avors on nicotine intake and pharmacology of e-cigarettes. Drug Alcohol Depend. 2017;178:391–8.

[60] St Helen G, Havel C, Dempsey DA, Jacob P III, Benowitz NL. Nicotine delivery, retention and pharmacokinetics from various electronic cigarettes. Addiction. 2016;111(3):535–44.

[61] Lopez AA, Hiler MM, Soule EK, Ramôa CP, Karaoghlanian N, Lipato T et al. Effects of electronic cigarette liquid nicotine concentration on plasma nicotine and pus topography in tobacco cigarette smokers: a preliminary report. Nicotine Tob Res. 2016;18(5):720–3.

[62] Ramôa CP, Hiler MM, Spindle TR, Lopez AA, Karaoghlanian N, Lipato T et al. Electronic cigarette nicotine delivery can exceed that of combustible cigarettes: a preliminary report. Tob Control. 2016;25(e1):e6–9.

[63] Marsot A, Simon N. Nicotine and cotinine levels with electronic cigarette: a review. Int J Toxicol. 2016;35(2):179–85.

[64] Hajek P, Pittaccio K, Pesola F, Myers Smith K, Phillips-Waller A et al. Nicotine delivery and users' reactions to Juul compared with cigarettes and other e-cigarette products. Addiction. 2020;115(6).

[65] Voos N, Kaiser L, Mahoney MC, Bradizza CM, Kozlowski LT, Benowitz NL et al. Randomized within-subject trial to evaluate smokers' initial perceptions, subjective effects and nicotine delivery across six vaporized nicotine products. Addiction. 2019;114(7):1236–48.

[66] Duell AK, Pankow JF, Peyton DH. Nicotine in tobacco product aerosols: "It's déjà vu all over again". Tob Control, 2019;29(6):656–62.

[67] Duell AK, Pankow JF, Peyton DH. Free-base nicotine determination in electronic cigarette liquids by 1H NMR spectroscopy. Chem Res Toxicol. 2018;31(6):431–4.

[68] Stepanov I, Fujioka N. Bringing attention to e-cigarette pH as an important element for research and regulation. Tob Control. 2015;24(4):413.

[69] Farsalinos KE, Spyrou A, Stefopoulos C, Tsimopoulou K, Kourkoveli P, Tsiapras D et al. Corrigendum: Nicotine absorption from electronic cigarette use: comparison between experienced consumers (vapers) and naive users (smokers). Sci Rep. 2015;5:13506.

[70] Lee YO, Morgan-Lopez AA, Nonnemaker JM, Pepper JK, Hensel EC, Robinson RJ. Latent class analysis of e-cigarette use sessions in their natural environments. Nicotine Tob Res. 2018;21(10):1408–13.

[71] Coleman BN, Rostron B, Johnson SE, Ambrose BK, Pearson J, Stanton CA et al. Electronic cigarette use among US adults in the Population Assessment of Tobacco and Health (PATH) study, 2013–2014. Tob Control. 2017;26(e2):e117–26.

[72] Vogel EA, Cho J, McConnell RS, Barrington-Trimis JL, Leventhal AM. Prevalence of electronic cigarette dependence among youth and its association with future use. JAMA Netw Open. 2020;3(2):e1921513.

[73] Lee YO, Nonnemaker JM, Brad#eld B, Hensel EC, Robinson RJ. Examining daily electronic cigarette puff topography among established and nonestablished cigarette smokers in their natural environment. Nicotine Tob Res. 2017;20(10):1283–8.

[74] St Helen G, Nardone N, Addo N, Dempsey D, Havel C, Jacob P III et al. Differences in nicotine intake and effects from electronic and combustible cigarettes among dual users. Addiction. 2019;115(4):757–67.

[75] Chen A, Krebs NM, Zhu J, Muscat JE. Nicotine metabolite ratio predicts smoking topography: The Pennsylvania Adult Smoking Study. Drug Alcohol Depend. 2018;190:89–93.

[76] Holford TR, Levy DT, Meza R. Comparison of smoking history patterns among African American and white cohorts in the United States born 1890 to 1990. Nicotine Tob Res. 2016;18(suppl 1):S16–29.

[77] Ross KC, Gubner NR, Tyndale RF, Hawk LW Jr, Lerman C, George TP et al. Racial differences in the relationship between rate of nicotine metabolism and nicotine intake from cigarette smoking. Pharmacol Biochem Behav. 2016;148:1–7.

[78] Preventing tobacco use among youth and young adults: a report of the Surgeon General. Atlanta (GA): Department of Health and Human Services, Centers for Disease Control and Prevention; 2012.

[79] Hoffman SJ, Tan C. Overview of systematic reviews on the health-related effects of government tobacco control policies. BMC Public Health. 2015;15:744.

[80] Hopkins DP, Razi S, Leeks KD, Kalra GP, Chattopadhyay SK, Soler RE et al. Smokefree policies to reduce tobacco use: a systematic review. Am J Prev Med. 2010;38(2):S275–89.

[81] Alcalá HE, Shimoga SV. It is about trust: Trust in sources of tobacco health information, perceptions of harm, and use of e-cigarettes. Nicotine Tob Res. 2020;22(5):822–6.

[82] Vasiljevic M, Petrescu DC, Marteau TM. Impact of advertisements promoting candy-like "avoured e-cigarettes on appeal of tobacco smoking among children: an experimental study. Tob Control. 2016;25(e2):e107–12.

[83] Farrelly MC, Duke JC, Crankshaw EC, Eggers ME, Lee YO, Nonnemaker JM et al. A randomized trial of the effect of e-cigarette TV advertisements on intentions to use e-cigarettes. Am J Prev Med. 2015;49(5):686–93.

[84] Camenga D, Gutierrez KM, Kong G, Cavallo D, Simon P, Krishnan-Sarin S. e-Cigarette advertising exposure in e-cigarette naïve adolescents and subsequent e-cigarette use: A longitudinal cohort study. Addict Behav. 2018;81:78–83.

[85] Zeller M. Evolving "The real cost" campaign to address the rising epidemic of youth e-cigarette

[86] Leavens ELS, Stevens EM, Brett EI, Hébert ET, Villanti AC, Pearson JL et al. JUUL electronic cigarette use patterns, other tobacco product use, and reasons for use among ever users: Results from a convenience sample. Addict Behav. 2019;95:178–83.

[87] Ickes M, Hester JW, Wiggins AT, Rayens MK, Hahn EJ, Kavuluru R. Prevalence and reasons for Juul use among college students. J Am Coll Health. 2019;68(5):1–5.

[88] Kong G, Bold KW, Morean ME, Bhatti H, Camenga DR, Jackson A et al. Appeal of JUUL among adolescents. Drug Alcohol Depend. 2019;205:107691.

[89] Leavens ELS, Stevens EM, Brett EI, Le%ngwell TR, Wagener TL. JUUL in school: JUUL electronic cigarette use patterns, reasons for use, and social normative perceptions among college student ever users. Addict Behav. 2019;99:106047.

[90] Nardone N, St Helen G, Addo N, Meighan S, Benowitz NL. JUUL electronic cigarettes: Nicotine exposure and the user experience. Drug Alcohol Depend. 2019;203:83–7.

[91] Patel M, Cuccia AF, Willett J, Zhou Y, Kierstead EC, Czaplicki L et al. JUUL use and reasons for initiation among adult tobacco users. Tob Control. 2019;28(6):681–4.

[92] Keamy-Minor E, McQuoid J, Ling PM. Young adult perceptions of JUUL and other pod electronic cigarette devices in California: a qualitative study. BMJ Open. 2019;9(4):e026306.

[93] Hrywna M, Bover Manderski MT, Delnevo CD. Prevalence of electronic cigarette use among adolescents in New Jersey and association with social factors. JAMA Netw Open. 2020;3(2):e1920961.

[94] McKelvey K. Halpern-Felsher B. How and why California young adults are using different brands of pod-type electronic cigarettes in 2019: Implications for researchers and regulators. J Adolesc Health. 2020;67(1):46–52.

[95] Kavuluru R, Han S, Hahn EJ. On the popularity of the USB "ash drive-shaped electronic cigarette Juul. Tob Control. 2019;28(1):110–2.

[96] Ramamurthi D, Chau C, Jackler RK. JUUL and other stealth vaporisers: hiding the habit from parents and teachers. Tob Control. 2018;28(6):610–6.

[97] Farsalinos K, Poulas K, Voudris V. Changes in pu%ng topography and nicotine consumption depending on the power setting of electronic cigarettes. Nicotine Tob Res. 2017;20(8):993–7.

[98] Dawkins L, Cox S, Goniewicz M, McRobbie H, Kimber C, Doig M et al. "Real-world" compensatory behaviour with low nicotine concentration e-liquid: subjective effects and nicotine, acrolein and formaldehyde exposure. Addiction. 2018;113(10):1874–82.

[99] Audrain-McGovern J, Rodriguez D, Pianin S, Alexander E. Initial e-cigarette "avoring and nicotine exposure and e-cigarette uptake among adolescents. Drug Alcohol Depend. 2019;202:149–55.

[100] Melstrom P, Sosnoff C, Koszowski B, King BA, Bunnell R, Le G et al. Systemic absorption of nicotine following acute secondhand exposure to electronic cigarette aerosol in a realistic social setting. Int J Hyg Environ Health. 2018;221(5):816–22.

[101] St Helen G, Shahid M, Chu S, Benowitz NL. Impact of e-liquid "avors on e-cigarette vaping behavior. Drug Alcohol Depend. 2018;189:42–8.

[102] Robinson RJ, Hensel EC, al-Olayan AA, Nonnemaker JM, Lee YO. Effect of e-liquid "avor on electronic cigarette topography and consumption behavior in a 2-week natural environment

switching study. PLoS One. 2018;13(5):e0196640.

[103] Schneller LM, Bansal-Travers M, Goniewicz ML, McIntosh S, Ossip D, O'Connor RJ. Use of "avored electronic cigarette re#ll liquids among adults and youth in the US – Results from wave 2 of the Population Assessment of Tobacco and Health Study (2014–2015). PLoS One. 2018;13(8):e0202744.

[104] Villanti AC, Johnson AL, Ambrose BK, Cummings KM, Stanton CA, Rose SW et al. Flavored tobacco product use in youth and adults: Findings from the #rst wave of the PATH study (2013–2014). Am J Prev Med. 2017;53(2):139–51.

[105] Chen R, Ahererra A, Isichei C, Olmedo P, Jarmul S, Cohen JE et al. Assessment of indoor air quality at an electronic cigarette (vaping) convention. J Expo Sci Environ Epidemiol. 2018;28(6):522–9.

[106] Melstrom P, Koszowski B, Thanner MH, Hoh E, King B, Bunnell R et al. Measuring PM2.5, ultra#- ne particles, nicotine air and wipe samples following the use of electronic cigarettes. Nicotine Tob Res. 2017;19(9):1055–61.

[107] Schober W, Fembacher L, Frenzen A, Fromme H. Passive exposure to pollutants from conventional cigarettes and new electronic smoking devices (IQOS, e-cigarette) in passenger cars. Int J Hyg Environ Health. 2019;222(3):486–93.

[108] Soule EK, Maloney SF, Spindle TR, Rudy AK, Hiler MM, Cobb CO. Electronic cigarette use and indoor air quality in a natural setting. Tob Control. 2017;26(1):109–12.

[109] Volesky KD, Maki A, Scherf C, Watson L, Van Ryswyck K, Fraser B et al. The in"uence of three e-cigarette models on indoor #ne and ultra#ne particulate matter concentrations under real-world conditions. Environ Pollut. 2018;243(Pt B):882–9.

[110] Quintana PJE, Hoh E, Dodder NG, Matt GE, Zakarian JM, Anderson KA et al. Nicotine levels in silicone wristband samplers worn by children exposed to secondhand smoke and electronic cigarette vapor are highly correlated with child's urinary cotinine. J Expo Sci Environ Epidemiol. 2019;29(6):733–41.

[111] Visser FW, Klerx WN, Cremers HWJM, Ramlal R, Schwillens PL, Talhout R. The health risks of electronic cigarette use to bystanders. Int J Environ Res Public Health. 2019;16(9):1525.

[112] Tzortzi A, Teloniatis S, Matiampa G, Bakelas G, Tzavara C, Vyzikidou VK et al. Passive exposure of non-smokers to e-cigarette aerosols: Sensory irritation, timing and association with volatile organic compounds. Environ Res. 2020;182:108963.

[113] Luck W, Nau H, Hansen R, Steldinger R. Extent of nicotine and cotinine transfer to the human fetus, placenta and amniotic "uid of smoking mothers. Dev Pharmacol Ther. 1985;8:384–95.

[114] Cairns NJ, Wonnacott S. (3H)(-)Nicotine binding sites in fetal human brain. Brain Res. 1988;475(1):1–7.

[115] Suter MA, Aagaard KM. The impact of tobacco chemicals and nicotine on placental development. Prenatal Diagn. 2020;40(9): 10.1002/pd.5660.

[116] Holbrook BD. The effects of nicotine on human fetal development. Birth Defects Res C: Embryo Today Rev. 2016;108(2):181–92.

[117] Wikström AK, Cnattingius S, Stephansson O. Maternal use of Swedish snuff (snus) and risk of stillbirth. Epidemiology. 2010;21(6):772–8.

[118] Gupta PC, Sreevidya S. Smokeless tobacco use, birth weight, and gestational age: pop-

ulation based, prospective cohort study of 1217 women in Mumbai, India. Br Med J. 2004;328(7455):1538.

[119] Gupta PC, Subramoney S. Smokeless tobacco use and risk of stillbirth: a cohort study in Mumbai, India. Epidemiology. 2006;17(1):47–51.

[120] Steyn K, De Wet T, Salojee Y, Nel H, Yach D. The in"uence of maternal cigarette smoking, snuff use and passive smoking on pregnancy outcomes: the Birth To Ten Study. Paediatric Perinatal Epidemiol. 2006;20(2):90–9.

[121] Hickson C, Lewis S, Campbell KA, Cooper S, Berlin I, Claire R et al. Comparison of nicotine exposure during pregnancy when smoking and abstinent with nicotine replacement therapy: systematic review and meta-analysis. Addiction. 2019;114(3):406–24.

[122] Forinash AB, Pitlick JM, Clark K, Alstat V. Nicotine replacement therapy effect on pregnancy outcomes. Ann Pharmacother. 2010;44(11):1817–21.

[123] Bahl V, Lin S, Xu N, Davis B, Wang Y, Talbot P. Comparison of electronic cigarette re#ll "uid cytotoxicity using embryonic and adult models. Reprod Toxicol. 2012;34(4):529–37.

[124] Balster RL, Schuster CR. Fixed-interval schedule of cocaine reinforcement: effect of dose and infusion duration. J Exp Anal Behav. 1973;20(1):119–29.

[125] Carter LP, Stitzer ML, Henning#eld JE, O'Connor RJ, Cummings KM, Hatsukami DK. Abuse liability assessment of tobacco products including potential reduced exposure products. Cancer Epidemiol Biomarkers Prev. 2009;18(12):3241–62.

[126] Jensen KP, Valentine G, Gueorguieva R, Sofuoglu M. Differential effects of nicotine delivery rate on subjective drug effects, urges to smoke, heart rate and blood pressure in tobacco smokers. Psychopharmacology (Berl), 2020;237(5):1359–69.

[127] Djordjevic MV, Fan J, Ferguson S, Hoffmann D. Self-regulation of smoking intensity. Smoke yields of the low-nicotine, low-"tar" cigarettes. Carcinogenesis. 1995;16(9):2015–21.

[128] Henning#eld JE, London ED, Benowitz NL. Arterial–venous differences in plasma concentrations of nicotine after cigarette smoking. J Am Med Assoc. 1990;263(15):2049–50.

[129] Henning#eld JE, Stapleton JM, Benowitz NL, Grayson RF, London ED. Higher levels of nicotine in arterial than in venous blood after cigarette smoking. Drug Alcohol Depend. 1993;33(1):23–9.

[130] Talih S, Balhas Z, Salman R, El-Hage R, Karaoghlanian N, El-Hellani A et al. Transport phenomena governing nicotine emissions from electronic cigarettes: model formulation and experimental investigation. Aerosol Sci Technol. 2017;51(1):1–11.

[131] Breland A, Balster RL, Cobb C, Fagan P, Foulds J, Koch JR et al. Answering questions about electronic cigarettes using a multidisciplinary model. Am Psychol. 2019;74(3):368–79.

8. 电子烟碱／非烟碱传输系统和加热型烟草制品相关个体风险的量化方法探索：对人体健康和监管的影响

Frank Henkler-Stephani, German Federal Institute for Risk Assessment, Department of Chemicals and Product Safety, Berlin, Germany

Yvonne Staal, National Institute for Public Health and the Environment, Centre for Health Protection, Bilthoven, Netherlands

摘 要

电子烟加热和雾化电子烟烟液以供吸入。许多烟液含有烟碱（电子烟碱传输系统，ENDS），而有些烟液则不含烟碱（电子非烟碱传输系统，ENNDS）。虽然ENDS或ENNDS的基本设计相似或相同，但该设备在使用、操作温度或性能标准方面差异很大。健康风险不仅取决于设备的特性，还取决于电子烟烟液的组成。目前，几千种口味的烟液不仅能被买到，也可以在家里自制。

初步评估表明，单个ENDS的毒理学风险取决于设备和电子烟烟液，但其中与大功率设备相关的风险可能相对较高。由于这些风险来自单个化合物或释放物中的混合物，因此，本章总结了可用于量化使用ENDS、ENNDS和加热型烟草制品（HTP）相关健康风险的方法。大多数方法都需要大量有关释放物和有害物质的数据，其中只有部分数据可用。目前，定量风险评估方法不能被用于监管，尽管最有希望的方法是基于化合物在释放物中的相对效应。由于ENDS和ENNDS的多样性，风险评估仍然是一项挑战，结果无法推广到全部设备。烟液和设备种类繁多，表明应对烟液和设备的不同组合以及单个产品进行不同的健康风险评估。可对高风险的相关指标进行描述，例如特定成分或特定设备设置。

8.1 背 景

烟草使用是全世界引起过早死亡的主要原因。每年约800万的人死于与烟草

8. 电子烟碱/非烟碱传输系统和加热型烟草制品相关个体风险的量化方法探索：对人体健康和监管的影响

有关的疾病，其中包括估算的暴露于二手烟的120万非吸烟者[1]。尽管卷烟是最常见的烟草制品，尤其是在发达国家，但其他烟草制品和替代产品也构成严重的健康风险。在印度，每年有超过35万人死于咀嚼和口用烟草[2]。

虽然成瘾是维持这种危险行为的驱动力，但烟碱不会引发与烟草消费相关高死亡率的主要毒理效应，而高死亡率是致癌物、其他有害的燃烧产物和烟草成分造成的。通过对烟草使用致癌性中单个化合物的贡献评估[3,4]，确定了主要致癌物，并根据其诱发肿瘤的能力对烟气成分进行了排序。类似的方法也用于心血管和其他健康风险评估。

已经提出了降低吸烟者暴露有害成分的策略，其中包括强制性限制卷烟烟气中最具相关性的有害成分[5-7]。然而，降低燃烧和燃烧产物的技术限制导致限制传统卷烟毒性水平的策略没有成功[8]。加热型烟草制品通过电加热和其他放热方式，使烟草材料产生由保润剂、烟碱和其他烟草成分和热解产物组成的气溶胶。加热型烟草制品产生的气溶胶通常比卷烟烟气中有害和潜在有害化合物含量少[9]，然而，目前还不清楚减少暴露有害成分是否显著降低了健康风险。

电子烟是ENDS最常见的形式，这两个术语经常作为同义词使用。与加热型烟草制品相反，ENDS不再将烟碱消费与烟草使用联系起来。虽然ENDS被明确定义为烟碱传输系统，但"电子非烟碱传输系统"（ENNDS）这一名称仅指不含烟碱。根据目前的理解[10]，ENDS和ENNDS是除了是否含有烟碱之外的同类产品。甘油和/或丙二醇通常是形成气溶胶的烟液主要成分。烟液还含有各种口味和其他成分，来提高气溶胶的可口性和吸引力[11-13]。第一代电子烟的形状像传统卷烟，由电池、烟液储液池和雾化室组成，其中含有各种材料细丝制成的灯芯，用于加热烟液[14]。生成的气溶胶被引导至烟嘴，并且可以被吸入。无论是一次性使用或可重新填充的烟弹，均可提供烟液。初代设备在向使用者递送烟碱方面的效率并不高，然而，这一基本功能已逐步得到改善。可以重新填充和重复使用的复杂设备，如开放式系统已经出现。其他改进包括增加电池容量、可变电源、去除锡的焊料接头和灯丝线圈。最近的发展包括低欧姆雾化装置，由于电阻低，这些单位可以在更高的可变电压下工作[14]。现代雾化器每次抽吸的烟液量比原来的电子烟设备高得多[15]。这些系统具有很强的适应性。例如，使用者可以搭建他们自己的线圈或自定义性能参数，例如功率、气流，间接地改变烟碱递送。然而，先进系统的运行也越来越复杂，已经开发更容易使用的产品模块，包括预填充的烟液储液器和一次性使用的雾化器[16]（参见第6章）。这些产品被称为"豆荚"系统，可能与咖啡胶囊类似。含有烟碱的"豆荚"可以连接到一个棍子或笔形装置上，里面装着低功率电池和烟嘴。JUUL作为美国领先的电子烟品牌，是一个很流行的例子[16,17]。虽然"豆荚"系统缺乏先进的可重新填充式电子烟的灵活性，但其烟碱的递送和致瘾性程度可与燃烧型卷烟相媲美[18]，其浓度可达到60 mg/mL，如美国版的JUUL。这比欧洲产品中允许的烟碱上限高出3倍。大多数"豆荚"系统含有烟碱盐，如烟碱苯甲酸盐或水杨酸盐，以限制碱性烟碱的刺激性。

直到最近，电子烟还主要被视为传输烟碱的装置，并没有考虑到它们造成的危害。ENNDS被一些人视为可接受的不含烟碱的燃烧型卷烟，然而，这个观点具有一定局限性。许多新技术最初保持其原始的应用，而随着时间的推移，也可能应用于其他方面。使用改型电子烟用于大麻和其他非法化合物的消费[19]很可能是始料未及的产品应用早期指标。最近，大麻二酚的使用已成为一种商业新应用，与烟碱或烟草的使用关系不大。大麻二酚已经被定义了药理学特性[20]，并声称有许多有益的效果。据报道，一些商业产品包含有其他大麻素[21]。因此，将大麻二酚烟液描述为ENNDS消耗品具有误导性，因为它们根本不含有烟碱，而诸如"电子大麻二酚吸入系统"或"电子吸入系统"等术语可能更合适。越来越多的电子传输产品现在超出了烟碱和烟草管制的传统范围，在一些国家，它们越来越多地（往往不受管制地使用）受人关注。

8.1.1　量化风险的挑战

从风险评估的角度来看，ENDS和传统卷烟之间的重要区别在于系统本身、使用方式、操控性和电子烟液等。然而，上述技术发展并非主要从健康角度出发。制造商将改善烟碱的传输、产品的口感和其他决定消费者接受和使用的特性作为考虑目标。这个目标是一把双刃剑。电子烟的可接受性可能是使用这些系统戒烟的先决条件，但它们的吸引力，特别是对儿童和青少年的吸引力，又会增加本来不吸烟的人使用电子烟的风险。

ENDS的变化如此迅速，导致电子烟毒理学风险评估越来越难。早期产品的风险评估假设温度最高达100℃，而新的高功率ENDS产品温度可以达到250℃，将促进某些成分化学降解。然而，其他技术特征，例如过热控制或者替换含锡或其他金属部件，可能会减少使用者毒性暴露[14]。技术特征在改变风险方面模棱两可的作用也在"豆荚"系统（如JUUL）中得到了说明，这些系统具有影响其毒理学特征的独特系统。例如，封闭系统很难篡改，防止皮肤或口腔意外暴露烟液，而且由于瓦数低，只能形成极少量加热产生的有害物质。然而，特别是在美国销售的"豆荚"系统对年轻人构成了更高的风险，因为它们有着更高的致瘾性和吸引力，而且烟碱含量很高[16,18]。

下面讨论与电子烟烟液成分相关的更多风险。然而，产品快速变化使其很难被归纳。除了如"豆荚"系统或低欧姆设备这种差异很大的产品之外，大多数ENDS产品快速多样化也导致了信息的滞后，当研究报告发表的时候，市场已经更新迭代了。风险评估和风险沟通面临的一个日益严峻的挑战是，如何将主要提供另一种烟碱输送方式的产品与含有非法或娱乐性药物或其他生理活性化合物的非常规液体的产品区分开来。尽管对某些产品可能很难进行区分，例如，将 Δ^9-四氢大麻酚（THC）或合成大麻素的任何不良影响归因于ENDS或ENNDS本身是误导性的，即使它们是通过电子传输系统输送的。应考虑限制ENDS将烟碱与其他药理或生理活性化合物结合使用。因为不含烟碱不足以作为定义这一新兴产品类别的标准，因此应对ENNDS及其烟液进行具体定义。而要对含有大麻二酚、THC或其他药物的产品制定替代评估框架。原则上，应考虑用于输送生理活性化合物或烟碱以外药物的产品是否在烟草管制范围内。

基于个体或人群的风险评估还取决于电子烟的使用和吸烟行为，包括双重使用或多重使用（即与卷烟或其他烟草制品同时使用）的流行率（另见第5章）。在美国，尽管目前使用电子烟的成年人吸烟率已经下降，从2015年的56.9%下降到2018年的40.8%[22]，但双重使用很常见，虽然ENDS可以减少有害成分暴露，但如果吸烟者继续吸烟，任何假设的健康益处都将是有限的。即使重度吸烟者（每天>15支）将吸烟量减少超过50%，肺癌的发病率也仅下降27%，其患肺癌风险仍然比不吸烟者高出7倍以上[23]。由于与烟草使用相关的风险仍然存在，逐步用ENDS替代烟草的价值是有限的。不幸的是，由于数据有限，很难对持续的、只使用电子烟的健康影响进行流行病学研究。此外，电子烟的流行率在吸烟者和戒烟者中最高[22]，而且很难将使用电子烟相关的疾病与以前烟草消费的持续影响区分开来。

8.2 使用 ENDS 和 ENNDS 相关的风险评估与量化

定量风险评估是评估卷烟、ENDS、ENNDS或HTP对人体健康影响的有力工具。例如，肿瘤效力建模[4]可以评估与卷烟烟气的单个成分或整个烟气相关的致癌风险，如下所述。建模和风险量化已经证实，卷烟与ENDS、ENNDS或HTP相比有非常严重的不良影响[24]。尽管也可以对ENDS和ENNDS进行建模，但很难涵盖整个产品范围。然而，可以识别出显著增加健康风险的相关危害。监管机构应意识到ENDS类型和类别之间的差距越来越大。因此，管控应针对对风险影响最大的成分、释放物和技术特征。

总体来说，ENDS比传统卷烟释放的有害成分更少，但有两个例外值得注意。第一个是烟碱，这是一种剧毒化合物，添加到电子烟烟液中的浓度高达60 mg/mL，在所谓的"烟碱冲击"中甚至更高，具体取决于司法管辖区。摄入10 mg符合欧盟法规的烟液相当于成人2.8 mg/kg bw（体重70 kg）或儿童摄入量为20 mg/kg（体重10 kg）。估计最低潜在致死剂量为6.5~13 mg/kg体重[25]，婴幼儿的致死剂量可<5 mg/kg bw[26]。由于灌装容器可能含有数百毫克，因此意外中毒的风险很高。开放系统和可灌装容器的使用者面临皮肤暴露的风险，还可能发生意外的口服中

毒，尽管口服吸收烟碱往往受到呕吐效应的限制。2018年报告了一些轻度或中度中毒病例[27]，但致命的中毒事件是非常罕见的[28]。与传统卷烟相比，第二个例外是污染物和有意添加的化合物，包括精油、草本提取物和某些香料，如双乙酰和乙酰丙酰，它们会导致严重的（急性）肺损伤。

符合欧洲、美国和其他国家规定的商业产品不应构成此类风险。已报告了一例肺脂性肺炎病例，可能是由被脂肪油污染的甘油引起的[29]。许多使用者自己制备电子烟烟液，有时还会自己配制配料，可能是因为他们没有意识到风险。此类风险是可以避免的，如果监管得当，则不应用于商业产品。然而，在最糟糕的情况下，使用非常规、被操纵或有问题的产品可能会产生急性和致命的毒性影响，而与吸烟有关的疾病需要数年时间才能发展起来。下面讨论电子烟的典型危害。

8.2.1　与特定成分或无意添加的物质相关的健康风险

儿童或青少年吸入ENDS或ENNDS气溶胶时，直接或间接暴露于有害和潜在有害成分可能会对健康造成的影响概述如下。实际发生的影响将取决于吸入的化合物的数量。

烟碱

吸烟会增加动脉硬化、心肌梗死、中风和其他心血管疾病风险。这不仅是因为烟碱，还因为其他化合物，如一氧化碳、氮氧化合物、金属和颗粒物。发病机制通常与炎症有关[30,31]。烟碱的某些作用，如血压升高和冠状动脉灌注减少，可能会导致心血管损伤。然而，关于ENDS中烟碱对心血管的影响的数据有限。无烟烟草的使用研究表明，没有燃烧产物的烟碱对心血管的毒性低于卷烟烟气[30]。例如，瑞典口含烟没有增加中风或梗死的风险[32]，尽管烟碱被有效吸收，但这还需要进一步的研究[33]。对瑞典和美国的11项研究进行的综合分析表明，无烟烟草使用者心肌梗死和中风的风险增加[34]。INTERHEART对52个国家27089名参与者的数据进行的研究也报告了无烟烟草使用会增加心肌梗死的概率[35]。

现在电子烟可以经过优化，以达到传统卷烟的烟碱水平[15]。在开放系统中，烟碱水平通常也可以随着功率设置而调整，并随着每口抽吸雾化烟液的量而增加。吸入时，雾化烟碱可能导致炎症和血管损伤[36]。然而，发病机制也依赖于炎症辅助因子[31]，例如活性氧，而这些因子是否由电子烟产生目前尚不清楚。怀孕期间吸烟会影响胚胎发育，降低出生体重，并增加早产、死胎或婴儿猝死等并发症的风险。此外，母亲吸烟会损害肺功能和肺发育。然而，又很难将烟气和燃烧产物的不利影响与烟碱的不利影响区分开来。据报道，烟碱通过收缩脐带和子宫的血管来限制胎儿获得氧气[37]。产前暴露于烟碱会干扰大脑的发育[38]，并与神经行为损伤有关，包括过度活跃、焦虑和认知功能损伤。在一项对大鼠的研究中，烟碱对神经系统有致畸作用[39]，其他实验研究表明，在发育过程中暴露于烟碱会产生长期的不良影响，如生育能力下降、高血压、肥胖和呼吸功能障碍[40]。

烟碱可能引起其他不良反应，例如胰岛素抵抗，从而增加了患2型糖尿病的风险[41]，尽管这个问题还需要进一步的研究。有人认为烟碱可以抑制肺中的黏膜纤毛的清除作用[42]，这可能会增加对有害物质的暴露。

ENDS和ENNDS的有害成分和释放物

甘油和丙二醇。甘油和丙二醇是最常用的雾化溶剂，也是电子烟液的主要成分。虽然吸入后有轻微的副作用，如刺激性[43,44]，但这两种化合物都被认为是相对安全的。然而，每口抽吸雾化烟液的量是不同的，并且在低欧姆设备中非常高[45]。目前还没有报告根据气溶胶的密度或吸入物质的总质量来比较电子烟气溶胶的毒性特性，特别是长期而言。其他溶剂如乙二醇已被使用[12]，它具有较高的毒理学风险[46]。然而，目前尚不清楚市场上是否存在包含其他雾化剂作为主要成分的ENDS。

香料和其他成分。有些香料，包括双乙酰，令人担忧，因为它会导致闭塞性细支气管炎，这是一种罕见但严重的肺部疾病。虽然早在5年前，在许多地区禁止在电子烟中使用双乙酰，但在接受测试的烟液中仍检测到了很大比例的双乙酰[11]。甜味剂三氯蔗糖是一种卤化二糖，当设备温度达到200℃以上时，可以在电子烟液中降解，这会产生潜在的有害有机氯[47]。普通电子烟液中的其他化合物可能包括敏化剂，例如，柠檬香浓缩液中发现的氧化柠檬烯[13]，被认为是一种重要的接触过敏原[48]。已有罕见的过敏性肺炎病例报告，可能与使用ENDS和ENNDS有关。然而，目前还没有在电子烟液中鉴定出特定的过敏原[49]。2019年，英国又报告了一例过敏性肺炎[50]。这一病例与电子烟和雾化相关的肺损伤（EVA-LI）无关（见HSP2），但表明吸入电子烟气溶胶可引起过敏反应，尽管这种现象很少见。因此，应阐明这些病例的原因，并监测未来的发展。电子烟烟液的种类越来越繁多，应该被视为潜在危险，因为许多成分都会产生热降解产物并进行化学转化。≥250℃的线圈温度促进了不确定的化学反应[51]。

羰基化合物和热降解产物。羰基化合物，包括甲醛、乙醛和丙烯醛，被认为是ENDS商业用途中最相关的有害释放物。它们源于甘油和丙二醇的降解，具体取决于设备温度。在低功率设备中，羰基化合物主要产生在干吸的条件下，因为在没有烟液的情况下电线会过热[12,52]。在高瓦数的设备中，羰基化合物是根据所应用的功率形成的。Talih等[45]证明了在大功率单线圈常规ENDS和低欧姆设备中羰基化合物形成增强，这与线圈表面和消耗的烟液量有关。研究表明，每15次抽吸的总羰基化合物释放量≤400 μg。气溶胶中的羰基化合物含量可以接近传统卷烟的含量，但变化很大。例如，它的含量可能会被调味品极大地增加[53]，抽吸模式可以解释甲醛水平变化，从20 ng/口到255 ng/口[45]。一般来说，ENDS和ENNDS中的羰基化合物含量从几乎无法检出到每口几微克。乙醛不仅有致癌和其他危险的特性，而且通过抑制单胺氧化酶增加烟碱的致瘾性[54]。因此，适度的羰基化合物含量可能会使产品更具吸引力，从而提高消费者的满意度。

与不吸烟的青少年相比，成年吸烟者不太可能会使用JUUL作为卷烟的替代品，尽管这种设备可以提供高水平的烟碱[55]。将瓦数增加到可变的、更高水平的趋势可能部分是为了增加使用者的满意度，因为更高水平的羰基化合物可以增强烟碱的效果及其致瘾性。应该对不同组的ENDS和ENNDS产品进行毒理学评估。对于在≥15 W条件下工作的ENDS，可以考虑作为一个特定类别，因为这些设备比低功率设备产生更浓的气溶胶和更高水平的有害物质。有报道称，高功率ENDS和ENNDS会增加肺损伤的风险，包括暂时性炎症，并干扰气体交换[56]。监管机构应该意识到，对高容量开放式系统的ENDS和ENNDS很难设定成分含量的上限，因为可以通过切换到更高的功率进行补偿[57]。例如，当更大数量的烟液被雾化时，烟碱的摄入量会增加，从而增加健康风险。

金属。ENDS和ENNDS的气溶胶中出现的金属通常来自设备本身[58]。水平低于毒理学关注的限值[59,60]，但在高功率设置下会增加[61]。由于氧化还原活性金属可能会增加活性氧的水平，心血管风险可能会增加，特别是在烟碱水平高的情况下。Haddad等[62]表明，高功率的ENDS可以产生与燃烧型卷烟相当的活性氧。

总之，危害分析证实，烟碱、羰基化合物和金属是危险相关因素，这意味着使用电子烟可能增加患心血管疾病的风险。高烟碱含量、高功率和低质量标准的应用也影响到释放物中的金属水平。由于气溶胶可能含有刺激物，还应评估呼吸道疾病的风险。ENDS产生的致癌物水平低于典型的烟气致癌物水平，但羰基化合物除外[63,64]。这一发现最近在一项企业赞助的研究中被证实，为低欧姆和高功率设备[65]，但应该在独立研究中得到证实。

8.2.2　ENDS和ENNDS的潜在健康影响

吸烟可能会增加动脉硬化、心肌梗死、中风和其他心血管疾病的风险。危险因素包括许多化学物质，如一氧化碳、氮氧化物、金属和颗粒物。发病机制通常与炎症有关，一些烟碱相关的影响，如血压升高和冠状动脉灌注减少，可加重心血管损伤[30,31]。然而，对于使用电子烟是否会增加心血管风险，仍然存在一些争论。一项关于电子烟与心血管疾病研究的系统综述表明，电子烟可能会增加血栓和动脉粥样硬化的风险[66]。

短期暴露于电子烟释放物会引发血管氧化应激和功能障碍[67]。即使是使用一次无烟碱电子烟，也会对内皮功能产生短暂的影响[68]。在一项荟萃分析中，Skotsimara等[69]发现电子烟使用与内皮损伤、动脉僵硬和冠状动脉事件的长期风险之间存在关联。然而，这些发现是在单个研究中得出的，并没有在其他研究中得到证实。对血管功能的短期影响不一定发展为临床相关疾病。对2016年行为风险因素监测系统收集的数据进行分析显示，在7万名报告使用电子烟的受访者中，中风、心肌梗死和冠状动脉疾病的风险大幅增加[70,71]。此外，国家健康访谈调查（2014~2016年）的分析证实，在调整吸烟和其他风险因素后，心肌梗死的风险更高[72]。对2016年和2017年收集的风险因素监测系统数据的汇总分析证实从不吸烟的电子烟使用者心血管风险增加[73]。目前还没有关于可能增加这种风险的设备的具体数据。关于电子烟对心血管和一般健康影响的临床前研究结果和流行病学证据的全面审查尚无定论，需要更多数据。作者得出结论，尽管认为电子烟比燃烧型烟草制品危害更小是合理的，但不吸烟比使用电子烟更好[70]。一份报告显示，卷烟抽吸者转向电子烟后，内皮功能和血管健康得到了显著改善[74]。

其他健康影响包括呼吸道，主要是上呼吸道刺激[75]。对肺功能的影响也有报道，电子烟使用者的肺功能比非使用者低，可能会增加肺阻力[76,77]。此外，电子烟使用者对感染的反应可能会降低[78]。对2016~2017年行为风险因素监测系统数据的分析表明，电子烟增强了肺毒性。呼吸道疾病增加也已得到其他方面的证实[79]。另一项研究表明，吸烟的慢性阻塞性肺疾病患者在完全转用电子烟后的健康状况有所改善，包括更好的效果和逆转了吸烟造成的一些危害[80]。这说明了电子烟健康风险对于吸烟者和非吸烟者的不同观点。有报道称电子烟的使用与哮喘之间存在联系。然而，在成年哮喘患者组别中，电子烟使用的高流行率可能与尝试戒烟有关[73]。

2019年夏季，美国报告了一系列与使用雾化器或电子烟产品相关的严重肺部损伤病例（另见第12章）。与电子烟使用相关的病例事件数量增加，并报告了无肺部疾病史的死亡[81]。这种疾病被命名为"电子烟和雾化器相关肺损伤"（EVALI），主要局限在美国。截至2020年2月18日，已报告2807例住院病例，包括68例EVALI相关死亡[82,83]。患者呈现相似的临床特征，例如呼吸困难、咳嗽、低血氧，胸部影像学表现为双侧混浊[84]。由于这些症状与其他呼吸系统疾病类似，这些病例与使用特定类型的电子烟没有直接联系[85]。许多（但不是全部）患者都使用过含有四氢大麻酚的烟液，进一步研究表明，在含有四氢大麻酚的电子烟液中加入维生素E醋酸酯作为增稠剂可能是呼吸道损伤的原因[86-88]。研究发现，呼吸系统疾病患者比非患者更经常使用含有四氢大麻酚的电子烟液，更重要的是，据报道，他们更经常从非正规渠道获取产品[89]。然而，也有病例发生在不使用含四氢大麻酚烟液的患者中。患者在吸入含维生素E醋酸酯后几天至几周内出现呼吸系统后果，这有助于确定疾病的原因。然而，要建立与长期健康影响之间的联系则困难得多。在一项证实了这种关联性的研究中[88]，从EVALI患者收集的产品中发现了极高水平的维生素E醋酸酯。报道浓度为31%~88%，而THC含量常常低于广告[90]。需要注意的是，这些产品与普通电子烟几乎没有共同之处。风险评估和公共教育都是必要的，以应对电子烟日益激增的非常规且往往是非法地使用电子烟来递送烟碱以外的有害物质。风险评估将受益于这些产品和常规ENDS之间的明确区分，但在实践中很难实现。

8.2.3 非使用者的风险

非使用者可能会暴露于使用者使用ENDS、ENNDS或HTP后呼出的释放物。他们的实际暴露很大程度上取决于房间的大小和通风情况。由于暴露于丙二醇和甘油，非使用者可能会受到呼吸道刺激。如果使用含有烟碱的电子烟烟液，还包括心悸和收缩压升高，也可能会出现烟碱的全身效应。由于某些电子烟烟液中存在烟草特有亚硝胺，在最坏的情况下，非使用者可能会增加肿瘤的风险[91]。另一项研究得出结论，电子烟使用对非使用者的健康不太可能造成威胁[92]。对于非使用者暴露于电子烟和传统卷烟的研究表明，与二手电子烟气溶胶相关的健康风险低于二手卷烟烟气相关的健康风险[93,94]。其他研究表明，尽管非使用者会从二手烟中摄取烟碱，但使用电子烟时空气中烟碱的浓度比传统卷烟低得多[95,96]。与使用电子烟相比，使用HTP会导致非使用者暴露更多[97]。非使用者暴露于二手HTP或电子烟释放物是否会产生不利影响，在很大程度上尚不清楚。一项研究表明，二手电子烟暴露与过去12个月报告的哮喘发作之间存在关联[98]。这表明，呼吸道疾病患者和幼儿等人群，更容易受到二手烟释放物的不利影响。

8.3 量化风险的方法

烟草制品不仅种类不同，而且在释放物和使用方面也不同，从而造成不同的健康风险。即使是同类产品，如ENDS和ENNDS，对健康也有不同的影响。希望转换到潜在危害较小的产品的传统卷烟吸烟者需要这类产品的相对风险信息。对吸烟者潜在危害较小的产品对非吸烟者的危害更大。定量风险特征（包括剂量或浓度-反应关系）可用于确定在人口水平对健康的潜在影响，前提是有关于使用者人数及其使用模式的信息。然而，鉴于某类产品之间和使用者之间的差异，概括烟草制品的健康风险在科学上是不合适的[99]。对ENDS和ENNDS的风险评估应该是对于一组设备，甚至单个设备或烟液分别进行。在估计人暴露于电子烟的情况时，还必须考虑广泛的电子烟使用参数。

量化化学混合物的风险在本质上是困难的，而新型烟碱和烟草制品的出现增加了产品种类、成分和使用多样性的挑战。产品之间的差异使得暴露和危害的量化不确定。为了测量暴露，需要了解电子烟使用模式、吸入的释放物以及（根据使用的方法）颗粒物的大小和分布，这些信息决定了呼吸道中的沉积物，从而决定了局部暴露和影响[100,101]。还需要提供有关释放物的化学成分的信息，以确定

8.电子烟碱/非烟碱传输系统和加热型烟草制品相关个体风险的量化方法探索：对人体健康和监管的影响

使用者暴露的化合物。烟草制品的释放物包含许多不同的化合物，这些化合物取决于电子烟使用模式和设备设置，如温度和功率[62,102]。为了评估这些产品的风险，必须对释放物中的化合物进行表征和量化[9]。遗憾的是，仅关于成分的信息是不够的，因为它们可能在雾化过程中降解或燃烧，设备也可能会带来风险。

关于危害的信息可从化合物的毒理学研究中获得，最好是通过吸入给药，这是ENDS、ENNDS和烟草制品最相关的暴露途径。无法获得这些产品中所有化合物的危害信息。电子烟液通常含有调味品，已有经口服给药测试其用于食品的毒性，但没有关于吸入产生的毒性的信息。这些信息是必要的，因为在毒性研究中，一些化合物，如双乙酰和肉桂醛，吸入时可对呼吸道产生局部影响[103,104]。此外，ENDS、ENNDS和HTP产品的使用者暴露于可能发生或不发生生物相互作用的混合物中。

与混合物相关的风险可以通过结合单个化合物的危害数据和产品释放物中存在的数量来量化。另外，可以在毒性研究中逐个病例量化ENDS、ENNDS或HTP的风险。然而，实验动物研究可能无法为评估烟草制品对人的危害提供有意义的结果。此外，实验研究中的暴露时间一般为6小时/天，5天/周，这与使用ENDS、ENNDS或HTP的7天/周的不规则峰值暴露无法相比。替代模型（如细胞模型）的开发和使用正在迅速增加，但还不能对混合物进行危害特性表征。

已有对混合物的毒理学效应进行的一些研究[105]。在2019年欧盟完成的Euro-Mix项目（https://www.euromixproject.eu/）中[106]，根据化合物的靶器官和作用机制进行了分类。具有相似和不同作用模式的化合物在针对靶器官的特异性测试中进行评估，以确定混合物的效果是否与单个化合物的效果不同。该项目形成了一个实用指南手册，概述了暴露场景和测试策略工具箱。与EuroMix项目一样，大多数混合物毒性研究都设计二元组合。然而，即使对于这些相对简单的混合物，在没有至少一些实验数据输入的情况下，也没有可靠的方法来定量预测混合物中两种化合物任何组合的效果。很难预测混合物中的化合物是否会相互作用，这可能会被生物反应的变化所掩盖。二元混合物通常具有相加效应，因此可以将单个化合物的效力调整剂量相加，以预测它们的联合效应。数学模型可用来确定混合物的毒性是否来源于化合物之间的相互作用。常用的两种数学模型是剂量-浓度加和模型和独立作用-反应加和模型[107-109]。剂量-浓度加和模型基于这样的假设：所有化合物都具有类似的作用模式，但可能具有不同的诱导效应的效力。一旦确定了一种化合物相对于另一种化合物的效力，这两种化合物的浓度就可以相对于其中一种化合物的浓度来表示并作为参考，当浓度相加时可以用来确定二元混合物对参比化合物剂量-效应曲线的影响[110]。在浓度加和模型中，假定有类似的作用模式，而独立作用模式可以用来确定由于不同机制或作用模式而产生的化合物效应[111]。数学模型可用于确定化合物是否有相互作用，是协同作用或拮抗作用。

这些模型不能用来预测烟草烟气等复杂混合物的影响。

相较于其他复杂化合物，如石油衍生产品和水泥类产品的评估来说，烟草制品、ENDS和ENNDS释放物的风险评估更为复杂。一般来说，评估此类产品风险的方法是评估具有代表性的混合物样本的整体毒性，主要是对实验动物的毒性。类似的方法可用于评估ENDS和ENNDS的危害，并区分产品类别。虽然实验动物研究在伦理上是有争议的，但各种烟草制品已经被测试。然而，将结果转化为人类效应并评估一个类别内的不同产品仍然是一个挑战。

8.3.1 毒理学关注的阈值

评估暴露于复杂混合物的一种方法是毒理学关注阈值（TTC）[112]。这种方法最初是为评估致癌性而开发的，通过结构-活性关系和交叉参考（read-across）来识别的混合物中毒性最大的化合物。TTC值（以µg/人或每天µg/kg bw为单位）已根据结构要素定义为三个级别（Cramer Ⅰ~Ⅲ级），但仅在口腔暴露后。Cramer Ⅲ级表示最高致癌风险，因此TTC值最低[113]。然后，通过比较暴露于这些化合物的风险（单独或合计）与适当的TTC值来评估混合物的风险。该方法已被应用于复杂混合物，如植物提取物[114]、香料混合物[115]和吸入有害成分[116,117]。该方法并不表明对健康的风险，但表明如果化合物超过TTC阈值，则需要进一步测试。否则，存在健康风险概率较低。当没有可用的危害性数据时，可使用该方法。众所周知，烟草制品对健康有不利影响，因此不能使用TTC风险评估方法来量化健康影响。然而，对于ENDS和ENNDS，它可能被用来识别构成健康风险的化合物，并对它们优先进行进一步的测试。

8.3.2 基于单个化合物的风险评估

有关单个成分暴露和危害的信息可用于评估产品的整体风险。对于卷烟，可以根据其潜在的危险选择化合物[6,118,119]。对于ENDS，释放物的化合物数量是有限的，而且它们不一定与已知的烟草有害成分重叠。对于ENDS和ENNDS，还应考虑热降解产生的化合物，如醛类。该方法已用于评估电子烟对使用者和非使用者的有害性[91]。这种方法中所使用的危害数据来自为设定安全暴露水平而进行的研究。在烟草制品的释放物中，某些情况下也包括来自ENDS和ENNDS的释放物，其浓度通常高于安全暴露水平，因此，这些信息可以用作潜在关注的指标，但不能用于量化风险。在实验动物身上看到的影响可能不会直接反映在人类类似暴露水平上，因为实验研究中的暴露模式与人的暴露模式不同。为了估计复杂混合物的风险，而相加单个化合物的风险可能会导致低估健康风险，因为混合化合物之间的相互作用被忽略了。由于许多实验研究的设计都是基于单个化合物的风险评估而不能用于比较不同（烟草）产品的风险。为了比较影响的严重程度，有必要提供详细信息，说明暴露与健康影响之间的关系，以及如何将其外推到对人类的影响。

8.3.3 相对风险方法

还开展了评估烟草制品作为整体和相对于参比卷烟的致癌效力的研究[3,24]。在这种方法中，通过使用暴露水平与诱发肿瘤数量之间的模型线性关系，使用啮齿类动物致癌性研究的数据来确定化合物的致癌性。例如，基准剂量方法可用于确定导致肿瘤发生率为10%的有效剂量的置信区间下限（通常为95%）。可根据技术指南文件，如美国加利福尼亚州环境健康危害评估办公室发布的技术指南文件[120]，使用致癌效力定量数据来计算每种化合物的相对致癌效力。混合物或气溶胶的总相对致癌效力可以通过相加单个化合物的值来计算。在这种方法中，假定在整个剂量范围内的相对效力因子是相等的，其机制是可比的。虽然致癌物的作用机制不同，但结果可能表明致癌风险，并可用于比较不同产品的致癌风险。在方法有效期间评估假设的有效性。

这种相对风险方法取决于是否有关于释放物和致癌性的数据。需要尽可能多的有关化合物释放的数据，因为包含更多的化合物可以改进相对风险的计算。用于分析两种产品的释放物的方法应该是相似的，必须有关于释放物变化的信息，以确定相对风险的不确定性。关于致癌性的信息应从啮齿动物研究中获得。或者，如果有关于剂量-反应的数据，也可以使用其他来源的数据，这些数据是致癌效力的指标。目前，这种方法似乎是量化产品相对风险的最佳方法。但是，对于系统应用，应提供更多关于释放物和危害水平的数据。类似的方法可能适用于其他健康问题，如心血管疾病或慢性阻塞性肺病。

8.3.4 暴露边界值法

暴露边界值法是基于暴露水平与未发生影响的剂量或出现预先确定的不良影响剂量（例如基准剂量水平）的比例。化合物暴露的边界值越大，风险可能越高。该方法已应用于烟草制品中的化合物[121,122]，用于确定烟气释放物中应降低的化合物的优先顺序，并评估ENDS和ENNDS释放物中的单个化合物。根据有关吸入的释放物的数据和有关危害的信息计算每种化合物的暴露边界值，并且暴露边界值取决于数据的质量。该方法不能产生定量风险特征。其主要目标是确定某一特定暴露是否值得关注。暴露边界值的大小并不是衡量风险的标准。

8.3.5 非致癌作用

许多方法涉及致癌效应的风险评估，这比非致癌效应更容易建模，因为癌症终点更容易比较，致癌物的剂量-反应曲线也是可比较的。从本质上讲，非致癌作用是多种多样的。应区分局部效应和全身效应，不同的化合物对不同的器官有作用，即使化合物对同一器官有作用，其机制和结果也可能不同。非致癌效应的量化可能必须在作用机制层面进行[106]。然而，这种方法需要的有关化合物作用机制的剂量-反应数据有限，但已为卷烟烟气对心血管和呼吸系统的影响生成了非癌症风险指数[4]。

或者，可使用释放物的毒理学分析来表征产品的风险。如上所述，实验动物的生物测试可能不是优选的，并且细胞测试的结果很难转化为对人的影响。此外，不仅应将效应（read-out参数）外推到人体效应，而且暴露应该类似于人体暴露。这包括抽吸轮廓，在肺部模型的情况下，还包括呼吸道中的沉淀。该领域正在发展，但用于表征ENDS、ENNDS和HTP非致癌作用的体外分析仍在研究。

8.3.6　评估框架

可以根据适当的评估框架评估ENDS和ENNDS对健康的影响。在这种方法中，专家判断被用来对产品的各个方面进行评分，以识别最重要的风险，例如有害物质[123]。荷兰国家公共卫生和环境研究所建立了一个烟草制品评估框架，总结了可能影响产品吸引力、致瘾性和有害性的所有因素，可用于确定知识差距或对特定产品进行优先研究[124]。这些模型允许在数据有限的情况下对产品进行评估。随着对产品知识的增加，评估也会得到改善，但这种方法仅限于基于专家判断的定性结果。

上述方法可用于比较不同产品的危害。表8.1总结了这些方法、数据要求及其应用。应用这些方法的可行性取决于这些信息是可测的，而不是产品本身提供的。目前还没有全面风险表征所需的信息。在某些方法中，对于不同质量的数据可以采用加权概率法。许多风险量化方法还需要释放物数据，这是人类暴露的指标。烟气和ENDS或ENNDS的释放物是动态的。当释放物从加热元件传到设备出口时会冷却，导致挥发性化合物的冷凝和颗粒物的凝聚。此外，吸入的空气（包括释放物）在上呼吸道中湿度增加。这些进程同时发生，并决定了呼吸道中的局部沉积，这可能会导致呼吸道中特定位置出现高剂量，并产生特定部位的不良反应。正在对烟草烟气和ENDS或ENNDS释放物的呼吸道沉积进行建模[100,125]。定性风险评估的结果取决于剂量学质量，而剂量学质量是有限的。

表8.2总结了本章描述的方法的局限性和优点。应该注意的是，所有方法都是为了评估对使用者的风险，如果有关于非使用者暴露的信息，也可以使用类似的方法来评估风险。

表8.1 根据单个化合物或其混合物（1）、所需数据（2和3）及其应用（4~7）对ENDS和ENNDS进行定量风险评估的方法总结

方法	1.是单个化合物还是混合物？	2.是否取决于释放特性？	3.是否依赖于可用的危害性数据？	4.是否允许量化单个化合物的风险？	5.是否允许量化产品的风险？	6.是否允许量化产品作为一个整体对弱势群体的风险？	7.改良产品所需数据
毒理学关注阈值	单个化合物	是	否	否，允许识别潜在的有害化合物	否	否	释放物数据
基于单个化合物的风险评估	单个化合物	是	是	是	是，但只有在所有化合物的数据可用的情况下	是，但前提是所有化合物的数据和关于弱势群体的具体危害性数据都是可用的	释放物数据
相对风险法	混合物	是	是	是	是，但只有在所有化合物的数据可用的情况下	是，但前提是所有化合物的数据和关于弱势群体的具体危害性数据都是可用的	释放物数据
暴露边界值法	单个化合物	是	是	是	否	否	否	释放物数据
非致癌作用的生物测试	混合物	否	否	否	是	是，但前提是有关于弱势群体危害的具体数据	毒性生物测试
评估框架	混合物	否	是	否	否，只有非定量风险	否，只有在弱势群体的数据可用的情况下才有非定量的风险	重新评估，例如释放物数据

注：第一列，该方法输出对单个化合物或作为混合物释放物的适用性；第二列，对释放物（理想情况下）中所有化合物的定量数据要求；第三列，与剂量-反应相关的危害性数据的要求；第四列，量化暴露释放物中一种化合物风险的方法适用性；第五列，量化暴露产品风险的方法适用性；第六列，量化易感人群（如婴儿）暴露产品风险的方法适用性；第七列，量化轻微变化的产品风险所需的数据，如电子烟烟液中的新口味或设备的技术改变

表8.2 量化ENDS和ENNDS健康风险的每种方法的局限性和优势

方法	主要局限性	主要优势	对ENDS、ENNDS和HTP的潜在应用
毒理学关注阈值	无法评估完整的产品风险，无法量化风险	有关暴露的可能的风险信息	为进一步测试确定化合物的优先级
基于单个化合物的风险评估	无法评估完整的产品风险	能识别具有最高健康风险的化合物	基于现有数据的健康风险评估
相对风险法	目前只针对致癌物	允许比较产品之间的风险	根据现有数据进行健康风险评估，并允许对产品进行比较
暴露边界值法	无法评估完整的产品风险	与健康有关的暴露信息	为进一步测试确定化合物的优先级
非致癌作用的生物测试	需要进行广泛的测试，并外推至人的暴露和结果	不需要释放物或危害性数据	基于现有生物分析的健康风险评估
评估框架	最主观的方法	需要有限的数据；更多的数据将改善结果	非定量健康风险评估，可用于确定优先事项

8.4 加热型烟草制品

用于评估ENDS和ENNDS风险的方法原则上适用于加热型烟草制品（HTP）。目前，HTP变化性小于电子传输系统，并且可以获得一些关于气溶胶成分的可靠、独立的数据[9]。企业数据必须经过独立核实，才能用于风险评估。之前的调查针对的是卷烟烟气中常见的有害成分，需要进一步的独立研究来全面分析释放物和健康风险。用于分析卷烟烟气的方法和标准化的吸入模式仍然必须采用，来获得使用者代表性的测量结果。此外，一些有害成分可能与卷烟烟气无关，但可能会出现在HTP气溶胶的优先级中。这个问题已经通过非靶向筛查方法[126-128]得到了解决，企业科学家也使用了这些方法[129]。一些关注成分，包括缩水甘油（国际癌症研究机构工作组将其列为2A组）和糠醇（被分类为2B组）已经在HTP气溶胶中被识别出，特定HTP的数据可用于进行风险建模和比较或定量评估。Stephens[24]对卷烟、ENDS和一个HTP装置产生的气溶胶的致癌力进行了建模，并改进了比较建模方法[3]，以确定单个化合物和产品释放物的相对癌症效力及置信区间。然后可以用概率方法计算两种产品累积暴露的比例或变化。对于HTP，所选化合物的累积暴露变化比卷烟烟气低10~25倍。有了人体剂量-反应的相关信息，累积暴露的变化就可以转化为每个装置的相关健康影响。这种方法最初用于HTP气溶胶和卷烟烟气中的8种致癌物，但应该扩展到HTP气溶胶中比卷烟烟气含量更高的化合物。

这些计算说明了HTP气溶胶和卷烟烟气成分中可能会影响健康风险的组成差异。随着HTP的不断发展，关键有害成分的性能标准和上限应被认为是最有用、最可行的监管选择。上述ENDS和ENNDS的定量风险评估方法是以释放物和危害性数据为基础，因此，如果数据可以测量到，那么这种方法也适用于HTP。

8.5 对监管的影响

目前关于ENDS、ENNDS和烟草制品健康风险的量化方法还不足以用于监管。然而，可以使用一些方法来获得产品对人体绝对健康风险的表征。这些方法是基于与释放物中特定或独特化合物相关的风险，这取决于数据的可用性。应识别释放物中的化合物并量化，最好是在代表性使用者环境中。此外，需要这些化合物的人体毒理学数据来量化健康风险。然而，这两个参数都缺乏数据。本章描述的风险评估方法可考虑用于监管。然而，目前由于缺乏数据，还达不到这一阶段，因为只有当模型和数据的质量都满足要求才行。

健康风险量化可以用在人群水平的健康风险评估模型中。虽然还没有这样做，但已经探索了对人群健康影响进行建模的可行性[130]。在监测ENDS、ENNDS或HTP的流行和使用情况时，可评估其对吸烟者、非吸烟者和既往吸烟者的健康影响，并可将其作为限制使用该产品的理由或作为公众教育的基础。输出结果在很大程度上取决于输入数据，在这种情况下，输入数据还将包括流行病学数据。在应用人群模型时，应考虑到有关产品使用和产品间转换的流行病学数据不一致的观察结果。

建模表明，从卷烟转向HTP可能会影响人体健康。然而，其效果取决于设备是否明显降低了先前已证明的释放物中有害成分的水平。监管机构应该考虑对羰基化合物、一氧化碳和其他关键有害成分设定强制性上限，以确保设备符合技术和性能标准。

有效监管ENDS和ENNDS及其特性也有助于限制其健康风险。应优先考虑不容易被消费者改装的封闭系统，如"豆荚"类装置或封闭的烟弹。应禁止有问题的添加物和成分（如双乙酰、三氯蔗糖、精油和所有致癌、致突变和致畸化合物），还可以考虑设定封闭系统ENDS释放物中的烟碱上限。虽然一般情况下，对有害化合物的禁令应适用于烟液，但可灵活操作、高功率的开放式系统的释放物很难监管。因此，重点可能会转移到技术特性上，例如过热控制、最大瓦数或温度。进一步监管将需要对ENDS和ENNDS子类别进行详细评估，特别是高功率产品，包括气溶胶化学、毒理学以及设备的设计和性能。

ENNDS的模糊术语与定义也是令人担忧的问题。例如，不含烟碱的设备可以用来吸入大麻二酚等化合物，人们声称大麻二酚对健康有一些好处。然而，认为某些ENNDS可能有益于健康的印象可能会分散和混淆当前的评估，应该尽量避免。应禁止ENDS将烟碱与其他药理或生理活性化合物结合使用，因为这可能增加烟碱消费的吸引力。应规定ENNDS的术语和定义，以将其确定为与ENDS产品相同但不含烟碱的产品，且应该同时适用于设备和消耗品。"ENNDS"一词仍将包括传统的无烟碱电子烟烟液。ENDS和ENNDS的监管框架可能仍会包括其他电子吸入系统的产品，从而同样地定义所有设备。然而，出于几个原因，将烟碱和烟草替代品与其他电子吸入产品明确分开是有益的。第一，可以对用于吸入其他物质或材料的ENDS、ENNDS和其他电子吸入系统采用特定规则。第二，需要特定的风险信息说明，以防止产生误导性结论，比如在EVALI事件早期所观察到的那种情况。第三，将来监管机构在处理新出现的吸入系统时可能会获得更大的灵活性。

应当指出，健康风险的量化不是一种静态结果，而是基于可获得的知识的评估。关于ENDS和ENNDS的信息正在增加，这对ENDS和ENNDS来说可能更加重要，因为设备、使用者设置和电子烟烟液种类繁多，会影响健康风险。向公众提供关于ENDS和ENNDS对健康影响的信息是一项挑战。很难向公众传达随着时间的推移因设备或信息的差异而发生的相对风险的改变。此外，企业可利用有关健康风险信息不适当地推广替代性产品。

本章描述的几种方法有望用于评估ENDS、ENNDS和HTP的风险，尽管可能需要多个模型来进行全面评估。目前，还没有足够的科学数据来进行最终评估。由于许多方法需要大量的危害和暴露数据，我们可以通过收集这些数据并进行标准化的分析来为未来做准备，以便将结果输入数据库以供未来使用。非靶向筛选可以用来识别特定产品的化合物，它们的危害性可以从数据库中获得。关于人体实际暴露与发生不良反应之间相关性的信息对于风险表征是必要的。风险评估模型的开发应该继续进行，并且在某些时候，应该用人体数据进行验证。在风险评估中还应开发呼吸道沉积模型，因为这是释放物量化和危害特性之间的关键步骤。最终，只需要对一种新产品进行化学分析，并结合沉积模型和风险评估，就可以确定对健康的影响。

8.6 讨 论

因为暴露和产生影响之间的时间很短，使用ENDS或ENNDS与急性影响（短期健康风险）之间的因果关系通常更容易确定。在许多情况下，当使用者停止使用该产品时，不利影响会逆转。长期电子烟使用者的健康影响数据将有助于对ENDS和ENNDS健康风险的评估。遗憾的是，目前还没有这方面的数据，因为电子烟使用还没有足够的时间产生癌症等慢性健康影响。此外，目前ENDS和ENND使用者往往是以前的卷烟使用者。因此，如果ENDS和ENND使用者患病，可能是抽吸卷烟的延迟效应，而不一定与ENDS和ENND的使用有关。评估健康风险最可靠的数据是不曾抽吸卷烟的ENDS和ENND使用者的数据。各种各样的设备和电子烟烟液持续快速发展，而且目前使用产品的信息缺乏，使人们无法得出关于整个ENDS和ENND产品类别构成的风险的结论。随着此类产品在市场上存在时间的延长，或如果更多的非吸烟者开始使用ENDS和ENND，那么长期数据和非吸烟者信息的缺乏可能会随着时间的推移而改变。预计产品的变化不会停止，相反，越来越多的产品将进入市场。和其他国家一样，在荷兰的市场上已经有超过两万种不同的电子烟烟液可供选择。如前所述，一个务实的做法是确定电子烟液或设备设置对健康有不利影响的化合物。单个化合物的健康风险可以表征产品整体的健康风险，可以为有关许可成分、设备设计特性和某些成分水平的政策提供信息。

风险建模、流行病学研究和对单个化合物的评估已经使一些专家达成共识，即未掺杂质的ENDS的危害比传统卷烟小[131]。然而，正如WHO所承认并总结的那样，仍然存在一些健康风险[10]。虽然定量评估仍然很困难[70]，但是，正如模型所证实的那样，使用ENDS可以显著降低致癌风险[24]。然而，ENDS与烟碱致瘾、呼吸道疾病和其他不利影响的风险增加相关，尤其是对儿童、青少年和非吸烟者。在美国，使用"豆荚"系统的人数迅速增加令人担忧[16]。然而，对于那些使用ENDS代替卷烟的吸烟者来说，潜在的好处也应该被考虑。例如，换用ENDS后，慢性阻塞性肺炎患者的病情有所改善[80]。这里有一些证据表明ENDS对戒烟有用[132]，但作为一个普遍的结论是不科学的，因为ENDS和ENNDS是各种各样的、不断变化的产品[99]。健康风险信息交流应避免不准确的概括，特别是因为产品的多样性及其可能的影响。

在风险评估和立法方面，ENDS和ENNDS仍经常被作为一个类别的产品。然而，监管机构和ENDS使用者应认识到风险的高度差异性，风险范围可能从非常低到与设备温度≥250 ℃的HTP相当。即使是单个化合物的健康风险也是高度可变的，这取决于设备和性能设置。因此，应将各主要种类（即"豆荚"系统、低欧姆）进行分类，以确保在风险评估和监管方面有更具体的术语范围。

导致形成有害化合物的有害添加成分或设备设置也可以进行。虽然从健康角度来看这是非常有益的，但吸入途径电子烟液组成成分的暴露毒理学信息非常有限。根据毒理学信息管制或禁止特定成分可能是一个有用的选择。如上所述，需要灵活的策略来覆盖不同产品组、开放系统和封闭系统。重要的是，仅对烟液的限制并不意味着其他成分是安全的，特别是它们的吸入风险还不知道。

此外，应根据其药理或生理作用和反应将ENDS和ENNDS与用于传递大麻二酚和其他物质的吸入装置区分开来。例如，非常规烟液成分和可使油或蜡雾化或用于吸入非法药物的改装装置所造成的健康风险和伤害，不能被视为电子烟或雾化器使用后的不良后果。应禁止在ENDS中将烟碱与其他生理或药理活性化合物结合使用，因为这可能增加烟碱的吸引力。烟草控制应包括根据这些新挑战而定义和规范ENDS和ENNDS的评估框架。

8.7 建 议

混合物的风险评估非常重要，不仅对ENDS和ENNDS如此，对烟草制品也是如此。有几个模型可以用来评估混合物的风险，尽管大多数模型用于表征致癌效应。缺乏数据限制了模型的实行，这也意味着这些模型还不能对人使用的数据进行验证。下面列出了一些基于本次审查结果的建议：

- ■ 为应用定量风险评估的方法，应收集关于ENDS、ENNDS和HTP对暴露人群的释放物、有害性、使用和影响的数据。
- ■ 有害成分的特性应包括非靶向筛查方法，以识别通常在烟草气溶胶中测量不到的产品特有化合物。

- 对长期不良反应和产品转换应进行适当的实验动物研究和流行病学研究。
- 应核实用于证明任何声称HTP及其营销具有积极健康影响的模型。

每个产品和产品的变化可能会导致风险的变化。此外，使用者可以根据他们的需求灵活地调整系统，这可能会改变他们在风险评估中使用的暴露程度，并改变他们的健康风险。其他供监管机构考虑的建议如下：

- 限制产品的变化。
- 监管机构应定义电子烟和无烟碱电子烟产品的子类或类别和术语，作为差异化风险评估的基础。不同的产品类别，如"一次性"产品、开放式、可填充和高功率系统，可能需要特定的技术标准或监管方法，以最大限度地降低健康风险和致瘾。
- 除了电子烟中含有烟碱外，电子烟和无烟碱电子烟都不应含有任何可能介导药物相关效应或潜在导致健康影响的化合物。
- "ENNDS"一词不应适用于含有大麻二酚或大麻油等药理活性成分的电子"雾化"产品。

8.8 结　论

尽管电子烟和无烟碱电子烟具有许多相同功能，但他们仍是高度可变的产品类别。电子烟烟液和设备的多样性使得无法评估这类产品的健康风险。因此，应该对每个单独的设备、烟液和使用进行风险评估。或者，可以突出风险标志，例如可以应用于许多产品的特定成分或特定设置。

有几种方法被用来量化烟草制品的健康风险，无论是绝对风险还是相对于卷烟的风险。目前，最有希望的方法是那些基于化合物在释放物中相对效力的方法。然而，这些方法的实用性取决于数据的可用性，而数据往往是有限的。没有一种方法可以用于监管，尽管在适当的时候可能会这样做，但在向公众传达健康风险时应谨慎使用，以确保信息明确。

8.9 参　考　文　献

[1] Tobacco. Geneva: World Health Organization; 2019 (https://www.who.int/news-room/fact-sheets/detail/tobacco, accessed 10 January 2021).

[2] Sinha DN, Palipudi KM, Gupta PC, Singhal S, Ramasundarahettige C, Jha P et al. Smokeless tobacco use: a meta-analysis of risk and attribu表 mortality estimates for India. Indian J Cancer. 2014;51(Suppl 1):S73–7.

[3] Slob W, Soeteman-Hernández LG, Bil W, Staal YCM, Stephens WE, Talhout R. A method for comparing the impact on carcinogenicity of tobacco products: a case study on heated tobacco versus cigarettes. Risk Anal. 2020;40(7):1355–66.

[4] Fowles J, Dybing E. Application of toxicological risk assessment principles to the chemical constituents of cigarette smoke. Tob Control. 2003;12(4):424–30.

[5] The scientific basis of tobacco product regulation. Second report of a WHO study group (WHO Technical Report Series, No. 951). Geneva: World Health Organization; 2008:277 (https://www.who.int/tobacco/global_interaction/tobreg/publications/9789241209519.pdf?ua=1, acces-sed 10 January 2021).

[6] Burns DM, Dybing E, Gray N, Hecht S, Anderson C, Sanner T et al. Mandated lowering of toxicants in cigarette smoke: a description of the World Health Organization TobReg proposal. Tob Control. 2008;17(2):132–41.

[7] Work in progress in relation to Articles 9 and 10 of the WHO FCTC. Geneva: World Health Organization; 2014:21 (https://apps.who.int/gb/fctc/PDF/cop4/FCTC_COP4_ID2-en.pdf, accessed 10 January 2021).

[8] Hoffmann D, Hoffmann I, El-Bayoumy K. The less harmful cigarette: a controversial issue. a tribute to Ernst L. Wynder. Chem Res Toxicol. 2001;14(7):767–90.

[9] Mallock N, Pieper E, Hutzler C, Henkler-Stephani F, Luch A. Heated tobacco products: A review of current knowledge and initial assessments. Front Public Health. 2019;7:287.

[10] Electronic nicotine delivery systems and electronic non-nicotine delivery systems (ENDS and ENNDS). Geneva: World Health Organization; 2016.

[11] Allen JG, Flanigan SS, LeBlanc M, Vallarino J, MacNaughton P, Stewart JH et al. Flavoring chemicals in e-cigarettes: Diacetyl, 2,3-pentanedione, and acetoin in a sample of 51 products, including fruit-, candy-, and cocktail-flavored e-cigarettes. Environ Health Perspect. 2016;124(6):733–9.

[12] Hutzler C, Paschke M, Kruschinski S, Henkler F, Hahn J, Luch A. Chemical hazards present in liquids and vapors of electronic cigarettes. Arch Toxicol. 2014;88(7):1295–308.

[13] Noel JC, Rainer D, Gstir R, Rainer M, Bonn G. Quantification of selected aroma compounds in e-cigarette products and toxicity evaluation in HUVEC/Tert2 cells. Biomed Chromatogr. 2020;34(3):e4761.

[14] Williams M, Talbot P. Design features in multiple generations of electronic cigarette atomizers. Int J Environ Res Public Health. 2019;16(16):e2904.

[15] Yingst J, Foulds J, Zurlo J, Steinberg MB, Eissenberg T, Du P. Acceptability of electronic nicotine delivery systems (ENDS) among HIV positive smokers. AIDS Care. 2020;32(10):1224–8.

[16] Fadus MC, Smith TT, Squeglia LM. The rise of e-cigarettes, pod mod devices, and JUUL among youth: Factors influencing use, health implications, and downstream effects. Drug Alcohol Depend. 2019;201:85–93.

[17] Walley SC, Wilson KM, Winickoff JP, Groner J. A public health crisis: Electronic cigarettes, vape, and JUUL. Pediatrics. 2019;143(6):e20182741.

[18] Goniewicz ML, Boykan R, Messina CR, Eliscu A, Tolentino J. High exposure to nicotine among adolescents who use Juul and other vape pod systems ("pods"). Tob Control. 2019;28(6):676–7.

[19] Giroud C, de Cesare M, Berthet A, Varlet V, Concha-Lozano N, Favrat B. e-Cigarettes: A review of new trends in cannabis use. Int J Environ Res Public Health. 2015;12(8):9988–10008.

[20] Pisanti S, Malfitano AM, Ciaglia E, Lamberti A, Ranieri R, Cuomo G et al. Cannabidiol: State of the art and new challenges for therapeutic applications. Pharmacol Ther. 2017;175:133–50.

[21] Poklis JL, Mulder HA, Peace MR. The unexpected identification of the cannabimimetic, 5F-ADB, and dextromethorphan in commercially available cannabidiol e-liquids. Forensic Sci Int. 2019;294:e25-7.

[22] Owusu D, Huang J, Weaver SR, Pasch KE, Perry CL. Patterns and trends of dual use of e-cigarettes and cigarettes among US adults, 2015-2018. Prev Med Rep. 2019;16:101009.

[23] Godtfredsen NS, Prescott E, Osler M. Effect of smoking reduction on lung cancer risk. JAMA. 2005;294(12):1505-10.

[24] Stephens WE. Comparing the cancer potencies of emissions from vapourised nicotine products including e-cigarettes with those of tobacco smoke. Tob Control. 2018;27(1):10-7.

[25] Mayer B. How much nicotine kills a human? Tracing back the generally accepted lethal dose to dubious self-experiments in the nineteenth century. Arch Toxicol. 2014;88(1):5-7.

[26] Seo AD, Kim DC, Yu HJ, Paulus MC, van Heel DAM, Bomers BHA et al. Accidental ingestion of e-cigarette liquid nicotine in a 15-month-old child: an infant mortality case of nicotine in-toxi-cation. Korean J Pediatr. 2016;59(12):490-3.

[27] Wang B, Liu S, Persoskie A. Poisoning exposure cases involving e-cigarettes and e-liquid in the United States, 2010-2018. Clin Toxicol. 2020;58(6):488-94.

[28] Morley S, Slaughter J, Smith PR. Death from ingestion of e-liquid. J Emerg Med. 2017;53(6):862-4.

[29] Viswam D, Trotter S, Burge PS, Walters GI. Respiratory failure caused by lipoid pneumonia from vaping e-cigarettes. BMJ Case Rep. 2018;2018:bcr2018224350.

[30] Benowitz NL, Burbank AD. Cardiovascular toxicity of nicotine: Implications for electronic ciga-rette use. Trends Cardiovasc Med. 2016;26(6):515-23.

[31] Messner B, Bernhard D. Smoking and cardiovascular disease: mechanisms of endothelial dys-function and early atherogenesis. Arterioscler Thromb Vasc Biol. 2014;34(3):509-15.

[32] Hansson J, Galanti MR, Hergens MP, Fredlund P, Ahlbom A, Alfredsson L et al. Use of snus and acute myocardial infarction: pooled analysis of eight prospective observational studies. Eur J Epidemiol. 2012;27(10):771-9.

[33] MacDonald A, Middlekauff HR. Electronic cigarettes and cardiovascular health: What do we know so far? Vasc Health Risk Manag. 2019;15:159-74.

[34] Boffetta P, Straif K. Use of smokeless tobacco and risk of myocardial infarction and stroke: Sys-tematic review with meta-analysis. BMJ. 2009;339:b3060.

[35] Teo KK, Ounpuu S, Hawken S, Pandey MR, Valentin V, Hunt D et al. Tobacco use and risk of myocardial infarction in 52 countries in the INTERHEART study: a case–control study. Lancet. 2006;368:647-58.

[36] Babic M, Schuchardt M, Tölle M, van der Giet M. In times of tobacco-free nicotine consump-tion: The influence of nicotine on vascular calcification. Eur J Clin Invest. 2019;49(4):e13077.

[37] The 2004 United States Surgeon General's report: The health consequences of smoking. New S Wales Public Health Bull. 2004;15(5-6):107.

[38] Dwyer JB, McQuown SC, Leslie FM. The dynamic effects of nicotine on the developing brain. Pharmacol Ther. 2009;122(2):125-39.

[39] Joschko MA, Dreosti IE, Tulsi RS. The teratogenic effects of nicotine in vitro in rats: A light and electron microscope study. Neurotoxicol Teratol. 1991;13(3):307-16.

[40] Bruin JE, Gerstein HC, Holloway AC. Long-term consequences of fetal and neonatal nicotine

exposure: a critical review. Toxicol Sci. 2010;116(2):364–74.

[41] Bergman BC, Perreault L, Hunerdosse D, Kerege A, Playdon M, Samek AM et al. Novel and reversible mechanisms of smoking-induced insulin resistance in humans. Diabetes. 2012;61(12):3156–66.

[42] Hess R, Bartels MJ, Pottenger LH. Ethylene glycol: an estimate of tolerable levels of exposure based on a review of animal and human data. Arch Toxicol. 2004;78(12):671–80.

[43] Chung S, Baumlin N, Dennis JS, Moore R, Salathe SF, Whitney PL et al. Electronic cigarette vapor with nicotine causes airway mucociliary dysfunction preferentially via TRPA1 receptors. Am J Respir Crit Care Med. 2019;200(9):1134–45.

[44] Pisinger C, Døssing M. A systematic review of health effects of electronic cigarettes. Prev Med. 2014;69:248–60.

[45] Wieslander G, Norbäck D, Lindgren T. Experimental exposure to propylene glycol mist in aviation emergency training: Acute ocular and respiratory effects. Occup Environ Med. 2001;58(10):649–55.

[46] Talih S, Salman R, Karaoghlanian N, El-Hellani A, Saliba N, Eissenberg T et al. "Juice monsters": sub-ohm vaping and toxic volatile aldehyde emissions. Chem Res Toxicol. 2017;30(10):1791–3.

[47] Duell AK, McWhirter KJ, Korzun T, Strongin RM, Peyton DH. Sucralose-enhanced degradation of electronic cigarette liquids during vaping. Chem Res Toxicol. 2019;32(6):1241–9.

[48] Matura M, Goossens A, Bordalo O, Garcia-Bravo B, Magnusson K, Wrangsjö A et al. Oxidized citrus oil (R-limonene): a frequent skin sensitizer in Europe. J Am Acad Dermatol. 2002;47(5):709–14.

[49] Sommerfeld CG, Weiner DJ, Nowalk A, Larkin A. Hypersensitivity pneumonitis and acute respiratory distress syndrome from e-cigarette use. Pediatrics. 2018;141(6):e20163927.

[50] Gallagher J. Vaping nearly killed me, says British teenager. BBC News, 12 November 2019 (https://www.bbc.com/news/health-50377256).

[51] Chen W, Wang P, Ito K, Fowles J, Shusterman D, Jaques PA et al. Measurement of heating coil temperature for e-cigarettes with a "top-coil" clearomizer. PLoS One. 2018;13(4):e0195925.

[52] Farsalinos KE, Voudris V, Spyrou A, Poulas K. e-Cigarettes emit very high formaldehyde levels only in conditions that are aversive to users: A replication study under verified realistic use conditions. Food Chem Toxicol. 2017;109(1):90–4.

[53] Gillman IG, Pennington ASC, Humphries KE, Oldham MJ. Determining the impact of flavored e-liquids on aldehyde production during vaping. Regul Toxicol Pharmacol. 2020;112:104588.

[54] Talhout R, Opperhuizen A, van Amsterdam JG. Role of acetaldehyde in tobacco smoke addiction. Eur Neuropsychopharmacol. 2007;17(10):627–36.

[55] Patel M, Cuccia A, Willett J, Zhou Y, Kierstead EC, Czaplicki L et al. JUUL use and reasons for initiation among adult tobacco users. Tob Control. 2019;28(6):681–4.

[56] Chaumont M, van de Borne P, Bernard A, Van Muylem A, Deprez G, Ullmo J et al. Fourth generation e-cigarette vaping induces transient lung inflammation and gas exchange disturbances: results from two randomized clinical trials. Am J Physiol Lung Cell Mol Physiol. 2019;316(5):L705-19.

[57] Gotts JE. High-power vaping injures the human lung. Am J Physiol Lung Cell Mol Physiol. 2019;316(5):L703–4.

[58] Gray N, Halstead M, Gonzalez-Jimenez N, Valentin-Blasisni L, Watson C, Pappas RS. Analysis of toxic metals in liquid from electronic cigarettes. Int J Environ Res Public Health. 2019;16(22):4450.

[59] Farsalinos KE, Rodu B. Metal emissions from e-cigarettes: A risk assessment analysis of a recently-published study. Inhal Toxicol. 2018;30(7–8):321–6.

[60] Olmedo P, Goessler W, Tanda S, Grau-Perez M, Jarmul S, Aherrera A et al. Metal concentrations in e-cigarette liquid and aerosol samples: The contribution of metallic coils. Environ Health Perspect. 2018;126(2):027010.

[61] Zhao D, Navas-Acien A, Ilievski V, Slavkovich V, Olmedo P, Adria-Mora B et al. Metal concentrations in electronic cigarette aerosol: Effect of open-system and closed-system devices and power settings. Environ Res. 2019;174:125–34.

[62] Haddad C, Salman R, El-Hellani A, Talih S, Shihadeh A, Saliba NA. Reactive oxygen species emissions from supra- and sub-ohm electronic cigarettes. J Anal Toxicol. 2019;43(1):45–50.

[63] Goniewicz ML, Knysak J, Gawron M, Kosmider L, Sobczak A, Kurek J et al. Levels of selected carcinogens and toxicants in vapour from electronic cigarettes. Tob Control. 2014;23(2):133–9.

[64] Goniewicz ML, Smith DM, Edwards KC, Blount BC, Caldwell KL, Feng J et al. Comparison of nicotine and toxicant exposure in users of electronic cigarettes and combustible cigarettes. JAMA Netw Open. 2018;1(8):e185937.

[65] Belushkin M, Tafin Djoko D, Esposito M, Korneliou A, Jeannet C, Lazzerini M et al. Selected harmful and potentially harmful constituents levels in commercial e-cigarettes. Chem Res Toxicol. 2020: doi:10.1021/acs.chemrestox.9b00470.

[66] Kennedy CD, van Schalkwyk MCI, McKee M, Pisinger C. The cardiovascular effects of electronic cigarettes: A systematic review of experimental studies. Prev Med. 2019;127:105770.

[67] Kuntic M, Oelze M, Steven S, Kröller-Schön S, Stamm P, Kalinovic S et al. Short-term e-cigarette vapour exposure causes vascular oxidative stress and dysfunction: Evidence for a close connection to brain damage and a key role of the phagocytic NADPH oxidase (NOX-2). Eur Heart J. 2019;41(26):2472–83.

[68] Caporale A, Langham MC, Guo W, Johncola A, Chatterjee S, Wehrli SW. Acute effects of electronic cigarette aerosol inhalation on vascular function detected at quantitative MRI. Radiology. 2019;293(1):97–106.

[69] Skotsimara G, Antonopoulos AS, Oikonomou E, Siasos G, Ioakeimidis N, Tsalamandris S et al. Cardiovascular effects of electronic cigarettes: A systematic review and meta-analysis. Eur J Prev Cardiol. 2019;26(11):1219–28.

[70] D'Amario D, Migliaro S, Borovac JA, Vergallo R, Galli M, Restivo A et al. Electronic cigarettes and cardiovascular risk: Caution waiting for evidence. Eur Cardiol. 2019;14(3):151–8.

[71] Ndunda PM, Muutu TM. Electronic cigarette use is associated with a higher risk of stroke. Stroke. 2019;50(Suppl 1):A9.

[72] Alzahrani T, Pena I, Temesgen N, Glantz SA. Association between electronic cigarette use and myocardial infarction. Am J Prev Med. 2018;55(4):455–61.

[73] Osei AD, Mirbolouk M, Orimoloye OA, Dzaye O, Iftekhar Uddin SM, Dardari ZA et al. The association between e-cigarette use and asthma among never combustible cigarette smokers: Behavioral risk factor surveillance system (BRFSS) 2016 & 2017. BMC Pulmon Med.

2019;19(1):180.

[74] George J, Hussain M, Vadiveloo T, Ireland S, Hopkinson P, Struthers AD et al. Cardiovascular effects of switching from tobacco cigarettes to electronic cigarettes. J Am Coll Cardiol. 2019;74(25):3112–20.

[75] Jordt SE, Jabba S, Ghoreshi K, Smith GJ, Morris JB. Propylene glycol and glycerin in e-cigarettes elicit respiratory irritation responses and modulate human sensory irritant receptor function. Am J Respir Crit Care Med. 2019;201:A4169.

[76] Meo SA, Ansary MA, Barayan FR, Almusallam AS, Almehaid AM, Alarifi NS et al. Electronic cigarettes: Impact on lung function and fractional exhaled nitric oxide among healthy adults. Am J Mens Health. 2019;13(1):1557988318806073.

[77] Stockley J, Sapey E, Gompertz S, Edgar R, Cooper B. Pilot data of the short-term effects of e-cigarette vaping on lung function. Eur Respir J. 2018;52(Suppl 62):PA2420.

[78] Rebuli ME, Glista-Baker E, Speen AM, Hoffmann JR, Duffney PF, Pawlak E et al. Nasal mucosal immune response to infection with live-attenuated influenza virus (LAIV) is altered with exposure to e-cigarettes and cigarettes. Am J Respir Crit Care Med. 2019;199(9):A4170.

[79] Wills TA, Pagano I, Williams RJ, Tam EK. e-Cigarette use and respiratory disorder in an adult sample. Drug Alcohol Depend. 2019;194:363–70.

[80] Polosa R, Morjaria JB, Prosperini U, Russo C, Pennisi A, Puleo R et al. Health effects in COPD smokers who switch to electronic cigarettes: A retrospective-prospective 3-year follow-up. Int J Chron Obstruct Pulmon Dis. 2018;13:2533–42.

[81] Layden JE, Ghinai I, Pray I, Kimball A, Layer M, Tenforde MW et al. Pulmonary illness related to e-cigarette use in Illinois and Wisconsin – Preliminary report. N Engl J Med. 2020;382(10):903–16.

[82] Moritz ED, Zapata LB, Lekiachvili A, Glidden E, Annor FB, Werner AK et al. Update: Characteristics of patients in a national outbreak of e-cigarette, or vaping, product use-associated lung injuries – United States, October 2019. Morb Mortal Wkly Rep. 2019;68(43):985–9.

[83] Outbreak of lung injury associated with e-cigarette use, or vaping. Atlanta (GA): Centers for Disease Control and Prevention; 2020 (https://www.cdc.gov/tobacco/basic_information/e-cigarettes/severe-lung-disease.html).

[84] Kalininskiy A, Bach CT, Nacca NE, Ginsberg G, Marraffa J, Navarette KA et al. e-Cigarette, or vaping, product use associated lung injury (EVALI): case series and diagnostic approach. Lancet Respir Med. 2019;7(12):1017–26.

[85] Blagev DP, Harris D, Dunn AC, Guidry DW, Grissom CK, Lanspa MJ. Clinical presentation, treatment, and short-term outcomes of lung injury associated with e-cigarettes or vaping: A prospective observational cohort study. Lancet. 2019;394(10214):2073–83.

[86] Blount BC, Karwowski MP, Morel-Espinosa M, Rees J, Sosnoff C, Cowan E et al. Evaluation of bronchoalveolar lavage fluid from patients in an outbreak of e-cigarette, or vaping, product use-associated lung injury – 10 states, August–October 2019. Morb Mortal Wkly Rep. 2019;68(45):1040–1.

[87] Boudi FB, Patel S, Boudi A, Chan C. Vitamin E acetate as a plausible cause of acute vaping-related illness. Cureus. 2019;11(12):e6350.

[88] Blount BC, Karwowski MP, Shields PG, Morel-Espinosa M, Valentin-Blasini L, Gardner M

et al. Vitamin E acetate in bronchoalveolar-lavage fluid associated with EVALI. N Engl J Med. 2020;382(8):697–705.

[89] Navon L, Jones CM, Ghinai I, King BA, Briss PA, Hacker KA et al. Risk factors for e-cigarette, or vaping, product use-associated lung injury (EVALI) among adults who use e-cigarette, or vaping, products – Illinois, July–October 2019. Morb Mortal Wkly Rep. 2019;68(45):1034–9.

[90] Lewis N, McCaffrey K, Sage K, Cheng CJ, Green J, Goldstein L et al. e-Cigarette use, or vaping, practices and characteristics among persons with associated lung injury – Utah, April–October 2019. Morb Mortal Wkly Rep. 2019;68(42):953–6.

[91] Visser WF, Klerx WN, Cremers HWJM, Ramlal R, Schwillens PL, Talhout R. The health risks of electronic cigarette use to bystanders. Int J Environ Res Public Health. 2019;16(9):1525.

[92] McAuley TR, Hopke PK, Zhao J, Babaian S. Comparison of the effects of e-cigarette vapor and cigarette smoke on indoor air quality. Inhal Toxicol. 2012;24(12):850–7.

[93] Avino P, Scungio M, Stabile L, Cortelessa G, Buonnano G, Manigrasso M. Second-hand aerosol from tobacco and electronic cigarettes: Evaluation of the smoker emission rates and doses and lung cancer risk of passive smokers and vapers. Sci Total Environ. 2018;642:137–47.

[94] Hess IM, Lachireddy K, Capon A. A systematic review of the health risks from passive exposure to electronic cigarette vapour. Public Health Res Pract. 2016;26(2):e2621617.

[95] Czogala J, Goniewicz ML, Fidelus B, Zielinska-Danch W, Travers MJ, Sobczak A. Secondhand exposure to vapors from electronic cigarettes. Nicotine Tob Res. 2014;16(6):655–62.

[96] Ballbe M, Martínez-Sánchez JM, Sureda X, Fu M, Pérez-Ortuña R, Pascual JA et al. Cigarettes vs. e-cigarettes: Passive exposure at home measured by means of airborne marker and biomarkers. Environ Res. 2014;135:76–80.

[97] Protano C, Manigrasso M, Avino P, Vitali M. Second-hand smoke generated by combustion and electronic smoking devices used in real scenarios: Ultrafine particle pollution and age-related dose assessment. Environ Int. 2017;107:190–5.

[98] Bayly JE, Bernat D, Porter L, Choi K. Secondhand exposure to aerosols from electronic nicotine delivery systems and asthma exacerbations among youth with asthma. Chest. 2019;155(1):88–93.

[99] Smoking cessation: A report of the Surgeon General. Rockville (MD): Department of Health and Human Services; 2020 (https://www.hhs.gov/sites/default/files/2020-cessation-sgr-full-report.pdf).

[100] Kane DB, Asgharian B, Price OT, Rostami A, Oldham M. Effect of smoking parameters on the particle size distribution and predicted airway deposition of mainstream cigarette smoke. In-hal Toxicol. 2010;22(3):199–209.

[101] Son Y, Mainelis G, Delnevo C, Wackowski OA, Schwander S, Meng Q. Investigating e-cigarette particle emissions and human airway depositions under various e-cigarette-use conditions. Chem Res Toxicol. 2020;33(2):343–52.

[102] Beauval N, Verriele M, Garat A, Fronval I, Dusautoir R, Anthérieu S et al. Influence of puffing conditions on the carbonyl composition of e-cigarette aerosols. Int J Hyg Environ Health. 2019;222(1):136–46.

[103] Hubbs AF, Kreiss K, Cummings KJ, Fluharty KL, O'Connell R, Cole A et al. Flavorings-related lung disease: A brief review and new mechanistic data. Toxicol Pathol. 2019;47(8):1012–6.

[104] Clapp PW, Lavrich KS, van Heusden CA, Lazarowski ER, Carson JL, Jaspers I. Cinnamaldehyde in flavored e-cigarette liquids temporarily suppresses bronchial epithelial cell ciliary motility by dysregulation of mitochondrial function. Am J Physiol Lung Cell Mol Physiol. 2019;316(3):L470–86.

[105] Bopp SK, Barouki R, Brack W, Dalla Costa S, Dorne JCM, Drakvik PE et al. Current EU research activities on combined exposure to multiple chemicals. Environ Int. 2018;120:544–62.

[106] EFSA Scientific Committee, More SJ, Bampidis V, Benford D, Bennekou SH, Bragard C et al. Guidance on harmonised methodologies for human health, animal health and ecotoxicological risk assessment of combined exposure to multiple chemicals. EFSA J. 2019;17(3):5634.

[107] Ukić S, Sigurnjak M, Cvetnić M, Markić M, Novak Stankov M et al. Toxicity of pharmaceuticals in binary mixtures: Assessment by additive and non-additive toxicity models. Ecotoxicol Environ Saf. 2019;185:109696.

[108] Gao Y, Feng J, Kang L, Xu X, Zhu L. Concentration addition and independent action model: Which is better in predicting the toxicity for metal mixtures on zebrafish larvae. Sci Total Environ. 2018;610-611:442–50.

[109] Jonker MJ, Svendsen C, Bedaux JJ, Bongers M, Kammenga JE. Significance testing of synergistic/antagonistic, dose level-dependent, or dose ratio-dependent effects in mixture dose-response analysis. Environ Toxicol Chem. 2005;24(10):2701–13.

[110] Altenburger R, Backhaus T, Boedeker W, Faust M, Scholze M, Grimme LH. Predictability of the toxicity of multiple chemical mixtures to Vibrio fischeri: Mixtures composed of similarly acting chemicals. Environ Toxicol Chem. 2000;19(9):2341–7.

[111] Baas J, Van Houte BP, Van Gestel CA, Kooijman SALM. Modeling the effects of binary mixtures on survival in time. Environ Toxicol Chem. 2007;26(6):1320–7.

[112] Leeman WR, Krul L, Houben GF. Complex mixtures: Relevance of combined exposure to substances at low dose levels. Food Chem Toxicol. 2013;58:141–8.

[113] EFSA Scientific Committee. Scientific opinion on exploring options for providing advice about possible human health risks based on the concept of threshold of toxicological concern (TTC). EFSA J. 2012;10(7).

[114] Kawamoto T, Fuchs A, Fautz R, Morita O. Threshold of toxicological concern (TTC) for botanical extracts (botanical-TTC) derived from a meta-analysis of repeated-dose toxicity studies. Toxicol Lett. 2019;316:1–9.

[115] Rietjens I, Cohen SM, Eisenbrand G, Fukushima S, Gooderham NJ, Guengerich FP et al. FEMA GRAS assessment of natural flavor complexes: Cinnamomum and Myroxylon-derived flavoring ingredients. Food Chem Toxicol. 2020;135:110949.

[116] Tluczkiewicz I, Kühne R, Ebert RU, Batke M, Schüürmann G, Mangelsdorf I et al. Inhalation TTC values: A new integrative grouping approach considering structural, toxicological and mechanistic features. Regul Toxicol Pharmacol. 2016;78:8–23.

[117] Schüürmann G, Ebert RU, Tluczkiewicz I, Escher SE, Kühne R. Inhalation threshold of toxicological concern (TTC) – Structural alerts discriminate high from low repeated-dose inhalation toxicity. Environ Int. 2016;88:123–32.

[118] The scientific basis of tobacco product regulation. Report of a WHO Study Group (WHO Technical Report Series, No. 945). Geneva: World Health Organization; 2007 (https://www.who.int/

tobacco/global_interaction/tobreg/who_tsr.pdf, accessed 10 January 2021).

[119] Wagner KA, Flora JW, Melvin MS, Avery KC, Ballentine RM, Brown AP et al. An evaluation of electronic cigarette formulations and aerosols for harmful and potentially harmful constituents (HPHCs) typically derived from combustion. Regul Toxicol Pharmacol. 2018;95:153–60.

[120] Air Toxics Hot Spots Program risk assessment guidelines. Part II. Technical support document for cancer potency factors: Methodologies for derivation, listing of available values, and adjustments to allow for early life stage exposures. Sacramento (CA): California Environmental Protection Agency, Office of Environmental Health Hazard Assessment, Air Toxicology and Epidemiology Branch; 2008 (https://oehha.ca.gov/media/downloads/crnr/tsd062008.pdf).

[121] Cunningham FH, Fiebelkorn S, Johnson M, Meredith C. A novel application of the margin of exposure approach: segregation of tobacco smoke toxicants. Food Chem Toxicol. 2011;49(11):2921–33.

[122] Soeteman-Hernández LG, Bos PM, Talhout R. Tobacco smoke-related health effects induced by 1,3-butadiene and strategies for risk reduction. Toxicol Sci. 2013;136(2):566–80.

[123] Risicobeoordeling lachgas [Risk assessment on nitrous oxide]. Bilthoven: Coördinatiepunt Assessment en Monitoring nieuwe drugs; 2019 (https://www.vvgn.nl/wp-content/uploads/2019/12/risicobeoordelingsrapport-lachgas-20191209-beveiligd.pdf).

[124] Staal YCM, Havermans A, van Nierop L et al. Evaluation framework to qualitatively assess the health effects of new tobacco and related products [submitted 2020].

[125] Sosnowski TR, Kramek-Romanowska K. Predicted deposition of e-cigarette aerosol in the human lungs. J Aerosol Med Pulm Drug Deliv. 2016;29(3):299–309.

[126] Savareear B, Escobar-Arnanz J, Brokl M, Saxton MJ, Wright C, Liu C et al. Non-targeted analysis of the particulate phase of heated tobacco product aerosol and cigarette mainstream tobacco smoke by thermal desorption comprehensive two-dimensional gas chromatography with dual flame ionisation and mass spectrometric detection. J Chromatogr A. 2019;1603:327–37.

[127] Salman R, Talih S, El-Hage R, Haddad C, Karaoghlanian N, El-Hellani A et al. Free-base and total nicotine, reactive oxygen species, and carbonyl emissions from IQOS, a heated tobacco product. Nicotine Tob Res. 2019;21(9):1285–8.

[128] Cancelada L, Sleiman M, Tang X, Russell ML, Montesinos VN, Litter MI et al. Heated tobacco products: Volatile emissions and their predicted impact on indoor air quality. Environ Sci Te-chnol. 2019;53(13):7866–76.

[129] Bentley MC, Almstetter M, Arndt D, Knorr A, Martin E, Pospisil P et al. Comprehensive chemical characterization of the aerosol generated by a heated tobacco product by untargeted screening. Anal Bioanal Chem. 2020;412(11):2675–85.

[130] Apelberg BJ, Feirman SP, Salazar E, Corey CG, Ambrose BK, Paredes A et al. Potential public health effects of reducing nicotine levels in cigarettes in the United States. N Engl J Med. 2018;378(18):1725–33.

[131] Chen J, Bullen C, Dirks K. A comparative health risk assessment of electronic cigarettes and conventional cigarettes. Int J Environ Res Public Health. 2017;14(4):382.

[132] Hajek P, Phillips-Waller A, Przulj D, Pesola F, Myers Smith K, Bisal N et al. A randomized trial of e-cigarettes versus nicotine-replacement therapy. N Engl J Med. 2019;380(7):629–37.

9. 新型烟草制品的风味

Ahmad El-Hellani, Department of Chemistry, Faculty of Arts and Sciences, American University of Beirut, Beirut, Lebanon; Center for the Study of Tobacco Products, Virginia Commonwealth University, Richmond (VA), USA

Danielle Davis, Department of Psychiatry, Yale School of Medicine, Yale University, New Haven (CT), USA

Suchitra Krishnan-Sarin, Department of Psychiatry, Yale School of Medicine, Yale University, New Haven (CT), USA

摘 要

烟碱和烟草制品包含特征性风味来掩盖刺激性，使其便于使用并提高可接受性。最近一项估算表明，调味物质的使用很普遍，目前电子烟中已有数千种调味物质可供使用。非传统口味，如水果和糖果，对年轻人特别有吸引力，而且这些口味在成年人中，特别是在试图戒烟的成年吸烟者中的使用也在增加。所有烟碱和烟草制品中的调味物质不仅增加了产品的吸引力和对它的首次使用率，而且还可能增加使用程度、从尝试使用到常规使用和依赖的发展。另外令人担忧的是调味物质的化学物质可能会增加产品的有害性。在各种烟碱和烟草制品中，或在各国之间，对调味物质的规定并不统一。有些国家禁止所有产品中的所有口味，有些国家只禁止某些口味（如禁止薄荷醇），而有些国家只在法规中包括某些产品（如卷烟和无烟烟草等传统产品）。这份关于烟碱和烟草制品中调味物质的报告呼吁采用烟碱和烟草制品中调味物质的通用术语，并考虑减少市场上风味烟碱和烟草制品供应的政策，因为有明确证据表明这将有助于吸烟者戒掉传统卷烟。

9.1 背景和引言

在烟草制品中添加调味物质可追溯到19世纪，当时在无烟烟草制品中添加了水果味的调味物质[1]。现在，几乎所有烟碱和烟草制品，包括卷烟和雪茄等传统卷烟、鼻烟和口含烟、加热型烟草制品（HTP）和电子烟碱传输系统（ENDS）

等较新的产品都有多种口味,这增加了它们的吸引力、使用流行性和伤害减少的认知[2,3]。调味物质掩盖了烟碱和烟草的刺鼻味,使其易于使用,也减少了刺鼻气味的二次暴露,从而提高了它们的可接受性,并有可能导致人们对这些产品依赖性的发展和维持。调味物质还增强了烟碱和烟草制品对新手使用者和易感人群的吸引力,从而增加了对这些产品的使用[4]。调味物质会产生双重作用,吸引新手使用者,特别是年轻人,促使其开始使用烟碱和烟草制品,并使当前使用者延长使用时间。公共卫生界应解决这个问题,并且卫生机构应制定最适用的法规确保解决烟碱和烟草制品中含有调味物质的问题[5-8]。

全世界都在使用风味烟碱和烟草制品。关于它们在全球的可用性和使用情况,系统性证据有限,对这些产品的偏好往往是特定于某些国家和地区的。例如,含有丁香及其精油的传统卷烟(*kretek*)在印度尼西亚很受欢迎,风味无烟烟草(*pan masala, gutka*)包含烟草和香精香料、精油、调味品、槟榔和其他成分,在印度使用非常广泛。水烟加热添加浓重香味、甜味的烟草,起源于印度和中东,并在欧洲和北美的年轻人中越来越流行。

对烟碱和烟草制品中使用调味物质的限制因国家而异。一些国家,如巴西、智利、埃塞俄比亚、摩尔多瓦共和国和加拿大一些省份,限制烟碱和烟草制品中的所有口味,包括薄荷醇,然而加拿大允许使用具有波特酒、葡萄酒、朗姆酒或威士忌口味的产品[9]。2020年5月,欧盟和28个成员国实施了一项禁止薄荷醇卷烟和自卷烟的法令[10]。巴西、加拿大、智利和埃塞俄比亚将短支雪茄和小雪茄等非卷吸烟产品纳入风味限制范围内,而其他国家,包括欧盟、摩尔多瓦共和国和土耳其,不将限制扩大到卷烟和自卷烟以外的产品[11]。是否限制薄荷醇也存在差异。例如,虽然巴西已禁止所有烟草制品中的所有特征香味(定义为除烟草外,在烟草消费之前、期间或之后可识别的具有味道或香气的调味物质)[12],包括薄荷醇。在美国大多数地方,除了薄荷醇以外的特征风味,仅在卷烟和自卷烟中被禁止[13],但是有一些地方(例如旧金山、加利福尼亚)已禁止了烟草制品中包括薄荷醇[14]在内所有的特征风味。在欧盟,与摩尔多瓦共和国、土耳其和美国一样,该法令不适用于其他烟草制品,如雪茄、小雪茄、短支雪茄和无烟烟草制品[15]。

由于ENDS不含烟草,因此对ENDS的监管可能与烟草制品不同。虽然世界卫生组织(WHO)不认为它们是烟草制品[16],但一些国家在现有烟草制品法规中把这些产品包括在内,而另一些国家则将ENDS与烟草制品分开考虑。

ENDS在30个国家被禁止销售,一些国家有法规限制调味物质的应用[17],并限制最大烟碱浓度。2019年末,美国开始限制在烟弹式ENDS中添加调味物质,但许多州对风味烟碱和烟草制品的销售实行地方限制[18]。包括澳大利亚和日本在内的一些国家禁止销售含有烟碱的电子烟[19]。不含烟碱的电子烟被称为电子非烟碱传输系统(ENNDS),并且根据法规,仍然可以提供不同的口味。在有ENDS

的国家,报告显示ENNDS的使用率较低[20],因此,在风味电子系统的评估中,通常包括ENNDS。在这篇综述中,在对电子烟使用的描述中尽可能区分了两者。

自2003年ENDS进入全球市场以来,人们对调味物质的作用重新产生了兴趣。它在全球范围内越来越受欢迎,可以说在加拿大、欧洲和美国最受欢迎[21]。ENDS有越来越多的可定制口味的电子烟烟液。2013~2014年对在电子烟市场的一项研究发现了7764种独特的电子烟烟液口味[22],2018年的后续调查显示,这一数字翻了一番,达到15586种口味[23]。电子烟的使用在公共卫生领域引发了关于使用这些设备帮助吸烟者停止使用更有害的传统吸烟来减少危害的辩论。这场辩论已经演变成了一场关于电子烟中调味物质的风险与益处的辩论。一些研究人员和提倡者认为,在电子烟中添加调味物质是有益的,因为它可能有助于吸烟者戒烟[24-26],而其他人则认为,调味物质的存在只会增强这些产品对年轻人的吸引力,并导致对其使用和依赖性的增加[26-29]。

电子烟并不是市场上相对较新的唯一产品。加热型烟草制品具有独特性,但它属于烟草制品最新的迭代产品,在2013年出现。这些产品没有电子烟的多种口味,但在市场上的销售情况与电子烟类似,因为它们被宣传为传统烟草制品的潜在替代品。

鉴于加热型烟草制品在市场上的出现相对较新,且可能作为一种低风险产品进行销售,因此了解风味如何有助于提高其吸引力和使用非常重要,尤其是对非使用者和年轻人[30]。

本报告是对世界卫生组织烟草制品管制研究小组第七次报告[31]的更新,描述了风味烟草制品的流行病学、对吸引力的影响、烟碱和烟草制品的相关试验和持续使用影响、最常用的调味物质、对健康的影响以及当前对这些产品中调味物质的监管。

9.2 风味电子烟的流行病学(使用频率、模式和原因,按社会人口统计变量)

虽然风味烟碱和烟草制品在全球使用,但关于其使用的系统信息有限。大部分信息是针对美国的,调查显示美国的使用率很高。2013~2014年的全国成年人烟草调查表明,估计有1020万电子烟使用者(68.2%),610万水烟使用者(82.3%),410万雪茄抽吸者(36.2%),400万无烟烟草制品使用者(50.6%)在过去30天[32]使用了风味产品。在同一项调查中,吸烟者使用含薄荷醇的卷烟(调查时美国卷烟中唯一的特征风味)的比例相对较高,占使用卷烟的39%[33]。2014~2015年的一项基于人口的调查报告了类似的结果,报告表明41.4%的烟碱和烟草使用者使用了风味电子烟,根据产品类型变化,雪茄吸烟者中占比28.3%,水烟使用者中占比87.2%[34]。

关于使用风味烟碱和烟草制品的全球数据有限,因此迫切需要评估全球调味物质使用范围,应提供世界其他地方关于使用风味烟碱和烟草制品的信息。虽然美国的数据提供了一个合理的估计,但使用的风味烟碱和烟草制品的类型以及这些产品的监管制度因国家而异,可能会影响不同调味物质的使用。风味电子烟的使用模式可能因其可用性和受欢迎程度而异。然而,青少年和年轻人的使用率高于老年人。不同烟草制品中调味物质使用的流行病学证据如下所述。

9.2.1 电子烟碱/非烟碱传输系统

由于可用调味物质数量的动态增长[23]及其在青少年和年轻人中的流行[8],调味物质越来越多地用于ENDS。烟草与健康人口评估(PATH)是在美国进行的一项纵向全国性调查,该调查表明,当前吸烟者使用风味电子烟的人在12~17岁的青少年中最为普遍(97%),其次是18~24岁的年轻人(97%)和成年人≥25岁(81%)。在使用电子烟的起始阶段也观察到类似的特征,报告显示93%的青少年、84%的年轻人和55%的成年人在起始阶段使用风味电子烟[35]。其他研究[36-38]和最新的综述[39,40]证实了风味电子烟的使用特征。这些研究表明,相较于老年人,风味电子烟的使用在青少年和年轻人中更流行。2019年,美国一项关于中学生和高中生使用烟草制品的全国烟草调查报告表明,当前电子烟使用者中使用风味电子烟的人在非西班牙裔白人青少年中最为普遍(77%),男性(71%)和女性(69%)情况相似[41]。

青少年和年轻人使用非传统调味物质(即除烟草或薄荷醇外)的情况更为普遍,然而,在所有年龄组中,似乎都出现了口味偏好向非传统口味转变的现象。PATH第2轮研究表明,水果口味是青少年最常用的类别。而薄荷醇或薄荷作为传统上在烟草制品中出现的口味,显示是成年人最常使用的类别[42]。后续第3轮研究表明,水果口味是年轻人和成年人使用者最常用的口味,甜味或糖果口味在年轻人和成年人使用者中也非常流行[29]。这些发现与最近的其他研究一致,具体来说,有证据表明水果味的电子烟在各年龄组(包括青少年和成年人)[36,43,44]都是最受欢迎的口味类别。美国一项长期使用风味电子烟的研究发现,随着时间的推移,人们对烟草、薄荷醇或薄荷的偏好逐渐降低,对水果味的偏好保持稳定,对甜味的偏好也逐渐增加[45]。

此外,还对当前或既往使用其他烟草制品(如卷烟)的电子烟使用者的口味偏好进行了研究。这一点很重要,特别是在欧洲和美国,关于口味是否以及如何降低电子烟对年轻人的吸引力,或者增强它们对烟草制品使用者的吸引力,如希望改用电子烟的吸烟者,存在着争论。无论使用何种烟草制品,水果口味似乎都很受欢迎。一项全球互联网调查发现，尽管水果和甜味是最受欢迎的（分别为69%和61%），但与既往吸烟者相比，当前吸烟者更倾向使用烟草味，而使用水果味和甜味的可能性更小[24]。在新西兰的一项研究中，水果口味是仅使用电子烟者、既往吸烟者和当前吸烟者的优选[46]。尽管所有年龄组的口味偏好似乎都在向非传统口味转变，但值得注意的是，在电子烟中，年轻人使用非传统口味的流行度似乎仍然最高[47]。

许多使用风味电子烟的人表示，他们使用多种口味（例如水果、甜味或糖果、薄荷醇或薄荷）。一项对美国电子烟使用者的研究表明，青少年（46%）比成年人（32%）更普遍地使用多种口味[29]。在印度研究显示，成年人电子烟使用者使用多种口味的比例相对较高（65%）[48]。全球互联网调查显示，使用电子烟[24]的当前和既往成年吸烟者在不同口味之间进行转换是常见的，68%的人表示至少每天更换一次，16%的人每周更换一次，10%的人不到一周更换一次。

尽管关于使用风味无烟碱电子烟的证据有限，但一项针对日本青少年和年轻人（18~29岁）的研究表明，4.3%的人使用了这些产品，这些产品是禁止使用的。虽然没有直接研究调味物质的使用，但据报道使用该产品的主要原因是水果口味，这表明调味物质是使用无烟碱电子烟的一个原因[49]。

9.2.2 传统卷烟和无烟烟草制品

尽管到目前为止，电子烟是所有烟碱和烟草制品中口味最多的，但其他烟碱和烟草制品也使用了调味物质，包括传统卷烟和无烟烟草制品。对于电子烟，青少年和年轻人使用风味烟草制品的情况比成年人更为普遍。在一个具有全国代表性的加拿大年轻人样本中，52%的人报告使用了烟草制品（卷烟、烟斗、雪茄、小雪茄、bidis、无烟烟草、水烟、blunts、自卷烟）[50]。同样，波兰的一项全国性调查显示，年轻吸烟者更有可能使用风味卷烟[51]。在美国PATH研究中，当前使用小雪茄和过滤嘴雪茄的所有年龄段的使用者中，约有一半人表示使用了风味电子烟，而当前雪茄吸烟者使用风味电子烟的比例相对较低，成年人中占24%、青少年中占27%，年轻人中占36%。在同一项研究中，调查显示，小雪茄烟使用者首次使用风味产品在青少年中的比例为68%、年轻人中为63%，成年人中占42%；滤嘴雪茄使用者中首次使用的就是风味烟草制品的人在青少年中占56%，年轻人中占54%，成年人中占40%。在传统雪茄使用者中，风味卷烟总体流行率较低，但年龄的分级效应仍然明显，39%的青少年、35%的年轻人和22%的成年人表示首次使用产品是添加了调味物质的[35]。

2013~2014年美国全国成年人烟草调查显示，多种传统烟草制品中添加了调味物质。无烟烟草中有薄荷醇或薄荷（77%）；水烟中有水果味（74%）；雪茄、小雪茄、滤嘴小雪茄中有水果味（52%），糖果、巧克力和其他甜味（22%），酒精（14%）；烟斗中有水果（57%），糖果、巧克力和其他甜味（26%），薄荷醇或薄荷（25%）[32]。

卷烟中嵌入爆珠是为了让它具有传统卷烟烟草和薄荷醇以外的其他口味，爆珠卷烟在全球市场所占份额似乎日益增长[52]。爆珠烟有不同风味，薄荷醇、绿茶、威士忌以及其他风味。爆珠卷烟很受青少年和年轻吸烟者的欢迎。澳大利亚超过一半的青少年吸烟者报告使用过爆珠卷烟[53]，英国和美国的年轻人表示更喜欢这些爆珠卷烟[53,54]。这种现象可能会改变，因为自2020年春季起在巴西和欧盟风味爆珠已被禁止使用[55]。

9.2.3 加热型烟草制品

在美国自20世纪60年代起、全球自80年代起，HTP就已经发展起来了，然而，早期的产品并不成功。自2013年起，新一代HTP进入市场[30]，并且目前已在约50个国家上市，包括加拿大、以色列、意大利、日本、韩国，最近，在美国也上市了[56,57]。HTP中含有调味物质，可能会增加其吸引力和使用[58]。当加热烟草或激活含有烟草的设备时，这些设备会产生含有烟碱和有害化学物质的气溶胶，使用者吸吮或抽吸过程中会吸入气溶胶。这些产品已作为燃烧型卷烟的更安全替代品[59]上市。与电子烟相比，口味选择较少，主要有烟草或薄荷醇，其他口味包括"含薄荷醇的水果味"和咖啡味。

日本最近一直在测试这些产品的市场。2018年收集的日本国际烟草管制调查第1轮数据表明HTP的使用流行率仅为2.7%。在HTP使用者和双重使用者中，风味的受欢迎程度似乎相似，薄荷醇味最受欢迎，其次是烟草和含薄荷醇的水果味[60]。有证据表明，口味会增加其他烟碱和烟草制品的吸引力，尤其是对不吸烟的青少年和年轻人而言。鉴于此，人们发出了警报，认为这些产品可能因其风味而具有吸引力[58]。随着HTP市场扩展到其他国家，应就这些产品中调味物质的使用进行全面研究。

9.3 调味物质对吸引力、尝试使用、使用和持续使用的影响

9.3.1 青少年和年轻人

吸引力

由于使用烟草制品的青少年和年轻人似乎比成年人更频繁地使用风味烟草制品和ENDS[36,51]，因此了解这些产品对青少年的吸引力以及与成年使用者吸引力的不同之处非常重要。按年龄对产品使用情况进行比较，结果表明风味雪茄、卷

烟和水烟对年轻人的吸引力大于成年人[61,62]。年轻人表示口味是开始和持续使用ENDS的主要原因[63,64]，而且甜味和水果味的电子烟烟液比不是甜味（如烟草味）的电子烟烟液更吸引青少年和年轻人[65,66]。年轻人表示口味也是他们使用小雪茄的主要原因[67]。

值得注意的是，风味烟草制品对人的吸引力很可能在使用风味ENDS之前就开始了。烟草制品和ENDS中的味道可能会吸引人，这不仅是因为与口味相关的积极感受[68-70]，还因为它们降低了这些产品的风险[71-75]与危害感知。ENDS不仅提供非传统卷烟草口味，还凭借其色彩鲜艳的图像和吸引人的口味描述进行广告宣传。功能磁共振成像显示，在易使用ENDS的年轻人中，仅观看显示水果、薄荷和甜味ENDS的广告比烟草味的广告更能增加中枢神经的活动。当参与者观看不是水果、薄荷和甜味电子烟的广告时，中枢神经活动也会增加，这表明非传统风味的吸引力可能在ENDS体验之前就开始了，广告也可能会导致人们开始去使用电子烟[76]。

关于电子烟的尝试使用、使用和持续使用

ENDS的口味可能对青少年和年轻人尝试和持续使用发挥重要作用。因此，最初使用风味电子烟不仅与连续使用有关，还与使用天数有关。这表明随着时间的推移，使用量会增加[77]。在另一项研究中显示，年轻人对特定口味的偏好和使用的口味总数与使用电子烟的天数有关，在成年人中却不是这样[28]，这表明口味偏好在青少年与成年人使用中所起的作用不同。甜味这种非传统风味尤其吸引年轻人，并可能导致年轻人使用电子烟[28,37,63,78,79]。在一项针对年轻人的研究中，风味电子烟的使用与非传统风味（即非烟草、薄荷或薄荷醇）和连续使用电子烟有关，但与使用电子烟的天数无关，这表明使用非传统口味可能会导致电子烟的持续使用[37]。这可能部分是由于非传统风味的感官效应。在一项实验室研究中，采用市场上可买到的6种口味，水果味被认为是最甜的，烟草味被认为是最苦的。当口味被认为是凉爽和刺激时，甜度和凉爽度与喜好度呈正相关，而刺激性和苦味与喜好度呈负相关[80]。

烟碱和烟草制品中的调味物质可能会增加其致瘾性，从而使人们持续使用[81]。据报道，电子烟中的青苹果烟草味会改变吸烟行为，这可能与烟碱乙酰胆碱受体的上调有关[82]，薄荷醇改变吸烟行为的生物学研究表明，其中的机制包括加强与烟碱相关的感官诱因，上调烟碱乙酰胆碱受体并改变烟碱代谢以提高其生物利用度[83]。

如年轻人初始使用的是风味电子烟，会导致他们继续大量使用，这种特征也可能出现在其他烟草制品和年龄群体中。烟草与健康人群评估研究的纵向结果表明，青少年、年轻人和成年吸烟者持续使用卷烟与他们最初使用风味（即薄荷醇）

卷烟有关，并且在年轻人和成年人中，这种特征会扩展到其他烟草制品。因此，人们对风味电子烟、雪茄、小雪茄、滤嘴雪茄、水烟或无烟烟草的持续使用与最初使用风味电子烟有关[38]。

9.3.2 成年人

吸引力

在成年人中，调味物质似乎也有助于提高烟草制品的吸引力。和年轻人一样，风味电子烟尤其具有高吸引力。在对荷兰、新西兰和美国的电子烟使用者的研究中，电子烟的吸引力与调味物质的可获得性相关[46,84]。一项针对亚太地区大学生（从本科生到博士生）的研究发现，34%的电子烟使用者使用该产品是因为其提供的口味[85]。口味可以提高成年人对产品的积极预期[86]，以及使用者和非使用者对产品的总体正面认知[87]，从而提高产品对成年人的吸引力。

关于电子烟的尝试使用、使用和持续使用

在青少年和年轻人中，甚至在老年人中，口味对产品的使用都起非常重要作用。在一份成人样本（≥25岁）中，目前经常使用风味卷烟、雪茄、小雪茄、滤嘴雪茄、水烟、无烟烟草和电子烟的人与他们首次使用风味电子烟有关，而与其首次使用非风味电子烟无关，在青少年和年轻人中也是这样[38]。无烟烟草使用者也有类似的使用特征，即那些开始使用无味产品的人可能会转而使用风味电子烟，而那些开始使用风味电子烟的人可能会继续使用风味电子烟[88]。

有证据表明，成年人使用风味烟草制品会对其产生更大的依赖性，这会导致对风味电子烟连续使用。2014~2015年美国的一项调查显示，对烟草制品依赖的两个标志，每日使用量和早晨首次使用时间，与风味电子烟的使用有关。同一项研究表明，每天使用风味电子烟比使用无味产品的可能性更大。据报道，早上醒来后30分钟内首次使用风味雪茄（大雪茄、雪茄和小雪茄）的人数多于首次使用无味雪茄的人数[89]。

使用风味电子烟后可能会持续使用，因为它们会影响奖赏机制并降低人对烟碱的厌恶性，而烟碱是这些产品中产生依赖性的成分[90]。有关燃烧型卷烟中烟碱和薄荷醇相互作用的大量文献表明，薄荷醇可作为烟碱感官效应的提示线索[91]，并增强烟碱的奖赏机制[92]和戒断症状[93]。这一发现可能揭示了为什么在总体使用量减少的情况下，薄荷醇卷烟的使用量在燃烧型卷烟的吸烟者中不断增加[94]，以及为什么薄荷醇卷烟的吸烟者戒烟更困难[95]。电子烟的水果味道已被证明能增强烟碱的奖赏和强化作用[78,96]，并抑制其厌恶作用[97]。薄荷醇口味能改善电子烟烟液的味道，并使较高浓度的烟碱不那么令人厌恶，也更令人愉悦[96-98]。在燃烧型

卷烟产品中也观察到类似的特征。烟草行业相关文件表明，风味雪茄产品通过减少对喉咙的刺激和易于释放物的吸入，从而增加了对初始使用者的吸引力[99]。

另一种连续使用电子烟的可能机制是，人们普遍认为，风味烟草制品，包括电子烟，比无味烟草制品对人体伤害更少。电子烟的口味不仅给使用者也给非使用者带来了错误的安全感知[100]。然而这种态度正在改变，越来越多的人呼吁在指定的无烟室内使用电子烟，因为风味电子烟产生的二手气溶胶会在空气中留下刺鼻的气味[101]。对风味电子烟在室内和公共场所的二手气溶胶的化学和毒理学评估研究落后于它们的过度使用现象[102,103]。

关于调味物质使人从传统卷烟转向其他产品（如电子烟或加热型烟草制品）中的作用，仍存在疑问。目前还没有关于加热型烟草制品、口味和转换行为之间关系的研究报告。电子烟中的口味可能会吸引传统卷烟的使用者，使用风味电子烟可能会增大戒掉传统卷烟的可能性[104,105]。虽然许多人描述了青少年和年轻人以及非燃烧型产品使用者滥用电子烟的可能性，但尚未对电子烟风味在传统卷烟转换中的具体作用进行实验研究。需要证据来确定电子烟口味在产品转换中的作用（另见第3章）。

9.4　常见调味物质特性、健康效应及其对公众健康的影响

9.4.1　电子烟和烟草制品中常见的调味物质

荷兰国家公共卫生与环境研究所出版了一本"调味物质库"，收录了添加到卷烟和自卷烟产品中的调味物质[106]。调味物质分为8类：水果、香料、草本、酒精、薄荷醇、甜味、花香和其他。一年后，发表了一份类似的报告，将电子烟中的口味分为13类：烟草、薄荷、水果、甜点、酒精、坚果、香辛料、糖果、咖啡/茶、饮料、甜味、其他和无添加口味[107]。如上所述，近年来，市场上有独特风味的电子烟数量急剧增加，从2013~2014年的7764种增加到2016~2017年的15586种[22,23]。使用者通常在可再充填的电子烟中混合和搭配口味[108]，有时添加非法物质[109]，这是电子烟特有的，"自己动手"是常见的做法。研究人员也应考虑这些添加剂对电子烟使用的影响。

9.4.2　风味电子烟中常见调味物质的化学和物理性质

卷烟中的调味物质

上述调味物质库[106]包括卷烟、自卷烟风味的化学分析。这补充了卷烟烟丝中的烟草基质总量的数据分析。卷烟中调味成分的鉴定和热解评估研究已被报道[110]。

ENDS中的调味物质

研究已报道了风味电子烟烟液[111,112]的化学图谱,表明添加到电子烟基质中的调味物质[113]的复杂性。虽然没有对数千种可能的调味物质进行全面研究,但对市售调味物质的分析表明,一些化学物质通常用于多种调味物质中[114]。本报告的一位作者最近发表了对已报道文献[115-117]的荟萃分析,其中指出了某些化学物质的出现频率,如乙基麦芽酚(47%)、香兰素(37%)、薄荷醇(29%)、乙基香兰素(23%)、芳樟醇(23%)、苯甲醛(22%)、苯甲醇(21%)、麦芽酚(20%)、肉桂醛(20%)、丁酸乙酯(19%)和羟基丙酮(16%)。这项工作还描述了这些风味添加剂对电子烟产生的气溶胶有害性的可能诱因[115]。调味物质要么被完全蒸馏到气溶胶中(它们的有害性取决于性质和释放水平)[118],与电子烟载体(丙二醇和甘油)反应,形成具有独特毒理学性质的缩醛化合物[119],或经过热降解生成有害物质,如羰基化合物[120]、活性氧[121]和挥发性有机化合物[122]。测定了几种风味成分的气粒相分配系数,发现它们与产品的毒性评估有关,因为可以确定这些化学物质进入体内的吸收位置[123]。

添加到电子烟的调味物质不仅赋予其特殊的口味,还增加了甜味(例如蔗糖)[124],就像在其他产品(包括无烟产品[125]和雪茄[126])中观察到的那样。甜味剂,包括人造甜味剂(如三氯蔗糖)在内,可以增加烟碱和烟草制品的吸引力[65],尽管表明甜味剂直接影响力的证据有限[127]。

其他烟草制品中的调味物质

有两项关于无烟烟草制品中调味物质的研究[128,129],对比迪烟[130]、丁香烟[131]和风味水烟[132]中的调味物质进行了定性和定量研究。

9.4.3 调味物质的毒性

电子烟中的调味物质会强烈增加气溶胶的一般毒性[133]。包括对坚果味的电子烟烟液中双乙酰的定量检测[134]在内,对电子烟烟液的几项针对性分析发现吸入该化学品的释放物会增加闭塞性细支气管炎或"爆米花肺"的疾病风险[135]。一些樱桃味的电子烟烟液会让使用者暴露于苯甲醛,尽管苯甲醛浓度很低[136]。为了评估疾病风险,通常将电子烟气溶胶中这些有害物质的水平与工作场所暴露限值进行比较。在细胞系和动物中对电子烟烟液和气溶胶中的调味物质进行毒理学评估表明,调味物质会以各种方式增加电子烟气溶胶的有害性[121,137]。一份报告显示,电子烟烟液基质中的调味物质加合物比它们的本体化合物更具有细胞毒性[119]。

调味物质还可能增加有害物质的释放,从而增加电子烟气溶胶的有害性。例如，烟液中的调味物质会增加羰基化合物和其他已知或可能致癌的化合物的释放量[120]。调味物质还通过扰乱体内的氧化平衡来增加毒性，因为它们增加了电子烟气溶胶中自由基和活性氧的存在，而不是仅由载体组成的普通烟液产生的自由基和活性氧[138]。这种导致气溶胶有害性增加的途径取决于设备操作参数，如功率输入和烟液成分[139]，因为功率越高，加热线圈的温度越高，导致风味化合物的降解越充分。

目前，除了电子烟外，无法确定调味物质对其他烟草和烟碱产品释放物毒性的净影响。

9.5 风味产品的监管

9.5.1 全球风味电子烟监管

调味物质在寻求戒烟的吸烟者对电子烟的接受度和满意度以及对年轻人尝试和持续使用的吸引力的影响问题有高度两极分化和激烈的争议[81,140,141]。这一争议反映在不同国家的公共卫生机构用来应对电子烟流行的不同方法上：调味物质是被禁止、被限制还是被允许（表9.1），这取决于公共卫生机构对现有信息的评估以及对辩论双方论点的看法[142,143]。目前，大约有100个国家通过新的法规或主要通过调整其他烟草制品的法规来规范电子烟的使用[144]。大约30个国家禁止电子烟的销售[145,146]。2020年6月，在美国巴尔的摩约翰·霍普金斯大学布隆伯格公共卫生学院全球烟草管制研究所的网站上搜索发现，35个国家存在涉及配料或调味物质的电子烟政策[147]。大多数都专注于标签和确保使用高质量的配料，只有五个国家（加拿大、芬兰、卢森堡、沙特阿拉伯和美国）对电子烟的调味物质有具体的规定[148]。芬兰禁止在所有电子烟中使用特征风味（如水果味、糖果味）；卢森堡的政策禁止会影响电子烟使用者健康感知的添加剂；沙特阿拉伯允许水果风味和薄荷醇，但禁止其他特征风味（如可可、香草、咖啡、茶、香辛料、糖果、口香糖、可乐和酒精）[149]，而美国食品药品监督管理局（FDA）发布了一项政策，禁止所有风味电子烟烟弹（除了烟草和薄荷醇风味的产品），因为有证据表明，调味物质强烈地影响年轻人对电子烟的使用，特别是非常受欢迎的烟弹产品（如JUUL）[150]。它还禁用电子烟[150,151]中使用的风味烟液。加拿大限制销售可能会吸引年轻人的调味物质，最近，新斯科舍省禁止了风味电子烟，其他省份也在考虑这样做[152,153]。

表9.1 各国对风味电子烟的监管办法（截至2020年6月）

国家或地区	管制规则
阿根廷、澳大利亚、阿塞拜疆、巴林、巴巴多斯、巴西、文莱达鲁萨兰国、柬埔寨、智利、哥伦比亚、哥斯达黎加、厄瓜多尔、埃及、萨尔瓦多、斐济、冈比亚、格鲁吉亚、洪都拉斯、匈牙利、冰岛、印度、印度尼西亚、伊朗、以色列、牙买加、日本、约旦、科威特、老挝、黎巴嫩、马来西亚、马尔代夫、毛里求斯、墨西哥、尼泊尔、新西兰、尼加拉瓜、挪威、阿曼、帕劳、巴拿马、巴拉圭、菲律宾、卡塔尔、韩国、摩尔多瓦、塞舌尔、新加坡、阿拉伯叙利亚共和国、塔吉克斯坦、泰国、东帝汶、多哥、土耳其、土库曼斯坦、乌干达、乌克兰、阿拉伯联合酋长国、乌拉圭、委内瑞拉玻利瓦尔共和国、越南	没有具体规定
奥地利、比利时、保加利亚、克罗地亚、塞浦路斯、捷克、丹麦、爱沙尼亚、法国、德国、希腊、爱尔兰、意大利、拉脱维亚、立陶宛、马耳他、荷兰、波兰、葡萄牙、罗马尼亚、斯洛伐克、斯洛文尼亚、西班牙、瑞典、英国（英格兰、北爱尔兰、威尔士）	电子烟烟液不应含有某些添加剂（未指明）。在电子烟的制造中应使用高质量的原料。只有加热或未加热状态下不会对人体健康构成危险的成分才能使用
加拿大	禁止销售含有某些添加剂的电子烟（未指明）。限制销售可能吸引年轻人的风味剂（包括建议的风味剂：糖果、甜点、大麻、软饮料和能量饮料）
芬兰	电子烟烟液不应包含某些添加剂和特征性风味（如糖果或水果口味）。在电子烟的制造中应该使用高质量的原料。只有加热或未加热状态下不会对人体健康构成危险的成分才能使用
沙特阿拉伯	电子烟烟液中的部分调味物质被禁止。水果口味和薄荷醇是允许的，但可可、香草、咖啡、茶、香辛料、糖果、口香糖、可乐和酒精口味是被禁止的
美国	除烟草和薄荷醇口味的产品外，所有含调味物质的烟弹都是被禁止的。在开放系统电子烟使用的风味液除外
卢森堡	禁止使用可能给人电子烟产品有益健康或降低健康风险印象的添加剂（如维生素）。咖啡因、牛磺酸和其他与能量或活力相关的兴奋剂是被禁止的。禁止添加任何增加颜色、改变释放物性质或促进吸入或烟碱摄取的添加剂。只有加热或未加热状态下不会对人体健康构成危险的成分才能使用。具有致癌、致突变或生殖毒性的未燃烧形式的添加剂也是被禁止的。在电子烟的制造中应该使用高质量的原料

9.5.2　全球风味烟草制品监管

对于电子烟，各国对其他风味烟草制品的监管也有所不同。表9.2列出了根据无烟草儿童运动[154]对法规的分析得出的不同国家和地区的监管方法。概括如下，有些地方没有相关法规，部分地区禁止特定种类的调味物质（含或不含薄荷醇），还有些地区全面禁止所有特征性调味物质（定义可能因司法管辖区而异），其中可能还包括薄荷醇。

表9.2　国家和地区对烟草制品（包括调味物质）含量的监管方法（截至2020年6月）

国家或地区	管制规则
阿富汗、阿根廷、阿塞拜疆、孟加拉国、白俄罗斯、贝宁、不丹、博茨瓦纳、布基纳法索、布隆迪、柬埔寨、喀麦隆、佛得角、乍得、中国、哥伦比亚、科摩罗、刚果、科特迪瓦、吉布提、埃及、萨尔瓦多、厄立特里亚、埃斯瓦蒂尼、斐济、法国、加蓬、德国、加纳、几内亚、冰岛、印度、伊朗、以色列、牙买加、日本、约旦、哈萨克斯坦、老挝、拉脱维亚、黎巴嫩、利比里亚、马达加斯加、马拉维、马来西亚、马尔代夫、马里、毛里求斯、墨西哥、缅甸、挪威、阿曼、巴基斯坦、巴拿马、秘鲁、菲律宾、波兰、卡塔尔、沙特阿拉伯、塞舌尔、新加坡、所罗门群岛、南非、西班牙、苏里南、瑞典、阿拉伯叙利亚共和国、东帝汶、多哥、土库曼斯坦、阿拉伯联合酋长国、委内瑞拉玻利瓦尔共和国、越南、包括东耶路撒冷在内的巴勒斯坦被占领土	法律没有授权管制卷烟的成分
阿尔及利亚、文莱达鲁萨兰国、智利、哥斯达黎加、刚果民主共和国、厄瓜多尔、冈比亚、格鲁吉亚、危地马拉、圭亚那、洪都拉斯、印度尼西亚、伊拉克、肯尼亚、卢旺达、泰国、乌克兰、坦桑尼亚联合共和国、乌拉圭	法律授权对卷烟的成分进行管制；但是，还没有发布任何规定
澳大利亚	卷烟的成分和添加剂不受国家层面的监管；然而，水果味和糖果味的卷烟在所有州和地区都是被禁止的。至少有一个州禁止薄荷醇
亚美尼亚、俄罗斯	法律规定了卷烟的特定成分，包括禁止薄荷和一些草本，以及其他未指明的风味品
巴西、加拿大[a]、爱尔兰[a]、意大利、毛里塔尼亚、摩尔多瓦、尼日尔[b]、罗马尼亚[a]、塞内加尔、斯洛文尼亚[a]、斯里兰卡、土耳其、乌干达、英国（英格兰、北爱尔兰、威尔士）[a]	法律规定了卷烟的特定成分，包括禁止使用糖和甜味剂、特征香精香料、薄荷和留兰香、香辛料和草本、促进烟碱摄取的成分、给人留下有益健康印象的成分、其他风味物质、与能量和活力有关的成分以及着色剂
埃塞俄比亚、尼日利亚	法律规定了卷烟的特定成分，包括禁止具有特征性的香精香料、给人留下健康益处印象的成分，以及与能量和活力相关的成分

a. 自2020年5月20日起，薄荷醇作为特征香味物质被禁止
b. 不禁止使用薄荷醇

2009年，美国食品药品监督管理局禁止卷烟中除薄荷醇以外的所有特征性调味物质[13]。2014年，欧盟出台了一系列政策，禁止在卷烟中加入薄荷醇以外的其他调味物质[28,155]。而且在2020年5月，《欧盟烟草制品指令》禁止在卷烟中添加薄荷醇[156,157]，然而这些规定并不适用于其他烟草制品[28]。其他的国家也已经在包括卷烟的烟草制品中禁止添加调味物质[158]，这些法规有的仍处于立法层面，如巴西[159]和乌干达[160]，有的最近开始实施，如土耳其[161]。2014年，新加坡禁止在水烟中加入水果味调味物质[63]。2010年，加拿大禁止销售所有风味卷烟和小雪茄，但不禁止薄荷醇味的卷烟，并禁止其他烟草制品中的所有风味剂，包括水烟、无烟烟草和比迪烟。2017年，艾伯塔省和安大略省禁止销售薄荷醇卷烟[161]。

在加拿大一项对法规反应的研究中，观察在薄荷醇产品被禁用后，那些每天、"偶尔"或"从不"使用薄荷醇卷烟的人（即非薄荷醇卷烟的使用者）的戒烟行为，

发现"每天"和"偶尔"使用薄荷醇卷烟的人比使用非薄荷醇产品的人尝试或已经戒烟的人数更多[162]。在一项针对美国旧金山居民的研究中，在禁止薄荷醇口味后，年轻人（18~24岁）和年长一些的人（25~34岁）对薄荷醇产品和电子烟的使用减少，但青少年的使用人数增加了，并且这项调查中有65%的人不相信禁令会在全市范围内得到统一执行[14]。埃塞俄比亚为严格的烟草管制战略提供了一个理想的例子，禁止销售所有风味烟草制品，包括那些含有薄荷醇的产品，并且全面禁止所有烟草制品，以保护公众健康[163]。该国的烟草制品使用量低于其他人类发展指数较低的国家[164]。

9.5.3 常用方法的优缺点

烟草制品的调味物质增加了它们的吸引力以及使用者和非使用者的安全性认知[100,165,166]。因此，国家监管部门已试图减少使用这些产品对公众健康产生的影响。如上所述，只有少数国家禁止在烟碱和烟草制品中添加特征调味物质，这是为了履行世界卫生组织《烟草控制框架公约》规定的义务，以及为了保护青少年和公众健康[158]。

在全球范围内，对调味物质的不同管制方法呈现出一系列政策，可以根据它们对公众健康的估计效益进行权衡。有些没有涉及风味成分、烟草成分或一般成分，这种缺乏具体内容的做法可能会留下一些漏洞，烟草行业可以借此解决年轻人的烟草使用问题。部分禁止某些调味物质或禁止某些烟碱和烟草制品中的调味物质也为烟草行业宣传其他调味物质或调味物质替代型的烟碱和烟草制品留下了广阔的空间，也因此挑战了监管机构的工作。全面禁止所有口味的烟碱和烟草制品似乎是遏制青少年使用烟草制品的一种强有力的方法，然而监管机构应该考虑到一个潜在的论点，即调味物质可能是一种工具，来使人们适应从传统卷烟到其他产品的转变。监管机构应确保撤出市场上可用于在开放系统电子烟等产品中递送烟碱的定制产品，否则，使用者将会在他们的产品中添加非传统和非法的添加剂[167]。由于相关风险超过其潜在的公共卫生益处，因此也有人呼吁从电子烟中去除调味物质[168,169]。所有法规都应通过广泛的宣传和提高认识活动来解决供给和需求问题，以寻求公众的支持来实施[170-172]。

9.5.4 对烟碱产品特定风味监管产生的影响

由于大多数关于风味烟草制品的法规是最近出台的，关于其影响力的信息有限[173-175]。尽管如此，我们还是可以从限制烟草制品调味物质的政策对其消费的影响中吸取经验[176-178]。例如，加拿大限制销售风味雪茄（<1.4 g）对其他雪茄销售的影响的评估显示，总体上有所下降。此外，在禁止使用含薄荷醇的燃烧型烟草制品后，更多的含薄荷醇的吸烟者戒烟或尝试戒烟，而吸不含薄荷醇产品的人

没有此现象[162]，尽管作者指出，禁止特定调味物质和产品类型可能降低政策的有效性[179]。类似地，在美国纽约对禁止添加调味物质的烟草制品（如雪茄、小雪茄、自卷烟）的效果进行评估显示，青少年使用任何烟草制品[180]的概率都显著降低（28%）。而在美国旧金山禁止薄荷醇的效果显示，年轻人中的电子烟使用量而不是卷烟使用量减少[14]。另一份报告指出，降低风味烟草制品的可获得性和可负担性有助于减少所有烟草制品的使用[181]。

由于没有或只有相对较新的电子烟调味物质法规，因此没有电子烟或其他烟草制品中使用的纵向数据[182]。已使用影响分析和建模来估计法规对电子烟中调味物质使用的可能影响[143,183,184]，包括对所有烟草制品使用的净影响[185]。一份报告显示，尽管重新评估的数据表明对这两种产品监管的净效果有利于公众健康[185]，对电子烟调味物质的限制性法规可能会增加年轻人使用卷烟以及同时使用电子烟和燃烧型卷烟的意愿[183]。对电子烟中所有调味物质、卷烟中的薄荷醇或卷烟中所有调味物质禁令影响的评估表明，禁止这两种产品中的所有调味物质的措施将降低燃烧型卷烟和电子烟的吸烟率，然而卷烟的使用率仍比现状高2.7%[169]。由于加热型烟草制品被视为烟草制品，并且在某些政策中将加热型烟草制品包括在内，因此对这些产品中的调味物质没有具体规定[186]。已经证明，电子烟中的调味物质会增加这些产品的滥用风险，特别是对于青少年和年轻人，没有证据表明电子烟中的特定调味物质会帮助吸烟者戒烟。在考虑管制措施时，有必要考虑无味电子烟在支持戒烟和减少滥用燃烧型烟草方面是否具有与风味电子烟相似的效果。

9.5.5 调味物质的未来监管

调味物质增加了烟碱和烟草制品的吸引力、持续使用、使用范围、依赖性和有害性，并增加了新一代烟碱和烟草成瘾者的风险。烟碱成瘾和暴露与释放的其他有害物质对公众健康造成重大负担[187]。证明调味物质对公众健康有益的唯一方式就是证明其在烟草制品中有害性更小，风险更小，并有助于减少燃烧型烟草的使用[81]。尽管如此，应鼓励此类替代风味烟草制品的使用者停止使用，以戒除烟碱成瘾，并避免再次使用烟草或烟碱产品。

世界卫生组织烟草制品管制研究小组在关于风味烟草和烟碱产品的第七份报告[31]中指出，研究重点是系统监测常规风味、传统及新型烟草制品的全球流行病学，鉴别调味物质如何促进这些产品对人的吸引力，以及调味物质的有害性及其对健康的影响。这份报告证实调味物质在烟碱和烟草制品中仍然普遍存在，电子烟的受欢迎程度有所提高，虽然对烟碱和烟草制品中调味物质的可用性有规定，但各国的情况差别很大，应系统地监测风味烟碱和烟草制品使用的全球流行病学情况。

这份报告还对风味电子烟对青少年和年轻人的吸引力表示担忧。在这些人群中，电子烟是最常用的烟碱和烟草制品，而且在所有年龄组中，风味产品使用量最高。因此，电子烟的口味可能对青少年和年轻人具有独特的吸引力。一种解释是可能因为非传统口味的广泛存在，这种口味在这些群体中比在老年人中更受欢迎。研究表明，电子烟的水果味会增强对使用者的奖赏和强化作用[24,91]。

为了更好地了解电子烟口味对使用的影响，将其分为几类，类似于燃烧型烟草[107]。例如，蓝莓味和青苹果味被认为是水果口味，而松饼和纸杯蛋糕口味被归类为甜点味。这种分类有助于解释具有越来越多口味选择的产品的结果。

使用风味电子烟与使用其他烟草制品的可能性关联性更大[188]，尤其是在年轻人中[189]。因此，减少所有烟草使用的政策必须基于实际使用特征[190]。从监管的角度来看，调味物质对电子烟吸引力和使用性的影响是有争议的，因为关于调味物质的可获得性是否有助于从燃烧型卷烟转向电子烟的使用，增加年轻人对电子烟的使用存在激烈的争论。已有证据表明，电子烟有不同口味是吸烟者接受这些产品的重要考虑因素[24,191,192]。

烟碱和烟草制品的口味，特别是电子烟的口味，通过味道和感官体验的生动描述进行营销[193]。也许不足为奇，绝大多数电子烟使用者（青少年、年轻人和成年人）支持"吸进来的味道我喜欢"作为使用电子烟的理由[35]。而且电子烟使用者将口味和独特的风味作为两个最重要的因素来选择电子烟[157]。多种口味的电子烟可能会使使用者更有可能找到一种对他们有吸引力的产品，也可能解释了为什么调味物质在电子烟中比在其他烟草和烟碱产品中使用得更多。由于调味物质的吸引力和促进持续使用的特点普遍存在于其他产品中，例如无烟烟草和燃烧型卷烟以及加热型烟草制品等较新的烟草制品。随着全球烟草监管日益关注电子烟，为减少其使用量，监管机构应继续监测其他烟碱和烟草制品的使用。

对电子烟的监管始于主要的烟草公司开始生产后的一段时间[194]。一些倡导团体批评美国FDA没有采取足够快的行动来阻止青少年使用电子烟，这可能导致美国青少年对电子烟的高使用率，卫生组织赢得了一场诉讼，迫使美国FDA将电子烟上市前申请中成分安全性研究的提交截止日期从2022年提前到2020年5月12日[195-198]。对包括卷烟在内的其他烟草制品中的调味物质的监管也落后于它们在人群中的传播速度[37]。19世纪无烟烟草中引入了调味物质，而卷烟中的调味物质的使用仅仅在几十年前才兴盛起来[1,4,199]。调味物质广泛用于其他烟草制品，如水烟和比迪烟[2,32]，众多的调味物质选择有助于它们在人群中的流行，尤其是在年轻人中[27,61]。调味物质不仅会增加对烟草制品的上瘾程度，还会成倍增加有害物质的释放量。因此，调味物质的监管是烟草监管中减少危害和预防措施的交汇点[200]。阻止烟草流行的最佳监管方法是对所有风味烟草制品制定补充政策，最终终止烟草流行[201,202]。

9.6 讨 论

在烟碱和烟草制品中使用调味物质是有争议的，因为它们已被明确证明会增加这些产品的使用和吸引力，特别是在年轻人中。电子烟仍然是一个主要的关注点，因为它们越来越受欢迎。它们吸引人的一个主要特点就是口味多样，这促进了人们尝试使用和长期使用。此外，新的证据表明，调味物质可能会以特有的方式增加电子烟等新产品的有害性。口味的原因使烟草和烟碱的使用增加，增加了公共卫生负担。然而，调味物质也可能被用来减轻负担，因为一些成年吸烟者表示，像电子烟等产品中的调味物质有助于他们尝试戒烟或减少吸烟。决策者在监管烟草制品口味时应该考虑这一点。在限制烟碱和烟草使用的传播和发展以及减少燃烧型烟草制品的使用的所有监管方法中，烟草制品中调味物质的监管应该是一个优先事项。

9.7 研究差距、优先事项和问题

对风味烟碱和烟草制品，尤其是电子烟，有必要从以下方面进行研究：
- 风味烟碱和烟草制品使用的全球流行病学监测研究；
- 对电子烟使用特性、使用原因、长期和持续使用的风味的纵向研究；
- 科学分类，提供对风味烟草制品及其化学成分进行分类的方法；
- 就"特征风味"的当前定义达成共识；
- 对新型烟碱和烟草制品（尤其是电子烟）中风味的法规、限制和禁令进行影响分析，包括在假设情景和任务中对风味相关政策的反应建模；
- 生物医学和行为学研究，按年龄组和烟草使用状况，研究风味对烟碱和烟草制品使用奖赏体验的影响；
- 烟碱和烟草制品中的个别风味成分和化学物质以及新形成的组合体的有害性。

全球优先事项是：
- 建立证据，证明调味物质对不同年龄组使用烟碱和烟草制品以及不同国家使用不同产品的影响；
- 确定烟碱和烟草制品的口味对数十年来减少人群中烟碱和烟草使用的影响；
- 在各国之间交流烟草制品调味物质的监管经验。

向决策者和世界卫生组织等国际组织提出的问题包括：
- 如何快速解决烟碱和烟草制品中使用调味物质的问题来防止新一代使用

者对烟碱和烟草产生依赖？
- 可以做些什么来走在企业前端，解决企业利用调味物质来提高产品的吸引力和使用率的问题？
- 能否建立一个稳健、有发展前景、可持续的监管模式来解决新型或风险降低的烟草制品的类似问题？

9.8 政策建议

非系统性地监管风味烟碱和烟草制品不会扭转烟草流行的趋势。将各种政策工具与所有烟碱和烟草制品使用的全景相结合，有助于卫生机构解决风味电子烟的问题[203,204]。电子烟可以作为一个切入口，进而增加对所有烟草制品的监管[205]，以实现未来无烟碱和无烟草的最终目标[206]。新型烟碱和烟草制品的调味物质政策应包括以下内容：

- 在调味物质未被禁止的地方，对烟碱和烟草制品的监管应在全球范围内保持一致。也就是说，应该对所有烟碱和烟草制品而不是每种产品的调味物质供应进行类似的监管。
- 应研究像电子烟或加热型烟草制品这样的产品中的特征风味在帮助吸烟者戒烟方面的可能作用。

9.9 参考文献

[1] Kostygina G, Ling PM. Tobacco industry use of flavourings to promote smokeless tobacco products. Tob Control. 2016; 25(Suppl 2):ii40–9.

[2] Ben Taleb Z, Breland A, Bahelah R, Kalan ME, Vargas-Rivera M，Jaber R et al. Flavored versus nonflavored waterpipe tobacco: A comparison of toxicant exposure, puff topography, subjective experiences, and harm perceptions. Nicotine Tob Res. 2018；21(9):1213–9.

[3] Meernik C, Baker HM, Kowitt SD, Ranney LM, Goldstein AO. Impact of non-menthol flavours in e-cigarettes on perceptions and use: an updated systematic review. BMJ Open. 2019；9(10):e031598.

[4] Carpenter CM, Wayne GF, Pauly JL, Koh HK, Connolly GN. New cigarette brands with flavors that appeal to youth: tobacco marketing strategies. Health Aff (Millwood). 2005；24(6):1601–10.

[5] Gendall P, Hoek J. Role of flavours in vaping uptake and cessation among New Zealand smokers and non-smokers: a cross-sectional study. Tob Control. 2020: 055469.

[6] Yao T, Jiang N, Grana R, Ling PM, Glantz SA. A content analysis of electronic cigarette manufacturer websites in China. Tob Control. 2016；25(2):188–94.

[7] He G, Lin X, Ju G, Chen Y. Mapping public concerns of electronic cigarettes in China. J Psychoactive Drugs. 2019:52(1):1–7.

[8] Cullen KA, Gentzke AS, Sawdey MD, Chang JT, Anic GM, Wang TW et al. e-Cigarette use among youth in the United States, 2019. JAMA. 2019；322(21):2095–2103.

[9] Prohibited additives. In: Tobacco and Vaping Products Act (S.C. 1997, c. 13). Ottawa: Department of Justice; 1997 (https://laws.justice.gc.ca/eng/acts/T-11.5/page-12.html).

[10] Hiscock R, Silver K, Zatoński M, Gilmore AB. Tobacco industry tactics to circumvent and undermine the menthol cigarette ban in the UK. Tob Control. 2020；055769.

[11] Farrelly MC, Loomis BR, Kuiper N, Han B, Gfroerer J, Caraballo RS et al. Are tobacco control policies effective in reducing young adult smoking? J Adolesc Health. 2014；54(4):481–6.

[12] Talhout R, van de Nobelen S, Kienhuis AS. An inventory of methods sui table to assess additive-induced characterising flavours of tobacco products. Drug Alcohol Depend. 2016；161:9–14.

[13] Food and Drug Administration. Enforcement of general tobacco standard special rule for cigarettes. Fed Reg. 2009；E9:23144 (https://www.federalregister.gov/documents/2009/09/25/E9-23144/enforcement-of-general-tobacco-standard-special-rule-for-cigarettes, accessed 18 February 2020).

[14] Yang Y, Lindblom EN, Salloum RG, Ward KD. The impact of a comprehensive tobacco product flavor ban in San Francisco among young adults. Addict Behav Rep. 2020；11:100273.

[15] Family Smoking Prevention and Tobacco Control. Division A. Public Law. 22 June 2009; 111-31 (https://www.govinfo.gov/content/pkg/PLAW-111publ31/pdf/PLAW-111publ31.pdf, accessed 30 November 2020).

[16] Tobacco. Geneva: World Health Organization; 2020 (https://www.who.int/news-room/factsheets/detail/tobacco, accessed 18 August 2020).

[17] Kennedy RD, Awopegba A, De León E, Cohen JE. Global approaches to regulating electronic cigarettes. Tob Control. 2017；26(4):440–5.

[18] FDA finalizes enforcement policy on unauthorized flavored cartridge-based e-cigarettes that appeal to children, including fruit and mint. Silver Spring (MD): Food and Drug Administration; 2020 (https://www.fda.gov/news-events/press-announcements/fda-finalizes-enforcement-policy-unauthorized-flavored-cartridge-based-e-cigarettes-appeal-children, accessed 23 June 2020).

[19] Drope J, Schulger NW, editors. The Tobacco Atlas, sixth edition. Atlanta (GA): American Cancer Society; 2018.

[20] Dawkins L, Turner J, Roberts A, Soar K. "Vaping" profiles and preferences: an online survey of electronic cigarette users. Addiction. 2013；108(6):1115–25.

[21] WHO global report on trends in prevalence of tobacco smoking 2015. Geneva: World Health Organization; 2015.

[22] Zhu SH, Sun JY, Bonnevie E, Cummins SE, Gamst A, Yin L et al. Four hundred and sixty brands of e-cigarettes and counting: implications for product regulation. Tob Control. 2014; 23(suppl 3):iii3–9.

[23] Hsu G, Sun JY, Zhu SH. Evolution of electronic cigarette brands from 2013–2014 to 2016–2017: Analysis of brand websites. J Med Internet Res. 2018;20(3):e80.

[24] Farsalinos KE, Romagna G, Tsiapras D, Kyrzopoulos S, Spyrou A, Voudris V. Impact of flavour variability on electronic cigarette use experience: an internet survey. Int J Environ Res Public Health. 2013; 10(12):7272–82.

[25] Jones DM, Ashley DL, Weaver SR, Eriksen MP. Flavored ENDS use among adults who have used cigarettes and ENDS, 2016–2017. Tob Regul Sci. 2019; 5(6):518–31.

[26] Bold KW, Krishnan-Sarin S. e-Cigarettes: Tobacco policy and regulation. Curr Addict Rep. 2019; 6(2):75–85.

[27] Ambrose BK, Day HR, Rostron B, Conway KP, Borek N, Hyland A et al. Flavored tobacco product use among US youth aged 12–17 years, 2013–2014. JAMA. 2015; 314(17):1871–3.

[28] Morean ME, Butler ER, Bold KW, Kong G, Camenga DR, Simon P et al. Preferring more e-cigarette flavors is associated with e-cigarette use frequency among adolescents but not adults. PLoS One. 2018; 13(1):e0189015.

[29] Schneller LM, Bansal-Travers M, Goniewicz ML, McIntosh S, Ossip D, O'Connor RJ. Use of flavored e-cigarettes and the type of e-cigarette devices used among adults and youth in the US – Results from wave 3 of the Population Assessment of Tobacco and Health Study (2015–2016). Int J Environ Res Public Health. 2019; 16(16):2991.

[30] Heated tobacco products (HTPs) market monitoring information sheet 2018. Geneva: World Health Organization; 2018 (https://apps.who.int/iris/bitstream/handle/10665/273459/WHON-MH-PND-18.7-eng.pdf?ua=1, accessed 18 August 2020).

[31] Krishnan-Sarin S, Green BG, Jordt SE, O'Malley SS. The science of flavour in tobacco products. In: WHO Study Group on Tobacco Product Regulation, Seventh report (WHO Technical Report Series No. 1015). Geneva: World Health Organization; 2019.

[32] Bonhomme MG, Holder-Hayes E, Ambrose BK, Tworek C, Feirman SP, King BA et al. Flavoured non-cigarette tobacco product use among US adults: 2013–2014. Tob Control. 2016; 25(Suppl 2):ii4–13.

[33] Villanti AC, Mowery PD, Delnevo CD, Niaura RS, Abrams DB, Giovino GA. Changes in the prevalence and correlates of menthol cigarette use in the USA, 2004-2014. Tob Control. 2016; 25(Suppl 2):ii14–20.

[34] Odani S, Armour B, Agaku IT. Flavored tobacco product use and its association with indicators of tobacco dependence among US adults, 2014–2015. Nicotine Tob Res. 2020; 22(6):1004–15.

[35] Rostron BL, Cheng YC, Gardner LD, Ambrose BK. Prevalence and reasons for use of flavored cigars and ENDS among US youth and adults: Estimates from wave 4 of the PATH study, 2016–2017. Am J Health Behav. 2020; 44(1):76–81.

[36] Harrell MB, Weaver SR, Loukas A, Creamer M, Marti CN, Jackson CD et al. Flavored e-cigarette use: Characterizing youth, young adult, and adult users. Prev Med Rep. 2017; 5:33–40.

[37] Leventhal AM, Goldenson NI, Cho J, Kirkpatrick MG, McConnell RS, Stone MD et al. Flavored e-cigarette use and progression of vaping in adolescents. Pediatrics. 2019; 144(5):e20190789.

[38] Villanti AC, Johnson AL, Glasser AM, Rose SW, Ambrose BK, Conway KP et al. Association of flavored tobacco use with tobacco initiation and subsequent use among US youth and adults, 2013–2015. JAMA Netw Open. 2019;2(10):e1913804.

[39] Goldenson NI, Leventhal AM, Simpson KA, Barrington-Trimis JL. A review of the use and appeal of flavored electronic cigarettes. Curr Addict Rep. 2019;6(2):98–113.

[40] Zare S, Nemati M, Zheng Y. A systematic review of consumer preference for e-cigarette attributes: Flavor, nicotine strength, and type. PLoS One. 2018; 13(3):e0194145.

[41] Wang TW, Gentzke AS, Creamer MR, Cullen KA, Holder-Hayes E, Sawdey MD et al. Tobacco

product use and associated factors among middle and high school students – United States, 2019. Morbid Mortal Wkly Rep. 2019; 68(12):1–22.

[42] Schneller LM, Bansal-Travers M, Goniewicz ML, McIntosh S, Ossip D, O'Connor RJ. Use of flavored electronic cigarette refill liquids among adults and youth in the US – Results from wave 2 of the Population Assessment of Tobacco and Health Study (2014–2015). PLoS One. 2018; 13(8):e0202744.

[43] Russell C, McKeganey N, Dickson T, Nides M. Changing patterns of first e-cigarette flavor used and current flavors used by 20,836 adult frequent e-cigarette users in the USA. Harm Reduct J. 2018; 15(1):33.

[44] Berg CJ. Preferred flavors and reasons for e-cigarette use and discontinued use among never, current, and former smokers. Int J Public Health. 2016; 61(2):225–36.

[45] Du P, Bascom R, Fan T, Sinharoy A, Yingst J, Mondal P et al. Changes in flavor preference in a cohort of long-term electronic cigarette users. Ann Am Thorac Soc. 2020; 17(5):573–81.

[46] Gendall P, Hoek J. Role of flavours in vaping uptake and cessation among New Zealand smokers and non-smokers: a cross-sectional study. Tob Control. 2020; doi.org/10.1136/tobaccocontrol-2019-055469.

[47] Soneji SS, Knutzen KE, Villanti AC. Use of flavored e-cigarettes among adolescents, young adults, and older adults: Findings from the Population Assessment for Tobacco and Health Study. Public Health Rep. 2019; 134(3):282–92.

[48] Sharan RN, Chanu TM, Chakrabarty TK, Farsalinos K. Patterns of tobacco and e-cigarette use status in India: a cross-sectional survey of 3000 vapers in eight Indian cities. Harm Reduct J. 2020; 17(1):21.

[49] Okawa S, Tabuchi T, Miyashiro I. Who uses e-cigarettes and why? e-Cigarette use among older-adolescents and young adults in Japan: JASTIS study. J Psychoactive Drugs. 2020; 52(1):37–45.

[50] Minaker LM, Ahmed R, Hammond D, Manske S. Flavored tobacco use among Canadian students in grades 9 through 12: prevalence and patterns from the 2010–2011 youth smoking survey. Prev Chronic Dis. 2014; 11:E102.

[51] Kaleta D, Usidame B, Szosland-Faltyn A, Makowiec-Dabrowska T. Use of flavoured cigarettes in Poland: data from the global adult tobacco survey (2009–2010). BMC Public Health. 2014;14:127.

[52] Moodie C, Thrasher JF, Cho YJ, Barnoya J, Chaloupka FJ. Flavour capsule cigarettes continue to experience strong global growth. Tob Control. 2019;28(5):595–6.

[53] White V, Williams T. Australian secondary school students' use of tobacco in 2014. Victoria: Centre for Behavioural Research in Cancer, Cancer Council Victoria; 2014.

[54] Emond JA, Soneji S, Brunette MF, Sargent JD. Flavour capsule cigarette use among US adult cigarette smokers. Tob Control. 2018;27(6):650–5.

[55] Directive 2014/40/EU of the European Parliament and of the Council of 3 April 2014 on the approximation of the laws, regulations and administrative provisions of the Member States concerning the manufacture, presentation and sale of tobacco and related products and repealing Directive 2001/37/EC. Offic J Eur Union. 2014; L127:1–38.

[56] Kim M. Philip Morris International introduces new heat-not-burn product, IQOS, in South Korea. Tob Control. 2018;27(e1):e76–8.

[57] Valinsky J. A new non-vaping, non-smoking way to get nicotine has come to America. CNN Business, 4 October 2019.

[58] McKelvey K, Popova L, Kim M, Chaffee BW, Vijayaraghaven M, Ling P et al. Heated tobacco products likely appeal to adolescents and young adults. Tob Control. 2018;27(Suppl 1):s41–7.

[59] Sustainability. Lausanne: Philip Morris International; 2017 (https://www.pmi.com/sustainability, accessed 1 November 2017).

[60] Sutanto E, Miller C, Smith DM, O'Connor RJ, Quah ACK, Cummings KM et al. Prevalence, use behaviors, and preferences among users of heated tobacco products: Findings from the 2018 ITC Japan survey. Int J Environ Res Public Health. 2019;16(23):10.3390.

[61] Klein SM, Giovino GA, Barker DC, Tworek C, Cummings KM, O'Connor RJ. Use of flavored cigarettes among older adolescent and adult smokers: United States, 2004–2005. Nicotine Tob Res. 2008;10(7):1209–14.

[62] King BA, Tynan MA, Dube SR, Arrazola R. Flavored-little-cigar and flavored-cigarette use among US middle and high school students. J Adolesc Health. 2014;54(1):40–6.

[63] Kong G, Morean ME, Cavallo DA, Camenga DR, Krishnan-Sarin S. Reasons for electronic cigarette experimentation and discontinuation among adolescents and young adults. Nicotine Tob Res. 2015;17(7):847–54.

[64] Kong G, Bold KW, Morean ME, Bhatti H, Camenga DR, Jackson A et al. Appeal of JUUL among adolescents. Drug Alcohol Depend. 2019;205:107691.

[65] Goldenson NI, Kirkpatrick MG, Barrington-Trimis JL, Pang RD, McBeth JF, Pentz MA et al. Effects of sweet flavorings and nicotine on the appeal and sensory properties of e-cigarettes among young adult vapers: Application of a novel methodology. Drug Alcohol Depend. 2016;168:176–80.

[66] Jackson A, Green B, Erythropel HC, Kong G, Cavallo DA, Eid T et al. Influence of menthol and green apple e-liquids containing different nicotine concentrations among youth e-cigarette users. Exp Clin Psychopharmacol. 2020:10.1037.

[67] Kong G, Bold KW, Simon P, Camenga DR, Cavallo DA, Krishnan-Sarin S. Reasons for cigarillo initiation and cigarillo manipulation methods among adolescents. Tob Regul Sci. 2017;3(2 Suppl 1):S48–58.

[68] Sharma E, Clark PI, Sharp KE. Understanding psychosocial aspects of waterpipe smokingamong college students. Am J Health Behav. 2014;38(3):440–7.

[69] Choi K, Fabian L, Mottey N, Corbett A, Forster J. Young adults' favorable perceptions of snus, dissolvable tobacco products, and electronic cigarettes: findings from a focus group study. Am J Public Health. 2012;102(11):2088–93.

[70] Moodie C, Ford A, Mackintosh A, Purves R. Are all cigarettes just the same? Female's perceptions of slim, coloured, aromatized and capsule cigarettes. Health Educ Res. 2015;30(1):1–12.

[71] Griffiths MA, Harmon TR, Gilly MC. Hubble bubble trouble: The need for education about and regulation of hookah smoking. J Public Policy Marketing. 2011;30(1):119–32.

[72] Roskin J, Aveyard P. Canadian and English students' beliefs about waterpipe smoking: A qualitative study. BMC Public Health. 2009 ; 9:10.

[73] Hammal F, Wild TC, Nykiforuk C, Abdullahi K, Mussie D, Finegan BA. Waterpipe (hookah) smoking among youth and women in Canada is new, not traditional. Nicotine Tob Res.

2016;18(5):757-62.

[74] Kowitt SD, Meernik C, Baker HM, Osman A, Huang LL, Goldstein AO. Perceptions and experiences with flavored non-menthol tobacco products: A systematic review of qualitative studies. Int J Environ Res Public Health. 2017;14(4):338.

[75] Sterling KL, Fryer CS, Fagan P. The most natural tobacco used: A qualitative investigation of young adult smokers' risk perceptions of flavored little cigars and cigarillos. Nicotine Tob Res. 2016;18(5):827-33.

[76] Garrison KA, O'Malley SS, Gueorguieva R, Krishnan-Sarin S. A fMRI study on the impact of advertising for flavored e-cigarettes on susceptible young adults. Drug Alcohol Depend. 2018; 186:233-41.

[77] Audrain-McGovern J, Rodriguez D, Pianin S, Alexander E. Initial e-cigarette flavoring and nicotine exposure and e-cigarette uptake among adolescents. Drug Alcohol Depend. 2019; 202:149-55.

[78] Audrain-McGovern J, Strasser AA, Wileyto EP. The impact of flavoring on the rewarding and reinforcing value of e-cigarettes with nicotine among young adult smokers. Drug Alcohol Depend. 2016; 166:263-7.

[79] Krishnan-Sarin S, Morean ME, Camenga DR, Cavallo DA, Kong G. e-Cigarette use among high school and middle-school adolescents in Connecticut. Nicotine Tob Res. 2015;17(7):810-8.

[80] Kim H, Lim J, Buehler SS, Brinkman MC, Johnson NM, Wilson L et al. Role of sweet and other flavours in liking and disliking of electronic cigarettes. Tob Control. 2016;25(Suppl 2):ii55-61.

[81] Landry RL, Groom AL, Vu THT, Stokes AC, Berry KM, Kesh A et al. The role of flavors in vaping initiation and satisfaction among US adults. Addict Behav. 2019;99:106077.

[82] Avelar AJ, Akers AT, Baumgard ZJ, Cooper SY, Casinelli GP, Henderson BJ. Why flavored vape products may be attractive: Green apple tobacco flavor elicits reward-related behavior, upregulates nAChRs on VTA dopamine neurons, and alters midbrain dopamine and GABA neuron function. Neuropharmacology. 2019;158:107729.

[83] Wickham RJ. How menthol alters tobacco-smoking behavior: A biological perspective. Yale J Biol Med. 2015;88(3):279-87.

[84] Romijnders KA, Krüsemann EJ, Boesveldt S, Graaf K, Vries H, Talhout R. e-Liquid flavor preferences and individual factors related to vaping: A survey among Dutch never-users, smokers, dual users, and exclusive vapers. Int J Environ Res Public Health. 2019;16(23):4661.

[85] Wipfli H, Bhuiyan MR, Qin X, Gainullina Y, Palaganas E, Jimba M et al. Tobacco use and e-cigarette regulation: Perspectives of university students in the Asia-Pacific. Addict Behav. 2020 ; 107:106420.

[86] Ashare RL, Hawk LW Jr, Cummings KM, O'Connor RJ, Fix BV, Schmidt WC. Smoking expectancies for flavored and non-flavored cigarettes among college students. Addict Behav. 2007 ; 32(6):1252-61.

[87] Feirman SP, Lock D, Cohen JE, Holtgrave DR, Li T. Flavored tobacco products in the United States: A systematic review assessing use and attitudes. Nicotine Tob Res. 2016;18(5):739-49.

[88] Oliver AJ, Jensen JA, Vogel RI, Anderson AJ, Hatsukami DK. Flavored and nonflavored smokeless tobacco products: rate, pattern of use, and effects. Nicotine Tob Res. 2013;15(1):88-92.

[89] Odani S, Armour B, Agaku IT. Flavored tobacco product use and its association with indicators

of tobacco dependence among US adults, 2014–2015. Nicotine Tob Res. 2020;22(6):1004–15.

[90] Lynn WR, Davis RM, Novotny TE, editors. The health consequences of smoking: Nicotine addiction: A report of the Surgeon General. Atlanta (GA): Center for Health Promotion and Education, Office on Smoking and Health; 1988.

[91] DeVito EE, Valentine GW, Herman AI, Jensen KP, Sofuoglu M. Effect of menthol-preferring status on response to intravenous nicotine. Tob Regul Sci. 2016;2(4):317–28.

[92] Henderson BJ, Wall TR, Henley BM, Kim CH, McKinney S, Lester HA. Menthol enhances nicotine reward-related behavior by potentiating nicotine-induced changes in nAChR function, nAChR upregulation, and DA neuron excitability. Neuropsychopharmacology. 2017;42(12):2285–91.

[93] Alsharari SD, King JR, Nordman JC, Muldoon PP, Jackson A, Zhu AZX et al. Effects of menthol on nicotine pharmacokinetic, pharmacology and dependence in mice. PLoS One. 2015;10(9):e0137070.

[94] Villanti AC, Johnson AL, Ambrose BK, Cummings KM, Stanton CA, Rose SW et al. Flavored tobacco product use in youth and adults: Findings from the first wave of the PATH study (2013–2014). Am J Prev Med. 2017;53(2):139–51.

[95] Ahijevych K, Garrett BE. The role of menthol in cigarettes as a reinforcer of smoking behavior. Nicotine Tob Res. 2010;12(Suppl 2):S110–6.

[96] DeVito EE, Jensen KP, O'Malley SS, Gueorguieva R, Krishnan-Sarin S, Valentine G et al. Modulation of "protective" nicotine perception and use profile by flavorants: Preliminary findings in e-cigarettes. Nicotine Tob Res. 2020;22(5):771–81.

[97] Leventhal AM, Goldenson NI, Barrington-Trimis JL, Pang RD, Kirkpatrick MG. Effects of non-tobacco flavors and nicotine on e-cigarette product appeal among young adult never, former, and current smokers. Drug Alcohol Depend. 2019;203:99–106.

[98] Krishnan-Sarin S, Green BG, Kong G, Cavallo DA, Jatlow P, Gueorguieva R et al. Studying the interactive effects of menthol and nicotine among youth: An examination using e-cigarettes. Drug Alcohol Depend. 2017; 180:193–9.

[99] Kostygina G, Glantz SA, Ling PM. Tobacco industry use of flavours to recruit new users of little cigars and cigarillos. Tob Control. 2016; 25(1):66–74.

[100] Romijnders KAGJ, van Osch L, de Vries H, Talhout R. Perceptions and reasons regarding e-cigarette use among users and non-users: A narrative literature review. Int J Environ Res Public Health. 2018; 15(6):1190.

[101] Wilson N, Hoek J, Thomson G, Edwards R. Should e-cigarette use be included in indoor smoking bans? Bull World Health Organ. 2017; 95(7):540–1.

[102] Shearston J, Lee L, Eazor J, Meherally S, Park SH, Vilcassim MR et al. Effects of exposure to direct and secondhand hookah and e-cigarette aerosols on ambient air quality and cardiopulmonary health in adults and children: protocol for a panel study. BMJ Open. 2019; 9(6):e029490.

[103] Visser WF, Klerx WN, Cremers HWJM, Ramlal R, Schwillens PL, Talhout R. The health risks of electronic cgarette use to bystanders. Int J Environ Res Public Health. 2019; 16(9):1525.

[104] Hajek P, Phillips-Waller A, Przulj D, Pesola F, Myers Smith K, Bisal N et al. A randomized trial of e-cigarettes versus nicotine-replacement therapy. N Engl J Med. 2019; 380(7):629–37.

[105] Glasser A, Vojjala M, Cantrell J, Levy DT, Giovenco DP, Abrams D et al. Patterns of e-cigarette

use and subsequent cigarette smoking cessation over two years (2013/2014 to 2015/2016) in the Population Assessment of Tobacco and Health (PATH) study. Nicotine Tob Res. 2020: doi: 10.1093/ntr/ntaa182.

[106] Krüsemann EJ, Visser WF, Cremers JW, Pennings JL, Talhout R. Identification of flavour additives in tobacco products to develop a flavour library. Tob Control. 2018; 27(1):105–11.

[107] Krüsemann EJZ, Boesveldt S, de Graaf K, Talhout R. An e-liquid flavor wheel: A shared vocabulary based on systematically reviewing e-liquid flavor classifications in literature. Nicotine Tob Res. 2019; 21(10):1310–9.

[108] Cooper M, Harrell MB, Perry CL. A qualitative approach to understanding real-world electronic cigarette use: Implications for measurement and regulation. Prev Chronic Dis. 2016; 13:E07.

[109] Cox S, Leigh NJ, Vanderbush TS, Choo E, Goniewicz ML, Dawkins L. An exploration into "do-ityourself" (DIY) e-liquid mixing: Users' motivations, practices and product laboratory analysis. Addict Behav Rep. 2019; 9:100151.

[110] Baker RR, Bishop LJ. The pyrolysis of tobacco ingredients. J Anal Appl Pyrolysis. 2004; 71(1): 223–311.

[111] Aszyk J, Kubica P, Kot-Wasik A, Namieśnik J, Wasik A. Comprehensive determination of flavouring additives and nicotine in e-cigarette refill solutions. Part I: Liquid chromatography–tandem mass spectrometry analysis. J Chromatogr A. 2017;1519:45–54.

[112] Aszyk J, Woźniak MK, Kubica P, Kot-Wasik A, Namieśnik J, Wasik A. Comprehensive determination of flavouring additives and nicotine in e-cigarette refill solutions. Part II: Gas-chromatography–mass spectrometry analysis. J Chromatogr A. 2017;1517:156–64.

[113] Aszyk J, Kubica P, Woźniak MK, Namieśnik J, Wasik A, Kot-Wasik A. Evaluation of flavour profiles in e-cigarette refill solutions using gas chromatography–tandem mass spectrometry. J Chromatogr A. 2018; 1547:86–98.

[114] Krüsemann EJZ, Havermans A, Pennings JLA, de Graaf K, Boesveldt S, Talhout R. Comprehensive overview of common e-liquid ingredients and how they can be used to predict an e-liquid's flavour category. Tob Control. 2020:055447.

[115] Hua M, Omaiye EE, Luo W, McWhirter KJ, Pankow JF, Talbot P. Identification of cytotoxic flavor chemicals in top-selling electronic cigarette refill fluids. Sci Rep. 2019; 9(1):2782.

[116] Omaiye EE, McWhirter KJ, Luo W, Tierney PA, Pankow JF, Talbot P. High concentrations of flavor chemicals are present in electronic cigarette refill fluids. Sci Rep. 2019; 9(1):2468.

[117] Salam S, Saliba NA, Shihadeh A, Eissenberg T, El-Hellani A. Flavor–toxicant correlation in e-cigarette: A meta-analysis. Chem Res Toxicol. 2020.

[118] El-Hage R, El-Hellani A, Salman R, Talih S, Shihadeh A, Saliba NA. Fate of pyrazines in the flavored liquids of e-cigarettes. Aerosol Sci Technol. 2018;52(4):377–84.

[119] Erythropel HC, Jabba SV, DeWinter TM, Mendizabal M, Anastas PT, Jordt SE et al. Formation of flavorant–propylene glycol adducts with novel toxicological properties in chemically unstable e-cigarette liquids. Nicotine Tob Res. 2019; 21(9):1248–58.

[120] Khlystov A, Samburova V. Flavoring compounds dominate toxic aldehyde production during e-cigarette vaping. Environ Sci Technol. 2016; 50(23):13080–5.

[121] Lerner CA, Sundar IK, Yao H, Gerloff J, Ossip DJ, McIntosh S et al. Vapors produced by electronic cigarettes and e-juices with flavorings induce toxicity, oxidative stress, and inflammatory

response in lung epithelial cells and in mouse lung. PLoS One. 2015; 10(2):e0116732.

[122] Pankow JF, Kim K, McWhirter KJ, Luo W, Escobedo JO, Strongin RM et al. Benzene formation in electronic cigarettes. PLoS One. 2017; 12(3):e0173055.

[123] Pankow JF, Kim K, Luo W, McWhirter KJ. Gas/particle partitioning constants of nicotine, selected toxicants, and flavor chemicals in solutions of 50/50 propylene glycol/glycerol as used in electronic cigarettes. Chem Res Toxicol. 2018; 31(9):985–90.

[124] Kubica P, Wasik A, Kot-Wasik A, Namieśnik J. An evaluation of sucrose as a possible contaminant in e-liquids for electronic cigarettes by hydrophilic interaction liquid chromatography-tandem mass spectrometry. Anal Bioanal Chem. 2014; 406(13):3013–8.

[125] Miao S, Beach ES, Sommer TJ, Zimmerman JB, Jordt SE. High-intensity sweeteners in alterna-tive tobacco products. Nicotine Tob Res. 2016; 18(11):2169–73.

[126] Erythropel HC, Kong G, deWinter TM, O'Malley SS, Jordt SE, Anastas PE et al. Presence of high-intensity sweeteners in popular cigarillos of varying flavor profiles. JAMA. 2018; 320(13):1380–3.

[127] DeVito EE, Jensen KP, O'Malley SS, Gueorguieva R, Krishnan-Sarin S, Valentine G et al. Modulation of "protective" nicotine perception and use profile by flavorants: Preliminary findings in e-cigarettes. Nicotine Tob Res. 2020; 22(5):771–81.

[128] Lisko JG, Stanfill SB, Watson CH. Quantitation of ten flavor compounds in unburned tobacco products. Anal Meth. 2014; 6(13):4698–704.

[129] Chen C, Isabelle LM, Pickworth WB, Pankow JF. Levels of mint and wintergreen flavorants: Smokeless tobacco products vs. confectionery products. Food Chem Toxicol. 2010; 48(2):755–63.

[130] Stanfill SB, Brown CR, Yan X, Watson CH, Ashley DL. Quantification of flavor-related compounds in the unburned contents of bidi and clove cigarettes. J Agric Food Chem. 2006; 54(22):8580–8.

[131] Polzin GM, Stanfill SB, Brown CR, Ashley DL, Watson CH. Determination of eugenol, anethole, and coumarin in the mainstream cigarette smoke of Indonesian clove cigarettes. Food Chem Toxicol. 2007; 45(10):1948–53.

[132] Schubert J, Luch A, Schulz TG. Waterpipe smoking: Analysis of the aroma profile of flavored waterpipe tobaccos. Talanta. 2013; 115:665–74.

[133] Behar RZ, Luo W, McWhirter KJ, Pankow JF, Talbot P. Analytical and toxicological evaluation of flavor chemicals in electronic cigarette refill fluids. Sci Rep. 2018;8(1):8288.

[134] Allen JG, Flanigan SS, LeBlanc M, Vallarino J, MacNaughton P, Stewart JH et al. Flavoring chemicals in e-cigarettes: Diacetyl, 2,3-pentanedione, and acetoin in a sample of 51 products, including fruit-, candy-, and cocktail-flavored e-cigarettes. Environ Health Perspect. 2016; 124(6):733–9.

[135] Kreiss K, Gomaa A, Kullman G, Fedan K, Simoes EJ, Enright PL. Clinical bronchiolitis obliterans in workers at a microwave-popcorn plant. NEJM. 2002; 347(5):330–8.

[136] Kosmider L, Sobczak A, Prokopowicz A, Kurek J, Zaciera M, Knysak J et al. Cherry-flavoured electronic cigarettes expose users to the inhalation irritant, benzaldehyde. Thorax. 2016; 71(4):376–7.

[137] El-Hage R, El-Hellani A, Haddad C, Salman R, Talih S, Shihadeh A et al. Toxic emissions result-

ing from sucralose added to electronic cigarette liquids. Aerosol Sci Technol. 2019; 53(10):1197–203.

[138] Muthumalage T, Prinz M, Ansah KO, Gerloff J, Sundar IK, Rahman I. Inflammatory and oxidative responses induced by exposure to commonly used e-cigarette flavoring chemicals and flavored e-liquids without nicotine. Front Physiol. 2018; 8:1130.

[139] Sleiman M, Logue JM, Montesinos VN, Russell ML, Litter MI, Gundel LA et al. Emissions from electronic cigarettes: Key parameters affecting the release of harmful chemicals. Environ Sci Technol. 2016; 50(17):9644–51.

[140] Wagener TL, Meier E, Tackett AP, Matheny JD, Pechacek TF. A proposed collaboration against Big Tobacco: Common ground between the vaping and public health community in the United States. Nicotine Tob Res. 2015; 18(5):730–6.

[141] Gartner C. How can we protect youth from putative vaping gateway effects without denying smokers a less harmful option? Addiction. 2018; 113(10):1784–5.

[142] Kennedy RD, Awopegba A, De León E, Cohen JE. Global approaches to regulating electronic cigarettes. Tob Control. 2017; 26(4):440–5.

[143] Doan TTT, Tan KW, Dickens BSL, Lean YA, Yang Q, Cook AR. Evaluating smoking control policies in the e-cigarette era: a modelling study. Tob Control. 2019:054951.

[144] Country laws regulation e-cigarettes: A policy scan. Baltimore (MD): Johns Hopkins Bloomberg School of Public Health, Institute for Global Control; 2020 (https://www.globaltobaccocontrol.org/e-cigarette/country-laws/view?field_policy_domains_tid%5B%5D=119, accessed 17 February 2020).

[145] Villanti AC, Byron MJ, Mercincavage M, Pacek LR. Misperceptions of nicotine and nicotine reduction: The importance of public education to maximize the benefits of a nicotine reduction standard. Nicotine Tob Res. 2019; 21(Suppl 1):S88–90.

[146] Country laws regulation e-cigarettes: Policy domains: Sale. Baltimore (MD): Johns Hopkins Bloomberg School of Public Health, Institute for Global Tobacco Control; 2020 (https://www.globaltobaccocontrol.org/e-cigarette/sale, accessed 15 June 2020).

[147] Ingredients/flavors. Baltimore (MD): Johns Hopkins Bloomberg School of Public Health, Institute for Global Tobacco Control; 2020 (https://www.globaltobaccocontrol.org/category/ policy-domains/ingredientsflavors, accessed 15 June 2020).

[148] Ollila E. See you in court: Obstacles to enforcing the ban on electronic cigarette flavours and marketing in Finland. Tob Control. 2019:055260.

[149] Country laws regulating e-cigarettes: Saudi Arabia. Baltimore (MD): Johns Hopkins Bloomberg School of Public Health, Institute for Global Tobacco Control; 2020 (https://www.globaltobaccocontrol.org/e-cigarette/saudi-arabia, accessed 15 June 2020).

[150] Enforcement priorities for electronic nicotine delivery systems (ENDS) and other deemed products on the market without premarket authorization. Silver Spring (MD): Food and Drug Administration; 2020 (https://www.fda.gov/media/133880/download, accessed 17 February 2020).

[151] Tanne JH. FDA bans most flavoured e-cigarettes as lung injury epidemic slows. BMJ. 2020; 368:m12.

[152] Doucette K. Nova Scotia first province to ban flavoured e-cigarettes and juices. CTV News Atlantic, 5 December 2019 (https://atlantic.ctvnews.ca/nova-scotia-first-province-to-banfla-

voured-e-cigarettes-and-juices-1.4716489, accessed 17 February 2020).

[153] Ontario considering ban on flavoured vape products: health minister. The Canadian Press, 5 December 2019 (https://www.cbc.ca/news/canada/toronto/ontario-considering-ban-onflavoured-vape-products-health-minister-1.5386305, accessed 17 February 2020).

[154] Tobacco control laws. Legislation. Washington (DC): Campaign for Tobacco-Free Kids; 2020 (https://www.tobaccocontrollaws.org/legislation, accessed 17 June 2020).

[155] Vardavas CI, Bécuwe N, Demjén T, Fernández E, McNeill A, Mons U et al. Study protocol of European Regulatory Science on Tobacco (EUREST-PLUS): Policy implementation to reduce lung disease. Tob Induced Dis. 2018; 16(2):2.

[156] Zatoński M, Herbeć A, Zatoński W, Przewozniak K, Janik-Koncewicz K, Mons U et al. Characterising smokers of menthol and flavoured cigarettes, their attitudes towards tobacco regulation, and the anticipated impact of the Tobacco Products Directive on their smoking and quitting behaviours: The EUREST-PLUS ITC Europe Surveys. Tob Induced Dis. 2018; 16:A4.

[157] Sussman S, Garcia R, Cruz TB, Baezconde-Garbanati L, Pentz MA, Unger JB. Consumers' perceptions of vape shops in Southern California: an analysis of online Yelp reviews. Tob Induced Dis. 2014; 12(1):22.

[158] Erinoso O, Clegg Smith K, Iacobelli M, Saraf S, Welding K, Cohen JE. Global review of tobacco product flavour policies. Tob Control. 2020:055404.

[159] Oliveira Da Silva AL, Bialous SA, Albertassi PGD, Arquete DADR, Fernandes AMMS, Moreira JC. The taste of smoke: tobacco industry strategies to prevent the prohibition of additives in tobacco products in Brazil. Tob Control. 2019; 28(e2):e92–101.

[160] Agaku IT, Odani S, Armour BS, King BA. Adults' favorability toward prohibiting flavors in all tobacco products in the United States. Prev Med. 2019; 129:105862.

[161] Soule EK, Lopez AA, Guy MC, Cobb CO. Reasons for using flavored liquids among electronic cigarette users: A concept mapping study. Drug Alcohol Depend. 2016; 166:168–76.

[162] Chaiton MO, Nicolau I, Schwartz R, Cohen JE, Soule E, Zhang B et al. Ban on menthol-flavoured tobacco products predicts cigarette cessation at 1 year: a population cohort study. Tob Con-trol. 2020; 29(3):341–7.

[163] Tobacco Control Directive. Addis Ababa: Ethiopian Food, Medicine and Healthcare Administration and Control Authority. 2015 (https://www.tobaccocontrollaws.org/files/live/Ethiopia/Ethiopia%20-%20Tobacco%20Ctrl.%20Dir.%20No.%2028_2015%20-%20national.pdf, accessed 18 February 2020).

[164] Palandri F, Breccia M, Bonifacio M, Polverelli N, Elli EM, Benevolo G et al. Life after ruxolitinib: Reasons for discontinuation, impact of disease phase, and outcomes in 218 patients with myelofibrosis. Cancer. 2020; 126(6):1243–52.

[165] Gorukanti A, Delucchi K, Ling P, Fisher-Travis R, Halpern-Felsher B. Adolescents' attitudes towards e-cigarette ingredients, safety, addictive properties, social norms, and regulation. Prev Med. 2017; 94:65–71.

[166] Huang LL, Baker HM, Meernik C, Ranney LM, Richardson A, Goldstein AO. Impact of non-menthol flavours in tobacco products on perceptions and use among youth, young adults and adults: a systematic review. Tob Control. 2017; 26:709–19.

[167] Eissenberg T, Soule E, Shihadeh A. "Open-system" electronic cigarettes cannot be regulated ef-

fectively. Tob Control. 2020:055499.

[168] Drazen JM, Morrissey S, Campion EW. The dangerous flavors of e-cigarettes. N Engl J Med. 2019; 380:679–80.

[169] Buckell J, Marti J, Sindelar JL. Should flavours be banned in cigarettes and e-cigarettes? Evidence on adult smokers and recent quitters from a discrete choice experiment. Tob Control. 2018:054165.

[170] Keklik S, Gultekin-Karakas D. Anti-tobacco control industry strategies in Turkey. BMC Public Health. 2018; 18(1):282.

[171] McKee M. Evidence, policy, and e-cigarettes. NEJM. 2016; 375(5):e6.

[172] Jongenelis MI, Kameron C, Rudaizky D, Pettigrew S. Support for e-cigarette regulations among Australian young adults. BMC Public Health. 2019; 19(1):67.

[173] Courtemanche CJ, Palmer MK, Pesko MF. Influence of the flavored cigarette ban on adolescent tobacco use. Am J Prev Med. 2017; 52(5):e139–46.

[174] Yang Y, Lindblom EN, Salloum RG, Ward KD. The impact of a comprehensive tobacco product flavor ban in San Francisco among young adults. Addict Behav Rep. 2020; 11:100273.

[175] Chaiton M, Schwartz R, Cohen JE, Soule E, Eissenberg T. Association of Ontario's ban on menthol cigarettes with smoking behavior 1 month after implementation. JAMA Intern Med. 2018; 178(5):710–1.

[176] Lidón-Moyano C, Martín-Sánchez JC, Saliba P, Graffelman J, Martínez-Sánchez JM. Correlation between tobacco control policies, consumption of rolled tobacco and e-cigarettes, and intention to quit conventional tobacco, in Europe. Tob Control. 2017; 26(2):149–52.

[177] Harrell MB, Loukas A, Jackson CD, Marti CN, Perry CL. Flavored tobacco product use among youth and young adults: What if flavors didn't exist? Tob Regul Sci. 2017; 3(2):168–73.

[178] Chaiton MO, Nicolau I, Schwartz R, Cohen JE, Soule E, Zhang B et al. Ban on menthol-flavoured tobacco products predicts cigarette cessation at 1 year: a population cohort study. Tob Control. 2019:054841.

[179] Chaiton MO, Schwartz R, Tremblay G, Nugent R. Association of flavoured cigar regulations with wholesale tobacco volumes in Canada: an interrupted time series analysis. Tob Control. 2019; 28(4):457–61.

[180] Farley SM, Johns M. New York City flavoured tobacco product sales ban evaluation. Tob Control. 2017; 26(1):78–84.

[181] Agaku IT, Odani S, Armour B, Mahoney M, Garrett BE, Loomis BR et al. Differences in price of flavoured and non-flavoured tobacco products sold in the USA, 2011–2016. Tob Control. 2019; 29:537–47.

[182] Glantz SA, Bareham DW. e-Cigarettes: Use, effects on smoking, risks, and policy implications. Annu Rev Public Health. 2018; 39:215–35.

[183] Pacek LR, Rass O, Sweitzer MM, Oliver JA, McClernon FJ. Young adult dual combusted cigarette and e-cigarette users' anticipated responses to hypothetical e-cigarette market restrictions. Substance Use Misuse. 2019; 54(12):2033–42.

[184] Brock B, Carlson SC, Leizinger A, D'Silva J, Matter CM, Schillo BA. A tale of two cities: exploring the retail impact of flavoured tobacco restrictions in the twin cities of Minneapolis and Saint Paul, Minnesota. Tob Control. 2019; 28(2):176–80.

[185] Glantz SA. Net effect of young adult dual combusted cigarette and e-cigarette users' anticipated responses to hypothetical e-cigarette marketing restrictions. Substance Use Misuse. 2020:55(6):1028–30.

[186] Krishnan-Sarin S, Jackson A, Morean M, Kong G, Bold KW, Camenga DR et al. e-Cigarette devices used by high-school youth. Drug Alcohol Depend. 2019; 194:395–400.

[187] Jankowski M, Krzystanek M, Zejda JE, Majek P, Lubanski J, Lawson JA et al. e-Cigarettes are more addictive than traditional cigarettes – A study in highly educated young people. Int J Environ Res Public Health. 2019; 16(13):2279.

[188] Pacek LR, Villanti AC, McClernon FJ. Not quite the rule, but no longer the exception: Multiple tobacco product use and implications for treatment, research, and regulation. Nicotine Tob Res. 2020; 22(11):2114–7.

[189] Harrell PT, Naqvi SMH, Plunk AD, Ji M, Martins SS. Patterns of youth tobacco and polytobacco usage: The shift to alternative tobacco products. Am J Drug Alcohol Abuse. 2017; 43(6):694–702.

[190] Horn K, Pearson JL, Villanti AC. Polytobacco use and the "customization generation"– New perspectives for tobacco control. J Drug Educ. 2016 ;46(3-4):51–63.

[191] Shiffman S, Sembower MA, Pillitteri JL, Gerlach KK, Gitchell JG. The impact of flavor descriptors on nonsmoking teens' and adult smokers' interest in electronic cigarettes. Nicotine Tob Res. 2015; 17(10):1255–62.

[192] Tackett AP, Lechner WV, Meier E, Grant DM, Driskill LM, Tahirkheli NN et al. Biochemically verified smoking cessation and vaping beliefs among vape store customers. Addiction. 2015; 110(5):868–74.

[193] Soule EK, Sakuma K-LK, Palafox S, Pokhrel P, Herzog TA, Thompson N et al. Content analysis of internet marketing strategies used to promote flavored electronic cigarettes. Addict Behav. 2019; 91:128–35.

[194] Cox E, Barry RA, Glantz S. e-Cigarette policymaking by local and state governments: 2009-2014. Milbank Q. 2016; 94(3):520–96.

[195] Printz C. US Food and Drug Administration considers comments on proposed nicotine product regulations. Cancer. 2018; 124(20):3959–60.

[196] Printz C. FDA launches public education effort to prevent youth e-cigarette use. Cancer. 2018; 124(7):1313–4.

[197] Jenssen BP, Walley SC. e-Cigarettes and similar devices. Pediatrics. 2019; 143(2):e20183652.

[198] Jaffe S. Will Trump snuff out e-cigarettes? Lancet. 2019;394(10213):1977–8.

[199] Carpenter CM, Wayne GF, Pauly JL, Koh HK, Connolly GN. New cigarette brands with flavors that appeal to youth: Tobacco marketing strategies. Health Affairs. 2005;24(6):1601–10.

[200] Green SH, Bayer R, Fairchild AL. Evidence, policy, and e-cigarettes – Will England reframe the debate? NEJM. 2016; 374(14):1301–3.

[201] McDaniel PA, Smith EA, Malone RE. The tobacco endgame: a qualitative review and synthesis. Tob Control. 2016; 25(5):594–604.

[202] Malone RE. Tobacco endgames: what they are and are not, issues for tobacco control strategic planning and a possible US scenario. Tob Control. 2013; 22(Suppl 1):i42–4.

[203] Balaji S. Electronic cigarettes and its ban in India. Indian J Dental Res. 2019; 30:651.

[204] Guliani H, Gamtessa S, Çule M. Factors affecting tobacco smoking in Ethiopia: evidence from the demographic and health surveys. BMC Public Health. 2019; 19(1):938.
[205] Hall W, Gartner C, Forlini C. Nuances in the ethical regulation of electronic nicotine delivery systems. Addiction. 2015; 110(7):1074-5.
[206] Beaglehole R, Bonita R, Yach D, Mackay J, Reddy KS. A tobacco-free world: a call to action to phase out the sale of tobacco products by 2040. Lancet. 2015; 385(9972):1011-8.

10. 新型烟草制品的全球营销和推广及其影响

Meagan Robichaud, Caleb Clawson and Ryan David Kennedy, Institute for Global Tobacco Control, Department of Health, Behavior and Society, Johns Hopkins Bloomberg School of Public Health, Baltimore (MD), USA

摘　要

过去几十年来，全球烟草的使用有所减少，很大程度上是因为公共卫生界通过循证烟草控制战略成功阻止了烟草的使用。最近，烟碱和烟草制造商开发了新的产品，包括含电子烟碱传输系统（ENDS）、电子非烟碱传输系统（ENNDS）和加热型烟草制品（HTP）。这些产品的引入使全球烟草控制的进展变得更复杂。在许多市场中，这些产品在青少年和年轻人中特别受欢迎。尽管有证据表明使用烟草和烟碱有潜在的危害，还是有许多使用者和非使用者认为这些产品是无害的。这些产品的营销引发了对它们的尝试使用，包括从未使用过烟草的青少年和年轻人。因此，对产品的广告宣传、营销、推广和使用进行强有力的全球监管至关重要。对电子烟碱传输系统、电子非烟碱传输系统和加热型烟草制品营销的持续监管，包括传统媒体中的广告、直接面向消费者的营销、销售点营销、在线营销（包括社交媒体）、跨境营销和赞助，被认为可能是防止使用新型烟碱和烟草制品、减轻全球烟草公共卫生负担的一个有价值的综合战略。

10.1 背　景

许多国家在减轻烟草使用的公共卫生负担方面取得了进展，这在很大程度上是由于烟草使用的减少。全球≥15岁人群的烟草使用率从2000年的33.3%下降到2015年的24.9%，预计到2025年将下降至20.9%[1]。对烟草使用的监测显示，2018年是一个重大转折点，这是观察到男性烟草使用量减少的第一年，在2015年时，全球烟草使用者中男性约占81%[1]。2000年，估计有10.5亿男性使用烟草制品，这一数字在2000~2005年期间增加了2200万,在2005~2010年期间增加了1300万,在2010~2015年期间增加了700万。2018年男性烟草使用者人数为10.93亿,预计到2020年这一数字将减少200万,到2025年将再减少400万[1]。

2018年,全球15岁以上人口中有23.6%使用烟草,18.9%使用燃烧型烟草,16.1%使用卷烟[1]。因此,2018年全球约80%的烟草使用者使用燃烧型烟草制品。2018年,全球卷烟销量超过5.3万亿支[2]。虽然预计到2023年,全球卷烟销量将下降约7%[2],而新型烟碱和烟草制品,如ENDS、ENNDS和HTP的出现引发了人们的担忧。据估计,2018年全球1.2%的成年人是ENDS当前使用者,不同国家、世界卫生组织区域和人群的ENDS使用情况存在显著差异[3]。儿童和青年(包括一些以前从未使用过烟草的人)开始使用这些产品,尤其令人担忧。有证据表明,使用ENDS与以后使用燃烧型烟草制品有相关性,这引起了人们的担忧,即在几十年来阻止烟草使用的工作后,ENDS和HTP都可能促使吸烟重新"正常化"[4,5]。

在美国,儿童和年轻人对ENDS的使用越来越多,导致近几十年来烟草使用总量首次出现增长[6](图10.1)。2019年美国全国青少年烟草调查的结果显示,青少年使用ENDS的比例持续大幅上升[7]。美国高中生当前使用电子烟的比例从2018年的20.8%上升到2019年的27.5%,而初中生的这一比例从2018年的4.9%上升到2019年的10.5%[6,7]。青少年对ENDS的使用增加促使联邦和州的决策者更仔细地研究降低这些产品吸引力的策略[8]。在撰写本报告时,2020年的调查结果尚未公布。

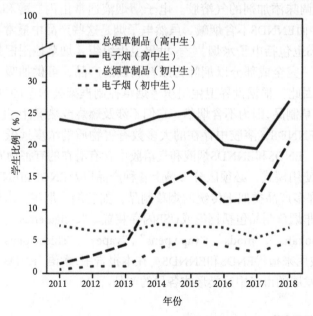

图10.1 2011~2018年美国初中生和高中生当前使用电子烟和总烟草制品的比例[6]

世界各地的决策者正在考虑和实施各种政策,将ENDS和HTP纳入现有的烟草预防框架,或对这些产品进行专门的监管或禁令。这类产品在市场的不断发展以及ENDS在儿童和年轻人中的流行,使得分享关于这类政策效果的信息和证据变得紧迫。大量证据表明,烟草和烟碱产品的营销与年轻人和成年人更容易使用这些产品以及产品使用率的提高有相关性。因此,更加重视营销在促进烟碱和烟草使用方面的作用,对于确保采取有效的公共卫生措施减少烟草使用至关重要。

本报告扩展了全球烟草监管者论坛第三次会议准备的背景文件[1],在该次会议上,讨论了ENDS的使用和营销,还有ENNDS和HTP的营销与推广。本报告还区分了ENDS、ENNDS和HTP的营销策略。响应世界卫生组织《烟草控制框架公约》第七届和第八届缔约方大会(FCTC/COP7和FCTC/COP8)的决议,继续监测和报告ENDS、ENNDS和HTP的市场发展,包括广告宣传和促销[9,10]。

10.2 电子烟碱传输系统和电子非烟碱传输系统

10.2.1 引言

电子烟碱传输系统(ENDS)和电子非烟碱传输系统(ENNDS)的烟液加热后会产生含有调味添加剂的气溶胶。电子烟烟液通常由丙二醇和/或甘油组成。ENDS含烟碱,而ENNDS不含烟碱。虽然电子烟是这些产品中最常见的类型,但ENDS和ENNDS也包括电子水烟[11,12]。世界卫生组织《烟草控制框架公约》将烟草制品定义为"完全或部分以烟叶为原料制成的产品,可供抽吸、吮吸、咀嚼或鼻吸"[13]。因此,根据世界卫生组织《烟草控制框架公约》的规定,ENDS和ENNDS不是烟草制品,因为不含烟草。它们不涉及燃烧或热解,因此不会产生"烟雾"。ENDS和ENNDS气溶胶中存在的大多数有害物质的浓度远低于烟草烟雾中的浓度。然而,ENDS和ENNDS烟液和气溶胶中含有潜在的有害物质[14]。

为了理解使用模式,必须区分市场上各种产品以及ENDS和ENNDS设备的非标准命名[4]。许多产品类似于传统的烟草制品,如卷烟、烟斗、水烟和雪茄,还有一些类似于非烟草制品包括钢笔或USB闪存装置。"e-cigarettes"、"e-cigs"、"cigalikes"、"e-hookahs"、"mods"、"vape pens"、"vapes"、"shisha pens"和"tank systems"等术语被用来指代ENDS和ENNDS。在本报告中,术语"ENDS"和"ENNDS"用于指代快速发展的市场中这类形式各异的产品。

---

1 Kennedy RD, Clawson C. Global landscape of electronic nicotine delivery system (ENDS) marketing and promotion. Paper prepared for the Third Meeting of the Global Tobacco Regulators Forum, Geneva, 11–12 September 2019(未发表)。

近年来，主要的跨国烟草公司广泛营销和销售ENDS和ENNDS，欧洲和北美青少年对这两种产品的使用量激增，让公共卫生专家、家长和政府官员感到震惊[15]。根据欧睿2017年发布的市场研究报告，2011~2016年间，"电子雾化"产品的消费增长了818%[16]。2011~2014年间，在一些市场ENDS产品的营销支出增长了近10倍（例如在美国从1200万美元增加到超过1.25亿美元），刺激了许多国家[17,18]ENDS使用的急剧增长。本节将重点关注ENDS的主要市场、产品和战略、ENDS在全球的使用情况、营销监管以及管制ENDS广告、促销和赞助的措施。

10.2.2 电子烟市场、产品和营销策略

在过去二十年里，全球燃烧型烟草市场见证了制造商整合为强大的跨国烟草公司的历程。2001年，菲利普·莫里斯国际公司、英美烟草公司、日本烟草国际公司、利是美公司和阿塔迪斯公司这五大跨国烟草公司的燃烧型卷烟销售市场份额为43%[19]。到2017年，中国烟草总公司、菲利普·莫里斯国际公司、英美烟草公司、日本烟草公司（JTI母公司）和帝国烟草集团这五家最大公司的市场份额已经增长到81%[2]。除了提供各种ENDS产品外，菲利普·莫里斯国际公司、英美烟草公司、日本烟草公司和帝国烟草集团还占据了2019年燃烧型卷烟零售量前六大市场位置中的四个[2]。新进入ENDS市场的公司，如JUUL实验室公司，获得了跨国烟草公司的大量投资，以保持其影响力并进入全球市场[20]。这些公司对ENDS创新方面的大量投资可能会使阻止烟草和烟碱的使用以及实现公众健康目标的工作变得更加复杂。

市场参与者、产品和市场份额

JUUL实验室公司（奥驰亚拥有35%的股份）。目前，JUUL凭借其广受欢迎的烟碱盐变体及小巧、符合人体工程学的设备设计，占据了ENDS全球市场的最大份额（26.2%）[16]。2019年，JUUL宣布计划在印度和菲律宾推出产品线[21]。然而，印度政府禁止ENDS的生产、制造、进口、出口、销售和分销，因为担心有引起年轻人使用的趋向[22]。JUUL目前正在测试一款应用程序，该程序可记录使用者的烟碱消费量，并允许该公司追踪新制造产品的二次销售[23]。2018年10月，JUUL实验室以7500万美元[24]收购了V2及其母公司VMR产品有限责任公司，V2 Cigs于2018年11月1日永久关闭。JUUL实验室目前在20个国家销售其产品[25]。

奥驰亚。2018年12月，菲利普·莫里斯美国公司的母公司奥驰亚宣布，决定将工作重点重新放在创新产品上，包括停止生产和分销所有Nu Mark系列ENDS产品，如MarkTen和Green Smoke产品，这些产品近年来在加拿大和美国市场占有重要的市场份额[26]。此外，2018年12月20日，奥驰亚通过全资子公司以128亿美元的价格购买了JUUL实验室公司无表决权转换股，相当于JUUL 35%的经济权益。

此后，奥驰亚同意至少在6年[20]内不与JUUL在ENDS营销方面竞争。

菲利普·莫里斯国际公司（PMI）。PMI官方网站关于其"无烟"产品线表示该公司正在"探索新的电子蒸汽产品"，并表示"我们还在开发受2011年收购的技术启发的产品……我们的科学家继续开发这项技术，以复制不用烟草和不经燃烧进行吸烟的感觉和仪式。其中一种正在开发的产品叫作STEEM……这与电子烟不同……以烟碱盐的形式产生含烟碱的蒸汽"[27]。欧睿公司报告了PMI其他ENDS品牌的销售情况，包括Solaris在西班牙和Nicocig、Vivid Vapor和MESH在英国的销售情况[28]。

日本烟草国际公司（JTI）。JTI的2018年度报告指出，"风险降低产品是我们烟草业务增长战略的关键支柱之一，我们将优先向该类别分配资源。"[29]Logic是该公司的旗舰电子烟品牌，产品在26个国家上市，包括英国和美国[30]。2018年9月17日，JTI在英国推出了一款口袋大小的设备Logic Compact。它的设计与近年来在ENDS市场占有巨大份额的JUUL（一款电子烟产品）有着惊人的相似之处。Logic Compact已在25个国家[31]上市。JTI还生产另一种ENDS产品"E-lites"，这种产品在保加利亚和德国都有销售。

英美烟草公司（BAT）。英美烟草声称，它"处于开发和销售一系列潜在风险降低产品的最前沿，这些产品在不燃烧烟草的情况下提供了吸烟的乐趣"[32]。其不断增长的所谓"潜在风险降低产品"组合包括一系列ENDS产品。BAT在2013年推出了他们的旗舰品牌Vype[33]。2017年，英美烟草收购了2013年推出Vuse的雷诺蒸汽公司。虽然BAT此后收购了多个ENDS品牌，包括Ten Motives（英国）、Chic（波兰）和VIP（英国），但该公司于2019年11月28日宣布，将在2020年期间尽可能将其ENDS品牌迁移到Vuse，以简化其"新品类产品组合"，其他新品类产品包括"现代口腔产品"Velo和HTP产品glo[34]。截至2019年12月底，英美烟草公司的ENDS产品已在27个市场上市[33]。

雷诺美国公司（现为英美烟草公司所有）。Vuse是2016年美国便利店销量第一的电子烟产品[35]。该公司指出，"Vuse和其他包括Vuse Vibe在内的RJR雾化电子烟未来是否成功将取决于不断发展的替代烟草制品的创新能力"。Vuse产品在美国销售。

帝国烟草公司（IMB）。该公司在2018年的年度报告中表示，"通过我们不断增长的下一代产品组合，我们正在为成年吸烟者提供一系列危害性较低的卷烟替代品，特别是雾化类产品"。该公司将"在专业电子烟渠道和网络上建立影响力"作为优先事项[36]。它的旗舰电子烟品牌是blu，并在2018年[36]推出了myblu和myblu加强版，作为该品牌的封闭式电子烟，带有预装好的烟弹。Myblu加强版是一种烟碱盐变体，该公司声称它"更接近地复制了吸烟的体验和满足感"[37]。myblu和myblu加强版都有不含烟碱的变体[36,38]。blu品牌还包括一款开放系统产品

blu pro，帝国烟草集团在2018年推出了另一款开放系统产品blu ACE，此后一直处于停产状态[36,39,40]。2019年底，blu在16个市场上市[38]。

促进电子烟（含或不含烟碱）销售的营销策略

烟碱和烟草公司使用各种各样的策略来营销ENDS和ENNDS。这些营销策略已显示出向青少年进行积极营销的趋势，青少年越来越多地接触到来自各种渠道的ENDS广告[41-43]。此外，市场上含有烟碱的电子烟烟液有成千上万种口味，包括对年轻人有吸引力的糖果和水果口味。除非对ENDS和ENNDS的营销进行监管，否则它们的使用可能会使烟碱和烟草的使用重新正常化[44]。以下是ENDS和ENNDS的一般营销策略。

- 广告[45]：

线上，包括社交媒体（如Facebook、Instagram、Twitter）[46]和利用在社交媒体上有影响力的人；电视、电影院[42]；广播[42]；印刷媒体(如报纸、杂志)[47]；广告牌和海报[47]；销售点的展示和广告[48]。

- 赞助：

体育、文化和艺术活动[49,50]；活动，包括学校节目[51]。

- 面向青少年的营销策略：

卡通人物的使用[52]；口味，特别是糖果、水果和其他甜味[53]；学校附近的营销[54]；针对学校和青少年营地的营销[55]；利用流行的网络或手机游戏进行营销（例如PokémonGo）[56]。

- 美化产品使用：

名人代言[45]；在"魅力"活动中的促销，例如纽约时装周上的免费传单[57]。

- 定价策略：

优惠券、折扣、折扣码、回扣[58]；"多买"，例如买一送一[59]；免费样品[60]。

- 产品创新[61]。

- 产品设计：

易于隐藏(尤其是针对青少年)[62]；颜色和图案定制[63]；时尚、现代的设计[64]。

- 产品使用的性别化[49,65]。

- 健康声明或危害降低声明[66-68]。

- ENDS品牌推销[69]。

- 为前沿团体提供资金，包括[70]：

智库（如欧洲政策信息中心）；公关公司（如蓝星策略）。

- 游说并雇佣他人代表行业进行游说[71]。

- 履行企业社会责任和开展慈善事业提升行业形象[72]。

常见策略

在不只生产ENDS和ENNDS的六个主要市场参与者中，有五个打算扩大其产品企业的创新和/或生产。ENDS在某些地区的使用量突然大幅增加，这促使了新型烟碱和烟草制品的开发，最近推出的许多产品提供了比前几代ENDS更高的烟碱浓度，例如，JUUL和NJOY提供的含有5%和6%烟碱的产品[73]。而且，在2018年，JUUL的制造商声称，一个含5%烟碱的JUUL烟弹含有的烟碱量大约与20支卷烟相同[74]。在2013~2015年供应的ENDS中，烟碱最高浓度为4.9%[73]。此外，随着市场转向JUUL推广的烟碱盐变体，许多产品比前几代更有效地提供高浓度的烟碱[75]。这些品牌包括菲利普·莫里斯国际公司的STEEM、日本烟草国际公司的Logic Compact、帝国烟草集团的Myblu以及影响力较小的品牌，如深圳IVPS科技的Smok Nord[64]。主要市场参与者正在开发的产品都是封闭系统，使用者不能用电子烟烟液重新填充设备，而是必须购买烟弹或胶囊。既往吸烟者和试图戒烟的人表现出了对开放系统的偏好[76]，但这可能会随着含烟碱盐封闭系统的日益流行而改变[77]。烟草企业正在购买竞争公司的股票和/或多数股权，奥驰亚、JUUL实验室和VMR的情况就是如此。开拓新市场一直是当务之急，帝国烟草集团、日本烟草国际公司和JUUL都明确宣布了各自的扩展计划。

电子烟公司也使用间接营销策略来吸引消费者，包括青少年。通常通过前沿团体来实现，这些团体被定义为自称独立但实际上是"为其他方或利益服务，其赞助者被隐藏或很少被提及"的组织[70]。值得注意的前沿团体包括自2019年以来完全由菲利普·莫里斯国际公司资助的无烟世界基金会，以及享受吸烟权利自由组织（Forest），该组织反对修订欧盟烟草制品指令，该指令要求对超过一定水平烟碱含量[78,79]的电子烟发放许可证。前沿团体还包括智库、公关公司和游说团体[70]。

电子烟制造商还利用承担企业社会责任战略来提升自己的公众形象，推广自己的品牌。企业社会责任"指的是自愿的企业行动，声称通过将社会目标放在首位来为公共利益服务"[72]。这些公司经常将慈善事业作为其承担企业社会责任的部分证明，包括向以青年为导向的组织和与受烟草使用影响过大的其他群体相关的事业提供慈善捐款，包括LGBTQ+社区以及少数民族群体[72,80]。例如，2018年，奥驰亚向非裔美国人历史文化国家博物馆和美国男生女生俱乐部提供了慈善捐款[80]。

10.2.3　电子烟的全球使用情况和使用流行程度

ENDS在全球的销量正在快速增长。2014年全球市场销售额达到27.6亿美元[81]，2017年达到93.9亿美元，预计到2026年将达到583.2亿美元[82]。

2018年，世界上大约有三分之一（32.4%）的男性和5.5%的女性吸烟[1]。据估计，2018年，全球有1.2%的成年人使用ENDS，其中男性占1.7%，女性占0.7%[3]。世界卫生组织各地区对燃烧型烟草和ENDS的使用各不相同。欧洲地区的吸烟率最高（26.2%），西太平洋地区的ENDS使用率最高（2.4%）（表10.1）。

表10.1 2018年世界卫生组织地区燃烧型烟草和电子烟的使用率[3]

国家和地区	卷烟使用（%）			电子烟使用（%）		
	合计	男性	女性	合计	男性	女性
非洲地区（46个国家中的3个代表国家）						
阿尔及利亚	18.4	33.5	3.2	3.8	7.0	0.6
尼日利亚	10.8	16.1	5.4	0.0	0.0	0.0
南非	18.9	31.1	7.3	0.4	0.6	0.2
美洲地区（35个国家中的9个代表国家）						
加拿大	14.8	17.1	12.6	3.5	3.7	3.3
智利	31.8	36.7	27.2	0.5	0.5	0.5
哥伦比亚	11.6	17.0	6.6	0.1	0.1	0.0
哥斯达黎加	12.5	17.0	8.0	0.2	0.3	0.2
多米尼加	9.3	10.4	8.3	0.1	0.1	0.1
厄瓜多尔	15.2	21.8	8.7	0.1	0.1	0.1
危地马拉	6.8	11.4	2.5	0.2	0.3	0.2
秘鲁	11.4	19.3	3.8	0.4	0.4	0.3
美国	13.7	15.5	11.9	3.8	4.1	3.5
地中海东部地区（22个国家中的5个代表国家）						
埃及	30.6	54.9	5.3	0.7	1.3	0.1
摩洛哥	20.3	37.6	3.7	0.7	1.4	0.0
巴基斯坦	21.0	34.6	6.9	0.0	0.0	0.0
突尼斯	32.1	54.4	11.0	0.6	1.1	0.1
沙特阿拉伯	29.8	39.2	15.5	0.3	0.4	0.3
欧洲地区（53个国家中的38个代表国家）						
奥地利	26.2	28.1	24.4	1.2	1.3	1.1
阿塞拜疆	26.0	33.1	19.3	0.3	0.5	0.1
白俄罗斯	24.8	46.2	7.0	2.1	3.3	1.1
比利时	22.0	23.0	21.0	4.4	4.8	4.0
波斯尼亚和黑塞哥维那	38.2	44.5	32.3	1.3	1.3	1.3
保加利亚	32.2	36.6	28.1	1.1	1.9	0.4
克罗地亚	27.5	35.5	20.3	1.2	1.6	0.8
捷克	33.4	37.0	30.0	5.7	6.9	4.5
丹麦	21.1	21.0	21.1	4.8	4.8	4.8
爱沙尼亚	23.4	31.0	17.0	1.6	2.2	1.1
芬兰	12.2	12.8	11.6	2.0	2.8	1.3
法国	26.2	30.5	22.3	4.3	5.0	3.6
格鲁吉亚	28.5	54.1	6.2	1.4	2.9	0.1

续表

国家和地区	卷烟使用（%）			电子烟使用（%）		
	合计	男性	女性	合计	男性	女性
德国	21.4	24.0	19.0	5.3	6.6	4.1
希腊	42.1	52.5	32.6	2.4	3.2	1.7
匈牙利	28.6	34.5	23.4	2.1	2.9	1.3
爱尔兰	19.5	19.8	19.3	5.4	5.4	5.3
以色列	23.5	31.3	16.0	0.4	0.5	0.3
意大利	21.1	25.5	17.0	1.5	1.8	1.3
哈萨克斯坦	30.3	42.0	20.0	4.0	8.2	0.3
拉脱维亚	27.0	42.5	14.4	1.1	2.0	0.4
立陶宛	26.3	37.4	17.1	1.6	1.9	1.3
荷兰	22.5	25.7	19.3	3.6	4.0	3.3
北马其顿	31.0	33.0	29.0	1.0	1.1	0.9
挪威	11.0	11.5	10.5	1.5	1.5	1.5
波兰	33.1	34.1	32.2	5.4	7.7	3.3
葡萄牙	19.1	26.4	12.8	2.2	3.2	1.3
罗马尼亚	30.2	39.8	21.2	3.3	5.0	1.8
大不列颠联合王国	14.7	16.5	13.0	6.1	7.7	4.6
俄罗斯	33.3	44.4	24.2	1.5	2.2	1.0
塞尔维亚	32.3	35.3	29.5	0.5	0.6	0.5
斯洛伐克	31.4	45.5	18.3	1.9	2.2	1.6
斯洛文尼亚	23.4	26.1	20.9	0.9	1.5	0.2
西班牙	25.5	28.7	22.4	1.7	1.8	1.7
瑞典	10.1	11.0	9.2	1.1	1.2	1.0
瑞士	25.0	28.5	21.6	1.8	2.5	1.2
乌克兰	28.8	36.3	22.7	2.0	2.5	1.6
乌兹别克斯坦	11.3	19.9	3.1	0.5	0.2	0.7
东南亚地区（11个国家中的2个代表国家）						
印度	3.8	6.9	0.6	0.1	0.1	0.0
印度尼西亚	36.4	67.9	5.0	0.5	1.1	0.0
西太平洋地区（27个国家中的7个代表国家）						
澳大利亚	13.7	15.5	11.9	0.8	1.0	0.6
中国	27.8	51.9	2.8	0.2	0.5	0.0
日本	18.2	29.0	8.1	0.2	0.3	0.1
马来西亚	21.5	40.1	1.5	2.7	4.3	1.0
新西兰	14.9	16.2	13.6	4.1	2.0	6.1
菲律宾	23.3	42.0	4.8	0.3	0.6	0.1
韩国	22.0	36.7	7.5	4.1	7.1	1.1

ENDS使用的总流行率介于0.0%~6.1%之间，男性为0.0%到8.2%，女性为0.0%

到6.1%。2018年的国家数据显示，欧洲地区的国家、美洲地区的加拿大和美国、西太平洋地区的新西兰和韩国以及非洲地区的阿尔及利亚的ENDS使用率更高。这些数字与早期公布的数据一致[83-85]。报告显示尼日利亚和巴基斯坦没有人使用ENDS。与燃烧型烟草的使用趋势一致，男性的ENDS使用率通常高于女性。世界卫生组织六大地区的ENDS的使用情况如下：

- 非洲地区。该地区46个国家中只有3个国家提供了关于ENDS的使用流行率的信息：阿尔及利亚、尼日利亚和南非。阿尔及利亚男性的ENDS使用率非常高，可达7.0%。
- 美洲地区。有9个国家提供了ENDS使用数据。报告显示ENDS的使用流行率最高的是美国（3.8%），其次是加拿大（3.5%）。该地区其他国家的比例低于1.0%。除了加拿大和美国之外，其他国家的男女使用率相似。
- 地中海东部地区。埃及、摩洛哥、巴基斯坦、沙特阿拉伯和突尼斯提供了ENDS使用率数据。尽管该地区几个国家的吸烟率仍然很高，但ENDS尚未以任何显著的方式渗透到市场中，在所有有数据可查的国家中，ENDS使用率仍低于1%。
- 欧洲地区。有38个国家ENDS的流行率的信息。流行率最高的是英国，为6.1%。捷克、爱尔兰和波兰的比例≥5.0%，男性使用率相对较高的有哈萨克斯坦（8.2%）、法国（5.0%）、德国（6.6%）和罗马尼亚（5.0%）。与吸烟的人口统计数据一致，北欧国家男性和女性中ENDS流行率相似。有趣的是，在乌兹别克斯坦女性中的流行率（0.7%）高于男性（0.2%）。
- 东南亚地区。只有印度和印度尼西亚的ENDS的使用数据。虽然这两个国家都报告了高吸烟率，尤其是在男性中，但ENDS的使用几乎不存在。
- 西太平洋地区。有7个国家的ENDS数据。韩国男性的使用率特别高(7.1%)。在新西兰，男性比女性更可能吸烟，但女性（6.1%）比男性（2.0%）更可能使用ENDS。

虽然这些数据显示了成年人使用ENDS的普遍程度，但在一些国家，18~24岁的年轻人比25岁及以上的成年人（既往或当前使用ENDS的人）人数更多，而且这一比例近年来一直在稳步上升[84-86]。2016年美国18~24岁成年人中既往使用电子烟的比例为23.5%（875例），2017年欧盟该年龄段成年人中既往使用电子烟的比例为28%，加拿大20~24岁成年人中既往使用电子烟的比例为29%[84-86]。2017~2018年的数据表明，加拿大和美国市场推出的含有烟碱盐的可再填充或一次性烟弹（pod mods）的新一代产品，导致高中生（在过去30天中使用ENDS的人）数量最近大幅增加（2019年为28%）[87,88]。加拿大和美国也被报告，2017~2019年间年轻人使用ENDS的比例比英国（英格兰）有更大的增长，因为英国有更全面的政策来监管电子烟的获取和分销[75]。

10.2.4 电子烟的广告、促销和赞助趋势

电子烟制造商如何通过社交媒体在市场上做广告

社交媒体平台是ENDS广告的重要渠道,网站可以在全球范围内推广这些产品。在Instagram和Twitter上浏览ENDS的广告和促销内容,检索条件包括带有特定ENDS品牌名称(如#JUUL)或与ENDS使用相关的通用术语(如#vape)的话题标签,以及世界卫生组织成员国的名称(如#白俄罗斯)。由于Instagram的检索功能只允许使用一个检索词,因此检索内容是电子烟品牌名称或检索词与世界卫生组织成员国的名称组合在一起(如#VapeCanada),而Twitter的检索功能可使用多个检索词和布尔运算符(如#JUUL和#Canada)。

调查显示,至少有149个(77%)世界卫生组织成员国在社交媒体平台Twitter上和Instagram上销售ENDS。对每个国家最多六个广告(Twitter上三个、Instagram上三个)内容进行分析揭示了世界卫生组织六大地区使用的几种常见广告策略(表10.2)。

表10.2　2019年世界卫生组织六大地区常见的ENDS广告策略

世界卫生组织	国家数量	可用数据数量(%)	广告数量	健康警示数量(%)	烟碱词汇数量(%)	风味词汇数量(%)	设计特点词汇数量(%)	设备图像数量(%)	烟液图像描述数量(%)
非洲地区	46	26(57)	84	1(1)	24(29)	63(75)	15(18)	22(26)	61(73)
美洲地区	35	26(74)	135	3(2)	14(10)	55(41)	72(53)	86(64)	59(44)
东南亚地区	11	9(82)	42	0	3(7)	19(45)	21(50)	25(60)	19(45)
欧洲地区	53	53(100)	267	5(2)	44(16)	141(53)	117(44)	142(53)	148(55)
地中海东部地区	22	21(95)	102	1(1)	17(16)	50(49)	47(46)	51(50)	50(49)
西太平洋地区	27	14(52)	70	3(4)	11(16)	34(49)	36(51)	42(60)	35(50)
合计	194	149(77)	700	13(2)	113(16)	362(52)	308(44)	368(53)	372(53)

此次检索发现的广告中,很少有健康警示的。广告通常包括设备或电子烟烟液图像和味道的描述或其他涉及特征的词汇内容。

COVID-19对电子烟营销策略的影响

在COVID-19大流行期间,一些ENDS制造商、零售商和使用者将其在社交媒体上的营销和推广策略与大流行和遏制策略相关的信息进行了统一。Instashop和Twitter上的电子烟营销中有几个主题。

应对无聊和孤独。例如,一家电子烟液制造商的Twitter广告称"COVID-19

封锁让你感到沮丧？我们有解药。蓝树莓和芒果冰完美同步——适合你的烟斗、烟弹或最喜欢的'豆荚'系统"。

网购。例如，来自Vuse中东的一篇Instagram上的帖子称，"使用Instashop在线订购！这些天我们都在附近，所以当你待在家里的时候，你的Vuse订单会马上到。

居家办公。零售商和制造商利用这一主题鼓励用户"在家工作"时在线购买产品。电子烟使用者在他们的家庭办公室张贴了他们使用的ENDS的照片。

获得防护设备和用品。一些制造商发布了广告，提供防护设备和用品，如口罩和洗手液。例如，电子烟品牌MOTI美洲的一则Instagram上的广告称，"与卷烟相比，#vapes#对肺部的伤害降低了95%。在#COVID-19大流行期间，我们建议使用MOTI，为了你的健康，可以替代卷烟。[你能得到什么？]2个一次性外科口罩"[66]。

保持健康（尤其是促进肺部健康）。来自MOTI美洲的上述例子也表明，在COVID-19大流行期间，品牌商也利用了降低伤害的主题[66]。

支持受疫情影响的企业。制造商和零售商在其广告中发布了支持受疫情影响的企业的信息。例如，ENDS的制造商和出口商INNOPHASE在Twitter上的一则广告中说，"目前，我们有一个VPOD的大促销活动，以帮助我们的合作伙伴度过COVID-19的困难时期"。

10.2.5　控制电子烟的广告、促销和赞助的措施

目前许多国家限制ENDS和ENNDS的广告、促销或赞助。然而，各国的监管策略差异很大。例如，有8个国家（哥斯达黎加、厄瓜多尔、格鲁吉亚、日本、墨西哥、新西兰、帕劳和摩尔多瓦）监管ENDS的营销，但不监管ENNDS的营销，因为对广告的限制仅适用于"含有烟碱的或作为药品监管的电子烟"[89]。在各欧盟成员国，对独特品牌元素实行禁令旨在降低广告的影响力，一些国家通过减少零售销售和包装上的品牌出现机会，进一步降低广告的影响力。在美国，食品药品监督管理局在2016年通过了几项关于电子烟营销和推广的法规，包括禁止在商店内免费提供电子烟烟液样品[90]。美国食品药品监督管理局还通过了禁止虚假或误导性广告的条令（例如使用"淡"、"温和"或"低"等描述词），并要求制造商提交"风险降低的烟草制品"的授权申请，在产品可以作为风险降低的烟草制品进行营销之前，对其营销对人群健康的影响进行全面的科学审查[90,91]。

有几个国家特别关注面向青少年的电子烟营销。由于青少年使用率的增加尤其令人担忧，美国食品药品监督管理局在2020年1月发布了一项政策，优先执行关于风味"烟弹式电子烟"（不包括薄荷醇和烟草调味物质）的法规，试图限制青少年获取某些风味电子烟[92]。加拿大已经禁止了所有会对青少年有吸引力的表明风味和设计属性的营销和包装元素[93,94]。

鉴于有证据表明，接触烟草和烟碱产品的营销、广告宣传和促销与易使用ENDS和ENNDS有相关性，因此对针对青少年的营销和其他任何类型的营销进行监管都很重要。人们注意到，社交媒体影响者被用来推广电子烟。在大多数社交媒体平台上注册账户都有最低年龄要求，一些平台不接受或发布烟草制品广告。然而，目前还不清楚这些政策是否会限制社交媒体影响者推广ENDS产品。英国已经讨论并确定了监管社交媒体影响者的可能解决方案，包括定义为"影响者"的最低追随者人数，并要求网络影响者公开为产品代言的费用。

接触烟草广告和获得价格促销的机会增加与成年人更易使用ENDS和燃烧型烟草制品有相关性[95]，接触ENDS广告的机会增加，尤其与成年人和青少年使用ENDS的易感性和当前使用ENDS的可能性增加有关[96,97]。对电子烟广告的接受度也随着广告接触的增加而增加[96]。

ENDS广告的某些方面似乎与青少年报告对使用ENDS的兴趣和以后使用ENDS的可能性增加有关。带有社交信息而非健康信息的ENDS广告和在社交媒体平台上看到的广告分别与使用ENDS的兴趣和ENDS使用量的增加有关[98,99]。励志人物或名人在广告中的代言也与使用的可能性增加有关[100]。与这些广告策略以及糖果和水果口味产品营销相关的新一代高科技ENDS设备很快在青少年和年轻人中流行起来[43,52,101]。

10.2.6 建议

对监测电子烟和无烟碱电子烟的营销、广告、促销和赞助的建议

1) 更好地监测电子烟和无烟碱电子烟的营销，关注社交媒体、销售点营销和赞助

社交媒体。为了让决策者充分了解行业的营销策略，监测传统和社交媒体广告渠道并关注它们随时间的动态变化很重要。监测可以在政府部门内部独立进行，也可以利用媒体和互联网监测服务、行业报告和人口调查进行。报告应强调青少年接触营销的程度。

销售点营销。烟草企业长期以来一直将销售点营销作为促销活动的契机。这种策略可以通过实地检查来监测。

赞助。赞助仍然是烟碱和烟草行业营销活动不可或缺的一部分。应监测所有活动的赞助情况以及在广告中使用证明材料的情况，以了解行业如何将赞助作为促销策略，以及接触此类营销的人群。

2）政府间合作监测营销趋势

跨境广告，包括通过媒体进入边境司法管辖区，将需要政府之间的合作。

3）监测年轻人接触直接营销的途径

直接面向消费者的营销（通过邮寄和电子邮件）是烟草行业的一项关键战略。可以实施政策以确保营销活动的材料只给成年人，而且只给那些同意接收这些材料的人。人群层面的调查可以帮助监管机构监督这类广告。

4）监测全球电子烟监管政策

对于不同产品和产品特性的政策和营销差异如何影响人们对与使用电子烟有关的风险或危害的认知，以及使用烟草和烟碱产品的差异，目前缺乏相关证据。有些国家对ENNDS产品的监管不同于ENDS产品，这增加了对新型产品监管的复杂性和模糊性[89]。更有力的政策发展报告将使世界卫生组织《烟草控制框架公约》所有缔约方更及时、更有效地制定策略。

5）监测含烟碱电子烟和无烟碱电子烟营销的差异

鉴于烟草行业历来针对特定人群（如低收入社区、少数民族群体以及性少数群体）[102,103]进行定向广告，应监测社区、印刷和数字媒体中ENDS和ENNDS广告在数量和内容上的差异。

对监管机构的建议

1）考虑支持州、省和地方对含烟碱电子烟和无烟碱电子烟的监管

州、省和地方卫生部门以及地方联盟在推进烟草管制和减少烟草使用负担方面发挥着重要作用[104-106]。关键的信息提供者（例如当地公共卫生中心主任）和当地联盟已经改变了烟草使用的社会规范，建立了对烟草管制政策的支持，并实施了烟草管制措施[105]，这些当地的监管者可以用来监管ENDS和ENNDS的营销。

2）考虑保护烟草管制不受行业干扰的策略和政策

电子烟公司使用各种策略破坏对电子烟的监管，以及阻止青少年使用这些产品的努力，包括在赞助学校的预防计划[51]，游说反对监管电子烟的政策以及企业社会责任和慈善活动[107,108]。监管机构可以采取措施实施世界卫生组织《烟草控制框架公约》第5.3条，该条要求在制定和实施与烟草管制相关的公共卫生政策时，缔约方应根据国家法律采取行动，保护这些政策不受烟草行业的商业利益和其他

既得利益的影响[109]。

这些步骤包括避免与生产电子烟的公司以及由这些公司资助的倡议组织建立伙伴关系（例如无烟世界基金会，该基金会从PMI获得了初始资金）[110]，拒绝行业捐款（资金或其他），禁止行业赞助活动，特别是面向青年的活动[107,108]。

3）继续关注循证的吸烟预防策略

各国政府和卫生组织应继续使用循证措施来减少吸烟（如世界卫生组织《烟草控制框架公约》所述），不应因电子烟等新型产品的推广和营销而偏离这些领域的行动。

4）考虑具有成本效益的反营销策略

鉴于电子烟在社交媒体上的推广覆盖全球，政府和社交媒体平台都难以有效监管此类内容，尤其是由社交媒体用户发布的内容。因此，反营销可能是最可行的选择。反营销策略可以采取各种形式，包括社交媒体活动，这可能比传统媒体活动更具成本效益。除了劝阻烟碱和烟草的使用，反营销还可包括教育公众，使其了解行业活动。

5）在可能的情况下禁止所有烟草广告、促销和赞助。

纵观其历史，烟草行业几乎规避了所有对广告、促销和赞助的限制，以接触到包括年轻人在内的消费者。因此，可能有必要全面禁止烟草营销来尽量减少青少年接触烟碱和烟草制品营销的机会。这样的禁令可以确保关于这些产品的大部分信息是来自国家和地方政府以及公共卫生机构的。

6）促进各国政府和政府部门在考虑、实施和执行电子烟营销法规方面的合作

跨境广告，包括通过媒体的跨境广告，将需要政府之间的合作。

7）及时了解电子烟行业营销策略，尤其是针对年轻人的策略

为了确保关于电子烟营销的监管政策的有效性，监管机构必须了解如何识别行业营销这些产品所使用的策略，包括定向广告、赞助和价格促销。

10.2.7 电子烟的研究差距

（1）有必要对电子烟营销作进一步研究，特别是在社交媒体上的营销，以便监管机构了解企业和零售商使用的营销和促销策略。

由于社交媒体平台上的广告与人们对使用电子烟的兴趣增加有关，研究人员

应继续收集电子烟制造商和零售商在社交媒体上使用的营销策略的证据[98,99]。

（2）应专门就ENNDS的营销及其对风险认知的影响开展进一步的研究。

由于在某些司法管辖区，ENNDS与ENDS的监管方式不同，因此它们的营销方式也可能不同。许多人错误地认为烟碱是卷烟中的主要致癌物，如在一项研究中，许多参与者认为极低烟碱含量的卷烟比目前市面上的卷烟致癌性更低[111]。因此，应研究消费者对与ENNDS和ENDS相关风险的认知，以及ENNDS的营销和推广如何影响消费者认知。

（3）应就用户在社交媒体上发布的有关电子烟的内容及其对风险认知、产品使用和营销监管有效性的潜在影响作进一步研究。

社交媒体用户发布的信息增加了这些产品的存在感。例如，即使在JUUL停止发布自己的产品信息之后[112]，与JUUL相关的信息仍继续在社交媒体上被广泛发布。鉴于用户发表的信息已被电子烟公司用作重要营销策略，因此应对其进行监测，以了解其对风险认知、产品使用和营销法规有效性的影响。

10.3 加热型烟草制品

10.3.1 引言

加热型烟草制品（HTP）是"当烟草被加热或含有烟草的装置被激活时，会产生含有烟碱和有害化学物质的气溶胶的烟草制品"[113]。HTP和ENDS的区别在于ENDS直接传输从烟草中提取的烟碱，而HTP是通过加热烟草将烟碱传递给使用者[5,114]。它们还含有非烟草添加剂，并且通常是调味剂。HTP模拟传统的吸烟行为，有些专门设计为含有烟草的卷烟用于加热。虽然HTP技术自20世纪80年代就已存在，但早期产品并不成功（图10.2）[115]。

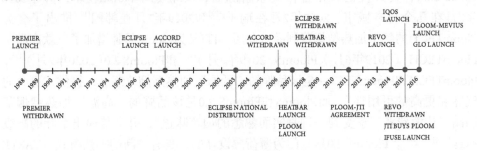

图10.2 加热型烟草制品时间表[115]

10.3.2 市场参与者、产品和策略

根据2016年烟草销售的数据和趋势，欧睿国际预测燃烧型卷烟在烟草销售总

额占比将继续下降,但会被新型烟碱和烟草制品(如加热型烟草制品)带来的市场收益所填补[116]。预计加热型烟草制品的销售额将快速增长,到2024年全球市值将从2018年的63亿美元[117]增长至220亿美元。2017年ENDS的全球市场价值为93.9亿美元,预计到2026年将达到583.2亿美元[82]。尽管预计ENDS和加热型烟草制品将持续快速增长,但全球燃烧型卷烟市场仍远高于两者,2018年为8880亿美元,预计到2024年将达到11240亿美元[118]。

市场参与者、产品和市场份额

目前加热型烟草制品市场由三家制造商主导:PMI、JTI和BAT[115]。如上所述,这三家跨国烟草公司2019年[119]燃烧型卷烟零售额也位列前六,并在全球范围内大举投资ENDS的生产,以扩大其产品组合。包括ENDS和加热型烟草制品在内的产品多元化生产,可显著巩固市场占有率,并使抵制烟草使用的努力复杂化。

菲利普·莫里斯国际公司(PMI)

2014年底,PMI在日本推出IQOS。截至2020年6月30日,IQOS已在57个国家(地区)上市[120]。PMI发布的2020年第二季度报告及其网站显示,该公司不仅在投资下一代IQOS产品,还在投资Teeps等新型加热型烟草制品,这类加热型烟草制品采用碳源加热烟草[120,121]。2020年第二季度,IQOS使用者总数估计达到1540万[120]。PMI的IQOS的利润率比传统卷烟[115]高30%~50%,IQOS现在占PMI销量的10%以上[120]。

日本烟草国际公司(JTI)

JTI在2013年推出了第一款新一代HTP产品Ploom,该产品是与一家美国公司Pax Labs合资开发的。在合作关系解除后,JTI收购了Ploom技术,并于2016年3月在几个日本城市、2017年7月在瑞士[122]和2019年7月在韩国[123]推出了名为PloomTECH的新型加热型烟草制品。此后,JTI又为Ploom品牌增加了三款产品:PloomTECH+(2019年6月)、PloomS(2019年8月)[124]和PloomS 2.0(2020年7月)[125]。PloomTECH和PloomTECH+是该品牌的"低温"加热型烟草制品,提供"更少的气味和更高的可用性",而PloomS和PloomS 2.0是该品牌的"高温"加热型烟草制品,让消费者"享受到一种真实而熟悉的烟草味道",并"传递出烟叶的高级味道"[126,127]。PloomS 2.0是专门为薄荷醇设计的,具有"新的加热模式'TASTE ACCEL',与目前的PloomS相比,它延长了峰值加热温度的持续时间"[127]。

英美烟草公司(BAT)

BAT是第三家进入新一代加热型烟草制品市场的公司,于2015年在罗马尼亚推出了iFuse[115]。2016年，BAT在日本开发并推出了glo，此后又推出了glo品牌下的其他产品[128]。2019年，该公司推出了glo pro，不同于之前glo的"双区加热室"，该产品具有感应加热功能，用以提高"消费者满意度和感官体验"[129,130]。BAT还推出了设备更轻薄的glo nano，以及glo sens，后者是一种"将蒸汽技术与真正的烟草相结合"的混合产品[131]。2020年，该公司在德国、意大利、日本、罗马尼亚和俄罗斯联邦推出了glo Hyper。glo Hyper旨在配合该公司的"Neo demi-slim系列"产品，这些产品"烟草含量比现有的Neo sticks多[132,133]"。截至2019年底，BAT的加热型烟草制品已在17个市场上市[130]。其声称的"潜在风险降低产品"组合不断增长，包括五个子公司名下的一系列加热型烟草制品[32]。

韩国烟草与人参公司（KT&G）

KT&G于2017年第四季度在韩国推出lil品牌，进入了加热型烟草制品市场。KT&G是韩国领先的卷烟生产商，见证了市场上卷烟使用者向加热型烟草制品的快速转化。lil旨在打造国内市场[115]。2018年，该公司以lil品牌推出了三款采用烟草加热技术的产品：lil plus、lil mini和lil hybrid[134]。lil plus和lil mini完全采用HTP技术，而lil hybrid同时采用HTP和ENDS技术[135]。2020年1月，KT&G和PMI达成协议，允许PMI分销KT&G的无烟产品，包括lil品牌下的HTP和lil的ENDS产品Vapour[135]。

促进加热型烟草制品销售的营销策略

烟草公司使用了广泛的营销策略来推广加热型烟草制品，通常这些营销针对青少年和年轻人[136,137]。加热型烟草制品的营销包括以下策略。

■ 广告：

线上，包括社交媒体(如Facebook、Instagram、Twitter)[138]；电视[138]；广播[138]；报纸和杂志[138]；广告牌和海报[138]；销售点的展示和广告[139]；加热型烟草制品专用零售店[115]；酒吧和酒馆[138]。

■ 强调与卷烟的相似性[115]。

■ 承认卷烟的危害，同时将加热型烟草制品作为"更清洁的替代品"[140]。

在美国，制造商试图利用这一说法，通过申请将加热型烟草制品指定为"风险降低的烟草制品"[92]来规避严格的广告用语规定。例如，2020年7月，PMI成功从FDA获得了IQOS系统及其三款万宝路烟弹的"暴露降低"命令，该命令"允许销售带有某些声明的产品"。

■ 利用品牌"大使"（现场和在社交媒体上）和演示[151,141]。

■ 产品设计：

时尚、高科技的外观[138,142]；快速充电[115]；气味更少[142]；二手烟释放量更少[141]；

定制颜色和限量版设计[115]。

■ 赞助[141]：

体育赛事；艺术表演；音乐会；美食美酒节。

■ 定价策略：

"饵钩"定价：设备的折扣价格和特殊设计的再填充物或插入物的经常性成本[115]；免费样品[141]。

■ 客户服务：

呼叫中心支持[115]；专门的品牌零售店和网站[115]；帮助客户定位附近商店并排除设备故障的应用程序[141,143]。

■ 面向青少年的营销：

在销售点面向青少年的商品附近放置加热型烟草制品[139]；赞助面向青少年的活动（例如特拉维夫的TLV学生日）[141]。

■ 资助前沿团体（例如无烟世界基金会）[70]。

■ 游说[144]。

■ 履行企业社会责任以提升企业形象[72]。

常见策略

最新一代的加热型烟草制品不仅针对特定的烟草使用者细分群体，还以非传统方式进行营销和分销。由于预期销量会增加，烟草公司大力投资增加其HTP组合。例如，英美烟草公司正在为其glo牌加热型烟草制品创建更多的功能，包括下一代设备、其他的口味和混合技术。PMI的网站显示，它正在采用类似的策略，不仅投资下一代IQOS产品，还投资像Teeps这样的新型加热型烟草制品，这种加热型烟草制品有一种加热烟草的替代热源[121]。

加热型烟草制品的制造商还使用间接营销来接触包括青少年在内的消费者。通常会利用包括无烟世界基金会在内的前沿团体，或者游说的形式来营销[70]。加热型烟草制品公司还利用企业社会责任战略来提升其公众形象进而推广其品牌[72]，例如将慈善事业作为其企业社会责任的证据，经常向环境事业、防止童工组织和扩大受教育机会的组织进行慈善捐赠[145]。

10.3.3 加热型烟草制品的全球使用和流行情况

虽然目前没有全球监测加热型烟草制品流行趋势的可靠、公开的数据，但已有某些国家和区域趋势的数据。目前亚太地区的报告显示，18~39岁年龄组的人群贡献了加热型烟草制品销售和使用的最大收入份额。2018年，日本占全球加热型烟草制品市场收入的85%[146]，加热型烟草制品收入增长最快的是韩国[147]。2018年在日本的调查结果显示2.7%的成年人在过去30天内使用过加热型烟草制品，

1.7%的成年人每天使用；几乎所有接受调查的加热型烟草制品使用者都是燃烧型卷烟的当前或既往使用者[148]。

在韩国自IQOS于2017年进入市场以来，19~24岁的年轻人中HTP既往或当前使用者一直在快速增长，5.7%的受访者表示在IQOS上市3个月后就使用过，3.5%的人表示目前正在使用[149]。2017年，加热型烟草制品占卷烟销量的2.2%，到2018年，这一比例达到9.6%。加热型烟草制品上市1年后，2.8%的12~18岁韩国青少年表示曾使用过[150]。

2019~2025年的市场预测显示，由于克罗地亚、德国、意大利、波兰和俄罗斯联邦加热型烟草制品的增长，制造商进行了大量投资，预计欧洲市场的销售也将保持强劲[151]。2019年，欧睿报道意大利是亚太地区之外最大的加热型烟草制品市场。2018~2019年市场快速增长的其他国家有捷克、德国、罗马尼亚、俄罗斯和乌克兰[152]。2017年，美国只有0.7%的成年人报告曾使用过加热型烟草制品（占当前燃烧型烟草抽吸者的2.7%），然而，这一比例仅在1年内就大幅上升，2018年成年人使用率达到2.4%（占目前吸烟者的6.7%）[124]。欧睿的数据也被用来预测2021年加热型烟草制品的市场价值，预测德国、日本、韩国、土耳其和美国为加热型烟草制品市场价值最高的一些国家[153]。

虽然许多国家开始调查和报告成年人使用加热型烟草制品的趋势，但缺乏关于年轻人使用这些产品的情况以及成年使用者偏好和使用模式的证据。这两个都是未来研究的关键领域。

10.3.4 加热型烟草制品的广告、促销和赞助趋势

加热型烟草制品如何通过社交媒体在市场上做广告

社交媒体平台是加热型烟草制品广告的新兴渠道，因为世界各地的成年人和青少年都会使用Twitter和Instagram等网站。浏览Instagram和Twitter上加热型烟草制品的广告和促销活动，检索条件包括带有加热型烟草制品品牌名称（如#IQOS）和世界卫生组织成员国名称（如#白俄罗斯）的话题标签。由于Instagram的检索功能只允许使用一个检索词，因此将HTP品牌名称或检索词与世界卫生组织成员国名称相结合（如#IQOS Canada），而Twitter的检索功能可使用多个搜索词和布尔运算符（如#IQOS和#Canada）。检索显示，至少有95个世界卫生组织成员国在Twitter和Instagram上销售加热型烟草制品（表10.3）。对每个国家最多6个广告（3个在Twitter广告、3个在Instagram广告）的内容进行分析表明，世界卫生组织六大地区使用了类似的广告策略。

表10.3　2020年世界卫生组织六大地区常见的HTP社交媒体广告策略

WHO地区	国家数量	可用数据数量（%）	广告数量	健康警示数量（%）	烟碱词汇数量（%）	风味词汇数量（%）	设计特点词汇数量（%）	设备图像数量（%）	烟液图像数量（%）
非洲地区	46	3(7)	9	1(11)	3(33)	0	0	7(78)	4(44)
美洲地区	35	12(34)	42	17(40)	3(7)	12(29)	9(21)	38(90)	12(29)
东南亚地区	11	6(56)	16	11(69)	3(19)	3(19)	4(25)	14(89)	4(25)
欧洲地区	53	48(91)	186	86(46)	21(11)	41(22)	22(12)	146(78)	30(16)
东中海东部地区	22	15(68)	62	12(19)	17(27)	17(27)	4(60)	55(89)	18(29)
西太平洋地区	27	11(41)	39	13(33)	2(50)	8(21)	13(33)	38(97)	8(21)
合计	194	95(49)	354	140(40)	49(14)	81(23)	52(15)	298(84)	76(21)

大约一半的世界卫生组织成员国有加热型烟草制品广告，其中大多数广告包括设备的图像，只有不到一半的加热型烟草制品社交媒体广告中包含健康警示，约四分之一的广告提到了风味。

COVID-19对加热型烟草制品营销策略的影响

在COVID-19大流行期间，一些加热型烟草制品制造商、零售商和使用者在社交媒体上发布了与COVID-19相关的信息以及相关遏制策略作为营销和推广策略。Instagram和Twitter上加热型烟草制品营销中出现的与大流行和遏制有关的主题包括以下内容：

（1）应对无聊和孤独。制造商将其产品作为让使用者在封锁措施期间在家安全享受的一种手段进行营销。例如，glo全球账户在Instagram上发布的一条帖子中写道："左脑说待在家里，右脑想出去。我们很喜欢这些新工具，让你在家聚会，不用妥协。#BreakBinary#NetflixParty#Discoveryglo#Myglo"[66]。

（2）储备必需品。社交媒体用户暗示应该"储备"喜爱的加热型烟草制品，为封锁或隔离做好准备，将这些产品等同于封锁期间的必需品。例如，一名用户发布了几个加热型烟草制品设备（主要是IQOS）的照片以及几包HEETS（插入IQOS设备的加热烟草单元），并声明"我准备好隔离了！"另一位用户在Instagram和Twitter上都发布了一张IQOS设备和三包HEETS的照片，并配文"#封锁#必需品和/或#隔离#成瘾。感谢你的礼物@iqos-it"。

（3）"足不出户"活动。一些加热型烟草制品品牌利用社交媒体上的"足不出户"活动和信息来宣传产品。例如，glo希腊举办了一场"居家挑战"，参与者可以赢得奖品[66]。这篇帖子利用了Instagram上的#menoumespiti信息，这是一个热门的话题标签，在希腊语中意思是"我们待在家里"。

（4）获得防护设备和用品。一些制造商在购买其产品时会提供品牌口罩和洗手液。例如，glo哈萨克斯坦公司的一则Instagram上的广告显示，一名妇女戴着glo品牌的口罩，并配文"哪儿都找不到口罩？我们会给你[眨眼表情符号]。在3月26日星期四之前，你可以在纳扎尔巴耶夫100G的glo空间免费获得口罩。此外，你还可以通过在网站或我们的销售点购买设备来获得相同的口罩……"[66]

10.3.5 加热型烟草制品广告、促销和赞助的控制措施

在对烟草制品的广告、促销或赞助进行监管的国家中，许多国家通过将加热型烟草制品纳入现有法规进行监管，有些将它归为新型烟草制品，有些将它归为其他类别，还有一些国家则对这些产品制定了专门的法规。此外，一些国家把加热型烟草制品当作烟草制品，而不是将其单独命名，以便它们受所有现行国家烟草制品法规的约束[113, 154]。继续监测这些监管的趋势及其对烟草使用的影响，将为世界各地正在观察加热型烟草制品使用趋势的决策者提供宝贵见解，并为设计全面的监管框架或更新现有法规，以减少与烟草使用有关的危害提供宝贵见解。

虽然还没有经过同行评议的证据表明加热型烟草制品的广告与促销和使用之间存在关联，研究人员指出，青少年友好的HTP广告也被用于营销ENDS，比如高科技、新颖的设计特征，声称加热型烟草制品比燃烧型产品的危害更小，以及加热型烟草制品可能比燃烧型产品更容易被社会接受[136,150]。鉴于有证据表明成功地使用这些策略来宣传ENDS产品，应密切监测加热型烟草制品广告、促销信息和媒体，以及儿童和青少年对使用加热型烟草制品的易感性的趋势。

10.3.6 对监测加热型烟草制品的广告、促销和赞助趋势的建议

更好地监测加热型烟草制品的使用和销售趋势

为了增加以下建议的影响，应在各国收集关于加热型烟草制品使用趋势的更可靠的数据，包括人群统计数据和产品偏好。虽然新一代加热型烟草制品在烟草制品市场上相对较新，但ENDS的迅速流行表明，更快速地监测加热型烟草制品的使用和销售并报告国家政策对于将其纳入现有控制框架至关重要。

更好地监督加热型烟草制品的营销，特别要注意社交媒体上的营销

替代型烟草制品在青少年友好的社交媒体平台上的强劲营销表明，有必要加强对营销趋势的监测，以更好地了解人们接触此类营销是否与对加热型烟草制品的态度和使用情况有关。这应包括对加热型烟草制品公司、零售商和社交媒体用户发布的内容进行监控。

全面搜集全球政府如何监管加热型烟草制品，包括其广告、促销和赞助

应收集有关加热型烟草制品监管、营销和促销的更多信息，以了解全球情况。更好地报告世界各地的关键政策发展将有助于世界卫生组织《烟草控制框架公约》所有缔约方制定及时、有效的策略。

10.3.7 对监管机构的建议

继续关注循证的吸烟预防策略

各国政府和卫生组织应继续关注减少吸烟的循证措施（如世界卫生组织《烟草控制框架公约》所述），不应因加热型烟草制品等新型产品的推广和营销而分散行动的注意力。

考虑更具成本效益的反营销策略

考虑到加热型烟草制品广告在社交媒体上的广告覆盖全球，无论是政府还是社交媒体平台都无法有效地监管此类内容，尤其是由社会媒体用户发布的内容。反营销可能是最可行的选择。可以使用各种形式，包括社交媒体活动，这可能比传统媒体活动更具成本效益。反营销除了劝阻烟碱和烟草的使用，还可以包括关于行业活动的教育。

应用从电子烟监管中吸取的相关经验教训

由于加热型烟草制品公司采用了许多与推广电子烟相同的营销策略，监管机构可以将从电子烟营销中吸取的经验教训应用到监管加热型烟草制品的营销和推广中。

促进各国政府和政府部门在实施和执行加热型烟草制品的营销法规方面的合作

跨境广告，包括向邻国的媒体传播，应在政府一级协同处理。

在可能的情况下禁止所有烟草广告、促销和赞助

烟草行业几乎规避了所有对广告、促销和赞助的限制，以接触到包括青少年在内的消费者。因此，可能有必要全面禁止烟草营销，以尽量减少对烟碱和烟草制品营销的接触。这样的禁令将确保有关这些产品的大部分信息来自国家、地方政府以及公共卫生机构。

及时了解加热型烟草制品行业营销策略，尤其是面向青少年的策略

为了使控制加热型烟草制品营销的政策有效，监管机构必须了解如何识别行业使用的策略，包括使用健康相关的声明、赞助和价格促销。

10.3.8 加热型烟草制品研究的研究差距

（1）需要更多的证据来了解加热型烟草制品的监管、产品风险认知和产品使用之间的关系。

关于新型烟碱和烟草制品的许多证据表明了政策和使用趋势之间的关联。然而，针对加热型烟草制品的证据不足，有必要进一步评估不同的政策和产品特性如何改变人们（特别是青少年）对风险和/或产品使用的看法。

（2）有必要对加热型烟草制品市场营销进行更多的研究，特别要关注社交媒体，使监管机构了解加热型烟草制品公司和零售商使用的营销和促销策略。

社交媒体平台是在全球范围内宣传和推广加热型烟草制品的重要渠道。由于在社交媒体平台上的广告与人们对使用烟草和烟碱产品的兴趣增加及其实际使用有关，因此应研究HTP在这些平台上的营销和推广的影响[98,99]。

（3）应对社交媒体用户发布的有关HTP的内容及其对风险认知、产品使用和营销法规有效性的影响进行更多研究。

社交媒体用户发布的内容已被用作电子烟公司的重要营销策略。因此，应监测与加热型烟草制品相关的用户发布内容的趋势，并研究其对风险认知和产品使用以及营销法规有效性的影响。

10.4 总　　结

ENDS、ENNDS和HTP在全球范围内通过传统和新兴渠道（如社交媒体）进行营销。本报告收集的证据表明，ENDS和HTP在Twitter上和Instagram上被大力推广。各国监管这些产品营销和推广的策略差异很大，有些国家禁止某类产品，有些国家只监管含有烟碱的产品，还有一些国家对风味、包装和广告加以限制。建立NDS、ENNDS和HTP营销的全球监测体系是非常有必要的，可了解这些产品是如何做广告的。还需要更多的证据来了解这些产品的营销如何影响产品的认知和使用（尤其是在青少年和年轻人中）。如果没有此类法规的话，各级政府应规范ENDS、ENNDS和HTP的广告、促销和赞助。

10.5 参考文献

[1] WHO global report on trends in prevalence of tobacco use 2000–2025. Third edition. Geneva: World Health Organization; 2019.

[2] The global cigarette industry. Washington (DC): Campaign for Tobacco-Free Kids; 2019.

[3] Smoking population – number of adult smokers. London: Euromonitor International Ltd; 2019 (http://www.portal.euromonitor.com.proxy1.library.jhu.edu/portal/statisticsevolution/ index).

[4] e-Cigarette use among youth and young adults: A report of the Surgeon General. Washington (DC): Department of Health and Human Services; 2016.

[5] 6 important things to know about IQOS, the new heated cigarette product. Washington (DC): Truth Initiative; 2019 (https://truthinitiative.org/research-resources/emerging-tobaccoproducts/6-important-things-know-about-iqos-new-heated, accessed 18 December 2020).

[6] Cullen KA, Ambrose BK, Gentzke A, Apelberg BJ, Jamal A, King BA. Notes from the field: Use of electronic cigarettes and any tobacco product among middle and high school students – United States, 2011–2018. Morb Mortal Wkly Rep. 2018;67:629–33.

[7] Youth tobacco use: Results from the National Youth Tobacco Survey. Silver Spring (MD): Food and Drug Administration; 2020 (https://www.fda.gov/tobacco-products/youth-and-tobacco/youth-tobacco-use-results-national-youth-tobacco-survey#1, accessed 18 August 2020).

[8] Trump administration combating epidemic of youth e-cigarette use with plan to clear market of unauthorized, non-tobacco-flavored e-cigarette products. Silver Spring (MD): Food and Drug Administration; 2019 (https://www.fda.gov/news-events/press-announcements/trump-administration-combating-epidemic-youth-e-cigarette-use-plan-clear-market-unauthorizednon#:~:text=%E2%80%9CThe%20Trump%20Administration%20is%20making,Human%20Services%20Secretary%20Alex%20Azar, accessed 10 January 2021).

[9] Conference of the Parties to the WHO Framework Convention on Tobacco Control. Decision FCTC/COP7(9) Electronic nicotine delivery systems and electronic non-nicotine delivery systems. Geneva: World Health Organization; 2016 (https://www.who.int/fctc/cop/cop7/FCTC_COP7_9_EN.pdf?ua=1, accessed 10 January 2021).

[10] Conference of the Parties to the WHO Framework Convention on Tobacco Control. Decision FCTC/COP8(22) Novel and emerging tobacco products. Geneva: World Health Organization; 2018 (https://www.who.int/fctc/cop/sessions/cop8/FCTC_COP8(22).pdf?ua=1, accessed 13 August 2020).

[11] Public health consequences of e-cigarettes. Washington (DC): National Academies of Sciences Engineering and Medicine; 2018. doi:10.17226/24952.

[12] Williams M, Bozhilov K, Ghai S, Talbot P. Elements including metals in the atomizer and aerosol of disposable electronic cigarettes and electronic hookahs. PLoS One. 2017;12(4):e0175430.

[13] World Health Assembly resolution 56.1. Geneva: World Health Organization; 2003 (https://www.who.int/tobacco/framework/final_text/en/). https://www.who.int/tobacco/framework/final_text/en/index3.html, accessed August 16, 2020.

[14] Conference of the Parties to the WHO Framework Convention on Tobacco Control. Electronic nicotine delivery systems and electronic non-nicotine delivery systems (ENDS/ENNDS): Re-

[15] Stone E, Marshall H. Tobacco and electronic nicotine delivery systems regulation. Transl Lung Cancer Res. 2019;8(S1):S67–76.

[16] MacGuill S. Growth in vapour products. London: Euromonitor International; 2017 (https://blog. euromonitor.com/growth-vapour-products/, accessed 13 December 2020).

[17] Kim AE, Arnold KY, Makarenko O. e-Cigarette advertising expenditures in the US, 2011–2012. Am J Prev Med. 2014;46(4):409–12.

[18] Electronic cigarettes: An overview of key issues. Washington (DC): Campaign for Tobacco-Free Kids; 2020 (https://www.tobaccofreekids.org/assets/factsheets/0379.pdf).

[19] Tobacco companies. In: Mackay J, Eriksen M, editors. The tobacco atlas. Geneva: World Health Organization; 2002:51 (https://www.who.int/tobacco/en/atlas18.pdf?ua=1, accessed 13 August 2020).

[20] Exhibit 99.1. Altria makes $12.8 billion minority investment in Juul to accelerate harm reduction and drive growth. Washington (DC): United States Securities Exchange Commission; 2019 (https://www.sec.gov/Archives/edgar/data/764180/000119312518353970/d660871dex991. htm, accessed 13 August 2020).

[21] Kalra A, Kirkham C. Exclusive: Juul plans India e-cigarette entry with new hires, subsidiary. Reuters, 30 January 2019 (https://www.reuters.com/article/us-juul-india-exclusive-idUSKCN-1PO0VV, accessed 11 November 2020).

[22] Gupta S. India is banning all e-cigarettes over fears about youth vaping. CNN, 18 September 2019 (https://edition.cnn.com/2019/09/18/health/india-e-cigarette-ban-intl/index.html#:~:text=India%20is%20banning%20all%20e%2Dcigarettes%20over%20fears%20about%20youth%20vaping&text=Electronic%20cigarette%20devices%20on%20display,risk%2C%20especially%20to%20young%20people).

[23] Kastrenakes J, Garun N. Juul launches a Bluetooth e-cigarette that tracks how much you vape. The Verge, 6 August 2019 (https://www.theverge.com/2019/8/6/20754655/juul-c1-bluetoothe-cigarette-vape-monitor-consumption-age-restriction, accessed 13 August 2020).

[24] Turning Point Brands, Inc. announces third quarter 2018 results. Business Wire, 7 November 2018 (https://www.businesswire.com/news/home/20181107005281/en/Turning-Point-Brands-Announces-Quarter-2018-Results, accessed 13 August 2020).

[25] Find your country's online store. San Francisco (CA): JUUL Labs (https://www.juul.com/global, accessed 16 August 2020).

[26] 2018. Altria Group, Inc. Annual Report. Richmond (VA): Altria Group, Inc.; 2018 (https://www.annualreports.com/HostedData/AnnualReportArchive/a/NYSE_MO_2018.pdf, accessed 10 January 2021).

[27] STEEM: Exploring new e-vapor products. Lausanne: Philip Morris International (https://www.pmi.com/smoke-free-products/steem-creating-a-vapor-with-nicotine-salt, accessed 13 August 2020).

[28] Brand shares: E-Vapour products. London: Euromonitor International Inc.; 2018 (https://www-portal-euromonitor-com.proxy1.library.jhu.edu/portal/statisticsevolution/index, accessed 13 August 2020).

[29] Japan Tobacco Inc. Annual Report 2018. Tokyo: Japan Tobacco Inc.; 2019 (https://www.jt.com/investors/results/annual_report/pdf/2018/annual.fy2018_E2.pdf, accessed 13 August 2020).

[30] 2019 earnings report. Geneva: Japan Tobacco International; 2020 (https://www.jti.com/sites/default/files/press-releases/documents/2020/jt-group-2019-financial-results-and-2020-forecast.pdf, accessed 13 August 2020).

[31] JTI launches Logic Compact in the UK – the first market to launch this new premium e-cigarette: The moment vaping "just clicks" has arrived. Geneva: Japan Tobacco International; 2018 (https://www.jti.com/jti-launches-logic-compact-uk-first-market-launch-new-premium-ecigarette, accessed 13 August 2020).

[32] British American Tobacco plc. Transforming tobacco: Annual report and Form 20-F 2018. London; 2018 (https://www.bat.com/group/sites/UK__9D9KCY.nsf/vwPagesWebLive/DOAWWG-JT/$file/Annual_Report_and_Form_20-F_2018.pdf, accessed 16 December 2020).

[33] Vapour products. London: British American Tobacco plc; 2020 (https://www.bat.com/ecigarettes, accessed 13 August 2020).

[34] British American Tobacco to focus on three global new category brands to further accelerate their growth. London: British American Tobacco plc; 2019 (https://www.bat.com/group/sites/UK__9D9KCY.nsf/vwPagesWebLive/DOBJBMSE, accessed 16 August 2020).

[35] Form 10-K. Reynolds American Inc. Securities registered pursuant to Section 12(b) of the Securities Exchange Act of 1934; Washington (DC): United States Securities Exchange Commission; 2017 (https://www.sec.gov/Archives/edgar/data/1275283/000156459017001245/rai-10k_20161231.htm, accessed 13 August 2020).

[36] Something better: Annual report and accounts 2018. London; Imperial Brands plc; 2018 (https://www.imperialbrandsplc.com/content/dam/imperial-brands/corporate/investors/annual-report-and-accounts/2018/annual-report-and-accounts-2018.pdf, accessed 13 August 2020).

[37] Imperial Brands CAGNY presentation 21 02 2019. London: Imperial Brands plc; 2019 (https://www.imperialbrandsplc.com/content/dam/imperial-brands/corporate/investors/presentations/conferences/2019/CAGNY Script.pdf.downloadasset.pdf, accessed 13 August 2020).

[38] Imperial Brands plc. Annual report and accounts 2019. London: Imperial Brands plc; 2019 (https://www.imperialbrandsplc.com/content/dam/imperial-brands/corporate/investors/annual-report-and-accounts/2019/Annual Report 2019.pdf, accessed 13 August 2020).

[39] What happened to the blu ACE device? Amsterdam: Fontem Ventures BV; 2020 (https://support-uk.blu.com/hc/en-gb/articles/360015460600-What-happened-to-the-blu-ACE-device-, accessed 13 August 2020).

[40] Imperial Brands plc interim results for the six months ended 31 March 2018. London: Imperial Brands plc; 2018 (https://www.imperialbrandsplc.com/content/dam/imperial-brands/corporate/investors/results-centre/2018/2018-05-09 Interims 18 script.pdf, accessed 13 August 2020).

[41] Padon AA, Lochbuehler K, Maloney EK, Cappella JN. A randomized trial of the effect of youth appealing e-cigarette advertising on susceptibility to use e-cigarettes among youth. Nicotine Tob Res. 2018;20(8):954–61.

[42] Mantey DS, Cooper MR, Clendennen SL, Pasch KE, Perry CL. e-Cigarette marketing exposure is associated with e-cigarette use among US youth. J Adolesc Health. 2016;58(6):686–90.

[43] Jackler RK, Chau C, Getachew BD, Whitcomb MM, Lee-Heidenreich J, Bhatt AM et al. JUUL

advertising over its first three years on the market. Berkeley (CA): Stanford Research into the Impact of Tobacco Advertising; 2019 (http://tobacco.stanford.edu/tobacco_main/publications/JUUL_Marketing_Stanford.pdf).

[44] Sebastian M. e-Cig marketing budgets growing by more than 100% year over year. AdAge, 14 April 2014 (https://adage.com/article/media/e-cig-companies-spent-60-million-ads-year/292641).

[45] Duke JC, Lee YO, Kim AE, Watson KA, Arnold KY, Nonnemaker JM et al. Exposure to electronic cigarette television advertisements among youth and young adults. Pediatrics. 2014;134(1):e29–36.

[46] Gregory A. Tobacco companies "pushing e-cigs on youngsters via Facebook and Twitter". The Mirror, 27 November 2013 (https://www.mirror.co.uk/lifestyle/health/tobacco-compa-ni-es-pushing-e-cigs-youngsters-2854699, accessed 13 August 2020).

[47] Chen-Sankey JC, Unger JB, Bansal-Travers M, Niederdeppe J, Bernat E, Choi K. e-Cigarette marketing exposure and subsequent experimentation among youth and young adults. Pediatrics. 2019;144(5):e20191119.

[48] Carpenter CM, Wayne GF, Pauly JL, Koh HK, Connolly GN. New cigarette brands with flavors that appeal to youth: Tobacco marketing strategies. Health Aff. 2005;24(6):1601–10.

[49] 7 ways e-cigarette companies are copying big tobacco's playbook. Washington (DC): Campaign for Tobacco-Free Kids; 2013 (https://www.tobaccofreekids.org/blog/2013_10_02_ecigarettes, accessed August 16, 2020).

[50] 4 marketing tactics e-cigarette companies use to target youth. Washington (DC): Truth Initiative; 2018 (https://truthinitiative.org/research-resources/tobacco-industry-marketing/4-market-ing-tactics-e-cigarette-companies-use-target, accessed 16 August 2020).

[51] Evidence brief: Tobacco industry sponsored youth prevention programs in schools. Atlanta (GA): Centers for Disease Control and Prevention; 2019 (https://www.cdc.gov/tobacco/basic_ infor-mation/youth/evidence-brief/index.htm, accessed 16 August 2020).

[52] Jackler RK, Ramamurthi D. Unicorns cartoons: marketing sweet and creamy e-juice to youth. Tob Control. 2017;26(4):471–5.

[53] Ashley DL, Spears CA, Weaver SR, Huang J, Eriksen MP. e-Cigarettes: How can they help smokers quit without addicting a new generation? Prev Med. 2020:106145.

[54] Giovenco DP, Casseus M, Duncan DT, Coups EJ, Lewis MJ, Delnevo CD. Association between electronic cigarette marketing near schools and e-cigarette use among youth. J Adolesc Health. 2016;59(6):627–34.

[55] Kaplan S. Juul targeted schools and youth camps, House Panel on Vaping Claims. The New York Times, 25 July 2019 (https://www.nytimes.com/2019/07/25/health/juul-teens-vaping.html, accessed 13 August 2020).

[56] Kirkpatrick MG, Cruz TB, Goldenson NI, Allem JP, Chu KH, Pentz MA et al. Electronic cigarette retailers use Pokémon Go to market products. Tob Control. 2017;26(E2):e147.

[57] Cardellino C. Electronic cigarettes available for free at Fashion Week. Cosmopolitan. 5 September 2013 (https://www.cosmopolitan.com/style-beauty/fashion/advice/a4736/nyfw-2013-ciga-rettes/, accessed 13 August 2020).

[58] Ali FRM, Xu X, Tynan MA, King BA. Use of price promotions among US adults who use elec-

tronic vapor products. Am J Prev Med. 2018;55(2):240-3. 0.

[59] D'Angelo H, Rose SW, Golden SD, Queen T, Ribisl KM. e-Cigarette availability, price promotions and marketing at the point-of sale in the contiguous United States (2014–2015): National estimates and multilevel correlates. Prev Med Rep. 2020;19:101152.

[60] Wadsworth E, McNeill A, Li L, Hammond D, Thrasher JF, Yong HH et al. Reported exposure to e-cigarette advertising and promotion in different regulatory environments: Findings from the International Tobacco Control Four Country (ITC-4C) survey. Prev Med. 2018;112:130–7.

[61] De Andrade M, Hastings G, Angus K. Promotion of electronic cigarettes: tobacco marketing reinvented? BMJ. 2013;347:f7473.

[62] Lee SJ, Rees VW, Yossefy N, Emmons KM, Tan ASL. Youth and young adult use of pod-based electronic cigarettes from 2015 to 2019: A systematic review. JAMA Pediatr. 2020;174(7):714–20.

[63] Bach L. JUUL and youth: Rising e-cigarette popularity. Washongton DC: TobaccoFree Kids; 2019 (https://www.tobaccofreekids.org/assets/factsheets/0394.pdf, accessed 16 August 2020).

[64] Hammond D, Wackowski OA, Reid JL, O'Connor RJ. Use of JUUL e-cigarettes among youth in the United States. Nicotine Tob Res. 2018; 22(5):827–32.

[65] Jivanda T. "Put it in my mouth": Viewers outraged by apparent reference to oral sex in VIP e-cig advert. Independent, 4 December 2013 (https://www.independent.co.uk/news/media/advertising/put-it-my-mouth-viewers-outraged-apparent-reference-oral-sex-vip-e-cig-advert-8982918.html, accessed 10 January 2021).

[66] Hickman A. "Big tobacco" using COVID-19 messaging and influencers to market products. PR Week, 15 May 2020 (https://www.prweek.com/article/1683314/big-tobacco-using-covid-19-messaging-influencers-market-products, accessed 13 August 2020).

[67] Yao T, Jiang N, Grana R, Ling PM, Glantz SA. A content analysis of electronic cigarette manufacturer websites in China. Tob Control. 2016; 25(2):188–94.

[68] Klein EG, Berman M, Hemmerich N, Carlson C, Htut S, Slater M. Online e-cigarette marketing claims: A systematic content and legal analysis. Tob Regul Sci. 2016; 2(3):252–62.

[69] Hrywna M, Bover Manderski MT, Delnevo CD. Prevalence of electronic cigarette use among adolescents in New Jersey and association with social factors. JAMA Netw Open. 2020; 3(2):e1920961.

[70] Tobacco tactics. Front groups. Bath: University of Bath; 2019 (https://tobaccotactics.org/wiki/front-groups, accessed 14 October 2020).

[71] Tobacco tactics. Lobby groups. Bath: University of Bath; 2019 (https://tobaccotactics.org/topics/lobby-groups/, accessed 14 October 2020).

[72] Tobacco tactics. CSR strategy. Bath: University of Bath; 2020 (https://tobaccotactics.org/wiki/csr-strategy/, accessed 15 October 2020).

[73] Romberg AR, Miller Lo EJ, Cuccia AF, Willett GJ, Xiao H, Hair EC et al. Patterns of nicotine concentrations in electronic cigarettes sold in the United States, 2013–2018. Drug Alcohol Depend. 2019; 203:1–7.

[74] How much nicotine is in JUUL? Washington (DC): Truth Initiative; 2019 (https://truthinitiative.org/research-resources/emerging-tobacco-products/how-much-nicotine-juul, accessed December 2020).

[75] Hammond D, Rynard VL, Reid JL. Changes in prevalence of vaping among youths in the United

States, Canada, and England from 2017 to 2019. JAMA Pediatr. 2020; 174(8):797–800.

[76] Chen C, Zhuang YL, Zhu SH. e-Cigarette design preference and smoking cessation. Am J Prev Med. 2016;51(3):356–63.

[77] e-Cigarettes: Facts, stats and regulations. Washington (DC): Truth Initiative; 2019 (https://truth-initiative.org/research-resources/emerging-tobacco-products/e-cigarettes-facts-stats-and-regu-lations#:~:text=Between%202012%20and%202013%2C%20 2.4,among%20adults%20aged%20 45%2D64).

[78] Tobacco tactics. Forest. Bath: University of Bath; 2020 (https://tobaccotactics.org/wiki/forest/, accessed 15 October 2020).

[79] Tobacco tactics. Foundation for a Smoke-Free World. Bath: University of Bath (https://tobacco-tactics.org/wiki/foundation-for-a-smoke-free-world/, accessed 10 January 2021).

[80] 2018 recipients of chari table contributions from the Altria family of companies. Richmond (VA): Altria; 2018 (https://web.archive.org/web/20200320145936/https:/www.altria.com/-/ media/ Project/Altria/Altria/responsibility/investing-in-communities/2018Grantees.pdf, accessed 15 October 2020).

[81] Statistics and market data on consumer goods & FMCG: Tobacco. New York City (NY): Statista; 2020.

[82] Electronic cigarette – global market outlook (2017–2026). Dublin: Research and Markets; 2019 (https://www.researchandmarkets.com/reports/4827644/electronic-cigarette-global-mar-ket-outlook, accessed 10 January 2021).

[83] Bao W, Xu G, Lu J, Snetselaar LG, Wallace RB. Changes in electronic cigarette use among adults in the United States, 2014–2016. JAMA. 2018; 319(19):2039.

[84] Wang T, Asman K, Gentzke A, Cullen KA, Holder-Hayes E, Reyes-Guzman C et al. Tobacco product use among adults — United States, 2017. Morb Mortal Wkly Rep. 2017; 67(44):1225–32.

[85] Canadian Tobacco, Alcohols and Drugs Survey (CTADS): summary of results for 2017. Ottawa: Health Canada, Statistics Canada; 2019 (https://www.canada.ca/en/health-canada/services/ ca-nadian-tobacco-alcohol-drugs-survey/2017-summary.html, accessed 6 December 2020).

[86] Special Eurobarometer 458: Attitudes of Europeans toward tobacco and electronic cigarettes. Brussels: European Commission; 2017.

[87] Schoenborn CA, Clarke TC. QuickStats: Percentage of adults who ever used an e-cigarette and percentage who currently use e-cigarettes, by age group – National Health Interview Survey, United States, 2016. Morb Mortal Wkly Rep. 2017; 66(33):892.

[88] Hammond D, Reid JL, Rynard VL, Fong GT, Cummings KM, McNeill A et al. Prevalence of va-ping and smoking among adolescents in Canada, England, and the United States: repeat nation-al cross sectional surveys. BMJ. 2019; 365:l2219.

[89] Country laws regulating e-cigarettes. Advertising, promotion and sponsorship. Baltimore (MD): Institute for Global Tobacco Control; 2020 (https://www.globaltobaccocontrol.org/e-cigarette/advertising-promotion-and-sponsorship, accessed 16 August 2020).

[90] e-Cigarettes at the point of sale. CounterTobacco.Org; 2020 (https://countertobacco.org/re-sources-tools/evidence-summaries/e-cigarettes-at-the-point-of-sale/, accessed 16 August 2020).

[91] Modified risk tobacco product applications. Silver Spring (MD): Food and Drug Administra-tion, Center for Tobacco Products; 2020.

[92] FDA finalizes enforcement policy on unauthorized flavored cartridge-based e-cigarettes that appeal to children, including fruit and mint. Silver Spring (MD): Food and Drug Administration; 2020 (https://www.fda.gov/news-events/press-announcements/fda-finalizes-enforcement-policy-unauthorized-flavored-cartridge-based-e-cigarettes-appeal-children, accessed 17 August 2020).

[93] Notice of intent – Potential measures to reduce the impact of vaping product advertising on youth and non-users of tobacco products. Ottawa: Health Canada; 2019 (https://www.canada.ca/en/health-canada/programs/consultation-measures-reduce-impact-vaping-products-advertising-youth-non-users-tobacco-products/notice-document.html).

[94] Canada Gazette, Part I, Volume 153, Number 25: Vaping Products Labelling and Packaging Regulations. Ottawa: Health Canada; 2019 (http://gazette.gc.ca/rp-pr/p1/2019/2019-06-22/html/reg4-eng.html).

[95] Nicksic NE, Snell LM, Rudy AK, Cobb CO, Barnes AJ. Tobacco marketing, e-cigarette susceptibility, and perceptions among adults. Am J Health Behav. 2017; 41(5):579–90.

[96] Nicksic NE, Snell LM, Barnes AJ. Does exposure and receptivity to e-cigarette advertisements relate to e-cigarette and conventional cigarette use behaviors among youth？ Results from wave 1 of the Population Assessment of Tobacco and Health Study. J Appl Res Child. 2017; 8(2):3.

[97] Cho YJ, Thrasher JF, Reid JL, Hitchman S, Hammond D. Youth self-reported exposure to and perceptions of vaping advertisements: Findings from the 2017 International Tobacco Control Youth Tobacco and Vaping Survey. Prev Med. 2019; 126:105775.

[98] Pokhrel P, Fagan P, Herzog TA, Chen Q, Muranaka N, Kehl L et al. e-Cigarette advertising exposure and implicit attitudes among young adult non-smokers. Drug Alcohol Depend. 2016; 163:134–40.

[99] Camenga D, Gutierrez KM, Kong G, Cavallo D, Simon P, Krishnan-Sarin S. e-Cigarette advertising exposure in e-cigarette naïve adolescents and subsequent e-cigarette use: A longitudinal cohort study. Addict Behav. 2018; 81:78–83.

[100] Phua J, Jin SV, Hahm JM. Celebrity-endorsed e-cigarette brand Instagram advertisements: Effects on young adults' attitudes towards e-cigarettes and smoking intentions. J Health Psychol. 2018; 23(4):550–60.

[101] Behind the explosive growth of JUUL. Washington (DC): Truth Initiative; 2019.

[102] Marketing tobacco to LGBT communities. Washington (DC): National Institutes of Health; 2020(https://smokefree.gov/marketing-tobacco-lgbt-communities, accessed 17 August 2020).

[103] Tobacco industry marketing. Atlanta (GA): Centers for Disease Control and Prevention; 2020 (https://www.cdc.gov/tobacco/data_statistics/fact_sheets/tobacco_industry/marketing/index.htm, accessed 17 August 2020).

[104] Berg CJ, Dekanosidze A, Torosyan A, Grigoryan L, Sargsyan Z, Hayrumyan V et al. Examining smoke-free coalitions in Armenia and Georgia: Baseline community capacity. Health Educ Res. 2019; 34(5):495–504.

[105] Berg CJ. Local coalitions as an underutilized and understudied approach for promoting tobacco control in low- and middle-income countries. J Glob Health. 2019;9(1):10301.

[106] Restricting tobacco advertising. Counter Tobacco; 2020 (https://countertobacco.org/policy/restricting-tobacco-advertising-and-promotions/, accessed 17 August 2020).

[107] Public health groups and leaders worldwide urge rejection of Philip Morris International's new

foundation. Washington (DC): Campaign for Tobacco-Free Kids; 2020 (https://www.tobaccofreekids.org/what-we-do/industry-watch/pmi-foundation/compilation, accessed 17 August 2020).

[108] Spinning a new tobacco industry: How Big Tobacco is trying to sell a do-gooder image and what Americans think about it. Washington (DC): Truth Initiative; 2019 (https: //truthinitiative. org/research-resources/tobacco-industry-marketing/spinning-new-tobacco-industry-howbig-tobacco-trying, accessed 17 August 2020).

[109] Guidelines for implementation of Article 5.3 of the WHO Framework Convention on Tobacco Control. Geneva: World Health Organization; 2011 (https://www.who.int/fctc/guidelines/article_5_3.pdf? ua=1, accessed 17 August 2020).

[110] Van der Eijk Y, Bero LA, Malone RE. Philip Morris International-funded "Foundation for a Smoke-Free World": analysing its claims of independence. Tob Control. 2019; 28:712–8.

[111] Byron MJ, Jeong M, Abrams DB, Brewer NT. Public misperception that very low nicotine cigarettes are less carcinogenic. Tob Control. 2018;27(6):712–4.

[112] Most JUUL-related Instagram posts appeal to youth culture and lifestyles, study finds. Washington (DC): Truth Initiative; 2019 (https://truthinitiative.org/research-resources/emerging-tobacco-products/most-juul-related-instagram-posts-appeal-youth-culture, accessed 15 October 2020).

[113] Heated tobacco products. Information sheet. 2nd edition. Geneva: World Health Organization; 2020 (https://www.who.int/publications/i/item/WHO-HEP-HPR-2020.2).

[114] Heated tobacco products. Atlanta (GA): Centers for Disease Control and Prevention; 2020 (https://www.cdc.gov/tobacco/basic_information/heated-tobacco-products/index.html, accessed 10 January 2021).

[115] Heated tobacco products (HTPs) market monitoring information sheet. Geneva: World Health Organization; 2018 (https://apps.who.int/iris/bitstream/handle/10665/273459/WHO-NMH-PND-18.7-eng.pdf? ua=1).

[116] Cigarettes to record US$7.7 billion loss by 2021 as heated tobacco grows 691 percent. Euromonitor International, 24 June 2017 (https://blog.euromonitor.com/cigarettes-record-loss-heated-tobacco-grows-691-percent/).

[117] Heat not burn tobacco products market – forecasts from 2019 to 2024. Dublin: Research and Markets; 2019 (https://www.researchandmarkets.com/reports/4849645/heat-not-burn-tobacco-products-market-forecasts).

[118] Cigarette market: Global industry trends, share, size, growth, opportunity and forecast 2019–2024. Dublin: Research and Markets; 2019 (https://www.researchandmarkets.com/reports/4752264/cigarette-market-global-industry-trends-share#:~:text=The%20global%20cigarette%20market%20was,4%25%20during%202019%2D2024).

[119] The global cigarette industry. Washington (DC): Campaign for Tobacco-Free Kids; 2019 (https://www.tobaccofreekids.org/assets/global/pdfs/en/Global_Cigarette_Industry_pdf.pdf, accessed 13 August 2020).

[120] 2020 second-quarter & year-to-date highlights. Lausanne: Philip Morris International; 2020 (https://philipmorrisinternational.gcs-web.com/static-files/b88146c3-a4e9-421e-b2aab470e74fcdf2, accessed 13 August 2020).

[121] Carbon heated tobacco product: TEEPS. Lausanne: Philip Morris International; 2020.

[122] Cigarettes in Japan: Country report. London: Euromonitor International; 2020 (https://www.euromonitor.com/cigarettes-in-japan/report, accessed 13 December 2020).

[123] Korea – Ploom Tech. Geneva: Japan Tobacco International; 2020 (https://www.jti.com/node/5242, accessed 13 August 2020).

[124] Japan Tobacco Inc. Integrated report 2019. Tokyo: Japan Tobacco Inc.; 2020 (https://www.jti.com/sites/default/files/global-files/documents/jti-annual-reports/integrated-report-2019v.pdf, accessed 13 August 2020).

[125] Japan Tobacco International. 2020 second quarter results. Lausanne; 2020 (https://www.jt.com/investors/results/forecast/pdf/2020/Second_Quarter/20200731_07.pdf, accessed 13 August 2020).

[126] JT launches two new tobacco vapor products under its Ploom brand. Tokyo: Japan Tobacco Inc.; 2019 (https://www.jti.com/sites/default/files/press-releases/documents/2019/JT-launchestwo-new-tobacco-vapor-products-under-its-ploom-brand_0.pdf, accessed 13 August 2020).

[127] JT launches Ploom S 2.0, an evolved heated tobacco device, in Japan on July 2nd and introduces two new Camel menthol tobacco stick products. Tokyo: Japan Tobacco Inc., 2020 (https://www.jt.com/media/news/2020/0601_01.html, accessed 13 August 2020.

[128] Tobacco heating products: Innovating to lead the transformation of tobacco. London: British American Tobacco plc; 2020 (https://www.bat.com/group/sites/UK__9D9KCY.nsf/vwPagesWebLive/DOAWUGNJ, accessed 17 August 2020).

[129] BAT Science: Introduction to tobacco heating products. London: British American Tobacco plc; 2020 (https://www.bat-science.com/groupms/sites/BAT_B9JBW3.nsf/vwPagesWebLive/ DOBB-3CEX, accessed 17 August 2020).

[130] Delivering for today & investing in the future: Annual report and form 20-F 2019. London: British American Tobacco plc; 2020 (https://www.bat.com/ar/2019/pdf/BAT_Annual_Report_and_Form_20-F_2019.pdf, accessed 14 August 2020).

[131] News release: British American Tobacco expands its new category portfolio with the launch of three innovative new tobacco heating products. London: British American Tobacco plc; 2019 (https://www.bat.com/group/sites/UK__9D9KCY.nsf/vwPagesWebLive/DOBFPCWP, accessed 14 August 2020.

[132] British American Tobacco plc half-year report to 30 June 2020. London: British American Tobacco plc; 2020 (https://www.bat.com/group/sites/uk__9d9kcy.nsf/vwPagesWebLive/ DO72T-JQU/$FILE/medMDBRZQL4.pdf?openelement, accessed 17 August 2020).

[133] Introduction to tobacco heating products. London: British American Tobacco plc; 2020 (https://www.bat-science.com/groupms/sites/BAT_B9JBW3.nsf/vwPagesWebLive/DOBB3CEX, accessed 14 August 2020).

[134] lil: a little is a lot. Seoul: Korea Tobacco & Ginseng Corp., 2019 (http://en.its-lil.com/brand/main_brand. Published 2019, accessed 14 August 2020).

[135] Philip Morris International Inc. announces agreement with KT&G to accelerate the achievement of a smoke-free future. Lausanne: Philip Morris International; 2020 (https://www.pmi.com/media-center/press-releases/press-release-details/?newsId=21786, accessed 14 August 2020).

[136] McKelvey K, Popova L, Kim M, Chaffee BW, Vijayaraghavan M, Ling P et al. Heated tobacco

products likely appeal to adolescents and young adults. Tob Control. 2018;27(Suppl 1):s41–7.

[137] Heated tobacco products. Washington (DC): Campaign for Tobacco-Free Kids; 2020 (https://www.tobaccofreekids.org/what-we-do/global/heated-tobacco-products, accessed 13 August 2020).

[138] Gravely S, Fong GT, Sutanto E, Loewen R, Ouimet J, Xu SS et al. Perceptions of harmfulness of heated tobacco products compared to combustible cigarettes among adult smokers in Japan: Findings from the 2018 ITC Japan survey. Int J Environ Res Public Health. 2020;17(7):2394.

[139] Bar-Zeev Y, Levine H, Rubinstein G, Khateb I, Berg CJ. IQOS point-of-sale marketing strategies in Israel: A pilot study. Isr J Health Policy Res. 2019;8(1):11.

[140] Bialous SA, Glantz SA. Heated tobacco products: Another tobacco industry global strategy to slow progress in tobacco control. Tob Control. 2018;27(Suppl 1):s111–7.

[141] Jackler RK, Ramamurthi D, Axelrod AK, Jung JK, Louis-Ferdinand NG, Reidel JE et al. Global marketing of IQOS: The Philip Morris campaign to popularize "heat not burn". Berkeley (CA): Stanford Research into the Impact of Tobacco Advertising; 2020 (http://tobacco.stanford.edu/tobacco_main/publications/IQOS_Paper_2-21-2020F.pdf, accessed 17 August 2020).

[142] Hair EC, Bennett M, Sheen E, Cantrell J, Briggs J, Fenn Z et al. Examining perceptions about IQOS heated tobacco product: Consumer studies in Japan and Switzerland. Tob Control. 2018;27(Suppl 1):s70–3.

[143] IQOS App. Lausanne: Philip Morris International Management SA; 2020 (https://play.google.com/store/apps/details?id=com.pmi.store.PMIAPPM06278, accessed 14 August 2020).

[144] IQOS in the US. Washington (DC): Truth Initiative; 2020 (https://truthinitiative.org/research-resources/emerging-tobacco-products/iqos-us, accessed 15 October 2020).

[145] 2018 charitable contributions at a glance. Lausanne: Philip Morris International; 2018 (https://www.pmi.com/resources/docs/default-source/our_company/transparency/charitable-2018.pdf?sfvrsn=d97d91b5_2, accessed 15 October 2020).

[146] Uranaka T, Ando R. Philip Morris aims to revive Japan sales with cheaper heat-not-burn tobacco. Reuters, 23 October 2018.

[147] Asia Pacific heat-not-burn tobacco product market (2019–2025): Market forecast by product type, by demography, by sales channels, by countries, and competitive landscape. Dublin: Research and Markets; 2019 (https://www.researchandmarkets.com/reports/4911922/asia-pacific-heat-not-burn-tobacco-product-market).

[148] Sutanto E, Miller C, Smith DM, O'Connor RJ, Quah ACK, Cummings KM et al. Prevalence, use behaviors, and preferences among users of heated tobacco products: Findings from the 2018 ITC Japan survey. Int J Environ Res Public Health. 2019;16(23):4630.

[149] Kim J, Yu H, Lee S, Paek YJ. Awareness, experience and prevalence of heated tobacco product, IQOS, among young Korean adults. Tob Control. 2018;27(Suppl 1):s74–7.

[150] Kang H, Cho S. Heated tobacco product use among Korean adolescents. Tob Control. 2020;29:466–8.

[151] Heated tobacco products market size, share & trends analysis report by product (stick, leaf), by distribution channel (online, offline), by region, and segment forecasts, 2019–2025. San Francisco (CA): Grand View Research; 2019 (https://www.grandviewresearch.com/industry-analysis/heated-tobacco-products-htps-market).

[152] Tobacco tactics. Heated tobacco products. Bath: University of Bath; 2020 (https://tobaccotactics.org/topics/heated-tobacco-products/, accessed 10 January 2021).

[153] MacGuill S, Genov I. Emerging and next generation nicotine products (webinar). London: Euromonitor International; 2017 (http://go.euromonitor.com/rs/805-KOK-719/images/Emerging-and-Next-Generation-Nicotine-Products-Euromonitor.pdf, accessed 6 December 2020).

[154] Countries that regulate heated tobacco. Baltimore (MD): Institute for Global Tobacco Control; 2020.

11. 烟草中的烟碱形态、化学修饰及其对电子烟碱传输系统的影响

Nuan Ping Cheah, Pharmaceutical, Cosmetics and Cigarette Testing Laboratory, Health Sciences Authority, Singapore

Najat Saliba, Department of Chemistry, Faculty of Arts and Sciences, American University of Beirut, Beirut, Lebanon; Center for the Study of Tobacco Products, Virginia Commonwealth University, Richmond (VA), USA

Ahmad El-Hellani, Department of Chemistry, Faculty of Arts and Sciences, American University of Beirut, Beirut, Lebanon; Center for the Study of Tobacco Products, Virginia Commonwealth University, Richmond (VA), USA

摘　要

烟碱形态的影响，即科学家和监管者一直在争论的烟草制品和电子烟碱传输系统（ENDS）中的游离态烟碱或烟碱盐。本章简要讨论了改变烟草制品和ENDS中烟碱形态比例的各种方法。主要关注ENDS烟液中烟碱形态的分配，特别是最近引入的烟弹式ENDS。讨论了各种参数对ENDS烟碱释放的影响，例如烟碱形态、烟碱盐阴离子、设备输出功率和使用者抽吸模式。建议避免将ENDS烟液中烟碱浓度上限设定作为监管ENDS烟碱释放的唯一措施，并采用"烟碱释放量"作为监管工具，考虑影响ENDS烟碱释放的所有参数。强调在构建烟碱释放量模型时考虑烟碱形态的重要性，以及最大限度减少使用者对ENDS的定制可能。尽管仍有必要研究测试ENDS烟液和气溶胶中烟碱形态的方法，以及在有无调味物质的情况下身体对不同形态烟碱的吸收，建议世界卫生组织敦促各国将烟碱释放量和形态纳入ENDS的监管中，以便更好地向使用者通报其设备输送烟碱的情况。

11.1 背　景

本章是世界卫生组织烟草制品监管研究小组第十次会议的背景文件，也是讨论和审议ENDS中烟碱问题的平台。

烟碱既可以作为游离态存在，也可以与有机酸结合成为各种烟碱盐。本报告简要回顾了已发表的有关烟碱的科学信息，烟碱在烟草制品中的存在、在制造前的修饰及其在烟草成品中的形态。还讨论了ENDS中烟碱的形态对吸引力、致瘾性和健康的影响。还介绍了烟碱形态对烟草控制、监管和研究的影响。此外，还简要讨论了烟碱释放量与ENDS烟碱释放监管的相关性，认为烟碱形态可以纳入释放量模型。

烟碱是烟草中的主要生物碱[1]，是33个品种中含量最高的吡啶类生物碱；降烟碱在24个烟草种类中含量最高，假木贼碱在两个烟草种类（N. glauca和N. debneyi）中含量最高，新烟草碱在一个烟草种类（N. otophora）中含量最高。烟碱主要存在于51个种类中，降烟碱存在于2个种类中（N. alata和N. africana），假木贼碱存在于7个种类中（N. glauca、N. solanifolia、N. benavidesi、N. cordifolia、N. debneyi、N. maritima和N. hesperis）[2]。烟碱是在烟草的根部通过酶促途径合成的[3]，由烟酸（吡啶环）和N-甲基-Δ^1-吡咯啉鎓阳离子（吡咯烷环）缩合而成[4]。烟草种类（Nicotiana）中烟碱和其他三种主要吡啶类生物碱的含量见表11.1。

表11.1 烟草中烟碱和主要生物碱的含量[2]

亚属，组	物种	含量（mg/g干重）		占总量的比例（%）							
				烟碱		降烟碱		假木贼碱		新烟草碱	
		叶	根	叶	根	叶	根	叶	根	叶	根
黄花烟草圆锥烟草组	光烟草	8872	5246	12.5	35.5	1.5	2.8	85.1	51.3	0.9	10.4
	茄叶烟草	848	9326	3.2	27.7	81.4	10.0	15.4	60.3	痕量	2.0
	贝纳末特氏烟草	2166	14666	82.7	44.9	1.3	0.8	14.8	48.1	1.2	6.2
	心叶烟草	789	13435	58.4	26.4	6.1	2.5	29.0	64.4	6.5	6.7
黄花烟草组	黄花烟草	7752	8439	96.4	81.6	0.9	1.7	1.1	6.6	1.6	10.1
绒毛烟草组	耳状烟草*	377	7924	6.9	61.3	32.9	27.0	痕量	0.6	60.2	11.1
红花烟草	红花烟草	11462	2176	94.8	81.3	3.0	6.0	0.3	1.7	1.9	11.0
碧冬烟	花烟草	26	1998	100	37.7	痕量	46.4	—	痕量	—	15.8
香甜烟草组	迪勒纳氏烟草	2457	3038	31.1	34.7	15.8	1.4	46.0	53.2	7.1	10.7
	海滨烟草	608	14030	7.2	20.8	70.4	30.0	15.8	44.5	6.6	4.6
	西烟草	4108	1930	52.1	22.1	0.4	1.2	44.3	74.9	3.2	1.8
	非洲烟草*	6776	7698	4.7	45.0	92.4	45.1	0.3	1.0	2.6	8.9

*物种不开花

栽培烟草（N. tabacum L.和N. rustica L.）中生物合成烟碱的含量已可通过基因修饰[5]和工业应用的靶向基因操作[6-8]实现。N. tabacum L.和N.rustica L.是烟草制

11. 烟草中的烟碱形态、化学修饰及其对电子烟碱传输系统的影响

品制造中使用的主要烟草品种，因为这些品种中存在大量烟碱[9,10]。典型的烟碱浓度范围为每克烟草15~35 mg[3]，总生物碱浓度高达每克烟草79 mg[11]。

烟碱的结构包括连接到吡啶环的吡咯烷环。烟碱的游离态、单质子化或双质子化三种形态，取决于烟叶中天然存在的酸对两个氮中心的质子化（图11.1）。质子化形态，也称为烟碱盐，主要存在于未加工的烟叶中。然而，烟草制品有不同的游离态烟碱与盐的比例，在pH值为7、8和9时，溶液中各存在9%、49%和90.5%的游离态烟碱，相应地，烟碱盐分别为91%、51%和9.5%。

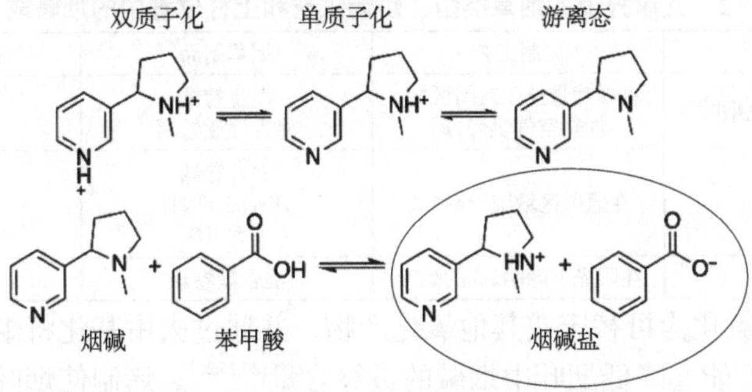

图11.1　游离态烟碱和单、双质子化烟碱盐[12,13]
图的下半部分显示了烟碱与苯甲酸的质子化，苯甲酸可能在烟叶或烟草制品制造过程中自然产生

烟碱的刺激和致瘾作用归因于吡啶类生物碱对大脑神经元烟碱乙酰胆碱受体的作用。目前，吸烟是烟碱传递的最有效形式。燃烧烟草的主流烟气中的烟碱迅速被肺部吸收，并在7秒内到达大脑[14]。虽然几十年来，吸烟一直是烟碱摄入的最普遍形式，但最近引入市场的替代型烟草制品（如电子烟碱传输系统、加热型烟草制品）在全球范围内越来越流行，导致烟碱和烟草制品的总使用量全面增加，尤其是青少年等易感群体。因此，尽管在烟草管制和预防方面取得了重大进展，但烟碱和烟草制品的使用继续增加。本章介绍了市场最广泛、最受欢迎的烟碱传输产品之一的电子烟碱传输系统（ENDS）的烟碱滥用倾向。电子烟碱传输系统销售组合多种多样，比其他烟碱和烟草制品更能满足使用者的个性化需求[14]，实施一套规范的监管是一项挑战，因此，正在开展研究烟碱形态（游离态烟碱与烟碱盐）对这些装置传输能力的影响[15]。

11.2　烟碱的化学修饰及其对烟碱传输的影响

11.2.1　烘烤对烟碱的影响

烟碱和烟草制品依赖性的主要决定因素是其快速释放并达到烟碱药理活性水平的能力[16,17]。制造商仔细控制烟碱剂量以确保其能够产生预期效果，如放松和

精神敏锐度，同时最大限度地降低不良影响的风险，如恶心和中毒[18]。烟碱占烤烟干重的2%~8%，某些烟草种类的烟碱含量范围更广[19]。商品卷烟中使用的三种主要烟叶类型是烤烟、白肋烟和香料烟（表11.2）[20]。传统卷烟是由这些烟草的混合物制成的。不同国家主要的烟草混合物有所不同，但商品卷烟中使用量最大的是烤烟和白肋烟。大多数卷烟主要含有烤烟（例如在加拿大）或主要含有烤烟和白肋烟的混合物，并添加少量香料烟（美国混合型）。

表11.2 文献报道的烟草类型、烘烤工艺和上杆位置中的烟碱成分[21]

烟草类型	调制工艺	烟草制品	烟碱含量（mg/g）
弗吉尼亚（或淡色烟叶）	将烟叶悬挂在封闭区域，加热空气烘烤1周	混合卷烟 弗吉尼亚卷烟	6.52~60.4
白肋烟	在通风区域晾干4~8周	混合卷烟 Kretek卷烟 （丁香风味）	35.6~47.73
香料烟	在阳光下晾晒2周的烟叶	混合型卷烟	1.80~12.6

由于烟碱氧化为可替宁或其他氧化产物，并通过去甲基化将烟碱转化为降烟碱[22-24]，晾制可能会降低烟叶中烟碱的最终含量[23,25,26]。烤制使烟叶中的糖分含量较高，由于糖分是烟草烟气中有机酸的前体，烟气中含有小比例的游离态烟碱[27]。

11.2.2 碱改性

在烟草加工过程中，使用铵类化合物（如磷酸二铵）和其他一些物质（如碳酸钙）可以提高烟草和"烟气pH"。在卷烟滤嘴中加入碳酸钙和碳酸钠以提高"烟气pH"，可能消除了在烟丝中添加碱的情况[28,29]。自20世纪60年代人们意外发现pH升高有助于烟碱吸收，氨水一直被用于烟草制造，尽管被企业否认[33,34]，但卷烟烟气和烟草制品中的游离态烟碱增加[30-32]。氨还会与烟草中的天然有机羟基化合物发生反应，改善烟气质量，产生更顺滑、"巧克力般"、酸味更小的口感[35,36]。

其他常用的增加烟气pH和改善烟气口感的碱性物质包括尿素、磷酸二铵、乙醇胺和碳酸盐[37,38]。在碱性或高pH环境中，非质子化（游离态）形式的烟碱通过黏膜迅速被吸收，然而，这种快速通量会对使用者产生刺激性。

11.2.3 酸改性

当卷烟烟气被认为太刺激时，吸烟者吸入的深度会减少[39]。乙酰丙酸等添加剂通过降低游离态烟碱的比例，使烟气在上呼吸道更平滑，因此，烟气更容易被吸入肺部[40]。据报道，乙酰丙酸也能增加烟碱的释放量[41]。研究发现，添加硝酸镁等无机盐可以降低烟碱向烟气的转移[42]。在酸性条件下，烟碱被离子化（质子化），因此穿过生物膜的速度要慢得多，刺激性也要小得多[43]。表11.3总结了在烟草制品中添加酸的研究结果。

表11.3　向烟草制品中添加酸的研究结果

酸的类型	目的
乳酸	减少粗糙感和苦味，使口感更甜、更顺滑[44]
柠檬酸	降低粗糙感，改善风味，降低烟气pH，"中和"烟碱的影响，增强再造烟草薄片形成[45,46]
酒石酸	与乳酸相似，降低烟气的pH[46,47]
苹果酸	在普通的制造和储存过程中没有促进迁移[48,49]
甲酸	增加烟碱释放，但有明显的酸味，不能改善主观表现[50,51]
乙酰丙酸	减少粗糙感，但不会减少烟气中烟碱的释放，并且没有令人不快的味道。乙酰丙酸烟碱盐也增加了烟气中烟碱的释放[41,52]
苯甲酸和山梨酸	减少粗糙感，增加烟碱释放。与烟碱形成盐
丙酮酸和月桂酸	通常添加以形成烟碱盐[53,54]

11.3　对电子烟碱传输系统产品和多样性的影响

11.3.1　电子烟碱传输系统中游离态烟碱与烟碱盐

上一节表明，制造商可在烟草烘烤或者加工过程中控制燃烧型卷烟中烟碱形态和剂量。在上文中报告，烟草烟气中不同形态烟碱的分布影响烟碱的可吸入性。游离态烟碱很容易在上呼吸道被吸收，而烟碱盐则被输送到支气管肺泡区[55]。虽然不同形态烟碱的吸收部位是已知的，但关于该部位如何影响烟碱向大脑递送的速率仍存在争议[56]。

电子烟碱传输系统允许使用者在剂量和形态方面对他们的烟碱体验进行更大的定制。尽管最流行的电子烟碱传输系统（例如烟弹式ENDS）的气溶胶具有低pH和高水平的烟碱盐，烟液中烟碱形态的比例可以在整个pH范围变化（5.3~9.3）[57]。添加酸，如乙酰丙酸，形成单质子态和双质子态烟碱，使从电子烟碱传输系统吸入的气溶胶在喉咙和上呼吸道更顺畅。用于形成烟碱盐的其他常见酸有乳酸、苯甲酸、山梨酸、丙酮酸、水杨酸、苹果酸、月桂酸和酒石酸[54]。一份报告指出，酚类、香兰素和乙基香兰素等风味添加剂可作为电子烟液中的质子化剂[58]。一项基于随机对照试验[59]的研究表明，ENDS中烟碱盐的存在与传统卷烟一样，在一定程度上降低了对烟瘾的渴望。药代动力学和主观数据表明，乳酸烟碱盐通过肺部途径快速吸收烟碱，尽管其最大烟碱水平不超过传统卷烟，且还显示出可接受的主观满意度和吸烟欲望的缓解[59]。

三种形态的烟碱被认为是烟草制品中主要的致瘾性物质。到了20世纪90年代，越来越多的人认为不含烟碱的烟草制品不会持续致瘾性[16,60]。因此，应重点关注使用者对电子烟碱传输系统气溶胶的感知中口味和烟碱之间的相互作用。调味物质可以减少电子烟碱传输系统气溶胶中高烟碱水平对上呼吸道的刺激，或有助于低烟碱水平气溶胶的感官影响,如薄荷醇[61]。此外,公布的数据显示,一些口味,如苹果味,可能会增加电子烟碱传输系统气溶胶中烟碱的强化作用[62],正如卷烟烟气中的薄荷醇的情况一样[63]。

烟碱盐如苯甲酸烟碱盐是单质子化盐。据报道,这些产品在使用者中产生了很高的满意度,受欢迎的JUUL产品[64]证明了这一点,JUUL是一种含苯甲酸盐[54]和类似电子烟碱传输系统[65]的专利产品。在雾化过程中,烟碱盐蒸发分解,产生游离态烟碱和酸,这些烟碱和酸在与环境空气暴露时重新结合,凝结成可吸入的气溶胶[54]。色谱法可鉴定电子烟烟液或气溶胶中来自添加酸(如水杨酸盐、酒石酸盐、乙酰丙酸盐和苹果酸盐)的阴离子[66,67]。

11.3.2 有效浓度和滥用倾向

有研究者已开发了测定电子烟碱传输系统烟液和气溶胶中烟碱含量的分析方法[68-70]。电子烟碱传输系统烟液和预灌装烟弹中的烟碱浓度范围很广,从无烟碱烟弹到一些"DIY"烟液中浓度约为130 mg/mL[71-73]。在电子烟碱传输系统流行的最初几年,烟弹式电子烟碱传输系统(封闭系统)的烟碱浓度低于开放系统[71]。然而,最近推出的烟弹式电子烟碱传输系统烟碱水平非常高,含量超过60 mg/mL[69,74]。电子烟碱传输系统的吸引力不仅与烟碱浓度有关,还与烟碱的形态(游离态或烟碱盐)有关[75-77]。到目前为止,对电子烟碱传输系统烟液中烟碱形态的分析报告表明,pH和烟碱形态比率范围很广[68,78,79]。在这些研究中使用了多种方法,包括pH测量,然后使用Henderson-Hasselbalch方程估计烟碱形态的比,或者用有机溶剂提取溶解在水中的电子烟碱传输系统烟液的烟碱,通过气相色谱法测量不同形态的烟碱[68]或者通过质子核磁共振波谱测定不同形态烟碱的比[80]。此外,烟液中存在的烟碱被有效地转移到气溶胶中进而影响主观效果[67]。本报告作者最近的研究表明,烟碱形态并不影响气溶胶中烟碱的总释放量[81]。

有几个因素促成了电子烟碱传输系统的吸引力和持续使用,包括口味、高"可定制性"和烟碱释放[82,83]。在美国,与过去十年中任何其他非卷烟烟草制品不同,电子烟碱传输系统的使用在儿童和青少年中已经超过了抽吸卷烟[84]。某些电子烟碱传输系统,操作参数(即功率、烟液组成、抽吸曲线)无限组合的可能性允许烟碱释放量从微量到高于燃烧型卷烟的数量级[71,85,86]。这种广泛的烟碱递送可能会增加使用者的滥用倾向和烟碱依赖风险[87]。与其他药物一样,烟碱致瘾性与给药剂量和给药速度有关[88]。抽吸卷烟是一种高效、快速的烟碱递送方式,因此具有致瘾性[89]。同样的道理也适用于电子烟碱传输系统,最近的大脑成像研究表明,电子烟碱传输系统可以将烟碱以类似于卷烟的速度输送到大脑[90,91]。

值得注意的是,电子烟碱传输系统和加热型烟草制品通常被用作吸烟的补充,而不是替代品,特别是在无烟环境中[92]。因此,电子烟碱传输系统和卷烟的双重使用是维持烟碱依赖的常见做法[93]。纵向研究还表明，电子烟碱传输系统使用者同时抽吸卷烟，可能是因为对烟碱的依赖性更大[94,95]。因此，电子烟碱传输系统与卷烟的双重使用呈增长趋势[96,97]。

11.3.3 对产品刺激性可能的掩饰

烟碱释放是使用电子烟碱传输系统的主要原因之一[98]。另一个促成电子烟碱传输系统流行的因素是独特风味的广泛可用性[99,100]，其数量近年来急剧增长[101]。最常见的口味是烟草、薄荷醇或薄荷、水果、糖果或甜点和饮料[102-104]。风味可以掩盖与吸入的游离态烟碱相关的刺激感，尤其是如果它们能引起凉爽的感觉，如薄荷和薄荷醇[105]。因此，一个有趣的研究领域是口味选择与新手和有经验的使用者使用的烟碱形态之间的相关性。

11.3.4 健康影响和监管考虑

电子烟碱传输系统通常被认为比抽吸卷烟更安全、致瘾性更低，这可能是电子烟碱传输系统越来越受欢迎的原因[106,107]。在过去十年中，全球对电子烟碱传输系统的使用迅速增加，特别是在年轻人中，有证据表明这一人群中存在烟碱依赖[81,108-113]。早期电子烟碱传输系统装置被认为在向使用者输送烟碱方面效率低下[114]。然而，随着设备的发展和使用经验的增加，烟碱的递送效率大大提高[115-117]。

电子烟碱传输系统产品具有明显的多样性，在材料、结构、电功率输出、溶剂和组成方面存在差异[118]。因此，烟碱释放量取决于多种变量的组合，如装置功率、烟液组成和使用者的抽吸行为[84,119]。如上所述，电子烟碱传输系统为使用者提供了前所未有的机会，通过制备自己的烟液和修改操作参数，定制烟碱浓度和形态，从而确定烟碱剂量。

电子烟碱传输系统的烟碱释放是科学家和决策者争论的话题。一些人认为，如果电子烟碱传输系统产品释放烟碱的剂量和速率与卷烟相当，电子烟碱传输系统可能有助于吸烟者减少吸烟或戒烟，从而减少其暴露于相关危害的程度[120]。然而，相当有效的烟碱输送可能会使电子烟碱传输系统使用者，包括以前未接触烟碱的人，对烟碱更上瘾[110,121,122]。最近的一项研究表明，对电子烟碱传输系统烟液的烟碱浓度施加限制，不足以确保烟碱释放量低于卷烟[123]，因为使用者可以增加装置的功率以获得卷烟中的烟碱水平。此外，转用高功率设备的使用者吸入更多的气溶胶，从而吸入更多的有害物质，带来意外的健康后果[86]。

大多数国家尚未修订其立法，包括关于电子烟碱传输系统的规定。主要是欧洲的研究人员和监管机构认为，降低风险的第一个监管措施可能是限制电子烟碱传输系统烟液的烟碱含量，以控制烟碱的释放。自2014年以来，这一政策在欧盟一直有效，电子烟碱传输系统烟液中的烟碱浓度限制为20 mg/mL。其他司法管辖区也可考虑采取类似做法。然而，该政策并未考虑电子烟碱传输系统产品特性的变化，如功率和抽吸曲线。例如，在烟碱浓度受到限制的地区，使用者可以通过选择功率更大的装置来规避法规的目的，以获得超过卷烟的烟碱释放量。Shihadeh和Eisenberg[124]提出了"烟碱释放量"的测量方法，即单位时间内从烟嘴输送的烟碱量（mg/s），作为一种合适的监管工具，它包含了影响递送给使用者的烟碱速率和剂量的电子烟碱传输系统的所有相关操作参数。该小组目前的工作重点是将烟碱形态纳入烟碱释放量结构。对电子烟碱传输系统致瘾性和滥用倾向的临床研究将确定烟碱形态对输送速度的影响。烟碱形态也可能影响有害性，Pankow等[125]证明了这一点，在制备烟碱盐时使用苯甲酸可能会导致释放物中苯的形成。

11.4 讨　　论

烟草行业已经使用了许多方法来提高卷烟中烟碱的递送效率，包括控制卷烟烟丝中烟碱形态的比例。尽管如此，为了平衡烟碱的释放和产生烟气的刺激性，传统卷烟的生产受到了限制。最近在电子烟碱传输系统设计中采用了类似的方法[126]。由于添加了有机酸，市场份额最大的电子烟碱传输系统使用的烟液pH较低，这掩盖了雾化过程中大量烟碱的刺激。然而，烟碱盐可能由于阴离子的降解而导致电子烟气溶胶的毒性[125]。

11.5 研究差距、优先事项和问题

证据表明，各种形态的烟碱，即游离态烟碱和单、双质子烟碱盐已在烟碱和烟草制品市场上销售，并正在迅速推广到其他产品。最近，烟草企业开始生产合成烟碱，其生产成本比以前的技术要低，已经应用于电子烟烟液[127]。这可能对某些司法管辖区的监管构成挑战，因为是非烟草来源的烟碱。

研究和开发需求：

■ 制定和/或验证电子烟烟液和电子烟碱传输系统气溶胶中游离态烟碱和游离态与质子化和双质子化烟碱比率的测定标准方法；

■ 电子烟碱传输系统烟液和气溶胶中不同烟碱形态、总量和有机酸[68,81]；

■ 在有/无混杂因素的情况下进行研究，确定烟碱形态对烟碱传输系统使用者烟碱传递的影响和潜在依赖性，包括成瘾的维持；

■ 研究使用有机酸改变烟碱药代动力学对健康的影响[128]及对有害性的影响；

■ 验证烟碱通量和烟碱形态，作为调节电子烟碱传输系统烟碱释放的工具。

如果这些空白领域得到解决，全球优先事项应是：
- 确保电子烟碱传输系统中烟碱形态的比例有助于抽吸卷烟者戒烟，并且不会导致新手吸烟者对烟碱成瘾；
- 基于严格的证据限制制造商操纵烟碱浓度和形态。

鉴于电子烟碱传输系统烟碱输送的重要性以及电子烟碱传输系统烟液中烟碱形态和浓度对其吸引力和致瘾性可能产生的综合影响，应考虑在将来要求提交一份完整文献。

11.6 建　议

如果该主题被视为优先事项，并且有足够的信息，则应考虑为今后的会议编写一份关于烟碱形态的完整文件。

11.7 考 虑 因 素

在各国加强烟草监管框架时，首要目标应该是减少对烟碱的暴露，烟碱是烟草中最重要的致瘾物质。正如世界卫生组织《烟草控制框架公约》第1条所述，许多国家有权管制烟草制品或烟草衍生产品。到目前为止，对电子烟碱传输系统烟碱释放的唯一监管是欧盟的监管，该监管限制了电子烟碱传输系统烟液中烟碱的含量。在上文中提出，这可能无法解决这一新类别烟碱传输产品的所有功能，因为它没有反映出电子烟碱传输系统烟碱释放量是许多因素结果的事实，如烟碱浓度、功率输出和使用者抽吸方式。烟碱通量可能是监管考虑的一个选项。

此外，烟碱形态对其释放、药代动力学和药效学的影响还没有得到很好的研究，需要进行更多的研究。监管机构还应考虑尽可能减少使用者对电子烟碱传输系统的烟碱载量、形态和其他特征如风味的定制。

为了公众健康的利益，各国可能必须审查其立法，全面监管含烟碱的产品，而不管烟碱的来源如何。如果烟碱传输产品不受管制，各国和世界卫生组织在减少烟草使用和烟碱滥用倾向方面的工作可能会受到影响。最后，世界卫生组织应考虑在世界卫生组织《烟草控制框架公约》中包含"电子烟碱传输系统"或在公约内提供解决这些产品特定问题的规定。

11.8 参 考 文 献

[1] Wernsman E. Time and site of nicotine conversion in tobacco. Tob Sci. 1968;12:226–8.
[2] Saitoh F, Noma M, Kawashima N. The alkaloid contents of sixty Nicotiana species. Phyto-che-mistry. 1985;24(3):477–80.

[3] Collins W, Hawks S. Principles of flue-cured tobacco production. Raleigh (NC): HarperCollins Publishers; 1993.

[4] Sun B, Tian YX, Zhang F, Chen Q, Zhang Y, Luo Y et al. Variations of alkaloid accumulation and gene transcription in Nicotiana tabacum. Biomolecules. 2018;8(4):114.

[5] Liu H, Kotova TI, Timko MP. Increased leaf nicotine content by targeting transcription factor gene expression in commercial flue-cured tobacco(NicotianatabacumL.). Genes. 2019;10(11):10.3390/genes10110930.

[6] Saunders JW, Bush LP . Nicotine biosynthetic enzyme activities in Nicotiana tabacum L. geno-types with different alkaloid levels. Plant Physiol. 1979;64(2):236–40.

[7] Moghbel N, Ryu B, Ratsch A, Steadman KJ. Nicotine alkaloid levels, and nicotine to nor-nicotine conversion, in Australian Nicotiana species used as chewing tobacco. Heliyon. 2017;3(11):e00469.

[8] Griffith R, Valleau W, Stokes G. Determination and inheritance of nicotine to nornicotine con-version in tobacco. Science. 1955;121(3140):343–4.

[9] Chaplin JF, Burk L. Agronomic, chemical, and smoke characteristics of flue-cured tobacco lines with different levels of total alkaloids 1. Agron J. 1984;76(1):133–6.

[10] Chaplin JF, Burk L. Genetic approaches to varying chemical constituents in tobacco and smoke. Beitr Tabakforsch. 1977;9(2):102–6.

[11] Tso TC. Production, physiology, and biochemistry of tobacco plant. Beltsville (MD): Ideals; 1990.

[12] Baxendale IR, Brusotti G, Matsuoka M, Ley SV. Synthesis of nornicotine, nicotine and other functionalised derivatives using solid-supported reagents and scavengers. J Chem Soc Perkin Trans. 2002;2002(2):143–54.

[13] Domino EF, Hornbach E, Demana T. The nicotine content of common vegetables. N Engl J Med. 1993;329(6):437.

[14] Maisto SA, Galizio M, Connors GJ, editors. Drug use and abuse. Seventh edition. Belmont (CA): Cengage Learning; 2014.

[15] Gholap VV, Kosmider L, Golshahi L, Halquist MS. Nicotine forms: Why and how do they matter in nicotine delivery from electronic cigarettes? Expert Opin Drug Delivery. 2020;1–10.

[16] Benowitz NL, Henningfield JE. Establishing a nicotine threshold for addiction – the implicati-ons for tobacco regulation. New Engl J Med. 1994;331(2):123–5.

[17] Henningfield J, Benowitz N, Slade J, Houston T, Davis R, Deitchman S. Reducing the addictive-ness of cigarettes. Tob Control. 1998;7(3):281.

[18] Carpenter CM, Wayne GF, Connolly GN. Designing cigarettes for women: new findings from the tobacco industry documents. Addiction. 2005;100(6):837–51.

[19] Gorrod JW, Jacob P III, editors. Analytical determination of nicotine and related compounds and their metabolites. Amsterdam: Elsevier Science; 1999.

[20] Paschke T, Scherer G, Heller WD. Effects of ingredients on cigarette smoke composition and biological activity: a literature overview. Beitr Tabakforsg. 2002;20(3):107–244.

[21] Djordjevic MV, Gay SL, Bush LP , Chaplin JF. Tobacco-specific nitrosamine accumulation and distribution in flue-cured tobacco alkaloid isolines. J Agric Food Chem. 1989;37(3):752–6.

[22] Burton H, Bush L, Hamilton J. Effect of curing on the chemical composition of burley tobacco.

Rec Adv Tob Sci. 1983;9:91–153.

[23] Wahlberg I, Karlsson K, Austin DJ, Junker N, Roeraade J, Enzell C et al. Effects of flue-curing and ageing on the volatile, neutral and acidic constituents of Virginia tobacco. Phytochemistry. 1977;16(8):1217–31.

[24] Burton HR, Kasperbauer M. Changes in chemical composition of tobacco lamina during se-nescence and curing. 1. Plastid pigments. J Agric Food Chem. 1985;33(5):879–83.

[25] Bush L. Physiology and biochemistry of tobacco alkaloids. Recent Adv Tob Sci. 1981;7:75–106.

[26] Frankenburg WG, Gottscho AM, Mayaud EW, Tso TC. The chemistry of tobacco fermentation. I. Conversion of the alkaloids. A. The formation of 3-pyridyl methyl ketone and of 2,3´-dipyridyl. J Am Chem Soc. 1952;74(17):4309–14.

[27] Elson L, Betts T, Passey R. The sugar content and the pH of the smoke of cigarette, cigar and pipe tobaccos in relation to lung cancer. Int J Cancer. 1972;9(3):666–75.

[28] Leffingwell JC. Tobacco, production, chemistry and technology. Oxford: Blackwell;. 1999.

[29] Douglas C. Tobacco manufacturers manipulate nicotine content of cigarettes to cause and sustain addiction. Tobacco: the growing epidemic. Berlin: Springer; 2000:209–11.

[30] Brambles Australia Ltd (Brambles) v British American Tobacco Australia Services Ltd(-BA-TAS).2007.BatesNo.:proctorr20071203b (https://www. industrydocuments.ucsf.edu/docs/tmyg0225, accessed 18 March 2020).

[31] Fant RV, Henningfield JE, Nelson RA, Pickworth WB. Pharmacokinetics and pharmacodyna-mics of moist snuff in humans. Tob Control. 1999;8(4):387–92.

[32] Henningfield JE, Radzius A, Cone EJ. Estimation of available nicotine content of six smokeless tobacco products. Tob Control. 1995;4(1):57.

[33] Stevenson T, Proctor RN. The secret and soul of Marlboro. Am J Public Health. 2008;98(7):1184–94.

[34] Hind JD, Seligman RB. Tobacco sheet material. US Patent. 3,353,541. 1967. Filed 16 June 1966 and issued 21 November 1967.

[35] Teague CE Jr. Modification of tobacco stem materials by treatment with ammonia and ot-her substances. Winston-Salem (NC): RJ Reynolds; 1954. Bates no. 504175083-5084 (http://legacy.library.ucsf.edu /tid/gpt58d00, accessed 9 November 2020).

[36] Pankow JF, Mader BT, Isabelle LM, Luo W, Pavlick A, Liang C. Conversion of nicotine in tobacco smoke to its volatile and available free-base form through the action of gaseous ammonia. Environ Sci Technol. 1997;31(8):2428–33.

[37] Seligman R.The use of alkalis to improve smoke flavor.PhilipMorris.1965. Batesno.2026351158-2026351163(https://www.industrydocuments.ucsf.edu/docs/tfpv0109, acces-sed 4 December 2020).

[38] Armitage AK, Dixon M, Frost BE, Mariner DC, Sinclair NM. The effect of tobacco blend ad-ditives on the retention of nicotine and solanesol in the human respiratory tract and on subsequent plasma nicotine concentrations during cigarette smoking. Chem Res Toxicol. 2004;17(4):537–44.

[39] Henningfield JE, Fant RV, Radzius A, Frost S. Nicotine concentration, smoke pH and whole tobacco aqueous pH of some cigar brands and types popular in the United States. Nicotine Tob Res. 1999;1(2):163–8.

[40] Scientific Committee on Emerging and Newly Identified Health Risks. Addictiveness and at-tractiveness of tobacco additives. Brussels: European Commission, Directorate-General forHealth and Consumers; 2010 (https://ec.europa.eu/health/scientific_committees /emerging/ docs/scenihr_o_029.pdf, accessed 9 November 2020).

[41] Keithly L, Ferris Wayne G, Cullen DM, Connolly GN. Industry research on the use and effects of levulinic acid: a case study in cigarette additives. Nicotine Tob Res. 2005;7(5):761–71.

[42] Kobashi Y, Sakaguchi S. The influence of inorganic salts on the combustion of cigarette and on the transfer of nicotine on to smoke. Agric Biol Chem. 1961;25(3):200–5.

[43] van Amsterdam J, Sleijffers A, van Spiegel P, Blom R, Witte M, van de Kassteele J et al. Effect of ammonia in cigarette tobacco on nicotine absorption in human smokers. Food Chem Toxicol. 2011;49(12):3025–30.

[44] Meyer L. Lactic acid sprayed cellulose acetate filters. 1963, Bates No.: 1003105149-1003105150 (https://www.industrydocuments.ucsf.edu/docs/qjwd0107, accessed 5 April 2020).

[45] Morris P. Citric acid. 1989. Bates No.: 2028670128-2028670129 (https://www.industrydocu-ments.ucsf.edu/tobacco/docs/#id=nzfm0111, accessed 5 April 2020).

[46] Winterson WD, Cochran TD, Holland TC, Torrence KM, Rinehart S, Scott GR. Method of making pouched tobacco product. US Patent no. US 7,980,251 B2; 2011.

[47] JD Backhurst. The effects of ameliorants on the smoke from Burley tobacco. 1968. Bates No.:302075334-302075350(https://www.industrydocuments. ucsf.edu/tobacco/docs/#id=-jn-vj0189, accessed 5 April 2020).

[48] Sudholt MA. Increased impact of cellulose–malic acid filter containing migrated nicotine.1985. Bates No.: 81122186-81122188 (https://www.industrydocuments.ucsf.edu /tobacco/docs/#id=-jnfk0037, accessed 5 April 2020).

[49] Sudholt MA. Continuation of study of malic acid treated cellulose filters. 1985. Bates No.:80551009-80551010(https://www.industrydocuments.ucsf.edu/tobacco/ docs/#id=km-fk0037, accessed 5 April 2020).

[50] Cipriano JJ, Kounnas CN, Spielberg HL. Manipulation of nicotine delivery by addition of acids To filler. 1979. BatesNo.:3990037328-3990037339https: //www.industrydocuments. ucsf.edu/tobacco/docs/#id=kpxp0180, accessed 5 April 2020).

[51] Cipriano JJ, Kounnas CN, Spielberg HL. Manipulation of nicotine delivery by addition of acids to filter. 1975. Bates No.: 3990093180-3990093188 (https://www.industrydocuments.ucsf.edu / tobacco/docs/#id=zpgp0180, accessed 5 April 2020).

[52] Levulinic acid. RJ Reynolds Collection. 1991. Bates No.: 512203269-512203295 (https://www.industrydocuments.ucsf.edu/tobacco/docs/#id=rymk0089, accessed 5 April 2020).

[53] Bowen A, Xing C. Nicotine salt formulations for aerosol devices and methods thereof. US Patent no. USS 9,215,895 B2; 2015.

[54] Bowen A, Xing C. Inventors; JUUL Labs, Inc., assignee (2016). Nicotine salt formulations for aerosol devices and methods thereof. US patent 20,160,044,967. 28 October 2015.

[55] Armitage A, Dollery C, Houseman T, Kohner E, Lewis P, Turner D. Absorption of nicotine by man during cigar smoking. Br J Pharmacol. 1977;59(3):493P.

[56] Pankow JF. A consideration of the role of gas/particle partitioning in the deposition of ni-cotine and other tobacco smoke compounds in the respiratory tract. Chem Res Toxicol.

2001;14(11):1465–81.

[57] El-Hellani A, Salman R, El-Hage R, Talih S, Malek N, Baalbaki R et al. Nicotine and carbonyl emissions from popular electronic cigarette products: correlation to liquid composition and design characteristics. Nicotine Tob Res. 2018;20(2):215–23.

[58] Pankow JF, Duell AK, Peyton DH. Free-base nicotine fraction αfb in non-aqueous versus aqu-eous solutions: electronic cigarette fluids without versus with dilution with water. Chem Res Toxicol. 2020;33(7):1729–35.

[59] O'Connell G, Pritchard JD, Prue C, Thompson J, Verron T, Graff D et al. A randomised, open-la-bel, cross-over clinical study to evaluate the pharmacokinetic profiles of cigarettes and e-ciga-ret-tes with nicotine salt formulations in US adult smokers. Intern Emerg Med. 2019;14(6):853–61.

[60] Henningfield J, Pankow J, Garrett B. Ammonia and other chemical base tobacco additives and cigarette nicotine delivery: issues and research needs. Nicotine Tob Res. 2004;6(2):199–205.

[61] Rosbrook K, Green BG. Sensory effects of menthol and nicotine in an e-cigarette. Nicotine Tob Res. 2016;18(7):1588–95.

[62] Avelar AJ, Akers AT, Baumgard ZJ, Cooper SY, Casinelli GP, Henderson BJ. Why flavored vape products may be attractive: Green apple tobacco flavor elicits reward-related behavior, upre-gu-lates nAChRs on VTA dopamine neurons, and alters midbrain dopamine and GABA neuron function. Neuropharmacology. 2019;158:107729.

[63] Ahijevych K, Garrett BE. The role of menthol in cigarettes as a reinforcer of smoking behavior. Nicotine Tob Res. 2010;12(Suppl 2):S110–6.

[64] Fadus MC, Smith TT, Squeglia LM. The rise of e-cigarettes, pod mod devices, and Juul among youth: Factors influencing use, health implications, and downstream effects. Drug Alcohol Depend. 2019;201:85–93.

[65] Barrington-Trimis JL, Leventhal AM. Adolescents' use of "pod mod" e-cigarettes – urgent concerns. New Engl J Med. 2018;379(12):1099–102.

[66] Tracy M, Liu X. Analysis of nicotine salts by HPLC on an anion-exchange/cation-exchange / reversed-phase tri-mode column. Sunnyvale (CA): DionexCorp.;2011(https://www.chroma-tographyonline.com/view/analysis-nicotine-salts-hplc-anion-exchangecation-exchange-re-versed-phase-tri-mode-column, accessed 7 March 2020).

[67] Sádecká J, Polonský J. Determination of organic acids in tobacco by capillary isotachophore-sis. J Chromatogr A. 2003;988(1):161–5.

[68] El-Hellani A, El-Hage R, Baalbaki R, Salman R, Talih S, Shihadeh A et al. Free-base and protona-ted nicotine in electronic cigarette liquids and aerosols. Chem Res Toxicol. 2015;28(8):1532–7.

[69] Omaiye EE, McWhirter KJ, Luo W, Pankow JF, Talbot P. High-nicotine electronic cigarette products: toxicity of JUUL fluids and aerosols correlates strongly with nicotine and some flavor chemical concentrations. Chem Res Toxicol. 2019;32(6):1058–69.

[70] Duell AK, Pankow JF, Peyton DH. Free-base nicotine determination in electronic cigarette liquids by 1H NMR spectroscopy. Chem Res Toxicol. 2018;31(7):431–4.

[71] El-Hellani A, Salman R, El-Hage R, Talih S, Malek N, Baalbaki R et al. Nicotine and carbonyl emissions from popular electronic cigarette products: correlation to liquid composition and de-

sign characteristics. Nicotine Tob Res. 2018;20(2):215–23.

[72] Grana RA, Ling PM. "Smoking revolution": a content analysis of electronic cigarette retail websites. Am J Prev Med. 2014;46(4):395–403.

[73] Davis B, Dang M, Kim J, Talbot P. Nicotine concentrations in electronic cigarette refill and do-it-yourself fluids. Nicotine Tob Res. 2015;17(2):134–41.

[74] Talih S, Salman R, El-Hage R, Karam E, Karaoghlanian N, El-Hellani E et al. Characteristics and toxicant emissions of JUUL electronic cigarettes. Tob Control. 2019;28(6):678–80.

[75] St Helen G, Ross KC, Dempsey DA, Havel CM, Jacob P, Benowitz NL. Nicotine delivery and vaping behavior during ad libitum e-cigarette access. Tob Regul Sci. 2016;2(4):363–76.

[76] Kong G, Bold KW, Morean ME, Bhatti H, Camenga DR, Jackson A et al. Appeal of JUUL among adolescents. Drug Alcohol Depend. 2019;205:107691.

[77] Seeman JI, Fournier JA, Paine JB, Waymack BE. The form of nicotine in tobacco. Thermal transfer of nicotine and nicotine acid salts to nicotine in the gas phase. J Agric Food Chem. 1999;47(12):5133–45.

[78] Stepanov I, Fujioka N. Bringing attention to e-cigarette pH as an important element for research and regulation. Tob Control. 2015;24(4):413–4.

[79] Lisko JG, Tran H, Stanfill SB, Blount BC, Watson CH. Chemical composition and evaluation of nicotine, tobacco alkaloids, pH, and selected flavors in e-cigarette cartridges and refill solutions. Nicotine Tob Res. 2015;17(10):1270–8.

[80] Gholap VV, Heyder RS, Kosmider L, Halquist MS. An analytical perspective on determination of free base nicotine in e-liquids. J Anal Meth Chem. 2020;2020:6178570.

[81] Talih S, Salman R, El-Hage R, Karaoghlanian N, El-Hellani A, Saliba N et al. Effect of free-base and protonated nicotine on nicotine yield from electronic cigarettes with varying power and liquid vehicle. Sci Rep.2020;10:16263.

[82] Cobb CO, Lopez AA, Soule EK, Yen MS, Rumsey H, Lester Scholtes R et al. Influence of electronic cigarette liquid flavors and nicotine concentration on subjective measures of abuse liability in young adult cigarette smokers. Drug Alcohol Depend. 2019;203:27–34.

[83] Camenga DR, Morean M, Kong G, Krishnan-Sarin S, Simon P, Bold K. Appeal and use of customizable e-cigarette product features in adolescents. Tob Regul Sci. 2018;4(2):51–60.

[84] Youth and tobacco use. Atlanta (GA): Centers for Disease Control and Prevention; 2019 (https://www.cdc.gov/tobacco/data_statistics/fact_sheets/youth_ data/tobacco_use/index.htm, accessed 21 January 2020).

[85] Talih S, Balhas Z, Salman R, El-Hage R, Karaoghlanian N, El-Hellani A et al. Transport phenomena governing nicotine emissions from electronic cigarettes: Model formulation and experimental investigation. Aerosol Sci Technol. 2017;51(1):1–11.

[86] Voos N, Goniewicz ML, Eissenberg T. What is the nicotine delivery profile of electronic cigarettes? Expert Opin Drug Delivery. 2019;16(11):1193–203.

[87] Dawkins L, Cox S, Goniewicz M, McRobbie H, Kimber C, Doig M et al. "Real-world" compensatory behaviour with low nicotine concentration e-liquid: subjective effects and nicotine,acrolein and formaldehyde exposure. Addiction. 2018;113(10):1874–82.

[88] Allain F, Minogianis EA, Roberts DCS, Samaha AN. How fast and how often: The pharmacokinetics of drug use are decisive in addiction. Neurosci Biobehav Rev. 2015;56:166–79.

[89] Berridge MS, Apana SM, Nagano KK, Berridge CE, Leisure GP, Boswell MV. Smoking produces rapid rise of [11C]nicotine in human brain. Psychopharmacology. 2010;209(4):383-94.

[90] Baldassarri SR, Hillmer AT, Anderson JM, Jatlow P, Nabulsi N, Labaree D et al. Use of electronic cigarettes leads to significant beta2-nicotinic acetylcholine receptor occupancy: Evidence from a PET imaging study. Nicotine Tob Res. 2017;20(4):425-33.

[91] Solingapuram Sai KK, Zuo Y, Rose JE, Garg PK, Garg S, Nazih R et al. Rapid brain nicotine uptake from electronic cigarettes. J Nuclear Med. 2019;10.2967.

[92] Shi Y, Cummins SE, Zhu SH. Use of electronic cigarettes in smoke-free environments. Tob Control. 2017;26(e1):e19-22.

[93] Owusu D, Huang J, Weaver SR, Pechacek TF, Ashley DL, Nayak P et al. Patterns and trends of dual use of e-cigarettes and cigarettes among US adults, 2015-2018. Prev Med Rep. 2019;16:101009.

[94] Stanton CA, Sharma E, Edwards KC, Halenar MJ, Taylor KA, Kasza KA et al. Longitudinal trans-itions of exclusive and polytobacco electronic nicotine delivery systems (ENDS) use among youth, young adults and adults in the USA: findings from the PATH Study waves 1-3 (2013-2016). Tob Control. 2020;29(Suppl 3):s147-54.

[95] Aleyan S, Hitchman SC, Ferro MA, Leatherdale ST. Trends and predictors of exclusive e-cigaret-te use, exclusive smoking and dual use among youth in Canada. Addict Behav. 2020:106481.

[96] Piper ME, Baker TB, Benowitz NL, Jorenby DE. Changes in use patterns over 1 year among smokers and dual users of combustible and electronic cigarettes. Nicotine Tob Res. 2020;22(5):672-80.

[97] Roh EJ, Chen-Sankey JC, Wang MQ. Electronic nicotine delivery system (ENDS) use patterns and its associations with cigarette smoking and nicotine addiction among Asian Americans: Findings from the national adult tobacco survey (NATS) 2013-2014. J Ethnicity Subst Abuse. 2020:1-19.

[98] Kong G, Morean ME, Cavallo DA, Camenga DR, Krishnan-Sarin S. Reasons for electronic ciga-rette experimentation and discontinuation among adolescents and young adults. Nicotine Tob Res. 2014;17(7):847-54.

[99] Soule EK, Lopez AA, Guy MC, Cobb CO. Reasons for using flavored liquids among electronic cigarette users: A concept mapping study. Drug Alcohol Depend. 2016;166:168-76.

[100] Romijnders KAGJ, van Osch L, de Vries H, Talhout R. Perceptions and reasons regarding e-ci-garette use among users and non-users: a narrative literature review. Int J Environ Res Public Health. 2018;15(6):1190.

[101] Soule EK, Sakuma KLK, Palafox S, Pokhrel P, Herzog TA, Thompson N et al. Content analysis of internet marketing strategies used to promote flavored electronic cigarettes. Addict Behav. 2019;91:128-35.

[102] Krüsemann EJZ, Boesveldt S, de Graaf K, Talhout R. An e-liquid flavor wheel: a shared voca-bulary based on systematically reviewing e-liquid flavor classifications in literature. Nicotine Tob Res. 2019;21(10):1310-9.

[103] Krüsemann EJZ, Havermans A, Pennings JLA, de Graaf K, Boesveldt S, Talhout R. Compre-hensive overview of common e-liquid ingredients and how they can be used to predict an e-liquid's flavour category. Tob Control. 2020: 055447.

[104] E-Cigarette market – growth, trends and forecast (2019–2024). Hyderabad: MordorIntelligence; 2018.

[105] Wickham RJ. How menthol alters tobacco-smoking behavior: a biological perspective. Yale J Biol Med. 2015;88(3):279–87.

[106] Huang J, Feng B, Weaver SR, Pechacek TF, Slovic P, Eriksen MP. Changing perceptions of harm of e-cigarette vs cigarette use among adults in 2 US national surveys from 2012 to 2017. JAMA Network Open. 2019;2(3):e191047.

[107] Amrock SM, Lee L, Weitzman M. Perceptions of e-cigarettes and noncigarette tobacco products among US youth. Pediatrics. 2016;138(5):e20154306.

[108] Cullen KA, Gentzke AS, Sawdey MD, Chang JT, Anic GM, Wang TW et al. e-Cigarette use among youth in the United States, 2019. JAMA. 2019;322(21):2095–103.

[109] Cullen KA, Ambrose BK, Gentzke AS, Apelberg BJ, Jamal A, King BA. Notes from the field: use of electronic cigarettes and any tobacco product among middle and high school students–United States, 2011–2018. Morb Mortal Wkly Rep. 2018;67:1276–7.

[110] Soule EK, Lee JGL, Egan KL, Bode KM, Desrosiers AC, Guy MC et al. "I cannot live without my vape": Electronic cigarette user-identified indicators of vaping dependence. Drug Alcohol Depend. 2020:107886.

[111] Gentzke AS, Creamer M, Cullen KA, Ambrose BK, Willis G, Jamal A et al. Vital signs: tobacco product use among middle and high school students – United States, 2011–2018. Morbid Mortal Wkly Rep. 2019;68(6):157–64.

[112] Dai H, Leventhal AM. Prevalence of e-cigarette use among adults in the United States, 2014–2018. JAMA. 2019;322(18):1824–7.

[113] WHO global report on trends in prevalence of tobacco smoking. Geneva: WorldHealthOrga-nization;2015(https://apps.who.int/iris/bitstream/handle/10665/156262/9789241564922_eng.pdf?sequence=1&isAllowed=y, accessed 9 November 2020).

[114] Eissenberg T. Electronic nicotine delivery devices: ineffective nicotine delivery and craving suppression after acute administration. Tob Control. 2010;19(1):87–8.

[115] Vansickel AR, Eissenberg T. Electronic cigarettes: effective nicotine delivery after acute admi-nistration. Nicotine Tob Res. 2013;15(1):267–70.

[116] Dawkins L, Corcoran O. Acute electronic cigarette use: nicotine delivery and subjective effects in regular users. Psychopharmacology. 2014;231(2):401–7.

[117] Dawkins L, Kimber C, Puwanesarasa Y, Soar K. First- versus second-generation electronic cigarettes: predictors of choice and effects on urge to smoke and withdrawal symptoms. Addiction. 2015;110(4):669–77.

[118] Breland A, Soule E, Lopez A, Ramôa C, El-Hellani A, Eissenberg T. Electronic cigarettes: what are they and what do they do? Ann NY Acad Sci. 2017;1394(1):5–30.

[119] Talih S, Balhas Z, Eissenberg T, Salman R, Karaoghlanian N, El-Hellani A et al. Effects of user puff topography, device voltage, and liquid nicotine concentration on electronic cigarette nicotine yield: measurements and model predictions. Nicotine Tob Res. 2015;17(2):150–7.

[120] Hajek P, Phillips-Waller A, Przulj D, Pesola F, Myers Smith K, Bisal N et al. A randomized trial of e-cigarettes versus nicotine-replacement therapy. New Engl J Med. 2019;380(7):629–37.

[121] Jankowski M, Krzystanek M, Zejda JE, Majek P, Lubanski J, Lawson JA et al. E-cigarettes are

more addictive than traditional cigarettes – A study in highly educated young people. Int J Environ Res Public Health. 2019;16(13):2279.

[122] Vogel EA, Prochaska JJ, Ramo DE, Andres J, Rubinstein ML. Adolescents' e-cigarette use: in-creases in frequency, dependence, and nicotine exposure over 12 months. J Adolesc Health. 2019;64(6):770–5.

[123] Talih S, El-Hage R, Karaoghlanian A, Saliba NA, Salman R, Karam EE et al. Might limiting liquid nicotine concentration result in more toxic electronic cigarette aerosols? Tob Control. 2020: 10.1136/tobaccocontrol-2019-055523.

[124] Shihadeh A, Eissenberg T. Electronic cigarette effectiveness and abuse liability: predicting and regulating nicotine flux. Nicotine Tob Res. 2015;17(2):158–62.

[125] Pankow JF, Kim K, McWhirter KJ, Luo W, Escobedo JO, Strongin RM et al. Benzene formation in electronic cigarettes. PLoS One. 2017;12(5):e0173055.

[126] Duell AK, Pankow JF, Peyton DH. Nicotine in tobacco product aerosols: "It's déjà vu all over again". Tobacco Control. 2019;29(6):656–62.

[127] Process for preparing racemic nicotine. United States Patent US 8,884,021 B2; 2014 (https://patentimages.storage.googleapis.com/d1/cd/24/4141350fb210fc/US8884021.pdf).

[128] El-Hellani A, El-Hage R, Salman R, Talih S, Shihadeh A, Saliba NA. Carboxylate countera-nions in electronic cigarette liquids: Influence on nicotine emissions. Chem Res Toxicol. 2017;30(8):1577–81.

12. 电子烟碱传输系统或雾化产品使用相关肺损伤

Farrah Kheradmand, Professor of Medicine, Immunology and Pathology at Baylor College of Medicine and Staff Physician at Michael E DeBakey VA, Houston (TX), USA

Laura E. Crotty Alexander, Associate Professor of Medicine, University of California at San Diego and Section Chief of Pulmonary and Critical Care at Veterans Administration San Diego Healthcare System, San Diego (CA), USA

摘 要

在过去的十年里，无论是学术界还是新闻界，都对电子烟碱传输系统（ENDS，电子烟）对肺部健康的有益或有害影响进行了激烈的辩论。2019年夏天，有报道称几个电子烟使用者集体出现急性呼吸衰竭，导致住院治疗，在某些情况下甚至死亡，使这场辩论朝着新的方向发展。这里描述了电子烟或雾化产品使用相关的肺损伤（EVALI）的暴发，有关其临床特征和肺病理的最新资料，以及对受影响个人使用的许多商业和/或非法产品中查明的致病化学物质的调查。尽管新闻报道强调了美国EVALI的几个聚类性病例，但在包括欧洲国家在内的其他地区也有与电子烟相关的呼吸衰竭的孤立病例。关于暴露于电子烟和雾化产品气溶胶中的化学物质的急性和长期影响，仍有许多未回答的问题。有必要在流行病学、转化研究和基础研究的指导下有必要采取综合方法来评估吸入这些在世界上许多国家都在使用的产品释放物所产生的风险。

12.1 背 景

12.1.1 与电子烟和雾化产品相关的呼吸影响

电子烟碱传输系统（ENDS）通常被称为电子烟或雾化装置，其使用相对较新，但在许多国家的所有年龄段人群中，尤其是在加拿大和美国等一些国家的年轻人中，正在迅速发展。雾化装置通过雾化含有亲水性溶剂、丙二醇和植物甘油的烟液载体将烟碱输送到肺部。然而加热这些化学物质组合的味道不吸引人，因此>99%的电子烟液含有化学香料。虽然这些装置的长期呼吸和/或全身影响尚不清楚，但其使用与对肺部的急性和亚急性影响有关，包括嗜酸性肺炎、过敏性肺炎、类脂肺炎、急性呼吸窘迫综合征和弥漫性肺泡出血[1,2]。"Vaping"是一个非正式术语，用于指这些产品的使用，也有报道称会加剧先前存在的肺部疾病，尤其是哮喘患者的呼吸道高敏性和咳嗽[3]。本报告提供了2019年与电子烟产品相关的肺损伤暴发的信息。

注意，"vaping"一词可能有积极的隐含意义，因为它与"水蒸气"有关，这可能暗示着产品是无风险的，然而，电子烟并非无害。

12.1.2 电子烟碱传输系统或雾化产品使用相关肺损伤（EVALI）

2019年夏天，美国发现了几起使用雾化器或电子烟导致的肺部损伤。EVALI一词由美国疾病控制和预防中心（US CDC）于2019年10月11日首次使用。美国的流行病学分析已证实，2019年之前不存在特定疾病实体EVALI。尽管2019年之前曾报道过电子烟或雾化装置引起的肺部疾病，但EVALI被认为是由不同的化学暴露通过不同的病理机制引起的[4]。EVALI在2019年9月达到流行水平，然而，尽管与EVALI相关的急诊就诊次数有所减少，全国仍有病例发生。截至2020年2月，已确认需要住院治疗的病例超过2800例，其中68例死亡。已经确定了各种类型的肺部病理学，包括弥漫性肺泡损伤[5]。

12.1.3 与EVALI有关的产品和化学制品

EVALI的幸存者报告说使用了各种电子烟、雾化装置、电子烟烟液和调味物质，并且没有任何特定品牌或设备是所有病例共同使用的。超过80%受影响的个体报告称其电子烟烟液中存在大麻和其他大麻素（Δ^9-四氢大麻酚，THC），一半同时吸食THC和烟碱。所有含有THC的电子烟烟液均被确定为潜在危险。含有维生素E醋酸盐的电子烟烟液也与该病有关，因为在受影响患者使用的大多数电子烟烟液以及支气管肺泡灌洗液中都检测到这种物质[6]。尽管许多品牌的电子烟液与这种疾病有关，但经销商经常用"自制混合"的电子烟烟液填充空烟弹。因此，任何电子设备或电子烟烟液都不能被认为是安全的。

由于超过80%的EVALI患者报告使用THC，并且他们使用的大多数电子烟液对THC呈阳性反应，因此，含有THC的电子烟或雾化装置被认为是此次暴发的主要原因。尽管14%的EVALI患者仅吸食烟碱，但专家认为他们有其他形式的电子烟引起的肺损伤，而不是EVALI。这一结论得到了广泛的、非特异性的EVALI定义的支持，而且仅使用烟碱的队列中的大多数患者是老年女性。这些产品所使用的运营商，特别那些实体或在线经销的运营商受到极大影响。在确定的152种含THC的不同品牌产品中，Dank Vapes（在许多地点出售的含有不明来源THC烟液的烟弹）在美国东北部和南部最常见。TKO和Smart Cart品牌在西部有报道，Rove品牌在中西部各州也有报道。这些发现表明EVALI与含THC的产品有关，可能不是由单一品牌造成的。公共卫生机构和美国食品药品监督管理局都已确定维生素E醋酸盐是与EVALI最密切相关的化学物质，因为从EVALI患者获得的51份支气管肺泡灌洗液样本中，有48份检测到了维生素E醋酸盐[6]。

12.2 EVALI

12.2.1 详细说明和历史记录

2019年7月在美国伊利诺伊州和威斯康星州发现EVALI首次爆发[7,8]。EVALI病例特征被确定为：症状出现的90天内使用过电子烟，胸部X光片或胸部计算机轮廓图显示双侧肺浸润，并且没有或不太可能有引起感染的证据。美国疾病控制和预防中心随后将EVALI病例归类为疑似病例（微生物研究呈阳性但不太可能导致临床表现）或确诊病例（如果临床医生可以排除呼吸道感染）[9]。例如，如果金黄色葡萄球菌是从痰培养物中生长出来的，则该病例可能被认为是EVALI，金黄色葡萄球菌只是一种定植菌，而不是症状的原因。如果最常见的呼吸道微生物检测结果为阴性（例如流感、其他呼吸道病毒，包括SARS-CoV2、痰革兰氏染色和培养、肺炎链球菌尿液抗原和肺炎军团菌尿液抗原），则该病例符合确诊EVALI的标准。

经美国疾病控制和预防中心分类后，美国各地又有数百例确诊病例[10,11]。疫情高峰期出现在2019年9月，此后报告和确诊病例数量有所下降。随后减少的可能原因包括以下几点。

- 媒体对EVALI的强烈兴趣可能导致电子烟使用者放弃电子烟，从可靠来源购买电子烟，尤其是停止使用含THC的产品。
- 报告病例的医疗保健专业人员较少，因为这种诊断并不是唯一的或新的。
- 电子烟或雾化液制造商可能已经停止在电子烟液中添加与EVALI相关的化学物质。

12.2.2 症状

同时出现胃肠道和呼吸道症状是EVALI最明确的症状。EVALI最常见的症状包括腹痛、恶心、呕吐或腹泻伴随呼吸急促、咳嗽、运动时气急或胸痛。不太典型的症状包括发烧、不适、疲劳和体重减轻[12,13]。

12.2.3 临床表现

一些患者在首次出现症状后数小时内得到临床护理，但另一些患者在首次出现病况前数周至数月就有症状，这使对疾病发病变化的理解变得复杂。半数住院EVALI患者缺氧，需要进入重症监护室，约一半患者需要机械通气和/或体外膜肺氧合。患者需要2~6 L氧气的中度病例非常常见。人们越来越多地认识到轻度病例，这些患者不需要补充氧气来维持血氧饱和度>94%，但发现其症状与中度和重度患者相似。据报道，气胸和纵隔气肿很常见，这增加了肺实质损伤导致支气管肺瘘的可能性增加[12,13]。最近，有报道称，一些EVALI患者在出院后2~3天死亡，可能是由于突发性气胸。

12.2.4 报告病例

美国疾控中心仅报告了2020年1月以来的住院患者人数，反映出对报告中度至重度病例的偏倚，并于2020年2月25日停止报告EVALI病例数。加拿大、日本、墨西哥、英国和其他一些国家也报告了电子烟相关肺损伤的病例。有些报告在有EVALI描述之前就有了。大多数患者需要住院治疗，其疾病影像学和临床描述反映了EVALI的情况。众所周知，在EVALI出现之前，电子烟会导致各种肺部疾病，而吸入性肺部疾病也有类似的表现，因此这些病例很可能与THC或维生素E醋酸酯无关，也不是EVALI。其他国家尚未报告对电子烟引发的肺部疾病病例、趋势、流行率或病例群体进行系统回顾。

12.3 EVALI 的识别

如果EVALI有一个病因，那么就可以建立一个通用定义，就像国际肺专家就急性呼吸窘迫综合征定义的全球共识一样。任何定义都必须比目前的定义[9]更具体和详细，目前的定义包括90天内吸过电子烟、双侧肺浸润且无明确感染原因的人。理想情况下，EVALI的定义将排除其他与电子烟相关的肺部疾病以及与电子烟无关的疾病。增加特异性检测以排除特发性或与电子烟相关但与EVALI无关的肺部疾病将是有帮助的。EVALI的病理特征广泛，与急性间质性肺炎、超敏性肺炎、弥漫性肺泡出血、类脂性肺炎和成人呼吸系统疾病综合征的病理特征重叠。

12.4 对 EVALI 的监测

12.4.1 国家监测机制

很难收集美国的所有病例，因为美国疾病控制和预防中心及食品药品监督管理局依赖于当地和联邦公共卫生部门，这些部门是稳健可靠的，但需根据当地优先事项和州健康隐私法要求分享其数据。另一个困难是教育医疗保健提供者，因为大多数人没有明确询问电子烟、雾化装置或THC专用装置的使用情况。有人建议为患者和医疗保健提供者建立一个开放的报告门户来识别更多病例[14]。隐私问题是主要的妨碍因素，对其他疾病也是同样的情况，如铅中毒和传染病。

12.4.2 地区监测机制

最近形成了一个地区监测机制[11]。地区监测可以帮助每个医疗系统准确检测和跟踪所有EVALI病例，然后数据可以与区域公共卫生部门、美国CDC和FDA共享。英国有一个称为"黄牌计划"的系统，可在药品和保健品管理局网站上查阅[15]。该计划允许报告与药物相关的不良或疑似不良事件的信息，并规定报告使用电子烟的任何副作用或与这些产品或其电子烟烟液相关的安全问题。英国公众和卫生保健人员都可以通过该系统提交报告。

12.4.3 国际监测机制和验证

监测方面的一个挑战是在医疗中心和农村地区通过相似的调查和确认方法来识别病例，以控制和应对每个病例。世界卫生组织发起了一个由250多个机构组成的全球网络，致力于应对和提高对急性公共卫生事件的认识，即全球疫情警报和应对网络[16]。

另一个困难是定义EVALI，因为当前的定义很宽泛，难以识别。为了改善监测，需要在国家和国际层面上制定标准（如病例定义）和培训，以识别EVALI。美国疾病控制和预防中心报告，近3%的EVALI患者需要重新住院，近七分之一的EVALI死亡发生在出院后，尤其是患有一种或多种慢性病的患者[17]。

12.5 讨 论

据估计，全球有3500万~4000万成年人和儿童会使用电子烟，这表明有大量人易受EVALI和其他雾化装置相关疾病的影响。电子烟的使用或雾化装置本身会带来健康风险，并可能损害肺部以外的健康。此外，在产品中添加药物和其他物质会增加风险，因此应对产品进行适当监管。

EVALI在美国的爆发突出了将电子烟有害性的定义扩展到吸烟之外的重要性，因为抽吸电子烟导致的疾病风险不同于与吸烟相关的疾病风险[18]。2019年新型冠状病毒感染（COVID-19）大流行期间的一个主要问题是呼吸衰竭患者EVALI识别和报告的延迟。例如，在2020年4月，8名因呼吸衰竭住院的患者符合美国疾病控制和预防中心对EVALI的病例定义，但医生在住院后1~8天（中位值第3天）首次在鉴别诊断中考虑EVALI[19]。本报告强调了识别EVALI引起的呼吸衰竭的一些困难，因此，在COVID-19大流行期间，EVALI可能未得到充分诊断。

国家和国际层面的优先事项是建立研究人员和临床医生可访问的EVALI患者登记文档，以评估长期效果和临床结果。此外，采集和分析人类样本（例如全血、气管抽吸物、支气管肺泡灌洗液、肺活检样本、尿液和尸检样本）可以深入了解EVALI的病理生理学特征。应进行更多的机理研究，以了解电子烟产品在肺部的毒性作用。具体而言，目前尚不清楚在暴露于维生素E醋酸盐之前吸入丙二醇或植物甘油是否以及如何导致肺损伤。此外，雾化温度，尤其是在大功率设备中，在肺损伤中发挥作用[20]。EVALI的动物模型有助于研究与电子烟和THC产品使用相关肺部毒性的潜在原因[14]。

12.6 考虑事项

建议对EVALI疾病谱的许多方面及其对其他国家肺损伤的潜在影响进行全面报告。应解决几个关键问题，以提高目前对EVALI和呼吸衰竭患者护理的理解。

- 一个重大挑战是确保医生识别EVALI病例，特别是在COVID-19大流行期间。应让医生了解EVALI的风险，并询问电子产品中烟碱或THC的使用情况，作为呼吸衰竭患者常规病史的一部分。
- 必须更好地定义EVALI，以指导初级保健和儿科医师正确诊断。
- 研究应该解决使用电子烟是否也会增加成人和儿童感染SARS-CoV-2的可能性。电子烟产品使用的临床前模型提供了强有力的证据，证明长期使用会改变肺部对流感（一种常见病毒病原体）的防御免疫[18]。

12.7 建 议

12.7.1 主要建议

- 应建立EVALI患者和其他电子烟相关肺部疾病患者的国家和国际登记文档，以改进对病愈者长期临床结果的监测。

- 应组织有力的国际行动，提醒家长、儿童和青少年注意吸入电子烟产品气溶胶中所含化学物质的危害。
- 当获得更多信息后，应考虑撰写一份全面的关于EVALI的论文。

12.7.2 其他建议

- 为了更好地了解EVALI的病理生理学特征，应获取并分析血液、气管抽吸物、支气管肺泡灌洗液、肺组织和尿液等样本。
- 应开发EVALI的动物模型，以深入了解吸入烟碱和THC在肺部和全身引起毒性的细胞和分子机制，以及这些产品的装置和风味化学物质的健康影响。
- 应开展进一步研究以确定电子烟产品损害肺部的机制。

12.8 参 考 文 献

[1] Arter ZL, Wiggins A, Hudspath C, Kisling A, Hostler DC, Hostler JM. Acute eosinophilic pneu-monia following electronic cigarette use. Respir Med Case Rep. 2019;27:100825.

[2] Sommerfeld CG, Weiner DJ, Nowalk A, Larkin A. Hypersensitivity pneumonitis and acute re-spiratory distress syndrome from e-cigarette use. Pediatrics. 2018;141(6):20163927.

[3] Lappas AS, Tzortzi AS, Konstantinidi EM, Teloniatis SI, Tzavara CK, Gennimata SA et al. Short-term respiratory effects of e-cigarettes in healthy individuals and smokers with asthma. Re-spirology. 2018;23:291–7.

[4] McCauley L, Markin C, Hosmer D. An unexpected consequence of electronic cigarette use. Chest. 2012;141:1110–3.

[5] Butt YM, Smith ML, Tazelaar HD, Vaszar LT, Swanson KL, Cecchini MJ et al. Pathology of va-ping-associated lung injury. N Engl J Med. 2019;381(18):1780–1.

[6] Blount BC, Karwowski MP, Shields PG, Morel-Espinosa M, Valentin-Blasini L, Gardner M et al. Vitamin E acetate in bronchoalveolar-lavage fluid associated with EVALI. N Engl J Med. 2020;382:697–705.

[7] Layden JE, Ghinai I, Pray I, Kimball A, Layer M, Tenforde M et al. Pulmonary illness related to e-cigarette use in Illinois and Wisconsin – preliminary report. N Engl J Med. 2020;382:903–16.

[8] Ghinai I, Pray IW, Navon L, O'Laughlin K, Saathoff-Huber L, Hoots B et al. e-Cigarette product use, or vaping, among persons with associated lung injury – Illinois and Wisconsin, April–Sep-tember 2019. Morb Mortal Wkly Rep. 2019;68:865–9.

[9] 2019 lung injury surveillance. Primary case definitions. September 18, 2019. Atlanta (GA): Cen-ters for Disease Control and prevention; 2019 (https://www.cdc.gov/tobacco /basic_infor-ma-tion/e-cigarettes/assets/2019-Lung-Injury-Surveillance-Case-Definition-508.pdf, accessed 10 January 2021).

[10] Blagev DP, Harris D, Dunn AC, Guidry DW, Grissom CK, Lanspa MJ. Clinical presentation, tre-atment, and short-term outcomes of lung injury associated with e-cigarettes or vaping: a prospective observational cohort study. Lancet. 2019;394(10214):2073–83.

[11] Alexander LEC, Perez MF. Identifying, tracking, and treating lung injury associated with e-ci-garettes or vaping. Lancet. 2019;394(10214):2041–3.

[12] Zou RH, Tiberio PJ, Triantafyllou GA, Lamberty PE, Lynch MJ, Kreit JW et al. Clinical characte-rization of e-cigarette, or vaping, product use associated lung injury in 36 patients in Pittsburgh, PA. Am J Respir Crit Care Med. 2020;201(10):1303–6.

[13] Chatham-Stephens K, Roguski K, Jang Y, Cho P , Jatlaoui TC, Kabbani S et al. Characteristics of hospitalized and nonhospitalized patients in a nationwide outbreak of e-cigarette, or vaping, product use-associated lung injury – United States, November 2019. Morb Mortal Wkly Rep. 2019;68:1076–80.

[14] Crotty Alexander LE, Ware LB, Calfee CS, Callahan SJ, Eissenberg T, Farver C et al. NIH workshop report: E-cigarette or vaping product use associated lung injury: Developing a research agenda. Am J Respir Crit Care Med. 2020;202(6):795–802.

[15] Yellow card. London: Medicines and Healthcare Products Regulatory Agency; 2020 (https://yellowcard.mhra.gov.uk/the-yellow-card-scheme/#:~:text= The%20Yellow%20Card%20scheme%20is,by%20health%20professionals%20and%20patients, accessed 10 January 2021).

[16] Global Outbreak Alert and Response Network. Geneva: World Health Organization; 2020 (https://extranet.who.int/goarn/, accessed 10 January 2021).

[17] Evans ME, Twentyman E, Click ES, Goodman AB, Weissman DN, Kiernan E et al. Update: Inte-rim guidance for health care professionals evaluating and caring for patients with suspected E-cigarette, or vaping, product use-associated lung injury and for reducing the risk for rehos-pitalization and death following hospital discharge – United States, December 2019. Morb Mortal Wkly Rep.2020;68(5152):1189–94.

[18] Madison MC, Landers CT, Gu BH, Chang CY, Tung HY, You R et al. Electronic cigarettes dis-rupt lung lipid homeostasis and innate immunity independent of nicotine. J Clin Invest. 2019;129:4290–304.

[19] Armatas C, Heinzerling A, Wilken JA. Notes from the field: e-cigarette, or vaping, product use-associated lung injury cases during the COVID-19 response – California, 2020. Morb Mortal Wkly Rep. 2020;69:801–2.

[20] Chaumont M, Bernard A, Pochet S, Mélot C, El Khattabi C, Reye F et al. High-wattage e-ciga-rettes induce tissue hypoxia and lower airway injury: A randomized clinical trial. Am J Crit Care Med. 2018;198(1):123–6.

13. 总体建议

世界卫生组织烟草制品管制研究小组发表报告，为烟草制品管制提供科学依据。根据世界卫生组织《烟草控制框架公约》（FCTC）第9条和第10条，报告确定了对烟草制品成分、释放物和设计特征进行监管的循证方法。

研究小组第十次会议的审议情况、结果和建议均包含在本报告中，会议专门讨论了新型烟碱和烟草制品，包括电子烟碱传输系统（ENDS）和电子非烟碱传输系统（ENNDS）以及加热型烟草制品（HTP）。尽管如此，世界卫生组织《烟草控制框架公约》缔约方大会第八次会议（COP8）关于新型烟草制品的决议（FCTC/COP8(22)）[1]部分体现了这一重点，所有烟草制品都属于研究小组的职权范围。这使我们能够采取综合方法，向各国提供关于传统产品和新产品的证据，以应对烟草控制方面的挑战，这仍然是一个全球优先事项。

监管机构应该注意的是烟草每年导致800多万人死亡[2,3]，其中700多万人死于直接使用烟草，约130万非吸烟者死于二手烟暴露[4,5]。烟草最终会导致多达一半的使用者死亡，因此烟草仍然是一个全球卫生紧急事件[2]。在一些地区，制造商积极营销和推广新型烟碱和烟草制品，包括对儿童和青少年，使烟草管制更加复杂，并使监管机构分心。因此，本报告的建议应在更广泛的烟草控制背景下考虑，并补充研究小组在其他烟草制品管制报告（涉及卷烟、无烟烟草和水烟）[6-12]中提出的建议。

新型烟碱和烟草制品的积极营销和推广对烟草控制构成严重威胁。研究小组审议了各国关于监管这些产品的技术支持请求后得出结论，必须将重点放在更广泛的烟草控制上，监管机构不应因烟草和相关行业的策略以及对这些产品的积极推广而分散注意力。报告强调了以下方面的重要性：

■ 行业研究的良好科学性及可验证性；
■ 向监管机构全面披露产品信息；
■ 澄清研究资金来源，以确定不当影响；
■ 独立研究；
■ 所有烟草制品均适用于烟草控制法，不能有例外；
■ 监测烟草及相关行业的活动；
■ 保护政策不受烟碱和烟草行业的影响，特别是在世界卫生组织《烟草控制框架公约》第5.3条及其指南的背景下[13,14]。

报告第2~12章提供了科学信息、政策建议和指导，以弥补烟草控制方面的监管差距。报告还确定了需要进一步开展工作和研究的领域，重点放在各国的监管需求上，同时考虑到区域差异，从而为所有国家，特别是世界卫生组织成员国提供持续、有针对性的技术支持和策略。

13.1 主要建议

对决策者和所有其他相关方的主要建议如下：

- 确保继续关注世界卫生组织《烟草控制框架公约》中所述的减少烟草使用的循证措施，并力求避免被烟草行业推广新型烟草制品（如加热型烟草制品）的行动分散注意力；
- 使用烟草制品（包括装置）现行法规监管加热型烟草制品，并考虑扩大现有法规的范围，因为烟草行业可能会使用其中的监管漏洞，包括目前不能合法获得加热型烟草制品的国家；
- 考虑到对人体健康的高度保护，酌情在国家法律范围内对加热型烟草制品（包括装置）适用最严格的烟草管制法规；
- 禁止所有制造商和相关团体声称与其他产品相比，加热型烟草制品的危害更小，或将加热型烟草制品描述为戒烟的适当方法，并禁止在公共场所使用，除非出现有力的独立证据支持政策的改变；
- 确保公众充分了解与使用加热型烟草制品相关的风险，包括与传统卷烟和其他烟气型烟草制品双重使用以及在怀孕期间使用的风险，纠正错误观念，反驳错误信息，强调减少暴露并不一定意味着减少伤害；
- 依靠独立数据，支持就使用加热型烟草制品对公众健康的影响进行持续的独立研究，并对烟草行业资助的数据进行批判性分析和解释，包括关于加热型烟草制品的释放物和毒性以及对使用者和非使用者的相关暴露和影响的数据；
- 要求烟草制造商至少每年次向相关监管机构披露所有产品信息，包括产品设计、化学成分、总烟碱含量、烟碱形态、毒性、产品测试的其他结果和测试方法，对产品的任何修改都需要更新报告；
- 禁止电子烟碱传输系统、电子非烟碱传输系统和加热型烟草制品的所有商业营销，包括在社交媒体上以及通过由烟草行业资助或与烟草行业有关联的组织进行营销；
- 禁止销售使用者可以控制设备功能和烟液成分的电子烟碱传输系统和电子非烟碱传输系统（即开放式系统）；

■ 禁止销售比传统卷烟滥用倾向更高的电子烟碱传输系统，例如限制烟碱释放率或通量；
■ 禁止在电子烟碱传输系统和电子非烟碱传输系统中添加除烟碱以外的药理活性物质，如大麻和四氢大麻酚（在其合法的司法管辖区）。

敦促各国执行上述建议，因为有足够的关于烟碱和烟草制品的信息，各国可以采取行动保护本国公民，特别是年轻一代的健康。尽管该报告承认，对这些产品还需要更多的了解，并强调有必要继续进行独立研究，以进一步了解这些产品，包括其营销、特征、使用率和可用性，以及烟草和相关行业的促销策略，但仍有超过10亿的烟草使用者，数百万人使用这些新产品。因此，公共卫生界应响应继续加速实施循证政策和建议的呼吁，如WHO FCTC及MPOWER措施和缔约大会报告中的政策和建议。因此，各国应实施经过验证的政策措施，此外，还应考虑执行本报告中的建议。关于所考虑的每个主题的具体建议见第2.8、3.5.4、4.9、5.11、6.7、7.11、8.5、8.7、9.8、10.2.6、10.3、7、11.6和12.7节。

13.2　对公共卫生政策的意义

研究小组的报告为所有烟草制品的科学、研究和证据提供了有益的指导，包括卷烟、无烟烟草和水烟。最近，研究小组扩大了其工作范围，向监管机构提供了关于新型烟碱和烟草制品，特别是ENDS、ENNDS和HTP的成分、释放物、变化和特征的急需信息。该报告强调了这些产品及其特性对使用者和非使用者的公共卫生影响，包括：致瘾性，产品的认知和使用，吸引力，在初吸和戒烟方面的潜在作用，营销（包括促销策略）和影响，减少危害的声明，产品变化，对个体和人群健康风险的量化，各国监管情况和特定国家的经验，对烟草控制的影响以及研究差距。如上所述，研究小组的建议直接解决了某些成员国由于这些产品进入其市场而面临的一些独特的监管挑战。此外，该报告将帮助成员国更新其对新型烟碱和烟草制品的了解，并帮助其制定有效的烟碱和烟草制品监管策略。

研究小组是由监管、技术和科学方面的专家组成，具有独特性，能够对复杂的数据和研究进行分析和提炼，并将其综合为政策建议，为国家、区域和全球各级的政策制定提供信息。这份由多学科专家团队撰写的权威报告深入探讨了各国政府在新兴产品方面面临的关键挑战。该报告的性质意味着，监管机构、政府和相关方可以酌情依据所提供的科学和证据来反驳烟草和相关行业的论点。在烟碱和烟草制品的政策和研究中发现的差距表明了信息不足的领域。各国在制定研究议程时，可以将重点放在与其政策目标、目的和国情相关的领域。这是研究小组的一个关键作用，特别是对于资源或能力不足，无法掌握烟草制品管制技术信息的政府来说。

报告中提出的建议促进了监管工作的国际协调和产品监管最佳做法的采用，加强了世界卫生组织所有区域的产品监管能力，为成员国提供基于可靠科学的现成资源，并支持其缔约国实施世界卫生组织《烟草控制框架公约》。烟草制品管制补充了世界卫生组织《烟草控制框架公约》关于减少需求的其他规定。研究小组的建议如果得到有效实施，将有助于减少烟草使用，从而降低烟草使用的流行率，促进健康。

13.3 对WHO规划的影响

本报告履行了世界卫生组织烟草制品管制研究小组的职责，即向总干事提供关于烟草制品管制的科学合理、基于证据的建议[1]，烟草制品管制是烟草控制的一个高度技术性领域，成员国在这一领域面临着复杂的监管挑战。研究小组的审议结果和主要建议将促进成员国对ENDS、ENND和HTP的理解，以及在更广泛的烟草控制背景下这些产品在多国市场上扩散的影响。

该报告对产品监管知识体系的贡献将为世界卫生组织健康促进部烟草规划的工作提供信息，特别是在向成员国提供技术支持方面。它还将通过世界卫生组织全球烟草监管机构论坛会议和EZCollab网络共享信息，帮助成员国和监管机构更新信息。根据缔约方大会第八次会议的要求，世界卫生组织《烟草控制框架公约》缔约方将通过关于新型烟草制品研究和证据的综合报告获得最新情况[2]。综合报告将包括研究小组第八次报告的主要信息和建议。所有这些都将有助于实现可持续发展目标（即加强世卫组织《烟草控制框架公约》的实施）和世界卫生组织第十三个全球工作规划的三十亿目标。

本报告是世界卫生组织的全球公共卫生公益项目（即世界卫生组织制定或实施的一项有益于全球许多国家或地区的举措[15]），可供所有国家使用，以帮助在国家一级和全球范围内产生影响，减少烟草使用并改善总体公共卫生。

13.4 参考文献

[1] Novel and emerging tobacco products. Decision FCTC/COP8(22) of the Conference of the Parties to the WHO Framework Convention on Tobacco Control at its Eighth session. Geneva: World Health Organization; 2018 (https://www.who.int/fctc/cop/sessions/cop8/FCTC＿COP8(22).pdf?ua=1, accessed 19 December 2020).

[2] WHO report on the global tobacco epidemic, 2019. Geneva: World Health Organization; 2019

---

1 2003年11月，总干事将前烟草制品监管科学咨询委员会的地位从一个科学咨询委员会正式确定为一个研究小组。

2 见 FCTC/COP8(22), 2(a)。

[3] GBD 2019 Risk Factors Collaborators. Global burden of 87 risk factors in 204 countries and territories, 1990–2019: a systematic analysis for the Global Burden of Disease Study 2019. Lancet 2020;396:1223-49.

[4] Tobacco.Fact sheet. Geneva: World Health Organization; 2020 (https://www.who.int /newsroom /fact -sheets/detail/tobacco, accessed 23 December 2020).

[5] Findings from the Global Burden of Disease Study 2017: GBD Compare Seattle (WA): Institute for Health Metrics and Evaluation; 2018 (http://vizhub. healthdata.org/gbd-compare, accessed 19 December 2020).

[6] The scientific basis of tobacco product regulation. Report of a WHO study group (WHO Technical Report Series, No. 945). Geneva: World Health Organization; 2007 (https://www.who.int / tobacco/global_interaction/tobreg/who_ tsr.pdf, accessed 10 January 2021).

[7] The scientific basis of tobacco product regulation. Second report of a WHO study group (WHO Technical Report Series, No. 951). Geneva: World Health Organization; 2008 (https://apps. who.int/iris/bitstream/handle/10665/43997/TRS951_eng.pdf?sequence=1, accessed 10 January 2021).

[8] WHO Study Group on Tobacco Product Regulation. Report on the scientific basis of tobacco product regulation: Third report of a WHO study group (WHO Technical Report Series, No. 955). Geneva: World Health Organization; 2009(https://apps.who.int/iris/bitstream/handle/10665 /44213/9789241209557_eng.pdf?sequence=1, accessed 10 January 2021).

[9] WHO Study Group on Tobacco Product Regulation. Report on the scientific basis of tobacco product regulation: Fourth report of a WHO study group (WHO Technical Report Series, No. 967). Geneva: World Health Organization; 2012 (https://apps.who.int/iris/bitstream/handle/10665/44800/9789241209670_eng.pdf?sequence=1, accessed 10 January 2021).

[10] WHO Study Group on Tobacco Product Regulation. Report on the scientific basis of tobacco product regulation: Fifth report of a WHO study group (WHO Technical Report Series, No. 989). Geneva: World Health Organization; 2015 (https://apps.who.int/iris/bitstream/handle/10665/161512/9789241209892.pdf?sequence=1, accessed 10 January 2021).

[11] WHO Study Group on Tobacco Product Regulation. Report on the scientific basis of tobacco product regulation: Sixth report of a WHO study group (WHO Technical Report Series, No. 1001). Geneva: World Health Organization; 2017 (https://apps.who.int/iris/bitstream/handle/10 665/260245/9789241210010-eng.pdf?sequence=1, accessed 10 January 2021).

[12] WHO Study Group on Tobacco Product Regulation. Report on the scientific basis of tobacco product regulation: Seventh report of a WHO study group (WHO Technical Report Series, No. 1015). Geneva: World Health Organization; 2019 (https://apps.who.int/iris/bitstream/handle/10 665/329445/9789241210249-eng.pdf, accessed 10 January 2021).

[13] Article 5.3.WHO Framework Convention on Tobacco Control. Geneva:World Health Organization;2005(https://apps.who.int/iris/bitstream/handle/10665/42811/9241591013.pdf;jsessionid=1DE94B49AE5482D746B237CAC38D5658? sequence=1, accessed 19 December 2020).

[14] Guidelines for implementation of Article 5.3 of the WHO Framework Convention on Tobacco Control on the protection of public health policies with respect to tobacco control from commercial and other vested interests of the tobacco industry. Geneva: World Health Organiza-

tion;2008 (https://www.who.int/fctc/guidelines/article_5_3.pdf, accessed 19 December 2020).

[15] Thirteenth General Programme of Work 2019–2023. Geneva: World Health Organization; 2019(https://apps.who.int/iris/bitstream/handle/10665/324775/WHO- PRP-18.1-eng.pdf, accessed 29 December 2020).

1. Introduction

Effective tobacco product regulation is an essential component of a comprehensive tobacco control programme. It includes regulation of contents and emissions by mandated testing, disclosure of test results, setting limits as appropriate, disclosure of product information and imposing standards for product packaging and labelling. Tobacco product regulation is covered under Articles 9, 10 and 11 of the WHO Framework Convention on Tobacco Control (WHO FCTC) *(1)* and in the partial guidelines on implementation of Articles 9 and 10 of the WHO FCTC *(2)*. Other WHO resources, including the basic handbook on tobacco product regulation *(3)*, the handbook on building laboratory testing capacity *(4)* and the online modular courses based on the handbooks, which were published in 2018 *(5)*, support Member States in this respect.

The WHO Study Group on Tobacco Product Regulation (TobReg) was formally constituted by the WHO Director-General in 2003 to address gaps in the regulation of tobacco products. Its mandate is to provide evidence-based recommendations on policy for tobacco product regulation to the Director-General. TobReg is composed of national and international scientific experts on product regulation, treatment of tobacco dependence, toxicology and laboratory analyses of tobacco product ingredients and emissions. The experts are from countries in all six WHO regions *(6)*. As a formal entity of WHO, TobReg submits technical reports that provide the scientific basis for tobacco product regulation to the WHO Executive Board, through the Director-General, to draw the attention of Member States to WHO's work in this field. The reports, in the WHO Technical Report Series, include previously unpublished background papers that synthesize published scientific literature and have been discussed, evaluated and reviewed by TobReg. In accordance with Articles 9 and 10 of the WHO FCTC, relevant decisions of the Conference of the Parties (COP) to the WHO FCTC and relevant WHO reports submitted to the COP, the TobReg reports identify evidence-based approaches to regulating all forms of nicotine and tobacco products, including new and emerging products such as electronic nicotine delivery systems (ENDS), electronic non-nicotine delivery systems (ENNDS), heated tobacco products (HTPs) and nicotine pouches. These reports, now considered to be WHO global public health goods, respond to World Health Assembly resolutions WHA54.18 (2001), WHA53.17 (2000) and WHA53.8 (2000). "Global public health goods" are initiatives developed or undertaken by WHO that are of benefit either globally or to many countries in many regions *(7)*. This designation presents a unique opportunity for TobReg to speak directly to Member States and influence national, regional and global policy.

The 10th meeting of TobReg took place virtually on 28 September–2 October 2020, coordinated from WHO headquarters in Geneva. Over 50 participants, including the members of TobReg, discussed the scientific literature on the attractiveness, toxicity, appeal, variation, marketing, health effects and regulation of novel and emerging nicotine and tobacco products, with a focus on HTPs. These products, which have a long history but have attracted increased international interest in the past decade, present various regulatory challenges, as there is a lack of capacity to regulate them, they distract from evidence-based interventions, conflation of product categories, unsubstantiated health claims and opposition to and interference with effective tobacco control policies and marketing and promotional activities that target children and adolescents by tobacco and related industries. The meeting provided a platform for discussing nine background papers:

- toxicants in heated tobacco products, exposure, health effects and claims of reduced risk (section 2);
- the attractiveness and addictive potential of heated tobacco products: effects on perception and use and associated effects (section 3);
- variations among heated tobacco products, considerations and implications (section 4);
- use of heated tobacco products: product switching and dual or poly product use (section 5);
- regulations on HTPs, ENDS and ENNDS, with country approaches, barriers to regulation and regulatory considerations (section 6);
- estimation of exposure to nicotine from use of electronic nicotine delivery systems and from conventional cigarettes (section 7);
- exploration of methods for quantifying individual risks associated with electronic nicotine and non-nicotine delivery systems and heated tobacco products: impact on population health and implications for regulation (section 8).
- flavours in novel and emerging nicotine and tobacco products (section 9);
- global marketing and promotion of novel and emerging nicotine and tobacco products and their impacts (section 10);

TobReg also discussed two supplementary "horizon scanning" papers, on:

- forms of nicotine in tobacco plants, chemical modifications and implications for electronic nicotine delivery systems products (section 11); and
- EVALI: "e-cigarette or vaping product use-associated lung injury" (section 12).

The background papers and the horizon scanning papers were prepared by subject matter experts according to the terms of reference or outline drawn up by the WHO Secretariat for each paper and were reviewed and revised by expert reviewers and members of the Study Group. The horizon scanning papers, which are new additions to the report, are short papers on emerging issues in product regulation for members to decide whether further work on these topics is warranted, such as a full background paper for a future meeting of the Study Group. The papers on HTPs and novel and emerging tobacco products addressed the request in paragraph 2a of decision FCTC/COP8(22) *(8)* at the eighth session of the COP (COP8):

> 2. REQUESTS the Convention Secretariat to invite WHO and, as appropriate, the WHO Tobacco Laboratory Network (TobLabNet):
>
> (a) to prepare a comprehensive report, with scientists and experts, independent from the tobacco industry, and competent national authorities, to be submitted to the Ninth session of the COP on research and evidence on novel and emerging tobacco products, in particular heated tobacco products, regarding their health impacts including on non-users, their addictive potential, perception and use, attractiveness, potential role in initiating and quitting smoking, marketing including promotional strategies and impacts, claims of reduced harm, variability of products, regulatory experience and monitoring of Parties, impact on tobacco control efforts and research gaps, and to subsequently propose potential policy options to achieve the objectives and measures outlined in paragraph 5 of the present decision.

The request made to WHO via the Convention Secretariat on novel and emerging tobacco products was examined by WHO and was the basis for deciding the topics of the papers, with other emerging issues in nicotine and tobacco product regulation, including requests for technical assistance to WHO from Member States. The Secretariat, in consultation with the Study Group, invited experts who not only contributed to discussions but also provided the most recent empirical scientific evidence and regulations on nicotine and tobacco products in their background papers. The period of the literature search is indicated in each paper; for most, it was the second quarter of 2020. The papers were subject to several rounds of review before and after the meeting by independent technical experts, the WHO Secretariat, people in other relevant WHO departments, regional colleagues and members of the Study Group before compilation into the technical report. This eighth report of TobReg on the scientific basis of tobacco product regulation is designed to guide Member States in finding the most effective, evidence-based means to bridge regulatory gaps and address challenges in tobacco control and should lead to development of coordinated regulatory frameworks for nicotine and tobacco products. All experts and other participants in the meeting, including members of the Study Group, were required to complete a declaration of interests, which was evaluated by WHO to ensure independence from tobacco and related industries.

This report includes five papers on HTPs (sections 2–6), two on ENDS (sections 7 and 8) and two general reviews on novel and emerging nicotine and tobacco products, on the use of flavours in these products (section 9) and global marketing and promotion (section 10). It also includes two supplementary sections (11 and 12) that scan the horizon of studies on forms of nicotine in tobacco plants and on e-cigarette or vaping product use-associated lung injury (EVALI) and concludes with a section of overall recommendations, summarizing the recommendations in each section. The recommendations, which represent syntheses of complex research and evidence, promote international coordination of regulation and adoption of best practices in product regulation, strengthen capacity-building for product regulation in all WHO regions, represent a ready resource for Member States based on sound science and support implementation of the WHO FCTC by its States Parties.

This eighth report of the Study Group addresses ENDS, ENNDS and HTPs; however, it does not cover all aspects of these products, because many of the papers were written to meet the request of COP8, which was to review understanding on novel and emerging tobacco products (FCTC/COP8(22)). Continued research is necessary on these products; the Study Group will cover other products of interest (including traditional products, such as waterpipes, cigarettes and smokeless tobacco) in its next report, guided by countries' regulatory requirements and pertinent issues in tobacco product regulation. This will ensure continued, timely technical support to all countries and address all products, recognizing that their availability depends on the jurisdiction.

In summary, the outcomes of TobReg's deliberations and its recommendations will improve Member States' understanding of the evidence on ENDS, ENNDS and HTPs, contribute to the body of knowledge on product regulation, inform WHO's work, especially in providing technical support to Member States and keep Member States, regulators, civil society, research institutions and other interested parties up to date on product regulation through various platforms. States Parties to the WHO FCTC will be updated by a comprehensive report to be submitted to COP9, via the Convention Secretariat, on novel and emerging tobacco products, in line with decision FCTC/COP8(22), which will include the messages and recommendations in this report. Thus, the Study Group's activities will contribute to meeting target 3.a of the Sustainable Development Goals: strengthening implementation of the WHO FCTC (9).

References

1. WHO Framework Convention on Tobacco Control. Geneva: World Health Organization; 2003 (http://www.who.int/fctc/en/, accessed 10 January 2021).
2. Partial guidelines on implementation of Articles 9 and 10. Geneva: World Health Organization; 2012 (https://www.who.int/fctc/guidelines/Guideliness_Articles_9_10_rev_240613.pdf, accessed January 2019).
3. Tobacco product regulation: basic handbook. Geneva: World Health Organization; 2018 (https://www.who.int/tobacco/publications/prod_regulation/basic-handbook/en/, accessed 10 January 2021).
4. Tobacco product regulation: building laboratory testing capacity. Geneva: World Health Organization; 2018 (https://www.who.int/tobacco/publications/prod_regulation/building-laboratory-testing-capacity/en/, accessed 14 January 2019).
5. Tobacco product regulation courses. Geneva: World Health Organization (https://openwho.org/courses/TPRS-building-laboratory-testing-capacity/items/3S11LKUFGyoTksZ5RSZbID; https://openwho.org/courses/TPRS-tobacco-product-regulation-handbook/items/7zq-7S1jxAtpbUWiZdfH98I).
6. TobReg members. In: World Health Organization [website]. Geneva: World Health Organization (https://www.who.int/groups/who-study-group-on-tobacco-product-regulation/about, accessed 10 January 2021).
7. Feacham RGA, Sachs JD. Global public goods for health. The report of working group 2 of the Commission on Macroeconomics and Health. Geneva: World Health Organization; 2002 (https://apps.who.int/iris/bitstream/handle/10665/42518/9241590106.pdf?sequence=1, accessed 10 January 2021).
8. Decision FCTC/COP8(22). Novel and emerging tobacco products. Decision of the Conference of the Parties to the WHO Framework Convention on Tobacco Control, Eighth session. Geneva: World Health Organization; 2018 (https://www.who.int/fctc/cop/sessions/cop8/FCTC__COP8(22).pdf, accessed 7 November 2020).
9. Sustainable development goals. Geneva: World Health Organization (https://www.who.int/health-topics/sustainable-development-goals#tab=tab_2, accessed 10 January 2021).

2. Toxicants in heated tobacco products, exposure, health effects and claims of reduced risk

Irina Stepanov, Professor, Division of Environmental Health Sciences and Masonic Cancer Center, University of Minnesota, Minneapolis (MN), USA

Federica Mattioli, Mass Spectrometry Laboratory, Environmental Health Sciences Department, Istituto di Ricerche Farmacologiche Mario Negri, Milan, Italy

Enrico Davoli, Head, Mass Spectrometry Laboratory, Environmental Health Sciences Department, Istituto di Ricerche Farmacologiche Mario Negri, Milan, Italy

Contents
Abstract
2.1 Background
2.2 Toxicants in heated tobacco product (HTP) emissions
  2.2.1 Laboratory methods for measuring toxicants
  2.2.2 Nicotine
  2.2.3 Other toxicants
2.3 Exposure and effects of heated tobacco products in vitro and in laboratory animals
  2.3.1 In-vitro studies
  2.3.2 Studies in laboratory animals
2.4 Exposure of humans to toxicants in HTPs and implications for health
  2.4.1 Product use and topography
  2.4.2 Biomarkers of exposure and effect
  2.4.3 Passive exposure
  2.4.4 Impact on health outcomes
2.5 Review of the evidence for reduced risk or harm with use of HTPs
  2.5.1 Harm reduction in the context of tobacco products
  2.5.2 Claims of reduced risk
2.6 Summary and implications for public health
  2.6.1 Summary of data
  2.6.2 Implications for public health
2.7 Research gaps and priorities
2.8 Policy recommendations
2.9 References

Abstract

Increased marketing and the increasing popularity of novel and emerging heated tobacco products (HTPs) calls for urgent assessment of their potential impact on public health. In this report, we evaluate the published literature on the chemical composition of HTPs, exposure to these products and their effects in model toxicological systems (in vitro and in experimental animals) and in humans. Such evaluation is the key initial step in characterizing the addictive, toxic and carcinogenic potential of HTPs and comparing them with other tobacco products on the market. The data indicate that the chemical profile of HTP aerosols is substantially different from that of conventional cigarettes or e-cigarettes. In the absence of standard analytical methods, however, these findings should be interpreted with caution. Studies also suggests that nicotine intake from some HTPs may be similar to or greater than that from conventional cigarettes. Studies of exposure in vitro and in experimental animals and humans generally confirm reduced exposure to some combustion-derived toxicants; however, some studies raise concern about potential cardiopulmonary toxicity and hepatotoxicity in users and harm from second-hand exposure in non-users. The report highlights the paucity of independent, non-industry research and recommends research and regulatory priorities.

2.1 Background

Heated tobacco products (HTPs), also referred to by manufacturers and some regulators as "heat-not-burn" or tobacco heating products, produce aerosols by heating tobacco at lower temperatures than in conventional tobacco-burning cigarettes (generally < 600 °C) in battery-powered heating systems. Early versions of HTPs, such as the cigarettes Eclipse and Accord, were poorly accepted by smokers or did not gain a meaningful market share; however, the versions of HTPs introduced during the past decade have started to gain popularity in some parts of the world. As combustion of conventional cigarettes results in many harmful emissions, including numerous toxicants and carcinogens, the purported aim of the new HTPs is to offer smokers a "less harmful alternative" to conventional cigarettes. HTPs are thus marketed with claims of "reduced risk products", "cleaner alternatives to conventional cigarettes" and "smoke-free alternatives". Examples of currently marketed HTPs are IQOS (Philip Morris International, PMI), glo and iFuse (British American Tobacco, BAT) and Ploom TECH (Japan Tobacco). Although they all heat tobacco, they differ in their construction and composition and in the mechanisms for heating tobacco and generating aerosol. For example, IQOS and glo generate aerosol containing nicotine by heating cigarette-like tobacco sticks at 240–350 °C, while Ploom TECH and iFuse produce aerosols by heating a mixture of glycerol and propylene glycol, which is then passed through a capsule containing tobacco material.

There are several major challenges to assessing user and non-user exposure to emitted chemical constituents, many of which are important toxicants and carcinogens, and the effects associated with the currently marketed HTPs. The first is that products such as IQOS were introduced relatively recently, in 2015, and were initially available only on limited markets. The second is the diversity of

HTPs, outlined above, as the chemical composition of the aerosols produced by these devices may differ. While the concentrations of many combustion-associated chemicals in HTP aerosols may be lower than in conventional cigarettes, some are higher, and HTPs may emit unique harmful chemicals because of their distinctive characteristics and how they are used. A third issue is that some of the tobacco sticks manufactured for one device can be used with other devices and can be reused; the interchangeability and misuse of products make it difficult to determine the toxic emissions to which the user is exposed. Fourthly, most of the scientific data on currently marketed HTPs were generated and published by the tobacco industry or funded by affiliates of the industry, and there is a critical lack of independent academic literature on HTPs. Given the history of misleading marketing and misinterpretation of research data by some tobacco product manufacturers, their publications must be interpreted with caution and after rigorous review of the raw data and methods. Unpublished data reported to regulators by manufacturers should be reviewed in a similarly rigorous way.

This report was commissioned by WHO in response to the request made by the Conference of the Parties to the WHO Framework Convention on Tobacco Control at its eighth session (Geneva, Switzerland, 1–6 October 2018) to the Convention Secretariat to invite WHO to prepare a comprehensive report of research and evidence on novel and emerging tobacco products, in particular HTPs, and to propose policy options to achieve the objectives and measures outlined in the relevant decision (FCTC/COP8(22)). This report partly addresses that request. It covers a broad range of aspects of the health impacts of HTPs, including in non-users, their addictive potential, perception and use, attractiveness, potential role in initiating and quitting smoking, marketing, including promotional strategies and impacts, claims of reduced harm, variation among products, regulatory experience and monitoring by Parties, impact on tobacco control and research gaps.

In this paper, we review the current literature on the toxicity of and exposure to HTP emissions and associated health effects and evaluate claims that HTPs reduce risk. In particular, we cover:

- toxicants in HTP emissions and in other tobacco products;
- exposure to HTP emissions and their effects in model systems (in vitro and in experimental animals) and in humans and the implications for health;
- assessment of the basis of the claims that HTPs reduce risk or harm;
- implications for public health;
- research gaps and priorities; and
- relevant policy recommendations.

The literature search was based primarily on the PubMed database and the SciFinder search tool, which retrieves data from the Medline and CAplus databases. Relevant articles cited in publications obtained in the search were also included. In addition, we consulted manufacturers' websites and the websites of the US Centers for Disease Control and Prevention, the United States Food and Drug Administration (FDA) and other relevant organizations that provide information on the chemistry and health effects of HTPs.

2.2 Toxicants in heated tobacco product emissions

2.2.1 Laboratory methods for measuring toxicants

As outlined above, the HTP devices on the market vary in design, the main differences among them being how the tobacco is heated and the temperatures reached in the devices. The effect of temperature on the formation of harmful constituents in emissions of tobacco products, including conventional and e-cigarettes, is well documented. HTP devices produced by third parties are also available, mainly online, usually without product specifications. (A simple Google or Amazon Internet search with "IQOS sticks compatible" as keywords indicates a number of brands, like Uwoo, Luckten, Kacig, Hotcig, Uwell, Vaptio, G-taste and Smok Nord.) Reliable, analytically validated testing procedures are necessary to draw reliable inferences from assessments of the levels of harmful emissions from HTPs and to compare them with emissions from other products, such as conventional and e-cigarettes.

Most of the methods for analysing HTP emissions used to date by both the industry and independent research groups are adapted from methods used to analyse conventional cigarettes, mainly for determining common emissions and other parameters (e.g. nicotine and some other tobacco-derived constituents) *(1,2)*. To analyse compounds that may be specific to HTPs, some laboratories used complex analytical approaches, such as multidimensional gas chromatography coupled with mass spectrometry or liquid chromatography–tandem mass spectrometry *(1,3)*. In 2019, the Tobacco and Tobacco Products Technical Committee, TC126, of the International Organization for Standardization (ISO) created a working group on HTP, but they have not yet published HTP-specific methods.

While, in principle, methods used to analyse conventional cigarettes could be adapted for analysing HTPs, there are key differences between the two products. First, the tobacco material in HTPs may contain more humectant and water than that of conventional cigarettes *(4)*, which may have implications for analysis of emissions. At a minimum, studies should be conducted to determine the efficiency of the usual glass-fibre Cambridge filter pads in capturing toxicants found in both cigarette smoke and HTP-specific constituents derived mainly from thermal degradation of glycerol, propylene glycol and additives. Secondly, it is uncertain what puffing regimens (standardized puffing volume and frequency) should be used in the analysis of HTPs, as information on human topography in use of HTPs is lacking. The puffing parameters of HTPs are limited by HTP firmware and differ by device; manufacturers provide only specifications for "proper" product use and operation. As a result, use of different devices with different suggested puffing regimes *(5)* might result in large differences in desorption temperature and in toxicant emissions among studies. In addition, the higher intensity of HTP puffing may change the emission profile, because the kinetics of pyrolysis product generation is different from the kinetics of constituent desorption from tobacco material *(6)*. For example, increasing puffing intensity to generate more nicotine in the smoke, and thus the temperature at which the tobacco material is heated, is likely to increase the emissions of toxic pyrolysis products significantly. Finally, reference materials are not yet available, for either HTP devices or tobacco filler, obviating adequate analytical quality control.

In the absence of standard reference materials, analytical methods and puffing topography, accurate comparisons of HTPs and with other tobacco products cannot yet be made. For these reasons, generic statements of relative risk for users of these products are still preliminary and should be used carefully and cited only with recognition of this context.

2.2.2 Nicotine

As both conventional cigarettes and HTPs contain tobacco, both emit nicotine during use. Nicotine is the main known addictive constituent in tobacco products *(7)*; therefore, its levels in HTP emissions are of great importance. Table 2.1 lists the reported nicotine contents and emissions in two HTPs, IQOS and glo. Overall, the results of the analyses are consistent in both industry-supported and independent academic research. The corresponding nicotine values in reference cigarettes, which represent popular "full-flavour" (3R4F and CM6) and ventilated, formerly referred to as "light" (1R5F), filtered conventional cigarettes, are also shown in Table 2.1. The reported levels in the tobacco material (contents) of regular and menthol IQOS sticks were comparable to those of conventional cigarettes *(8,9)*; however, because HTP sticks are shorter and thinner than conventional cigarettes, they contain less tobacco material and, as a consequence, less total nicotine content per stick than a conventional cigarette. For example, Liu et al. *(10)* reported a nicotine content ranging from 1.9 to 4.6 mg per stick in three different HTPs.

Table 2.1. Representative reported nicotine contents and emissions in IQOS and glo HTPs

Product brand and variety	Range of reported average nicotine levels			References
	Contents (tobacco material) (mg/g)	Emissions		
		By ISO regimen[a]	By HCI regimen[a]	
IQOS				
Regular	15.2–15.7	0.4–0.77 mg/stick 0.3 mg/14 puffs[b]	1.1–1.5 mg/stick 1.1–1.4 mg/12 puffs	1,6,8,9,11–13
Menthol	15.6–17.1	0.43 mg/stick	1.2 mg/stick 1.38 mg/12 puffs	6,8,9
Mint	NR	0.32 mg/stick	1.2 mg/stick	6
Essence	NR	NR	1.14 mg/stick	4
Glo				
Regular	NR	0.07 mg/stick	0.27 mg/stick	6
Bright tobacco	NR	0.09–0.15 mg/stick	0.31–0.57 mg/stick	1,4,6
Fresh mix	NR	0.14 mg/stick	0.51 mg/stick	6
Intensely fresh	NR	0.13–0.15 mg/stick	0.36–0.51 mg/stick	1,4,6
Coolar green	NR	0.068 mg/stick	0.17 mg/stick	6
Coolar purple	NR	0.06 mg/stick	0.25 mg/stick	6
Reference conventional cigarettes				
3R4F	15.9–19.7 mg/g	0.73–0.76 mg/cigarette	1.4–2.1 mg/cigarette	1,6,12,14–16
1R5F	15.9–17.2 mg/g	0.12 mg/cigarette	1.0–1.1 mg/cigarette	6,8,15,16
CM6	18.7 mg/g	1.2 ± 0.13 mg/cigarette	2.6–2.73 mg/cigarette	6,15,16

NR: not reported.
[a] Smoking machine regimes used to puff HTPs for emission analyses: ISO, International Organization for Standardization (ISO) method; HCI, Health Canada method for testing tobacco products "Intense puffing regime".
[b] Some studies reported data based on the number of puffs taken to consume one HTP stick by using the corresponding machine-smoking regimen.

Data on HTP emissions also show significant differences among brands. For example, the nicotine levels in the aerosol of IQOS puffed under ISO or intense smoking conditions are comparable to those found in conventional cigarettes (Table 2.1). This suggests that the efficiency of nicotine desorption or transfer from the tobacco material to aerosol or smoke is much higher for IQOS than for conventional cigarettes. In contrast, the nicotine levels in the aerosol of glo were reported to be about 40% of those reported for IQOS and 23% of those for reference cigarettes *(17)*. Other studies show similar differences between the two products (Table 2.1). These observations indicate the importance of systematic surveillance and reporting of nicotine content and emissions in HTPs. Research should be conducted on the implications of these differences for the abuse liability and toxicity of the products.

2.2.3 Other toxicants

HTP emissions generally contain lower concentrations of toxic chemicals than conventional cigarettes because of the lower temperature at which they operate, which is the main source of many toxicants in the smoke of conventional cigarettes. The levels of most of the harmful and potentially harmful constituents measured in HTP aerosol are lower than those in reference cigarette smoke *(1,2,18–22)*, with the exception of glycidol, which is found at higher levels *(23)*. Nevertheless, both independent and manufacturer-funded studies show that, even if the temperatures reached by HTPs are not sufficient for combustion, they are still sufficient for the formation of harmful chemicals. Davis et al. *(24)* reported that IQOS tobacco filler appears to char without ignition and that charring increased when the IQOS is not cleaned after each use. Some signs of combustion were also identified in another study *(3)*. Auer et al. *(11)* found that the aerosol released by IQOS contains elements from pyrolysis and thermogenic degradation that are the same harmful constituents of conventional cigarette smoke.

A study by BAT researchers showed that nicotine and some cigarette smoke compounds were released at between 100 °C and 200 °C as a result of evaporative transfer or initial thermal decomposition from the tobacco blend. It is important to note that tobacco heated to 200 °C can generate emissions for a substantially longer time than a burning cigarette. With increments of temperature, the levels of some analytes increase gradually: between 180 °C and 200 °C, the amounts of carbon monoxide (CO), those of crotonaldehyde and methyl ethyl ketone double, and that of formaldehyde triples; between 120 °C and 200 °C, that of acetaldehyde increases 15 times; between 160 °C and 200 °C, that of acetone doubles and that of propionaldehyde triples; and between 140 °C and 200 °C, that of butyraldehyde doubles. These chemicals can be formed by pyrolytic decomposition of carbohydrates and tobacco structural polymers. Tobacco-specific nitrosamines (TSNAs) were quantifiable, but there was no consistent difference at different temperatures *(25)*.

Humectants contribute mainly to the total particulate matter of aerosols generated by HTPs. Much higher levels of humectants (e.g. glycerol) were found in HTP aerosol than in conventional cigarette smoke *(4)*. Independent studies found that the content of glycerol generated from HTPs in the HCI regimen was approximately 360 μg/stick for IQOS, 520 μg/stick for glo and 5900 μg/stick for Ploom TECH, as compared with 18 μg/cigarette from conventional cigarettes. HTPs generated fewer chemical compounds than conventional cigarettes, except for water, propylene glycol, glycerol and acetol *(5)*. The total particulate matter of a prototype HTP, THS 2.2, was composed mainly of glycerine (56.3%) and propylene glycol.

Although glycerine and propylene glycol may be safe for humans, they were found to produce harmful products when heated, including acrolein (a strong airway irritant) and glycidol (a carcinogen), in studies of IQOS and e-cigarette emissions. These carbonyls have been reported as by-products of propylene glycol and glycerine thermal decomposition in e-cigarettes. As the HTP stick contains a large amount of glycerine, its degradation by-products were present in IQOS emissions *(26)*.

Carbon monoxide

CO was found as a product of incomplete combustion in in HTPs emissions in both independent and tobacco industry studies.

Industry research

Forster et al. *(1)* reported that CO yields were below the reporting limits at temperatures < 180 °C, representing a reduction of > 99% from those in conventional cigarette emissions. Above this temperature (to 200 °C), the CO level increased with increasing temperature. The CO yield from an HTP was below the limit of detection, whereas that from a conventional cigarette was 31.2 mg/cigarette *(2)*. No correlation was found between the presence of flavours and CO levels: both the flavoured and unflavoured HTPs generated ≤ 0.22 mg/stick, while conventional cigarettes generated 32.8 mg/cigarette *(23)*.

Academic research

CO was found in IQOS aerosol *(11)*, even though the temperature was only 330 °C, as compared with 684 °C in smoke from a conventional cigarette, but at lower levels than in mainstream smoke of conventional cigarettes. The concentration of CO emitted by IQOS, measured with the official WHO TobLabNet method, was approximately one hundredth of that emitted by conventional cigarettes *(8)*. Similar results were found with the ISO and HCI regimes, THS 2.2 releasing 90% less CO than conventional cigarettes *(3)*.

Tobacco-specific nitrosamines

Manufacturers of HTPs claim that the levels of tobacco-specific nitrosamines (TSNA) are lower in HTPs. Several studies reported considerably lower levels of TSNA in e-liquids and in HTPs than in conventional cigarettes *(29)*. Independent studies reported less tar and more TSNA than the manufacturers did *(17)*.

Industry research

PMI and BAT reported a mean reduction of 90% in the levels of TSNAs in HTP aerosols as compared with mainstream smoke of conventional cigarettes. They claimed that the reduction is due to lower evaporating transfer and a lower working temperature, which reduce pyrosynthesis and pyrorelease *(27)*. TSNA emissions from a prototype HTP, THP1.0, were reported to be reduced by 80–98% from those in the smoke of conventional cigarettes *(1)*.

Academic research

Studies by non-industry researchers have also demonstrated lower levels of TSNA in the aerosol of HTPs than in smoke from conventional cigarettes *(30,31)*, the reductions of individual TSNA ranging from 8–22 times per puff in the HTP aerosol as compared with cigarette smoke; the reduction in HTP HeatStick aerosol was 7–17 times. TSNA yields per puff in IQOS aerosol were an order of magnitude lower than those in the smoke of conventional cigarettes but an order of magnitude higher than those in the aerosol of e-cigarettes *(32,33)*. Other independent studies confirmed that the levels of TSNA in tobacco material and mainstream smoke of IQOS were significantly lower than those of conventional cigarettes, although the transfer rates of N'-nitrosonornicotine (NNN), N'nitrosoanatabine (NAT) and 4-(methylnitrosamino)-1-(3-pyridyl)-1-butanone (NNK) in IQOS were slightly higher than those in conventional cigarettes *(8,13)*. Ratajczak et al. *(28)* reported that the concentration of TSNAs was one fifth that of conventional cigarettes. Li et al. *(3)* found that > 92% less NNN, NNK and NAT was released than from conventional cigarettes under both the ISO and the HCI regimen, and 72% less N'-nitrosoanabasine than 3R4F was released under both regimens.

Carbonyl compounds

HTPs emit toxic carbonyl compounds generated from thermal decomposition, and their levels increase gradually with temperature rising from 160 °C or 180 °C to 200 °C *(8,25)*. Formaldehyde, acetaldehyde, acrolein and other aldehydes form during heating of mixtures of glycerol and propylene glycol in e-cigarettes *(34,35)*. As these humectants are also present in HTPs, aldehydes may be derived from them *(36)*. In most cases, however, the levels are lower than in conventional cigarette smoke, according to both manufacturer-funded and independent studies.

Industry research

Aldehydes accounted for 41% of the total estimated concentration of constituents in THP1.0 emissions *(37)*. Crooks et al. *(23)* measured the levels of some aldehydes in aerosol from flavoured and unflavoured HTPs and in conventional cigarette smoke. They found formaldehyde at 1.52 µg/stick in aerosols from flavoured and 1.79 µg/stick in those from unflavoured HTPs and at 66.67 µg/stick in smoke from conventional cigarettes; and acetaldehyde at 35.48 µg/stick in aerosols from flavoured and 35.54 µg/stick in those from unflavoured HTPs and at 2164.73 µg/stick in smoke from conventional cigarettes. The reduction in formaldehyde in HTP aerosols as compared with smoke from conventional cigarettes was reported to be 90% *(2,18)*. The concentrations of formaldehyde (16.3 µg/m^3) and acetaldehyde (12.4 µg/m^3) fell within the range of the mean concentrations observed in residential and public environments *(19)*.

Academic research

Studies by non-industry researchers confirm the lower carbonyl production by IQOS than by conventional cigarettes, although it was higher than that from e-cigarettes *(26)*. The level of formaldehyde in HTP aerosols was 91.6% lower than in smoke from conventional cigarettes, and reductions of 84.9% were seen for acetaldehyde, 90.6% for acrolein, 89% for propionaldehyde and 95.3% for crotonaldehyde. At more intense puffing regimens, minimal differences in carbonyl emissions were observed between IQOS and conventional products, except that formaldehyde levels were increased three to four times over those with the HCI puffing regimen, from 6.4 to 17.1 µg/stick for regular IQOS. Carbonyl levels were higher in aerosols from HTPs than those from e-cigarettes *(38)*. Similar results were obtained by Ruprecht et al. *(39)*, who observed that the levels of aldehydes formed by use of IQOS were 2% higher than with conventional cigarettes for acrolein, 6% higher for acetaldehyde and 7% higher for formaldehyde; e-cigarettes generated only 1% of the amount of aldehydes generated by conventional cigarettes. Other authors observed that the yields of carbonyls were 80–96% lower than those from conventional cigarettes *(12,13,30,40,43)*.

Uchiyama et al. *(5)* compared acetaldehyde production by different HTP types under the HCI regimen and found that IQOS generated 210 µg/stick, glo™ generated 250 µg/stick and Ploom TECH 0.45 µg/stick, while conventional cigarettes generated 1300 µg/cigarette. Salman et al. *(41)* estimated reductions of 70% and 65% in the daily intake of formaldehyde and acetaldehyde, respectively, with use of IQOS instead of conventional cigarettes. Li et al. *(3)* observed reductions of 55.80% and 77.34% in formaldehyde and acetaldehyde, respectively, under the ISO regime.

Cyanidric formaldehyde can be formed from pyrolysis of the polymer filter in HTPs *(42)*. This thin plastic sheet melts during IQOS use, releasing formaldehyde cyanohydrin *(24)*.

Benzo[a]pyrene and other polycyclic aromatic hydrocarbons

Benzo[a]pyrene and other polycyclic aromatic hydrocarbons (PAH) are typical products of incomplete combustion. Their determination is important, as they are carcinogens *(13)*.

Industry research

Manufacturer-funded studies reported lower levels of PAH, including benzo[a]pyrene, in HTPs than in conventional cigarettes. The formation of PAH, aromatic amines, phenols and aldehydes was reduced by > 75% *(22)*, and acyclic, alicyclic and monocyclic aromatic hydrocarbons in HTP aerosol accounted for < 4%, as compared with 64% in mainstream smoke of conventional cigarettes *(37)*. Benzo[a]pyrene levels were very low in the aerosols of all products, in contrast to the yields found in the mainstream smoke of conventional cigarettes *(18)*.

Takahashi et al. *(2)* found < 0.531 ng benzo[a]pyrene in an HTP aerosol and 12.9 ng in conventional cigarette smoke, while Crooks et al. *(23)* found 0.44 ng/stick benzo[a]pyrene in the aerosol of flavoured HTPs, 0.41 ng/stick in that from unflavoured HTPs and 12.76 ng/stick in smoke from conventional cigarettes, indicating that the levels were not associated with the presence of flavours.

Pyrolysed menthol could be a precursor of benzo[a]pyrene in the smoke of a mentholated cigarette product, although no significant contribution of menthol to the yield of benzo[a]pyrene was observed in THS2.2 *(20)*.

Academic research

Auer et al. *(11)* found 0.8 ng/stick benzo[a]pyrene in HTP emissions and 20 ng/cigarette in smoke from conventional cigarettes; these levels are higher than those found in manufacturer-funded studies. HTP released higher levels of acenaphthene than conventional cigarettes. St Helen et al. *(12)* found levels similar to those in tobacco industry studies, with 0.736 ng/stick benzo[a]pyrene in HTP aerosols and 13.3 ng/stick in smoke from conventional cigarettes, indicating a reduction of 94%.

Other toxic chemicals

Most of the chemical constituents of the particulate phase of HTP aerosol and conventional cigarette smoke were oxygenated compounds, comprising 39% and 70% of the total estimated concentration of analytes in cigarette and HTP particulate phase, respectively. The levels of oxygenated compounds are higher in the HTP particulate phase probably because of the large amount of glycerine and other humectants *(43)*.

Industry research

Nitrogen-containing compounds. The levels of these compounds were 12% lower in HTP aerosols than in smoke from conventional cigarettes, accounting for 58% and 29% of the total estimated concentration of analytes in the cigarette and the HTP particulate phase, respectively *(43)*. The level of nitrogen oxide increased with time in conventional cigarette smoke but remained constant in HTP aerosol.

The levels found in HTP aerosol were 5.5–7.3% those of smoke from conventional cigarettes *(21)*.

Manufacturer-funded studies reported reductions of 25–50% in the levels of ammonia and some other toxicants in HTPs *(20)*. Ammonia was found at a level 88% lower than in conventional cigarette smoke in HTP emissions *(1)*. Crooks et al. *(23)* reported that the levels of ammonia, nitrogen oxides and o-cresol were higher in aerosol from flavoured than in that from unflavoured Neostik.

Metals. In one study, the level of mercury in HTP emissions was 69% lower than in cigarette mainstream smoke, while others reported reductions of 25–50% *(1,20)*. The levels of chromium, nickel and selenium were below the limits of detection in reference cigarette smoke *(2)*.

Volatile organic compounds. In some studies, the levels of certain volatile hydrocarbons in HTP emissions were 97–99% lower than in smoke from conventional cigarettes *(13,40)*. Industry researchers reported that the concentrations of Hoffmann list volatile compounds were significantly lower in the aerosol of prototype HTP product THP1.0 than in smoke from conventional cigarettes, with the level of toluene was reduced by 99% and that of 2-propanone by 91% *(37)*. After HTP use, the level of benzene in the aerosol was 0.93 µg/m^3, and that after e-cigarette use was lower. Toluene was not detectable in HTP aerosol but was present at 151.1 µg/m^3 after conventional cigarette use *(36)*.

Other chemicals. The levels of many constituents of HTP emissions were below the level of quantification or detection, except for formaldehyde, acetone and ammonia *(2)*. Aldehydes, ketones and heterocyclic compounds accounted for 41%, 32% and 10% of the total concentration of analytes, respectively. The level of 2-propanone was higher in the mainstream smoke of conventional cigarettes (152 µg/cigarette) than in HTP aerosol (13.3 µg/stick), whereas the levels of pyridine and dimethyl trisulfide in HTP aerosols were marginally higher *(37)*. Forster et al. *(1)* calculated a reduction of 96–99% in the levels of phenols (except for resorcinol, p-cresol and caffeic acid) and a 99% reduction in the levels of ethylene oxide and propylene oxide; however, acetoin and methylglyoxal were present at higher levels in HTP emissions than in conventional cigarette smoke.

Particulate matter. Manufacturer-funded studies found that the levels of particulate matter < 2.5 µm (PM2.5) in HTP emissions were 28 times lower than in conventional cigarette smoke *(19)*. The yield of total particulate matter from HTPs was approximately twice that of conventional cigarettes. The levels of water and humectants in total particulate matter were higher in HTPs than in conventional cigarettes: 90% (w/w) in HTP and 37% in conventional cigarettes *(2)*. PMI studies showed that the respirable fraction of particles is 90% lower in HTP aerosols than in conventional cigarette smoke *(44)*, equivalent to the concentration in background air; however, the limits were below the lower working range of the methods *(45)*.

Academic research

Reactive oxygen species. Independent studies reported that the level of total reactive oxygen species was 85% lower in HTP emissions than in those of conventional cigarettes *(41)*. Emission of reactive oxygen species during use of IQOS can be harmful *(46)*.

Metals. Independent studies found lower levels of metals in HTP emissions than in cigarette smoke *(13,42)*. Ruprecht et al. *(39)* observed that use of mentholated IQOS is associated with higher metal concentrations than use of IQOS without menthol. They found metals such as aluminium, titanium, strontium, molybdenum, tin and antimony in IQOS that are not present in conventional cigarettes, and metals such as nickel, copper, zinc, lanthanum and lead in conventional cigarettes that are not detected in IQOS. One article reported an unusual case of criminal mercury poisoning with use of HTPs; its addition to a tobacco stick caused the victim to inhale vaporized mercury *(47)*.

Volatile organic compounds. More than 70 volatile compounds were detected in mainstream emissions of IQOS, including isoprene, acrylonitrile, cresols, benzene, phenol, naphthalene, acetaldehyde, propanal, acrolein, formaldehyde, 2-butanone, acetone, crotonaldehyde and quinoline *(26)*. All these compounds are considered potentially harmful by the FDA. Some studies, however, found that the levels of some volatile hydrocarbons were 97–99% lower in HTP aerosols than in smoke from conventional cigarettes *(13,40)*. No PAHs were found in IQOS side-stream aerosol *(42)*. Heated tobacco materials contained more types of volatile compounds than conventional cigarette emissions *(48)*.

Black carbon. Ruprecht et al. *(39)* found the highest black carbon concentrations in conventional cigarettes, with 78 µg/m^3 for organic compounds and 2.3 µg/m^3 for elemental carbon. Elemental carbon was not detectable during puffing of IQOS, while the levels of organic compounds were 2.81–3.89% of those in conventional cigarettes.

Other chemicals. Results similar to those of manufacturer-funded studies were obtained in an independent study, with 90% lower levels of most toxicants, except for carbonyls, ammonia and N'-nitrosoanabasine. The level of ammonia, in particular, was 63.4% lower with the HCI regimen than in conventional cigarette smoke *(3)*.

Particulate matter. The concentration of particles in the mainstream aerosol of IQOS was lower than that in emissions from e-cigarettes and conventional cigarettes *(49)*, and use of HTPs resulted in the lowest levels of fine particulate matter *(36)*. The concentrations of PM>0.1 and PM>0.3 were significant in aerosol from conventional cigarettes and that from IQOS (for PM>0.3) but were trivial in aerosol from e-cigarettes as compared with conventional cigarettes *(39)*. Other studies found that the level of particulates in emissions from e-cigarettes and HTPs was about 25% that in cigarette smoke. The diameter of most particles in IQOS aerosol is < 1000 nm, which is considered safer than a lower mass. The respirable fraction of particles is higher in glo and conventional cigarette smoke *(50)*.

As these results show that the composition of IQOS HeatSticks is different from that of conventional cigarettes, including flavourings and additives, IQOS aerosol may contain other chemical constituents not present in tobacco smoke, which have not yet been investigated in untargeted analyses *(12)*.

2.3 Exposure and effects of HTPs in vitro and in laboratory animals

2.3.1 In-vitro studies

Industry research

Research groups at PMI and BAT have published several reports on the results of in-vitro toxicological studies with prototype HTPs that differed in the device and tobacco characteristics. Most reported substantially lower cytotoxicity of HTP aerosol than of the smoke of reference cigarettes.

PMI research. PMI published a series of papers on the cytotoxicity (measured in the neutral red uptake assay) and mutagenicity (*Salmonella* reverse mutation assay) of a prototype electrically heated cigarette smoking system puffed under ISO smoking conditions and one alternative puffing condition *(51–53)*. They reported that the aerosol from these devices was up to 40% less cytotoxic and up to 90% less mutagenic than the smoke of a reference 1R4F cigarette (comparisons based on total particulate matter yield). In a study published in 2012, PMI reported toxicological evaluation of the same prototype system with 25 additional smoking regimens that reflect human puffing behaviour *(54)*. While the overall biological activity of the HTPs tested was lower than that of a reference cigarette, increased smoking intensity led to substantial increases in cytotoxicity and nearly 36 times more bacterial mutagenicity than with HTP aerosols generated under low-intensity ISO conditions. They concluded that the increases were probably due to changes in the emissions of harmful constituents, suggesting that the results of ISO-based testing are not informative in terms of potential effects in humans.

PMI also reported on the biological effects of the aerosol from another HTP prototype, THS2.2. In one study, the ability of THS2.2 aerosol to inhibit monoamine oxidase activity was investigated, to assess the potential abuse liability of this product *(55)*. The authors reported significant inhibitory activity of a 3R4F reference cigarette aerosol but not by THS2.2. In another study, histology, cytotoxicity, secreted cytokines and chemokines, ciliary beating and genome-wide mRNA/miRNA profiles were assessed in human organotypic bronchial epithelial cultures at various times after exposure to THS2.2 aerosol or reference 3R4F cigarette smoke, with similar nicotine concentrations *(56)*. Cell fate, cell proliferation, cell stress and inflammatory network models 4 h after exposure to THS2.2 aerosol were only 7.6% of those observed after exposure to 3R4F smoke. No morphological changes were reported after exposure to THS2.2 aerosol, even at a nicotine concentration three times that in 3R4F smoke. Similar studies were conducted with another HTP prototype, CHTP1.2, in human organotypic cultures derived from buccal and gingival epithelia *(57)*, small airway and nasal epithelial cells *(58)* and endothelial cells *(59)*. Cells were exposed acutely (28 min) or repeatedly (28 min/day for 3 days) to CHTP1.2 aerosol or to 3R4F smoke. The authors reported lack of cytotoxicity, fewer pathophysiological alterations and less change in toxicological and inflammatory biomarkers after exposure to CHTP1.2 than after exposure to 3R4F smoke. Alterations in mRNA expression were, however, detected in small airway and nasal epithelial cultures exposed to CHTP1.2 aerosol.

BAT research. BAT researchers assessed nicotine delivery and cytotoxicity in H292 human bronchial epithelial cells of aerosols from IQOS and glo in comparison with tobacco smoke from 3R4F cigarettes; all products were smoked under HCI *(60)*. The authors reported more nicotine delivery but less cytotoxicity of HTPs than 3R4F cigarettes. They concluded that there was no difference between the HTPs; however, the figures in the report suggest that IQOS was more cytotoxic than the glo product. In another study, RNA sequencing was used to compare transcriptomic perturbations after acute exposure of 3D airway tissue to the aerosols of the same products *(61)*. Differential expression of 115 RNAs was reported with IQOS and of two RNAs with the glo product as compared with air, while 2809 RNAs were differentially expressed in response to 3R4F. Examination of the data and charts in the report suggests that the results depended on the thresholds set for data analyses (i.e. P values and fold change in expression) and that inflammatory and xenobiotic metabolizing gene expression may be affected by the HTPs tested. A subsequent study was conducted to complement these two assessments and to determine the effect of flavours on in-vitro responses to Neostik *(62)*. The authors concluded that the addition of flavours does not change the in-vitro baseline responses to unflavoured Neostik.

Academic research

Leigh et al. *(63)* used an in-vitro model of an air–liquid interface with human bronchial epithelial cells (H292) to investigate the toxic effects of inhaling emissions from IQOS and from e-cigarettes and conventional cigarettes. The number of puffs of each product was adjusted to achieve similar nicotine delivery to the cells, and cytotoxicity was measured in the neutral red uptake and trypan blue assays. The cytotoxicity of IQOS in the neutral red assay, but not in the trypan blue assay, was higher than that in air controls and lower than that with conventional cigarette smoke. In another study, the cytotoxicity of

IQOS aerosols was compared with that of smoke from Marlboro Red and 3R4F reference cigarettes *(64)* in three assays with eight different cell types. The results with transformed mouse NIH/3T3 fibroblasts were similar to those previously reported by PMI *(51–53)*; but assessments in other types of cells and with higher concentrations of aerosols showed comparable depression of mitochondrial and lysosomal activity by IOQS and cigarette smoke solutions. No adjustment was made in this study, however, for the levels of nicotine in the different products, and the puffing regimen used was described in insufficient detail.

In another study, the effect of exposure of culture medium to IQOS aerosol, e-cigarette aerosol and conventional cigarette smoke was studied on the viability and differentiation of 3T3-L1 pre-adipocytes to beige adipocytes *(65)*. The authors concluded that exposure to IQOS had limited or no effect as compared with air. Examination of the data in this report shows that expression of the adipogenic markers Ppar-γ and Resistin was statistically significantly decreased in IQOS-treated cells at the end of differentiation (10 days), an effect similar to that observed in cigarette smoke-treated cells. It was noted that the authors of this report have a history of industry-linked research funding.

2.3.2 Studies in laboratory animals
Industry research

Most reports on the effects of HTPs in laboratory animals have been published by tobacco industry research groups, namely PMI and Japan Tobacco, which generally report substantially less toxicity and carcinogenicity with HTP aerosols than with the smoke of reference tobacco cigarettes.

Japan Tobacco research. Two studies have been published on the effects of a prototype "heated" cigarette on dermal tumorigenicity and inhalation toxicity in mouse models. In the first study, the effect of HTP on dermal tumour promotion in female SENCAR mice was studied after topical application of 7,12-dimethylbenz[*a*]anthracene as a tumour initiator *(66)*. Condensates of HTP aerosol or cigarette smoke (3R4F cigarettes) generated under the modified HCI smoking regimen were applied repeatedly for 30 weeks at five doses up to 30 mg tar/application. At ≤ 15 mg tar, animals treated with HTP aerosol developed neoplasms with a longer latency and lower incidence and multiplicity and lower incidences of inflammation and squamous epithelial hyperplasia than animals treated with cigarette smoke. At the end of treatment, however, these effects were more prevalent than in untreated animals. At doses ≥ 22.5 mg of tar, the differences between HTP aerosol and 3R4F condensate were less clear.

In the second study, the same HTP was investigated in nose-only 5-week and 13-week inhalation studies *(67)*. Lesser histopathological changes were found in the respiratory tracts of HTP-treated animals (respiratory epithelial hyperplasia in the nasal cavity and accumulation of pigmented macrophages in alveoli) and less pulmonary inflammation (as measured by the percentage of neutrophils and activity of γ-glutamyl transpeptidase, alkaline phosphatase and lactic dehydrogenase in bronchoalveolar lavage fluids) than in cigarette smoke-treated animals. HTP treatment nevertheless had significant effects on the histopathological outcomes as compared with air treatment, as HTP-treated animals showed a 100% incidence of hyperplasia and hyperkeratosis in the larynx and epiglottis at all doses tested in both treatment regimens and substantial dose-related increases (≤ 100%) in the incidence of alterations in nasal epithelium, ventral pouch and lung tissues.

PMI research. In a 90-day study in rats exposed by inhalation through the nose only to the same or twice the nicotine concentration in the inhalation zone, CHTP1.2 aerosol resulted in significantly less exposure to harmful constituents and induced less respiratory tract irritation and systemic and pathological effects than 3R4F cigarettes *(68)*. The toxicology arm of this study, which included transcriptomics, proteomics and lipidomics analyses *(69)*, showed much weaker inflammatory and cellular stress responses in the respiratory nasal epithelium and lungs with CHTP1.2 aerosol than with 3R4F smoke. Many of these effects showed dose–response relations. CHTP1.2 aerosol also induced lower lipid concentrations in the serum of exposed animals.

PMI researchers also conducted 6-month exposure studies in the ApoE-/- mouse model to investigate the effects of THS 2.2 and CHTP1.2 on the cardiorespiratory system. One report, a systems toxicology approach with a combination of physiology, histology and molecular measurements, demonstrated a lower impact of 3R4F on the cardiorespiratory system, including little to no lung inflammation or emphysematous changes and reduced atherosclerotic plaque formation *(70)*. In another study, cardiovascular effects were investigated in echocardiographic, histopathological, immunohistochemical and transcriptomics analyses. The authors reported that continuous exposure to HTP aerosols did not affect atherosclerosis progression, heart function, left ventricular structure or the cardiovascular transcriptome *(71)*. Review of the data in these publications reveals, however, consistent increases in many of the outcome measures in animals treated with HTPs as compared with sham controls, although the increases did not reach statistical significance.

Academic research

In a study to investigate whether exposure to IQOS aerosol affects vascular endothelial function, which is known to be impaired by conventional cigarette smoke *(72)*, rats were exposed acutely through the nose only to IQOS aerosol generated by single HeatSticks, to mainstream smoke from single Marlboro Red cigarettes or to clean air. Arterial flow-mediated dilation, a measure of vascular endothelial function, was measured before and after exposure. The authors reported that flow-mediated dilation was impaired to comparable degrees by exposure to IQOS aerosol and to conventional cigarette smoke, and no effect was observed with clean air. Serum nicotine and cotinine levels were approximately 4.5 times higher in rats exposed to IQOS than in those exposed to cigarettes, even though the IQOS aerosol contained less nicotine than cigarette smoke. Brief exposure regimens that resulted in similar serum nicotine levels in groups exposed to IQOS and to cigarettes induced comparably impaired flow-mediated dilation. The authors concluded that IQOS use does not necessarily avoid the adverse cardiovascular effects of smoking cigarettes.

2.4 Exposure of humans to toxicants in HTPs and implications for health

2.4.1 Product use and topography

Davis et al. (24) evaluated the performance of IQOS under five conditions with two different protocols for cleaning devices. HeatSticks were inspected before and after use to determine any signs of tobacco material pyrolysis (charring) and melting of the polymer-film filter and to assess the effects of cleaning on charring. The results showed charring of the tobacco material after use, and gas chromatography–mass spectrometry headspace analysis of the polymer-film filter showed release of highly toxic formaldehyde cyanohydrin at 90 °C (a lower temperature than during normal IQOS usage). Increases in charring and formaldehyde cyanohydrin emissions were observed when the device was not cleaned between consecutive uses of HeatSticks. The authors concluded that limitations of the device (i.e. short tobacco sticks and shutting of the device after a certain number of puffs) may contribute to decreases in inter-puff intervals, potentially increasing users' intake of nicotine and other harmful chemicals.

Some industry research acknowledges the potential modifying effects of use topography and device characteristics on the exposure of human users. For example, researchers at BAT investigated the impact of puffing parameters on the volume of emissions (73) and found that the choice of puffing parameters affects the volume, with significant differences among types of HTP. They suggested that detailed real-world HTP puffing topography should be studied in order to identify the most appropriate puffing parameters for laboratory testing. In another study, PMI researchers presented a modelling approach, named "nicotine bridging", to estimate human exposure to HTP emissions (74). The approach involves determination of harmful constituents and in-vitro toxicity parameter-to-nicotine regressions for multiple machine-smoking protocols; the distribution of nicotine uptake is determined from 24-h excretion of nicotine metabolites in a clinical study. The approach was illustrated with data for a prototype HTP (54) that showed less exposure of HTP users than conventional cigarette smokers, with little or no overlap between the distribution curves. The authors proposed that the method could be used to extrapolate exposure distributions to smoke constituents for which there are no specific biomarkers. As this method relies on machine-smoking protocols that are not representative of human exposure, extrapolation may not be justified, and additional independent research is required.

2.4.2 Biomarkers of exposure and effect
Industry research

As for the studies with cell cultures and animal models, many of the reports on biomarkers are provided by industry, particularly PMI and BAT.

PMI research. PMI researchers reported on biomarkers of exposure and effect in smokers who switched to prototype HTPs, some of which have been tested in vitro and in laboratory animals, as summarized above. Prototype HTPs were developed and tested by PMI for more than a decade before the release of IQOS and are apparently precursors of contemporary HTPs *(75)*. A series of publications in 2012 reported on randomized, controlled, open-label, parallel-group, single-centre studies on the use of type-K prototype HTPs (EHCSS-K3 and EHCSS-K6). In a study in the United Kingdom, biomarkers of exposure to nine cigarette smoke constituents and urinary mutagenicity were measured in 160 male and female smokers of Marlboro cigarettes who switched to EHCSS-K3 or EHCSS-K6 for 8 days *(76)*. Statistically significant mean decreases between baseline and day 8 ($P \leq 0.05$) were found for all biomarker measures, with reductions in urinary mutagenicity in smokers assigned to both HTPs. In a similar study in the Republic of Korea, biomarkers of exposure to 12 selected constituents and urinary mutagenicity were measured in 72 male and female subjects who smoked low-yield Lark One cigarettes at baseline and switched to EHCSS-K3 for 8 days *(77)*. Statistically significant reductions in urinary mutagenicity and in 10 of 12 constituent biomarkers were found between baseline and day 8 for the EHCSS-K3 group. In a study in Japan, the same measurements were made in 128 male and female participants who smoked Marlboro cigarettes at baseline and switched to EHCSS-K3 or EHCSS-K6 for 8 days *(78)*. The mean decreases in all constituents and in urinary mutagenicity between baseline and day 8 were statistically significant for participants who switched to HTPs ($P \leq 0.05$). An additional report on Japanese smokers presented data on biomarkers of exposure to 12 constituents, urinary mutagenicity and serum club cell 16-kDa protein (CC16) in 102 male and female subjects who smoked Marlboro Ultra Light menthol cigarettes at baseline and switched to the menthol version of the HTP prototype EHCSS-K6(M) for 6 days *(79)*. Statistically significant decreases in exposure to 10 of the 12 cigarette smoke constituents and in urinary mutagenicity were found, as in the previous studies; however, serum CC16, an indicator of lung epithelial injury, was unchanged in all groups, including that assigned to use no product. Despite the reductions in exposure measures after switching to HTPs, the levels of biomarkers in all these studies remained substantially higher than in the group assigned to stop smoking and not to use any product (Table 2.2). Another PMI study, in Polish smokers, included measurement of a broad panel of biomarkers associated with cardiovascular risk, in addition to biomarkers of exposure *(80)*. In this study, 316 male and female smokers were randomized to continue smoking conventional cigarettes or to switch to smoking the EHCSS-K6 prototype HTP for 1 month. Most of the cardiovascular biomarkers did not change after use of the HTP device, although the biomarkers of exposure decreased, as in the previous studies. A substantial mean increase was seen in high-density lipoprotein cholesterol, from a baseline median of 36.7 ng/mL (95% CI 2.1, 3410) to 59.0 ng/mL (95% CI 2.1, 4535), after 1 month of using EHCSS-K6; however, reductions in red blood cell count, haemoglobin and haematocrit were observed in the EHCSS-K6 group.

Table 2.2. Reported levels of biomarkers of exposure in smokers who switched to HTPs and in smokers assigned to continuous smoking or abstinence from all tobacco products

Biomarker of exposure [source]	HTP prototype (duration of use)	Mean ± SD level after switching to HTP	Comparison (↑, higher, or ↓, lower) with biomarker levels in participants assigned to other conditions		Reference
			HTP vs continuous smoking	HTP vs abstinence	
SPMA, µg/24 h [benzene]	EHCSS-K3 (8 days)	1.26 ± 1.26	↓80%	↑473%	66
	EHCSS-K3 (8 days)	1.63 ± 0.55	↓66%	↑19%	77
	EHCSS-K3 (8 days)	0.48 ± 0.28	↓78%	↑41%	68
	EHCSS-K6 (8 days)	0.86 ± 0.81	↓86%	↑291%	66
	EHCSS-K6 (8 days)	0.57 ± 0.57	↓75%	↑68%	68
	EHCSS-K6^M (6 days)	0.35 ± 0.13	↓85%	↓34%	69
MHBMA, µg/24 h [1,3-butadiene]	EHCSS-K3 (8 days)	2.63 ± 2.78	↓49%	↑874%	66
	EHCSS-K3 (8 days)	0.59 ± 0.58	↓75%	↑111%	67
	EHCSS-K3 (8 days)	0.66 ± 0.69	↓57%	↑50%	68
	EHCSS-K6 (8 days)	1.54 ± 1.52	↓70%	↑470%	66
	EHCSS-K6 (8 days)	0.74 ± 0.56	↓51%	↑68%	68
	EHCSS-K6^M (6 days)	0.61 ± 1.06	↓58%	↑69%	69
3-HPMA, mg/24 h [acrolein]	EHCSS-K3 (8 days)	1.15 ± 0.74	↓37%	↑140%	66
	EHCSS-K3 (8 days)	2.41 ± 0.67	↓18%	↑44%	67
	EHCSS-K3 (8 days)	1.01 ± 0.36	↓26%	↑110%	68
	EHCSS-K6 (8 days)	1.28 ± 0.87	↓30%	↑167%	66
	EHCSS-K6 (8 days)	1.09 ± 0.51	↓20%	↓127%	68
	EHCSS-K6^M (6 days)	0.83 ± 0.32	↓30%	↑110%	69
3-HMPMA, mg/24 h [crotonaldehyde]	EHCSS-K3 (8 days)	2.59 ± 1.90	↓50%	↑49%	66
	EHCSS-K3 (8 days)	1.99 ± 1.07	↓18%	↑20%	67
	EHCSS-K3 (8 days)	0.61 ± 0.17	↓52%	↑30%	68
	EHCSS-K6 (8 days)	2.62 ± 1.29	↓49%	↑51%	66
	EHCSS-K6 (8 days)	0.63 ± 0.15	↓51%	↑34%	68
	EHCSS-K6^M (6 days)	0.19 ± 0.08	↓85%	↓60%	69

Biomarker of exposure [source]	HTP prototype (duration of use)	Mean ± SD level after switching to HTP	Comparison (↑, higher, or ↓, lower) with biomarker levels in participants assigned to other conditions		Reference
			HTP vs continuous smoking	HTP vs abstinence	
Total NNAL, ng/24 h [NNK]	EHCSS-K3 (8 days)	104.3 ± 55.3	↓65%	↑76%	66
	EHCSS-K3 (8 days)	80.4 ± 59.3	↓59%	↑78%	67
	EHCSS-K3 (8 days)	102 ± 48	↓53%	↑23%	68
	EHCSS-K6 (8 days)	100.6 ± 68.8	↓66%	↑70%	66
	EHCSS-K6 (8 days)	100 ± 65	↓54%	↑20%	68
	EHCSS-K6^M (6 days)	95 ± 53	↓49%	↑19%	69
1-HOP, ng/24 h [pyrene, PAH]	EHCSS-K3 (8 days)	73.1 ± 30.4	↓60%	↓3%	66
	EHCSS-K3 (8 days)	143.62 ± 76.08	↓40%	↑13%	67
	EHCSS-K3 (8 days)	59.0 ± 35.3	↓56%	↑40%	68
	EHCSS-K6 (8 days)	71.9 ± 38.8	↓60%	↓4%	66
	EHCSS-K6 (8 days)	56.0 ± 28.4	↓59%	↑33%	68
	EHCSS-K6^M (6 days)	38.4 ± 22.7	↓64%	↑2%	69

1-HOP: 1-hydroxypyrene; 3-HPMA: 3-hydroxypropyl mercapturic acid; MHBMA: monohydroxybutenyl mercapturic acid; NNAL: 4-(methylnitrosamino)-1-(3-pyridyl)-1-butanol; SPMA: S-phenylmercapturic acid.

PMI published a separate series of reports on studies in Polish smokers who switched to prototype HTPs. In one study, harmful exposures were assessed in 112 male and female adult smokers who switched to a prototype carbon-heated tobacco product, MD2-E7, for 5 days, continued smoking or abstained *(81)*. Puffing topography during HTP use was also studied. Smoking intensity and nicotine intake increased after the switch to the test HTP; however, despite the more intense puffing topography, switching to the prototype HTP decreased all the measured biomarkers of exposure and urinary mutagenicity, consistent with the previous reports. In a study of smokers who switched to the THS 2.1 prototype for 5 days, puff duration and frequency increased over the baseline patterns of smoking conventional cigarettes, and use of HTP tobacco sticks increased by 27% during the study period *(82)*. The total puff volume returned to baseline at the end of the study, however, and the mean exposure to nicotine was comparable to that at baseline. Increased product consumption and total puff volume were also observed in a larger study of the THS 2.2 prototype *(83)*. The average puff duration was about 32% longer, the total puff duration was about 37% longer, and puff frequency was about 32% higher on day 4 than smoking patterns for conventional cigarettes at baseline. While the authors claimed that there was no change in nicotine intake, review of the data reveals an approximately 23% increase in nicotine biomarkers (total nicotine equivalents) between baseline and day 4. Although reductions in exposure to harmful constituents were observed consistently in these studies, the higher-intensity HTP puffing topography indicates that additional, longer studies are necessary to accurately assess reductions in the exposure to toxicants of smokers who switch to HTPs.

A longer study of 984 adult smokers in the USA has been reported *(84)*. The participants were randomized to switch to a THS2.2 device or to continue smoking for 6 months, and changes in a panel of biomarkers of exposure and effect were assessed. Reductions in biomarkers of exposure and in four biomarkers of effect (high-density lipoprotein cholesterol, white blood cell count, forced expiratory volume in 1 s as percentage predicted and carboxyhaemoglobin) were reported in smokers who switched as compared with those who continued smoking. Approximately 30% of smokers assigned to the HTP became dual users of conventional cigarettes and HTPs. In the group who predominantly used HTPs (average use, 16.5 sticks and two conventional cigarettes per day), the reductions in biomarkers of exposure ranged from 16% to 49% of the baseline smoking level, which were not as substantial as in previous, shorter switching studies. While some differences in the study procedures could have contributed to the lesser impact of HTP switching, longitudinal adaptations in HTP puffing topography could have played a role. The report did not provide information on product consumption or intensity of use of the test HTP.

BAT research. BAT researchers reported the results of a randomized, controlled study in Japan on switching to two commercial HTPs, glo and IQOS, for 5 days *(85)*. The 180 Japanese smokers were randomized to either continue smoking conventional cigarettes, switch to glo (mentholated or non-mentholated), switch to a non-mentholated IQOS or abstinence. As in the PMI studies, switching to HTPs resulted in significant decreases in urinary biomarkers of exposure and exhaled CO, as compared with baseline smoking. The reductions were similar for the two HTPs. An increase in HTP consumption was observed during the study; however, a similar increase was seen in consumption of tobacco cigarettes, which the authors attributed to the typical escalating product use seen in confinement studies.

BAT has been conducting another long-term randomized, multi-centre, controlled clinical switching study since March 2018 *(86)*. Up to 280 smokers were randomized to switching to a commercially available HTP for 1 year or to continuing smoking. In addition, up to 190 participants who wished to quit smoking were enrolled in a smoking cessation arm, and 40 never smokers served as a control group. Biomarkers of exposure and effect to toxicants are being assessed, and the changes will be compared with those in the smoking cessation and never smoker cohorts.

Japan Tobacco. In 2014, Japan Tobacco researchers reported on a controlled, semi-randomized, open-label, residential clinical study of changes in levels of biomarkers of exposure to tobacco constituents in healthy Japanese male smokers who switched to a prototype HTP *(87)*. A total of 70 smokers were enrolled and randomized to either an HTP or continued smoking for four consecutive weeks. As in the studies by PMI and BAT researchers, measured urinary biomarkers of exposure and urinary mutagenicity were significantly reduced in the HTP group.

Academic research

Little independent research on the exposure of HTP users has been published. A randomized, cross-over laboratory behavioural trial of use of IQOS was conducted in Belgium *(88)*, in which 30 participants were randomized to use of conventional cigarettes, e-cigarettes or IQOS for 5 min. Using an IQOS resulted in a small but reproducible increase in exhaled CO (0.3 ppm), although the level was lower than that observed after smoking a conventional cigarette (4.7 ppm). A similar randomized cross-over study was conducted in Italy by academic researchers with a history of tobacco industry-related funding *(30)*. A total of 12 healthy smokers were recruited, who used IQOS in the clinic (10 puffs, two sessions separated by 5 min) and provided exhaled CO samples at several times after the first puff of the first puffing session. No increase in exhaled CO was reported after use of IQOS. In a study in the USA, nicotine delivery and exhaled CO were measured in smokers of IQOS, JUUL e-cigarettes and conventional cigarettes *(89)*. The researchers recruited 18 smokers with no experience of IQOS or JUUL to complete a within-subject, laboratory study of controlled (10 puffs, ~30-s inter-puff interval) and ad-libitum (90 min) use of the test products or their own brand of conventional cigarettes. The amount of exhaled CO did not increase after use of IQOS, while the mean plasma nicotine increased significantly, from 2.1 (0.2) ng/mL to 12.7 (6.2) ng/mL after 10 puffs and to 11.3 (8.0) ng/mL after use ad libitum. The increase in plasma nicotine was about half that after smoking conventional cigarettes.

2.4.3 Passive exposure

Non-industry researchers have conducted studies of the potential impact of passive exposure to HTP emissions. In a study of particle size distribution, the profiles of exposure to submicron particles (5.6–560 nm) in conventional cigarettes and in IQOS were evaluated to estimate their potential deposition in the human respiratory system *(90)*. IQOS aerosol contained approximately four times lower amounts of such particles than the smoke of conventional cigarettes, and the particles in the IQOS aerosol dissipated rapidly; however, approximately half of such particles are small enough to reach the alveolar region upon inhalation. In a follow-up modelling study, the same group estimated the deposition of ultrafine particles from IQOS, e-cigarettes and conventional cigarettes in the respiratory tract of people of different ages in a multiple-path particle dosimetry model *(91)*. IQOS delivered significantly lower doses of second-hand particles than conventional cigarettes, but the doses were 50–110% higher than those from e-cigarettes, suggesting that non-users may have meaningful second-hand exposure. In the same study, 60–80% of the particles deposited in the head region of a 3-month-old infant measured < 100 nm, suggesting that they could be translocated to the brain via the olfactory bulb. In another study, the concentrations of aerosol particles, carbonyl and nicotine were analysed in a model chamber during use of an unidentified HTP *(19)*. Use of the HTP resulted in statistically significant increases in the amounts of several analytes, including nicotine, acetaldehyde and PM2.5, and in particle number as compared with background measurements. As in the previous study, the authors reported that HTP particles dissipated or evaporated within seconds and that the levels of particles and individual constituents were significantly lower than in smoke from conventional cigarettes under the same conditions. The authors concluded that intensive use of HTPs in a confined space with limited ventilation can substantially increase the exposure of bystanders to second-hand aerosol emissions.

In a study in Germany, particle size and concentrations were also measured to compare potential passive exposure due to use of IQOS, e-cigarettes and conventional tobacco cigarettes in cars *(92)*. The results showed that use of an IQOS had almost no effect on the mean concentration of fine particles (> 300 nm) or on the PM2.5 concentration in the interior of the car, but the concentration of smaller particles (25–300 nm) increased in all vehicles. The nicotine concentrations obtained from IQOS and e-cigarette use were comparable and were both lower than those from conventional cigarette smoking. In a more comprehensive chemical analysis of IQOS emissions in the context of second-hand exposures, the HCI smoking regimen was used to generate IQOS aerosol in an environmental chamber, and 33 volatile organic compounds were analysed, including aldehydes and nitrogenated aromatic species, in mainstream and sidestream emissions *(26)*. As in the studies described above, the yields from IQOS were substantially lower than those from conventional cigarettes and sometimes higher than those from e-cigarettes. Acrolein at > 0.35 $\mu g/m^3$ was identified as an exposure of potential concern.

Passive exposure from HTP use has also been studied in population-based research in Japan. In one study, the weighted prevalence of HTP use in indoor public spaces was estimated from nationally representative data in the International Tobacco Control Japan Survey (February–March 2018) *(93)*. It was found that 15.6% of current tobacco users in Japan reported using HTPs in indoor public spaces. In a survey of perceived symptoms due to exposure to second-hand HTP aerosol *(94)*, 8240 individuals aged 15–69 years were followed up between 2015 and 2017 in a longitudinal Internet survey. Of the respondents who reported having been exposed to second-hand HTP aerosol, 37% reported having experienced at least one symptom as a result. The most common symptoms were generally feeling ill, eye discomfort and sore throat. Nearly half of never smokers who had been exposed to second-hand HTP aerosol reported at least one symptom.

In an industry study of second-hand emissions from a prototype HTP in simulated "office" and "hospitality" environments with different baseline indoor air quality *(47)*, smoking an HTP under ISO conditions gave significantly lower yields of many constituents than smoking a Marlboro cigarette: the levels of 24 of 29 smoke constituents were reduced by a mean of > 90%, and the concentrations of five smoke constituents were reduced by a mean of 80–90%. Nicotine emissions were on average 97% lower from HTPs than from smoking Marlboro cigarettes, and the total number of respirable suspended particles was reduced by 90%. These results are generally consistent with those of non-industry research, which show that HTPs are a weaker source of indoor pollution than conventional cigarettes. Their impact on indoor air quality and on passive exposures is, however, not negligible and is not well understood.

2.4.4 Impact on health outcomes

Little research has been conducted on exposure to and effects of HTPs in toxicological models or human participants, partly because extensive marketing and use of these products is relatively recent. Assessment of the health implications of HTP use is therefore limited, and the available data should be interpreted with caution.

The apparent reduction in exposure of smokers who switch to HTPs to various emissions of traditional smoked products is an important but not sufficient factor in assessing the health effects of HTP use. Highly efficient nicotine intake from HTPs is an emerging concern, and particular attention should be paid to the potential effects in vulnerable populations, such as individuals with pre-existing conditions, pregnant women and young people.

Nicotine intake. Nicotine is a powerful, addictive chemical and the main driver of users' exposure to the toxic and carcinogenic constituents present in tobacco products. Emerging indications from studies in vitro and in laboratory animals and humans that more nicotine may be absorbed from HTPs than from cigarette smoke raise concern about the potentially high abuse liability of these products. In addition, nicotine is an important reproductive and neurobehavioural toxicant and contributes to mortality from cardiovascular diseases. Such effects are expected to be seen in HTP users.

Cardiovascular disease. The evidence that many cardiovascular biomarkers are not reduced in smokers who switch to HTPs *(80,95)* suggests that HTPs may represent a similar risk for cardiovascular disease as smoking. While the reasons are not clear at this time, elevated levels of certain chemical constituents in HTPs could play a role. These observations suggest that switching to HTPs is not likely to reduce the risk of cardiovascular morbidity and mortality associated with tobacco use.

Chronic respiratory disease. Although some research suggests relief of respiratory symptoms in smokers with chronic obstructive pulmonary disease who switch to HTPs *(96)*, other studies and expert groups have raised concern about the association between e-cigarette use and respiratory diseases *(97–99)*.

HTPs contain higher levels of volatile respiratory toxicants than e-cigarettes (Table 2.2). Furthermore, data from studies in laboratory animals suggest an impact of HTP exposure on respiratory organs *(67)* and the expression of RNAs associated with pulmonary response to injury *(61)*. An industry trial showed no reduction in the levels of a biomarker of lung epithelial injury (CC16) in smokers who switched to HTPs *(78)*. Thus, addicted adult smokers who switch to HTPs may not reduce their risk of chronic respiratory disease associated with tobacco use, and the use of these products by non-smokers may increase their risk of pulmonary disorder, particularly if they have other health conditions.

Pregnant women and children. Given the adverse effect of nicotine on development in utero and on birth outcomes, such as impaired cardiorespiratory and pulmonary function in infants, and the negative cognitive and neurobehavioural effects of exposure to nicotine during adolescence, use of HTPs by pregnant women and children is of particular concern.

The above considerations are based on knowledge of the effects associated with specific constituents reported to be present in HTPs and indirect evidence from in-vitro, in-vivo and biomarker studies on HTPs. There are virtually no studies in which the association between HTPs and health outcomes was studied directly. The few relevant studies are summarized below.

Health effects in young people. A large survey was conducted in the Republic of Korea to evaluate the effects of cigarettes, HTPs and e-cigarettes on the prevalence of asthma and allergic rhinitis among 60 040 middle- and high-school students *(100)*. "Ever HTP use" was significantly associated with current asthma and allergic rhinitis in adjusted models, although the association was weaker than that with current use of conventional cigarettes. The odds ratio for current asthma was particularly increased in those who were dual users of HTP and/or e-cigarettes with conventional cigarettes. In an earlier study *(101)*, the association between use of tobacco products and the risk of allergic diseases was assessed from cross-sectional data on 58 336 students aged 12–18 years from the 2018 Korea Youth Risk Behavior Survey. Use of conventional cigarettes, HTPs and e-cigarettes was each significantly associated with increased risks of asthma, allergic rhinitis and atopic dermatitis.

Health effects in adults. A study is being conducted in Kazakhstan to evaluate the health outcomes in men and women aged 40–59 who use IQOS with HeatSticks, as compared with smokers of conventional cigarettes *(102)*. In this 5-year, single-centre cohort observational study, data on the frequency of exacerbated respiratory symptoms, intolerance of physical exercise, abnormal lung function and other parameters and comorbid conditions are analysed. The study also includes comprehensive clinical assessments at baseline and annually, continuous monitoring of chronic obstructive pulmonary disease and registration of exacerbation of acute respiratory conditions. The clinical assessments include spirometry, chest computed tomography, electrocardiography, physical examinations, laboratory testing of serum for biomarkers of inflammation and metabolic syndrome, anthropometry and the 6-min walk test. Recruitment began in December 2017; results were not available as of July 2020. Similar studies should be conducted in other countries in which IQOS and other HTPs are marketed, for early identification of the potential health risks associated with these products.

2.5 Review of the evidence for reduced risk or harm with use of HTPs

2.5.1 Harm reduction in the context of tobacco products

The health consequences of smoking conventional cigarettes are well documented – and devastating. Nicotine is a powerful addictive substance, however, and quitting smoking is very difficult. The concept of reducing the harm of tobacco, as described by some in the tobacco control community, is based on the idea that cigarette smokers who are unwilling or unable to quit nicotine intake should have a less harmful alternative to conventional combusted cigarettes *(103,104)*. Over the past two decades, novel products claimed to reduce exposure and risk have been introduced continuously, including smokeless tobacco products, e-cigarettes and now HTPs. As these novel and emerging products generally evolve rapidly and are sometimes limited to a few markets, it has been difficult to generate timely research, independent of the tobacco industry, on their potential to reduce harm. This is particularly true for HTPs, and most of the available research on these products has been conducted by their manufacturers. Analysis of the data and the claims made on the basis of those data is essential to inform the research agenda and regulatory considerations.

2.5.2 Claims of reduced risk

Industry claims

All the studies conducted by the industry are based on a similar strategy for generating evidence to support claims of "reduced risk" associated with HTPs. The model is based on standard in-vitro assays, toxicological modelling, studies in laboratory animals and measurement of human biomarkers of exposure and potential harm. Importantly, the conclusions of all the published reports are based on comparisons of HTPs with conventional cigarettes.

To support its claim of "reduced risk" associated with use of HTPs, PMI modelled the population health impact of introducing IQOS *(105)*. Using the PMI assessment method, the authors conducted various simulations to estimate the population health impact of introducing a "reduced-risk tobacco product" in Japan and modelled different situations over a 20-year period from 1990. They estimated that, if tobacco use was stopped completely at baseline, the overall reduction in tobacco-attributable deaths from lung cancer, ischaemic heart disease, stroke and chronic obstructive pulmonary disease for men and women combined would be 269 916 deaths; if smoking was completely replaced by the reduced-risk tobacco product at baseline, they estimated 167 041–232 519 fewer deaths; and, if the product was introduced at baseline, with uptake rates consistent with the known uptake of IQOS, the estimated reduction in the number of deaths was 65 126–86 885.

Analysis of claims

While industry-sponsored publications fully disclose experimental data, their interpretation and the conclusions often omit important observations or present the findings inadequately. Therefore, careful analysis of industry claims and data is critical.

Analysis of specific claims and concerns about their accuracy. Several independent studies have examined the claims and conclusions of industry-published research reports. For example, St Helen et al. *(12)* examined sections on aerosol chemistry and human exposure assessment in the PMI application to the United States Food and Drug Administration (FDA) for authorization of a "modified risk tobacco product" (MRTP) to assess the validity of the claims of reduced exposure and risk. The authors of the analysis noted that PMI reported the levels in IQOS aerosol of only 40 of the 93 harmful and potentially harmful constituents on the FDA list. They also noted that, while the levels of all 58 constituents on the PMI list were lower in IQOS aerosol than in 3R4F cigarette smoke, the levels of 56 other constituents (not on either the PMI or the FDA list) were higher in IQOS emissions. The levels of some of these constituents were strikingly increased: 22 were > 200% higher and seven were > 1000% higher in IQOS aerosol than in 3R4F smoke (Fig. 2.1). The full list of constituents reported is available on the FDA website (https://www.fda.gov/media/110668/download). The impact of these constituents at the levels found in IQOS emissions on the overall toxicity or harm of this product should be assessed independently, as they may increase the overall risk for disease due to use of IQOS.

Fig. 2.1. Constituents reported by PMI to be at levels at least 200% higher in IQOS aerosol than in the smoke of 3R4F cigarettes

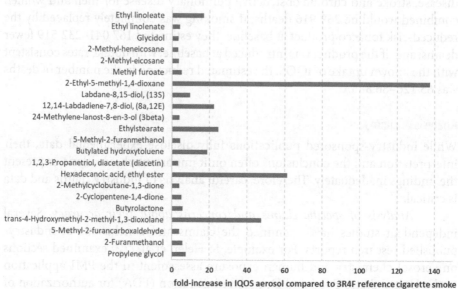

Data and chemical compound names as listed by PMI in its application to the FDA (https://www.fda.gov/media/110668/download, accessed 10 January 2021).

Data on the toxicity of IQOS in vitro and in vivo included by PMI in its MRTP application have also been evaluated. Moazed et al. *(106)* reviewed the pulmonary and immune toxicity of IQOS in studies in laboratory animals and humans. The data in the MRTP application provide evidence of pulmonary inflammation and immunomodulation after use of IQOS and no evidence of improvement in pulmonary inflammation or pulmonary function in cigarette smokers who switched to IQOS. Another group assessed the possible hepatotoxicity of IQOS *(107)* in PMI preclinical studies *(108)* and in studies of human IQOS use in PMI's MRTP application. They found that exposure to HTP increased liver weights, serum alanine aminotransferase activity and hepatocellular vacuolization in experimental animals, whereas these effects were not seen after exposure to conventional cigarettes. Clinical data showed increased alanine aminotransferase activity and plasma bilirubin in smokers who switched to IQOS, indicating hepatocellular injury. The potential impact of passive exposure to IQOS aerosol in non-users has not been considered.

Reduced exposure vs reduced risk. Assessment of the published studies on biomarkers after HTP use tend to support the claims that these products reduce exposure to many harmful constituents. For example, Drovandi et al. *(109)* conducted a meta-analysis of trials published between 2010 and 2019 to compare the levels of biomarkers of exposure with use of conventional combusted cigarettes and various HTP devices. The authors identified 10 non-blinded, randomized controlled trials involving a total of 1766 participants, all conducted by industry researchers. The analysis showed that the levels of 12 biomarkers of exposure to toxicants were significantly lower in participants assigned to an HTP than in those who smoked conventional cigarettes. In comparison with abstinence from smoking, the levels of eight biomarkers were similar and those of four biomarkers were significantly elevated in people who switched to HTPs. An analysis of the results for 24 biomarkers of potential harm in PMI studies of smokers who switched to IQOS or prototype HTPs *(95)* found no statistically detectable difference between users of IQOS and conventional cigarettes for the majority of the biomarkers, suggesting that reductions in exposure to tobacco constituents do not necessarily result in proportional reductions in risk for disease.

Analysis of the strategies used by the industry to support their claims of reduced risk. The overall approach of the industry to assessing the public health impact of HTPs, particularly PMI's modelling with regard to IQOS, has also been analysed. Max et al. *(110)* reviewed the "population health impact model" used by PMI in its application to the FDA to market IQOS as an MRTP and compared it with the FDA guidelines for MRTP applications; more general criteria for evaluating reduced-risk tobacco products were also considered. They found that the model addressed the impact of IQOS on mortality from four tobacco-attributable diseases but not morbidity, that it underestimates mortality rates, does not apply to tobacco products other than cigarettes, does not include FDA-recommended impacts on non-users and underestimates the impact on other population groups. Thus, the industry model underestimates the health impact of IQOS. Although even an improved model will have to rely on industry data and on a number of assumptions, the health outcomes and/or surrogate measures of health effects associated with HTP use must be assessed by independent researchers to provide regulatory agencies with information on the impact of HTPs on public health.

A review was reported of previously secret internal PMI documents on Accord, a precursor product similar to IQOS, with regard to public communications and the application for IQOS as an MRTP *(111)*. Like IQOS, Accord was marketed as a product that reduced users' exposure to harmful tobacco constituents; however, PMI consistently emphasized that such reductions did not render Accord safer. The review concluded that claims that use of IQOS reduces risks to health are not supported, given the similarity of the two products and the absence of consistent reductions in toxic emissions from IQOS aerosols as compared with Accord.

Even when there is no claim of reduced risk, however, a claim of reduced exposure may suggest the safety of currently marketed HTPs. A review *(112)* was conducted of PMI's qualitative and quantitative studies of perceptions of the claims of reduced risk submitted to the FDA in the MRTP application, which found that adult consumers perceived claims of reduced exposure as claims of reduced risk.

2.6 Summary and implications for public health

Although HTPs are not a new concept, the active marketing and uptake by users of the newer generation of these products around the world is relatively recent. Most of the information on HTP contents and emissions and the exposure and effects in HTP users was generated and published by the industry, although some more recent studies have been published by independent research groups. The current literature shows some agreement and some inconsistencies among publications. The industry-published reports tend to be biased towards favourable conclusions about the benefits of switching to HTPs, even when they contain clear data in tables and figures that do not fully support those conclusions.

2.6.1 Summary of data

The main conclusions from this review of publications on HTP toxicant emissions, exposures and effects in model systems and humans, and implications for health are outlined below.

Toxicant emissions and comparison with other tobacco products: The machine puffing regimens that have been used to test HTP emissions are based mainly on those used for conventional cigarettes, such as ISO and HCI. There may, however be important differences in how HTPs and conventional cigarettes are puffed by human users, which adds complexity to the known limitations of interpreting data generated by smoking machines. Nevertheless, in the absence of other data, approximate comparisons can be made of the emissions of HTPs and of conventional and e-cigarettes. Most publications, including non-industry studies, show that the levels of nicotine in HTPs and conventional cigarettes (on per-stick basis) are comparable. The levels of many harmful constituents that derive from the combustion process are consistently reported to be significantly lower in HTP aerosol than in conventional cigarette smoke. These include CO, PAH, some carbonyl compounds (formaldehyde, acetaldehyde) and other volatile toxicants, as well as components such as black carbon, nitrogen oxide and ammonia. The levels of TSNAs are also lower in HTP aerosols than in cigarette smoke. Some reports, however, indicate that the levels of other constituents, such as pyridine, dimethyl trisulfide, acetoin and methylglyoxal, may be comparable to or higher than those in the smoke of conventional cigarettes, and the levels of toxicants such as TSNA, CO, benzo[a]pyrene and carbonyls are higher in emissions of HTPs than in e-cigarettes.

Effects of HTPs in vitro: Studies in cell cultures can provide important mechanistic insights into any acute or chronic harmful effects associated with HTP use. Industry-published studies generally claim reduced cytotoxicity and mutagenicity and lower levels of a range of toxicological and inflammatory biomarkers after exposure in vitro to HTP aerosols as compared with conventional cigarette smoke. Increasing smoking intensity, however, results in substantial increases in these effects. Furthermore, more nicotine is delivered into cells exposed to HTPs than into those exposed to smoke from reference cigarettes. Both industry and independent publications show that cytotoxicity, mutagenicity and expression of certain RNAs are higher after exposure to HTP aerosol than after exposure to air. Differences in outcome measures have also been seen among HTPs, such as between IQOS and glo™.

Effects of HTPs in experimental animals: Industry studies of dermal tumorigenicity and acute and chronic inhalation toxicity in rodents reported that animals treated with HTP aerosol had lower tumour incidence and multiplicity, fewer inflammatory and cellular stress responses and fewer histological changes than animals treated with conventional cigarette smoke. Analysis of these publications, however, reveals dose–response relations for many of these effects and consistently greater responses in animals treated with HTPs than in air controls. Limited data suggest that exposure to HTPs delivers more nicotine than conventional cigarette smoke, consistent with the in-vitro results.

Exposure and effects in smokers who switch to HTPs and comparison with use of e-cigarettes or abstinence from tobacco: Industry publications report reductions in biomarkers of exposure to certain constituents, less urinary mutagenicity and reduction in some biomarkers of effect in smokers who switch to HTPs. Examination of the publications, however, shows substantially higher levels of biomarkers of exposure than in groups assigned to stop smoking and not use any product (Table 2.2). In addition, the levels of biomarkers of many cardiovascular and other diseases did not decrease and in some cases increased (CC16, alanine aminotransferase activity, plasma bilirubin) after a switch to HTPs over baseline levels. This suggests that HTPs have similar or greater cardiovascular toxicity than conventional cigarettes. Lastly, HTP consumption and nicotine intake clearly increased over time in the switching studies, suggesting increasing exposures to other aerosol constituents.

Passive exposure to HTPs and comparison with other tobacco products or clean air: Research on passive exposure to HTP aerosol has been limited. The results to date suggest that use of HTPs may expose bystanders to certain constituents at levels lower than with passive exposure to conventional cigarette smoke but at higher levels than exposure to clean air or e-cigarette aerosol.

2.6.2 Implications for public health

Real-world exposure and its effects on HTP users are not well characterized, and it is impossible at this time to accurately evaluate the long-term health outcomes in users who switch completely to HTPs or use them in combination with conventional cigarettes, e-cigarettes or other tobacco- or nicotine-containing products. Careful analysis of data from industry and academic studies calls attention to the following potential public health consequences of HTP use.

- **Addictive potential**. HTPs may be more efficient in delivering nicotine to users than other tobacco products, including conventional cigarettes. Therefore, in the absence of clear understanding of the health consequences of HTP use, the potential for addiction and subsequent long-term use of these products by various population subgroups, including young people and adults with comorbid conditions, HTPs are a public health concern.
- **Significant differences in exposure of users to toxicants**. Increasing the intensity of HTP puffing dramatically increases the yields of toxicants and the cytotoxic and mutagenic effects of HTP emissions in a dose–response manner. There could therefore be significant variation in toxic exposures and subsequent risks among individuals who use the same HTP, depending on the product type and use topography.
- **No reduction in the chronic disease burden among smokers who switch to HTPs**. Some smokers may choose to switch to HTPs to reduce harmful exposure without quitting tobacco use. As summarized above, the data indicate no improvement in several pulmonary and cardiovascular indicators and a high prevalence of dual use (with smoking) in participants in switching studies. Therefore, uptake of HTPs by smokers may not significantly reduce the prevalence of smoking-associated chronic diseases.
- **Increased risk of chronic diseases in non-smokers who initiate HTP use**. Studies conducted to date consistently show higher exposure and more effects with HTPs than with no exposure (such as sham controls in experimental studies and smoking abstinence in human trials). Uptake of HTPs by non-users of any tobacco product will therefore increase their risk for adverse outcomes such as respiratory, cardiovascular and potentially other diseases.
- **Unknown or unique toxic effects of HTPs**. Most research on HTP aerosols has been limited to analyses of key combustion and tobacco-specific constituents and comparisons with conventional cigarette smoke. HTP aerosols may contain unique harmful constituents that have not yet been identified or well characterized, as suggested by

indications of hepatocellular injury in response to HTPs but not to conventional cigarettes in experimental and clinical trials.
- **Second-hand exposure of non-users.** Exposure to particulates, nicotine and other components of HTP aerosols may pose risks to non-users.
- **Perceptions of safety among users.** Claims of reduced exposure from HTPs may be perceived by users and non-users of other tobacco products as claims of reduced risk, which may lead to uptake of HTPs by individuals who would otherwise quit smoking or not initiate tobacco use.

Lastly, there is sufficient variation among HTP brands to suggest that the technologies used in HTPs will continue to evolve. This may include changes in the chemical composition of aerosols produced by HTPs and in subsequent exposure and effects in users. Therefore, the limited data available on currently marketed HTPs, much of which was generated by industry, may not be directly applicable in the future.

2.7 Research gaps and priorities

This review of the toxicants in HTPs and reports on in vitro and in vivo toxicity and levels of exposure and effects in humans indicates the following areas for research:

- sound, more rigorous, innovative laboratory assessment of HTP emissions, such as non-targeted analyses to identify toxicants that may be responsible for the lack of improvement in biomarkers of potential harm to smokers who switch to HTPs;
- better characterization of human topography to understand how HTPs are used in the real world (as opposed to clinical confinement trials), including potential increases in product consumption over time and how they affect addictiveness, harmful exposures and health outcomes;
- better characterization of nicotine uptake by HTP users and the potential impact on abuse liability;
- use of toxicological models (e.g. cells or laboratory animals) to better understand nicotine absorption in cells and tissues and to predict unique or unknown toxicity of HTPs;
- identification of biomarkers of exposure and effect that are specific to HTPs to facilitate research on the amount of HTP use, associated exposure and potential health effects;

- the impact of dual or poly use of HTPs with other tobacco products, particularly conventional cigarettes;
- the effects of second-hand exposure on non-users, particularly vulnerable populations (those with pre-existing medical conditions, children, pregnant women); and
- monitoring of health outcomes in longitudinal population-based cohorts of HTP users and exposed non-users in various countries and comparison with non-users of any tobacco product.

2.8 Policy recommendations

The following recommendations are proposed for consideration by policy-makers, researchers and the public health community, as appropriate.

- Thoroughly examine industry research data on HTPs to ensure accurate interpretation.
- Prioritize and support independent research of the public health impact of HTPs.
- Develop standards for HTP surveillance and research, such as product-related terminology, standard testing procedures for HTP emissions, including puffing regimens, and HTP-specific reference products that could be used as quality controls.
- Conduct surveillance of the prevalence of use of the available HTPs and the potential associations with health outcomes, by country.
- Until further, independent evidence on the public health impact of HTPs is available, prohibit all manufacturers and associated groups from making claims about reduced harm as compared with other products, including advertisements and modifications to health warnings, and from portraying HTPs as appropriate for cessation of use of any tobacco product; and prohibit use of HTPs in public spaces, as they are tobacco products.
- Policy-makers should clearly communicate to the public that there is currently no evidence that HTPs reduce the risks associated with tobacco products.

2.9 References

1. Forster M, Fiebelkorn S, Yurteri C, Mariner D, Liu C, Wright C et al. Assessment of novel tobacco heating product THP1.0. Part 3: Comprehensive chemical characterisation of harmful and potentially harmful aerosol emissions. Regul Toxicol Pharmacol. 2018;93:14–33.
2. Takahashi Y, Kanemaru Y, Fukushima T, Eguchi K, Yoshida S, Miller-Holt J et al. Chemical analysis

and in vitro toxicological evaluation of aerosol from a novel tobacco vapor product: A comparison with cigarette smoke. Regul Toxicol Pharmacol. 2018;92:94–103.
3. Li X, Luo Y, Jiang X, Zhang H, Zhu F, Hu S et al. Chemical analysis and simulated pyrolysis of tobacco heating system 2.2 compared to conventional cigarettes. Nicotine Tob Res. 2019;21(1):111–8.
4. Gasparyan H, Mariner D, Wright C, Nicol J, Murphy J, Liu C et al. Accurate measurement of main aerosol constituents from heated tobacco products (HTPs): Implications for a fundamentally different aerosol. Regul Toxicol Pharmacol. 2018;99:131–41.
5. Uchiyama S, Noguchi M, Takagi N, Hayashida H, Inaba Y, Ogura H et al. Simple determination of gaseous and particulate compounds generated from heated tobacco products. Chem Res Toxicol. 2018;31(7):585–93.
6. Uchiyama S, Noguchi M, Sato A, Ishitsuka M, Inaba Y, Kunugita N. Determination of thermal decomposition products generated from e-cigarettes. Chem Res Toxicol. 2020;33(2):576–83.
7. Hukkanen J, Jacob P III, Benowitz NL. Metabolism and disposition kinetics of nicotine. Pharmacol Rev. 2005;57(1):79–115.
8. Bekki K, Inaba Y, Uchiyama S, Kunugita N. Comparison of chemicals in mainstream smoke in heat-not-burn tobacco and combustion cigarettes. J UOEH. 2017;39(3):201–7.
9. Farsalinos KE, Yannovits N, Sarri T, Voudris V, Poulas K. Nicotine delivery to the aerosol of a heat-not-burn tobacco product: Comparison with a tobacco cigarette and e-cigarettes. Nicotine Tob Res. 2018;20(8):1004–9.
10. Liu S, Li J, Zhu L, Liu W. Quantitative determination and puff-by-puff analysis on the release pattern of nicotine and menthol in heat-not-burn tobacco material and emissions. Paris: Cooperation Centre for Scientific Research Relative to Tobacco; 2020.
11. Auer R, Concha-Lozano N, Jacot-Sadowski I, Cornuz J, Berthet A. Heat-not-burn tobacco cigarettes: Smoke by any other name. JAMA Intern Med. 2017;177(7):1050–2.
12. St Helen G, Jacob P III, Nardone N, Benowitz NL. IQOS: examination of Philip Morris International's claim of reduced exposure. Tob Control. 2018;27(Suppl 1):s30–6.
13. Mallock N, Pieper E, Hutzler C, Henkler-Stephani F, Luch A. Heated tobacco products: a review of current knowledge and initial assessments. Front Public Health. 2019;7:287.
14. Counts ME, Morton MJ, Laffoon SW, Cox RH, Lipowicz PJ. Smoke composition and predicting relationships for international commercial cigarettes smoked with three machine-smoking conditions. Regul Toxicol Pharmacol. 2005;41(3):185–227.
15. Standard operating procedure for determination of nicotine and carbon monoxide in mainstream cigarette smoke under intense smoking conditions, WHO TobLabNet SOP10. Geneva: World Health Organization; 2016.
16. Standard operating procedure for determination of nicotine in cigarette tobacco filler, TobLabNet SOP 4. Geneva: World Health Organization; 2014.
17. Simonavicius E, McNeill A, Shahab L, Brose LS. Heat-not-burn tobacco products: a systematic literature review. Tob Control. 2019;28(5):582–94.
18. Jaccard G, Tafin Djoko D, Moennikes O, Jeannet C, Kondylis A, Belushkin M. Comparative assessment of HPHC yields in the tobacco heating system THS2.2 and commercial cigarettes. Regul Toxicol Pharmacol. 2017;90:1–8.
19. Meišutovič-Akhtarieva M, Prasauskas T, Čiužas D, Krugly E, Keratayté K, Martuzevičius D et al. Impacts of exhaled aerosol from the usage of the tobacco heating system to indoor air quality: A chamber study. Chemosphere. 2019;223:474–82.
20. Schaller JP, Keller D, Poget L, Pratte P, Kaelin E, McHugh D et al. Evaluation of the tobacco heating system 2.2. Part 2: Chemical composition, genotoxicity, cytotoxicity, and physical properties of the aerosol. Regul Toxicol Pharmacol. 2016;81(Suppl 2):S27–47.
21. Shein M, Jeschke G. Comparison of free radical levels in the aerosol from conventional ci-

garettes, electronic cigarettes, and heat-not-burn tobacco products. Chem Res Toxicol. 2019;32(6):1289–98.
22. Smith MR, Clark B, Lüdicke F, Schaller JP, Vanscheeuwijck P, Hoeing J et al. Evaluation of the tobacco heating system 2.2. Part 1: Description of the system and the scientific assessment program. Regul Toxicol Pharmacol. 2016;81(Suppl 2):S17–26.
23. Crooks I, Neilson L, Scott K, Reynolds L, Oke T, Meredith C et al. Evaluation of flavourings potentially used in a heated tobacco product: Chemical analysis, in vitro mutagenicity, genotoxicity, cytotoxicity and in vitro tumour promoting activity. Food Chem Toxicol. 2018;118:940–52.
24. Davis B, Williams M, Talbot P. iQOS: evidence of pyrolysis and release of a toxicant from plastic. Tob Control. 2019;28(1):34–41.
25. Forster M, Liu C, Duke MG, McAdam KG, Proctor CJ. An experimental method to study emissions from heated tobacco between 100–200°C. Chem Cent J. 2015;9:20.
26. Cancelada L, Sleiman M, Tang X, Russell ML, Montesinos VN, Litter MI et al. Heated tobacco products: volatile emissions and their predicted impact on indoor air quality. Environ Sci Technol. 2019;53(13):7866–76.
27. Jaccard G, Kondylis A, Gunduz I, Pijnenburg J, Belushkin M. Investigation and comparison of the transfer of TSNA from tobacco to cigarette mainstream smoke and to the aerosol of a heated tobacco product, THS2.2. Regul Toxicol Pharmacol. 2018;97:103–9.
28. Ratajczak A, Jankowski P, Strus P, Feleszko W. Heat not burn tobacco product – a new global trend: Impact of heat-not-burn tobacco products on public health, a systematic review. Int J Environ Res Public Health. 2020;17(2):409.
29. Konstantinou E, Fotopoulou F, Drosos A, Dimakopoulou N, Zagoriti Z, Niarchos A et al. Tobacco-specific nitrosamines: A literature review. Food Chem Toxicol. 2018;118:198–203.
30. Caponnetto P, Maglia M, Prosperini G, Busà B, Polosa R. Carbon monoxide levels after inhalation from new generation heated tobacco products. Respir Res. 2018;19(1):164.
31. Ishizaki A, Kataoka H. A sensitive method for the determination of tobacco-specific nitrosamines in mainstream and sidestream smokes of combustion cigarettes and heated tobacco products by online in-tube solid-phase microextraction coupled with liquid chromatography-tandem mass spectrometry. Anal Chim Acta. 2019;1075:98–105.
32. Glantz SA. Heated tobacco products: the example of IQOS. Tob Control. 2018;27(Suppl 1):s1–6.
33. Leigh NJ, Palumbo MN, Marino AM, O'Connor RJ, Goniewicz ML. Tobacco-specific nitrosamines (TSNA) in heated tobacco product IQOS. Tob Control. 2018;27(Suppl 1):s37–8.
34. Ogunwale MA, Li M, Ramakrishnam Raju MV, Chen Y, Nantz MN, Conklin DJ et al. Aldehyde detection in electronic cigarette aerosols. ACS Omega. 2017;2(3):1207–14.
35. Conklin DJ, Ogunwale MA, Chen Y, Theis WS, Nantz MH, Fu XA et al. Electronic cigarette-generated aldehydes: The contribution of e-liquid components to their formation and the use of urinary aldehyde metabolites as biomarkers of exposure. Aerosol Sci Technol. 2018;52(11):1219–32.
36. Kaunelienė V, Meišutovič-Akhtarieva M, Martuzevičius D. A review of the impacts of tobacco heating system on indoor air quality versus conventional pollution sources. Chemosphere. 2018;206:568–78.
37. Savareear B, Escobar-Arnanz J, Brokl M, Saxton MJ, Wright C, Liu C et al. Comprehensive comparative compositional study of the vapour phase of cigarette mainstream tobacco smoke and tobacco heating product aerosol. J Chromatogr A. 2018;1581–2.
38. Farsalinos KE, Yannovits N, Sarri T, Voudris V, Poulas K, Leischow SJ. Carbonyl emissions from a novel heated tobacco product (IQOS): comparison with an e-cigarette and a tobacco cigarette. Addiction. 2018;113(11):2099–106.
39. Ruprecht AA, Marco CD, Saffari A, Pozzi P, Mazza R, Veronese C et al. Environmental pollution and emission factors of electronic cigarettes, heat-not-burn tobacco products, and conventional cigarettes. Aerosol Sci Technol. 2017;51(6):674–84.

40. Mallock N, Böss L, Burk R, Danziger M, Welsch T, Hahn H et al. Levels of selected analytes in the emissions of "heat not burn" tobacco products that are relevant to assess human health risks. Arch Toxicol. 2018;92(6):2145–9.
41. Salman R, Talih S, El-Hage R, Haddad C, Karaoghlanian N, El-Hellani A et al. Free-base and total nicotine, reactive oxygen species, and carbonyl emissions From IQOS, a heated tobacco product. Nicotine Tob Res. 2019;21(9):1285–8.
42. De Marco C, Borgini A, Ruprecht AA, Veronese C, Mazza R, Bertoldi M et al. La formaldeide nelle sigarette elettroniche e nei riscaldatori di tabacco (HnB): facciamo il punto. [Formaldehyde in electronic cigarettes and in heat-not-burn products (HnB): let's make the point]. Epidemiol Prev. 2018;42(5–6):351–5.
43. Savareear B, Escobar-Arnanz J, Brokl M, Saxton MJ, Wright C, Liu C et al. Non-targeted analysis of the particulate phase of heated tobacco product aerosol and cigarette mainstream tobacco smoke by thermal desorption comprehensive two-dimensional gas chromatography with dual flame ionisation and mass spectrometric detection. J Chromatogr A. 2019;1603:327–37.
44. Tricker AR, Schorp MK, Urban HJ, Leyden D, Hagedorn HW, Engl J et al. Comparison of environmental tobacco smoke (ETS) concentrations generated by an electrically heated cigarette smoking system and a conventional cigarette. Inhal Toxicol. 2009;21(1):62–77.
45. Mitova MI, Campelos PB, Goujon-Ginglinger CG, Maeder S, Mottier N, Rouget EJR et al. Comparison of the impact of the tobacco heating system 2.2 and a cigarette on indoor air quality. Regul Toxicol Pharmacol. 2016;80:91–101.
46. Jankowski M, Brożek GM, Lawson J, Skoczyński S, Majek P, Zejda JE. New ideas, old problems? Heated tobacco products – a systematic review. Int J Occup Med Environ Health. 2019;32(5):595–634.
47. Hitosugi M, Tojo M, Kane M, Shiomi N, Shimizu T, Nomiyama T. Criminal mercury vapor poisoning using heated tobacco product. Int J Legal Med. 2019;133(2):479–81.
48. A study on production and delivery of volatile aroma substances from heat-not-burn products' tobacco material to emissions. Paris: Cooperation Centre for Scientific Research Relative to Tobacco; 2020.
49. Pacitto A, Stabile L, Scungio M, Rizza V, Buonanno G. Characterization of airborne particles emitted by an electrically heated tobacco smoking system. Environ Pollut. 2018;240:248–54.
50. Górski P. e-Cigarettes or heat-not-burn tobacco products – advantages or disadvantages for the lungs of smokers. Adv Resp Med. 2019;87(2):123–34.
51. Roemer E, Stabbert R, Veltel D, Müller BP, Meisgen TJ, Schramke H et al. Reduced toxicological activity of cigarette smoke by the addition of ammonium magnesium phosphate to the paper of an electrically heated cigarette: smoke chemistry and in vitro cytotoxicity and genotoxicity. Toxicol In Vitro. 2008;22(3):671–81.
52. Tewes FJ, Meisgen TJ, Veltel DJ, Roemer E, Patskan G. Toxicological evaluation of an electrically heated cigarette. Part 3: Genotoxicity and cytotoxicity of mainstream smoke. J Appl Toxicol. 2003;23(5):341–8.
53. Werley MS, Freelin SA, Wrenn SE, Gerstenberg B, Roemer E, Schramke H et al. Smoke chemistry, in vitro and in vivo toxicology evaluations of the electrically heated cigarette smoking system series K. Regul Toxicol Pharmacol. 2008;52(2):122–39.
54. Zenzen V, Diekmann J, Gerstenberg B, Weber S, Wittke S, Schorp MK. Reduced exposure evaluation of an electrically heated cigarette smoking system. Part 2: Smoke chemistry and in vitro toxicological evaluation using smoking regimens reflecting human puffing behavior. Regul Toxicol Pharmacol. 2012;64(2 Suppl):S11–34.
55. van der Toorn M, Koshibu K, Schlage WK, Majeed S, Pospisil P, Hoeng J et al. Comparison of monoamine oxidase inhibition by cigarettes and modified risk tobacco products. Toxicol Rep. 2019;6:1206–15.

56. Iskandar AR, Mathis C, Schlage WK, Frentzel S, Leroy P, Xiang Y et al. A systems toxicology approach for comparative assessment: Biological impact of an aerosol from a candidate modified risk tobacco product and cigarette smoke on human organotypic bronchial epithelial cultures. Toxicol In Vitro. 2017;39:29–51.
57. Zanetti F, Sewer A, Scotti E, Titz B, Schlage WK, Leroy P et al. Assessment of the impact of aerosol from a potential modified risk tobacco product compared with cigarette smoke on human organotypic oral epithelial cultures under different exposure regimens. Food Chem Toxicol. 2018;115:148–69.
58. Iskandar AR, Martin F, Leroy P, Schlage WK, Mathis C, Titz B et al. Comparative biological impacts of an aerosol from carbon-heated tobacco and smoke from cigarettes on human respiratory epithelial cultures: A systems toxicology assessment. Food Chem Toxicol. 2018;115:109–26.
59. Poussin C, Laurent A, Kondylis A, Marescotti D, van der Toorn M, Guedj E et al. In vitro systems toxicology-based assessment of the potential modified risk tobacco product CHTP 1.2 for vascular inflammation- and cytotoxicity-associated mechanisms promoting adhesion of monocytic cells to human coronary arterial endothelial cells. Food Chem Toxicol. 2018;120:390–406.
60. Jaunky T, Adamson J, Santopietro S, Terry A, Thorne D, Breheny D et al. Assessment of tobacco heating product THP1.0. Part 5: In vitro dosimetric and cytotoxic assessment. Regul Toxicol Pharmacol. 2018;93:52–61.
61. Haswell LE, Corke S, Verrastro I, Baxter A, Banerjee A, Adamson J et al. In vitro RNA-seq-based toxicogenomics assessment shows reduced biological effect of tobacco heating products when compared to cigarette smoke. Sci Rep. 2018;8(1):1145.
62. Godec TL, Crooks I, Scott K, Meredith C. In vitro mutagenicity of gas-vapour phase extracts from flavoured and unflavoured heated tobacco products. Toxicol Rep. 2019;6:1155–63.
63. Leigh NJ, Tran PL, O'Connor RJ, Goniewicz ML. Cytotoxic effects of heated tobacco products (HTP) on human bronchial epithelial cells. Tob Control. 2018;27(Suppl 1):s26–9.
64. Davis B, To V, Talbot P. Comparison of cytotoxicity of IQOS aerosols to smoke from Marlboro Red and 3R4F reference cigarettes. Toxicol In Vitro. 2019;61:104652.
65. Zagoriti Z, El Mubarak MA, Farsalinos K, Topouzis S. Effects of exposure to tobacco cigarette, electronic cigarette and heated tobacco product on adipocyte survival and differentiation in vitro. Toxics. 2020;8(1):9.
66. Tsuji H, Okubo C, Fujimoto H, Fukuda I, Nishino T, Lee KM et al. Comparison of dermal tumor promotion activity of cigarette smoke condensate from prototype (heated) cigarette and reference (combusted) cigarette in SENCAR mice. Food Chem Toxicol. 2014;72:187–94.
67. Fujimoto H, Tsuji H, Okubo C, Fukuda I, Nishino T, Lee KM et al. Biological responses in rats exposed to mainstream smoke from a heated cigarette compared to a conventional reference cigarette. Inhal Toxicol. 2015;27(4):224–36.
68. Phillips BW, Schlage WK, Titz B, Kogel U, Sciuscio D, Martin F et al. A 90-day OECD TG 413 rat inhalation study with systems toxicology endpoints demonstrates reduced exposure effects of the aerosol from the carbon heated tobacco product version 1.2 (CHTP1.2) compared with cigarette smoke. I. Inhalation exposure, clinical pathology and histopathology. Food Chem Toxicol. 2018;116(Pt B):388–413.
69. Titz B, Kogel U, Martin F, Kogel U, Sciuscio D, Martin F et al. A 90-day OECD TG 413 rat inhalation study with systems toxicology endpoints demonstrates reduced exposure effects of the aerosol from the carbon heated tobacco product version 1.2 (CHTP1.2) compared with cigarette smoke. II. Systems toxicology assessment. Food Chem Toxicol. 2018;115:284–301.
70. Phillips B, Szostak J, Titz B, Schlage WK, Guedj E, Leroy P et al. A six-month systems toxicology inhalation/cessation study in ApoE-/- mice to investigate cardiovascular and respiratory exposure effects of modified risk tobacco products, CHTP 1.2 and THS 2.2, compared with conventional cigarettes. Food Chem Toxicol. 2019;126:113–41.

71. Szostak J, Titz B, Schlage WK, Guedj E, Sewer A, Phillips B et al. Structural, functional, and molecular impact on the cardiovascular system in ApoE-/- mice exposed to aerosol from candidate modified risk tobacco products, carbon heated tobacco product 1.2 and tobacco heating system 2.2, compared with cigarette smoke. Chem Biol Interact. 2020;315:108887.
72. Nabavizadeh P, Liu J, Havel CM, Ibrahim S, Derakhshandeh R, Jacob P III et al. Vascular endothelial function is impaired by aerosol from a single IQOS HeatStick to the same extent as by cigarette smoke. Tob Control. 2018;27(Suppl 1):s13–9.
73. McAdam K, Davis P, Ashmore L, Eaton D, Jakaj B, Eldridge A et al. Influence of machine-based puffing parameters on aerosol and smoke emissions from next generation nicotine inhalation products. Regul Toxicol Pharmacol. 2019;101:156–65.
74. Urban HJ, Tricker AR, Leyden DE, Forte N, Zenzen V, Fueursenger A et al. Reduced exposure evaluation of an electrically heated cigarette smoking system. Part 8: Nicotine bridging – estimating smoke constituent exposure by their relationships to both nicotine levels in mainstream cigarette smoke and in smokers. Regul Toxicol Pharmacol. 2012;64(2 Suppl):S85–97.
75. Patskan G, Reininghaus W. Toxicological evaluation of an electrically heated cigarette. Part 1: Overview of technical concepts and summary of findings. J Appl Toxicol. 2003;23(5):323–8.
76. Tricker AR, Stewart AJ, Martin Leroy C, Lindner D, Schorp MK, Dempsey R. Reduced exposure evaluation of an electrically heated cigarette smoking system. Part 3: Eight-day randomized clinical trial in the UK. Regul Toxicol Pharmacol. 2012;64(2 Suppl):S35–44.
77. Tricker AR, Jang IJ, Martin Leroy C, Lindner D, Dempsey R. Reduced exposure evaluation of an electrically heated cigarette smoking system. Part 4: Eight-day randomized clinical trial in Korea. Regul Toxicol Pharmacol. 2012;64(2 Suppl):S45–53.
78. Tricker AR, Kanada S, Takada K, Martin Leroy C, Lindner D, Schorp MK et al. Reduced exposure evaluation of an electrically heated cigarette smoking system. Part 5: 8-Day randomized clinical trial in Japan. Regul Toxicol Pharmacol. 2012;64(2 Suppl):S54–63.
79. Tricker AR, Kanada S, Takada K, Martin Leroy C, Lindner D, Schorp MK et al. Reduced exposure evaluation of an electrically heated cigarette smoking system. Part 6: 6-Day randomized clinical trial of a menthol cigarette in Japan. Regul Toxicol Pharmacol. 2012;64(2 Suppl):S64–73.
80. Martin Leroy C, Jarus-Dziedzic K, Ancerewicz J, Lindner D, Kulesza A, Magnette J. Reduced exposure evaluation of an electrically heated cigarette smoking system. Part 7: A one-month, randomized, ambulatory, controlled clinical study in Poland. Regul Toxicol Pharmacol. 2012;64(2 Suppl):S74–84.
81. Lüdicke F, Haziza C, Weitkunat R, Magnette J. Evaluation of biomarkers of exposure in smokers switching to a carbon-heated tobacco product: A controlled, randomized, open-label 5-day exposure study. Nicotine Tob Res. 2016;18(7):1606–13.
82. Lüdicke F, Baker G, Magnette J, Picavet P, Weitkunat R. Reduced exposure to harmful and potentially harmful smoke constituents with the tobacco heating system 2.1. Nicotine Tob Res. 2017;19(2):168–75.
83. Haziza C, de La Bourdonnaye G, Skiada D, Ancerewicz J, Baker G, Picavet P et al. Evaluation of the tobacco heating system 2.2. Part 8: 5-Day randomized reduced exposure clinical study in Poland. Regul Toxicol Pharmacol. 2016;81 Suppl 2:S139–50.
84. Lüdicke F, Ansari SM, Lama N, Blanc N, Bosilkova M, Donelli A et al. Effects of switching to a heat-not-burn tobacco product on biologically relevant biomarkers to assess a candidate modified risk tobacco product: a randomized trial. Cancer Epidemiol Biomarkers Prev. 2019;28(11):1934–43.
85. Gale N, McEwan M, Eldridge AC, Fearon IM, Sherwood N, Bowen E et al. Changes in biomarkers of exposure on switching from a conventional cigarette to tobacco heating products: a randomized, controlled study in healthy Japanese subjects. Nicotine Tob Res. 2019;21(9):1220–7.
86. Newland N, Lowe FJ, Camacho OM, McEwan M, Gale N, Ebajemito J et al. Evaluating the effects

of switching from cigarette smoking to using a heated tobacco product on health effect indicators in healthy subjects: study protocol for a randomized controlled trial. Intern Emerg Med. 2019;14(6):885–98.
87. Sakaguchi C, Kakehi A, Minami N, Kikuchi A, Futamura Y. Exposure evaluation of adult male Japanese smokers switched to a heated cigarette in a controlled clinical setting. Regul Toxicol Pharmacol. 2014;69(3):338–47.
88. Adriaens K, Gucht DV, Baeyens F. IQOSTM vs. e-cigarette vs. tobacco cigarette: A direct comparison of short-term effects after overnight-abstinence. Int J Environ Res Public Health. 2018;15(12):2902.
89. Maloney S, Eversole A, Crabtree M, Soule E, Eissenberg T, Breland A. Acute effects of JUUL and IQOS in cigarette smokers. Tob Control. 2020: doi: 10.1136/tobaccocontrol-2019-055475.
90. Protano C, Manigrasso M, Avino P, Sernia S, Vitali M. Second-hand smoke exposure generated by new electronic devices (IQOS® and e-cigs) and traditional cigarettes: submicron particle behaviour in human respiratory system. Ann Ig. 2016;28(2):109–12.
91. Protano C, Manigrasso M, Avino P, Vitali M. Second-hand smoke generated by combustion and electronic smoking devices used in real scenarios: Ultrafine particle pollution and age-related dose assessment. Environ Int. 2017;107:190–5.
92. Schober W, Fembacher L, Frenzen A, Fromme H. Passive exposure to pollutants from conventional cigarettes and new electronic smoking devices (IQOS, e-cigarette) in passenger cars. Int J Hyg Environ Health. 2019;222(3):486–93.
93. Sutanto E, Smith DM, Miller C, O'Connor RJ, Hyland A, Tabuchi T et al. Use of heated tobacco products within indoor spaces: Findings from the 2018 ITC Japan survey. Int J Environ Res Public Health. 2019;16(23):4862.
94. Tabuchi T, Gallus S, Shinozaki T, Nakaya T, Kunugita N, Colwell B. Heat-not-burn tobacco product use in Japan: its prevalence, predictors and perceived symptoms from exposure to second-hand heat-not-burn tobacco aerosol. Tob Control. 2018;27(e1):e25–33.
95. Glantz SA. PMI's own in vivo clinical data on biomarkers of potential harm in Americans show that IQOS is not detectably different from conventional cigarettes. Tob Control. 2018;27(Suppl 1):s9–12.
96. Polosa R, Morjaria JB, Caponnetto P, Prosperini U, Russo C, Pennisi A et al. Evidence for harm reduction in COPD smokers who switch to electronic cigarettes. Respir Res. 2016;17(1):166.
97. Bhatta DN, Glantz SA. Association of e-cigarette use with respiratory disease among adults: A longitudinal analysis. Am J Prev Med. 2020;58(2):182–90.
98. Osei AD, Mirbolouk M, Orimoloye OA, Dzaye O, Iftekhar Uddin SM, Benjamin EJ et al. Association between e-cigarette use and chronic obstructive pulmonary disease by smoking status: behavioral risk factor surveillance system 2016 and 2017. Am J Prev Med. 2019;58(3):336–42.
99. Wills TA, Pagano I, Williams RJ, Tam EK. e-Cigarette use and respiratory disorder in an adult sample. Drug Alcohol Depend. 2019;194:363–70.
100. Chung SJ, Kim BK, Oh JH, Shim JS, Chang YS, Cho SH et al. Novel tobacco products including electronic cigarette and heated tobacco products increase risk of allergic rhinitis and asthma in adolescents: Analysis of Korean youth survey. Allergy. 2020;75(7):1640–8.
101. Lee A, Lee SY, Lee KS. The use of heated tobacco products is associated with asthma, allergic rhinitis, and atopic dermatitis in Korean adolescents. Sci Rep. 2019;9(1):17699.
102. Sharman A, Zhussupov B, Sharman D, Kim I, Yerenchina E. Lung function in users of a smoke-free electronic device with HeatSticks (iQOS) versus smokers of conventional cigarettes: Protocol for a longitudinal cohort observational study. JMIR Res Protoc. 2018;7(11):e10006.
103. Hatsukami DK, Slade J, Benowitz NL, Giovino GA, Gritz ER, Leischow S et al. Reducing tobacco harm: research challenges and issues. Nicotine Tob Res. 2002;4 Suppl 2:S89–101.
104. Warner KE. Tobacco harm reduction: promise and perils. Nicotine Tob Res. 2002;4(Suppl 2):S61–71.

105. Lee PN, Djurdjevic S, Weitkunat R, Baker G. Estimating the population health impact of introducing a reduced-risk tobacco product into Japan. The effect of differing assumptions, and some comparisons with the US. Regul Toxicol Pharmacol. 2018;100:92–104.
106. Moazed F, Chun L, Matthay MA, Calfee CS, Gotts J. Assessment of industry data on pulmonary and immunosuppressive effects of IQOS. Tob Control. 2018;27(Suppl 1):s20–5.
107. Chun L, Moazed F, Matthay M, Calfee C, Gotts J. Possible hepatotoxicity of IQOS. Tob Control. 2018;27(Suppl 1):s39–40.
108. Wong ET, Kogel U, Veljkovic E, Martin F, Xiang Y, Boue S et al. Evaluation of the tobacco heating system 2.2. Part 4: 90-day OECD 413 rat inhalation study with systems toxicology endpoints demonstrates reduced exposure effects compared with cigarette smoke. Regul Toxicol Pharmacol. 2016;81(Suppl 2):S59–81.
109. Drovandi A, Salem S, Barker D, Booth D, Kairuz T. Human biomarker exposure from cigarettes versus novel heat-not-burn devices: A systematic review and meta-analysis. Nicotine Tob Res. 2020;22(7):1077–85.
110. Max WB, Sung HY, Lightwood J, Wang Y, Yao T. Modelling the impact of a new tobacco product: review of Philip Morris International's population health impact model as applied to the IQOS heated tobacco product. Tob Control. 2018;27(Suppl 1):s82–6.
111. Elias J, Dutra LM, St Helen G, Ling PM. Revolution or redux? Assessing IQOS through a precursor product. Tob Control. 2018;27(Suppl 1):s102–10.
112. Popova L, Lempert LK, Glantz SA. Light and mild redux: heated tobacco products' reduced exposure claims are likely to be misunderstood as reduced risk claims. Tob Control. 2018;27(Suppl 1):s87–95.

3. The attractiveness and addictive potential of heated tobacco products: effects on perception and use and associated effects

Reinskje Talhout, National Institute for Public Health and the Environment, Centre for Health Protection, Bilthoven, Netherlands

Richard J. O'Connor, Department of Health Behavior, Roswell Park Comprehensive Cancer Center, Buffalo (NY), USA

Contents
Abstract
3.1 Background
3.2 Attractiveness of heated tobacco products (HTPs)
    3.2.1 Definitions of attractiveness in the context of Articles 9 and 10 of the WHO FCTC
    3.2.2 Attractive features of HTPs
    3.2.3 What we can learn from ENDS and ENNDS and relevance to HTPs
3.3 Addictiveness of HTPs
    3.3.1 Addictiveness
    3.3.2 What we can learn from ENDS and ENNDS and relevance to HTPs
    3.3.3 Overall abuse liability of HTPs
3.4 Effects of the attractiveness and addictiveness of ENDS, ENNDS and HTPs on perceptions of risk and harm and use
    3.4.1 Contributions of attractiveness and addictiveness to initiation, switching, complementing and quitting conventional tobacco products
    3.4.2 Learning from ENDS and ENNDS and application to HTPs
3.5 Discussion
    3.5.1 Behavioural implications of different patterns of use among different groups
    3.5.2 Implications for public health
    3.5.3 Research gaps, priorities and questions
    3.5.4 Policy recommendations
3.6 Conclusions
3.7 References
Annex 3.1. Menthol concentrations in IQOS, cigarettes and JUUL products

Abstract

Decision FCTC/COP8(22) requests a report on several aspects of novel and emerging tobacco products, in particular heated tobacco products (HTPs). This paper addresses the aspect of addictive potential, perception and use, attractiveness, potential role in initiating and quitting smoking, marketing including promotional strategies and impacts, and claims of "reduced harm". We reviewed the attractive and addictive features of HTPs and the effects of those features on consumer perception and use. The available literature on HTPs was complemented by information from the wider body of knowledge on e-cigarettes.

We searched the bibliographic database PubMed, with no restriction on time, up to January 2020. Studies on toxicity in users (i.e. toxicants in emissions, in-vitro studies, biomarkers of exposure) and environmental smoke and studies in a language other than English were excluded. We also included studies in the application of Philip Morris International (PMI) to the United States Food and Drug Administration (FDA) for their IQOS modified risk tobacco product (MRTP) were also used.

With regard to features that increase attractiveness, information was found on sensory attributes, ease of use, cost, reputation and image, and assumed risks and benefit. Little is known about how these different features affect consumer perception and use; a recent study reported that six important factors were health, cost, enjoyment and satisfaction, ease of use, use practices and social aspects. With regard to addictiveness, currently marketed HTPs deliver significant levels of nicotine in aerosol, and their pharmacokinetics and physiological and subjective effects are similar to those of contemporary ENDS products, suggesting comparable abuse liability.

HTPs have become popular in some markets, probably due to factors such as marketing as a "clean", modern, elegant, "reduced harm" product. Their sensory properties and ease of use are generally rated lower than those of conventional cigarettes but are directly correlated with their attractiveness, perceived risk and appeal, thus determining their uptake. The history of e-cigarettes shows that any new tobacco and related product that comes onto the market can quickly become popular. Knowledge of e-cigarettes indicates that the factors of concern for HTPs, and therefore potential regulatory targets, are nicotine levels and "throat hit", flavour variety, design of the device, marketing and perception of reduced harm.

Common regulatory principles for e-cigarettes include minimizing product appeal and thus potential uptake by young people, increasing product safety and minimizing false beliefs about health effects. A similar strategy could be followed for HTPs. Policy-makers are advised to monitor the HTP market, communicate the risks to the general public, limit marketing, consider regulating flavours and stimulate research, especially on perceptions and use.

3.1 Background

Since the 1980s, tobacco companies have tried to promote HTPs on the market as "healthier" than conventional cigarettes. Until recently, they failed *(1)*, but HTPs are now increasingly marketed as an alternative to smoking combustible products, primarily cigarettes, although controversy has surrounded the public health context of their marketing and use *(2)*. Since 2014, various new products have been introduced – including Philip Morris International (PMI)'s IQOS, and industry analysts predict that HTPs will absorb 30% of the regular cigarette market in the USA by 2025 *(3)*. Other examples are Ploom TECH from Japan Tobacco International, glo from British American Tobacco and PAX from PAX Labs *(4)*. Production of HTPs is expected to grow quickly *(5)*.

Like e-cigarettes, HTPs are rechargeable battery-powered devices that heat the product; however, the product consumed is tobacco *(6,7)*. According to WHO *(4)*, HTPs

> ... produce aerosols containing nicotine and toxic chemicals when tobacco is heated or when a device containing tobacco is activated. These aerosols are inhaled by users during a process of sucking or smoking involving a device. They contain the highly addictive substance nicotine as well as non-tobacco additives and are often flavoured. The tobacco may be in the form of specially designed cigarettes (e.g. "heat sticks" and "Neo sticks") or pods or plugs.

In decision FCTC/COP8(22) on novel and emerging tobacco products *(8)*, the Conference of the Parties to the WHO FCTC at its eighth session noted the evolution of HTPs, their marketing as "harm reduction" products and the resulting regulatory challenges, with "limited guidance to guide Parties on the classification and regulation of heated tobacco products". Hence, they requested a report to be submitted to the ninth session

> on novel and emerging tobacco products, in particular heated tobacco products, regarding their health impacts including on non-users, *their addictive potential, perception and use, attractiveness, potential role in initiating and quitting smoking, marketing including promotional strategies and impacts, claims of reduced harm* [italics added], variability of products, regulatory experience and monitoring of Parties, impact on tobacco control efforts and research gaps, and to subsequently propose potential policy options ....

The objectives of this paper address the italicized part of the request, as below, by reviewing current literature on:

- the attractive features of HTPs, in light of WHO's definition of attractiveness (section 3.2), including filling gaps in the available literature with expectations from studies on e-cigarettes, for which there is a larger body of information;
- the addictive features of HTPs, including nicotine delivery (section 3.3), also complemented by data on e-cigarettes and with an assessment of the overall abuse potential of HTPs; and
- the effects of attractiveness and addictiveness on the perception and

use of consumers (section 3.4), including the following, related constructs: awareness, attitude, knowledge, intention, reasons for use and risk perception. Consumer use includes prevalence, user behaviour (such as frequency, intensity and duration and place of use), user profiles, initiation, switching, complementing and quitting conventional tobacco products. Again, lessons learnt from e-cigarettes are used to hypothesize factors that could play a role.

The behavioural implications of different patterns of use among different groups and the implications for public health are covered in the discussion. We conclude with recommendations for research and policy.

A search was conducted of the bibliographic database PubMed, with no restrictions on time, up to January 2020, with the following (combinations of) keywords: heated tobacco products, heat-not-burn tobacco products, HTP, heat sticks and heatsticks and heets, tobacco sticks and IQOS. Studies on toxicity (toxicants in emissions, in-vitro data, biomarkers of exposure) and environmental smoke and studies in a language other than English were excluded. We also used data sent in 2016 by Philip Morris International (PMI) to the US Food and Drug Administration (FDA) seeking authorization to market its IQOS HTP system and flavoured "HeatSticks" in the USA as a modified risk tobacco product (MRTP), and the resulting files of the FDA assessment *(9–12)*.

3.2 Attractiveness of HTPs

3.2.1 Definition of attractiveness in the context of Articles 9 and 10 of the WHO FCTC

"Attractiveness" has been defined by WHO *(13)* as

> factors such as taste, smell and other sensory attributes, ease of use, flexibility of the dosing system, cost, reputation or image, assumed risks and benefits, and other characteristics of a product designed to stimulate use.

Data on all of these aspects are reviewed below. We will also cover marketing, including promotional strategies, and claims of reduced harm.

3.2.2 Attractive features of HTPs
Taste, smell and other sensory attributes
Few studies are available on the taste, smell and other sensory attributes of HTPs in humans. Three PMI studies *(14–16)*, report that smoking IQOS suppressed the urge to smoke to the same extent as smoking cigarettes but was consistently rated as providing less sensory and psychological satisfaction than cigarettes *(17)*. An independent study similarly showed that the HTP PAX was considered

significantly less satisfying, good tasting and calming than own-brand cigarettes but showed no significant effect of PAX on the urge to smoke *(17,18)*. Participants in focus groups in Japan and Switzerland also reported less satisfaction with the IQOS than with combustible cigarettes, a strange or unpleasant taste and smell, milder taste and reduced sensory cues, but less throat discomfort *(19)*.

Information on the availability and variability of flavours in HTPs is also important, as it is known that a variety of available flavours plays an important role in liking tobacco and related products such as e-cigarettes *(20)*. For example, in the Republic of Korea, IQOS HeatSticks (HEETS) are available in tobacco, menthol, bubble gum and lime flavours and glo Dunhill Neosticks in tobacco, menthol and lemon ginger, cherry and grape flavours *(7)*. KT&G "lil" sticks (Fiit) contain novel flavour capsules (menthol, mint, apple mint, bubble gum and apricot flavours) *(7,21)*. Like capsule cigarettes, capsule sticks contain menthol and other flavours that can mask the harshness of tobacco and may appeal to female and young non-smokers *(7)*. Perhaps in reaction to this novelty, British American Tobacco introduced Dunhill Neosticks containing capsules (strong menthol and tobacco/menthol), and PMI introduced Sienna Caps ("Sienna selection with a menthol capsule"). These flavoured HTPs are marketed only in countries outside the USA, where HEETS are considered a cigarette product, for which characterizing flavours are not allowed[1] *(22)*, and only menthol and tobacco may be sold *(11)*.

Ease of use

Focus group participants in Japan and Switzerland reported using IQOS indoors instead of combustible cigarettes, "because it creates no ash or odour" *(19)*. Many participants in both countries commented that the product felt unfamiliar and complicated to use and that using IQOS was cumbersome, as the charger and HeatSticks may be bulky, and the IQOS must be charged and cleaned. In the newer generation IQOS, the holder is integrated with the charger and the product can be used up to 10 times before recharging *(23)*. This could make use easier and thus increase its appeal.

Flexibility of the dosing system

Nicotine dosing is described in the sections on addictiveness, under the broader heading of abuse liability. No other information was found.

---

1   A cigarette or any of its component parts (including the tobacco, filter or paper) shall not contain, as a constituent (including a smoke constituent) or additive, an artificial or natural flavour (other than tobacco or menthol) or a herb or spice, including strawberry, grape, orange, clove, cinnamon, pineapple, vanilla, coconut, liquorice, cocoa, chocolate, cherry or coffee, that is a characterizing flavour of the tobacco product or tobacco smoke.

Cost

Unlike combustible cigarettes, which can be used directly from the package, use of an HTP generally requires the purchase of an external device. The price of such devices can far exceed the price of the consumables, for example, about 25 times the price of a pack of HEETS in the Republic of Korea *(3)*. While the excise tax on HTPs is generally lower than that on combustible cigarettes, HTPs were less expensive to use than combustible cigarettes in fewer than half the countries studied *(21,24)*. In Israel in 2018, HEETS were sold at prices on average 9.5% higher than those of cigarettes *(25)*. In Japan, younger non-users participating in a focus group commented that price could be a potential barrier, but overall, the price contributed to the cachet of the product as luxurious and prestigious *(19)*.

Reputation or image: marketing at point of sale, package and device

Advertising and promotion of HTPs are not always banned in the countries in which they are on the market *(6)*. A systematic Internet search for new tobacco and related products showed that common terms used in marketing or promoting HTPs include "reduced risk", "alternative", "clean", "smoke-free" and "innovative" *(26)*. Expert interviews and IQOS packaging and marketing analyses in Japan and Switzerland also showed that the product is marketed as a clean, chic, pure product *(19)*. In Israel, PMI promoted IQOS as part of its "Smoke-free Israel vision", focusing on "harm reduction" and stressing that the product was clean with less smell and no ash *(27)*. Retailers described the IQOS products as less harmful, a cessation device and not producing smoke *(25)*.

IQOS shops are situated prominently and strategically in selected cities as a core component of marketing. When HTPs were released in futuristic IQOS flagship stores across Italy, Japan and Switzerland, awareness and use of these products increased dramatically *(28)*. In Italy, the "IQOS embassy" and "IQOS boutique" are fancy concept stores where IQOS is promoted as a status symbol and people can try it for free *(6)*. Similarly, in Canada, IQOS was marketed in many tobacco retail outlets (1029 in Ontario) *(29)*. In IQOS boutiques, promotion activities include exchanging a pack of cigarettes or a lighter for an IQOS device, launch parties, "meet and greet" lunches and after-hours events. Promotional elements outside the shops are IQOS signs, sandwich-board signs reading "Building a smoke-free future" and sales representatives regularly smoking IQOS. In Ontario. however, the IQOS signage had to be taken down to comply with national and provincial laws on display and advertising *(30)*. In the Republic of Korea, IQOS flagship shops are located at prime locations in Seoul, again with shop design and product display giving a clean, refined look and feel to IQOS *(3)*. In the USA, IQOS was launched in Atlanta, Georgia, and IQOS shops are located in shopping malls in affluent areas *(31)*.

The IQOS name, device, packaging and shops resemble those of popular cell phones that attract children and adolescents; in combination with the purchasing process, this positions IQOS as a high-demand, upscale product for tech-savvy users, rather different from regular cigarettes *(3)*. Focus group participants in Japan and Switzerland found the product packaging appealing, and even non-users were intrigued, indicating that the product's sleek appearance compared well with that of tech devices *(19)*.

For stick packaging, in Israel, displays of HEETS packages were prominently placed close to consumers, in most cases near youth-oriented merchandise and pack colours indicating tobacco flavourings and strength *(25)*. While cigarette packs in many countries are required to feature graphic warning labels showing various negative consequences of smoking, with explicit colour pictures, to the best of our knowledge this is not required for HTPs in any country as yet. In the Republic of Korea, for example, HEETS packs have only a black-and-white warning label about nicotine addiction *(3)*. The visual design of tobacco products can influence consumers by implying product characteristics. Tests with three IQOS packages that decreasingly linked the product to the Marlboro brand but that were similarly noticeable showed that the packaging appeal, uniqueness and brand equity was significantly lower for the HEETS package than for the Marlboro package *(32)*; however, perceived safety of the Marlboro pack was lower than for the other two packs.

Assumed risks and benefits

HTPs are part of a long tradition in tobacco companies of developing and marketing products that they claim to be less dangerous than conventional cigarettes, beginning with so-called "safer cigarettes" in the 1960s *(33)*. Marketing and media accounts of HTPs explicitly or implicitly claim that they are safer than cigarettes, and some HTPs are claimed by the tobacco industry to help smokers to quit *(5,33)*. HTPs are often claimed to be less harmful than cigarettes because they expose users to lower levels of some toxicants *(33)*. While IQOS may contain lower levels of some toxicants, the data in the PMI application to the FDA for IQOS as an MRTP do not support claims of reduced risk *(34)*. The data do demonstrate, however, that adult consumers in the USA perceive claims of reduced exposure as claims of reduced risk *(35)*, as confirmed in an independent study of adults and adolescents in the USA *(36)*. Analysis of the first nine waves of a survey of the population aged ≥ 14 years in Germany also showed that the majority of 61 HTP users perceived HTPs as somewhat (41.0%) or much (14.8%) less harmful, and 37.7% perceived them as harmful as tobacco cigarettes *(37)*. In Italy, the pilot results of a questionnaire administered to 60 high-school students showed that 40 considered HTPs to be harmful to health, while 24 students said they would accept one of these products if offered by a friend *(38)*.

Several perceived benefits of IQOS use have been identified by focus group participants in Japan and Switzerland, including less throat discomfort, appealing packaging, cleanliness, lack of ash and smoke and greater social acceptability, but only few reported any health benefits of use as compared with combustible tobacco products *(19)*. Use in smoking cessation might be another perceived benefit. In Italy, 19 of 60 high-school students said they would recommend HTPs to a person who wished to stop smoking *(38)*.

Two interesting study protocols have been published, but, unfortunately, neither is independent of industry. PMI cross-sectional surveys are under way in Germany, Italy and the United Kingdom (Greater London) to estimate the prevalence and use patterns of IQOS and other tobacco- and nicotine-containing products *(39)*. The questionnaire also contains items on potential benefits (self-reported improvement in teeth colouring, breath smell, exercise capacity and skin appearance), use experience, perceived risk and experienced reinforcing effects, such as satisfaction, psychological rewards, aversion, enjoyment of respiratory tract sensations and reduced craving. The other protocol *(40)* is for a prospective study to compare changes in cigarette consumption and adoption rates among smokers randomized to HTPs or electronic cigarettes. Product acceptability, tolerability and their harm reduction potential will also be compared. There is, however a potential conflict of interest, as the research is supported by an Investigator-initiated Study award by Philip Morris Products SA, although the authors state that PMI "had no role in the design of the study protocol and will not have any role during its execution, analysis, data interpretation or writing of the manuscript". Further, some of the authors have undeclared conflicts of interest related to tobacco companies, both directly and through funding from tobacco industry front groups.

3.2.3 What we can learn from studies on ENDS and ENNDS and relevance to HTPs

The history of e-cigarettes should make us cautious about any new tobacco or related product coming onto the market, as e-cigarettes have attracted young people who may then proceed to cigarette use *(33)*. As many features of e-cigarettes are also found in HTP, knowledge about their attractiveness may be useful. For example, a study of e-cigarette users showed that the attractive characteristics were the variety of e-liquid flavours (69%), e-cigarette design (44%), ability to adjust e-liquid nicotine levels (31%), ability to adjust settings of device (25%), variety of e-cigarette design (21%), ability to do "cloud chasing" (16%) and price (13%) *(41)*. Fewer dual users, smokers and non-users found these product characteristics attractive, but in the same order. Similarly, analysis of self-reported data showed the importance of the following factors in the choice of an e-cigarette: flavour (39%), price (39%), amount of nicotine (27%), type of e-cigarette (22%), health claims (12%), design (10%), brand (9.4%) and packaging (3.7%) *(20)*.

With regard to sensory properties, users reported that they were dissatisfied with some aspects of e-cigarettes as compared with cigarettes, because many do not deliver nicotine into the bloodstream as quickly as cigarettes and lack the "throat hit" of cigarettes *(1)*. Although this observation was made before nicotine salt-containing e-liquids with a stronger nicotine hit came on the market, this may be one reason why smokers try HTPs instead of e-cigarettes. The sensory studies on HTPs described above show that smokers rate the taste, smell and throat hit are lower than for cigarettes. Flavours and flavour variety are considered the most attractive features of e-cigarettes *(20,41)*. Therefore, banning or restricting available HTP flavours might be helpful in decreasing their popularity among never users.

Although the ease of use of e-cigarettes and HTPs appears to be similar, e-cigarettes are much cheaper to use. Cost has been mentioned as a reason for using e-cigarettes *(20,41)* and could be a barrier for use of HTPs by some groups, especially children and adolescents.

Many e-cigarette users have accepted the marketing claim that non-combustible devices are safer than conventional cigarettes and may see HTPs as a means of enjoying an authentic tobacco taste with lower perceived risk *(1)*. Moreover, like e-cigarette users, HTP consumers may not understand that they must completely quit smoking cigarettes to achieve the claimed health benefits of HTPs and probably also wrongly believe that unsubstantiated claims of reduced risk by the tobacco industry mean that HTPs are risk-free *(33)*.

3.3 Addictiveness of HTPs

Addictiveness is a summary indicator of the abuse liability or abuse potential of a drug and its delivery system. In one model *(42)*, the abuse liability of tobacco products is considered to comprise the likelihood of repeated use (summarized as pharmacokinetics, drug effects and reinforcement) and the consequences of use (effects on functioning, physical dependence, adverse effects).

3.3.1 Addictiveness

The nicotine emissions under machine-smoking conditions from the three types of HTP for which there are the most published data can be summarized as follows (in μg/cigarette): Eclipse under ISO conditions, 0.18 *(43)*; an electrically heated cigarette smoking system under ISO conditions, 0.313 *(44)*; tobacco heating system (THS) 2.2 (IQOS) under Health Canada Intense (HCI) conditions, 1.32 *(45)*; and tobacco heating product (THP) 1.0 under HCI conditions, 0.462 *(46)*, keeping in mind that the data were published by the manufacturers. In general, the nicotine levels are lower than those in a comparison reference cigarette, 1R6F, which has certified values for many of the emissions of concern *(47)*, for which the values are 0.721 μg/per cigarette under ISO conditions and 1.90 μg/cigarette under HCI conditions. The levels for the HTPs have been largely replicated by studies inside *(48)* and outside *(49,50)* the industry. Note in particular the lower levels of nicotine for Eclipse and the electrically heated cigarette smoking system. At the meeting of the Society for Research on Nicotine and Tobacco in 2017, British American Tobacco presented two posters on its HTP (THP 1.0) *(46,51)* and in late 2017 published eight studies in a supplement to a journal *(52–60)*. The studies generally report lower nicotine emissions from HTP than from cigarettes smoked under standard machine conditions.

Bekki et al. *(50)* showed that the rate of transfer of nicotine from tobacco filler to aerosol in IQOS was comparable or slightly higher than in conventional cigarettes. Salman and colleagues *(61,62)* showed that IQOS and traditional cigarettes delivered similar quantities of total nicotine in aerosol under ISO conditions (0.77 and 0.80 mg/cigarette) and slightly less under HCI conditions (1.5 and 1.8 mg/cigarette). They also showed levels of 13% free nicotine in IQOS aerosol under ISO and 5.7% under HCI conditions. Meehan-Atrash and colleagues, using nuclear magnetic resonance, found a level of 0.53 mg/cigarette free nicotine *(63)*. Uchiyama et al. *(64)*, in an extensive investigation of the aerosols emitted by IQOS, glo and PloomTECH (Table 3.1), found that IQOS delivered by far the most nicotine in aerosol and also had more than three times the amount of nicotine in the rod than glo (5.2 mg vs 1.7 mg), although the transfer rates were comparable (23% for IQOS, 30% for glo).

Table 3.1. Nicotine concentrations emitted by IQOS, glo and PloomTECH with various flavours

HTP and flavour	ISO (μg/cig)	HCI (μg/cig)
IQOS tobacco	400	1200
IQOS menthol	430	1200
IQOS mint	320	1200
glo bright tobacco	150	570
glo fresh mix	140	510
glo intensely fresh	150	440
PloomTECH regular	70	270
PloomTECH green	68	170
PloomTECH purple	60	250

Source: reference *64*.

The FDA concluded from the data submitted by PMI that "nicotine pharmacokinetics (PK) in smokers who switched [to IQOS]… is similar to those who continued to smoke CC [conventional smoke, RT]" and that "IQOS is addictive and has nicotine delivery, addiction potential, and abuse liability similar to combustible cigarettes"; however, no comparisons with other substitutes, such as e-cigarettes, were available. The FDA concluded *(11)* that IQOS "provides nicotine at a high enough level to satisfy the withdrawal and craving symptoms of current smokers", but with no comparisons with ENDS. Fig. 3.1 shows the pharmacokinetics of nicotine delivery from IQOS and cigarettes in four studies conducted by PMI. The average C_{max} of IQOS in smokers in the USA was substantially lower than that in smokers in the other countries, although the C_{max} of cigarettes was remarkably consistent. No differences among products were found that explain this finding. Maloney et al. *(65)* found in a laboratory study that IQOS increased mean plasma nicotine significantly, from 2.1 to 12.7 ng/mL after 10 puffs and to 11.3 ng/mL after ad-libitum use, comparable to the rate with JUUL and somewhat lower than that with own-brand cigarettes.

Fig. 3.1. Pharmacokinetics of nicotine in four PMI studies of IQOS

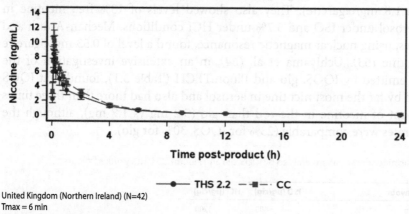

United Kingdom (Northern Ireland) (N=42)
Tmax = 6 min
Cmax = 9.6 ng/mL (THS), 12.4 ng/mL (CC)

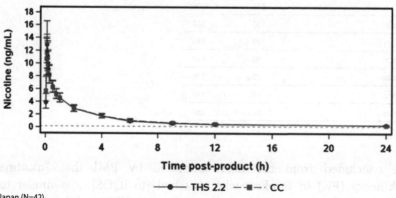

Japan (N=42)
Tmax = 6 min
Cmax = 14.3 ng/mL (THS), 13.8 ng/mL (CC)

Fig. 3.1. Pharmacokinetics of nicotine in four PMI studies of IQOS (continued)

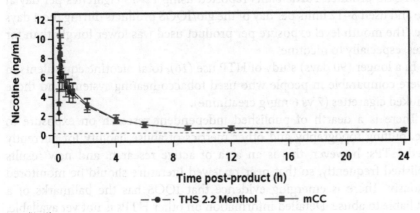

Japan (N=43)
Tmax = 6 min
Cmax = 10.7 ng/mL (THS), 12.1 ng/mL (CC)

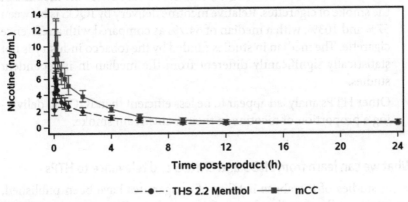

USA (N=41)
Tmax = 7 min (THS), 10 min (CC)
Cmax = 7.4 ng/mL (THS), 13.1 ng/mL (CC)

Source: reference 11.
THS 2.2, IQOS precursor or prototype; CC, conventional cigarette
Limit of quantification in all studies, 0.2 ng/mL

British American Tobacco has published one study on use patterns of its glo product *(58)*. Three groups took glo menthol, glo tobacco and glo + IQOS home for up to 14 days, with up to four laboratory visits. A fourth group used glo in the laboratory only. Laboratory measures included puffing topography, mouth level exposure and depth of mouth insertion. Participants who took the products home completed daily diaries of product use. Overall, the puff volume was about 60 mL with the glo product, with an average of 10–12 puffs per session, a duration of 1.8–2.0 s and a mean interval of 8.8 s. The volume per puff and total volume were significantly higher than with comparison cigarettes but comparable to those with the IQOS product. Participants reported using 12–15 cigarettes per day at baseline and used 8–12 units per day of the glo/IQOS products during the 4 days at home. The mouth level exposure per product used was lower for glo than for cigarettes, especially to nicotine.

In a longer (90 days) study of HTP use *(16)*, total nicotine equivalents in urine were comparable in people who used tobacco heating systems and those who smoked cigarettes (7 vs 6 mg/g creatinine).

There is a dearth of published independent research on exposure to nicotine, puffing topography and other metrics of abuse liability for currently marketed HTPs; however, this is an area of active research, and new results are published frequently, so that peer-reviewed literature should be monitored continuously. There is emerging evidence that IQOS has the hallmarks of a product liable to abuse. Detailed information on other HTPs is not yet available. Two conclusions can be reached.

- Mainstream aerosol from IQOS delivers about 70% of the nicotine in the smoke of cigarettes. Relative nicotine delivery by IQOS is between 57% and 103%, with a median of 64.7% as compared with a reference cigarette. The median in studies funded by the tobacco industry is not statistically significantly different from the median in independent studies.
- Other HTPs analysed appear to be less efficient than IQOS in delivering a proportion of nicotine in their mainstream smoke.

3.3.2 What we can learn from ENDS and ENNDS and relevance to HTPs

A number of studies of the abuse liability of e-cigarettes have been published, which indicate broadly that the design features that affect nicotine delivery should be considered in evaluating the abuse potential of ENDS *(66–69)*. As at least one HTP delivers nicotine and suppresses craving in the same way as an ENDS *(65)*, it is reasonable to hypothesize that the usage patterns observed for ENDS could be generalized to HTPs. Both HTP and ENDS are, however, broad classes of products, and the findings for one product might not be generalizable to others. More studies of direct comparisons of the use of leading HTPs and leading ENDS could clarify this issue.

3.3.3 Overall abuse liability of HTPs

As noted earlier, abuse liability comprises the likelihood of repeated use (summarized as pharmacokinetics), drug effects, reinforcement (rewarding effects that support future use) and consequences of use (effects on functioning, physical dependence, adverse effects). HTPs appear to deliver nicotine at least as well as cigarettes and suppress withdrawal and craving for nicotine. HTP users also show signs of nicotine dependence *(16,58,65)*. Thus, the abuse liability of at least some HTPs for which data are available is likely to be comparable to that of conventional cigarette. The liability may differ by HTP brand and type according to factors such as nicotine delivery, sensory properties and ease of use.

3.4 Effects of the attractiveness and addictiveness of ENDS, ENNDS and HTPs on perceptions of risk and harm and use

3.4.1 Contributions of attractiveness and addictiveness to initiation, switching, complementing and quitting conventional tobacco products

Few studies were found on awareness, use, user profile, initiation, switching, complementing or quitting conventional tobacco products in relation to the attractive aspects of the product. A few addressed perception and reasons for using HTPs. One study implied a direct link between marketing and initiation of IQOS. Google search query data showed that, in Japan, the largest Internet search volume for IQOS was in the week after a popular national television show introduced IQOS *(70)*. Furthermore, the prevalence of use of IQOS increased from 0.3% in January–February 2015 to 0.6% in January–February 2016 and up to 3.6% in January–February 2017, while the estimated rates of use of other HTP remained low in 2017. Respondents who had seen the TV programme in 2016 were more likely to have used IQOS than those who had not seen it (10.3% vs 2.7%).

An online survey of 228 young adults, including current, ever and non-users, in the Republic of Korea indicated that the reasons for using IQOS were a belief that they are less harmful or useful for stopping smoking *(71)*. A PMI premarket observational study showed that the main predictors of adoption were liking the smell, taste, aftertaste and ease of use *(72)*. Adoption was higher among participants who used both regular and menthol THS than among those who used only one variant. In wave 1 of the International Tobacco Control survey in Japan of 4684 adult participants, menthol was the most common flavour reported (41.5%) *(73)*. It has been reported[1] that the levels of menthol in IQOS vary markedly in different markets.

In an exploratory study among adults in London, United Kingdom *(74)*, with 22 current and eight ex-users of IQOS, the six main factors that influenced initiation and use of IQOS were health (wanting to reduce or quit smoking and perceptions of reduced harm), cost (high start-up cost but cheaper continuous cost than smoking), enjoyment and satisfaction (e.g. discretion, cleanliness, less smell, tactile qualities comparable to combustible cigarettes), ease of use (accessibility, shortcomings of maintenance or operation limited continuous use but increased use in smoke-free places), use practices (similar to smoking but new practices developed to charge and clean, new technology) and social aspects

---
1 Goniewicz ML, unpublished data. See Annex 3.1.

(improved social interactions with use of IQOS rather than smoking, but fewer shared social experiences for some).

Studies published by PMI on IQOS precursors *(75,76)* include information on subjective responses to the product, which are often important for understanding why it is used and how effective it might be as a substitute. The cigarette evaluation questionnaire consisted of 12 items in five categories *(77)*: smoking satisfaction, aversion, reduced craving, enjoyment of respiratory tract sensation and psychological reward. The questionnaire on smoking urges consisted of 10 items, with a single score on a seven-point scale *(78)*.

In a laboratory study of THS in Japan *(14)*, the mean satisfaction scores during the study decreased more for THS than for conventional cigarettes. In a similar study in Poland *(15)*, the observed differences in evaluation scales between THS and cigarettes were large and statistically significant. Broadly, cigarettes were rated more highly on satisfaction, craving reduction, sensation and reward and lower on aversion. An earlier study of THS 2.1 showed a similar pattern of results, with satisfaction scores on day 5 an average of 1.4 points lower for THS than for conventional cigarettes ($P < 0.001$) *(79)*. Significant differences were also seen on the reward, sensation and craving subscales, THS scoring lower than cigarettes in all cases. Adriaens and colleagues *(80)* conducted a small study of subjective responses to IQOS and to a tank-style ENDS in comparison with own-brand cigarettes. They found that IQOS and ENDS were equivalent in reducing craving for a cigarette but that both were less effective in suppressing craving than the own-brand cigarette. A study on longer-term use in Japan *(16)* suggested that the difference between cigarettes and HTP with regard to satisfaction fades with continued use over 90 days. These studies show that scores on questionnaires on smoking urges increase for THS over time, as do scores for withdrawal (as measured on the Minnesota nicotine withdrawal scale *(81)*), which may suggest some dissatisfaction with the product as a longer-term substitute for smoking.

3.4.2 Learning from ENDS and ENNDS and application to HTPs

As summarized above, few data are available on the effects of attractiveness and addictiveness on perceptions and reasons for using HTP and on initiation, switching, complementing and quitting conventional tobacco products. Below we summarize the available evidence on those factors in relation to the attractiveness and addictiveness of ENDS.

A review of reasons for e-cigarette use reported by e-cigarette users, cigarette smokers, dual users and non-users among both adults and young people showed that adults' perceptions and reasons for e-cigarette use are often related to smoking cessation, while the young like the novelty of the product *(82)*. Young non-users perceived e-cigarettes as a cool, fashionable product that mimics the smoking routine and is safe to use. In general, the perceived benefits included avoidance of smoking restrictions, the product being cool and fashionable, having health benefits, lower cost than cigarettes, positive experiences (mimics smoking routine, enjoyable taste, throat hit, weight control, increases concentration), safe to use, smoking cessation or reduction, social acceptability and perceived benefits for bystanders. Another review of studies in young adults *(83)* showed that their reasons for using e-cigarettes are more varied than only smoking cessation. Independently of smoking status, curiosity was the most frequently reported reason for initiating use of e-cigarettes. Continued use of e-cigarettes could be due to either replication of smoking habits or a different, personalized use of nicotine by inhalation. In Europe *(84)*, the most frequently mentioned reason (61%) for taking up e-cigarettes was to stop or reduce tobacco consumption. Other reasons included a perception of e-cigarettes as less harmful (31%) and lower cost (25%). Reducing tobacco consumption and being less harmful were cited more often by participants aged > 40 (76–78%) than those aged 15–24 (59%).

Flavours attract both young people and adults to e-cigarettes *(85)*. Flavours decrease the perception of harm and increase willingness to try and initiate use of e-cigarettes. Among adults, e-cigarette flavours increase product appeal and are a primary reason for using the product. "Pod mod" devices have become popular, especially among adolescents, due to their design, user-friendliness, less aversive vaping experiences, desirable flavours and discretion in places where smoking is forbidden *(86)*. Currently marketed HTPs share several of these characteristics, suggesting some generalizability of the ENDS experience.

3.5 Discussion

3.5.1 Behavioural implications of different patterns of use in different groups

Few studies are available on the perception and reasons for using HTPs among users, ex-users, smokers, dual users and never users of tobacco and related products (section 3.1). While the role of attractive or addictive aspects of HTPs is little known, studies are available on awareness, use, user profile, initiation, switching, complementing and quitting conventional tobacco products. Information on the patterns and prevalence of use in several groups is important for regulators, as the risk profile of smokers is different from that of never smokers.

An important question is whether HTPs are used primarily by smokers, or whether non-smokers also use the product. It would appear that, currently, most HTP users are smokers. For example, in Japan, virtually all HTP users were current (67.8%) or former smokers (25.0%), and only 1.0% were never smokers *(73)*. According to the premarket review by the FDA *(5)*,

> … although the data for IQOS uptake by never smokers, former smokers, and youth is limited, there are some data from countries where IQOS is

marketed – Italy and Japan – which show low uptake by youth and current non-smokers. In these countries, the likelihood of uptake is slightly higher in former smokers, but still low.

While most HTP users may be smokers, most are dual users of both HTPs and conventional cigarettes and are therefore exposed to emissions from both. PMI studies showed that most people who use IQOS, concurrently use cigarettes *(33)*. Studies in the Republic of Korea showing that 96% of current HTP users (2% of the study population) were dual users and that dual use with conventional cigarettes was not associated with an intention to quit cigarette smoking *(87)*. A survey of adolescents aged 12–18 years showed that 75.5% of the ever HTP users (2.8%) were current cigarette users, 45.6% were current e-cigarette users and 40.3% were concurrent users of cigarettes and e-cigarettes *(88)*. No difference in cigarette quit attempts was found with ever use of HTPs.

PMI and independent data suggest that IQOS will attract adolescent and young adult non-users to initiate tobacco use *(3)*. In Italy, marketing led to an increase in IQOS use, with an intent to use IQOS among non-smokers and long-term former smokers *(6,89)*. According to the FDA *(5)*,

> Certainly, the potential for rapid uptake of a novel tobacco product among youth exists. In the decade since e-cigarettes were introduced to the U.S. market, youth use rose rapidly but the limited flavour choices may reduce IQOS' appeal to youth. The limited options in terms of flavour choice and the price of the IQOS device may reduce the appeal to youth.

Evidence from Japan indicates that younger (< 30 years), wealthier people adopt IQOS *(90)*, and emerging evidence from Canada, United Kingdom (England) and the USA suggests that HTP have at least some appeal for young adults, including non-smokers *(91,92)*.

3.5.2 Implications for public health

HTPs have probably become popular in some markets because of factors such as their marketing as a "clean", modern, elegant, reduced-harm product. Factors that enhance their attractiveness appear to be sensory attributes, ease of use, cost, reputation and image and perceived risks and benefits. While HTPs suppress the urge to smoke, they are generally considered less satisfactory than cigarettes; however, different flavours are available, and this is known to be an attractive feature of tobacco and related products. Users find the fact that they create less ash and smell appealing and may use them indoors for that reason. Other attractive features include their marketing as an exclusive, modern and less harmful product suitable for smoking cessation. The product is, however, often considered to be less easy to use than cigarettes, and cost may be a barrier.

With regard to addictiveness, currently marketed HTPs deliver significant levels of nicotine in aerosol, and their pharmacokinetics, drug and subjective effects are similar to those of contemporary ENDS products, suggesting comparable abuse liability. Little information is available on how these features affect consumer perception and use, including initiation, switching, complementing and quitting conventional tobacco products. Industry studies show that the main predictors of adoption are liking of the smell, taste, aftertaste and ease of use. The only independent study that explored reasons for use and continuation and discontinuation of IQOS among smokers and ex-smokers found that the six important factors were: health, cost, enjoyment and satisfaction, ease of use, use practices and social aspects.

As described above, the vast majority of current IQOS users are current or ex-smokers, and most are dual users of HTPs and conventional cigarettes, with no intention to quit smoking cigarettes. Hence, the expected reduction in exposure to smoking-related toxicants will be much smaller than if they switch completely, and the reduction in risk, if any, will probably be much lower. For dual users, any risk reduction would be smaller than for people who switched completely, and reduced risk has not been proven even for complete switchers. Although few users of HTPs are never users or ex-users of cigarettes, their potential interest in HTPs is still a risk at population level, especially for never users. The history of e-cigarettes shows that any new tobacco or related product coming onto the market quickly becomes popular. Knowledge of e-cigarettes indicates that the factors of concern with regard to HTPs, and therefore potential regulation, are nicotine levels and "throat hit", flavour variety, marketing and perception of reduced harm. Furthermore, like e-cigarettes *(85)*, HTPs could be a gateway towards cigarette smoking.

3.5.3 Research gaps, priorities and questions

- There is almost no information on perceptions and reasons for using HTPs among users, ex-users, smokers, dual users and never users of tobacco and related products. Only one independent study addressed the reasons for use, continuation and discontinuation of IQOS among smokers and ex-smokers.
- There is also little information on how the attractiveness of HTPs affects consumer perception and use, including initiation, switching, complementing and quitting conventional tobacco products.
- There is no information on whether HTPs could be a gateway to use of combustible tobacco.

Most independent studies on HTPs have focused on emissions and toxicity rather than attractiveness and addictiveness. Instead of trying to reproduce PMI emission data and claims of assessing reduced harm, researchers should study the actual trajectory of smokers, never smokers and ex-smokers to initiating use of HTPs and the role of appealing product characteristics.

Additional independent research should be conducted on all the features that enhance the attractiveness and addictiveness of HTPs described in this paper. The studies should include:

- sensory studies, including the roles of flavourings and flavours and other attractiveness-enhancing content and additives, such as sugars and humectants, in the use of HTPs;
- behavioural studies, both qualitative and quantitative, on knowledge, attitude and risk perceptions, including health messages;
- studies on all the features that enhance attractiveness and addictiveness in relation to consumer perception and use, with, preferably, groups of smokers, dual users, HTP users and never users, ideally in longitudinal quantitative studies that include use of other types of tobacco and related products to establish the impact of several features of perception of different types of tobacco products and to study transitions from e.g. current or never smoking to HTP use or even to combustible tobacco products; and
- studies on the pharmacokinetics of use of HTPs other than IQOS, to establish patterns of use and nicotine delivery under real conditions.

3.5.4 Policy recommendations

Policy-makers might consider adopting regulatory principles that have been applied successfully to tobacco and related products in many jurisdictions in order to minimize product appeal and uptake among young people, increase product safety and minimize false beliefs about health with regard to HTPs. They should thus consider the following measures, with a focus on protecting young people and non-users.

- Ban sale to minors, price promotions, flavours that appeal to young people and flavour capsules; limit marketing at points of sale and elsewhere; and introduce plain packaging to minimize the appeal of HTPs and their uptake by young people.
- Ensure that the public is not misled by their appeal and by claims from the manufacturers but is well informed about the risks of HTPs, including the risk of dual use with cigarettes and use during

pregnancy; correct false perceptions, counter misinformation, and clarify that reduced exposure does not necessarily mean reduced harm.
- Monitor the prevalence of use and user profiles; establish or extend surveillance of the product and users, including demographics, use of other tobacco and related products, devices, brands, types and flavours used.

3.6 Conclusions

HTPs have become popular in some markets probably because of a combination of factors, including their marketing as a "reduced risk", "clean", modern, elegant product. Their sensory properties and ease of use are generally rated lower than those of conventional cigarettes but are important dimensions of their attractiveness, thus determining their uptake. Little is known about how these features affect consumer perceptions and use. Factors such as packaging, labelling, risk communication, price and smoke-free policies appear to influence initiation and use.

Currently marketed HTPs deliver significant levels of nicotine in aerosol, and their pharmacokinetics and drug and subjective effects are similar to those of contemporary ENDS products, suggesting comparable abuse liability. The history of e-cigarettes has shown that any new tobacco or related product coming onto the market may quickly become popular. Knowledge about e-cigarettes indicate that the factors of concern with regard to HTPs – and therefore potential aims of regulation – are nicotine levels and "throat hit", variety of flavours, design of the devices, marketing and a perception of reduced harm.

3.7 References

1. Caputi TL. Industry watch: heat-not-burn tobacco products are about to reach their boiling point. Tob Control. 2016;26(5):609–10.
2. Ratajczak A, Jankowski P, Strus P, Feleszko W. Heat not burn tobacco product – A new global trend: impact of heat-not-burn tobacco products on public health, a systematic review. Int J Environ Res Public Health. 2020;17(2):409.
3. Kim M. Philip Morris International introduces new heat-not-burn product, IQOS, in South Korea. Tob Control. 2018;27(e1):e76–8.
4. Heated tobacco products information sheet. Second edition. Geneva: World Health Organization; 2020 (https://apps.who.int/iris/bitstream/handle/10665/331297/WHO-HEP-HPR-2020.2-eng.pdf?sequence=1&isAllowed=y, accessed 10 January 2021).
5. Bialous SA, Glantz SA. Heated tobacco products: another tobacco industry global strategy to slow progress in tobacco control. Tob Control. 2018;27(Suppl 1):s111–7.
6. Liu X, Lugo A, Spizzichino L, Tabuchi T, Gorini G, Gallus S. Heat-not-burn tobacco products are getting hot in Italy. J Epidemiol. 2018;28(5):274–5.
7. Cho YJ, Thrasher JF. Flavour capsule heat-sticks for heated tobacco products. Tob Control. 2019;28(e2):e158–9.

8. Decision FCTC/COP8(22). Novel and emerging tobacco products. Geneva: World Health Organization; 2018 (https://www.who.int/fctc/cop/sessions/cop8/FCTC__COP8(22).pdf?ua=1, accessed 10 January 2021).
9. Philip Morris Products S.A. modified risk tobacco product (MRTP) applications. Silver Spring (MD): Food and Drug Administration; 2020 (https://www.fda.gov/tobacco-products/advertising-and-promotion/philip-morris-products-sa-modified-risk-tobacco-product-mrtp-applications, accessed 10 January 2021).
10. Marketing order. FDA submission tracking numbers (STNs): PM0000424-PM0000426, PM0000479. Silver Spring (MD): Food and Drug Administration; 2019 (https://www.fda.gov/media/124248/download, accessed 10 January 2021).
11. PMTA. Silver Spring (MD): Food and Drug Administration; 2019 (https://www.fda.gov/media/124247/download, accessed 10 January 2021).
12. The public health rationale for recommended restrictions on new tobacco product labeling, advertising, marketing, and promotion. Silver Spring (MD): Food and Drug Administration; 2019 (https://www.fda.gov/media/124174/download, accessed 10 January 2021).
13. Partial guidelines for implementation of Articles 9 and 10 of the WHO Framework Convention on Tobacco Control. Regulation of the contents of tobacco products and of tobacco product disclosures (FCTC/COP4(10)). Geneva: World Health Organization; 2012 (https://www.who.int/fctc/guidelines/Guideliness_Articles_9_10_rev_240613.pdf?ua=1, accessed 10 January 2021).
14. Haziza C, de la Bourdonnaye G, Merlet S, Benzimra M, Ancerewicz J, Donelli A et al. Assessment of the reduction in levels of exposure to harmful and potentially harmful constituents in Japanese subjects using a novel tobacco heating system compared with conventional cigarettes and smoking abstinence: A randomized controlled study in confinement. Regul Toxicol Pharmacol. 2016;81:489–99.
15. Haziza C, de la Bourdonnaye G, Skiada D, Ancerewicz J, Baker G, Picavet P et al. Evaluation of the tobacco heating system 2.2. Part 8: 5-day randomized reduced exposure clinical study in Poland. Regul Toxicol Pharmacol. 2016;81(Suppl 2):S139–50.
16. Lüdicke F, Picavet P, Baker G, Haziza C, Poux V, Lama N et al. Effects of switching to the tobacco heating system 2.2 menthol, smoking abstinence, or continued cigarette smoking on biomarkers of exposure: a randomized, controlled, open-label, multicenter study in sequential confinement and ambulatory settings (Part 1). Nicotine Tob Res. 2018;20(2):161–72.
17. Simonavicius E, McNeill A, Shahab L, Brose LS. Heat-not-burn tobacco products: a systematic literature review. Tob Control. 2019;28(5):582–94.
18. Lopez AA, Hiler M, Maloney S, Eissenberg T, Breland A. Expanding clinical laboratory tobacco product evaluation methods to loose-leaf tobacco vaporizers. Drug Alcohol Depend. 2016;169:33–40.
19. Hair EC, Bennett M, Sheen E, Cantrell J, Briggs J, Fenn Z et al. Examining perceptions about IQOS heated tobacco product: consumer studies in Japan and Switzerland. Tob Control. 2018;27(Suppl 1):s70–3.
20. Laverty AA, Vardavas CI, Filippidis FT. Design and marketing features influencing choice of e-cigarettes and tobacco in the EU. Eur J Public Health. 2016;26(5):838–41.
21. Lee J, Lee S. Korean-made heated tobacco product, "lil". Tob Control. 2019;28(e2):e156–7.
22. Reversing the Youth Tobacco Epidemic Act of 2019. 116th Congress, 1st session. H.R. 2339. Washington (DC): United States Government; 2019 (https://www.congress.gov/116/bills/hr2339/BILLS-116hr2339ih.xml, accessed 10 January 2021).
23. Our tobacco heating system. IQOS. Tobacco meets technology. Lausanne: Philip Morris International; 2020 (https://www.pmi.com/smoke-free-products/iqos-our-tobacco-heating-system, accessed 10 January 2021).
24. Liber AC. Heated tobacco products and combusted cigarettes: comparing global prices and

taxes. Tob Control. 2019;28(6):689–91.
25. Bar-Zeev Y, Levine H, Rubinstein G, Khateb I, Berg CJ. IQOS point-of-sale marketing strategies in Israel: a pilot study. Isr J Health Policy Res. 2019;8(1):11.
26. Staal YC, van de Nobelen S, Havermans A, Talhout R. New tobacco and tobacco-related products: early detection of product development, marketing strategies, and consumer interest. JMIR Public Health Surveill. 2018;4(2):e55.
27. Rosen LJ, Kislev S. IQOS campaign in Israel. Tob Control. 2018;27(Suppl 1):s78–81.
28. Tabuchi T, Kiyohara K, Hoshino T, Bekki K, Inaba Y, Kunugita N. Awareness and use of electronic cigarettes and heat-not-burn tobacco products in Japan. Addiction. 2016;111(4):706–13.
29. Mathers A, Schwartz R, O'Connor S, Fung M, Diemert L. Marketing IQOS in a dark market. Tob Control. 2019;28(2):237–8.
30. Yuen J. Health Canada orders IQOS tobacco storefront to remove its signs. Toronto Sun, 1 November 2018 (https://torontosun.com/news/local-news/health-canada-orders-iqos-tobacco-storefront-to-remove-its-signs, accessed 10 January 2021).
31. Churchill V, Weaver SR, Spears CA, Huang J, Massey ZB, Fairman RT et al. IQOS debut in the USA: Philip Morris International's heated tobacco device introduced in Atlanta, Georgia. Tob Control. 2020:doi: 10.1136/tobaccocontrol-2019-055488.
32. Lee JGL, Blanchflower TM, O'Brien KF, Averett PE, Cofie LE, Gregory KR. Evolving IQOS packaging designs change perceptions of product appeal, uniqueness, quality and safety: a randomised experiment, 2018, USA. Tob Control. 2019;28(e1):e52–5.
33. Glantz SA. Heated tobacco products: the example of IQOS. Tob Control. 2018;27(Suppl 1):s1–6.
34. Lempert LK, Glantz SA. Heated tobacco product regulation under US law and the FCTC. Tob Control. 2018;27(Suppl 1):s118–25.
35. Popova L, Lempert LK, Glantz SA. Light and mild redux: heated tobacco products' reduced exposure claims are likely to be misunderstood as reduced risk claims. Tob Control. 2018;27(Suppl 1):s87–95.
36. El-Toukhy S, Baig SA, Jeong M, Byron MJ, Ribisi KM, Brewer NT. Impact of modified risk tobacco product claims on beliefs of US adults and adolescents. Tob Control. 2018;27(Suppl 1):s62–9.
37. Kotz D, Kastaun S. E-Zigaretten und Tabakerhitzer: repräsentative Daten zu Konsumverhalten und assoziierten Faktoren in der deutschen Bevölkerung (die DEBRA-Studie) [E-cigarettes and heat-not-burn products: representative data on consumer behaviour and associated factors in the German population (the DEBRA study)]. Bundesgesundheitsblatt Gesundheitsforschung Gesundheitsschutz. 2018;61(11):1407–14.
38. La Torre G, Dorelli B, Ricciardi M, Grassi M, Mannocci A. Smoking E-CigaRette and HEat-no-T-burn products: validation of the SECRHET questionnaire. Clin Ter. 2019;170(4):e247–51.
39. Sponsiello-Wang Z, Langer P, Prieto L, Dobrynina M, Skiada D, Camille N et al. Household surveys in the general population and web-based surveys in IQOS users registered at the Philip Morris International IQOS user database: protocols on the use of tobacco- and nicotine-containing products in Germany, Italy, and the United Kingdom (Greater London), 2018–2020. JMIR Res Protoc. 2019;8(5):e12061.
40. Caponnetto P, Caruso M, Maglia M, Emma R, Saitta D, Busà B et al. Non-inferiority trial comparing cigarette consumption, adoption rates, acceptability, tolerability, and tobacco harm reduction potential in smokers switching to heated tobacco products or electronic cigarettes: Study protocol for a randomized controlled trial. Contemp Clin Trials Commun. 2020;17:100518.
41. Romijnders KA, Krüsemann EJ, Boesveldt S, Graaf K, de Vries H, Talhout R. e-Liquid flavor preferences and individual factors related to vaping: A survey among Dutch never-users, smokers, dual users, and exclusive vapers. Int J Environ Res Public Health. 2019;16:4661–76.
42. Carter LP, Stitzer ML, Henningfield JE, O'Connor RJ, Cummings KM, Hatsukami DK. Abuse liability assessment of tobacco products including potential reduced exposure products. Cancer

Epidemiol Biomarkers Prev. 2009;18(12):3241–62.
43. Slade J, Connolly GN, Lymperis D. Eclipse: does it live up to its health claims? Tob Control. 2002;11(Suppl 2):ii64–70.
44. Werley MS, Freelin SA, Wrenn SE, Gerstenberg B, Roemer E, Schramke H et al. Smoke chemistry, in vitro and in vivo toxicology evaluations of the electrically heated cigarette smoking system series K. Regul Toxicol Pharmacol. 2008;52(2):122–39.
45. Schaller JP, Keller D, Poget L, Pratte P, Kaelin E, McHugh D et al. Evaluation of the tobacco heating system 2.2. Part 2: Chemical composition, genotoxicity, cytotoxicity, and physical properties of the aerosol. Regul Toxicol Pharmacol. 2016;81(Suppl 2):S27–47.
46. Jakaj B, Eaton D, Forster M, Nicol T, Liu C, McAdam K et al. Characterizing key thermophysical processes in a novel tobacco heating product THP1.0(T). Poster. Society for Research on Nicotine and Tobacco, 7–11 March 2017, Florence, Italy; 2017.
47. Certificate of Analysis. 1R6F certified reference cigarette. Lexington (KY): University of Kentucky; 2016 (https://www.ecigstats.org/docs/research/CoA_1R6F.pdf, accessed 10 January 2021).
48. Li X, Luo Y, Jiang X, Zhang H, Zhu F, Hu S et al. Chemical analysis and simulated pyrolysis of tobacco heating system 2.2 compared to conventional cigarettes. Nicotine Tob Res. 2018;21(1):111–8.
49. Farsalinos KE, Yannovits N, Sarri T, Voudris V, Poulas K. Nicotine delivery to the aerosol of a heat-not-burn tobacco product: comparison with a tobacco cigarette and e-cigarettes. Nicotine Tob Res. 2018;20(8):1004–9.
50. Bekki K, Inaba Y, Uchuyama S, Kunugita N. Comparison of chemicals in mainstream smoke in heat-not-burn tobacco and combustion cigarettes. J UOEH. 2017;39(3):201–7.
51. Scott JK, Poynton S, Margham J, Forster M, Eaton D, Davis P et al. Controlled aerosol release to heat tobacco: product operation and aerosol chemistry assessment. Poster. Society for Research on Nicotine and Tobacco. 2–5 March 2016, Chicago, Illinois (http://www.researchgate.net/publication/298793405_Controlled_aerosol_release_to_heat_tobacco_product_operation_and_aerosol_chemistry_assessment, accessed 10 January 2021).
52. Eaton D, Jakaj B, Forster M, Nicol J, Mavropoulou E, Scott K et al. Assessment of tobacco heating product THP1.0. Part 2: Product design, operation and thermophysical characterisation. Regul Toxicol Pharmacol. 2018;93:4–13.
53. Forster M, Fiebelkorn S, Yurteri C, Mariner D, Liu C, Wright C et al. Assessment of novel tobacco heating product THP1.0. Part 3: Comprehensive chemical characterisation of harmful and potentially harmful aerosol emissions. Regul Toxicol Pharmacol. 2018;93:14–33.
54. Murphy J, Liu C, McAdam K, Gaça M, Prasad K, Camacho O et al. Assessment of tobacco heating product THP1.0. Part 9: The placement of a range of next-generation products on an emissions continuum relative to cigarettes via pre-clinical assessment studies. Regul Toxicol Pharmacol. 2018;93:92–104.
55. Taylor M, Thorne D, Carr T, Breheny D, Walker P, Proctor C et al. Assessment of novel tobacco heating product THP1.0. Part 6: A comparative in vitro study using contemporary screening approaches. Regul Toxicol Pharmacol. 2018;93:62–70.
56. Thorne D, Breheny D, Proctor C, Gaça M. Assessment of novel tobacco heating product THP1.0. Part 7: Comparative in vitro toxicological evaluation. Regul Toxicol Pharmacol. 2018;93:71–83.
57. Forster M, McAughey J, Prasad K, Mavropoulou E, Proctor C. Assessment of tobacco heating product THP1.0. Part 4: Characterisation of indoor air quality and odour. Regul Toxicol Pharmacol. 2018;93:34–51.
58. Gee J, Prasad K, Slayford S, Gray A, Nother K, Cunningham A et al. Assessment of tobacco heating product THP1.0. Part 8: Study to determine puffing topography, mouth level exposure and consumption among Japanese users. Regul Toxicol Pharmacol. 2018;93:84–91.
59. Jaunky T, Adamson J, Santopietro S, Terry A, Thorne D, Breheny D et al. Assessment of tobacco

heating product THP1.0. Part 5: In vitro dosimetric and cytotoxic assessment. Regul Toxicol Pharmacol. 2018;93:52–61.
60. Proctor C. Assessment of tobacco heating product THP1.0. Part 1: Series introduction. Regulatory toxicology and pharmacology : RTP 2017 doi: 10.1016/j.yrtph.2017.09.010 [published Online First: 2017/10/11]
61. Salman R, Talih S, El-Hage R, Haddad C, Karaoghlanian N, El-Hellani A et al. Free-base and total nicotine, reactive oxygen species, and carbonyl emissions from IQOS, a heated tobacco product. Nicotine Tob Res. 2019;21(9):1285–8.
62. Albert RE, Peterson HT Jr, Bohning DE, Lippmann M. Short-term effects of cigarette smoking on bronchial clearance in humans. Arch Environ Health. 1975;30(7):361–7.
63. Meehan-Atrash J, Duell AK, McWhirter KJ, Luo W, Peyton DH, Strongin RM. Free-base nicotine is nearly absent in aerosol from IQOS heat-not-burn devices, as determined by (1)H NMR spectroscopy. Chem Res Toxicol. 2019;32(6):974–6.
64. Uchiyama S, Noguchi M, Takagi N, Hayashida H, Inaba Y, Ogura H et al. Simple determination of gaseous and particulate compounds generated from heated tobacco products. Chem Res Toxicol. 2018;31(7):585–93.
65. Maloney S, Eversole A, Crabtree M, Soule E, Eissenberg T, Breland A. Acute effects of JUUL and IQOS in cigarette smokers. Tob Control. 2020:doi: 10.1136/tobaccocontrol-2019-055475.
66. Wagener TL, Floyd EL, Stepanov I, Driskill LM, Frank SG, Meier E et al. Have combustible cigarettes met their match? The nicotine delivery profiles and harmful constituent exposures of second-generation and third-generation electronic cigarette users. Tob Control. 2017;26(e1):e23–8.
67. Fearon IM, Eldridge AC, Gale N, McEwan M, Stiles MF, Round EK. Nicotine pharmacokinetics of electronic cigarettes: A review of the literature. Regul Toxicol Pharmacol. 2018;100:25–34.
68. Voos N, Goniewicz ML, Eissenberg T. What is the nicotine delivery profile of electronic cigarettes? Expert Opin Drug Deliv. 2019;16(11):1193–203.
69. Yingst JM, Foulds J, Veldheer S, Hrabovsky S, Trushin N, Eissenberg TT et al. Nicotine absorption during electronic cigarette use among regular users. PLoS One/ 2019;14(7):e0220300.
70. Tabuchi T, Gallus S, Shinozaki T, Nakaya T, Kunugita N, Colwell B. Heat-not-burn tobacco product use in Japan: its prevalence, predictors and perceived symptoms from exposure to secondhand heat-not-burn tobacco aerosol. Tob Control. 2018;27(e1):e25–33.
71. Kim J, Yu H, Lee S, Paek YJ. Awareness, experience and prevalence of heated tobacco product, IQOS, among young Korean adults. Tob Control. 2018;27(Suppl 1):s74–7.
72. Roulet S, Chrea C, Kanitscheider C, Kallischnigg G, Magnani P, Weitkunat R. Potential predictors of adoption of the tobacco heating system by US adult smokers: An actual use study. F1000Res. 2019;8:214.
73. Sutanto E, Miller C, Smith DM, O'Connor RJ, Quah ACK, Cummings KM et al. Prevalence, use behaviors, and preferences among users of heated tobacco products: Findings from the 2018 ITC Japan survey. Int J Environ Res Public Health. 2019;16(23):4630.
74. Tompkins CNE, Burnley A, McNeill A, Hitchman SC. Factors that influence smokers' and ex-smokers' use of IQOS: a qualitative study of IQOS users and ex-users in the UK. Tob Control. 2020:doi: 10.1136/tobaccocontrol-2019-055306.
75. Hanson K, O'Connor R, Hatsukami D. Measures for assessing subjective effects of potential reduced-exposure products. Cancer Epidemiol Biomarkers Prev. 2009;18(12):3209–24.
76. Rees VW, Kreslake JM, Cummings KM, O'Connor RJ, Hatsukami DK, Parascandola M et al. Assessing consumer responses to potential reduced-exposure tobacco products: a review of tobacco industry and independent research methods. Cancer Epidemiol Biomarkers Prev. 2009;18(12):3225–40.
77. Cappelleri JC, Bushmakin AG, Baker CL, Merikle E, Olufade AO, Gilbert DG. Confirmatory factor analyses and reliability of the modified cigarette evaluation questionnaire. Addict Behav.

2007;32(5):912–23.
78. Cox LS, Tiffany ST, Christen AG. Evaluation of the brief questionnaire of smoking urges (QSU-brief) in laboratory and clinical settings. Nicotine Tob Res. 2001;3(1):7–16.
79. Lüdicke F, Baker G, Magnette J, Picavet P, Weitkunat R. Reduced exposure to harmful and potentially harmful smoke constituents with the tobacco heating system 2.1. Nicotine Tob Res. 2017;19(2):168–75.
80. Adriaens K, Gucht DV, Baeyens F. IQOS(TM) vs. e-cigarette vs. tobacco cigarette: A direct comparison of short-term effects after overnight-abstinence. Int J Environ Res Public Health. 2018;15(12):2902.
81. Hughes JR, Hatsukami D. Signs and symptoms of tobacco withdrawal. Arch Gen Psychiatry. 1986;43(3):289–94.
82. Romijnders K, van Osch L, de Vries H, Talhout R. Perceptions and reasons regarding e-cigarette use among users and non-users: A narrative literature review. Int J Environ Res Public Health. 2018;15(6):1190.
83. Kinouani S, Leflot C, Vanderkam P, Auriacombe M, Langlois E, Tzourio C. Motivations for using electronic cigarettes in young adults: A systematic review. Subst Abus. 2020;41(3):315–22.
84. Attitudes of Europeans towards tobacco and electronic cigarettes (Special Eurobarometer 458, wave EB87.1). Brussels: European Commission; 2017.
85. Fadus MC, Smith TT, Squeglia LM. The rise of e-cigarettes, pod mod devices, and JUUL among youth: Factors influencing use, health implications, and downstream effects. Drug Alcohol Depend. 2019;201:85–93.
86. Meernik C, Baker HM, Kowitt SD, Ranney LM, Goldstein AO. Impact of non-menthol flavours in e-cigarettes on perceptions and use: an updated systematic review. BMJ Open. 2019;9(10):e031598.
87. Hwang JH, Ryu DH, Park SW. Heated tobacco products: Cigarette complements, not substitutes. Drug Alcohol Depend. 2019;204:107576.
88. Kang H, Cho SI. Heated tobacco product use among Korean adolescents. Tob Control. 2020;29(4):466–8.
89. Liu X, Lugo A, Spizzichino L, Tabuchi T, Pacifici R, Gallus S. Heat-not-burn tobacco products: concerns from the Italian experience. Tob Control. 2019;28(1):113–4.
90. Igarashi A, Aida J, Kusama T, Osaka K. Heated tobacco products have reached younger or more affluent people in Japan. J Epidemiol. 2020:doi: 10.2188/jea.JE20190260.
91. Czoli CD, White CM, Reid JL, O'Connor RJ, Hammond D. Awareness and interest in IQOS heated tobacco products among youth in Canada, England and the USA. Tob Control. 2020;29(1):89–95.
92. Dunbar MS, Seelam R, Tucker JS, Rodriguez A, Shih RA, D'Amico EJ. Correlates of awareness and use of heated tobacco products in a sample of US young adults in 2018–2019. Nicotine Tob Res. 2020:doi: 10.1093/ntr/ntaa007.

3. The attractiveness and addictive potential of heated tobacco products: effects on perception and use and associated effects

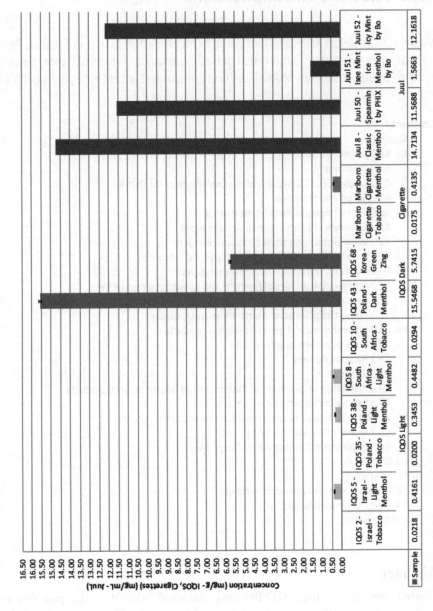

Annex 3.1. Menthol concentrations in IQOS, cigarettes and JUUL products

Source: Goniewicz ML, unpublished data, 2019.

4. Variations among heated tobacco products: considerations and implications

Maciej L. Goniewicz, Associate Professor of Oncology, Nicotine and Tobacco Product Assessment Resource, Department of Health Behavior, Division of Cancer Prevention and Population Studies, Roswell Park Comprehensive Cancer Center, Buffalo (NY), USA

Contents

Abstract
4.1 Background
    4.1.1 Overview
    4.1.2 Decision FCTC/COP8(22) of the Conference of the Parties
    4.1.3 Scope and objectives
4.2 Variations among products on the market
    4.2.1 Overview of product categories and types
    4.2.2 Variations among heated tobacco products
    4.2.3 Market distribution of product types and categories
4.3 Characteristics and design features of products
    4.3.1 Temperature profiles of products and operational capabilities
    4.3.2 Battery characteristics
    4.3.3 Properties of tobacco inserts, sticks and capsules
4.4 Content, emissions and general design of products
    4.4.1 Content and emissions
    4.4.2 Nicotine delivery
    4.4.3 Risk profiles
    4.4.4 Regulatory implications of the contents of heated tobacco products
    4.4.5 Regulatory implications of emissions
4.5 Variations among products, manufacturers and selling points
    4.5.1 Manufacturers and selling points
    4.5.2 Implications for customer pulling power
4.6 Discussion
4.7 Conclusions
4.8 Research gaps, priorities and questions
4.9 Policy recommendations
4.10 References

Abstract

Heated tobacco products (HTPs) have attracted interest in the past few years and are now available on approximately 50 markets. The interest is attributed to factors that include aggressive marketing by manufacturers, users' perception that HTPs represent a "safer alternative" to other smoked products, their capacity to deliver nicotine to users, the variety of flavours, including tobacco and menthol, technological advances and the variety of products and product features from which users can choose. The wide variation among the products, devices and their features result in the delivery of different levels of nicotine and toxicants, which has important regulatory implications. It is therefore important to understand how HTPs differ and how the differences affect the emissions of nicotine and other toxicants in order to formulate effective regulatory strategies and policies. The aim of this paper is thus to describe variations among the HTPs on the market in their characteristics and design features and how those features influence product toxicity, appeal and implications for regulations.

4.1 Background
4.1.1 Overview

HTPs are an emerging class of "potentially reduced exposure products" promoted by manufacturers as associated with "reduced risk", as "cleaner alternatives" and as "smoke-free" products. The concept of these products proceeds from the principle that most of the harm associated with tobacco smoking is from the combustion process. In a conventional tobacco cigarette, the temperature of the burning cone can reach up to 900 °C, with a median temperature across the rod of 600 °C *(1)*. This leads to myriad thermal reactions, including combustion, pyrolysis and pyrosynthesis, that result in the > 7000 compounds identified in tobacco smoke *(1)*. As burning tobacco is ultimately unnecessary to aerosolize nicotine (although it is efficient), alternative means for liberating nicotine from tobacco in an inhalable form have been explored; these might modify the toxicological risk associated with conventional tobacco cigarettes and influence consumer acceptability, which the earlier generation of these products failed to do.

4.1.2 Decision FCTC/COP8(22) of the Conference of the Parties

This paper was commissioned by WHO to address a component of the request to the Convention Secretariat in paragraph 2 of decision FCTC/COP8(22) by the Conference of the Parties to the WHO Framework Convention on Tobacco Control (WHO FCTC) at its eighth session, on novel and emerging tobacco products to address key topics related to those products and to submit a comprehensive report.

4.1.3 Scope and objectives

This report contains reviews of research on variations among novel and emerging HTPs and examination of the findings in the context of relevant aspects of the request. To the extent possible, the paper includes descriptions of the characteristics and design features of different types of HTPs, their contents and emissions, product diversity, market distribution and manufacturers and discussion of the evidence for implications for users, non-users and regulators. Research gaps and priorities, some key questions and some recommendations for policy-makers are identified.

4.2 Variations among products on the market

4.2.1 Overview of product categories and types

HTPs have a different operating system from those of electronic nicotine delivery systems (ENDS) and electronic non-nicotine delivery systems (ENNDS) (Table 4.1). HTPs contain a source of heat, which heats tobacco and vaporizes the tobacco constituents into an inhalable nicotine-containing aerosol, which contains other toxicants. Although HTPs contain tobacco material (as opposed to liquid in ENDS), they differ from conventional tobacco cigarettes, which must be combusted to create and deliver an aerosol to the user. HTPs do not achieve high temperatures during combustion and therefore aerosolize the nicotine from tobacco at a lower temperature (contemporary HTPs operate at < 350 °C) than conventional tobacco cigarettes (800 °C) *(2,3)*. Like ENDS, they are products that are purported to emit smaller amounts of toxicants than in the smoke from conventional cigarettes.

Table 4.1. Product performance characteristics and primary ingredients of heated tobacco products (HTPs), tobacco cigarettes and electronic nicotine delivery systems (ENDS)

Characteristic	Tobacco cigarettes	HTPs	ENDS
Nicotine	Yes	Yes	Yes[a]
Tobacco	Yes	Yes	No (nicotine derived mainly from tobacco)
Combustion	Yes	No (potential risk of incomplete combustion)	No (risk of thermal degradation of ingredients in the nicotine solution)
Temperature	Yes (very high during puffs)	Yes (generally lower than in tobacco cigarettes; may be overheated)	Yes (generally lower than in tobacco cigarettes; may be overheated)
Electronic system	No	Yes	Yes
Example of product			

[a] Electronic non-nicotine delivery systems (ENNDS) do not contain nicotine.

HTP systems have three common components: an insert (such as a stick, capsule or pod) that contains processed tobacco; a means for heating the tobacco (battery, carbon tip or aerosol); and a charger for electrically heated devices. Manufacturers have used four basic design approaches to HTPs that could serve as a useful basis for product classification. They differ in how the tobacco material is heated and whether it is separated from the heating element (Table 4.2).

Table 4.2. Classification of heated tobacco products according to the tobacco heating mechanism

Key characteristics	Examples of products
Integrated heating element	Premier, Eclipse, PMI "Platform 2" (TEEPS)
External heating element with specialized "cigarettes"	Accord, Heatbar, IQOS, glo
Hybrid devices: "vapour" plus tobacco; indirect heating	iFuse, PloomTECH
Heating chamber for loose tobacco material	PAX

The first and arguably the oldest type is a cigarette-like device with an embedded heat source that can be used to aerosolize nicotine (HTP type 1). The second approach is use of an external heat source to aerosolize nicotine from specially designed cigarettes (HTP type 2). This is the basic design of the Philip Morris International (PMI) IQOS (and its progenitors Accord and Heatbar) and the British American Tobacco (BAT) glo. For regulatory purposes, these represent two classes, one based on the heating device and the other on tobacco-containing sticks. Sticks generally meet the World Customs Organization classification of a cigarette (roll of tobacco in paper; harmonized system 2402.20). In the USA, the IQOS device is regulated as an accessory by the Food and Drug Administration (FDA) (much like ENDS, ENNDS and waterpipes), while the HeatSticks meet the statutory definition of cigarettes *(4)*. A third approach is to use ENDS technology to derive nicotine and tobacco flavour from small amounts of tobacco (HTP type 3). BAT's iFuse product appears to be a hybrid ENDS–tobacco product, in which the aerosol is passed over tobacco before reaching the user. Japan Tobacco International's (JTI) PloomTECH operates similarly, except that the solution from which the aerosol is created appears not to contain nicotine. A fourth approach is use of a heated sealed chamber to aerosolize nicotine directly from loose tobacco (HTP type 4). This class is represented by personal dry-herb vaporizers such as PAX, which have been marketed mainly for use with cannabis but can also aerosolize nicotine from tobacco. In the USA, since at least the 1970s, tobacco companies have been interested in marijuana and its legalization as both a potential and a rival product. Heating devices such as HTPs could be used with inserts containing marijuana and marketed for use with cannabis.

4.2.2 Variations among heated tobacco products

The concept of heating rather than burning tobacco emerged in the 1980s from the tobacco companies Philip Morris and RJ Reynolds in the USA, with Accord and Premier, respectively. These products and conceptually similar ones have continued to evolve and may now be poised to capture a significant market share. The introduction, aggressive marketing and growing popularity of ENDS may have set the conditions for these products to succeed, in part by changing social norms and perceptions of cigarette smoking and of devices that deliver nicotine. Strategies similar to those for promoting ENDS and ENNDS have been used for aggressive promotion and marketing of HTPs, and these products are now available in about 50 countries with plans for expansion to other markets.

4.2.3 Market distribution of product types and categories

Owing to declining sales of cigarettes, increased awareness of the health risk of smoked tobacco, increasing implementation of the WHO FCTC and the recent commercial success of ENDS and ENNDS, tobacco companies have reintroduced HTPs on the global market. Even as their usefulness for public health remains unclear, the marketing strategies of PMI, BAT and JTI are based on claims that the products reduce harm. Each of those companies has launched new-generation HTP brands in several countries since 2014 *(5)*. The international HTP market was valued at US$ 6.3 billion in 2018 *(6)*, with substantial market growth forecast over the next few years *(7)*. HTPs have attained a significant share of the tobacco market, particularly in Japan. Reports from market analysts indicate that Japan has the best-developed HTP market in the world, accounting for 85% of HTP sales in 2018 *(6)*, and tobacco inserts for PMI's primary HTP brand (IQOS) comprised 17% of all tobacco sales in July–September 2019 *(8)*. Table 4.3 shows product and pricing information for the three main HTPs in Japan. In the Republic of Korea, IQOS and domestic HTPs (KT&G, lil®) were launched simultaneously in 2017 *(9)*. HTPs are also gaining popularity elsewhere. BAT reported that its HTP brand, glo, had at least a 5% share of national tobacco markets in Poland, Romania and Serbia in June 2019 *(10)*. IQOS has been retailed online and in selected metropolitan shops in the United Kingdom (England) since December 2016 and in Canada since April 2017, while BAT launched glo in Vancouver, Canada, in May 2017. Early data suggest that, in both countries, awareness of HTPs was limited and uptake negligible 3–6 months after entry of HTPs to the market, and BAT halted glo sales in Canada in September 2019 *(11)*. Nevertheless, since 2018–2019, PMI has reported revenue increases of 92.5% in the United Kingdom (England) and 44.2% in Canada in their so-called "reduced risk" product line (including IQOS) *(7)*, suggesting that demand for HTPs is growing. In stark contrast, the sale of HTPs is effectively barred in Australia *(12)*, banned in a few other countries, including India, Saudi Arabia and Singapore, and, until recently, the USA prohibited sales of IQOS, glo and other contemporary HTPs. In April 2019, however, IQOS and three varieties of "HeatSticks" were authorized for sale as tobacco products *(13)*, with other HTP brands expected to follow suit. On 7 July 2020, the FDA authorized IQOS as a "modified risk tobacco product" after an assessment of scientific studies that indicated that switching completely from conventional cigarettes to the IQOS system significantly reduced exposure to harmful or potentially harmful chemicals *(14)*.

Table 4.3. Product and pricing information for the three most popular heated tobacco products in Japan

	IQOS	glo	Ploom TECH
Device image			
Manufacturer	Philip Morris International	British American Tobacco	Japan Tobacco International
Launched	November 2014	December 2016	March 2016
Type of tobacco insert	Stick	Stick	Capsule
Device generations	1st: IQOS 2nd: IQOS 2.4 3rd: IQOS 3 and IQOS 3 Multi	1st: glo 2nd: glo Series 2 and glo Series 2 Mini	1st: Ploom TECH 2nd: Ploom TECH+ and Ploom S[a]
Brand name of insert	Marlboro Heatsticks	Kent Neostick	Mevius for Ploom TECH
Flavours of inserts	Balanced Regular, Menthol, Mint, Purple Menthol, Smooth Regular	Bright Tobacco, Citrus Fresh, Dark Fresh, Fresh Mix, Intense Fresh, Refreshing Menthol, Regular, Rich Tobacco, Smooth Fresh, Spark Fresh, Strong Menthol	Brown Aroma, Cooler Green, Cooler Purple, Red Cooler, Regular
Price[b]	IQOS 2.4: ¥ 7980 (~US$ 76) IQOS 3 Multi: ¥ 8980 (~US$ 85) Marlboro Heatsticks: ¥ 500 (~US$ 4.73)	glo: ¥ 2980 (~US$ 28) glo Series 2: ¥ 2980 (~US$ 28) glo Series 2 Mini: ¥ 3980 (~US$ 38) Kent Neostick: ¥ 420 (~US$ 3.97)	Ploom TECH: ¥ 2980 (~US$ 28) Ploom TECH+: ¥ 4980 (~US$ 47) Ploom S: ¥ 7980 (~US$ 76) Mevius for Ploom TECH: ¥ 490 (~US$ 4.64)

Source: reference 15.
[a] Has a stick instead of a capsule.
[b] In comparison, the price of a pack of conventional cigarettes is approximately ¥ 500 (about US$ 4.73).

4.3 Characteristics and design features of products

4.3.1 Temperature profiles of products and operational capabilities

In PMI's IQOS, the tobacco stick ("HeatStick") is heated by a blade inserted into the end, so that the heat dissipates through the tobacco plug on a puff *(2)*. The aerosol then passes through a hollow acetate tube and a polymer-film filter on the way to the mouth. The product is designed not to exceed 350 °C, at which point the energy supplied to the blade is cut off. BAT describes its glo product as a heating tube consisting of two separately controlled chambers, which are activated by the user by a button on the device, and reach an operating temperature of 240 °C, within 30-40 s *(3)*. BAT's iFuse product appears to be a hybrid HTP, which has a liquid component but also contains tobacco; it passes the aerosol generated from the liquid over tobacco before it reaches the user. One study indicated a small heat loss in the aerosol when it is passed over the tobacco chamber (from 35 °C to 32 °C), implying some tobacco heating *(16)*. The delivery of toxicants under machine-smoking conditions without the tobacco chamber is reported to be nearly identical to that from an ENDS, implying a minimal contribution of the tobacco. JTI's PloomTECH operates in a similar manner, except that the solution used to create the aerosol appears not to contain nicotine.

4.3.2 Battery characteristics

Different sources of heat are used in different HTP devices, including electric energy from a battery and carbon tip that is lit with a match or lighter. Most HTPs have lithium ion batteries, which are rechargeable and are used in many products, such as laptops, mobile phones and electric cars. All lithium ion batteries operate in the same way: the ions flow, in a solvent, between two oppositely charged poles separated by a permeable thin sheet. The direction of the flow depends on whether the battery is charging or discharging. Generally, lithium ion batteries are considered safe, however, if the separator between the poles is breached, the poles short-circuit, causing an increase in temperature, which in turn causes the highly flammable electrolyte solvent to combust, with an explosion.

4.3.3 Properties of tobacco inserts, sticks and capsules

The sticks for PMI's IQOS (45 mm long, 7 mm diameter) contain approximately 320 mg of tobacco material, whereas a conventional cigarette is 84–100 mm long, 7.5–8.0 mm in diameter and contains about 700 mg of tobacco material *(17)*. The tobacco in IQOS appears not to be typical tobacco cut-filler but rather a reinforced web of cast-leaf tobacco (a type of reconstituted tobacco), which contains 5–30% by weight of aerosol-forming components such as polyols, glycol esters and fatty acids *(17)*. This composition is advantageous as an aerosol-forming substrate for use with a heating system. A tobacco insert for BAT's glo product is an 82-mm long, 5-mm diameter stick inserted into the heating chamber. The stick consists of a tobacco rod, a tubular cooling section, a filter and a mouthpiece. It contains approximately 260 mg of reconstituted sheet tobacco with 14.5% glycerol as the aerosolizing agent (3). BAT's iFuse product appears to be a hybrid, with a cartridge containing nicotine solution at a concentration of 1.86 mg/mL, with machine delivery of 20–40 μg/puff (16), and tobacco. It is, however, difficult to estimate the contribution, if any, of the tobacco in the iFuse device to the delivery of aerosolized nicotine, which the user inhales. JTI's PloomTECH operates in a similar manner, except that the liquid ENDS-like component appears not to contain nicotine (Fig. 4.1).

Fig. 4.1. Tobacco sticks and capsules in different heated tobacco products

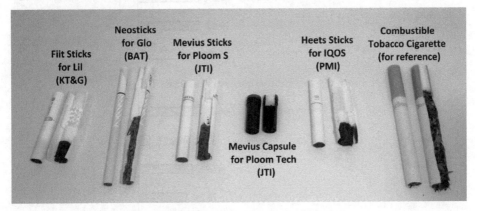

4.4 Content, emissions and general design of products

4.4.1 Content and emissions

Little is currently known about the individual ingredients and the emissions of HTPs products, and the health effects of many aerosolized constituents from tobacco and the other contents of the products, including humectants and additives, should be investigated to determine their effects on the health of users and non-users. HTP tobacco inserts often contain and emit toxicants, including cancer-causing chemicals, respiratory irritants and cardiovascular toxicants such as tobacco-specific nitrosamines, metals, volatile organic compounds, phenolic compounds, polycyclic aromatic hydrocarbons and minor tobacco alkaloids and organic solvents (18–21). Many of those chemicals are classified as carcinogens and as harmful and potentially harmful when inhaled. These toxicants are present in HTPs in various amounts, although typically at levels lower than those found in conventional cigarettes (Fig. 4.2).

Fig. 4.2. Percentage decrease in yields of selected harmful chemicals in emissions from IQOS from those in smoke from conventional cigarettes

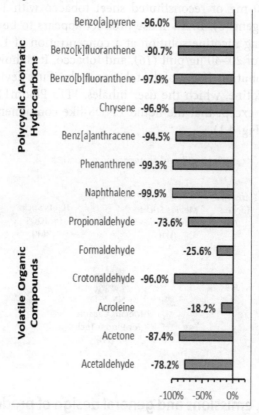

Source: reference 18.

A toxicological review by the FDA found lower levels of some harmful and potentially harmful constituents in IQOS aerosol than in smoke from 3R4F standard reference research cigarettes and commercially available conventional tobacco cigarettes; however, the FDA also found that 80 chemicals in HeatStick aerosols, including four that are possibly carcinogenic, are unique to IQOS or present at higher levels than in 3R4F smoke. Additionally, IQOS aerosol contains 15 other chemicals that are possibly genotoxic and 20 more compounds generally recognized as safe for ingestion that have potential adverse health effects (22,23). The FDA Technical Project Lead Review concluded that "although some of the chemicals are genotoxic or cytotoxic, these chemicals are present in very low levels and potential effects are outweighed by the substantial decrease in the number and levels of harmful and potentially harmful chemicals found in conventional cigarettes" and that the presence of these chemicals does "not raise significant concerns from a public health perspective" (24).

The risks associated with inhaling large doses of the humectants used in HTPs, e.g. propylene glycol and vegetable glycerine, are not well characterized, although they have been approved for use for other purposes. For example, propylene glycol is commonly used as an additive in foods and cosmetics, a solvent in pharmaceuticals, an antifreeze and as an ingredient of theatrical mist or fog. Studies of the health effects in theatre staff exposed to such mists concluded that massive, prolonged exposure results in irritation of the airways *(25,26)*. Vegetable glycerine, although widely used in the food and chemical industry as a non-toxic additive, may pose risks as used in HTPs because they can generate toxic aldehydes (including formaldehyde, acetaldehyde, acrolein and acetone) at high temperatures, some of which are classified as carcinogens.

Some HTPs contain flavouring agents, including menthol and fruit (Fig. 4.3). (See also section 6.) Although most of the flavourings are also commonly used in foods and indoor fragrances, little is known about their health effects when inhaled. Use of flavours in HTPs is widespread and is often cited as the main reason for using each product *(15)*. Flavours also reportedly play an important role in shaping consumer perceptions of emerging tobacco products, as their use is associated with experimentation or initiation of use *(27)*. A wide range of flavours is available on the market, which include tobacco, menthol, fruity and coffee. Certain chemicals (e.g. vanillin, limonene, isoamyl alcohol) and classes of chemicals that are used to provide the taste and odour of these flavours have been associated with respiratory toxicity *(28)*. Few studies have addressed the role of flavours in use of HTPs. There is also little information on whether smokers who initiate use of flavoured HTPs do so to substitute completely for conventional cigarette use or whether the same association is seen in countries with different regulatory environments for HTPs.

Fig. 4.3. Varieties of flavoured tobacco inserts, cartridges and capsules for heated tobacco products

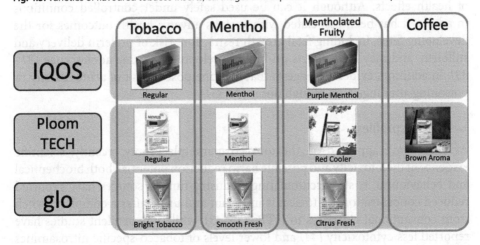

HTPs, as a product class, are exceptionally heterogeneous, with differences in the source of heat, heating elements and heating temperature. Each of these characteristics can influence the emissions of nicotine and non-nicotine toxicants and their delivery to users. Heating temperature in particular affects emissions of respiratory toxicants (including carbonyl compounds like formaldehyde, acetaldehyde and acrolein) that are generated by thermal degradation of humectants. Some HTPs may generate high levels of toxic chemicals.

4.4.2 Nicotine delivery

In most HTPs, nicotine is released from tobacco by heating the tobacco to a temperature that aerosolizes nicotine but may not achieve complete combustion of the plant material. In some hybrid HTPs, nicotine is present in the solution used to create inhalable aerosols. Volatilized nicotine that is not derived from combustion would, in principle, produce a less complex aerosol, with fewer toxic constituents than that from conventional tobacco cigarettes. Nicotine is not efficiently delivered as a gas; to deliver nicotine to a user's lungs, an aerosol-forming agent must be added to suspend nicotine on aerosol particles.

Nicotine is present in HTP aerosol in one of two forms: an unprotonated free base and a protonated salt. Nicotine is a weak base, and the fraction of any one form can be increased or decreased by altering the pH of the liquid. Free-base nicotine is volatile, is absorbed more readily than the monoprotonated form, produces enhanced electrophysiological and subjective responses in humans and may therefore be more addictive *(29)*. Free-base nicotine is, however, harsher on the throat than protonated nicotine when inhaled *(29)*. Laboratory studies have shown that nicotine in IQOS aerosol is present primarily in a protonated salt form *(20,30)*.

Nicotine is a pharmacologically active compound with a wide range of health effects. Although it can be used safely under controlled conditions in adults, it has been associated with various adverse health outcomes for the developing fetus, including fetal growth restriction, risk of preterm delivery and stillbirth, and may have effects on brain development during adolescence *(31–33)*. In addition, evidence suggests that nicotine poses a risk of acute toxicity or poisoning after ingestion of high doses *(34,35)*.

4.4.3 Risk profiles

As HTPs have appeared on the tobacco marketplace only recently, scientific evidence of their toxicity and possible harm reduction potential, both biochemical and behavioural, is still accumulating. Industry-funded sources have reported reduced concentrations of toxicants in serum and urine after a complete switch from conventional cigarettes to HTPs *(36–43)*, while independent studies have reported less cytotoxicity *(44)* and lower levels of tobacco-specific nitrosamines after use of HTPs than after smoking conventional cigarettes, although the levels are higher than with ENDS *(21)*. Unlike ENDS, HTPs contain tobacco and, therefore, if there is no combustion, are expected to expose users to numerous chemicals present in the tobacco material. Moreover, some tobacco constituents that would be completely burnt and decomposed during combustion may be present in emissions from HTPs. Thus, HTPs may deliver a unique chemical mixture with a distinct toxicity profile, and the potential benefits of reductions in exposure to selected toxicants might be replaced by new health risks. Side-stream emissions and second-hand exposure are also a concern with HTPs. Some tobacco manufacturers claim that certain HTPs result in minimal side-stream exposure, whereas other studies show more substantial levels *(45)*. This may depend in part on the design of each product.

In the studies of human exposure currently available (all conducted by manufacturers), IQOS appears to deliver fewer toxicants than cigarettes and may serve as an effective short-term substitute for cigarettes, as assessed by nicotine delivery and subjective effects. Published data on BAT's HTPs are limited, although a study protocol for a randomized trial was published recently, suggesting work in this area by BAT. No published studies are available on JTI's Ploom product. Recent studies have, however, suggested a high prevalence of concurrent use of HTPs with conventional cigarettes and/or ENDS *(46–48)*. Studies on concurrent use of ENDS and conventional cigarettes did not generally show any significant reduction in exposure to tobacco-specific toxicants *(49–51)*. Thus, it is unlikely that the exposure of dual users of HTPs and conventional cigarettes to toxicants is significantly decreased.

4.4.4 Regulatory implications of the contents of heated tobacco products

Introduction of HTPs into the current landscape of nicotine and tobacco products presents a regulatory challenge (see also section 3), as it is currently impossible to determine whether these products could play a role in reducing risk or harm to smokers. The variation in regulations on new and emerging nicotine and tobacco products among countries will probably affect nicotine content and emission, use of devices, uptake by non-smokers and substitution for conventional cigarettes among smokers. The differences in regulatory environments could also influence overall use and consumer behaviour by affecting the availability of these products. Thus, it is important to understand the nature of the emerging tobacco product market, including the diversity of HTPs and how they are used, to fully appreciate the impact of the regulatory environment, the potential implications of different types of devices on use and their potential harm.

HTPs are often not subject to mandatory manufacturing standards, resulting in minimal oversight to prevent inaccurate labelling of ingredients or the presence of toxic components. Although the commercial sensitivity of flavour recipes may limit disclosure of constituents, especially for ingredients of natural origin, governments and health authorities should require reporting of HTP constituents to appropriate government agencies. Several constituencies have enacted policies to regulate both the manufacture and marketing of flavoured tobacco products, including HTPs, to reduce their public appeal and health impact; these include the European Union, Canada, Ethiopia and the Republic of Moldova (52). Some countries consider HTPs to be tobacco products, and the provisions that apply to other tobacco products apply also to HTPs. Various product standards may be imposed on HTPs by a regulatory body, such as maximum nicotine content and yields and thresholds for tobacco-related toxicants, heavy metals, pesticides, residual solvents, mould, yeast, mycotoxins and other chemical and biological impurities. Product standards for HTPs may also specify chemicals that are prohibited, including certain flavourings, colouring agents and sweeteners.

Regulators should also be alert to changes in the design and composition of a product. In a presentation to the Consumer Analyst Group of New York (53), PMI claimed that IQOS had been modified from the original Japanese design in 2014–2017 in its aesthetics, blade self-cleaning technology, improved user interface, faster charging, Bluetooth connectivity, an accompanying mobile application and use of colours to increase the appeal of the device. Thus, a product may change after its introduction, as seen in the tobacco industry's practice of making minor adjustments to their cigarette products over time and among markets, such that a 2017 Marlboro is not necessarily identical to a 2010 Marlboro, and a Marlboro sold in France is not necessarily the same as one sold in the USA. In addition, research on a product may be done not with the product currently available to consumers but with a prototype (or even a series of different prototypes). While this practice is not in itself nefarious, any differences in design, function or presentation between the studied and the marketed product should be established and if and how such differences might affect consumer use. In the USA, the FDA follows a marketing authorization pathway in which changes to a product must be reported and can be monitored, with greater scrutiny and requirements for changes that impinge on public health (e.g. changes to delivery of harmful constituents; substantial design changes). European Union Member States, in the Tobacco Products Directive, have similar provisions. Regulators in other countries should consider a similar requirement for notification and justification of changes to products.

4.4.5 Regulatory implications of emissions

Given that HTPs have been found to contain various toxicants, product testing and constituent analysis can indicate the potential exposure of consumers to chemicals of concern for health. New tobacco products could be tested at independent laboratory facilities licensed for analytical testing. It is unclear whether the toxicants potentially emitted by HTPs differ among markets. For example, whether these products are covered by smoke-free legislation depends on the specific wording (53,54). The precautionary principle would support their inclusion in such regulations.

An additional concern with respect to emissions from HTPs is variation in the results reported by laboratories, which may result from use of differences in the analytical methods, the products tested and the puffing regimens used. For example, the puff duration for sampling aerosols may differ among laboratories, which will influence the toxicants emitted, as increasing the puff duration increases the mass of aerosol inhaled and exposes users to higher levels of toxicants. Other aspect of puffing protocols that may influence emissions include the number of puffs, flow rate and inter-puff intervals. Use of a standard puffing protocol designed for conventional cigarettes, such as those of the International Organization for Standardization or Health Canada (intense), could limit differences in results for emissions. A standardized puffing regimen may not, however, always be applicable, as the design of some products may limit it. For example, IQOS is designed to ensure the same puff duration and number of puffs as a cigarette, i.e. up to 14 puffs or 6 min of use. The puffing protocol may therefore be dictated by the device tested. This aspect of HTPs raises an additional challenge for designing regulations.

While the major ENDS and HTP companies sell their products worldwide, some smaller companies sell entire devices and device parts. For example, some tobacco inserts that are not manufactured by PMI can be used in IQOS devices (55,56). Tobacco sticks manufactured by Imperial Brands Plc in the United Kingdom for their Pulze HTP system also fit the IQOS device. Such combinations might yield toxicants different from those from the original product, due to changes in features such as electrical power. In general, combinations of products should be considered in designing regulations.

4.5 Variations among products, manufacturers and selling points

4.5.1 Manufacturers and selling points

Most current HTPs are manufactured and marketed by major transnational tobacco companies. While the distribution of HTPs is markedly unique, the inserts for HTPs are commonly sold at conventional retail outlets, and the devices are sold in specialty shops and online. The specialty shops therefore rely on a single product, and sales representatives explain the device, provide free cleaning of customers' devices and propose a free trial of the product (Fig. 4.4). HTP specialty shops usually have clean, sleek, modern designs (like the aesthetic of an Apple outlet).

Fig. 4.4. HTP shops in Japan and the Republic of Korea (2019)

4.5.2 Implications for customer pulling power

Cigarette smoking remains one of the leading causes of preventable morbidity and mortality worldwide, even as the prevalence of smoking continues to decrease in many countries. The main driver of the health consequences of tobacco smoking is inhalation of combustion by-products during use. Newer nicotine and tobacco products might therefore reduce health risks if smokers switch to exclusive use of products, with modified chemical and physical properties. The potential usefulness of HTPs for reducing harm has been examined. Industry data suggest that HTPs can serve as a long-term substitute in highly controlled settings *(36–41)*. As noted elsewhere in this report, however, the issues should be addressed by independent research. There is also concern that "real-world" concurrent use of cigarettes and HTPs might prolong smoking behaviour, whereby smokers initiate HTPs according to the situation, rather than switching from smoking completely *(57)*. Additionally, population studies in several Asian countries have found substantial use of other tobacco products with HTPs, raising concern about whether the products can replace conventional tobacco smoking or are rather used as complementary products *(15,46–48,58)*. Large tobacco companies use a variety of marketing strategies to promote HTPs to different sociodemographic groups, and differentiated marketing of cigarettes has disproportionately appealed to population subgroups such as adolescents and young adults, women, minorities and health-concerned smokers. (See section 10.)

Some concern has been raised about HTP users' privacy and security and how their personal information is collected and handled by HTP devices and by tobacco companies. HTPs are the first tobacco products that can harvest personal data on users' tobacco habits. PMI is already building a database of IQOS customers who register with the company *(59)*. Some HTP devices, including PMI's IQOS, are equipped with microcontroller chips that can store usage information and potentially transmit the information to the producer. The data could include details such as the number of puffs taken and how many times the user smoked the device in each day. The acquired data could potentially be used by tobacco industry in marketing. According to a statement by a PMI representative, the company extracts data from the device when investigating a malfunction *(59)*.

4.6 Discussion

HTPs are battery-operated devices that deliver nicotine aerosolized from tobacco, as well as other toxicants, to users. They are an emerging class of "potentially reduced exposure products" promoted by manufacturers as "reduced risk", "cleaner alternatives" and/or "smoke-free" products. HTPs have been marketed around the world with claims that they are less harmful than conventional cigarettes because they expose users to lower levels of some toxicants. There is, however, little evidence from independent studies on the chemistry of the aerosol produced by various types and brands of HTPs, the toxicology, effects on clinical measures, perceptions of the product and its packaging and behavioural factors.

HTPs present numerous regulatory challenges. The amounts of nicotine and non-nicotine toxicants depend on product features such as the composition of the tobacco insert, the temperature to which the heating element can rise and device design and characteristics. Understanding how these features influence important product characteristics such as temperature and emissions of nicotine and non-nicotine toxicants is essential for designing effective regulations and limiting the toxicity of these products. There is currently wide variation among the devices on the market, and users can control many of the devices' features that affect emissions, including those of numerous harmful chemicals, such as aldehydes, metals, volatile organic compounds and reactive oxygen species. As HTPs may emit chemicals that are not present in conventional cigarettes, chemical assessment of emissions from HTPs should go beyond those found in cigarette smoke. Additionally, the technology of HTPs is evolving rapidly, with new, more advanced devices constantly entering the market. These devices may have new features that could increase the levels of toxicants in emissions. Keeping track of the HTPs on the market and of any new features they may have will allow assessment of how the new features affect the emissions of aerosol toxicants and therefore users' health.

An important concern is that HTPs could increase concomitant use of other tobacco products. Although industry data suggest that HTPs could be used as a long-term substitute in highly controlled settings, there is concern that "real-world" concurrent use of cigarettes and HTPs might prolong smoking behaviour, whereby smokers initiate HTPs for use according to the situation rather than switching completely from smoking. Current smokers may not understand that "switching completely" means that they would have to quit conventional cigarettes to achieve any claimed health benefits of HTPs. According to WHO, however, "quitting" is complete cessation of tobacco use for at least 6 months with no use of cessation aids. As HTPs are tobacco products, conversion from use of conventional cigarettes to HTPs would not constitute cessation *(60)*. Currently, there is no information on how HTPs affect smokers' intentions to quit smoking.

4.7 Conclusions

- HTPs are an emerging class of "potentially reduced exposure products" promoted by manufacturers as "reduced risk", "cleaner alternatives" and/or "smoke-free" products.
- HTPs, as a product class, are exceptionally heterogeneous, differing in materials, configuration, content of tobacco inserts and temperature to which the heating element can rise. Each of these characteristics can influence emissions of nicotine and non-nicotine toxicants.
- HTPs contain and emit nicotine.
- HTPs emit numerous toxic chemicals, including tobacco-specific nitrosamines, aldehydes and metals, although exclusive users of those products appear to be exposed to lower levels of toxicants than cigarette smokers.
- Unlike ENDS, HTPs contain tobacco and are therefore expected to expose users to numerous chemicals present in the tobacco material.
- HTPs may emit chemicals that are not present in conventional cigarettes.
- While HTPs may expose users to lower levels of some toxicants than cigarettes, they might expose them to higher levels of other toxicants.
- As these products have been introduced recently into the tobacco marketplace, scientific evidence on their toxicity and long-term health effects is still accumulating.

- Although the public health utility of these new tobacco products remains unclear, tobacco companies are extensively using marketing strategies based on potential harm reduction.
- Although industry data suggest that HTPs could be used as a long-term substitute in highly controlled settings, independent population-based studies have raised concern that "real-world" concurrent use of cigarettes and HTPs might prolong smoking behaviour.

4.8 Research gaps, priorities and questions

Research conducted independently of the industry is required to inform product users, public health professionals and regulatory agencies about the potential public health impact of HTPs. As the tobacco marketplace continues to evolve in various regulatory environments, it is vital to assess trends in awareness and use of HTPs.

The chemical profile and toxicity of all emerging tobacco products, including HTPs, must be thoroughly investigated. It is important to understand where these products are positioned along the continuum of risk relative to conventional cigarettes, ENDS and other smoked tobacco products.

Studies of the prevalence of use and of substitution for tobacco cigarettes are limited. Many of these products were test-marketed in a single country or small geographical region. Thus, it can be difficult to predict the uptake of novel products, particularly among young people. The characteristics of current and potential users of new tobacco products should be considered in assessing their potential public health impact. The concept of a continuum of harm is often based on the toxicity profile of a product as compared with cigarette smoke, with less attention to the characteristics of users and other important factors.

It is important to understand whether HTPs could play a role in reducing the risks to smokers by reducing their exposure to certain toxic chemicals as compared with conventional tobacco smoke. They might either reduce the risk of smokers or impose serious risks to health. How this balance is viewed depends on the context.

4.9 Policy recommendations
Key recommendations

- Tobacco companies should be required to investigate the chemical profile and toxicity of all HTPs. HTP manufacturers should disclose the findings of product testing and provide detailed descriptions of the testing methods used to the appropriate regulatory agencies. The effects of the combination of factors in the product should be investigated.

- Policy-makers should continuously monitor the market to identify new HTPs and similar products and changed features of products and emissions. Changes to an existing product must be reported to regulatory agencies so they can be monitored, with greater scrutiny and requirements for changes that affect public health (e.g. changes to the delivery of harmful constituents; substantial design changes).
- New or modified products should be subjected to premarket review by appropriate regulatory agencies before marketing. All HTP components should be regulated as stringently as other tobacco products, including restrictions on labelling, advertising, sales to minors, price and taxation policies and smoke-free measures.

Other recommendations

- Research independent of the tobacco industry should be conducted to inform product users, public health professionals and regulatory agencies of the potential public health impact of HTPs and to assess trends in awareness and use of HTPs.
- The chemical profile and toxicity of all emerging tobacco products, including HTPs, should be established.
- Studies of the prevalence of use and of substitution for tobacco cigarettes should be conducted in several marketplaces, with consideration of the characteristics of current and potential users of newer tobacco products and assessment of the potential public health impact of the products.
- Carefully designed, independent studies should be conducted to understand the toxicity profile of the chemicals emitted from HTPs in order to assess the risks of smokers relative to those of non-smokers and smokers of conventional cigarettes. Such evaluations should address differences in prevalence, user behaviour and the population risk of other tobacco products.
- The privacy and security of HTP users should be protected. The collection and handling of personal information on HTP devices and by tobacco companies should be regulated.

4.10   References

1. Stedman RL. Chemical composition of tobacco and tobacco smoke. Chem Rev. 1968;68(2):153–207.
2. Smith MR, Clark B, Lüdicke F, Schaller JP, Vanscheeuwijck P, Hoeing J et al. Evaluation of the tobacco heating system 2.2. Part 1: Description of the system and the scientific assessment program. Regul Toxicol Pharmacol. 2016;81(Suppl 2):S17–26.
3. Eaton D, Jakaj B, Forster M, Nicol J, Mavropoulou E, Scott K et al. Assessment of tobacco heating product THP1.0. Part 2: Product design, operation and thermophysical characterisation. Regul Toxicol Pharmacol. 2018;93:4–13.
4. US Code: Federal Food, Drug, and Cosmetic Act, 21 U.S.C. §§ 301–392 (Suppl 3 1934). Washington (DC): US Congress; 1934 (https://www.loc.gov/item/uscode1934-005021009).
5. Bialous SA, Glantz SA. Heated tobacco products: another tobacco industry global strategy to slow progress in tobacco control. Tob Control. 2018;27(Suppl 1):s111–7.
6. Uranaka T, Ando R. Philip Morris aims to revive Japan sales with cheaper heat-not-burn tobacco. Reuters, 23 October 2018 (https://www.reuters.com/article/us-pmi-japan/philip-morris-aims-to-revive-japan-sales-with-cheaper-heat-not-burn-tobacco-idUSKCN1MX06E, accessed 29 August 2019).
7. Heat-not-burn tobacco products market by product and geography – Forecast and analysis 2020–2024. Toronto: Technavio; 2020.
8. Third-quarter results 2019. Lausanne: Philip Morris International; 2019 (https://www.pmi.com/investor-relations/reports-filings, accessed 10 January 2021).
9. Lee MH. KT&G's heat-not-burn cigar overcomes downsides of competitors. The Korea Times, 22 December 2017 (http://www.koreatimes.co.kr/www/tech/2017/12/133_241123.html, accessed 20 February 2019).
10. Half-year report for the six months to 30 June 2019. London: British American Tobacco; 2019 (https://www.bat.com/group/sites/UK__9D9KCY.nsf/vwPagesWebLive/DOBELLYE, accessed 10 January 2021).
11. glo is being discontinued. British American Tobacco (glo.ca).
12. Greenhalgh EC. Heated tobacco ("heat-not-burn") products. In: Tobacco in Australia: Facts and issues. Melbourne: Cancer Council Victoria; 2018.
13. FDA permits sale of IQOS Tobacco Heating System through premarket tobacco product application pathway. Silver Spring (MD): Food and Drug Administration; 2019.
14. FDA authorizes marketing of IQOS Tobacco Heating System with "reduced exposure" information. Press release; Silver Spring (MD): Food and Drug Administration; 2020 (https://www.fda.gov/news-events/press-announcements/fda-authorizesmarketing-IQOS-tobacco-heating-system-reduced-exposure-information, accessed 10 January 2021).
15. Sutanto E, Miller C, Smith DM, O'Connor RJ, Quah ACK, Cummings KM et al. Prevalence, use behaviors, and preferences among users of heated tobacco products: Findings from the 2018 ITC Japan survey. Int J Environ Res Public Health. 2019;16(23):4730.
16. Poynton S, Margham J, Forster M et al. Controlled aerosol release to heat tobacco: product operation and aerosol chemistry assessment. In: Society for Research on Nicotine and Tobacco, 2–5 March 2016, Chicago (IL) (https://cdn.ymaws.com/www.srnt.org/resource/resmgr/Conferences/2016_Annual_Meeting/Program/FINAL_SRNT_Abstract_WEB02171.pdf).
17. Batista RNM. Reinforced web of reconstituted tobacco. Neuchatel: Philip Morris Products SA; 2017.
18. Auer R, Concha-Lozano N, Jacot-Sadowski I, Cornuz J, Berthet A. Heat-not-burn tobacco cigarettes: Smoke by any other name. JAMA Intern Med. 2017;177(7):1050–2.
19. Davis B, Williams M, Talbot P. IQOS: evidence of pyrolysis and release of a toxicant from plastic.

Tob Control 2018;28(1).
20. Salman R, Talih S, El-Hage R, Karaoghlanian N, El-Hellani A, Saliba NA et al. Free-base and total nicotine, reactive oxygen species, and carbonyl emissions from IQOS, a heated tobacco product. Nicotine Tob Res. 2019;21(9):1285–8.
21. Leigh NJ, Palumbo MN, Marino AM, O'Connor RJ, Goniewicz ML. Tobacco-specific nitrosamines (TSNA) in heated tobacco product IQOS. Tob Control. 2018;27(Suppl 1):s37–8.
22. Lempert LK, Glantz S. Analysis of FDA's IQOS marketing authorisation and its policy impacts. Tob Control. 2020. doi: 10.1136/tobaccocontrol-2019-055585.
23. St. Helen G, Jacob III P, Nardone N, Benowitz NL. IQOS: examination of Philip Morris International's claim of reduced exposure. Tob Control. 2018;27:s30–6.
24. PMTA coversheet: Technical Project Lead Review (TPL), 29 April 2019. Silver Spring (MD): Food and Drug Administration; 2019 (https://www.fda.gov/media/124247/download, accessed 10 March 2020).
25. Varughese S, Teschke K, Brauer M, Chow Y, van Netten C, Kennedy SM. Effects of theatrical smokes and fogs on respiratory health in the entertainment industry. Am J Ind Med. 2005;47:411–8.
26. Teschke K, Chow Y, van Netten C, Varughese S, Kennedy SM, Brauer M. Exposures to atmospheric effects in the entertainment industry. J Occup Environ Hyg. 2005;2(5):277–84.
27. Meernik C, Baker HM, Kowitt SD, Ranney LM, Goldstein AO. Impact of non-menthol flavours in e-cigarettes on perceptions and use: an updated systematic review. BMJ Open. 2019;9(10):e031598.
28. Leigh NJ, Lawton RI, Hershberger PA, Goniewicz ML. Flavourings significantly affect inhalation toxicity of aerosol generated from electronic nicotine delivery systems (ENDS). Tob Control. 2016;25(Suppl 2):ii81–7.
29. Voos N, Goniewicz ML, Eissenberg T. What is the nicotine delivery profile of electronic cigarettes? Exp Opinion Drug Deliv. 2019:1–11.
30. Meehan-Atrash J, Duell AK, McWhirter KJ, Luo W, Peyton DH, Strongin RM. Free-base nicotine is nearly absent in aerosol from IQOS heat-not-burn devices, as determined by 1H NMR spectroscopy. Chem Res Toxicol. 2019;32(6):974–6.
31. Benowitz NL. Pharmacology of nicotine: addiction, smoking-induced disease, and therapeutics. Annu Rev Pharmacol Toxicol. 2009;49:57–71.
32. Dempsey DA, Benowitz NL. Risks and benefits of nicotine to aid smoking cessation in pregnancy. Drug Saf. 2001;24(4):277–322.
33. Benowitz NL. Toxicity of nicotine: implications with regard to nicotine replacement therapy. Prog Clin Biol Res. 1988;261:187–217.
34. Appleton S. Frequency and outcomes of accidental ingestion of tobacco products in young children. Regul Toxicol Pharmacol. 2011;61(2):210–4.
35. Solarino B, Rosenbaum F, Riesselmann B, Buschmann CT, Tsokos M. Death due to ingestion of nicotine-containing solution: case report and review of the literature. Forensic Sci Int. 2010;195(1–3):e19–22.
36. Martin Leroy C, Jarus-Dziedzic K, Ancerewicz J, Lindner D, Kulesza A, Magnette J. Reduced exposure evaluation of an electrically heated cigarette smoking system. Part 7: A one-month, randomized, ambulatory, controlled clinical study in Poland. Regul Toxicol Pharmacol. 2012;64(2 Suppl):S74–84.
37. Tricker AR, Kanada S, Takada K, Martin Leroy C, Lindner D, Schorp MK et al. Reduced exposure evaluation of an electrically heated cigarette smoking system. Part 6: 6-Day randomized clinical trial of a menthol cigarette in Japan. Regul Toxicol Pharmacol. 2012;64(2 Suppl):S64–73.
38. Tricker AR, Jang IJ, Martin Leroy C, Lindner D, Dempsey R. Reduced exposure evaluation of an electrically heated cigarette smoking system. Part 4: Eight-day randomized clinical trial in Korea. Regul Toxicol Pharmacol. 2012;64(2 Suppl):S45–53.

39. Urban HJ, Tricker AR, Leyden DE, Forte N, Zenzen V, Feuersenger A et al. Reduced exposure evaluation of an electrically heated cigarette smoking system. Part 8: Nicotine bridging – estimating smoke constituent exposure by their relationships to both nicotine levels in mainstream cigarette smoke and in smokers. Regul Toxicol Pharmacol. 2012;64(2 Suppl):S85–97. Erratum in: Regul Toxicol Pharmacol. 2015;71(2):185.
40. Tricker AR, Kanada S, Takada K, Leroy CM, Lindner D, Schorp MK et al. Reduced exposure evaluation of an electrically heated cigarette smoking system. Part 5: 8-Day randomized clinical trial in Japan. Regul Toxicol Pharmacol. 2012;64(2 Suppl):S54–63.
41. Tricker AR, Stewart AJ, Leroy CM, Lindner D, Schorp MK, Dempsey R. Reduced exposure evaluation of an electrically heated cigarette smoking system. Part 3: Eight-day randomized clinical trial in the UK. Regul Toxicol Pharmacol. 2012;64(2 Suppl):S35–44.
42. Schorp MK, Tricker AR, Dempsey R. Reduced exposure evaluation of an electrically heated cigarette smoking system. Part 1: Non-clinical and clinical insights. Regul Toxicol Pharmacol. 2012;64(2 Suppl):S1–10.
43. Gale N, McEwan M, Eldridge AC, Sherwood N, Bowen E, McDermott S et al. A randomised, controlled, two-centre open-label study in healthy Japanese subjects to evaluate the effect on biomarkers of exposure of switching from a conventional cigarette to a tobacco heating product. BMC Public Health. 2017;17(1):67.
44. Leigh NJ, Tran PL, O'Connor RJ, Goniewicz ML. Cytotoxic effects of heated tobacco products (HTP) on human bronchial epithelial cells. Tob Control. 2018;27(Suppl 1):s26–9.
45. Cancelada L, Sleiman M, Tang X, Russell ML, Montesinos VN, Litter Mi et al. Heated tobacco products: volatile emissions and their predicted impact on indoor air quality. Environ Sci Technol. 2019;53(13):7866–76.
46. Sutanto E, Miller C, Smith DM, Borland R, Hyland A, Cummings KM et al. Concurrent daily and non-daily use of heated tobacco products with combustible cigarettes: Findings from the 2018 ITC Japan Survey. Int J Environ Res Public Health. 2020;17(6):2098.
47. Hwang JH, Ryu DH, Park SW. Heated tobacco products: Cigarette complements, not substitutes. Drug Alcohol Depend. 2019;204:107576.
48. Kim J, Yu H, Lee S, Paek YJ. Awareness, experience and prevalence of heated tobacco product, IQOS, among young Korean adults. Tob Control. 2018;27(Suppl 1):s74–7.
49. Shahab L, Goniewicz ML, Blount BC, Brown J, McNeill A, Udeni Alwis K et al. Nicotine, carcinogen, and toxin exposure in long-term e-cigarette and nicotine replacement therapy users: A cross-sectional study. Ann Intern Med. 2017;166(6):390–400.
50. Goniewicz ML, Smith DM, Edwards KC, Blount BC, Caldwell KL, Feng J et al. Comparison of nicotine and toxicant exposure in users of electronic cigarettes and combustible cigarettes. JAMA Netw Open. 2018;1(8):e185937.
51. Czoli CD, Fong GT, Goniewicz ML, Hammond D. Biomarkers of exposure among "dual users" of tobacco cigarettes and electronic cigarettes in Canada. Nicotine Tob Res. 2019;21(9):1259–66.
52. How other countries regulate flavored tobacco products. Saint Paul (MN): Tobacco Control Legal Consortium; 2020 (https://www.publichealthlawcenter.org/sites/default/files/resources/tclc-fs-global-flavored-regs-2015.pdf, accessed 12 August 2020).
53. Event details. Philip Morris International Inc. presents at the Consumer Analyst Group of New York (CAGNY) conference. Philip Morris International, 19 February 2020 (http://www.pmi.com/2020cagny).
54. Sutanto E, Smith DM, Miller C, O'Connor RJ, Hyland A, Tabuchi T et al. Use of heated tobacco products within indoor spaces: Findings from the 2018 ITC Japan Survey. Int J Environ Res Public Health. 2019;16(23):4862.
55. Gretler C. Philip Morris has a Nespresso problem. Bloomberg, 21 June 2019 (https://www.bloomberg.com/news/articles/2019-06-20/philip-morris-arms-itself-to-battle-emerging-

iqos-knockoffs, accessed 12 August 2020).
56. Asun. IQOS sets to launch "cartridge recognition" technology to counter Imperial Tobacco. VapeBiz, 2 July 2019 (https://vapebiz.net/iqos-sets-to-launch-cartridge-recognition-technology-to-counter-imperial-tobacco/, accessed 12 August 2020).
57. Miller CR, Sutanto E, Smith DM, Hitchman SC, Gravely S, Yong HH et al. Awareness, trial, and use of heated tobacco products among adult cigarette smokers and ecigarettes users: Findings from the 2018 ITC Four Country Smoking & Vaping Survey. Tob Control. 2020. [In press.]
58. Kim SH, Kang SY, Cho HJ. Beliefs about the harmfulness of heated tobacco products compared with combustible cigarettes and their effectiveness for smoking cessation among Korean adults. Int J Environ Res Public Health. 2020;17(15):E5591.
59. Lasseter T, Wilson D, Wilson T, Bansal P. Every puff you take. Part 5. Philip Morris device knows a lot about your smoking habit. Reuters, 15 May 2018 (https://www.reuters.com/investigates/special-report/tobacco-iqos-device, accessed 12 August 2020).
60. WHO report on the global tobacco epidemic, 2019. Geneva: World Health Organization, 2019 (https://www.who.int/teams/health-promotion/tobacco-control/who-report-on-the-global-tobacco-epidemic-2019#:~:text=The%20%22WHO%20report%20on%20the,bans%20to%20no%20smoking%20areas, accessed 10 January 2021).

5. Use of heated tobacco products: product switching and dual or poly product use

Richard O'Connor, Department of Health Behavior, Roswell Park Comprehensive Cancer Center, Buffalo (NY), USA

Armando Peruga, Tobacco Control Group, Bellvitge Biomedical Research Institute, Barcelona, Spain; Centre for Epidemiology and Health Policy, School of Medicine Clínica Alemana, Universidad del Desarrollo, Santiago, Chile

Contents
    Abstract
    5.1 Introduction
    5.2 Information on HTP use at population level
    5.3 Dynamics of switching from conventional cigarettes to HTPs: Is dual or poly use a transitional or permanent state?
    5.4 Potential role of HTPs as a substitute for conventional cigarettes
    5.5 Exposure to nicotine and potential health risks among poly users
    5.6 Pharmacokinetics in animals
    5.7 Pharmacokinetics in people
    5.8 Subjective effects of use of HTPs and conventional cigarettes
    5.9 Discussion and implications
    5.10 Research gaps
    5.11 Policy recommendations
        5.11.1 Cessation policy
        5.11.2 Surveillance policy
        5.11.3 Research policy
        5.11.4 Cooperation and partnership policy
    5.12 References

Abstract

New-generation heated tobacco products (HTPs) are claimed by their manufacturers to assist smokers of conventional tobacco products to quit and switch completely to HTPs as a "safer alternative" source of nicotine. Concurrent use of two or more nicotine or tobacco products (poly use) consists of various types of behaviour, with heterogeneous frequencies of product use and different health risks, influenced by user characteristics. The capacity of newer-generation HTPs to substitute completely for conventional cigarette use is likely to depend on product features and on the characteristics of smokers, including experience, preparedness to switch to and use HTPs for a long time and perhaps the tobacco control regulatory environment. We reviewed literature on HTP use in both the laboratory and real-world contexts. Contemporary HTPs appear to deliver nicotine in pharmacologically meaningful doses in a similar manner to cigarettes or ENDS. Emerging independent studies indicate that poly use of cigarettes and HTPs is more common than implied by initial industry-sponsored studies. Little information is available on transitioning from cigarette smoking to HTPs, but the evidence suggests that smokers who use HTPs are more nicotine-dependent. Research is required on awareness about HTPs and the use behaviour in countries in which they have been introduced, including national surveys in the countries in which the products are available. Where their sale is permitted and the distinction is legally meaningful, HTPs should be considered as cigarettes rather than ENDS for the purposes of smoke-free laws, taxation, marketing and purchase.

5.1   Introduction

New-generation heated tobacco products (HTPs) are claimed by their manufacturers to assist smokers of conventional tobacco products to quit and switch to HTPs as a "safer, alternative" source of nicotine. This paper was commissioned by WHO to explore the potential role of HTPs in transitioning from conventional cigarettes (CCs) and other tobacco products. In particular, we addressed whether, as with use of electronic nicotine delivery systems (ENDS) and electronic non-nicotine delivery systems (ENNDS), the marketing of HTPs leads to significant concurrent use of HTPs and CCs or additional tobacco products and if dual or poly use with HTPs helps or prevents smokers of conventional tobacco products in switching completely to HTPs. As regulators are pressured by industry to apply more lenient regulations to HTPs than to other tobacco products, claims that these products are "reduced risk" or can help "smokers" to switch to other products must be carefully evaluated.

The difficulty in responding to these questions is due to the paucity of information on the use of HTPs at population level, whether the use is mainly dual or poly use and the health effects of such use. Data on exposure from exclusive or poly use are limited. Secondly, we found no empirical studies on whether HTPs can transition cigarette smokers completely from smoking (smoking cessation) or nicotine use (nicotine cessation); therefore, it is not possible to explore the relation between cigarette smoking cessation with HTPs among exclusive HTP users or poly users. Finally, no studies have been reported on cessation of HTPs per se.

5.2   Information on HTP use at population level

Table 5.1 summarizes the 13 studies found on the prevalence of HTP use (1–13), half of which describe experience of use of these products in Japan. Studies are available from only six countries between 2015 and 2019. All the data are from studies conducted at one time, except for one study in United Kingdom and one in Japan. Therefore, except for the United Kingdom, it is difficult to establish any trends in the use of HTPs in the general population. According to the most recent studies, in 2018 and 2019, about 3% of the adult population in Japan currently uses HTPs, with far lower numbers in the United Kingdom (England) and Poland. The studies do not provide any real sense of the frequency of dual use of HTPs and cigarettes or other smoking products (cigars, *bidis*, hookahs).

Table 5.1. Prevalence of HTP use (2015 – 2019): studies by country and year of data collection

Year	Germany	Italy	Japan	Poland	Republic of Korea	United Kingdom
2015			Adults aged 15-69[a] IQOS, current use, 0.3% Ploom, current use, 0.3%			
			Adults aged 16-69[b] HTP, ever use, 0.5% HTP, never smokers, 0.1% HTP, former smokers, 1.0% HTP, current smokers, 1.8%			
			Adult patients with NCDs aged 40–69[c] HTP, ever use, ♂1.7% ♀0.6% HTP, current HTP use, ♂0.8% ♀0%			
2016			Adults aged 15–69[a] IQOS, current use, 0.6% Ploom, current use, 0.3%			
2017	Current smokers and recent ex-smokers aged ≥ 14[d] Current HTP use, 0.3%	Adults aged ≥ 15[e] Ever tried IQOS, 1.4%	Adults aged 15–69[a] IQOS, current use[c] 3.6% Ploom, current use, 1.2% glo, current use, 0.8%		Young adults aged 19–24[g] IQOS, current use, 3.5%	Adults[h] HTP, ever use, 1.7%
			Adults[f] Current IQOS use, 1.8%			English adults[i] Current HTP use • Quarter 1, 0.1% • Quarter 2, 0.1% • Quarter 3, 0.1% • Quarter 4, 0.1%

Year	Germany	Italy	Japan	Poland	Republic of Korea	United Kingdom
2018			Adults[f] Current IQOS use, 3.2% Adult population ≥ 15[j] Current HTP use, 2.7% Daily HTP use, 1.7%		Secondary school students[k] HTP, ever use, ♂ 4.4% ♀ 1.2%	English adults[i] Current HTP use • Quarter 1, 0.1% • Quarter 2, 0.1% • Quarter 3, 0.1% • Quarter 4, 0.1%
2019				Adults aged ≥ 15[l] Current HTP use, 0.4%		English adults[i] Current HTP use • Quarter 1, 0% • Quarter 2, 0.1% • Quarter 3, 0.1% • Quarter 4, 0.2%

HTP: heated tobacco product; NCD: noncommunicable disease.
[a] Prevalence of current HTP use (i.e. use in the previous 30 days) was calculated from a longitudinal internet survey of a nationally representative sample of 8240 Japanese individuals (15–69 years old in 2015) followed up to 2017 *(1,2)*.
[b] Internet survey conducted between 31 January and 17 February 2015 among 7338 respondents aged 18–69 from a panel by Rakuten Research *(3)*.
[c] Among 4432 Japanese patients with chronic diseases aged 40–69 years from an Internet survey in 2015 *(4)*.
[d] Among 18 415 Germans over the age of 14 in a representative sample of people participating in a household survey between June 2016 and November 2017; 0.3% (95% CI = 0.09–0.64) of current tobacco smokers and new ex-tobacco smokers (< 12 months smoke-free) currently used HTPs. Use of HTPs increased with increasing education and income *(5)*.
[e] Among a sample of 3086 people selected by multistage sampling to be representative of the general Italian population aged ≥ 15 years and interviewed face-to-face *(6)*.
[f] Of 4878 Japanese adults in 2017, 1.8% used IQOS. Of 2394 Japanese adults in the first half of 2018, 3.2% used IQOS. Of "platform 1" registered users, 1.3% and 1.6% in 2017 and 2018, respectively, were never smokers, and 98% in 2016 and 98.6% in 2017 were dual users with tobacco *(7)*.
[g] Online survey in 2017 of 228 general young adults aged 19–24 years *(8)*.
[h] Among a nationally representative sample of 12 696 adults aged ≥ 17 interviewed by the market research company YouGov Plc in Great Britain in February–March 2017 *(9)*.
[i] From the Smoking Toolkit Study, a monthly household survey with a new representative sample of ~1800 English respondents aged > 15 each month. The fieldwork is conducted by the British Market Research Bureau. Cigarette smokers and recent ex-smokers (who smoked in the previous year) who agreed to be re-contacted are followed up 3 and 6 months later by postal questionnaire. The data on HTPs cover 63 499 adults since January 2017 *(10)*.
[j] Among 4684 Japanese adult participants in a nationally representative Internet survey conducted in February–March 2018; 2.7% used HTPs at least once a month and 1.7% daily. Among current smokers, 1.8% used HTPs at least once a month and about 50% daily. Among never smokers, 0.02% used HTPs at least once a month, all of them daily *(11)*.
[k] Among 60 040 adolescents attending secondary schools in the Republic of Korea as of April 2018, 4.4% of males and 1.2% of females ever used HTPs. About 6% were current smokers. Of these, 32.4% had ever used HTPs. Of the 86% never smokers, only 0.3% had ever used HTPs *(12)*.
[l] Among a representative nationwide sample of 1011 people aged ≥ 15 in Poland in September 2019; 0.4% used HTPs, all of whom were current smokers, representing 1.9% of current smokers *(13)*.

5.3 Dynamics of switching from conventional cigarettes to HTPs: Is dual or poly use a transitional or permanent state?

Concurrent use of two or more nicotine or tobacco products (poly use) comprises many types of behaviour and heterogeneous frequencies of product use and health risks, influenced by user characteristics. In the USA, poly users of tobacco tended to be male, use other drugs and be more nicotine-dependent *(14–23)*. Use of several tobacco products tends to be unstable over time *(24–27)*. Table 5.2 outlines some common definitions of poly use drawn from the published literature on ENDS and ENNDS.

Table 5.2. Types of poly use described in studies on tobacco and nicotine

Type of use	Reported use	Pros	Cons
Lifetime poly use	Ever use of two products or more in a lifetime	Broadest measure; captures a number of potential use patterns	Captures use that may have occurred years before or minimal experimentation that may have little impact on current behaviour or disease risk
Recent poly use	Use of two or more products in past 30 days	Captures contemporary use	Does not account for quantity or frequency of use; one use of two products is considered equivalent to daily use of each product
Predominantly poly user	Use of two or more products in past 30 days, used one of the products more than the other, daily or nearly daily	More comprehensive assessment of use pattern	Requires more questioning. Potentially subject to recall bias, particularly for the less frequently used product
Balanced poly user	Use of two or more products in past 30 days, in equivalent amounts, daily or nearly daily	More comprehensive assessment of use pattern	Requires more questioning. Potentially subject to recall bias, particularly for the less frequently used product
Intermittent poly user	Use of two or more products in past 30 days on at least some days, but no consistent pattern of use of any product	More comprehensive assessment of use pattern	Requires more questioning. Potentially subject to recall bias, particularly for the less frequently used product

While few studies are available on HTP poly use per se, studies of ENDS may be instructive. Borland and colleagues *(28)* analysed survey data from Australia, Canada, the United Kingdom (England) and the USA and described four subgroups of concurrent use of cigarettes and ENDS who differed in nicotine dependence, quitting behaviour and perceptions *(28)*: 1) dual daily users, 2) predominantly smokers (who used cigarette daily and ENDS less than daily), 3) predominantly vapers (who used ENDS daily and cigarette less than daily); and 4) concurrent non-daily users (who used both cigarette and ENDS less than daily). While many concurrent cigarette–ENDS users report trying to reduce smoking *(29,30)*, this claim tends not to be reflected in biomarkers of exposure *(31,32)*, and reductions in cigarettes per day may not meaningfully reduce the risk of mortality from smoking (33–36). A study by Baig and Giovenco (37) on dual use of ENDS and cigarettes suggests some probable transition pathways for different dual use behaviour. Broadly, dual users who had higher education or income were more likely to completely switch to e-cigarettes or to quit tobacco use over two years.

Currently, there is insufficient evidence to conclude that HTPs are less harmful than CCs. In fact, there is concern that, while they may expose users to lower levels of some toxicants than CCs, they expose users to higher levels of other toxicants (38–40). Studies indicate that up to 65% of HTP users in Japan and nearly all (96.2%) users in the Republic of Korea also smoked cigarettes (2,11,41–43). Sutanto and colleagues (44) analysed subgroups of poly users in Japan and found the overall distribution shown in Table 5.3.

Table 5.3. Proportions in four subgroups of concurrent users of HTPs in Japan, 2018

Weighted percentages (95% confidence interval)	Daily HTP user (n=594)	Non-daily HTP user (n=265)
	51.5 (46.7–56.3)[a]	48.5 (43.7–53.3)[a]
Daily smoker (n=3686) 94.4 (91.9–96.2)[b]	Dual daily user 51.0 (46.2–55.7)	Predominantly smoker 43.4 (38.6–48.4)
Non-daily smoker (n=213) 5.6 (3.8–8.1)[b]	Predominantly HTP user 0.5 (0.2–1.3)	Concurrent non-daily user 5.1 (3.4–7.6)

Source: reference 44.
[a] Values shown are the sum of the overall column.
[b] Values shown are the sum of the overall row.

In 2018, most HTP users in Japan concurrently smoked cigarettes, and most used both products every day (44). While there was no difference between exclusive daily smokers and dual daily users in the number of cigarettes per day, predominant smokers reported smoking more cigarettes per day than exclusive daily smokers, and predominant smokers used fewer tobacco-containing inserts per day than dual daily users; exclusive HTP users used more tobacco-containing inserts per day than dual daily users. Apart from greater frequency of use, this suggests that HTPs may not effectively substitute for cigarettes, consistent with the data from the Republic of Korea (41). Cigarette–HTP users were younger than exclusive smokers, while a study of actual use in the USA found greater interest among middle-aged smokers (45). Novel and emerging tobacco products often appeal to younger users for various reasons, including a perception of lower risk, marketing messaging and imagery and product appearance (29,46–49). Only about 10% of concurrent cigarette–HTP users planned to quit smoking in the next six months. This finding is in contrast to that of Borland et al. (28) that 50% of concurrent cigarette–ENDS users planned to quit but consistent with the findings of Baig et al. (37).

British American Tobacco (BAT) reported one study on use of its glo product *(50)*. Three groups took glo menthol, glo tobacco, glo and IQOS home for up to 14 days (with up to four laboratory visits). Participants reported using 12–15 cigarettes per day at baseline and used 8–12 units per day of the glo and IQOS products.

An application by Philip Morris International (PMI) to the US Food and Drug Administration for registration of IQOS as a "modified risk tobacco product" included a series of observational studies conducted by PMI on product switching in Germany, Italy, Japan, the Republic of Korea, Switzerland and the USA *(51)*. The study in the USA comprised 1106 current daily smokers who, after a 1-week baseline, were given access to IQOS (for free) for four weeks. The amount of product used (cigarettes and HeatSticks) was recorded in a diary. For this study, "switching" to IQOS was defined as using > 70% of total consumption as HeatSticks. About 15% of the participants met this definition by the end of the study, whereas 22% were dual users (30–70% of consumption as HeatSticks) and 63% were primarily smokers. In a second study, the product and its associated marketing were offered to 2089 daily smokers in Germany, Japan, Poland, the Republic of Korea and the USA. The prevalence of complete switching to a tobacco heating system (THS) ranged from 10% in Germany to 37% in the Republic of Korea, while dual use ranged from 32% in Japan to 39% in the Republic of Korea at the end of the 4-week trial. A 90-day use study on IQOS was conducted in the USA in 2013–2014, in which 88 of the 160 enrolled completed the study. It is noted that compliance with abstinence was substantially less in this study than in a similarly designed study in Japan, suggesting that experiences in one context cannot be generalized to others.

Limited post-market data are presented in the PMI application for IQOS. Those that are available are primarily from Japan, drawing on PMI's register of IQOS purchasers *(51)*. The proportion of exclusive IQOS use (> 95% of total consumption) increased from 52% to 65% between January and July 2016. Markov modelling of the transition in two cohorts of IQOS purchasers in Japan (in September 2015 and May 2016) suggested that those who transitioned to exclusive IQOS use are unlikely to transition back to exclusive cigarette use.

The Tobacco Products Scientific Advisory Committee of the US Food and Drug Administration expressed concern about some of the evidence on IQOS, and particularly the definition of "complete switching" and limitations of studies of consumer understanding of the claim of modified risk. The Committee offered qualified support only for a claim of exposure modification and expressed concern that the claims as worded would not be effective in communicating risk *(52)*.

In December 2017, the Committees on Toxicity, Carcinogenicity and Mutagenicity of Chemical Products in Food, Consumer Products and the Environment in the United Kingdom evaluated two HTPs on the market and concluded *(53)* that

> ... while there is a likely reduction in risk for smokers switching to "heat-not-burn" tobacco products, there will be a residual risk and it would be more beneficial for smokers to quit smoking entirely. This should form part of any long-term strategy to minimize risk from tobacco use.

As few studies are available specifically on exposure to and the health effects of contemporary HTP poly use and are short-term, studies on older HTPs were evaluated. A study of an early-generation HTP (Accord) *(54)* analysed concurrent use with use of subjects' own-brand cigarettes. After 6 weeks of use, Accord appeared to reduce cigarette smoking and exposure to carbon monoxide dose-dependently, i.e. more use of Accord was associated with fewer cigarettes smoked, and participants did not appear to increased their puff intensity when they reduced the number of cigarettes per day. A study of an early-generation HTP, Eclipse, was reported by Fagerstrom and colleagues *(55)*. After an initial 4-week randomized study, participants self-selected to use Eclipse (n=10), a nicotine inhaler (n=13) or cigarettes (n=13) for an additional eight weeks. At baseline, those who chose Eclipse smoked fewer cigarettes per day (18.0) on average than the other groups (20.4 for inhaler, 21.3 for cigarette). Over the 8 weeks, 30–60% of participants reported smoking no cigarettes at all, and an average of 2.6 cigarettes were smoked per day, with little change over time. Overall, the older literature suggests incomplete substitution of HTPs for cigarettes, which is generally not associated with a meaningfully lower health risk *(56–58)*, rather than the complete switching on which HTP's promotion of harm reduction is predicated.

5.4 Potential role of HTPs as a substitute for conventional cigarettes

PMI stated in January 2018 that "more than 3.7 million smokers outside the US have switched exclusively to IQOS in only two years. At the same time, non-smokers and former smokers show very little interest in the product" *(59)*. We were unable, however, to find any empirical study to substantiate this claim or any other with regards to use of new-generation HTPs to transition cigarette smokers from smoking. Some surveillance was conducted in the United Kingdom (England), where 0–1.4% of 4155 adults who smoked and tried to stop or who had stopped in the 12 months before the survey mentioned HTPs as a method for switching after 2016, depending on the quarter, with a median quarter prevalence of 0.4% *(10)*.

The capacity of the newer-generation HTPs to substitute for use of CCs probably depends on product features, the characteristics of smokers, including their experience and their preparedness to switch to HTPs for a prolonged period, and perhaps the characteristics of the tobacco control regulatory environment. In the absence of direct empirical evidence of the potential efficacy and effectiveness of HTPs in aiding switching, we used information on product features, such as their desirability and whether they deliver nicotine at a sufficient dose to reduce craving or withdrawal symptoms from CCs. The published studies of nicotine pharmacokinetics and evidence of the appeal of HTPs to smokers are described below, on the assumption that greater nicotine delivery and greater appeal might lead to greater substitution for cigarettes.

5.5 Exposure to nicotine and potential health risks among poly users

In this section, we compare use of HTPs with CCs among current smokers with regard to total nicotine delivered in mainstream emissions and key pharmacokinetics in plasma and urine.

Table 5.4 describes the 12 papers from 11 studies that we found up to January 2020 *(60–71)*, of which five were carried out or funded by the tobacco industry. The table updates and expands on that of Simonavicius et al. *(72)*. Comparison of the nicotine delivery in the aerosol of HTPs and the mainstream smoke of CCs is complicated by the variety of products and the methods used. The HTPs studied were IQOS (PMI), glo (BAT), iFuse (BAT) and a tobacco vaporizer. The reference products differed among studies. They included those developed for research, 3R4F 1R5F and 1R6F (most used 3R4F) and commercially available cigarettes, in which the yield of nicotine differs by brand, country and year. Mainstream emissions were obtained under the Health Canada Intense (HCI) regimen in most studies but with the International Standards Organization (ISO) regimen in others. It should be borne in mind that no machine-smoking regimen corresponds to human smoking and exposure, and their relevance to HTP use has not been validated. With these caveats, two conclusions can be reached from the studies.

- IQOS delivers about 70% of the nicotine contained in the smoke of CC. The relative nicotine delivery of IQOS is 40.7–102.8% (median 76.9%) when the reference is a commercial cigarette and 57–103% (median 64.7%) when the reference CC is one developed for research. The median in studies funded by the tobacco industry is not statistically significantly different from that in independent studies.
- The other HTPs studies appear to be less efficient (< 50%) than IQOS in delivering nicotine as compared with CC.

Table 5.4. Nicotine delivery in mainstream HTP aerosol and in mainstream smoke of conventional cigarettes

Reference	69	66	65	67	63	62	60	61	68	64	70	71
Year of publication	2019	2018	2018	2018	2018	2017	2017	2017	2017	2017	2016	2016
Tobacco industry affiliation or funding	No	No	No	No	Yes, BAT	No	No	No	Yes, BAT	Yes, PMI	Yes, PMI	Yes, PMI
Smoking machine	Custom-built	LM4E	SM450 linear	NR	LM	Custom-built	Custom-built	NR	NR	LM set to 12 puffs (1/30s)	LM SM405XR SM405RH LM20X	NR
Smoking regime	ISO, 6 puffs; HCI, 12 puffs	HCI, 12 puffs	ISO, HCI	HCI	HCI	HCI, 12 puffs	ISO 14 puffs	HCI 9–11 puffs	HCI for CC CORESTA 3–5 puff duration for HTP	HCI	HCI, ISO and five others based on observed HTP user puffing	NR
Conventional cigarettes — Product Type	Marlboro Red	48 commercial PM + 1R4F ISO, MDPH and HC smoking regimes[a]	3R4F	Marlboro Red, 1R6F	3R4F, 1R6F	Marlboro Regular	Lucky Strike Blue Lights	1R5F 3R4F	3R4F	3R4F	3R4F	3R4F
Conventional cigarettes — Nicotine mg/stick Total	ISO, 0.8; HCI, 1.8	Min. 1.07; max. 2.7	ISO, 0.71; HCI 1.9	MR: 1.07 1R6F: 0.65	1.86	1.99 mg/stick	0.361 mg/stick	1.0–1.7 mg/stick	1.8 mg/stick	1.86 mg/stick	1.89 mg/stick	1.88 mg/stick
Conventional cigarettes — Nicotine mg/stick FB	0.12–0.10	–	–	–	–	–	–	–	–	–	–	–

Reference			69	66	65	67	63	62	60	61	68	64	70	71
Heated tobacco product		Type	IQOS	IQOS	IQOS	IQOS	THS (IQOS) THP1.0 (glo)	IQOS	IQOS	IQOS	Disposable neopod (iFuse)	IQOS THS2.2 regular	IQOS THS2.2 regular and menthol	IQOS THS2.2 with 43 different tobacco blends
	Nicotine	Unit[b]	mg/stick	mg/stick	mg/stick	mg/stick		mg/12 puffs	mg/stick	mg/stick			mg/stick	Mean mg/stick
		Total	0.77–1.5	1.1	0.50–1.35	0.67	IQOS:1.16 glo: 0.462	1.4	0.301	1.1 (regular) 1.2 (menthol)	Average mg per puff of first 100 puffs 2.56	1.14	1.3 (regular) 1.2 (menthol)	Reference blend, 1.38 Range of 43 blends, 1.6–6
		FB	0.10–0.09											
	Relative to CC		83.3–96.3% 81.9–90.3%	40.7–102.8%[c]	70.4–71.1%	MR: 63% 1R6F: 103%	IQOS: 57% glo: 23%	70.4%	83.4%	64.7%	14.5%[d]	61%	70% (regular)[d] 64% (menthol)	Reference blend 73% Range of 43 blends: 87–33%

BAT: British American Tobacco; CC: conventional cigarette; CORESTA: Cooperation Centre for Scientific Research Relative to Tobacco; FB: Free-base; HCI: Health Canada Intensive; ISO: International Standards Organization; NR: not reported; PMI: Philip Morris International; THC: tobacco heating system
[a] As estimated in 2005 in a separate study (73).
[b] In comparison with the maximum and minimum nicotine yield of reference cigarettes. The authors report inconsistent nicotine delivery: the nicotine levels were initially < 50% of those found in the middle of the smoking procedure and therefore represent only 10–12% of the total nicotine yield. They argue that such inconsistency may influence consumer satisfaction, nicotine blood levels and adaptations of smoking behaviour.
[c] 14.5% is the authors' calculation; however, if the concentration in 14 puffs is the same as in other studies (2.56 * 0.14 = 0.358), the concentration is 19%.
[d] The authors reported a lower percentage in simulated Mediterranean, tropical and desert climate conditions, but did not report the exact figure.

5.6 Pharmacokinetics in animals

We found only one study in experimental animals. In this independent research, Nabavizadeh et al. *(67)* exposed three groups of eight rats each to IQOS aerosol from a single HeatStick, mainstream smoke from a single Marlboro Red cigarette or clean air. Exposure was for 1.5–5 min in a series of consecutive 30-s cycles, each cycle consisting of 5 or 15 s of exposure. After exposure, serum nicotine was about 4.5 times higher in rats exposed to IQOS than in those exposed to cigarettes, even though the IQOS aerosol contained about 63% the amount of the nicotine measured in smoke. When exposure to IQO emissions was shorter, the serum nicotine was similar in rats exposed to IQOS and cigarette emissions.

5.7 Pharmacokinetics in people

In most of these studies, values for biomarkers were reported after use of HTPs, ENDS and CCs. We report only the pharmacokinetics of nicotine after use of HTPs and CCs. The methods are summarized in Table 5.5.

Table 5.5. Methods used in studies on the pharmacokinetics of nicotine from HTPs and CCs in humans

Reference	Tobacco industry affiliation	Country	n	Participants	Design	Product HTPs	Product CCs	Exposure	Biomarker of exposure
74	PMI	Poland	160	Healthy white smokers aged 21–65 years with 3-year smoking history and smoked ≥ 10 CCs in past 4 weeks	Controlled, three-arm parallel, single-centre study in confinement with participants randomized to exclusive use of HTP, CC or abstinence	IQOS THS2.2	Own brand	Daily ad-libitum use of assigned product for 5 days	Urine NEQ
75	PMI	Japan	160	Healthy smokers aged 23–65 years with 3-year smoking history and ≥ 10 CCs in past 4 weeks	Controlled, three-arm parallel, single-centre study in confinement with participants randomized to exclusive use of HTP, CC or abstinence	IQOS THS2.2	Own brand	Daily ad-libitum of assigned product for 5 days	Urine NEQ
76	PMI	Japan	62	Healthy smokers aged 23–65 years with 3-year smoking history and ≥ 10 CCs in past 4 weeks	Open-label, cross-over study with participants randomized to exclusive use of HTP, CC or NRT gum	IQOS THS2.2 regular and menthol	Own-brand, regular and menthol	1 unit/day of assigned product (CC: 1 cigarette, HTTP: 1 stick used for 14 puffs or ≈ 6 min, Gum: 1 chewed for 35 ± 5 min) in sequences after 1-day washout. 1: THS-CC; 2: CC-THS; 3: THS-Gum; 4: Gum-THS	Plasma nicotine
77	PMI	United Kingdom	28	Healthy white cigarette smokers aged 23–65 years with 3-year smoking history and ≥ 10 non-menthol CCs in past 4 weeks. Excluded if used other tobacco products, ENDS or ENNDS	Open-label, randomized, two-period, two-sequence cross-over study with participants randomized 1:1 to exclusive use of HTP or CC	IQOS THS2.1	Own brand	HTTP: 1 stick and ad-libitum use daily in sequence of 1 day abstinence, 1 day single stick, 1 day ad libitum of one product; sequence repeated with the other product	Plasma nicotine
78, 79	PMI	Japan	160	Healthy smokers aged 23–65 years with BMI 18.5–32 kg/m², smoked ≥ 10/day menthol CCs in the past 4 weeks and reported smoking menthol CCs for ≥ 3 years	Controlled, three-arm parallel, single-centre study in confinement with participants randomized to exclusive use of HTP, CC or abstinence	IQOS THS2.2 menthol	Menthol own-brand	Limited ad-libitum use of assigned product for 5 days	Urine NEQ
80	PMI	Poland	112	Healthy white smokers with BMI 18.5–27.5kg/m²; aged 23–55 years, smoke 10–30 cigarettes per day, smoking for at least 5 consecutive years	Controlled, open-label, three-arm parallel group, single-centre confinement study with participants randomized 2:1:1 to exclusive use of HTP, CC or abstinence	CHTP prototype MD2-E7	Own brand	Ad-libitum use of assigned product for 5 days	Urine NEQ, cotinine
81	PMI	Poland	40	Healthy smokers who reported smoking for past 3 consecutive years and had smoked 10 commercially available non-menthol CCs daily for at least 4 weeks before the start of the study.	Controlled, open-label, two-arm parallel-group, single-centre confinement study with participants randomized 1:1 to exclusive use of HTP or CC	IQOS THS 2.1	Own brand	Ad-libitum use of assigned product for 5 days	Plasma nicotine, cotinine and urine NEQ

Reference	Tobacco industry affiliation	Country	n	Participants	Design	Product HTPs	Product CCs	Exposure	Biomarker of exposure
82	PMI	USA	160	Healthy smokers of at least 22 years of age who had a BMI 18.5–35 kg/m² and reported smoking for past three consecutive years and had smoked 10 commercially available non-menthol CCs daily for at least 4 weeks before the start of the study (verified on a urinary cotinine test). The subject did not plan to quit smoking within the next six months.	Subjects randomized (day 0) in a 2:1:1 ratio to the menthol THS, menthol CC, and abstention groups. Randomization was stratified by sex and daily mCC consumption quotas. The 5-day confinement period was followed by an 86-day ambulatory period and an additional 28-day safety follow-up period.	IQOS THS 2.2	Own brand	Ad-libitum use of assigned product for 5 days	Plasma nicotine, cotinine and urine NEQ
83	BAT	Japan	180	Healthy verified daily smokers of 10–30 cigarettes/day with a ≥ 3-year history of smoking; exclusion of regular users of other nicotine and tobacco products in the previous 2 weeks or planning to quit in the next 12 months	Randomized, controlled, parallel-group open-label, clinical 5-day confinement study	IQOS, glo, THP1.0 non-menthol and menthol	Own brand	Limited ad-libitum use of assigned product for 5 days	Urine NEQ
84	JTI	Japan		Healthy smokers aged 21–65 years who smoked an average of ≥ 11 manufactured cigarettes/day at screening and had smoked for ≥ 12 months before trial	Open-label, two-sequence, two-period, randomized cross-over, confinement study	Prototype tobacco vapour product	Commercial CC	HTP: 10 puffs for 3 min at approximately 20-s intervals CC: 10 puffs for 3 min at approximately 20s intervals Each product was used alone for 1 day and the other product the following day	Plasma nicotine
85	No	Italy	20	Healthy smokers	Cross-over randomized trial	IQOS THS2.2	Marlboro Gold	Used assigned product during six 1-day cycles, each rotating the assigned product with an inter-cycle washout of 1 week. The product used in each cycle was: CC: 1 cigarette with mean nicotine content of 0.60 mg ENDS: 9 puffs from a tobacco-flavoured e-liquid with a mean nicotine content of 16 mg, thus yielding 0.58 mg of nicotine content in 9 puffs HTP: 1 stick with a mean nicotine content of 0.50 mg per stick	Plasma cotinine
86	No	USA	15	Healthy smokers aged 18–55 years of ≥ 10 cigarettes per day. Had not used marijuana in past 30 days, had used ENDS ≤ 20 times and LLTV ≤ 4 times in lifetime		LLTV HTP (PAX) prefilled with 1 g LLT	Own brand	Three sessions of two daily bouts of 10 puffs per product with a 30-s inter-puff interval and an inter-bout period of 60 min, with ≥ 48 h intersession intervals with pre-session ≥ 2-h abstinence	Plasma nicotine

BAT: British American Tobacco; BMI: body mass index; CC: conventional cigarette; CHTP: carbon-heated tobacco product; END: electronic nicotine delivery; ENND: electronic non-nicotine delivery; HTP: heated tobacco product; JTI: Japan Tobacco International; LLTV: loose-leaf tobacco vaporizer; NEQ: nicotine equivalents; NRT: nicotine replacement therapy; PMI: Philip Morris International; THS: tobacco heating system.

Independent studies

Two independent studies were found, one conducted with IQOS and the other with a loose-leaf tobacco vaporizer from PAX. Biondi-Zoccai et al. *(85)* performed a randomized, cross-over trial to compare the effects on smokers of using one stick of IQOS, ENDS and one CC. Exposure to nicotine was evaluated by measuring serum cotinine before use, after a 1-week washout from any tobacco or nicotine product, and immediately after product use. Use of CC and IQOS increased cotinine plasma levels significantly: for CC, from 34.4 ng/mL (SD±19.3) before using a CC to 65.5 ng/mL (SD±10.2), afterwards, and, for IQOS, from 30.4 ±12.0 ng/mL to 61.0 ±16.7 ng/mL. In each case, the difference was statistically significant at $P < 0.001$, but no significant difference was found between products.

Lopez et al. *(86)* compared plasma nicotine in current smokers before and after use of a loose-leaf tobacco vaporizer HTP from PAX (Ploom), an ENDS and the participant's own brand of CC. Mean plasma nicotine concentration increased significantly immediately after each of two bouts of scheduled product use, from baseline (all $P < 0.025$) to 24.4 (SD±12.6) ng/mL after use of CC and 14.3 (±8.1) ng/mL after use of HTP in bout 1, and 23.7 (SD±14.5) ng/mL after use of CC and 16.4 (SD±11.3) ng/mL after use of an HTP in bout 2. The level of plasma nicotine attained immediately after each bout was higher with CCs than with HTPs; however, only the difference between nicotine levels in CCs and HTPs immediately after bout 1 was statistically significant (all $P < 0.017$). The mean plasma nicotine concentrations after use of CC were significantly higher than those after use of HTPs from the beginning to the end of experimental use.

Tobacco industry studies

Ten studies conducted by the tobacco industry were found: nine papers derived from eight studies by PMI *(74– 82)*, one by BAT *(83)* and one by Japan Tobacco International (JTI) *(84)* (see Table 5.5). All the PMI studies are randomized trials with allocation to use mainly of IQOS (a carbon-tip HTP was used in one study), the participant's regular brand of CC or, in five studies, abstinence or nicotine replacement therapy. Exposure was usually for about 5 days of ad-libitum use of the assigned product. The studies of BAT and JTI were randomized trials with participants assigned to use of an HTP (IQOS or glo in the case of BAT and Tobacco Vaporizer in the case of JTI) or a commercial CC. Participants were exposed to 5 days of ad-libitum use of the assigned product in the BAT trial and to use of one stick for 2 days in the JTI trial.

Levels of biomarkers in plasma

Three studies *(76,77,84)* compared pharmacokinetics after use of the products (Table 5.6), as the area under the plasma concentration versus time curve from time 0 to the last quantifiable concentration (AUC_{0-last}), an indicator of total exposure to nicotine on the assumption of equal clearance of the drug in all participants; the maximum observed plasma concentration attained (C_{max}), an indicator of the uptake of nicotine; the time to reach C_{max} (t_{max}), an indicator of the speed at which C_{max} is attained; and the half-life ($t_{1/2}$), an indicator of the duration of significant pharmacological effects. The values for IQOS THS2.2 were similar to those for CCs, with a t_{max} of 6 min. IQOS THS2.1 appeared to be less effective than CCs and IQOS THS2.2 in the uptake of nicotine and in providing the same maximal concentration. However, it presented a very similar t_{max} to CCs (8 min) and a slightly longer half-life than CCs. The tobacco vaporizer tested by Yuki et al. (84) reaching the t_{max} at the same time as CCs (3.8 min) but reached less than half of the C_{max} of CCs and generated < 70% of the nicotine taken up from CCs. Three studies in which IQOS, THS2.1 and THS2.2 were compared with CCs reported mean levels of nicotine and cotinine in plasma after 5 days of ad-libitum use of the HTP of about 85% for THS 2.1 and 100% for THS 2.2 of the levels reached after use of CCs.

Table 5.6. Pharmacokinetics of plasma nicotine after single use of IQOS and conventional cigarettes

Reference	76	76	77	84
Year of publication	2017	2017	2016	2017
Industry affiliation or funding	Yes	Yes	Yes	Yes
Reference CC	Own non-menthol brand	Own menthol brand	Own brand	CC1
HTP	THS2.2 IQOS	THS2.2 menthol IQOS	THS2.1	PNTV
HTP (single use) t to C_{max} (min)	6	6	8	3.8
AUC_{0-last} (ng*h/mL) Ratio of geometric LS means	96.3% 85.1–109.7%	98.1% 80.6–119.5%	77.4% 70.5–85.0[a]	68.3% 54.3–85.9%
C_{max} ng/mL Ratio of geometric least-squares means	103.5% 84.9–126.1%	88.5% 68.6–114%	70.3% 60.0–82.2%	45.7% 34.1–61.4%
t C_{max} (min) Median difference	0.04 -1.0–1.05	1.0 0.0–2.5	0.1 -1.0–2.0	-0.5 -1.1–0.03
$t_{1/2}$ (h) Ratio of geometric least-squares means	93.1% 84.6–102.4%	102.3% 85.3–122.7%	110.9% 101.7–120.9%	89.1% 78.2–102%

PNTV: prototype novel tobacco vapour. [a] 90% confidence interval.

Levels of biomarkers in urine

Six studies presented the value for nicotine equivalents in 24-h urine. The total with THS 2.2 and a carbon-tip HTP was ≥ 100% of that after use of CCs. For THS 2.1, the total was 87% of that of CCs. Use of glo resulted in urinary nicotine equivalents that were 57% and 74% of those of the comparable CCs, depending on whether they had menthol.

5.8 Subjective effects of use of HTPs and conventional cigarettes

We identified 10 studies in which the subjective effects of HTPs and CCs were compared *(74–77,79,81,82,86–88)*. IQOS was the HTP tested in all but one. The psychometric instruments most often used to assess the subjective impacts of HTPs and CCs were the brief questionnaire on smoking urges (QSU-brief) *(89)* and the modified cigarette evaluation questionnaire (mCEQ) *(90)*. The QSU-brief is a 10-item questionnaire, usually presented before use of the assigned product and then at the end of use to measure craving. The score may be reported as the total scale for its two components, desire to smoke with anticipation of pleasure from smoking and relief from nicotine withdrawal or negative affect with an urgent and overwhelming desire to smoke. The mCEQ assesses the reinforcing effects of product use, with three multidimensional domains, "smoking satisfaction", "psychological reward" and "aversion", and two single-item domains, "enjoyment of respiratory tract sensations" and "craving reduction".

The QSU-brief was used in eight studies, two independent *(86,88)* and six linked to industry *(74–77,79,82)*. All found, as expected, that the score for craving was high in all groups immediately before the start of the intervention. The score fell significantly immediately after use of IQOS or CC; however, the only independent study of IQOS reported that smoking resulted in lower craving scores than after use of IQOS (all $P < 0.001$) *(88)*, but the six industry studies did not. The least-squares mean differences between the IQOS and CC groups in the total QSU-brief scores in the industry studies, covering all times from the beginning to the end of use of the products, were generally small and none was statistically significant. The two independent studies *(86,88)*, which reported scores as the two-factor composition of the QSU-brief scale, found similar results over time. The study *(77)* that reported the QSU-brief scores for single-use and ad-libitum use found no difference in the mean total score (least square mean difference, 1.4 (95% CI: −1.0, 3.7) ad-libitum vs 0.2 (95% CI: −2.9, 5.3) for single use). This study, in which a loose-leaf tobacco vaporizer was compared with CCs, found that craving decreased significantly more after use of CC than after use of the vaporizer.

The mCEQ was used in seven studies *(74,75,77,79,81,82,88)*. In all the studies, the questionnaire was administered at the end of exposure, sometimes immediately. The only independent study *(88)* found that use of both products, IQOS and CCs, had a reinforcing effect on all subscales; however, CCs had a more substantial subjective effect than IQOS in terms of satisfaction, psychological reward, enjoyment of respiratory tract sensations and reduced craving. The industry studies tended to report significantly less reinforcing effect of IQOS use than of CC use on all or some of the mCEQ subscale scores, except aversion. Two studies *(75,79)* reported significant differences in all subscales at the beginning and one at the end of the exposure period, with the greatest difference for satisfaction and craving reduction. Another study *(81)* found that CC smokers had greater smoking satisfaction than IQOS users highlighted on the last day of exposure as compared with baseline. In another study *(82)*, the average results on the mCEQ for the entire 5-day exposure period were significantly lower for participants who switched to IQOS use than for participants who continued to smoke CC, after adjustment for baseline, smoking satisfaction, craving reduction, enjoyment of respiratory tract sensation and psychological reward. One study *(77)* indicated that the differences between the two products for smoking satisfaction, psychological reward, craving reduction and enjoyment of respiratory tract sensations were more significant after ad-libitum use. In an independent study, Biondi-Zoccai et al. *(85)* used neither the QSU-brief nor the mCEQ but a seven-question product satisfaction questionnaire *(91)*, which was administered after each session of product use. Satisfaction scores were higher for CCs than HTPs.

This limited body of research shows that HTPs overall deliver nicotine at a lower dose and more slowly than CCs. Of the HTPs analysed, only IQOS THS2.2 reaches the nicotine delivery of CCs. IQOS can reduce craving for smoking, perhaps to a lesser degree than CCs. Industry-linked studies showed little difference, and the one independent study showed significantly less reduction in craving than CCs. The sole study on other HTPs showed that a loose-leaf tobacco vaporizer was less able to quench smoking craving than CCs; however, IQOS and the vaporizer were perceived as less satisfying than CCs.

5.9 Discussion and implications

Little published information is available on use of HTPs at population level or whether the use is part of a poly use pattern or associated with cessation of use of CCs or nicotine. While research in this area is increasing and HTP use is being assessed in a number of surveys (e.g. International Tobacco Control, Japan Society and New Tobacco Internet Survey), publication will take months to years *(44,92–107)*. Independent studies do indicate that dual use of cigarettes and HTPs is more common than implied by initial industry-sponsored studies *(44,98,100)*. More information is required, however, on usage patterns in view of sociodemographic confounders *(95,99,106)*. There is little empirical support for the suggestion that new-generation HTPs overall help to transition smokers from CCs, and no studies on nicotine cessation have been published. Most of the available studies are industry-linked, and most studied IQOS. Laboratory studies suggested that only one of the HTPs analysed could deliver nicotine at a dose comparable to that of cigarettes *(76,77,84)*; however, other factors, including attractiveness and appeal, are often important in substitution behaviour. HTPs appear to reduce smoking craving subjectively, although perhaps not as significantly as CCs. It is, however, clear that HTPs are not as satisfying to smokers as CCs.

5.10 Research gaps

- Independent data on population-level usage patterns other than ever use. The key metrics include use with other products, amount used, daily or non-daily use and flavour preferences (where applicable) to determine the validity of claims of reduced risk.
- Studies on lil (KT&G), an HTP available in the Republic of Korea and now being marketed elsewhere by PMI, for which no published data were found.
- Independent studies of the pharmacokinetics of nicotine in HTPs other than IQOS, preferably between subjects to allow direct comparisons.
- Studies of the pharmacokinetics of nicotine delivery by HTPs, of leading ENDS products and/or of nicotine replacement therapy to compare the potential abuse liability of HTPs with that of products used to stop smoking.
- Independent studies of smoking cessation and use behaviour after adoption of HTPs expressly to cease smoking of conventional products.
- Studies of cessation of HTP use.

5.11 Policy recommendations

Policy-makers should consider the following recommendations for policy on the potential role of HTPs in cessation of smoking conventional tobacco products, particularly in the context of poly use of tobacco products.

5.11.1 Cessation policy

There is insufficient evidence that HTPs aid a switch from smoking. Therefore, claims should not be made to that effect. Even if future evidence supported HTPs as effective switching aids (i.e. substituting one tobacco product for another), they should never be considered as treatment for smoking cessation, which includes quitting nicotine use.

5.11.2 Surveillance policy

Surveillance of the prevalence and patterns of use of HTPs in various sociodemographic groups over time is rarely conducted at country level and should urgently be implemented. Understanding patterns of use among vulnerable populations (e.g. young people, racial and ethnic minorities, pregnant women) is of particular importance. The variables surveyed should include frequency of use (daily or non-daily use), amount used, concurrent use of other tobacco and nicotine products (poly use) and flavours used (where applicable). Surveillance systems might also include making it mandatory to record tobacco use, including HTPs, in medical notes.

5.11.3 Research policy

Studies should be conducted of consumers' use of HTPs to substitute completely for conventional cigarettes. Policy-makers are encouraged to prioritize funded research on ways to increase the reach, demand, quality, dissemination, implementation and sustainability of evidence-based smoking treatments.

5.11.4 Cooperation and partnership policy

Given the rapid dissemination of use of HTPs in the absence of the necessary scientific evidence of their effectiveness as aids for switching to conventional tobacco smoking, policy-makers are urged to share national experiences and to collaborate in developing an appropriate regulatory framework for HTPs.

5.12 References

1. Tabuchi T, Gallus S, Shinozaki T, Nakaya T, Kunugita N, Colwell B. Heat-not-burn tobacco product use in Japan: its prevalence, predictors and perceived symptoms from exposure to secondhand heat-not-burn tobacco aerosol. Tob Control. 2017;27(e1):e25–33.
2. Tabuchi T, Kiyohara K, Hoshino T, Bekki K, Inaba Y, Kunugita N. Awareness and use of electronic cigarettes and heat-not-burn tobacco products in Japan. Addiction. 2016;111(4):706–13.
3. Miyazaki Y, Tabuchi T. Educational gradients in the use of electronic cigarettes and heat-not-burn tobacco products in Japan. PloS One. 2018;13(1):e0191008.
4. Kioi Y, Tabuchi T. Electronic, heat-not-burn, and combustible cigarette use among chronic disease patients in Japan: A cross-sectional study. Tob Induc Dis. 2018;16:41.
5. Kotz D, Kastaun S. E-Zigaretten und Tabakerhitzer: repräsentative Daten zu Konsumverhalten und assoziierten Faktoren in der deutschen Bevölkerung (die DEBRA-Studie) [E-cigarettes and heat-not-burn products: representative data on consumer behaviour and associated factors in the German population (the DEBRA study)]. Bundesgesundheitsblatt Gesundheitsforschung Gesundheitsschutz. 2018;61(11):1407–14.
6. Liu X, Lugo A, Spizzichino L, Tabuchi T, Pacifici R, Gallus S. Heat-not-burn tobacco products: concerns from the Italian experience. Tob Control. 2019;28(1):113–4.
7. Langer P, Prieto L, Rousseau C. Tobacco product use after the launch of a heat-not-burn alternative in Japan: results of two cross-sectional surveys. Poster. Global Forum on Nicotine, 2019 (https://www.pmiscience.com/resources/docs/default-source/posters2019/langer-2019-tobacco-product-use-after-the-launch-of-a-heat-not-burn-alternative-in-japan.pdf?sfvrsn=460ed806_4, accessed 10 January 2021).
8. Kim J, Yu H, Lee S, Paek YJ. Awareness, experience and prevalence of heated tobacco product, IQOS, among young Korean adults. Tob Control. 2018;27(Suppl 1):s74–7.
9. Brose LS, Simonavicius E, Cheeseman H. Awareness and use of "Heat-not-burn" tobacco products in Great Britain. Tob Reg Sci. 2018;4(2):44–50.
10. West R, Beard E, Brown J. Trends in electronic cigarette use in England. Smoking Toolkit Study,

2020 (http://www.smokinginengland.info/sts-documents/, accessed 10 January 2021).
11. Sutanto E, Miller C, Smith DM, O'Connor RJ, Quah ACK, Cummings KM et al. Prevalence, use behaviors, and preferences among users of heated tobacco products: Findings from the 2018 ITC Japan survey. Int J Environ Res Public Health. 2019;16(23):4630.
12. Lee Y, Lee KS. Association of alcohol and drug use with use of electronic cigarettes and heat-not-burn tobacco products among Korean adolescents. PloS One. 2019;14(7):e0220241.
13. Pinkas J, Kaleta D, Zgliczynski WS, Lusawa A, Iwona Wrześniewska-Wal, Wierzba W et al. The prevalence of tobacco and e-cigarette use in Poland: A 2019 nationwide cross-sectional survey. Int J Environ Res Public Health. 2019;16(23):4820.
14. Berg CJ, Haardorfer R, Schauer G, Getachew B, Masters M, McDonald B et al. Reasons for polytobacco use among young adults: Scale development and validation. Tob Prev Cessation. 2016;2:69.
15. Bombard JM, Pederson LL, Koval JJ et al. How are lifetime polytobacco users different than current cigarette-only users? Results from a Canadian young adult population. Addictive behaviors 2009;34(12):1069-72. doi: 10.1016/j.addbeh.2009.06.009 [published Online First: 2009/08/04]
16. Bombard JM, Pederson LL, Nelson DE, Malarcher AM. Are smokers only using cigarettes? Exploring current polytobacco use among an adult population. Addict Behav. 2007;32(10):2411–9.
17. Fix BV, O'Connor RJ, Vogl L, Smith D, Bansal-Travers M, Conway KP et al. Patterns and correlates of polytobacco use in the United States over a decade: NSDUH 2002–2011. Addict Behav. 2014;39(4):768–81.
18. Horn K, Pearson JL, Villanti AC. Polytobacco use and the "customization generation" – new perspectives for tobacco control. J Drug Educ. 2016;46(3–4):51–63.
19. Lee YO, Hebert CJ, Nonnemaker JM, Kim AE. Multiple tobacco product use among adults in the United States: cigarettes, cigars, electronic cigarettes, hookah, smokeless tobacco, and snus. Prev Med. 2014;62:14–9.
20. Martinasek MP, Bowersock A, Wheldon CW. Patterns, perception and behavior of electronic nicotine delivery systems use and multiple product use among young adults. Respir Care. 2018;63(7):913–9.
21. Sung HY, Wang Y, Yao T et al. Polytobacco use of cigarettes, cigars, chewing tobacco, and snuff among US adults. Nicotine Tob Res. 2016;18(5):817-26. doi: 10.1093/ntr/ntv147 [published Online First: 2015/07/03].
22. Sung HY, Wang Y, Yao T, Lightwood J, Max W. Polytobacco use and nicotine dependence symptoms among US adults, 2012–2014. Nicotine Tob Res. 2018;20(Suppl 1):S88–98.
23. Wong EC, Haardorfer R, Windle M, Berg CJ. Distinct motives for use among polytobacco versus cigarette only users and among single tobacco product users. Nicotine Tob Res. 2017;20(1):117–23.
24. Hinton A, Nagaraja HN, Cooper S, Wewers ME. Tobacco product transition patterns in rural and urban cohorts: Where do dual users go? Prev Med Rep. 2018;12:241–4.
25. Miller CR, Smith DM, Goniewicz ML. Changes in nicotine product use among dual users of tobacco and electronic cigarettes: Findings from the Population Assessment of Tobacco and Health (PATH) study, 2013–2015. Subst Use Misuse. 2020:1–5. doi: 10.1080/10826084.2019.1710211.
26. Niaura R, Rich I, Johnson AL, Villanti AC, Romberg AR, Hair EC et al. Young adult tobacco and e-cigarette use transitions: Examining stability using multi-state modeling. Nicotine Tob Res. 2020;22(5):647–54.
27. Piper ME, Baker TB, Benowitz NL, Jorenby DE. Changes in use patterns over one year among smokers and dual users of combustible and electronic cigarettes. Nicotine Tob Res. 2020;22(5):672–80.
28. Borland R, Murray K, Gravely S, Fong GT, Thompson ME, McNeill A et al. A new classificati-

on system for describing concurrent use of nicotine vaping products alongside cigarettes (so-called "dual use"): findings from the ITC-4 Country Smoking and Vaping wave 1 survey. Addiction. 2019;114(Suppl 1):24–34.
29. Bhatnagar A, Whitsel LP, Blaha MJ, Huffman MD, Krishan-Sarin S, Maa J et al. New and emerging tobacco products and the nicotine endgame: The role of robust regulation and comprehensive tobacco control and prevention: A Presidential Advisory from the American Heart Association. Circulation. 2019;139(19):e937–58.
30. Farsalinos KE, Romagna G, Voudris V. Factors associated with dual use of tobacco and electronic cigarettes: A case control study. Int J Drug Policy. 2015;26(6):595–600.
31. Goniewicz ML, Smith DM, Edwards KC, Blount BC, Caldwell KL, Feng J et al. Comparison of nicotine and toxicant exposure in users of electronic cigarettes and combustible cigarettes. JAMA Netw Open. 2018;1(8):e185937.
32. Shahab L, Goniewicz ML, Blount BC, Brown J, McNeill A, Alwis KU et al. Nicotine, carcinogen, and toxin exposure in long-term e-cigarette and nicotine replacement therapy users: a cross-sectional study. Ann Intern Med. 2017;166(6):390–400.
33. Bjartveit K, Tverdal A. Health consequences of smoking 1–4 cigarettes per day. Tob Control. 2005;14(5):315–20.
34. Gerber Y, Myers V, Goldbourt U. Smoking reduction at midlife and lifetime mortality risk in men: a prospective cohort study. Am J Epidemiol. 2012;175(10):1006–12.
35. Hart C, Gruer L, Bauld L. Does smoking reduction in midlife reduce mortality risk? Results of 2 long-term prospective cohort studies of men and women in Scotland. Am J Epidemiol. 2013;178(5):770–9.
36. Tverdal A, Bjartveit K. Health consequences of reduced daily cigarette consumption. Tob Control. 2006;15(6):472–80.
37. Baig SA, Giovenco DP. Behavioral heterogeneity among cigarette and e-cigarette dual-users and associations with future tobacco use: Findings from the Population Assessment of Tobacco and Health study. Addict Behav. 2020;104:106263.
38. Farsalinos KE, Yannovits N, Sarri T, Voudris V, Poulas K. Nicotine delivery to the aerosol of a heat-not-burn tobacco product: comparison with a tobacco cigarette and e-cigarettes. Nicotine Tob Res. 2018;20(8):1004–9.
39. Farsalinos KE, Yannovits N, Sarri T, Voudris V, Poulas K, Leischow SJ. Carbonyl emissions from a novel heated tobacco product (IQOS): comparison with an e-cigarette and a tobacco cigarette. Addiction. 2018;113(11):2099–106.
40. Leigh NJ, Palumbo MN, Marino AM, O'Connor RJ, Goniewicz ML. Tobacco-specific nitrosamines (TSNA) in heated tobacco product IQOS. Tob Control. 2018;27(Suppl 1):s37–8.
41. Hwang JH, Ryu DH, Park SW. Heated tobacco products: Cigarette complements, not substitutes. Drug Alcohol Depend. 2019;204:107576.
42. Tabuchi T, Gallus S, Shinozaki T, Nakaya T, Kunugita N, Colwell B. Heat-not-burn tobacco product use in Japan: its prevalence, predictors and perceived symptoms from exposure to secondhand heat-not-burn tobacco aerosol. Tob Control. 2018;27(e1):e25–33.
43. Tabuchi T, Shinozaki T, Kunugita N, Nakamura M, Tsuji I. Study profile: The Japan "Society and New Tobacco" Internet Survey (JASTIS): A longitudinal Internet cohort study of heat-not-burn tobacco products, electronic cigarettes, and conventional tobacco products in Japan. J Epidemiol. 2019;29(11):444–50.
44. Sutanto E, Miller C, Smith DM, Borland R, Hyland A, Cummings KM et al. Concurrent daily and non-daily use of heated tobacco products with combustible cigarettes: Findings from the 2018 ITC Japan survey. Int J Environ Res Public Health. 2020;17(6):2098.
45. Roulet S, Chrea C, Kanitscheider C, Kallischnigg G, Magnani P, Weitkunat R. Potential predictors of adoption of the tobacco heating system by US adult smokers: An actual use study.

F1000Res. 2019;8:214.

46. Hair EC, Bennett M, Sheen E, Cantrell J, Briggs J, Fenn Z et al. Examining perceptions about IQOS heated tobacco product: consumer studies in Japan and Switzerland. Tob Control. 2018;27(Suppl 1):s70–3.

47. Mays D, Arrazola RA, Tworek C, Rolle IV, Neff LJ, Portnoy DB. Openness to using non-cigarette tobacco products among US young adults. Am J Prev Med. 2016;50(4):528–34.

48. McKelvey K, Popova L, Kim M, Chaffee BW, Vijayaraghavan M, Ling P et al. Heated tobacco products likely appeal to adolescents and young adults. Tob Control. 2018;27(Suppl 1):s41–7.

49. Soneji S, Sargent JD, Tanski SE, Primack BA. Associations between initial water pipe tobacco smoking and snus use and subsequent cigarette smoking: Results from a longitudinal study of US adolescents and young adults. JAMA Pediatr. 2015;169(2):129–36.

50. Gee J, Prasad K, Slayford S, Gray A, Nother K, Cunningham A et al. Assessment of tobacco heating product THP1.0. Part 8: Study to determine puffing topography, mouth level exposure and consumption among Japanese users. Regul Toxicol Pharmacol. 2018;93:84–81.

51. Philip Morris Products S.A. modified risk tobacco product (MRTP) applications. Silver Spring (MD): Food and Drug Administration; 2020 (https://www.fda.gov/tobacco-products/advertising-and-promotion/philip-morris-products-sa-modified-risk-tobacco-product-mrtp-applications, accessed 10 January 2021).

52. Tobacco Products Scientific Advisory Committee, 25 January 2018. Silver Spring (MD): Food and Drug Administration; 2018 (https://www.fda.gov/downloads/AdvisoryCommittees/CommitteesMeetingMaterials/TobaccoProductsScientificAdvisoryCommittee/UCM599235.pdf, accessed 10 January 2021).

53. Committees on Toxicity, Carcinogenicity and Mutagenicity of Chemical Products In Food, Consumer Products And The Environment (COT, COC and COM). Statement on the toxicological evaluation of novel heat-not-burn tobacco products. London: Food Standards Agency; 2017 (https://cot.food.gov.uk/sites/default/files/heat_not_burn_tobacco_statement.pdf, accessed 10 January 2021).

54. Hughes JR, Keely JP. The effect of a novel smoking system – Accord – on ongoing smoking and toxin exposure. Nicotine Tob Res. 2004;6(6):1021–7.

55. Fagerstrom KO, Hughes JR, Callas PW. Long-term effects of the Eclipse cigarette substitute and the nicotine inhaler in smokers not interested in quitting. Nicotine Tob Res. 2002;4(Suppl 2):S141–5.

56. Hackshaw A, Morris JK, Boniface S, Tang JL, Milenković M. Low cigarette consumption and risk of coronary heart disease and stroke: meta-analysis of 141 cohort studies in 55 study reports. BMJ. 2018;360:j5855.

57. Inoue-Choi M, Christensen CH, Rostron BL, Cosgrove CM, Reyes-Guzman C, Apelberg B et al. Dose–response association of low-intensity and nondaily smoking with mortality in the United States. JAMA Netw Open. 2020;3(6):e206436.

58. Inoue-Choi M, Hartge P, Liao LM, Caporaso N, Freedman ND. Association between long-term low-intensity cigarette smoking and incidence of smoking-related cancer in the National Institutes of Health–AARP cohort. Int J Cancer. 2018;142(2):271–80.

59. Gilchrist M. Presentation by Moira Gilchrist, PhD, Vice President Scientific and Public Communications, Philip Morris International, before the Tobacco Products Scientific Advisory Committee (TPSAC). Lausanne: Philip Morris International; 2018 (https://tobacco.ucsf.edu/sites/g/files/tkssra4661/f/wysiwyg/PMI-24Jan-TPSAC-Presentation.pdf).

60. Auer R, Concha-Lozano N, Jacot-Sadowski I, Cornuz J, Berthet A. Heat-not-burn tobacco cigarettes: Smoke by any other name. JAMA Intern Med. 2017;177(7):1050–2.

61. Bekki K, Inaba Y, Uchuyama S, Kunugita N. Comparison of chemicals in mainstream smoke in heat-not-burn tobacco and combustion cigarettes. J UOEH. 2017;39(3):201–7.

62. Farsalinos KE, Yannovits N, Sarri T, Voudris V, Poulas K. Nicotine delivery to the aerosol of a heat-not-burn tobacco product: comparison with a tobacco cigarette and e-cigarettes. Nicotine Tob Res. 2018;20(8):1004–9.
63. Forster M, Fiebelkorn S, Yurteri C, Mariner D, Liu C, Wright C et al. Assessment of novel tobacco heating product THP1.0. Part 3: Comprehensive chemical characterisation of harmful and potentially harmful aerosol emissions. Regul Toxicol Pharmacol. 2018;93:14–33.
64. Jaccard G, Tafin Djoko D, Moennikes O, Jeannet C, Kondylis A, Belushkin M. Comparative assessment of HPHC yields in the tobacco heating system THS2.2 and commercial cigarettes. Regul Toxicol Pharmacol. 2017;90:1–8.
65. Li X, Luo Y, Jiang X, Zhang H, Zhu F, Hus S et al. Chemical analysis and simulated pyrolysis of tobacco heating system 2.2 compared to conventional cigarettes. Nicotine Tob Res. 2019;21(1):111–8.
66. Mallock N, Boss L, Burk R, Dabziger M, Welsch T, Hahn H et al. Levels of selected analytes in the emissions of "heat not burn" tobacco products that are relevant to assess human health risks. Arch Toxicol. 2018;92(6):2145–9.
67. Nabavizadeh P, Liu J, Havel CM, Ibrahim S, Derakhshandeh R, Jacon P III et al. Vascular endothelial function is impaired by aerosol from a single IQOS HeatStick to the same extent as by cigarette smoke. Tob Control. 2018;27(Suppl 1):s13–9.
68. Poynton S, Sutton J, Goodall S, Margham J, Forster M, Scott K et al. A novel hybrid tobacco product that delivers a tobacco flavour note with vapour aerosol (Part 1): Product operation and preliminary aerosol chemistry assessment. Food Chem Toxicol. 2017;106(Pt A):522–32.
69. Salman R, Talih S, El-Hage R, Haddad C, Karaoghlanian N, El-Hellani A et al. Free-base and total nicotine, reactive oxygen species, and carbonyl emissions from IQOS, a heated tobacco product. Nicotine Tob Res. 2019;21(9):1285–8.
70. Schaller JP, Keller D, Poget L, Pratte P, Kaelin E, McHugh D et al. Evaluation of the tobacco heating system 2.2. Part 2: Chemical composition, genotoxicity, cytotoxicity, and physical properties of the aerosol. Regul Toxicol Pharmacol. 2016;81(Suppl 2):S27–47.
71. Schaller JP, Pijnenburg JP, Ajithkumar A, Tricker AR. Evaluation of the tobacco heating system 2.2. Part 3: Influence of the tobacco blend on the formation of harmful and potentially harmful constituents of the tobacco heating system 2.2 aerosol. Regul Toxicol Pharmacol. 2016;81(Suppl 2):S48–58.
72. Simonavicius E, McNeill A, Shahab L, Brose LS. Heat-not-burn tobacco products: a systematic literature review. Tob Control. 2019;28(5):582–94.
73. Counts M, Morton M, Laffoon S, Cox R, Lipowicz P. Smoke composition and predicting relationships for international commercial cigarettes smoked with three machine-smoking conditions. Regul Toxicol Pharmacol. 2005;41(3):185–227.
74. Haziza C, de la Bourdonnaye G, Skiada D, Ancerewicz J, Baker G, Picavet P et al. Evaluation of the tobacco heating system 2.2. Part 8: 5-day randomized reduced exposure clinical study in Poland. Regul Toxicol Pharmacol. 2016;81(Suppl 2):S139–50.
75. Haziza C, de la Bourdonnaye G, Merlet S, Benzimra M, Ancerewicz J, Donelli A et al. Assessment of the reduction in levels of exposure to harmful and potentially harmful constituents in Japanese subjects using a novel tobacco heating system compared with conventional cigarettes and smoking abstinence: A randomized controlled study in confinement. Regul Toxicol Pharmacol. 2016;81:489–99.
76. Brossard P, Weitkunat R, Poux V, Lama N, Haziza C, Picavet P et al. Nicotine pharmacokinetic profiles of the tobacco heating system 2.2, cigarettes and nicotine gum in Japanese smokers. Regul Toxicol Pharmacol. 2017;89:193–9.
77. Picavet P, Haziza C, Lama N, Weitkunat R, Lüdicke F. Comparison of the pharmacokinetics of nicotine following single and ad libitum use of a tobacco heating system or combustible

cigarettes. Nicotine Tob Res. 2016;18(5):557–63.
78. Lüdicke F, Picavet P, Baker G, Haziza C, Poux V, Lama N et al. Effects of switching to the menthol tobacco heating system 2.2, smoking abstinence, or continued cigarette smoking on clinically relevant risk markers: a randomized, controlled, open-label, multicenter study in sequential confinement and ambulatory settings (Part 2). Nicotine Tob Res. 2018;20(2):173–82.
79. Lüdicke F, Picavet P, Baker G, Haziza C, Poux V, Lama N et al. Effects of switching to the tobacco heating system 2.2 menthol, smoking abstinence, or continued cigarette smoking on biomarkers of exposure: a randomized, controlled, open-label, multicenter study in sequential confinement and ambulatory settings (Part 1). Nicotine Tob Res. 2018;20(2):161–72.
80. Lüdicke F, Haziza C, Weitkunat R et al. Evaluation of biomarkers of exposure in smokers switching to a carbon-heated tobacco product: a controlled, randomized, open-label 5-day exposure study. Nicotine Tob Res. 2016;18:1606–13. doi:10.1093/ntr/ntw022.
81. Lüdicke F, Baker G, Magnette J et al. Reduced exposure to harmful and potentially harmful smoke constituents with the tobacco heating system 2.1. Nicotine Tob Res. 2016;19:168-175. doi:10.1093/ntr/ntw164.
82. Haziza C, de la Bourdonnaye G, Donelli A, Poux V, Skiada D, Weitkunat R et al. Reduction in exposure to selected harmful and potentially harmful constituents approaching those observed upon smoking abstinence in smokers switching to the menthol tobacco heating system 2.2 for 3 months (Part 1). Nicotine Tob Res. 2020;22(4):539–48.
83. Gale N, McEwan M, Eldridge A, Fearon IM, Sherwood N, Bowen E et al. Changes in biomarkers of exposure on switching from a conventional cigarette to tobacco heating products: A randomized, controlled study in healthy Japanese subjects. Nicotine Tob Res. 2018;21:122–7.
84. Yuki D, Sakaguchi C, Kikuchi A, Futamura Y. Pharmacokinetics of nicotine following the controlled use of a prototype novel tobacco vapor product. Regul Toxicol Pharmacol. 2017;87:30–5.
85. Biondi-Zoccai G, Sciarretta S, Bullen C, Nocella C, Violo F, Loffredo L et al. Acute effects of heat-not-burn, electronic vaping, and traditional tobacco combustion cigarettes: the Sapienza University of Rome – Vascular Assessment of Proatherosclerotic Effects of Smoking (SUR-VAPES) 2 randomized trial. J Am Heart Assoc. 2019;8(6):e010455.
86. Lopez AA, Hiler M, Maloney S, Eissenberg T, Breland A. Expanding clinical laboratory tobacco product evaluation methods to loose-leaf tobacco vaporizers. Drug Alcohol Depend. 2016;169:33–40.
87. Newland N, Lowe FJ, Camacho OM, McEwan M, Gale N, Ebajemito J et al. Evaluating the effects of switching from cigarette smoking to using a heated tobacco product on health effect indicators in healthy subjects: study protocol for a randomized controlled trial. Intern Emergency Med. 2019;14(6):885–98.
88. Adriaens K, Gucht DV, Baeyens F. IQOS(TM) vs. e-cigarette vs. tobacco cigarette: A direct comparison of short-term effects after overnight-abstinence. Int J Environ Res Public Health. 2018;15(12):2902.
89. Cox LS, Tiffany ST, Christen AG. Evaluation of the brief questionnaire of smoking urges (QSU-brief) in laboratory and clinical settings. Nicotine Tob Res. 2001;3(1):7–16.
90. Cappelleri JC, Bushmakin AG, Baker CL, Merikle E, Olufade AO, Gilbert DG. Confirmatory factor analyses and reliability of the modified cigarette evaluation questionnaire. Addict Behav. 2007;32(5):912–23.
91. Shiffman S, Terhorst L. Intermittent and daily smokers' subjective responses to smoking. Psychopharmacology (Berl). 2017;234(19):2911–7.
92. Cruz-Jiménez L, Barrientos-Gutiérrez I, Coutiño-Escamilla L, Gallegos-Carrillo K, Arillo-Santillán E, Thrasher JF. Adult smokers' awareness and interest in trying heated tobacco products: perspectives from Mexico, where HTPs and e-cigarettes are banned. Int J Environ Res Public Health. 2020;17(7):2173.

93. Czoli CD, White CM, Reid JL, O'Connor RJ, Hammond D. Awareness and interest in IQOS heated tobacco products among youth in Canada, England and the USA. Tob Control. 2020;29(1):89–95.
94. Dunbar MS, Seelam R, Tucker JS, Rodriguez A, Shih RA, D'Amico EJ. Correlates of awareness and use of heated tobacco products in a sample of US young adults in 2018–2019. Nicotine Tob Res. 2020:doi: 10.1093/ntr/ntaa007.
95. Gravely S, Fong GT, Sutanto E, Loewen R, Ouimet J, Xu SS, et al. Perceptions of harmfulness of heated tobacco products compared to combustible cigarettes among adult smokers in Japan: Findings from the 2018 ITC Japan Survey. Int J Environ Res Public Health.2020;17(7):2394.
96. Hori A, Tabuchi T, Kunugita N. Rapid increase in heated tobacco product (HTP) use from 2015 to 2019: from the Japan "Society and New Tobacco" Internet Survey (JASTIS). Tob Control. 2020;5(5):652.
97. Igarashi A, Aida J, Kusama T, Tabuchi T, Tsuboya T, Sugiyama K et al. Heated tobacco products have reached younger or more affluent people in Japan. J Epidemiol. 2020:doi: 10.2188/jea.JE20190260.
98. Kang H, Cho SI. Heated tobacco product use among Korean adolescents. Tob Control. 2020;29(4):466–8.
99. Kim K, Kim J, Cho HJ. Gendered factors for heated tobacco product use: Focus group interviews with Korean adults. Tob Induc Dis. 2020;18:43.
100. Kim SH, Cho HJ. Prevalence and correlates of current use of heated tobacco products among a nationally representative sample of Korean adults: Results from a cross-sectional study. Tob Induc Dis. 2020;18:66.
101. Kim SH, Kang SY, Cho HJ. Beliefs about the harmfulness of heated tobacco products compared with combustible cigarettes and their effectiveness for smoking cessation among Korean adults. Int J Environ Res Public Health. 2020;17(15):5591.
102. Okawa S, Tabuchi T, Miyashiro I. Who uses e-cigarettes and why? e-Cigarette use among older adolescents and young adults in Japan: JASTIS Study. J Psychoactive Drugs. 2020;52(1):37–45.
103. Ratajczak A, Jankowski P, Strus P, Feleszko W. Heat not burn tobacco product – A new global trend: Impact of heat-not-burn tobacco products on public health, a systematic review. Int J Environ Res Public Health. 2020;17(2):0409.
104. Siripongvutikorn Y, Tabuchi T, Okawa S. Workplace smoke-free policies that allow heated tobacco products and electronic cigarettes use are associated with use of both these products and conventional tobacco smoking: the 2018 JASTIS study. Tob Control. 2020;doi: 10.1136/tobaccocontrol-2019-055465.
105. Sugiyama T, Tabuchi T. Use of multiple tobacco and tobacco-like products including heated tobacco and e-cigarettes in Japan: A cross-sectional assessment of the 2017 JASTIS study. Int J Environ Res Public Health. 2020;17(6):2161.
106. Tompkins CNE, Burnley A, McNeill A, Hitchman SC. Factors that influence smokers' and ex-smokers' use of IQOS: a qualitative study of IQOS users and ex-users in the UK. Tob Control. 2020:doi: 10.1136/tobaccocontrol-2019-055306.
107. Wu YS, Wang MP, Ho SY, Li HCW, Cheung YTD, Tabuchi T, et al. Heated tobacco products use in Chinese adults in Hong Kong: a population-based cross-sectional study. Tob Control. 2020;29(3):277–81.

6. Regulations on heated tobacco products, electronic nicotine delivery systems and electronic non-nicotine delivery systems, with country approaches, barriers to regulation and regulatory considerations

Ranti Fayokun, Moira Sy, Marine Perraudin and Vinayak Prasad
   No Tobacco Unit (Tobacco Free Initiative), Department of Health Promotion, Division of Universal Health Coverage and Healthier Populations, World Health Organization, Geneva, Switzerland

Contents
   Abstract
   6.1 Background
      6.1.1 Introduction and the request of the Conference of the Parties (FCTC/COP8(22))
      6.1.2 Scope and objectives
      6.1.3 Sources
   6.2 Regulatory mapping of novel and emerging nicotine and tobacco products
      6.2.1 Availability of HTPs
      6.2.2 Product classification
      6.2.3 Regulatory frameworks and measures to reduce tobacco demand
   6.3 Considerations and barriers to regulation, implementation and enforcement of policies
      6.3.1 Regulatory considerations in implementing policies
      6.3.2 Barriers to implementing and enforcing policies
      6.3.3 Other considerations and unintended consequences
   6.4 Discussion
   6.5 Conclusions
   6.6 Research gaps
   6.7 Policy recommendations
   6.8 References

Abstract

Heated tobacco products (HTPs) are increasingly marketed by the tobacco industry as part of a newer portfolio of products that are claimed to pose fewer risks to users and non-users than conventional tobacco products. These products have gained a considerable market share since they became available and are now found on about 50 markets worldwide. The new generation of HTPs, owing to their novelty, their unconventional technology and industry claims that they pose fewer risks to health than conventional tobacco products, are classified in various ways by different countries. The classifications have filtered through to the mechanisms adopted by countries to regulate these products, resulting in inconsistencies among countries, including the extent to which they apply the provisions of the WHO Framework Convention on Tobacco Control (WHO FCTC). This paper reviews the markets on which HTPs are available, the common classifications of these products and how the classifications affect regulatory outcomes. Further, we describe commonly used regulatory frameworks, barriers to regulation, considerations for regulations and unforeseen consequences. We also present guidance from WHO and the WHO FCTC to countries in formulating regulatory strategies for HTPs according to their national laws and ensuring strong protection of human health.

6.1 Background

6.1.1 Introduction and the request of the Conference of the Parties (FCTC/COP8(22))

HTPs produce aerosols containing nicotine and toxic chemicals when their tobacco material is heated or when a device containing tobacco is activated *(1)*. The tobacco may be in the form of specially designed cigarettes (e.g. "heat sticks" and "neo sticks"), pods or plugs. The resulting aerosols are inhaled by users after heating of tobacco in a device specifically designed for that purpose *(1)*. HTPs are aggressively marketed and promoted by the tobacco industry in a number of ways, including as "smoke-free", "cleaner alternatives", "safer alternatives" and "reduced risk" products relative to conventional cigarettes. They were available for sale legally in over 50 markets in all six WHO regions as of July 2020 *(2)*, examples of which are given in Table 6.1.

Table 6.1. Examples of markets on which HTPs are available

Device	Company	Markets
IQOS	Philip Morris International	Andorra, Albania, Armenia, Belarus, Bosnia and Herzegovina, Bulgaria, Canada, Colombia, Croatia, Cyprus, Czechia, Denmark, Dominican Republic, France (including La Réunion), Germany, Greece, Guatemala, Hungary, Israel, Italy, Japan, Kazakhstan, Latvia, Lithuania, Malaysia, Mexico, Monaco, Netherlands (including Curaçao), New Zealand, Poland, Portugal, Republic of Korea, Republic of Moldova, Romania, Russian Federation, Serbia, Slovakia, Slovenia, South Africa, Spain (including the Canary Islands), Sweden, Switzerland, Ukraine, United Arab Emirates, United Kingdom, USA and occupied Palestinian territory, including east Jerusalem
iFuse, glo	British American Tobacco	Canada, Italy, Japan, Republic of Korea, Romania, Russian Federation and Switzerland
Lil	KT&G	Japan, Republic of Korea, Russian Federation and Ukraine
Ploom	Japan Tobacco International	Italy, Japan, Republic of Korea, Russian Federation, Switzerland and United Kingdom

HTPs are one of the three broad categories of products, with electronic nicotine delivery systems (ENDS) and electronic non-nicotine delivery systems (ENNDS), that have become popular in several jurisdictions, especially in Japan and the Republic of Korea. While the technology for HTPs has been available since the

1980s, earlier attempts to introduce these products were unsuccessful, and the newer generations of these products became popular only in the past seven years (Fig. 6.1; see also sections 3 and 5).

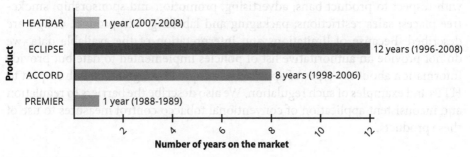

Fig. 6.1. Dates of launch and withdrawal from the market of early-generation heated tobacco products

Source: adapted from reference 3.

HTPs pose significant challenges to tobacco regulation, specifically because of their novel operating mechanisms and inadequate knowledge about their effects on health. The tobacco industry has exploited these challenges by using marketing tactics especially for "harm reduction" or "reduced risk" to facilitate their entry onto the market and have argued that HTPs should be categorized differently from tobacco products and specifically conventional cigarettes. The lack of an internationally agreed approach for assessing the risks of their use, because of insufficient knowledge, further complicates tobacco control. Novelty, misinformation and industry manipulation have resulted in disparate approaches to their classification and their regulation.

Decision FCTC/COP8(22) of the eighth session of the Conference of the Parties (COP) to the WHO FCTC on novel and emerging tobacco products) *(4)* requests the Convention Secretariat to invite WHO to prepare a comprehensive report on these products, covering several areas of research. WHO is expected to report to the ninth session of the COP on the regulatory experience and monitoring of Parties, effects on tobacco control and research gaps and to subsequently propose policy options for achieving the objectives and measures outlined in paragraph 5 of the decision. The aim of this paper is to map regulation of HTPs, review the regulatory experiences of WHO Member States with regard to HTPs, consider the impact of HTPs on tobacco control and identify research gaps.

6.1.2 Scope and objectives

We describe the regulatory experience of countries with HTPs after explaining common classifications for these products that may dictate specific regulatory pathways. For instance, classification in a certain category may result in a complete ban on HTPs, although the regulatory implications of a category differ by country and the classifications in a country determine the degree of application of tobacco control laws. An HTP classified as an "e-cigarette" product may result in a ban in one country and in regulation in another.

The common regulatory frameworks that follow from classifications with respect to product bans; advertising, promotion and sponsorship; smoke-free places; sales restrictions; packaging and labelling; and product design are described. Because of limitations and interpretation of the available data, we do not provide an authoritative list of policies implemented to date but provide information about the types of approaches adopted by regulators with respect to HTPs and examples of such regulation. We also describe the barriers to regulation and inconsistent application of conventional tobacco control measures to use of these products.

6.1.3 Sources

The summaries of legislation and policy were drawn from data collected by WHO in 2019–2020, the legislative database managed by the Campaign for Tobacco Free Kids *(5)*, desk research on specific countries and internal WHO correspondence with country regulators. The data are, however, limited, because they may depend on interpretation of domestic legislation to application of general tobacco control measures to these newer products.

6.2 Regulatory mapping of novel and emerging nicotine and tobacco products

6.2.1 Availability of HTPs

HTPs are currently available on over 50 markets *(2)* (see Table 6.1). The number is, however, increasing rapidly, as 15 markets have been added in only the past two years. The availability of these products has increased recently in some countries where HTPs are gaining a market share from conventional tobacco products, mainly cigarettes, although HTP sales revenue represents a small fraction of what the industry earns from cigarettes *(6)*. The tobacco industry plans to expand the market share and to increase the availability, visibility and access to these products globally *(7)*, in particular by maintaining their availability on the market and avoiding strict tobacco control measures resulting from Parties' obligations to the WHO FCTC. Regulatory exception would advance the tobacco industry's long-term objective of increasing the acceptability of a wide range of its newer tobacco products, undermining the WHO FCTC.

HTPs are not only sold legally in countries but are also traded illegally in some countries where they are banned and in countries where the products have not met regulatory requirements or have not had premarketing authorization.

6.2.2 Product classification

According to WHO guidance, HTPs should be classified as tobacco products *(8)*. At its eighth session, the COP to the WHO FCTC recognized these products as tobacco products and reminded Parties of their obligation to do so under the WHO FCTC. Some regulators have, however, classified HTPs into categories distinct from conventional cigarettes or tobacco products because of their unconventional characteristics. These distinct categories result mainly from industry arguments for weaker or no regulation of so-called "reduced-risk" products and insufficient knowledge about the products. To date, countries have classified HTPs into categories, including:

- tobacco products
- HTPs
- smokeless tobacco products
- novel, emerging, new or next-generation tobacco products and
- e-cigarettes *(9)*.

Table 6.2 lists examples of countries in which these categories are used.

Table 6.2. Examples of regulatory classification of HTPs

Classification	Example of countries (non-exhaustive lists)
Tobacco products	Republic of Korea, United Arab Emirates More than 110 other countries, where HTPs are classified as tobacco products by definition
ENDS	Brazil, India, Russian Federation, Saudi Arabia
Novel tobacco products	European Union countries, United Kingdom
Emerging and imitation tobacco products	Singapore
Smokeless tobacco products	New Zealand

ENDS: electronic nicotine delivery systems; HTPs: heated tobacco products.

HTPs may also fall into hybrid or exempt categories *(9)*, often resulting in more favourable treatment of HTPs than of conventional cigarettes. The nature of the device may also affect its classification: sticks (which contain tobacco) may be classified as tobacco products, while a heating device into which the sticks are inserted may be classified differently *(9)*.

The definitions of products are closely linked to their classification or categorization, as the definition of "tobacco products" under domestic law differs from one jurisdiction to another. Some countries define "tobacco products" as all tobacco-derived materials, including nicotine, as well as the way in which the product is consumed (e.g. sucked, smoked, chewed), and some may extend the definition to the way in which the products are presented or how the nicotine is derived. Some may include all nicotine-containing products. The way in which tobacco products are defined determines the extent to which existing tobacco control regulations and legislation are applicable to HTPs in that country, unless exceptions are specifically made for HTPs or specific laws that apply to HTPs.

Tobacco products

Many national laws and regulations in all regions of WHO define tobacco products broadly enough to include HTPs. If the tobacco control law includes no specific regulation, HTPs can be regulated through other applicable regulations, such as consumer protection or poisons laws. For example, Australia's Poison Standards Act classifies nicotine as a Schedule 7 poison, so that its sale and possession are largely illegal, although, in some states, a 3-month supply may be imported under the Therapeutic Goods Administration "personal importation scheme" with a medical prescription. Not surprisingly, this restrictive approach has been challenged by the tobacco industry. Philip Morris filed a regulatory application to the Advisory Committee on Chemicals and Poisons of the Therapeutic Goods Administration to seek an exemption from the poisons standard to allow legal sale of the nicotine in tobacco prepared and packed for "heating". The proposed amendment was rejected by the Therapeutic Goods Administration in June 2020 *(10)*.

e-Cigarettes

Some countries have classified HTPs as "e-cigarettes" or "electronic smoking devices" on the basis of legislative definitions. This may result in regulation similar to that for e-cigarettes or, in other countries, in a ban on the sale or importation of the entire category of products. For instance, in the Republic of Korea, although HTPs are primarily classified as tobacco products, under the law, tobacco products in which electronic devices are used to consume tobacco (e.g. by heating) are sub-categorized as e-cigarettes. This subcategory is thus applicable to HTPs, as they contain electronic devices, which heat the tobacco. In the country's regulatory context, this means that most tobacco product regulations, including on smoke-free areas, taxation, advertising, health warnings and labels, and prohibition of sale to minors, apply to HTPs; however, only 90% of the cigarette tax rate applies to HTPs. Depending on national legislation, classification of HTPs as e-cigarettes or electronic smoking devices could result in a range of measures, from a ban (Brazil) to regulation.

Novel tobacco products

In the European Union, HTPs are regulated as novel tobacco products according to the European Union Tobacco Product Directive 2014/40/EU *(11)*. "Novel tobacco products" are defined as tobacco products that are required to comply with the provisions of the Directive, including a ban on misleading elements foreseen by Article 13 and, notably, any suggestion that a particular tobacco product is less harmful than others *(12)*.

In line with Article 19 of the Directive on notification of novel tobacco products, tobacco manufacturers and importers are required to provide information and supporting documentation for all products that they intend to place on the national market and that fall into the category of novel tobacco products. Specifically, manufacturers and importers of these products are required to notify the competent authorities of Member States, in electronic form, 6 months before the products are placed on the market, accompanied by information about the products' ingredients and emissions.

Manufacturers and importers are also required to provide the competent authorities with:

(a) available scientific studies on toxicity, addictiveness and attractiveness of the novel tobacco product, in particular as regards its ingredients and emissions; (b) available studies, executive summaries thereof and market research on the preferences of various consumer groups, including young people and current smokers; and (c) other available and relevant information, including a risk/benefit analysis of the product, its expected effects on cessation of tobacco consumption, its expected effects on initiation of tobacco consumption and predicted consumer perception.

Additionally, they are required to submit to their competent authorities any new or updated information on the studies, research and other information referred to in a–c above and conduct additional tests or submit additional information as required by the competent authority in question. In addition to notification of these products to the relevant national authorities, Member States may introduce an authorization process, if deemed appropriate, and may charge manufacturers and importers proportional fees for authorization.

Countries including Poland (Article 11a of the Act of 9 November 1995 on the Protection of Health against Consequences of Consumption of Tobacco and Tobacco Products) and Spain (Royal Decree 579/2017) classify HTPs as novel tobacco products. France regulates HTPs as tobacco products under the French Public Health Code, Article L3512-1, 1° and as a novel tobacco product under Article L3512-1, 2°. French law does not establish an authorization system for products but includes a system for reporting product information (notification). HTP manufacturers are required to report the names, quantities and associated health effects of the ingredients to the national authorities. Use of HTPs in public places is banned under French Public Health Code Article L3512-8. Use of health claims, sale of HTPs to people under 18 and advertising, promotion and sponsorship of HTPs are also prohibited. Warning labels are required.

Emerging and imitation tobacco products

In Singapore, HTPs are treated in the same way as e-cigarettes that contain nicotine and electronic nicotine delivery systems, as both are considered to be emerging imitation tobacco products. Products that fall into this category include any device or article that:

- resembles, or is designed to resemble, a tobacco product;
- can be smoked;
- may be used in such a way as to mimic the act of smoking; or
- the packaging of which resembles, or is designed to resemble, the packaging commonly associated with tobacco products *(13)*.

It is prohibited to import, distribute, sell, purchase use or possess such products under the Tobacco (Control of Advertisement and Sale) Act, amended in 2011 *(13)*.

Smokeless tobacco

Article 2 of the European Union Tobacco Product Directive defines smokeless tobacco as "a tobacco product not involving a combustion process, including chewing tobacco, nasal tobacco and tobacco for oral use". Consequently, a number of countries in the European Union, such as Czechia (Section 2 (1) t) of Act No. 110/1997 Coll) and Portugal (Law 109/2015), classify HTPs as smokeless tobacco products.

In the Netherlands, HTPs are also regulated as smokeless tobacco under the Dutch Tobacco Act *(14)*, which is enforced by the Netherlands Food and Consumer Products Safety Authority. In line with the requirements of the European Union Directive for notification, the Dutch National Institute for Public Health and the Environment (RIVM) analyses and processes premarketing notification documents *(15)*. Requirements for HTPs include: warning labels similar to those for smokeless tobacco, a ban on health claims, a ban on sales to persons under the age of 18 and a ban on promotion and marketing, with few exceptions. HTPs are taxed at € 99.25/kg, like other smoked tobacco products. According to the Dutch Government, categorization of HTPs as smokeless tobacco could be changed if new evidence or information about their use becomes available. RIVM is currently analysing and conducting research on the contents and emissions of different IQOS HeatStick flavours and other HTPs for the Ministry of Health.

Next-generation products

In Italy, HTPs are regulated as next-generation products. Because of alleged belief that HTPs can reduce harm, these HTPs are exempt from the fiscal regimes of tobacco products. Rather, they are taxed under the category "inhalation product without combustion" under a specific excise structure *(16)*. Thus, these products enjoy the same tax reduction as electronic cigarettes, which is half the excise tax applied to conventional cigarettes *(17)*. Although the sale of HTPs to minors is banned (Decree n°6/2016 Chapter II Art. 24,3), enforcement of tobacco control regulations is only minimal for HTPs. Health warnings are required to cover only 30% of HTP packages (instead of 65% for conventional cigarettes), and they are not required to have pictorial images *(18)*. Comprehensive regulations prohibiting smoking in all public places and workplaces do not apply to HTPs. In addition, advertising and promotion of HTPs are not banned, and "IQOS embassies" and "IQOS boutiques", fancy concept stores in which people can try the products for free, are present in several strategic Italian cities. Therefore, for HTPs, the country has weakened the best-recognized tobacco control policies, i.e. price and tax increases, smoking bans, advertising bans and health warnings.

International approaches to the classification of HTPs

Discrepancies in national approaches to the classification of HTPs and legal challenges by the tobacco industry have raised concern about the lack of international standards for classifying HTPs. WHO is collaborating with experts and researchers on a classification tree for tobacco products, which will include HTPs, and will publish its findings once the project is finalized. A report will also be made to the next COP to the WHO FCTC, in November 2021, on appropriate classification of HTPs in accordance with paragraph 3(b) of decision FCTC/COP8(22), in which the COP requested the Convention Secretariat to advise on adequate classification of novel and emerging tobacco products such as HTPs *(19)*. The World Customs Organization is facilitating a revision of the "harmonized system code" to introduce new, specific customs codes for HTPs. The annex to the International Convention on the Harmonized Commodity Description and Coding System currently states that:

- heated tobacco units do not have a specific customs code and fall under the subheading of "other" (2403.99) in Chapter 24 of the International Convention on the Harmonized Commodity Description and Coding System, which addresses tobacco products; and
- devices used to heat tobacco units (i.e. HTPs) do not have a specific customs code and fall under the subheading of "other machines and apparatus" (8543.70) in Chapter 85, which concerns electrical machinery.

Mandatory changes have been made for 2022, to standardize these headings so that the products fall under heading 24.4:

Products containing tobacco, reconstituted tobacco, nicotine, or tobacco or nicotine substitutes, intended for inhalation without combustion; other nicotine-containing products intended for the intake of nicotine into the human body *(20)*.

Countries will be obliged to amend their domestic customs codes in 2020. The World Customs Organization will determine how disposable devices are to be classified later in the year.

The Harmonized System Code is not intended to affect domestic regulation of HTPs, but, as described in a WHO FCTC Secretariat information note *(21)*, in practice, these codes affect the entry and exit of goods at borders for the purposes of levying excise taxes and classification in domestic legislation. Such measures could be used by the tobacco industry to lobby for more favourable treatment of HTPs.

6.2.3 Regulatory frameworks and measures to reduce tobacco demand

As noted above, the classification of a tobacco product determines the regulations that are applicable. This in turn affects the availability and use of these products, as well as regulations on taxation, restrictions on advertising, promotion and sponsorship, use of the products in smoke-free places and packaging and labelling requirements. Some countries choose to ban the importation, sale or use of HTPs entirely, through a ban on an entire category of products, such as e-cigarettes or emerging and imitation tobacco products. For example, if a country has banned e-cigarettes and then makes a regulatory decision to classify HTPs as e-cigarettes, the classification will ensure that the HTPs do not enter the market (e.g. India). Nevertheless, the same product classification (i.e. "e-cigarettes") may result in a ban in one country and regulatory restrictions in another.

Various mechanisms have been adopted by WHO Member States to regulate HTPs. While many countries have used existing tobacco control laws, some have formulated specific provisions. Information held by WHO indicates the following common mechanisms.

Existing laws

HTPs could be defined in the same way as products that are already covered by law, such as in South Africa, where HTPs are considered tobacco products. As noted above (section 6.2.1), in Australia, HTPs are regulated under the Standard for the Uniform Scheduling of Poisons, with products containing nicotine categorized under Schedule 7, "dangerous poison". The Therapeutic Goods Administration recently refused to amend the Poisons Standard to allow sales of HTPs *(11)*.

Amendment of existing laws

Existing legislation can be amended to include HTPs if the definitions do not clearly cover these products. In Malaysia, an amendment to the Control of Tobacco Product Regulations 2004 changed the definition of "smoking" in 2015 to include use of HTPs *(22)*.

New legislation and other mechanisms

New legislation and other mechanisms may be used to regulate HTPs or to include them explicitly in existing legislation. The United Arab Emirates enacted an Electronic Nicotine Products (Equivalents of Traditional Tobacco Products) standard *(23)*, which regulates e-cigarettes, e-liquids with and without nicotine and HTPs, with a requirement to specify their production, import, retail and display. Consequently, these products now fall within the same regulatory framework as tobacco products, provided by Federal Law 15/2009. In the Philippines, the President issued an executive order in February 2020 regulating the commercialization and use of electronic cigarettes, HTPs and other novel tobacco products *(24)*. The executive order excludes HTPs from the definition of tobacco products but includes them in the definition of smoking.

As noted earlier, HTPs were recognized as tobacco products at COP8 (decision FCTC/COP8(22)) *(19)*. This decision reminded Parties of their commitments under the WHO FCTC when addressing the challenges of novel and emerging tobacco products such as HTPs and devices designed for consuming such products, to consider prioritizing specified tobacco control measures in accordance with the WHO FCTC and national law. These, listed in paragraph 5 of the decision, are:

(a) to prevent the initiation of novel and emerging tobacco products;

(b) to protect people from exposure to their emissions and to explicitly extend the scope of smoke-free legislation to these products in accordance with Article 8 of the WHO FCTC;

(c) to prevent health claims from being made about novel and emerging tobacco products;

(d) to apply measures regarding advertising, promotion and sponsorship of novel and emerging tobacco products in accordance with Article 13 of the WHO FCTC;

(e) to regulate the contents and the disclosure of the contents of novel and emerging tobacco products in accordance with Articles 9 and 10 of the WHO FCTC;

(f) to protect tobacco-control policies and activities from all commercial and other vested interests related to novel and emerging tobacco products, including interests of the tobacco industry, in accordance with Article 5.3 of the WHO FCTC;

(g) to regulate, including restrict, or prohibit, as appropriate, the manufacture, importation, distribution, presentation, sale and use of novel and emerging tobacco products, as appropriate to their national laws, taking into account a high level of protection for human health; and

(h) to apply, where appropriate, the above measures to the devices designed for consuming such products.

Advertising, promotion and sponsorship (Article 13)

Article 1 of the WHO FCTC provides a comprehensive definition of tobacco advertising, promotion and sponsorship. Tobacco advertising and promotion are defined as "any form of commercial communication, recommendation or action with the aim, effect or likely effect of promoting a tobacco product or tobacco use either directly or indirectly" and sponsorship as "any form of contribution to any event, activity or individual with the aim, effect or likely effect of promoting a tobacco product or tobacco use either directly or indirectly" *(25)*.

Although most countries do not specifically regulate the advertising, promotion and sponsorship of HTPs, the products should be covered by the bans on advertising, promotion and sponsorship applied to conventional tobacco products, in accordance with the guidance of WHO and the COP. If a distinction is made between HTP sticks and devices and if the definition of tobacco product covers only the sticks, advertising of the device may not be banned.

A comprehensive ban on tobacco advertising, promotion and sponsorship covers not only traditional forms of advertising such as television, radio and print but also "brand stretching", displays of products at points of sale and tobacco-industry-sponsored corporate social responsibility programmes, among others. Nevertheless, the fast-changing media landscape creates regulatory loopholes that allow tobacco product advertising in social media campaigns and by influencers, often targeting young people. For example, the tobacco industry engages in public relations and corporate social responsibility-related activities, sponsors events and uses social media and online platforms to promote HTPs, all of which have contributed to the proliferation of the products around the world. Early in 2020, the State Council in the Republic of Korea passed an amendment to the country's National Health Promotion Act banning any direct or indirect promotional activity by tobacco manufacturers to consumers. The Ministry of Health and Welfare plans to ban practices such as discounts on ENDS and HTPs and free distribution of these products, including the devices, during promotional events *(26)*.

Smoke-free spaces (Article 8)

HTPs are commonly referred to by some as "heat-not-burn" products, a term coined by the industry, which has positive connotations. Manufacturers suggest that the products are "ash-free", "smoke-free" and "cleaner alternatives" to conventional cigarettes, which may create confusion about their categorization. To reduce the confusion created by this terminology, especially in regulations on the application of smoke-free laws, WHO introduced the term "heated tobacco products". Philip Morris has tried to distinguish IQOS from conventional smoking by creating partnerships with hundreds of "IQOS-friendly" restaurants and bars in countries such as Romania and Ukraine (27), which may ban cigarettes but allow use of IQOS, undermining prohibitions on indoor smoking. Romania does not classify HTPs as "tobacco products for smoking" with regard to smoke-free policies on the grounds that these products do not generate smoke (28).

Packaging and labelling (Article 11)

Article 11 of the WHO FCTC states that regulators should ensure that tobacco product packaging and labelling

> do not promote a tobacco product by any means that are false, misleading, deceptive or likely to create an erroneous impression about its characteristics, health effects, hazards or emissions, including any term, descriptor, trademark, figurative or any other sign that directly or indirectly creates the false impression that a particular tobacco product is less harmful than other tobacco products. These may include terms such as "low tar", "light", "ultra-light" or "mild".

The aim of these prohibitions is to avoid misleading consumers into thinking that one tobacco product is healthier than another, an especially important aim with respect to HTPs. Currently, however, health warning requirements for HTPs tend to be less onerous than for those for conventional cigarettes. Even where health warnings are imposed, in some countries (Japan and Netherlands), they apply only to inserts and not to the devices.

Articles 9–12 of the European Union Tobacco Product Directive address health warnings and their dimensions. For novel tobacco products considered to be "smokeless tobacco", text (but not pictorial) health warnings must cover 30% of each of the two largest surface areas; for novel tobacco products intended for smoking, combined health warnings (graphic and text) must cover 65% of the two largest surfaces. Categorization as smokeless tobacco is therefore preferable for the tobacco industry. Article 13 of the Directive on product presentation prohibits labelling or packaging with any element or feature that creates an erroneous impression about the characteristics, health effects, risks or emissions of the product. Labelling or packaging may not include any information about the nicotine, tar or carbon monoxide content of the tobacco product; suggest that a particular tobacco product is less harmful than others, reduces the effects of any harmful components of smoke or has vitalizing, energizing, healing, rejuvenating, natural or organic properties or other health or lifestyle benefits; or refer to taste, smell, any flavourings or other additives or the absence thereof.

In the USA, the Food and Drug Administration (FDA) requires that all HTP package labels and advertisements include an additional warning about the addictiveness of nicotine, as well as the other warnings required for cigarettes. The aim of this requirement is to correct a misperception among users that IQOS pose a lower risk of addiction than conventional cigarettes. An application for designation as a modified risk tobacco product may be submitted to the FDA to allow a product to be marketed with a claim of reduced risk *(29)*, i.e. "any tobacco product that is sold or distributed for use to reduce harm or the risk of tobacco-related disease associated with commercially marketed tobacco products". In 2016, Philip Morris Products S.A. sought authorization to market IQOS with the claims that the product "can reduce the risks of tobacco-related diseases", "significantly reduce[s] your body's exposure to harmful or potentially harmful chemicals" and "presents less risk of harm than continuing to smoke cigarettes". The FDA concluded, however, that the company had not provided sufficient evidence that consumers would not be misled by those claims, and Philip Morris were consequently not allowed to market HTPs with a claim of reduced risk *(30)*. Under US law, the FDA may issue two types of order for a modified risk tobacco product: a "risk modification order" or an "exposure modification order". Philip Morris Products S.A. had requested both types of order. Although the FDA determined that the evidence did not support issuance of the first type, the evidence supported issuance of an exposure modification order for the IQOS device and the tobacco HeatSticks. The exposure modification order authorizes Philip Morris to make claims about how tobacco is heated and about the production of harmful and potentially harmful chemicals and exposure to those chemicals in advertising and marketing of the products. On 7 July 2020, the FDA authorized the marketing of an "IQOS tobacco heating system", which includes the IQOS device, Marlboro Heatsticks, Marlboro Smooth Menthol Heatsticks and Marlboro Fresh Menthol Heatsticks, as a modified risk tobacco product *(31)*. The FDA stressed that this authorization does not indicate that the products are safe or approved by the FDA, and it rejected claims that the company had adequately demonstrated that use of the products is less harmful than use of another tobacco product or reduces risks to health *(32)*. Philip Morris hailed the exposure authorization as a milestone for public health and cited it as "an important example of how governments and public health organizations can regulate smoke-free alternatives to differentiate them from cigarettes in order to promote the public health".

The Republic of Korea, which primarily regulates HTPs as tobacco products, requires graphic health warnings on HTP packages. The move by the Ministry of Health and Welfare to mandate use of graphic images of the consequences of tobacco use, such as cancer-ridden organs, and more concise written warnings with specific risk figures was part of a set of measures to deter smoking implemented in late 2018. The strengthened measures followed a one-year deliberation by a 13-member special committee comprising Government officials and private experts and a survey of 1500 smokers and non-smokers to gauge public opinion *(33)*. All the regulations that apply to tobacco products, such as taxation, smoke-free areas, advertising, package warnings and labels, also apply to HTPs *(1)*, in line with WHO recommendations that HTPs be subject to the same policy and regulatory measures as applied to all other tobacco products *(34)*.

In Canada, the Tobacco Products Regulations (Plain and Standardized Appearance), which came into force on 9 November 2019, apply to tobacco products, including devices necessary for the use of a product made in whole or in part of tobacco, such as HTPs, as they are defined as "tobacco products" under the Act *(35)*. Israel and New Zealand also require plain packaging for HTPs.

Sales restrictions

In most countries, the sales restrictions imposed on tobacco products are also applicable to HTPs. These include prohibition of certain methods of sale (e.g. from vending machines or the Internet), restricted locations, age restriction for purchasers and licensing or requirements for retailers. For example, Cyprus prohibits the sale of HTPs from automatic tobacco vending machines, sales to minors and free distribution of HTPs *(36)*. In Slovenia, the premises for the sale of tobacco, tobacco products and related products, including HTPs, must be registered under Article 35(1) of the Restriction of the Use of Tobacco Products Act 2017 *(37)*. The Act prohibits sales to minors and sale of HTPs at temporary and mobile points of sale, via the Internet, telecommunications or any other developing technology or cross-border distance sales and in single units, except in the manufacturer's original packaging (Article 30). The law in Saudi Arabia prohibits the sale of HTP sticks in packages of more than 20 sticks *(38)*.

All countries that ban the sale of tobacco products to minors implicitly extend the ban to HTPs; however, some countries apply different age limits. Japan for instance applies a sales ban to persons under 20 *(39)*, while Austria and Belgium prohibit HTP sales only to children under 16 *(40)*. Federal systems, such as those of Canada, Switzerland and the USA, may have different subnational limits. In Switzerland, an age restriction of either 16 or 18 years applies to the purchase of HTPs, depending on the canton, while, in Canada, the age restriction is between 18 and 19 years. In December 2019, the USA enacted a ban on sales of all tobacco products, including HTPs, to any person under the age of 21 years.

Contents and emissions (Articles 9 and 10)

Most countries in the European Union require manufacturers to report the names, quantities and health effects of ingredients, including flavours. Under the European Union Tobacco Product Directive, tobacco products with a characterizing flavour are prohibited. A "characterizing flavour" is defined in Article 1(25) of the Directive as:

> a clearly noticeable smell or taste other than one of tobacco, resulting from an additive or a combination of additives, including, but not limited to, fruit, spice, herbs, alcohol, candy, menthol or vanilla, which is noticeable before or during the consumption of the tobacco product.

The prohibition on characterizing flavours currently applies only to cigarettes and roll-your-own tobacco and not to HTPs. Article 7(2) notes, however, that the Commission may determine that a particular tobacco product is subject to this ban, either on the initiative of the European Commission or at the request of a Member State. Characterizing flavours in HTPs may therefore be banned in the European Union in the future.

Education, communication, training and public awareness (Article 12)

The tobacco industry, with its new portfolio of products, uses marketing and promotion mainly as "reduced harm", "reduced risk" and alternatives to conventional cigarette as a strategy to manipulate governments to open their markets to HTPs. These claims are, however, unsubstantiated, as these products have not been proven to be different from conventional cigarettes in terms of tobacco-related risk, and the claims have distracted attention from evidence-based tobacco control policy measures to reduce tobacco use and protect public health. Article 12 of the WHO FCTC states:

> Parties shall promote and strengthen public awareness of tobacco control issues, using all available communication tools, as appropriate, and adopt and implement effective legislative, executive, administrative or other measures to promote the following:
>
> (a) broad access to effective and comprehensive educational and public awareness programmes on the health risks including the addictive characteristics of tobacco consumption and exposure to tobacco smoke;
>
> (b) public awareness about the health risks of tobacco consumption and exposure to tobacco smoke, and about the benefits of the cessation of tobacco use and tobacco-free lifestyles as specified in Article 14.2;

(c) public access, in accordance with national law, to a wide range of information on the tobacco industry as relevant to the objective of this Convention;

(d) effective and appropriate training or sensitization and awareness programmes on tobacco control addressed to persons such as health workers, community workers, social workers, media professionals, educators, decision-makers, administrators and other concerned persons;

(e) awareness and participation of public and private agencies and non-governmental organizations not affiliated with the tobacco industry in developing and implementing intersectoral programmes and strategies for tobacco control; and

(f) public awareness of and access to information regarding the adverse health, economic, and environmental consequences of tobacco production and consumption.

These are evidence-based measures for sensitizing the public and raising awareness about the ill-effects of use of tobacco products. All countries, and not just Parties to the WHO FCTC, should consider prioritizing these measures to protect public health.

6.3 Considerations and barriers to regulation, implementation and enforcement of policies

6.3.1 Regulatory considerations in implementing policies

HTPs may be considered differently from conventional cigarettes because of insufficient knowledge about the products, tobacco industry lobbying, regulatory classification of smokeless products based on arguments that these products are "smoke-free", "ash-free" or "cleaner alternatives" than conventional cigarettes, and differential approaches to the devices and the inserts. The tobacco industry has aggressively marketed these products, lobbied governments for more lenient regulations and exerted substantial pressure on regulatory decisions concerning HTPs. This has resulted in only partial application of comprehensive tobacco control regulatory measures to HTPs, which will ultimately undermine existing tobacco control.

The authors of a study *(41)* on an IQOS campaign in Israel described ways in which the industry attempts to define a new product as part of a category not covered by existing tobacco laws, in this case by using the term "smoking" in the argument. When IQOS was launched in Israel in December 2016, Philip Morris International organized high-level meetings and other direct communications with the Israeli Ministries of Health and Finance to put pressure on the Government to exempt IQOS from existing tobacco regulations, which were reversed after three petitions to the Supreme Court. The authors warned that, in the absence of requirements for specific health warnings for HTPs, the industry may voluntarily place warnings on newer products, such as "research suggests that cigarettes cause addiction", which may introduce doubt about well accepted evidence regarding cigarettes.

The industry categorizes HTPs in the way that ensures the most favourable treatment under applicable national law. In New Zealand, HTPs were banned as a "smokeless product", but the ban was successfully challenged in court by Philip Morris International on the basis that HTPs are not "smokeless". New Zealand now applies all tobacco control laws for smoked products to HTPs, including plain packaging. In Romania, regulation of HTPs as smoked products was challenged in an industry submission on the basis of arguments of reduced harm, no combustion and therefore no smoke.

In determining the most appropriate approach to regulation of HTPs, countries should consider factors such as:

- the absolute and relative health risks to users and non-users;
- whether HTPs can be regulated continuously as scientific knowledge is gained on these products;
- the risk that tobacco use and smoking will be "renormalized";
- the risk of initiation by non-users of tobacco products, particularly young people;
- the possibility that smokers who have quit tobacco use, thereby improving their health, might switch to HTPs, although these are tobacco products and have not been proven to reduce tobacco-related risk;
- use with other nicotine and tobacco products, so that users are exposed to the emissions of two or more products; and
- capacity to assess industry claims regarding the relative harm of HTPs relative to conventional cigarettes and to prevent claims that could mislead consumers.

As noted previously, Parties to the WHO FCTC could go beyond its provisions in accordance with Article 2.1 of the Convention, which states that:

> In order to better protect human health, Parties are encouraged to implement measures beyond those required by this Convention and its protocols, and nothing in these instruments shall prevent a Party from imposing stricter requirements that are consistent with their provisions and are in accordance with international law.

6.3.2 Barriers to implementing and enforcing policies

In addition to lobbying, industry litigation threatens the passage, implementation and enforcement of policies. In New Zealand, a district court decision in 2018 (Philip Morris vs Ministry of Health) *(42)* overturned the previous classification of HTPs as "any tobacco product labelled or otherwise described as suitable for chewing, or for any other oral use (other than smoking)", which are banned under Section 29(2) of the Smokefree Environments Act 1990 *(43)*. Philip Morris Ltd was charged with violating the law by selling "Heets", the HTP inserts for IQOS. The holding found that, because the law was originally intended to control the sales of chewing tobacco and other tobacco products consumed orally, it should not apply to tobacco inserts for HTPs. Therefore, the district court ruled in favour of Philip Morris, and HTPs may be legally imported, sold, packed and distributed in New Zealand under the Act. Consequently, the Smokefree Environments Act regulations, including the ban on sales to minors and restrictions on advertising, apply to HTPs. This case highlights the challenges of regulating these products and the importance of legislation that can be adapted to the changing tobacco product landscape.

Much of the litigation is based on claims of combustion or non-combustion, whichever determines the most favourable treatment for the industry. As described earlier, regulation of HTPs as smoked products may be challenged by the industry on the basis of no combustion, while their regulation as smokeless products may be challenged on the basis that these products are not "smokeless".

6.3.3 Other considerations and unintended consequences

When countries regulate HTPs as smoked tobacco products, the health warnings for other smoked tobacco products apply. The same principle applies when HTPs are regulated as smokeless tobacco products. Many countries may be under the impression that these products require specific provisions, whereas they are already covered by their current tobacco control law. The Pan American Health Organization has made recommendations to countries in the Region of the Americas on regulation of HTPs under existing regulations for tobacco products. As the way in which tobacco products are defined in some regulations may make application of tobacco control laws difficult, regulations should be broadened to encompass novel and emerging nicotine and tobacco products. This would limit exploitation of regulatory loopholes by the tobacco industry. The WHO report on the global tobacco epidemic, 2019 *(44)* provides useful information and recommendations for countries.

- HTPs contain tobacco and should be regulated in the same way as tobacco products.

- HTPs produce toxic emissions, many of which are similar to those found in cigarette smoke.
- HTP users are exposed to toxic emissions from the products, and bystanders could also be exposed to toxic second-hand emissions.
- Although the levels of several toxicants in HTPs are generally lower than those in conventional cigarettes, the levels of others are higher. A lower level of a toxicant does not necessarily indicate a lower health risk.
- HTPs contain nicotine. Nicotine is highly addictive and is linked to harm, particularly in children, pregnant women and adolescents.
- The long-term health effects of HTP use and exposure to their emissions remain unknown. There is currently insufficient independent evidence on the relative and absolute risks. Independent studies should be conducted to determine the health risks they pose to users and bystanders.

This information includes important considerations for HTPs, as their availability on the market could have unintended consequences for public health, which should be considered in formulating policies and determining a regulatory path for HTPs.

6.4 Discussion

HTPs are tobacco products, defined in the WHO FCTC as "products entirely or partly made of the leaf tobacco as raw material which are manufactured to be used for smoking, sucking, chewing or snuffing". These products have gained a considerable market share in some countries and are now available on over 50 markets worldwide. Their unique characteristics, intensive industry lobbying, lack of clarity about their health risks and the absence of international approaches all pose challenges to regulators.

While limited data are available, as regulations depend on national interpretations of laws, which cannot be assessed independently, different countries clearly regulate HTPs in different ways, on the spectrum from bans to no regulation. Some countries consider HTPs to be in the same category as conventional tobacco products. Many countries already have domestic legislation and regulations with respect to basic tobacco control measures, including advertising, promotion and sponsorship, smoke-free spaces and packaging and labelling. A misconception is that regulating HTPs would be a new, resource-intensive initiative, when, in fact, these products are already covered by current tobacco control law.

The marketing of HTPs is, however, strategic, and its regulation presents challenges. The fact that devices and inserts are sold separately may exempt the devices (which do not contain tobacco) from, for instance, restrictions on advertising, promotion and sponsorship and even on sale to minors. The tobacco industry claims that there is no combustion in HTPs, and many European Union countries classify them as novel smokeless tobacco products, such that requirements for warnings and restrictions such as smoke-free areas may differ from those for conventional cigarettes. This paper poses a number of considerations for addressing regulation of these products.

6.5 Conclusions

In the past several years, the industry has significantly expanded its "reduced risk" portfolio with newer generation tobacco products, such as HTPs. The innovative technologies, design, marketing and health claims associated with these products have weakened tobacco control measures in some countries where there were relatively strong laws to regulate conventional cigarettes and attempts by the tobacco industry to reposition itself as a public health partner. Regulators were largely unprepared for these new products, especially their claims of "no combustion", "no smoke" and "no ash", which the industry has used to lobby governments for favourable regulatory treatment and in particular to circumvent smoke-free laws. As a result, the current regulations on HTPs are specific to each country. HTPs generate aerosols that contain toxicants, many at levels lower than those in conventional cigarette smoke, but in some cases higher. A lower level of a toxicant does not, however, necessarily mean lower risk. As the long-term effects on health of the use of and exposure to emissions from these products remain unknown and there is currently insufficient independent evidence on the relative and absolute risks, they should be fully subject to the provisions of the WHO FCTC, including a ban on their use in indoor spaces. The aim of this paper was to increase awareness about the inconsistent approaches used to regulate HTPs and to prepare regulators should a case be made by the industry to introduce HTPs and other novel and emerging tobacco products onto their markets.

Countries that are examining their legislative options can learn from regulatory successes and challenges in other countries. New tobacco control laws should anticipate not only HTPs but other emerging products, with definitions that are broad enough to encompass all innovative developments. Tobacco industry interference, including lobbying and misinformation, should be monitored and subjected to protection under WHO FCTC Article 5.3.

6.6 Research gaps

- Global surveillance of HTP products and their use to understand industry marketing strategies.
- Systematic monitoring of industry mechanisms to limit application of the WHO FCTC to HTPs and to undermine tobacco control.
- Comprehensive mapping of legislation on HTPs to identify regulatory loopholes that could be exploited by the industry and the level of implementation of existing policies, in order to improve it and to provide evidence on the regulatory approaches that promote maximum protection for public health.
- Effective surveillance for better understanding of the availability, marketing and use of HTPs.

6.7 Policy recommendations

As HTPs evolve and their availability spreads, regulators must address questions about these products in the face of industry pressure and scientific uncertainty. The varied approaches used by governments to classify and regulate these products reflect the absence of internationally agreed approaches. One thing is clear: HTPs are tobacco products. Therefore, policy-makers are urged to consider the following recommendations.

- Classify HTPs as tobacco products, except in countries where such classification would result in the more lenient regulations, undermine existing tobacco control provisions or allow market entry when similar products have been banned.
- Apply all the regulatory measures of the WHO FCTC to HTPs and especially those in Articles 5.3, 6, 8, 9 and 10, 11, 12, 13, 14 and 20. These include protecting tobacco control activities from all commercial and other vested interests, application of excise tax on these products, requiring reporting and disclosure of product information, requiring combined health warnings on HTPs and covering HTPs under smoke-free laws and bans and restrictions on tobacco advertising, promotion and sponsorship.
- In line with Article 13.4(a), prohibit "all forms of tobacco advertising, promotion and sponsorship that promote a tobacco product by any means that are false, misleading or deceptive or likely to create an erroneous impression about its characteristics, health effects, hazards or emissions".

- Use existing regulations for tobacco products to regulate HTPs, and broaden the scope of those regulations to ensure that regulatory loopholes cannot be exploited by the industry, even in countries in which these products are not currently (legally) on the market.
- Include HTPs in surveillance to understand their use and availability through existing channels, to inform regulation of these products and to ensure maximum protection of public health.
- Put the burden of proof on manufacturers to support claims about the products, and prohibit unsubstantiated claims about the relative risk or harmfulness of HTPs relative to other tobacco products.
- Monitor misinformation with respect to HTPs and claims about the risk or harm of these products relative to other products, and take appropriate regulatory action to curb such practices.
- Require premarket notification of novel and emerging tobacco products to enable the government to assess whether to authorize their sale.
- Define and classify these products to ensure that public health objectives are protected and to avoid regulatory loopholes. Given the variety of products on the market and under development, legal definitions must cover all product designs and be adaptable to product innovations.
- Closely monitor the products and their markets in the country, and institute effective measures to enforce adherence to relevant policies and regulations.
- Make clear regulatory distinctions among products and categories of products, and clearly define products and their components to ensure effective regulation.

6.8 References

1. Heated tobacco products: information sheet, 2nd edition. Geneva: World Health Organization; 2020 (https://www.who.int/publications/i/item/WHO-HEP-HPR-2020.2, accessed 10 January 2021).
2. Our tobacco heating system. IQOS. Lausanne: Philip Morris International; 2020 (https://www.pmi.com/smoke-free-products/iqos-our-tobacco-heating-system, accessed 10 January 2021).
3. Heated tobacco products (HTPs). Market monitoring information sheet (WHO/NMH/PND/18.7). Geneva: World Health Organization; 2018 (https://apps.who.int/iris/bitstream/handle/10665/273459/WHO-NMH-PND-18.7-eng.pdf, accessed 10 January 2021).
4. Novel and emerging tobacco products (FCTC/COP/8/22). Geneva: World Health Organization; 2018 (https://www.who.int/fctc/cop/sessions/cop8/FCTC__COP8(22).pdf?ua=1, accessed 28 December 2019).
5. Tobacco control laws. Washington (DC): Campaign for Tobacco-Free Kids; 2020 (https://www.tobaccocontrollaws.org/, accessed 10 January 2021).

6. Warner J. Vaping or cannabis: Where's the growth for Big Tobacco? IG, 17 February 2020 (https://www.ig.com/uk/news-and-trade-ideas/vaping-or-cannabis--wheres-the-growth-for-big-tobacco--200217, accessed 10 January 2021).
7. Bialous SA, Glantz SA. Heated tobacco products: another tobacco industry global strategy to slow progress in tobacco control. Tob Control. 2018;27:s111–7.
8. Preamble. Novel and emerging tobacco products (FCTC/COP/8/22). Geneva: World Health Organization; 2018 (https://www.who.int/fctc/cop/sessions/cop8/FCTC__COP8(22).pdf?ua=1, accessed 28 December 2019).
9. HTP information sheet. Geneva: World Health Organization; 2019 (https://www.who.int/tobacco/publications/prod_regulation/heated-tobacco-products/en/, accessed 10 January 2021).
10. Notice of final decisions to amend (or not amend) the current Poisons Standard 24 August 2020. Canberra: Department of Health, Therapeutic Goods Administration; 2020 (https://www.tga.gov.au/sites/default/files/public-notice-final-decisions-acms29-accs27-joint-acms-accs24-march-2020.pdf, accessed 10 January 2021).
11. Directive 2014/40/EU of the European Parliament and of the Council of 3 April 2014 on the approximation of the laws, regulations and administrative provisions of the Member States concerning the manufacture, presentation and sale of tobacco and related products. Strasbourg: European Parliament; 2014 (https://eur-lex.europa.eu/legal-content/EN/TXT/?uri=O-J%3AJOL_2014_127_R_0001, accessed 10 January 2021).
12. Parliamentary questions (P-009191/2016(ASW). Strasbourg: European Parliament; 2017 (https://www.europarl.europa.eu/doceo/document/P-8-2016-009191-ASW_ES.html).
13. Tobacco (Control of Advertisements and Sale) Act (as amended), Arts. 16(1)-(2). Singapore; 2011 (https://www.tobaccocontrollaws.org/.../Singapore%20-%20Control%20of%20Ads%20%26%20Sale%20-%20national.pdf, accessed 10 January 2021).
14. Act of April 26, 2016 Amending the Tobacco Act to Implement Directive 2014/40/EU on the Manufacture, Presentation and Sale of Tobacco and Related Products. The Hague: Staatsblad van het Koninkrijk der Nederlanden; 2016 (https://www.tobaccocontrollaws.org/files/live/Netherlands/Netherlands%20-%20Act of April 26%2C 2016%20-%20national.pdf, accessed 10 January 2021).
15. Information for suppliers. Bilthoven: National Institute for Public Health and the Environment; 2017 (https://www.rivm.nl/en/tobacco/information-for-suppliers, accessed 10 January 2021).
16. Gambaccini P. Blog. How should heated tobacco be taxed? Lausanne: Vapor Products Tax, 12 July 2017 (https://vaporproductstax.com/how-should-heated-tobacco-be-taxed/, accessed 10 January 2021).
17. Decreto legislativo, 15 dicembre 2014, n. 188. Disposizioni in materia di tassazione dei tabacchi lavorati, dei loro succedanei, nonche' di fiammiferi, a norma dell'articolo 13 della legge 11 marzo 2014, n. 23. Rome: Government of Italy; 2014 (http://www.governo.it/sites/governo.it/files/77443-9913.pdf, accessed 10 January 2021).
18. Legislative decree. Implementation of Directive 2014/40/EU on streamlining the legislative, regulatory and administrative provisions of the member states regarding the processing, presentation and sale of tobacco products and related products, which replaces directive 2001/37/EC. Rome: Government of Italy; 2016 (https://www.tobaccocontrollaws.org/.../Italy%20-%20Leg.%20Decree%20No. %206%20of%20Jan.%2012%2C%202016.pdf, accessed 10 January 2021).
19. Conference of the Parties to the WHO Framework Convention on Tobacco Control (eighth session), Decision FCTC/COP8(22). Novel and emerging tobacco products. Geneva: World Health Organization; 2018 (https://www.who.int/fctc/cop/sessions/cop8/FCTC__COP8(22).pdf?ua=1, accessed 10 January 2021).

20. International Convention on the Harmonized Commodity Description and Coding System (Brussels, 14 June 1983). Amendments to the nomenclature as an annex to the Convention accepted pursuant to the recommendation of 28 June 2019 of the Cistoms Co-operation Council (NG0262B1). Brussels: World Customs Organization; 2019:19 (http://www.wcoomd.org/-/media/wco/public/global/pdf/topics/nomenclature/instruments-and-tools/hs-nomenclature-2022/ng0262b1.pdf?db=web, accessed 10 January 2021).
21. Information note on classification of novel and emerging tobacco products. Geneva: World Health Organization; 2019 (https://untobaccocontrol.org/impldb/wp-content/uploads/Info-Note_Novel-Classification_EN.pdf, accessed 10 January 2021).
22. Federal Government Gazette. 28 December 2015. Control of Tobacco Product (Amendment) (No. 2) Regulations 2015. Kuala Lumpur: Federal Government; 2015 (https://www.tobaccocontrollaws.org/legislation/country/malaysia/laws, accessed 10 January 2021).
23. Notification detail. Draft of the UAE GCC Technical Regulation "Electronic nicotine products (equivalents of traditional tobacco products)" (G/TBTN/ARE/482). Brussels: European Commission; 2020 (https://ec.europa.eu/growth/tools-databases/tbt/en/search/?tbtaction=search.detail&num=482&Country_ID=ARE&dspLang=EN&BASDATEDEB=&basdatedeb=&basdatefin=&baspays=ARE&basnotifnum=482&basnotifnum2=482&bastypepays=ARE&baskeywords=, accessed 10 January 2021).
24. Executive order No. 106. Prohibiting the manufacture, distribution, marketing and sale of unregistered and/or adulterated electronic nicotine/non-nicotine delivery systems, heated tobacco products and other novel tobacco products, amending Executive Order No. 26 (S. 2017) and for other purposes. Manila: Malacañan Palace; 2020 (https://perma.cc/B7UY-BMR8, accessed 10 January 2021).
25. Guidelines for implementation of Article 13 of the WHO Framework Convention on Tobacco Control (Tobacco advertising, promotion and sponsorship). Geneva: World Health Organization; 2008 (https://www.who.int/fctc/guidelines/article_13.pdf?ua=1, accessed 10 January 2021).
26. Indirect promotional activities such as providing discount vouchers for electronic cigarette devices are prohibited. Sejong City: Health promotion Division; 2020 (http://www.mohw.go.kr/react/al/sal0301vw.jsp?PAR_MENU_ID=04&MENU_ID=0403&CONT_SEQ=352444, accessed 10 January 2021).
27. Jackler RK, Ramamurthi D, Axelrod AK, Jung JK, Louis-Ferdinand NG, Reidel JE et al. Global marketing of IQOS. The Philip Morris campaign to popularize "heat not burn" tobacco. Stanford (CA): Stanford Research into the Impact of Tobacco Advertising; 2020 (http://tobacco.stanford.edu/tobacco_main/publications/IQOS_Paper_2-21-2020F.pdf, accessed 10 January 2021).
28. Law No. 349 of June 6, 2002. On preventing the consumption of tobacco products and combating its effects. Bucharest: Government of Romania; 2002 (https://www.tobaccocontrollaws.org/files/live/Romania/Romania%20-%20Law%20No.%20349.pdf, accessed 10 January 2021).
29. Section 911 of the Federal Food, Drug, and Cosmetic Act – Modified Risk Tobacco Products. Silver Spring (MD): Food and Drug Administration; 2020 (https://www.fda.gov/tobacco-products/rules-regulations-and-guidance/section-911-federal-food-drug-and-cosmetic-act-modified-risk-tobacco-products accessed 10 January 2021).
30. McKelvey K, Popova L, Kim M, Kass Lempert L, Chaffee BW, Vijayaraghavan M et al. IQOS labelling will mislead consumers. Tob Control. 2018;27:s48–54.
31. Modified risk tobacco orders. FDA authorizes marketing of IQOS tobacco heating system with "reduced exposure" information. Press release, 7 July 2020. Silver Spring (MD): US Food and Drug Administration; 2020 (https://www.fda.gov/news-events/press-announcements/fda-authorizes-marketing-iqos-tobacco-heating-system-reduced-exposure-information,

accessed 10 January 2021).
32. WHO statement on heated tobacco products and the US FDA decision regarding IQOS. Geneva: World Health Organization; 2020 (https://www.who.int/news-room/detail/27-07-2020-who-statement-on-heated-tobacco-products-and-the-us-fda-decision-regarding-iqos, accessed 10 January 2021).
33. Lee KM. Gov't to mandate graphic warnings on heated tobacco products. The Korea Times, 2020 (https://www.koreatimes.co.kr/www/nation/2018/05/119_248951.html, accessed 10 January 2021).
34. Enforcement Decree of the National Health Promotion Act. Seoul: Ministry of Government Legislation; 2016 (https://www.tobaccocontrollaws.org/files/live/South%20Korea/South%20Korea%20-%20Enf.%20Decree%20of%20Nat'l%20Health%20Promotion%20Act.pdf, accessed 10 January 2021).
35. Tobacco product regulations: plain and standardized appearance. Ottawa: Government of Canada; 2020 (https://www.canada.ca/en/health-canada/services/health-concerns/tobacco/legislation/federal-regulations/products-regulations-plain-standardized-appearance.html, accessed 10 January 2021).
36. Law that provides for measures for the reduction of smoking. Nicosia: Government of Cyprus; 2002 (https://www.tobaccocontrollaws.org/files/live/Cyprus/Cyprus – Reduction of Smoking.pdf, accessed 10 January 2021).
37. The Restriction of the Use of Tobacco Products Act (Zoutpi). Ljubljana: National Assembly; 2017 (https://www.tobaccocontrollaws.org/.../Slovenia%20- %20TC%20Act%202017.pdf, accessed 10 January 2021).
38. Controls and requirements for electronic smoking devices. Version No. 1 (18/9/1440 AH). Riyadh: Saudi Food and Drug Authority; 2018 (https://www.tobaccocontrollaws.org/files/live/Saudi%20Arabia/Saudi%20Arabia%20-%20SFDA%20E-Cig%20Requirements.pdf, accessed 10 January 2021).
39. Act Prohibiting Smoking by Minors Act No. 33 on March 7, 1900. Last revision: Act No. 152 on December 12, 2001. Tokyo: Government of Japan; 2001 (https://www.tobaccocontrollaws.org/files/live/Japan/Japan%20-%20Act%20Prohibiting%20Smoking%20by%20Minors.pdf, accessed 10 January 2021).
40. Purchasing and consuming tobacco. Vienna: European Union Agency for Fundamental Rights; 2018 (https://fra.europa.eu/en/publication/2017/mapping-minimum-age-requirements/purchase-consumption-tobacco, accessed 10 January 2021).
41. Rosen L, Kislev S. The IQOS campaign in Israel. Tob Control 2018;27(Suppl1):s78–81.
42. Tobacco Control Laws. Litigation by country. New Zealand MOH v. PMI. Washington (DC: Campaign for Tobacco-Free Kids; 2020 (https://www.tobaccocontrollaws.org/litigation/decisions/nz-20180312-new-zealand-moh-v.-pmi, accessed 10 January 2021).
43. Smokefree Environments and Regulated Products Act 1990. Auckland: Ministry of Health; 1990 (https://www.tobaccocontrollaws.org/files/live/New%20Zealand/New%20Zealand%20-%20SF%20Act%201990%20-%20national.pdf, accessed 10 January 2021).
44. WHO report on the global tobacco epidemic, 2019. Geneva: World Health Organization; 2019.

7. Estimation of exposure to nicotine from use of electronic nicotine delivery systems and from conventional cigarettes

Anne Havermans, Centre for Health Protection, National Institute for Public Health and the Environment, Antonie van Leeuwenhoeklaan 9, 3721 MA Bilthoven, Netherlands

Thomas Eissenberg, Center for the Study of Tobacco Products, Department of Psychology, Virginia Commonwealth University, Richmond (VA), USA

Contents
Abstract
7.1 Background
7.2 Exposure to nicotine from ENDS
    7.2.1 ENDS nicotine emission
    7.2.2 Influence of ENDS electrical power on nicotine emission
    7.2.3 Contribution of the concentrations of nicotine and other compounds in ENDS liquids to nicotine emissions
7.3 Overview of exposure to accompanying substances
7.4 Nicotine delivery from ENDS
7.5 Behavioural patterns of exposure according to use
    7.5.1 Definition of user groups and user patterns
    7.5.2 Factors that influence behavioural patterns
7.6 Passive exposure to nicotine and other toxicants
7.7 Nicotine flux
7.8 Discussion
7.9 Conclusions
    7.10 Research gaps, priorities and questions
    7.11 Policy recommendations
    7.12 References

Abstract

Electronic nicotine delivery systems (ENDS) are a diverse class of products intended to deliver aerosolized nicotine. ENDS comprise a rapidly evolving range of technologies and a wide variety of types, from the first-generation "cig-a-like" devices to the currently popular "pod"-based devices. Factors such as device design, liquid ingredients and user behaviour all affect the content of nicotine and non-nicotine toxicants in ENDS aerosol. Although some evidence suggests that ENDS may help some smokers to replace conventional cigarettes, dual use of ENDS with combustible cigarettes and the increasingly common initiation of ENDS use among previously nicotine-naïve individuals raise clear public health concern. We reviewed the literature on nicotine emissions and delivery from ENDS and explored the factors that influence ENDS users' exposure to nicotine and non-nicotine toxicants. The review revealed that: ENDS are a heterogeneous product class that is evolving at a rate that outpaces regulation; flavoured ENDS liquids contribute to initiation and maintenance of their use by previously nicotine-naïve individuals; under certain circumstances, ENDS that deliver nicotine effectively might help some smokers to quit smoking combustible cigarettes; most ENDS users do not quit combustible smoking; regulation of ENDS nicotine emissions would be difficult because of the numerous inputs that control the emissions; and a regulatory focus on the rate of nicotine emission (e.g. nicotine "flux") might be useful, which would involve a requirement that only "closed-system" ENDS be marketed. In this context, future research needs and policy recommendations are proposed.

7.1 Background

Electronic nicotine delivery systems (ENDS) are a diverse class of products intended to deliver aerosolized nicotine to their users. They contain a battery-powered heating element known as a "coil" or "atomizer", which heats a liquid solution that contains nicotine, carrier liquids (e.g. propylene glycol, vegetable glycerine) and, usually, flavouring chemicals. The user inhales the resulting aerosol, which contains certain concentrations of nicotine and other toxicants. ENDS are a rapidly evolving class of products and include a wide variety of types, ranging from the first-generation "cig-a-like" devices to the currently popular "pod"-based devices in which a disposable cartridge holds the liquid *(1)*. Product design features and characteristics (such as wattage and coil dimensions), liquid constituents (such as carriers and nicotine concentration) and use behaviour (such as puff volume and duration) may be combined in numerous ways to affect the content (yield) of nicotine and other toxicants in the aerosol that the user inhales *(2)*.

Use of ENDS has risen substantially in some places during the past decade *(3,4)*. Use by children and adolescents has increased particularly rapidly in some countries, particularly in Europe, Canada and the USA, to the point that ENDS are now the most commonly used tobacco products in these age groups in the USA *(5,6)*. This raises concern, as ENDS emissions contain toxic chemicals that may be harmful to health *(7)* and also the dependence-producing drug nicotine. Nicotine is the primary addictive component of all tobacco products (e.g. combustible cigarettes, smokeless tobacco, heated tobacco) and in ENDS. In addition to causing dependence, nicotine can also have negative effects on health *(8)*. Children, adolescents and young adults are especially susceptible to the long-term neurocognitive effects of nicotine, as brain maturation continues into the early 20s *(9)*. It has been hypothesized that adolescents experience enhanced nicotine reward and reduced withdrawal via enhanced excitation and reduced inhibition of dopaminergic striatal cells, making them more vulnerable to long-term nicotine dependence than adults *(10)*. In addition, ENDS may serve as a "gateway" to smoking; several studies have found that their use is associated with an increased risk of initiating cigarette smoking among adolescents and young adults *(11)*.

Although some ENDS may help some smokers to replace cigarettes by providing nicotine in a similar amount and form (i.e. protonated state) *(12)*, initiation of ENDS use by young non-smokers raises clear concern *(11)*. Regulators might have to characterize and control the factors that influence nicotine delivery to users from ENDS in order to minimize their abuse liability and health impact while maximizing any opportunities to reduce the risk for cigarette smokers. As nicotine delivery is a combined result of product design, liquid composition and user behaviour *(2)*, however, it might be difficult for regulation to account for all these factors together. As nicotine delivery from ENDS is a function of so many variables (e.g. device characteristics and liquid constituents), it has been suggested that regulation focus on the rate at which ENDS emit nicotine and other toxicants, which would account simultaneously for all the device, liquid and user factors that control the emission rate. Nicotine flux – the rate at which ENDS emit nicotine – has thus been suggested as a regulatory target (e.g. *13*). As described in more detail below, regulation of ENDS nicotine flux (and also potentially the rate at which other toxicants are emitted) would have the advantage of directly controlling the factors that affect public health, rather than proxy factors (e.g. liquid nicotine concentration), which, when regulated individually, may not achieve public health goals.

This background paper provides a narrative review of the literature (as of March 2020) on emission and delivery of nicotine from ENDS and explores factors that influence users' exposure to nicotine and non-nicotine toxicants in ENDS emissions. We searched PubMed for relevant publications in the past five years using the search terms "ENDS" OR "E-cigarette" OR "electronic cigarette" AND "Nicotine" AND "exposure" OR "emission" OR "yield" OR "delivery". To find relevant literature about use patterns, an additional search was performed with the search terms "ENDS" OR "E-cigarette" OR "electronic cigarette" AND "topography" OR "behavior". Additional searches were performed for information about specific user groups, with the search terms "ENDS" OR "E-cigarette" OR "electronic cigarette" combined with terms related to specific hypothesized user groups, such as "race", "ethnicity", "gender", "male", "female" and "dual use". Relevant articles cited in publications obtained through the database search were also included (i.e. snowball method). As the aim of this document is to provide a narrative review, no formal selection criteria were applied to the results of these searches.

7.2 Exposure to nicotine from ENDS

7.2.1 ENDS nicotine emission

"Nicotine emission" can be defined as the amount of nicotine in the ENDS aerosol that leaves the device, in other words the nicotine yield. The nicotine yield can be analysed in the aerosol from a smoking machine with a predetermined puffing regime. The aerosol can be trapped on filter pads and extracted with suitable solvents, and the extract is analysed by chromatographic methods *(2)*. Studies with these methods and a variety of puffing regimes have shown various amounts of nicotine in ENDS aerosol, some showing yields below those generally obtained from combustible cigarettes and others showing yields equal to or exceeding that of combustible cigarettes (i.e. 1.76–2.20 mg/cigarette) *(2,14,15)*. Importantly, if machine puffing regimes do not mimic human puffing behaviour, they are not valid measures of human exposure. Arbitrarily chosen machine puffing regimens, however, allow valid comparisons when the regimen is applied equally to all products under study.

7.2.2 Influence of ENDS electrical power on nicotine emission

The amount of nicotine per puff in the aerosol is influenced by factors that include the electrical power flowing through the device, the nicotine concentration in the liquid aerosolized by the device and the puffing behaviour of the user *(2)*. Electrical power (W) is a function of battery voltage (V) and coil resistance (Ω), such that $W=V^2/\Omega$. The power of ENDS ranges from $\leq$ 10 W in early models to $\geq$ 250 W in currently marketed models *(16)*. Higher power is often achieved by integrating low resistance coils (i.e. < 1 Ω) into the device, colloquially referred to as "sub-ohm vaping" *(17)*. The voltage of the battery and default power settings differ widely among ENDS models, and more advanced devices often allow the user to adjust the power settings. Devices that cannot be adjusted in this manner, "closed-system" ENDS, often have lower power because they are smaller and more closely resemble combustible cigarettes, whereas "open-system" ENDS are larger and can thus contain larger batteries and lower-resistance heating elements *(1,18)*. An ENDS that is truly a "closed system" does not allow the user to alter any of the elements of the device or liquid that influence nicotine yield, e.g. battery voltage, coil resistance and liquid nicotine concentration; it may also limit user puffing behaviour (e.g. puff duration *(19)*).

Increasing the power flowing through the heating element that vaporizes the liquid can increase the amount of aerosol produced and may also lead the element to overheat, which can cause thermal degradation of the liquid, with resulting toxicant formation. The impact of electrical power on aerosol nicotine yield has not been studied extensively, but one study found that increasing the power output from 3 to 7.5 W increased the nicotine yield by four or five times *(2)*. Increased power can also increase the emissions of non-nicotine toxicants *(20)*.

7.2.3 Contribution of the concentrations of nicotine and other compounds in ENDS liquids to nicotine emissions

The nicotine-containing liquids used in ENDS come in refill bottles or prefilled cartridges or pods, with a wide range of nicotine concentrations, usually reported on the label in mg/mL or as a percentage of total volume. The maximum nicotine concentration may differ from country to country according to differences in regulations. For instance, the European Tobacco Products Directive states that liquids should not contain nicotine concentrations exceeding 20 mg/mL *(21)*. The rationale for this regulation, as described in the Directive, is that this concentration would allow delivery of nicotine at a concentration comparable to the permitted dose of nicotine from a standard cigarette during the time required to smoke the cigarette. The relation between the nicotine concentration in liquid and nicotine delivery to the ENDS user is not, however, straightforward, because of the interplay of factors in the device (e.g. electrical power), the composition of the liquid and user behaviour.

In the USA, up to about 2017, the nicotine concentration in commonly available liquids was usually 0–36 mg/mL *(1,22–25)*. Some newer products, however, contain nicotine at levels up to 67 mg/mL *(26,27)*, and there is concern that innovations in ENDS liquid formulations are spurring a "nicotine arms race" *(28)*. Furthermore, the nicotine concentrations in ENDS liquid often do not match the labelled content, with deviations of up to 52% *(15)*, and several studies have demonstrated measurable amounts of nicotine in some liquids labelled as not containing nicotine *(24,25,29)*.

Some studies have shown that the nicotine concentration in ENDS liquids directly influences nicotine yield, that is, higher liquid nicotine concentrations result in higher emissions of nicotine in the aerosol *(2,14)*. Power settings also play a role, as increasing the device power increases nicotine yields *(14,30)*. Furthermore, users of ENDS liquids with low nicotine strength can obtain the same amount of nicotine per puff as high-nicotine ENDS users by adjusting their puffing behaviour *(2,31)*. In this way, they may also be exposed to higher amounts of toxicants (see next section). Other chemicals in ENDS liquids also influence the nicotine yield in ENDS aerosols. For instance, liquids usually contain the solvents propylene glycol and/or vegetable glycerine in various ratios; higher levels of propylene glycol than vegetable glycerine result in higher nicotine yields at low device power settings *(30)*. This might be a consequence of the greater volatility of propylene glycol at relatively low temperatures, resulting in greater vaporization. As vegetable glycerine becomes more volatile at higher temperatures, the putative difference is thought to become less pronounced at higher power settings *(30)*.

7.3 Overview of exposure to accompanying substances

In addition to nicotine, ENDS emissions contain other toxicants, which are either present in the liquid or are formed by thermal breakdown of the liquid's ingredients. The toxicants present in liquids include propylene glycol, vegetable glycerine and various flavouring chemicals *(32,33)*. In addition, because the nicotine in ENDS is derived from tobacco plants, the liquid may contain tobacco-related toxicants such as tobacco-specific nitrosamines *(1)*. The flavouring agents used in ENDS liquids are "generally recognized as safe" when added to food, but their risk profiles when heated and inhaled are unknown *(34)*. Some flavouring chemicals such as diacetyl (buttery flavour) *(35,36)*, benzaldehyde (fruity flavour) *(37,38)* and cinnamaldehyde (cinnamon flavour) *(36,38–40)* are known to be toxic when inhaled *(41,42)*. Moreover, findings from the Population Assessment of Tobacco and Health study indicate that users of fruit-flavoured ENDS have significantly higher concentrations of the biomarker for the carcinogen acrylonitrile than users of other flavours *(43)*. Toxicants present after heating ENDS liquid include carbonyls, volatile organic compounds and polycyclic aromatic hydrocarbons, which are also present in tobacco smoke. Toxicant production from ENDS is affected by factors such as user behaviour and the type and power settings of the device *(2,44)*. For instance, more intensive puffing patterns can increase the production of carbonyls such as formaldehyde, acetaldehyde and acetone *(14,44)*, which have been correlated with pulmonary disease in smokers *(45)*.

ENDS emissions may also contain substances that potentiate the addictive effects of nicotine. For example, menthol is a common component of both ENDS and combustible cigarettes, and it is present in many ENDS liquids even when they are not labelled as containing menthol or mint flavour *(46)*. Menthol can enhance the reinforcing properties of nicotine in various ways, e.g. by facilitating inhalation and by acting on relevant nicotinic acetylcholine receptor subtypes in the brain *(46)*. Other examples are the popular ENDS flavouring agents vanillin and ethyl vanillin, which have been found to act as monoamine oxidase inhibitors and reinforce the brain's response to nicotine *(47)*. The green apple flavouring chemical farnesene can cause reward-related behaviour by stimulating nicotinic acetylcholine receptors and the potency of nicotine for activating those receptors *(48)*. Other compounds that may potentiate the effects of nicotine and affect its metabolism are alcohol and the minor tobacco alkaloid nicotyrine. The interaction of alcohol and nicotine in ENDS emissions has not been studied, but one study has shown that high levels of alcohol in ENDS liquid can acutely impact psychomotor function *(49)*. In addition, alcohol and tobacco are commonly used together *(50)*, and alcohol drinking can increase smoking *(46)*. Nicotyrine is a thermal reaction product of nicotine and is present in ENDS emissions at levels 2–63 times higher per unit of nicotine than in emissions from tobacco cigarettes *(51)*. It inhibits nicotine metabolism in vivo and may thereby increase nicotine delivery from ENDS *(1,46,51)*.

7.4 Nicotine delivery from ENDS

ENDS vary in their ability to deliver nicotine to users' blood and brain. Evaluation of the nicotine delivery profile of ENDS is important, as ENDS that deliver nicotine as effectively as a combustible cigarette are probably more effective substitutes for combustible cigarettes (52). The nicotine delivery profile of ENDS is also influenced by the combination of device type and power, the composition of the liquid and user behaviour (15,17,53). For instance, higher-wattage ENDS models deliver nicotine more effectively than lower-wattage models (16,54,55), higher liquid nicotine concentrations deliver more nicotine, especially in experienced users (17,56), and liquids with a higher propylene glycol than vegetable glycerine content increase nicotine delivery (probably due to the lower threshold of propylene glycol for evaporation and/or smaller particles that are more likely to reach users' lungs) (57).

One study showed that cherry and menthol flavours increase nicotine delivery (i.e. maximum concentration of nicotine in the blood) as compared with tobacco flavour (58). Another showed that more nicotine is delivered from a strawberry-flavoured liquid than from a tobacco-flavoured one, even though similar amounts of nicotine are inhaled, which may be related to differences in the pH of the liquids (59). Overall, substantial variation is seen in the nicotine delivery from different devices and liquids, some not increasing plasma nicotine concentrations and others delivering nicotine at a level approaching that of a tobacco cigarette (i.e. 10–30 ng/mL) (15,16,58,60–65).

Nicotine delivery to ENDS users may also depend on the bioavailability of nicotine in the liquid or aerosol. Thus, at a higher pH, a larger proportion of nicotine is in unprotonated form (free-base), which causes more irritation and increases the unpleasant taste of nicotine (28,66). At lower pH, more nicotine is present in protonated form, which reduces absorption in the upper respiratory tract and also reduces harshness and unpleasant taste, allowing users to inhale more deeply without experiencing discomfort, so that a larger portion of the aerosol reaches the lower lungs with enhanced absorption of nicotine. Originally, with very few exceptions, ENDS liquids contained only free-base nicotine. New liquids have been introduced onto the market, however, to which acids are added to increase the proportion of protonated nicotine (i.e. nicotine salts) (67). Liquids with a high nicotine concentration and small proportion of free-base nicotine are thought to be more likely to provide effective "cigarette-like" delivery of nicotine (66). In line with this notion, one study showed that the concentrations of urinary cotinine (a major metabolite of nicotine) in adolescents using "pod"-system ENDS containing nicotine salts were higher than those of adolescents who regularly smoked conventional tobacco cigarettes (26). The pH of ENDS liquids also varies widely, not only with brand and nicotine concentration but also within the same brand and nicotine concentration (68). ENDS liquids that have the same nicotine concentration and the same device characteristics, including electrical power but that differ in pH may have differing nicotine delivery profiles as well as differing sensory effects when the aerosols are inhaled. All other things being equal, protonated nicotine aerosol would be less harsh; however, this notion has not yet been tested empirically, as no studies have yet been reported in which liquid pH was manipulated systematically when all other variables were held constant.

7.5 Behavioural patterns of exposure according to use

7.5.1 Definition of user groups and user patterns

An important factor in exposure to nicotine is user behaviour, or puff topography. User puff topography includes variables such as the number, duration and volume of puffs and inter-puff interval and is highly individual. Various factors influence the way ENDS are used, such as the experience of the user and the composition of the liquid. Exposure to nicotine may also be affected by individual characteristics, and various user groups might be distinguished by the way in which they use ENDS. For example, experienced ENDS users typically take longer, larger puffs than ENDS-naïve users, resulting in higher nicotine delivery *(15,56,69)*. A study of "naturalistic" puffing topography identified three types of users: one that almost exclusively had "light" sessions (i.e. low puff volume (59.9 mL), flow rate (28.7 mL/s) and puff duration (2 s)), one with mainly "heavy" sessions (i.e. high puff volume (290.9 mL), flow rate (71.5 mL/s) and puff duration (4.4 s)) and a third with mainly "light" sessions (75%) and some "heavy" sessions (25%) *(70)*.

While some people use only ENDS, many ENDS users use tobacco cigarettes concurrently. In the USA, almost 70% of adult ENDS users also currently smoke cigarettes *(71)*, while the percentage of dual users among young people is lower, at 33% *(72)*. One study showed that cigarette smokers had longer puff duration and larger puff volume when using ENDS than non-smoking ENDS users *(73)*. Two other studies in users of both ENDS and tobacco cigarettes showed lower plasma nicotine concentrations after short-term ENDS use than after cigarette smoking in standardized laboratory settings *(55,74)*; however, the values were not compared in the same study to those for exclusive ENDS users. Other individual characteristics, such as gender and race, have been shown to affect exposure to nicotine from cigarette smoking *(75–77)* but have not been investigated for ENDS.

7.5.2 Factors that influence behavioural patterns

Various lifestyle and social factors also encourage or discourage ENDS use, potentially influencing exposure to nicotine. (See also section 3.) For example, local or national policies or regulations may prohibit the use of ENDS in certain enclosed public spaces (e.g. prohibition under smoke-free laws), and companies and institutions may ban ENDS use on their property, so that users have to restrict their use to home or outdoors. As smoke-free policies have reduced the social acceptability of smoking *(78)* and smoking *(79,80)*, a similar effect might be seen on ENDS use if it was included in such policies.

Advertisements and other information to which people are exposed through public channels may also influence their perception and use of ENDS *(81)*. Several studies have shown that e-cigarette advertising can increase interest in, purchase of and use of e-cigarettes *(82–84)*. Policy measures such as health warnings and public education campaigns may discourage people, especially children and adolescents, from initiating use of these products. For example, in the USA, an education campaign called "The real cost" has been highly successful in preventing young people from initiating smoking and has been extended to ENDS *(85)*. Such information may influence non-users' and users' knowledge and beliefs about the risks and benefits of ENDS use and thereby the likelihood of sustained use among users and uptake by non-users.

The social networks of ENDS users also play a role in uptake of the product *(86–91)*. Especially among young users, ENDS use tends to take place in the presence of peers *(92–94)*.

Design and characteristics of ENDS devices

Other factors that influence ENDS use and exposure to nicotine are the design and characteristics of the ENDS device. For example, several newer ENDS models are similar in appearance to a USB stick, which facilitates concealed use in schools and other public places *(95,96)*. They are also "smart"-looking and hence appeal to the e-generation. Other ENDS models are highly customizable, so that users can change power settings and use liquids with different nicotine concentrations and flavours, factors that are known to influence use. For example, power settings influence puff behaviour, such that higher power reduces the puff number and duration *(97)*. This change in response to device power may reflect users' attempt to titrate nicotine and/or the sensory effects of the inhaled aerosol. Use patterns also are correlated with the nicotine concentration in the liquid, such that lower concentrations of nicotine are associated with larger, longer puffs *(17,98)*. The first use of nicotine-containing ENDS may increase exposure to nicotine throughout life, as one study showed that adolescents who initially used an ENDS with nicotine tended to use ENDS on more days during the first year of high school than adolescents who initially used an ENDS without nicotine *(99)*.

Solvents (propylene glycol and vegetable glycerine)

Higher ratios of propylene glycol to vegetable glycerine in ENDS liquid have been related to reduced puff duration and size but increased nicotine delivery *(57)*. Liquids with higher propylene glycol ratios were also rated as less "pleasant" and less "satisfying" by participants in the same study. This may be because pure propylene glycol liquids produce little to no visible exhaled aerosol, which is usually considered a positive aspect by users and may be a conditioned reinforcer for nicotine. Another study showed that liquids with more vegetable glycerine were preferred to those with more propylene glycol and that "good taste" was the most important consideration when using and purchasing liquids *(100)*.

Flavours (See also section 6.)

ENDS liquid flavours have also been shown to affect puffing behaviour. For instance, one study showed that smokers took significantly longer puffs from flavoured ENDS (vanilla, cherry, menthol, espresso or tobacco flavours) than from tobacco cigarettes and tended to puff less frequently on vanilla- than on tobacco-flavoured ENDS *(58)*. In another study, experienced ENDS users took longer puffs from a strawberry-flavoured liquid than from a tobacco-flavoured one; however, they took even larger and more puffs when using their usual brand of ENDS liquid *(101)*. A third study found that ENDS liquid flavours influenced puff flow rate and puff volume but not puff duration *(102)*, although the direction of the effect was unclear. Flavours not only affect use behaviour but are also an important reason for initiating and continuing ENDS use, particularly for adolescents and young adults *(99,103,104)*.

7.6 Passive exposure to nicotine and other toxicants (See also section 8.)

ENDS users are exposed directly to nicotine and other toxicants by inhaling the aerosol emitted by their device. Non-users may be exposed to nicotine and other toxicants by "second-hand" exposure (also known as environmental exposure) or "third-hand" exposure to emissions that have settled onto surfaces, from skin contact or by ingestion of nicotine-containing liquid *(46)*. A growing body of literature suggests that ENDS use has a negative effect on indoor air quality *(105–109)*, supporting the idea that non-users may be exposed to toxicants exhaled by ENDS users when they share the same indoor space. Several studies have reported effective methods for assessing second- and third-hand exposure *(110)*. One study showed delivery of various levels of nicotine to non-users after acute second-hand exposure to ENDS aerosol in a real social setting *(100)*. Another, of exhaled breath of ENDS users, concluded that bystanders may experience systemic effects of nicotine, including increased heart rate and higher systolic blood pressure *(111)*. A further study confirmed that 30 min of second-hand exposure to ENDS aerosol caused sensory irritation and respiratory symptoms, which were related to the concentration of volatile organic compounds in the emissions *(112)*.

In pregnant women, nicotine readily crosses the placenta *(113)* and binds to nicotine acetylcholine receptors in the fetal brain, which play a critical role in brain development *(114)*. Early activation and desensitization of these receptors by nicotine can disrupt development, with long-term consequences *(9)*. Although there are no published studies on how ENDS use affects pregnancy outcomes or fetal development, nicotine is considered to contribute substantially to a range of adverse effects of maternal smoking, and CO is thought to be a cause of low birthweight. Neonates exposed prenatally to nicotine and tobacco have a lower birthweight and earlier gestational age, have a higher risk of lung and cardiorespiratory problems and are more prone to asthma and allergy in childhood *(9,115)*. They are also at higher risk of neurocognitive effects that can lead to poor academic performance and significant behavioural problems throughout life, including attention deficit hyperactivity disorder, aggressive behaviour and future substance abuse *(116)*. Although it is difficult to conclude that these effects are caused specifically by nicotine or by other components of tobacco smoke, nicotine is considered to be the substance mainly responsible for most of the adverse effects on fetuses from maternal smoking *(9,116)*. Studies of pregnant women who use smokeless nicotine-containing products have also found associations with preterm birth, stillbirth and orofacial cleft defects *(117–120)*. Use of nicotine replacement therapy during pregnancy is associated with lower exposure to nicotine *(121)* and a lower risk of preterm delivery and low birth weight than with smoking *(122)*. As some ENDS deliver nicotine in amounts comparable to those in combustible cigarettes, some of the adverse effects of maternal smoking also may occur after exposure to nicotine from maternal ENDS use. It should be noted that ENDS emissions contain other possibly harmful compounds, with effects on fetal development that have not been thoroughly studied. For example, one study showed that flavouring agents in ENDS refill solutions are cytotoxic to human embryonic stem cells *(123)*.

7.7 Nicotine flux

ENDS nicotine "flux" is the rate at which an ENDS device emits nicotine, or the ENDS nicotine yield per unit time (e.g. µg/s). The rate of drug delivery has long been a relevant metric for understanding drug abuse, as faster drug delivery leads to greater abuse potential *(124,125)*. When nicotine was delivered to cigarette smokers intravenously at different rates, faster delivery was considered to give more rewarding positive effects *(126)*. Combustible cigarettes are used by millions of people worldwide, and, generally, they emit nicotine at approximately 100 µg/s (calculated from data obtained by Djordjevic et al. *(127)*) and deliver nicotine to blood and brain very quickly *(128,129)*. While combustible cigarette nicotine flux is generally stable for all combustible cigarette brands, similar stability is not seen for ENDS, mainly because of the heterogeneity of the product class. When all possible combinations of device power, construction, liquid and nicotine and other ingredients are accounted for, ENDS fluxes may range from 0 µg/s (i.e. no nicotine emission) to > 100 µg/s. This vast range of nicotine flux explains the considerable variation in ENDS nicotine delivery profiles, with low-power devices and liquids with a low nicotine concentration delivering little or no nicotine and higher-power devices and liquids with higher concentrations delivering as much as or more nicotine than a combustible cigarette in the same number of puffs *(1)*. Importantly, nicotine flux is independent of user behaviour (e.g. longer puff durations do not alter flux), but flux and behaviour combined determine nicotine yield and exposure and the amount of drug delivered to the blood and brain. For example, a flux of 100 µg/s and a 1-s puff duration yields 100 µg nicotine, but a 4-s puff yields 400 µg nicotine. This explains why longer puffs result in greater nicotine delivery even when flux is controlled *(56)*. Longer puffs deliver a larger inhaled nicotine dose to the user.

While the heterogeneity of ENDS devices and liquids makes it difficult to measure flux in all possible combinations, ENDS flux can be predicted mathematically in a physics-based model *(130)*. As described elsewhere *(131)*, the model accounts for the time it takes for a coil to heat up after electricity begins to flow, cooling of the coil between puffs and the various ways in which heat can be transferred from the coil. Inputs to the model include the length, diameter, electrical resistance and thermal capacitance of the coil, the composition and thermodynamic properties of the liquid (including nicotine concentration), the ambient temperature and user behaviour (puff velocity and duration, inter-puff interval). In a validation study, model predictions were generated, and actual nicotine flux was measured in 100 conditions, in which power, device type, liquid composition and user behaviour were varied. The model accounted for 72% of the variation in nicotine flux under the conditions tested. This model could be used to predict the nicotine flux of any ENDS on the market today (open or closed system) as well as of ENDS that are being designed. Thus, mathematical modelling of nicotine flux presents a potential tool for policy-makers who wish to regulate ENDS nicotine emissions.

As has been noted *(18)*, if the goal of regulation is to decrease the likelihood that ENDS will be abused by a population such as non-smoking young people, an effective way may be to decrease ENDS nicotine flux. As the nicotine flux is a result of all of the ENDS characteristics (construction, wattage, liquid nicotine content), regulators can focus on a single product performance target – nicotine emission rate (i.e. nicotine flux), and manufacturers can choose the device and liquid characteristics that fall safely within that range. The flux target is not necessarily a single value but a range of allowable nicotine flux conditions (a nicotine emission rate no less than X and no greater than Y), allowing for a range of products designed to minimize abuse and maximize any potential benefit for smokers seeking to quit smoking combustible cigarettes and an eventual end to nicotine dependence, if ENDS can be demonstrated to provide such a therapeutic benefit. In sum, a mathematical model of nicotine flux allows regulators to examine an array of products efficiently to determine whether they meet or fall outside a specified nicotine flux range. Unfortunately, such a regulatory approach cannot succeed if users have control over key parameters such as device power and liquid nicotine concentration. Therefore, policy-makers also may wish to consider the extent to which "open-system" devices are amenable to effective regulation *(18)*.

7.8 Discussion

ENDS are a diverse, evolving product class with growing global popularity, particularly among children, adolescents and young adults. Some ENDS users were former cigarette smokers who used ENDS to quit smoking combustible cigarettes, and there is some empirical evidence from randomized clinical trials that ENDS assist smoking cessation, although results are inconsistent. Many ENDS users are dual users, who continue to use ENDS with other tobacco products, in particular conventional cigarettes. Others were nicotine-naïve before using ENDS and may be at risk for subsequent initiation of conventional cigarette smoking. The myriad flavours of ENDS liquids available on the market may help some smokers to quit smoking, may encourage dual or poly use and almost certainly encourage nicotine-naïve young people to initiate ENDS use. The proportion of naïve ENDS users who were potential smokers and of those who would have remained non-smokers is a potential confounder in such analyses.

Some ENDS can deliver as much or more nicotine than a combustible cigarette in the same number of puffs. Some ENDS also deliver much less nicotine than a combustible cigarette. The extent to which ENDS deliver or do not deliver nicotine depends on a variety of device characteristics (e.g. electrical power, coil dimensions), liquid constituents (e.g. nicotine concentration, ratio of propylene glycol to vegetable glycerine) and user behaviour (e.g. puff duration). These same factors influence the extent to which ENDS emit non-nicotine toxicants that may be injurious to users' health. A recent influence on ENDS nicotine delivery is the marketing of liquids that contain protonated nicotine (nicotine salts). The aerosol formed from a protonated liquid is less harsh to inhale than aerosol formed from free-base nicotine, so that manufacturers can increase the nicotine concentration of the liquid without making the resulting aerosol unpalatable.

In view of all the factors that influence nicotine and non-nicotine emissions from ENDS, regulation of this product class may be difficult. There is a temptation to focus on single factors when regulating ENDS nicotine delivery, such as liquid nicotine concentration; however, such an approach may drive users to obtain more nicotine by manipulating unregulated factors such as using higher-powered devices and/or increasing puff duration. Such behaviour could reduce the effectiveness of regulation, such that nicotine delivery remains higher than intended by the regulators, while also exposing users to more aerosol that may be more toxic than if they used lower-powered devices and/or took shorter puffs.

Thus, it has been suggested that regulators focus on the rate at which nicotine is emitted from ENDS, the nicotine flux, as a regulatory target. This focus would also require that ENDS products not allow users to access many of the device, liquid and user behaviour characteristics that influence nicotine flux, such as "closed-system" ENDS with built-in limits on puff duration. These devices exist in some markets and are therefore clearly feasible. Exactly which nicotine flux parameters are conducive to promoting smoking cessation by current cigarette smokers while limiting abuse liability in nicotine-naïve young people are yet to be determined.

7.9 Conclusions

The data reviewed lead to the following conclusions.

- ENDS are a heterogeneous product class that continues to evolve at a speed that outpaces current regulatory efforts.
- ENDS performance characteristics are also heterogeneous, some users being exposed to very low levels of nicotine and other toxicants and others being exposed to much higher levels.
- ENDS use by previously nicotine-naïve individuals is inconsistent with public health goals.
- Flavoured ENDS liquids contribute to initiation and maintenance of ENDS use among previously nicotine-naïve individuals. They may also be attractive for smokers who want to quit cigarettes.
- Under certain circumstances, such as in the context of intensive behavioural counselling, ENDS that deliver nicotine effectively might help some smokers to quit combustible smoking, with positive public health effects. Most of these individuals, however, continue to use ENDS, with uncertain individual health consequences and thus an uncertain public health impact.
- Most ENDS users do not quit smoking combustible cigarettes but rather use both ENDS and combustible cigarettes, which, at the least, maintains the substantial health risks associated with cigarette smoking and may increase their health risks.
- Regulation of the emissions of nicotine and other toxicants from ENDS is complicated by the numerous inputs to emissions.
- Regulation of the emissions of nicotine and non-nicotine toxicants from ENDS may be necessary. This would require that marketed ENDS be constructed so that users cannot alter important characteristics such as device power and liquid constituents.

- The profile of nicotine emission and delivery from ENDS that would be most likely to achieve cessation of conventional smoking, ideally while also reducing the abuse liability of ENDS among nicotine-naïve individuals, is not known. Identification of that profile, if it exists, will require careful empirical work similar to that conducted for other pharmacological compounds that are used therapeutically even though, in some forms or via some routes, they can also be abused (e.g. opioids).

7.10 Research gaps, priorities and questions

The data reviewed here raise many research questions, including those listed below.

- Studies to determine the range of nicotine flux, if any, that will reduce the abuse liability of ENDS products and limit initiation by young people while helping smokers to eliminate their use of cigarettes and other smoked products with ENDS.
- Studies to compare fetal exposure to nicotine and other toxicants from ENDS with that from maternal cigarette smoking.
- Which way of achieving a given ENDS nicotine flux poses the least risk for users? Many different combinations of device power and liquid nicotine concentration can achieve the same flux, and some may lead to more non-nicotine toxicants than others.
- To what extent are flavoured ENDS liquids required for ENDS-facilitated cigarette cessation? Could unflavoured ENDS liquids, made only of nicotine, propylene glycol and vegetable glycerine, when paired with a device that emits nicotine as effectively as a combustible cigarette, facilitate cigarette smoking cessation in a current smoker?
- To what extent is intensive behavioural counselling a requirement for ENDS-facilitated cigarette smoking cessation?
- Which smokers are most likely to achieve smoking cessation with an ENDS product that delivers nicotine effectively? Is cessation more likely to be facilitated by ENDS in some smokers than others?
- Given the diversity of regulatory approaches to ENDS (i.e. different countries have different approaches), which policies are most effective in protecting public health with respect to ENDS use?
- To what extent does the availability of ENDS at numerous retail outlets facilitate initiation among nicotine-naïve individuals? Would restricting ENDS sales to regulated venues (perhaps where individu-

alized smoking cessation counselling is available) be more consistent with public health goals?
- What is the effect of increased levels of protonated nicotine in ENDS liquids on toxicity (additional acid ingredients) and abuse liability (by making inhalation of high doses of nicotine less harsh)?

7.11   Policy recommendations

The data reviewed here support the following three recommendations.

- Regulators should not permit ENDS in which users can control device features and liquid ingredients (i.e. open-system ENDS) and should ensure that the ENDS that are permitted do not appeal to young people.
- Regulators should focus on nicotine emission rate or flux (i.e. outcome) as a regulatory target, instead of any single input variable (e.g. liquid nicotine concentration or device power).
- ENDS should not have a higher abuse liability than combustible cigarettes. Thus, the ENDS nicotine emission rate or flux should not be higher than the emission rate of combustible cigarettes.

7.12   References

1. Breland A, Soule E, Lopez A, Romôa C, El-Hellani A, Eissenberg T. Electronic cigarettes: what are they and what do they do? Ann NY Acad Sci. 2017;1394(1): 5–30.
2. Talih S, Balhas Z, Eissenberg T, Salman R, Karaoghlanian N, El-Hellani A et al. Effects of user puff topography, device voltage, and liquid nicotine concentration on electronic cigarette nicotine yield: measurements and model predictions. Nicotine Tob Res. 2015;17(2):150–7.
3. Filippidis FT, Laverty AA, Gerovasili V, Vardavas CI. Two-year trends and predictors of e-cigarette use in 27 European Union member states. Tob Control. 2017;26(1):98.
4. McMillen RC, Gottlieb MA, Whitmore Schaefer RM, Winickoff JP, Klein JD. Trends in electronic cigarette use among US adults: Use is increasing in both smokers and nonsmokers. Nicotine Tob Res. 2015;17(10):1195–202.
5. Jamal A, Gentzke A, Hu SS, Cullen KA, Apelberg BJ, Homa DM et al. Tobacco use among middle and high school students – United States, 2011–2016. Morbid Mortal Wkly Rep. 2016;66:597–603.
6. Cullen KA, Ambrose BK, Gentzke AS, Apelberg BJ, Jamal A, King BA. Notes from the field: Use of electronic cigarettes and any tobacco product among middle and high school students – United States, 2011–2018. Morbid Mortal Wkly Rep. 2018;67:1276–7.
7. Hutzler C, Paschke M, Kruschinski S, Henkler F, Hahn J, Luch A. Chemical hazards present in liquids and vapors of electronic cigarettes. Arch Toxicol. 2014;88(7):1295–308.
8. Benowitz NL, Burbank AD. Cardiovascular toxicity of nicotine: Implications for electronic cigarette use. Trends Cardiovasc Med. 2016;26(6):515–23.
9. England LJ, Bunnell RE, Pechacek TE, Tong VT, McAfee TA. Nicotine and the developing human: a neglected element in the electronic cigarette debate. Am J Prev Med. 2015;49(2):286–93.

10. O'Dell LE. A psychobiological framework of the substrates that mediate nicotine use during adolescence. Neuropharmacology. 2009;56(Suppl 1):263–78.
11. Soneji S, Barrington-Trimis JL, Wills TA, Levanthal AM, Unger JB, Gibson LA et al. Association between initial use of e-cigarettes and subsequent cigarette smoking among adolescents and young adults: a systematic review and meta-analysis. JAMA Pediatrics. 2017;171(8):788–97.
12. Rahman MA, Hann N, Wilson A, Mnatzaganian G, Worrall-Carter L. e-Cigarettes and smoking cessation: evidence from a systematic review and meta-analysis. PLoS One. 2015;10(3):e0122544.
13. Shihadeh A, Eissenberg T. Electronic cigarette effectiveness and abuse liability: predicting and regulating nicotine flux. Nicotine Tob Res. 2015;17(2):158–62.
14. El-Hellani A, Salman R, El-Hage R, Talih S, Malek N, Baalbaki R et al. Nicotine and carbonyl emissions from popular electronic cigarette products: Correlation to liquid composition and design characteristics. Nicotine Tob Res. 2018;20(2):215–23.
15. Voos N, Goniewicz ML, Eissenberg T. What is the nicotine delivery profile of electronic cigarettes? Expert Opin Drug Deliv. 2019;16(11):1193–203.
16. Wagener TL, Floyd EL, Stepanov I, Driskill LM, Frank SG, Meier E et al. Have combustible cigarettes met their match? The nicotine delivery profiles and harmful constituent exposures of second-generation and third-generation electronic cigarette users. Tob Control. 2017;26(e1):e23–8.
17. Hiler M, Karaoghlanian N, Talih S, Maloney S, Breland A, Shihadeh A et al. Effects of electronic cigarette heating coil resistance and liquid nicotine concentration on user nicotine delivery, heart rate, subjective effects, puff topography, and liquid consumption. Exp Clin Psychopharmacol, 2020;28(5):527–39.
18. Eissenberg T et al. "Open-system" electronic cigarettes cannot be regulated effectively. Tob Control, 2020 (in press).
19. Talih S, Salman R, El-Hage R, Karam E, Karaoghlanian N, El-Hellani A et al. Characteristics and toxicant emissions of JUUL electronic cigarettes. Tob Control. 2019;28(6):678.
20. Talih S, Salman R, El-Hage R, Karam EE, Karaoghlanian N, El-Hellani A et al. Might limiting liquid nicotine concentration result in more toxic electronic cigarette aerosols? Tob Control. 2020:10.1136/tobaccocontrol-2019-055523.
21. The European Parliament and the Council of the European Union, Directive 2014/40/EU on the approximation of the laws, regulations and administrative provisions of the Member States concerning the manufacture, presentation and sale of tobacco and related products and repealing Directive 2001/37/EC. Off J Eur Union. 2014;L127/1:127.
22. Nicotine salts e-liquid overview. Singapore: Wingle Group Electronics Ltd; 2018 (https://vape.hk/wp-content/uploads/2019/05/NICOTINE_SALTS_E-LIQUID.pdf, accessed 10 January 2021).
23. El-Hellani A, El-Hage R, Baalbaki R, Salman R, Talih S, Shihadeh A et al. Free-base and protonated nicotine in electronic cigarette liquids and aerosols. Chem Res Toxicol. 2015;28(8):1532–7.
24. Goniewicz ML, Gupta R, Lee YH, Reinhardt S, Kim S, Kim B et al. Nicotine levels in electronic cigarette refill solutions: A comparative analysis of products from the US, Korea, and Poland. Int J Drug Policy. 2015;26(6):583–8.
25. Raymond BH, Collette-Merrill K, Harrison RG, Jarvis S, Rasmussen RJ. The nicotine content of a sample of e-cigarette liquid manufactured in the United States. J Addict Med. 2018;12(2):127–31.
26. Goniewicz ML, Boykan R, Messina CR, Eliscu A, Tolentino J. High exposure to nicotine among adolescents who use Juul and other vape pod systems ("pods"). Tob Control. 2019;28(6):676–7.
27. Omaiye EE, McWhirter KJ, Luo W, Pankow JF, Talbot P. High-nicotine electronic cigarette products: toxicity of JUUL fluids and aerosols correlates strongly with nicotine and some flavor chemical concentrations. Chem Res Toxicol. 2019;32(6):1058–69.
28. Jackler RK, Ramamurthi D. Nicotine arms race: JUUL and the high-nicotine product market.

Tob Control, 2019;28(6):623–8.
29. Omaiye EE, Cordova I, Davis B, Talbot P. Counterfeit electronic cigarette products with mislabeled nicotine concentrations. Tob Regul Sci. 2017;3(3):347–57.
30. Kosmider L, Spindle TR, Gawron M, Sobczak A, Goniewicz ML. Nicotine emissions from electronic cigarettes: Individual and interactive effects of propylene glycol to vegetable glycerin composition and device power output. Food Chem Toxicol. 2018;115:302–5.
31. Robinson RJ, Hensel EC. Behavior-based yield for electronic cigarette users of different strength eliquids based on natural environment topography. Inhal Toxicol. 2019;31(13-14):484–91.
32. Madison MC, Landers CT, Gu BH, Chang CY, Yung HY, You R et al., Electronic cigarettes disrupt lung lipid homeostasis and innate immunity independent of nicotine. J Clin Invest. 2020;129(10):4290–304.
33. Chaumont M, Bernard A, Pochet S, Mélot C, El Khattabi C, Ree F et al. High-wattage e-cigarettes induce tissue hypoxia and lower airway injury: a randomized clinical trial. Am J Resp Critical Care Med. 2018;198(1):123–6.
34. Safety assessment and regulatory authority to use flavors – Focus on electronic nicotine delivery systems and flavored tobacco products. Washington (DC): Flavor and Extract Manufacturers Association of the United States; 2018.
35. Farsalinos KE, Kistler KA, Gillman G, Voudris V. Evaluation of electronic cigarette liquids and aerosol for the presence of selected inhalation toxins. Nicotine Tob Res. 2015;17(2):168–74.
36. Muthumalage T, Prinz M, Ansah KO, Gerloff J, Sundar IK, Rahman I. Inflammatory and oxidative responses induced by exposure to commonly used e-cigarette flavoring chemicals and flavored e-liquids without nicotine. Front Physiol. 2017;8:1130.
37. Kosmider L, Sobczak A, Prokopowicz A, Kurek J, Zaciera M, Knysak J et al. Cherry-flavoured electronic cigarettes expose users to the inhalation irritant, benzaldehyde. Thorax. 2016;71(4):376–7.
38. Hickman E, Herrera CA, Jaspers I. Common e-cigarette flavoring chemicals impair neutrophil phagocytosis and oxidative burst. Chem Res Toxicol. 2019;32(6):982–5.
39. Behar RZ, Davis B, Wang Y, Bahl V, Lin S, Talbot P. Identification of toxicants in cinnamon-flavored electronic cigarette refill fluids. Toxicol In Vitro. 2014;28(2):198–208.
40. Clapp PW, Lavrich KL, van Heusden CA, Lazarowski ER, Carson JL, Jaspers I. Cinnamaldehyde in flavored e-cigarette liquids temporarily suppresses bronchial epithelial cell ciliary motility by dysregulation of mitochondrial function. Am J Physiol Lung Cell Mol Physiol. 2019;316(3):L470–86.
41. Barrington-Trimis JL, Samet JM, McConnell R. Flavorings in electronic cigarettes: An unrecognized respiratory health hazard? JAMA. 2014;312(23):2493–4.
42. Tierney PA, Karpinski CD, Brown JE, Luo W, Pankow JF. Flavour chemicals in electronic cigarette fluids. Tob Control. 2016;25(e1):e10–5.
43. Smith DM, Schneller LM, O'Connor RJ, Goniewicz ML. Are e-cigarette flavors associated with exposure to nicotine and toxicants? Findings from wave 2 of the Population Assessment of Tobacco and Health (PATH) study. Int J Environ Res Public Health. 2019;16(24):5055.
44. Kosmider L, Kimber CF, Kurek J, Corcoran O, Dawkins LE. Compensatory puffing with lower nicotine concentration e-liquids increases carbonyl exposure in e-cigarette aerosols. Nicotine Tob Res. 2017;20(8):998–1003.
45. Talhout R, Schulz T, Florek E, van Benthern J, Wester P, Opperhuizen A. Hazardous compounds in tobacco smoke. Int J Environ Res Public Health. 2011;8(2):613–28.
46. DeVito EE, Krishnan-Sarin S. e-Cigarettes: Impact of e-liquid components and device characteristics on nicotine exposure. Curr Neuropharmacol. 2018;16(4):438–59.
47. Truman P, Stanfill S, Heydari A, Silver E, Fowles J. Monoamine oxidase inhibitory activity of

flavoured e-cigarette liquids. NeuroToxicology. 2019;75:123–8.
48. Cooper SY, Akers AT, Henderson BJ. Green apple e-cigarette flavorant farnesene triggers reward-related behavior by promoting high-sensitivity nAChRs in the ventral tegmental area. eNeuro. 2020;7(4):0172.
49. Valentine GW, Jatlow PI, Coffman M, Nadim H, Gueorgiueva R, Sofuoglu M. The effects of alcohol-containing e-cigarettes on young adult smokers. Drug Alcohol Depend; 2016;159:272–6.
50. Van Skike CE, Maggio SE, Reynolds AR, Casey EM, Bardo MT, Dwoskin LP et al. Critical needs in drug discovery for cessation of alcohol and nicotine polysubstance abuse. Prog Neuropsychopharmacol Biol Psychiatry. 2016;65:269–87.
51. Son Y, Wackowski O, Weisel C, Schwander S, Mainelis G, Delnevo et al. Evaluation of e-vapor nicotine and nicotyrine concentrations under various e-liquid compositions, device settings, and vaping topographies. Chem Res Toxicol. 2018;31(9):861–8.
52. Vansickel AR, Cobb CO, Weaver MF, Eissenberg TE. A clinical laboratory model for evaluating the acute effects of electronic "cigarettes": nicotine delivery profile and cardiovascular and subjective effects. Cancer Epidemiol Biomarkers Prev. 201019(8):1945–53.
53. Blank MD, Pearson J, Cobb CO, Felicione NJ, Hiler MM, Spindle TR et al. What factors reliably predict electronic cigarette nicotine delivery? Tob Control. 2019;29(6):644–51.
54. Yingst JM, Foulds J, Veldheer S, Hrabovsky S, Trushin N, Eissenberg T et al. Nicotine absorption during electronic cigarette use among regular users. PLoS One; 2019;14(7):e0220300.
55. Hajek P, Przulj D, Phillips A, Anderson R, McRobbie H. Nicotine delivery to users from cigarettes and from different types of e-cigarettes. Psychopharmacology. 2017;234(5):773–9.
56. Hiler M, Breland A, Spindle T, Maloney S, Lipato T, Karaoghlanian N et al. Electronic cigarette user plasma nicotine concentration, puff topography, heart rate, and subjective effects: Influence of liquid nicotine concentration and user experience. Exp Clin Psychopharmacol. 2017;25(5):380–92.
57. Spindle TR, Talih S, Hiler MM, Karaoghlanian N, Halquist MS, Breland AB et al. Effects of electronic cigarette liquid solvents propylene glycol and vegetable glycerin on user nicotine delivery, heart rate, subjective effects, and puff topography. Drug Alcohol Depend. 2018;188:193–9.
58. Voos N, Smith D, Kaiser L, Mahoney MC, Bradizza CM, Kozlowski LT et al. Effect of e-cigarette flavors on nicotine delivery and puffing topography: results from a randomized clinical trial of daily smokers. Psychopharmacology (Berl). 2020;237(2):491–502.
59. St Helen G, Dempsey DA, Havel CM, Jacob P III, Benowitz NL. Impact of e-liquid flavors on nicotine intake and pharmacology of e-cigarettes. Drug Alcohol Depend. 2017;178:391–8.
60. St Helen G, Havel C, Dempsey DA, Jacob P III, Benowitz NL. Nicotine delivery, retention and pharmacokinetics from various electronic cigarettes. Addiction. 2016;111(3):535–44.
61. Lopez AA, Hiler MM, Soule EK, Ramôa CP, Karaoghlanian N, Lipato T et al. Effects of electronic cigarette liquid nicotine concentration on plasma nicotine and puff topography in tobacco cigarette smokers: a preliminary report. Nicotine Tob Res. 2016;18(5):720–3.
62. Ramôa CP, Hiler MM, Spindle TR, Lopez AA, Karaoghlanian N, Lipato T et al. Electronic cigarette nicotine delivery can exceed that of combustible cigarettes: a preliminary report. Tob Control. 2016;25(e1):e6–9.
63. Marsot A, Simon N. Nicotine and cotinine levels with electronic cigarette: a review. Int J Toxicol. 2016;35(2):179–85.
64. Hajek P, Pittaccio K, Pesola F, Myers Smith K, Phillips-Waller A et al. Nicotine delivery and users' reactions to Juul compared with cigarettes and other e-cigarette products. Addiction. 2020;115(6).
65. Voos N, Kaiser L, Mahoney MC, Bradizza CM, Kozlowski LT, Benowitz NL et al. Randomized within-subject trial to evaluate smokers' initial perceptions, subjective effects and nicotine delivery across six vaporized nicotine products. Addiction. 2019;114(7):1236–48.

66. Duell AK, Pankow JF, Peyton DH. Nicotine in tobacco product aerosols: "It's déjà vu all over again". Tob Control, 2019;29(6):656–62.
67. Duell AK, Pankow JF, Peyton DH. Free-base nicotine determination in electronic cigarette liquids by 1H NMR spectroscopy. Chem Res Toxicol. 2018;31(6):431–4.
68. Stepanov I, Fujioka N. Bringing attention to e-cigarette pH as an important element for research and regulation. Tob Control. 2015;24(4):413.
69. Farsalinos KE, Spyrou A, Stefopoulos C, Tsimopoulou K, Kourkoveli P, Tsiapras D et al. Corrigendum: Nicotine absorption from electronic cigarette use: comparison between experienced consumers (vapers) and naive users (smokers). Sci Rep. 2015;5:13506.
70. Lee YO, Morgan-Lopez AA, Nonnemaker JM, Pepper JK, Hensel EC, Robinson RJ. Latent class analysis of e-cigarette use sessions in their natural environments. Nicotine Tob Res. 2018;21(10):1408–13.
71. Coleman BN, Rostron B, Johnson SE, Ambrose BK, Pearson J, Stanton CA et al. Electronic cigarette use among US adults in the Population Assessment of Tobacco and Health (PATH) study, 2013–2014. Tob Control. 2017;26(e2):e117–26.
72. Vogel EA, Cho J, McConnell RS, Barrington-Trimis JL, Leventhal AM. Prevalence of electronic cigarette dependence among youth and its association with future use. JAMA Netw Open. 2020;3(2):e1921513.
73. Lee YO, Nonnemaker JM, Bradfield B, Hensel EC, Robinson RJ. Examining daily electronic cigarette puff topography among established and nonestablished cigarette smokers in their natural environment. Nicotine Tob Res. 2017;20(10):1283–8.
74. St. Helen G, Nardone N, Addo N, Dempsey D, Havel C, Jacob P III et al. Differences in nicotine intake and effects from electronic and combustible cigarettes among dual users. Addiction. 2019;115(4):757–67.
75. Chen A, Krebs NM, Zhu J, Muscat JE. Nicotine metabolite ratio predicts smoking topography: The Pennsylvania Adult Smoking Study. Drug Alcohol Depend. 2018;190:89–93.
76. Holford TR, Levy DT, Meza R. Comparison of smoking history patterns among African American and white cohorts in the United States born 1890 to 1990. Nicotine Tob Res. 2016;18(suppl 1):S16–29.
77. Ross KC, Gubner NR, Tyndale RF, Hawk LW Jr, Lerman C, George TP et al. Racial differences in the relationship between rate of nicotine metabolism and nicotine intake from cigarette smoking. Pharmacol Biochem Behav. 2016;148:1–7.
78. Preventing tobacco use among youth and young adults: a report of the Surgeon General. Atlanta (GA): Department of Health and Human Services, Centers for Disease Control and Prevention; 2012.
79. Hoffman SJ, Tan C. Overview of systematic reviews on the health-related effects of government tobacco control policies. BMC Public Health. 2015;15:744.
80. Hopkins DP, Razi S, Leeks KD, Kalra GP, Chattopadhyay SK, Soler RE et al. Smokefree policies to reduce tobacco use: a systematic review. Am J Prev Med. 2010;38(2):S275–89.
81. Alcalá HE, Shimoga SV. It is about trust: Trust in sources of tobacco health information, perceptions of harm, and use of e-cigarettes. Nicotine Tob Res. 2020;22(5):822–6.
82. Vasiljevic M, Petrescu DC, Marteau TM. Impact of advertisements promoting candy-like flavoured e-cigarettes on appeal of tobacco smoking among children: an experimental study. Tob Control. 2016;25(e2):e107–12.
83. Farrelly MC, Duke JC, Crankshaw EC, Eggers ME, Lee YO, Nonnemaker JM et al. A randomized trial of the effect of e-cigarette TV advertisements on intentions to use e-cigarettes. Am J Prev Med. 2015;49(5):686–93.
84. Camenga D, Gutierrez KM, Kong G, Cavallo D, Simon P, Krishnan-Sarin S. e-Cigarette advertising exposure in e-cigarette naïve adolescents and subsequent e-cigarette use: A longitudi-

nal cohort study. Addict Behav. 2018;81:78–83.
85. Zeller M. Evolving "The real cost" campaign to address the rising epidemic of youth e-cigarette use. Am J Prev Med. 2019;56(2 Suppl 1):S76–8.
86. Leavens ELS, Stevens EM, Brett EI, Hébert ET, Villanti AC, Pearson JL et al. JUUL electronic cigarette use patterns, other tobacco product use, and reasons for use among ever users: Results from a convenience sample. Addict Behav. 2019;95:178–83.
87. Ickes M, Hester JW, Wiggins AT, Rayens MK, Hahn EJ, Kavuluru R. Prevalence and reasons for Juul use among college students. J Am Coll Health. 2019;68(5):1–5.
88. Kong G, Bold KW, Morean ME, Bhatti H, Camenga DR, Jackson A et al. Appeal of JUUL among adolescents. Drug Alcohol Depend. 2019;205:107691.
89. Leavens ELS, Stevens EM, Brett EI, Leffingwell TR, Wagener TL. JUUL in school: JUUL electronic cigarette use patterns, reasons for use, and social normative perceptions among college student ever users. Addict Behav. 2019;99:106047.
90. Nardone N, St Helen G, Addo N, Meighan S, Benowitz NL. JUUL electronic cigarettes: Nicotine exposure and the user experience. Drug Alcohol Depend. 2019;203:83–7.
91. Patel M, Cuccia AF, Willett J, Zhou Y, Kierstead EC, Czaplicki L et al. JUUL use and reasons for initiation among adult tobacco users. Tob Control. 2019;28(6):681–4.
92. Keamy-Minor E, McQuoid J, Ling PM. Young adult perceptions of JUUL and other pod electronic cigarette devices in California: a qualitative study. BMJ Open. 2019;9(4):e026306.
93. Hrywna M, Bover Manderski MT, Delnevo CD. Prevalence of electronic cigarette use among adolescents in New Jersey and association with social factors. JAMA Netw Open. 2020;3(2):e1920961.
94. McKelvey K, Halpern-Felsher B. How and why California young adults are using different brands of pod-type electronic cigarettes in 2019: Implications for researchers and regulators. J Adolesc Health. 2020;67(1):46–52.
95. Kavuluru R, Han S, Hahn EJ. On the popularity of the USB flash drive-shaped electronic cigarette Juul. Tob Control. 2019;28(1):110–2.
96. Ramamurthi D, Chau C, Jackler RK. JUUL and other stealth vaporisers: hiding the habit from parents and teachers. Tob Control. 2018;28(6):610–6.
97. Farsalinos K, Poulas K, Voudris V. Changes in puffing topography and nicotine consumption depending on the power setting of electronic cigarettes. Nicotine Tob Res. 2017;20(8):993–7.
98. Dawkins L, Cox S, Goniewicz M, McRobbie H, Kimber C, Doig M et al. "Real-world" compensatory behaviour with low nicotine concentration e-liquid: subjective effects and nicotine, acrolein and formaldehyde exposure. Addiction. 2018;113(10):1874–82.
99. Audrain-McGovern J, Rodriguez D, Pianin S, Alexander E. Initial e-cigarette flavoring and nicotine exposure and e-cigarette uptake among adolescents. Drug Alcohol Depend. 2019;202:149–55.
100. Melstrom P, Sosnoff C, Koszowski B, King BA, Bunnell R, Le G et al. Systemic absorption of nicotine following acute secondhand exposure to electronic cigarette aerosol in a realistic social setting. Int J Hyg Environ Health. 2018;221(5):816–22.
101. St Helen G, Shahid M, Chu S, Benowitz NL. Impact of e-liquid flavors on e-cigarette vaping behavior. Drug Alcohol Depend. 2018;189:42–8.
102. Robinson RJ, Hensel EC, al-Olayan AA, Nonnemaker JM, Lee YO. Effect of e-liquid flavor on electronic cigarette topography and consumption behavior in a 2-week natural environment switching study. PLoS One. 2018;13(5):e0196640.
103. Schneller LM, Bansal-Travers M, Goniewicz ML, McIntosh S, Ossip D, O'Connor RJ. Use of flavored electronic cigarette refill liquids among adults and youth in the US – Results from wave 2 of the Population Assessment of Tobacco and Health Study (2014–2015). PLoS One. 2018;13(8):e0202744.

104. Villanti AC, Johnson AL, Ambrose BK, Cummings KM, Stanton CA, Rose SW et al. Flavored tobacco product use in youth and adults: Findings from the first wave of the PATH study (2013–2014). Am J Prev Med. 2017;53(2):139–51.
105. Chen R, Ahererra A, Isichei C, Olmedo P, Jarmul S, Cohen JE et al. Assessment of indoor air quality at an electronic cigarette (vaping) convention. J Expo Sci Environ Epidemiol. 2018;28(6):522–9.
106. Melstrom P, Koszowski B, Thanner MH, Hoh E, King B, Bunnell R et al. Measuring PM2.5, ultrafine particles, nicotine air and wipe samples following the use of electronic cigarettes. Nicotine Tob Res. 2017;19(9):1055–61.
107. Schober W, Fembacher L, Frenzen A, Fromme H. Passive exposure to pollutants from conventional cigarettes and new electronic smoking devices (IQOS, e-cigarette) in passenger cars. Int J Hyg Environ Health. 2019;222(3):486–93.
108. Soule EK, Maloney SF, Spindle TR, Rudy AK, Hiler MM, Cobb CO. Electronic cigarette use and indoor air quality in a natural setting. Tob Control. 2017;26(1):109–12.
109. Volesky KD, Maki A, Scherf C, Watson L, Van Ryswyck K, Fraser B et al. The influence of three e-cigarette models on indoor fine and ultrafine particulate matter concentrations under real-world conditions. Environ Pollut. 2018;243(Pt B):882–9.
110. Quintana PJE, Hoh E, Dodder NG, Matt GE, Zakarian JM, Anderson KA et al. Nicotine levels in silicone wristband samplers worn by children exposed to secondhand smoke and electronic cigarette vapor are highly correlated with child's urinary cotinine. J Expo Sci Environ Epidemiol. 2019;29(6):733–41.
111. Visser FW, Klerx WN, Cremers HWJM, Ramlal R, Schwillens PL, Talhout R. The health risks of electronic cigarette use to bystanders. Int J Environ Res Public Health. 2019;16(9):1525.
112. Tzortzi A, Teloniatis S, Matiampa G, Bakelas G, Tzavara C, Vyzikidou VK et al. Passive exposure of non-smokers to e-cigarette aerosols: Sensory irritation, timing and association with volatile organic compounds. Environ Res. 2020;182:108963.
113. Luck W, Nau H, Hansen R, Steldinger R. Extent of nicotine and cotinine transfer to the human fetus, placenta and amniotic fluid of smoking mothers. Dev Pharmacol Ther. 1985;8:384–95.
114. Cairns NJ, Wonnacott S. (3H)(-)Nicotine binding sites in fetal human brain. Brain Res. 1988;475(1):1–7.
115. Suter MA, Aagaard KM. The impact of tobacco chemicals and nicotine on placental development. Prenatal Diagn. 2020;40(9): 10.1002/pd.5660.
116. Holbrook BD. The effects of nicotine on human fetal development. Birth Defects Res C: Embryo Today Rev. 2016;108(2):181–92.
117. Wikström AK, Cnattingius S, Stephansson O. Maternal use of Swedish snuff (snus) and risk of stillbirth. Epidemiology. 2010;21(6):772–8.
118. Gupta PC, Sreevidya S. Smokeless tobacco use, birth weight, and gestational age: population based, prospective cohort study of 1217 women in Mumbai, India. Br Med J. 2004;328(7455):1538.
119. Gupta PC, Subramoney S. Smokeless tobacco use and risk of stillbirth: a cohort study in Mumbai, India. Epidemiology. 2006;17(1):47–51.
120. Steyn K, De Wet T, Salojee Y, Nel H, Yach D. The influence of maternal cigarette smoking, snuff use and passive smoking on pregnancy outcomes: the Birth To Ten Study. Paediatric Perinatal Epidemiol. 2006;20(2):90–9.
121. Hickson C, Lewis S, Campbell KA, Cooper S, Berlin I, Claire R et al. Comparison of nicotine exposure during pregnancy when smoking and abstinent with nicotine replacement therapy: systematic review and meta-analysis. Addiction. 2019;114(3):406–24.
122. Forinash AB, Pitlick JM, Clark K, Alstat V. Nicotine replacement therapy effect on pregnancy outcomes. Ann Pharmacother. 2010;44(11):1817–21.

123. Bahl V, Lin S, Xu N, Davis B, Wang Y, Talbot P. Comparison of electronic cigarette refill fluid cytotoxicity using embryonic and adult models. Reprod Toxicol. 2012;34(4):529–37.
124. Balster RL, Schuster CR. Fixed-interval schedule of cocaine reinforcement: effect of dose and infusion duration. J Exp Anal Behav. 1973;20(1):119–29.
125. Carter LP, Stitzer ML, Henningfield JE, O'Connor RJ, Cummings KM, Hatsukami DK. Abuse liability assessment of tobacco products including potential reduced exposure products. Cancer Epidemiol Biomarkers Prev. 2009;18(12):3241–62.
126. Jensen KP, Valentine G, Gueorguieva R, Sofuoglu M. Differential effects of nicotine delivery rate on subjective drug effects, urges to smoke, heart rate and blood pressure in tobacco smokers. Psychopharmacology (Berl), 2020;237(5):1359–69.
127. Djordjevic MV, Fan J, Ferguson S, Hoffmann D. Self-regulation of smoking intensity. Smoke yields of the low-nicotine, low-"tar" cigarettes. Carcinogenesis. 1995;16(9):2015–21.
128. Henningfield JE, London ED, Benowitz NL. Arterial–venous differences in plasma concentrations of nicotine after cigarette smoking. J Am Med Assoc. 1990;263(15):2049–50.
129. Henningfield JE, Stapleton JM, Benowitz NL, Grayson RF, London ED. Higher levels of nicotine in arterial than in venous blood after cigarette smoking. Drug Alcohol Depend. 1993;33(1):23–9.
130. Talih S, Balhas Z, Salman R, El-Hage R, Karaoghlanian N, El-Hellani A et al. Transport phenomena governing nicotine emissions from electronic cigarettes: model formulation and experimental investigation. Aerosol Sci Technol. 2017;51(1):1–11.
131. Breland A, Balster RL, Cobb C, Fagan P, Foulds J, Koch JR et al. Answering questions about electronic cigarettes using a multidisciplinary model. Am Psychol. 2019;74(3):368–79.

8. Exploration of methods for quantifying individual risks associated with electronic nicotine and non-nicotine delivery systems and heated tobacco products: impact on population health and implications for regulation

Frank Henkler-Stephani, German Federal Institute for Risk Assessment, Department of Chemicals and Product Safety, Berlin, Germany

Yvonne Staal, National Institute for Public Health and the Environment, Centre for Health Protection, Bilthoven, Netherlands

Contents
Abstract
8.1 Background
    8.1.1 Challenges to quantifying risk
8.2 Risk assessment and quantification of risks associated with use of ENDS and ENNDS
    8.2.1 Health risks associated with specific ingredients or unintentionally added substances
    8.2.2 Potential health effects of ENDS and ENNDS
    8.2.3 Risks for bystanders
8.3 Methods for quantifying risk
    8.3.1 Threshold of toxicological concern
    8.3.2 Risk assessment based on individual compounds
    8.3.3 Relative risk approaches
    8.3.4 Margin-of-exposure approach
    8.3.5 Non-carcinogenic effects
    8.3.6 Evaluation frameworks
8.4 Heated tobacco products
8.5 Implications for regulation
8.6 Discussion
8.7 Recommendations
8.8 Conclusions
8.9 References

Abstract

e-Cigarettes heat and aerosolize e-liquids for inhalation. Many of the liquids contain nicotine (electronic nicotine delivery systems, ENDS), while others do not (electronic non-nicotine delivery systems, ENNDS). Although the basic design of ENDS or ENNDS is similar or the same, the devices that are used to heat liquids vary widely in details of their use, operation temperature or performance standards. The health risk depends not only on the properties of the device but also on the composition of the e-liquids, of which several thousand flavour varieties are commercially available or could be prepared at home.

Initial assessments suggest that the toxicological risks of individual ENDS depend on the devices and e-liquids, but the risk associated with high-powered devices may be relatively high. In this paper, we summarize methods that can be used to quantify the health risks associated with the use of ENDS, ENNDS and heated tobacco products (HTPs), due to either the individual compounds or the mixture in emissions. Most methods require substantial data on both emissions and hazards, only some of which are available. Currently, quantitative risk assessment methods cannot be used in regulation, although the most promising approaches are based on the relative potency of the compounds in emissions. Because of the diversity of ENDS and ENNDS, risk assessments remain a challenge, and the results cannot be generalized to the entire spectrum of devices. The wide variety of both liquids and devices indicates that different health risk assessments should be conducted for different combinations of liquid and device and for individual products. Relevant indicators of high risk could be characterized, such as specific ingredients or specific device settings.

8.1 Background

Tobacco use is the major cause of premature death worldwide. Each year, about 8 million people die from tobacco-related diseases, including an estimated 1.2 million non-smokers who were exposed to second-hand smoke (1). Although cigarettes are the most common tobacco product, especially in developed countries, other tobacco products and replacement products also pose serious health risks. In India, more than 350 000 deaths are attributed to use of chewing and oral tobacco each year (2).

While addiction is the driving force for maintenance of this hazardous behaviour, nicotine does not trigger the major toxic effects associated with the high mortality rate associated with tobacco consumption, which is due to carcinogenic and otherwise hazardous combustion products and tobacco constituents. The contributions of individual compounds to the carcinogenicity of tobacco use have been estimated (3,4), leading to identification of the major carcinogens and ranking of smoke constituents by their potency in inducing tumours. Similar approaches have been used for cardiovascular and other health risks.

Strategies have been proposed to reduce the exposure of smokers to toxicants, including mandatory limits on the most relevant toxicants in cigarette smoke (5-7). Attempts to restrict toxicant levels in conventional cigarettes have not, however, been successful because of technical limits to reducing combustion and combustion products (8). HTPs contain electrical heating and other exothermic processes for generating an aerosol from tobacco material that consists of humectants, nicotine and other tobacco constituents and pyrolysis products. Aerosols from HTPs have generally been found to contain fewer hazardous and potentially hazardous compounds than cigarette smoke *(9)*; however, it is not yet known whether reduced exposure to toxicants markedly reduces health risks.

e-Cigarettes are the most common form of ENDS, and the two terms are often used synonymously. In contrast to HTPs, ENDS no longer link nicotine consumption to tobacco use. While ENDS are clearly defined as nicotine delivery systems, the name "electronic non-nicotine delivery systems" (ENNDS) refers only to the absence of nicotine. According to current understanding *(10)*, ENDS and ENNDS are similar products, except for nicotine. Glycerol and/or propylene glycol are usually the major components of the liquids that form the aerosol. Liquids also contain various flavours and other ingredients to increase the palatability and attractiveness of the aerosol *(11–13)*. First-generation e-cigarettes are shaped like conventional cigarettes and consist of a battery, a liquid reservoir and an aerosolizing chamber containing a filament for heating liquid through wicks made of various materials *(14)*. The generated aerosol is directed to the mouthpiece and can be inhaled. Liquids are also provided in cartridges, either designed for single use or refillable. The initial devices were not very efficient at delivering nicotine to the user; however, this basic function has been gradually improved. Open systems have emerged, which are sophisticated devices that can be refilled and reused. Other improvements include increased battery capacity, variable electrical power, removal of tin solder joints and coiled filaments. Recent developments include sub-ohm-atomizing units that can be operated at higher variable voltages because of low electrical resistance *(14)*. Modern atomizers can evaporate a much higher volume of liquid per puff than the original cigarette-like devices *(15)*. These systems are highly adaptable. For example, users can build their own coils or customize performance parameters such as wattage, airflow and, indirectly, nicotine delivery. The operation of advanced systems is, however, also increasingly complicated, and products have been developed that are easier to use, with modules that contain prefilled liquid reservoirs and atomizers designed for single use *(16)*. (See also section 6.) These products have been termed "pod" systems, possibly as an analogy to coffee capsules. A nicotine-containing pod may be attached to a stick- or pen-shaped device containing a low-powered battery and a mouthpiece. A popular example is JUUL, which has become a leading e-cigarette brand in the USA *(16,17)*. Although pod systems lack the flexibility of advanced refillable e-cigarettes, their nicotine delivery and addictiveness are comparable to those of combustible cigarettes *(18)* due to very high nicotine concentrations in the liquid, which can reach 60 mg/mL, as in the US version of JUUL. This is three times higher than the upper limit of nicotine allowed in European products. Most pod systems contain nicotine salts, such as nicotine benzoate or salicylate, to limit the harshness of alkaline nicotine.

Until recently, e-cigarettes were considered mainly as devices for delivering nicotine, with no consideration of the harm they impose. ENNDS were regarded by some as an acceptable nicotine-free version of combustible cigarettes. This may have been a limited perspective. Many novel technologies initially retain their original applications, while more applications are explored over time. Use of modified e-cigarettes to consume cannabis and other illicit compounds *(19)* may well be seen as an early indicator of an initially unintended use of these products. Recently, use of cannabidiol has emerged as a commercially relevant novel application that shows little relation to nicotine or tobacco use. Cannabidiol has defined pharmacological properties *(20)* and is claimed to have many beneficial effects. Some commercial products have been reported to contain other cannabinoids *(21)*. It is therefore misleading to describe cannabidiol liquids as consumables for ENNDS because they do not contain nicotine, and terms such as "electronic cannabidiol inhalation system" or "electronic inhalation system" might be more appropriate. A growing number of electronic delivery products are now beyond the traditional reach of nicotine and tobacco control, and their increasing, often unregulated use in some countries is a cause of concern.

8.1.1 Challenges to quantifying risk

From the perspective of risk assessment, the important distinction between ENDS and combustible cigarettes is that the adverse health effects of e-cigarettes depend on factors such as the system itself, the way it is used, manipulation and the e-liquid. The technical developments described above were not, however, made primarily for health considerations. The aim of the manufacturers was to improve the delivery of nicotine and the palatability and other properties of the products that determine consumer acceptance and use. This aim is a double-edged sword. Acceptance of e-cigarettes is a prerequisite for potential use of these systems in smoking cessation, but their attractiveness, especially to children and adolescents, increases the risk of use by people who would otherwise not have done so.

It is increasingly difficult to assess the toxicological risks of the properties of ENDS, because they are diversifying so rapidly. Initial assessments of early products postulated a temperature range up to 100 °C. Novel high-powered systems can reach temperatures up to 250 °C, which could facilitate chemical degradation of some ingredients; however, other technical features, such as overheating control or replacement of parts containing tin or other metals, might have decreased the exposure of users to toxicants *(14)*. The ambiguous role of technical features in modifying risk is also illustrated by pod systems, such as JUUL, with unique systems that influence their toxicological profiles. For example, closed systems are difficult to manipulate, prevent incidental dermal or oral contact with the liquids and form only minimal amounts of heat-dependent toxicants because of the low wattage. Pod systems pose a higher risk for young people, however, because of their addictiveness and attractiveness and the high nicotine content, especially in products sold in the USA *(16,18)*.

Further risks related to the ingredients and constituents of e-liquids are discussed below. Again, however, rapid product diversity makes it difficult to generalize, except for categories such as pod systems or sub-ohm devices, with very different properties. The rapid diversification also results in an information lag, in that, by the time studies are published, the market has moved on. An increasing challenge for both risk assessment and risk communication is distinguishing those products intended primarily to provide another means of delivering nicotine from those of unconventional liquids containing illicit or recreational drugs or other physiologically active compounds. Although it might be difficult to make this distinction for some products, it would, for example, be misleading to attribute any adverse effects of Δ^9-tetrahydrocannabinol (THC) or synthetic cannabinoids to ENDS or ENNDS per se, even when they are delivered by electronic inhalation systems. Restrictions should be considered on use of ENDS to combine nicotine with other pharmacologically or physiologically active compounds. ENNDS and their liquids should be defined specifically, as the absence of nicotine is insufficient as a criterion for defining this emerging product group. Alternative assessment frameworks are required for products that contain cannabidiol, THC or other drugs. In principle, consideration should be given to whether products that are intended for the delivery of physiologically active compounds or drugs other than nicotine are within the scope of tobacco control.

Individual and population-based risk assessments also depend on e-cigarette use and smoking behaviour, including the prevalence of dual or poly use (i.e. parallel use with tobacco cigarettes or other tobacco products). (See also section 5.) In the USA, dual use is common, although the prevalence of smoking has decreased among adults who currently use e-cigarettes, from 56.9% in 2015 to 40.8% in 2018 *(22)*. Although ENDS can reduce exposure to toxicants, any putative health benefits will be limited if users continue to smoke cigarettes. Even when heavy smokers (> 15 cigarettes per day) reduced their smoking by > 50%, the incidence of lung cancer decreased by only 27%, and their lung cancer risk remained more than seven times higher than that of non-smokers *(23)*. Gradual substitution of tobacco cigarettes with ENDS is of only limited value, as the risks associated with tobacco use remain. Unfortunately, epidemiological studies of the health effects of consistent, exclusive e-cigarette use are difficult to conduct because of limited data. In addition, the prevalence of e-cigarette smoking is highest among cigarette smokers and ex-smokers *(22)*, and it is difficult to separate diseases putatively associated with e-cigarette use from continuing effects of previous tobacco consumption.

8.2 Risk assessment and quantification of risks associated with use of ENDS and ENNDS

Quantitative risk assessment is a powerful tool for assessing the impact of cigarettes, ENDS, ENNDS or HTPs on human health. For example, modelling of tumour potency *(4)* allows estimates of the carcinogenic risk associated with individual constituents of cigarette smoke or the entire smoke, as described below. Modelling and risk quantification have confirmed the exceptionally high adverse effects of tobacco cigarettes in relation to ENDS, ENNDS and HTP *(24)*. Although modelling might also be conducted for ENDS and ENNDS, it would be difficult to cover the entire product spectrum. Nevertheless, relevant hazards can be identified that substantially increase health risks. Regulators should be aware of the growing differences among types and categories of ENDS. Therefore, regulation should address the ingredients, emissions and technical features that have the strongest effects on risk.

In general, ENDS emit fewer toxicants than conventional cigarettes, with two notable exceptions. The first is nicotine, a highly toxic compound added to e-liquids at concentrations up to 60 mg/mL, depending on the jurisdiction, or even higher in so-called "nicotine shots". Intake of 10 mL of a liquid that complies with European regulations corresponds to a dose of 2.8 mg/kg body weight for an adult (70 kg) or 20 mg/kg body weight for a small child (10 kg); the minimum potentially lethal dosage is estimated to be 6.5–13 mg/kg body weight *(25)*, and the lethal dose for infants and small children can be < 5 mg/kg body weight *(26)*. As refill containers may contain several hundred milligrams, the risk of accidental or intentional poisoning is high. Users of open systems and refill containers are at risk of dermal exposure, and incidental oral poisoning can occur, although oral absorption of nicotine is often limited by its emetic effects. A number of weak or moderate cases of intoxication were reported in 2018 *(27)*, but fatal poisonings are very rare *(28)*. The second exception to emission of fewer toxicants than conventional cigarettes is contaminants and intentionally added compounds, including essential oils, herbal extracts and certain flavours such as diacetyl and acetyl propionyl, which can cause serious (acute) lung injury.

Commercial products that comply with regulations in Europe, the USA and other countries should not pose such risks. Single cases of lipid pneumonia have been reported, probably caused by glycerol contaminated with fatty oils *(29)*. Many users prepare their own e-liquids and sometimes make their own ingredients, perhaps because they are not aware of the risks. Such risks are avoidable and should not apply to commercial products if appropriately regulated. In the worst case, however, use of irregular, manipulated or faulty products could have acute and fatal toxic effects, while smoking-associated diseases take years to develop. Typical hazards of e-cigarettes are discussed below.

8.2.1 Health risks associated with specific ingredients or unintentionally added substances

The health effects that could occur from direct or indirect exposure to harmful and potentially harmful constituents of ENDS or ENNDS aerosols when inhaled by children or adolescents are summarized below. The actual occurrence of effects would depend on the quantity of the compound that is inhaled.

Nicotine

Tobacco smoking increases the risks for arteriosclerosis, myocardial infarction, stroke and other cardiovascular diseases. These are due not only to nicotine but also to other compounds, such as carbon monoxide, nitrogen oxides, metals and particulate matter. The pathogenesis is usually associated with inflammation *(30,31)*. Some effects of nicotine, such as increased blood pressure and decreased perfusion of coronary vessels, can contribute to cardiovascular injuries. Limited data are available, however, on the cardiovascular effects of nicotine in ENDS. The cardiovascular toxicity of nicotine is lower without combustion products than in cigarette smoke *(30)*, as indicated by studies on use of smokeless tobacco. For example, Swedish snus did not increase the risks of stroke or infarction *(32)*, despite efficient nicotine absorption, although this requires further study *(33)*. A meta-analysis of 11 studies in Sweden and the USA indicated that smokeless tobacco users had increased risks for myocardial infarction and stroke *(34)*. Increased odds for myocardial infarction with smokeless tobacco use were also reported in the INTERHEART study of data on 27 089 participants in 52 countries *(35)*.

Modern e-cigarettes can be optimized to deliver nicotine at levels comparable to those in combustible cigarettes *(15)*. In open systems, nicotine levels can usually also be adjusted with the power setting and increase with the amount of aerosolized liquid per puff. On inhalation, aerosolized nicotine might contribute to inflammation and vascular injuries *(36)*; however, pathogenesis also depends on inflammatory co-factors *(31)* such as reactive oxygen species, and it is not known whether these factors are generated by e-cigarettes. Smoking during pregnancy can affect embryonic development, reduce birth weight and increase the risk for complications, such as premature delivery, stillbirth or sudden infant death. Further, both lung function and development can be impaired by maternal smoking. It is again, however, difficult to distinguish the adverse effects of smoke and combustion products from those of nicotine. Nicotine has been reported to limit the availability of oxygen to the fetus by constricting blood vessels in the umbilical cord and uterus *(37)*. Prenatal exposure to nicotine interferes with development of the brain *(38)* and is associated with neurobehavioural impairment, including hyperactivity, anxiety and impaired cognitive function. Nicotine was confirmed as teratogenic to the nervous system in a study in rats

(39), and other experimental studies have suggested that developmental exposure to nicotine has long-term adverse effects such as impaired fertility, hypertension, obesity and respiratory dysfunction *(40)*.

Nicotine may induce other adverse effects, such as insulin resistance, thus increasing the risk for diabetes type 2 *(41)*, although the question should to be addressed in additional studies. One suggested that nicotine can inhibit mucociliary clearance in the lung *(42)*, which could increase exposure to toxicants.

Hazardous constituents and emissions of ENDS and ENNDS

Glycerol and propylene glycol. Glycerol and propylene glycol are the most commonly used solvents for aerosolization and are the major constituent of e-liquids. Although mild adverse effects such as irritation have been described after inhalation *(43,44)*, both compounds are considered relatively safe. The amount of evaporated liquid per puff varies, however, and is extremely high in sub-ohm devices *(45)*. No comparison has yet been reported of the toxic properties of e-cigarette aerosols according to the density of the aerosol or the total mass of inhaled material, particularly in the long term. Other solvents, such as ethylene glycol, have been used *(12)*, which have higher toxicological risks *(46)*; however, it is not clear whether ENDS containing other nebulizing agents as major components are currently on the market.

Flavours and other ingredients. Some flavours, including diacetyl, are of concern, as they can cause bronchiolitis obliterans, a rare but serious lung disease. Diacetyl was detected in a large proportion of liquids tested *(11)*, albeit 5 years ago, before the ban on its use in e-cigarettes in many jurisdictions. The sweetener sucralose, a halogenated disaccharide, can be degraded in e-liquids when devices reach temperatures > 200 °C, which can generate potentially harmful organochlorines *(47)*. Other compounds in regular e-liquids include sensitizers. For example, limonene oxide, found in lemon-flavoured concentrates *(13)*, is considered an important contact allergen *(48)*. Rare cases of hypersensitivity pneumonitis have been reported that might be related to use of ENDS and ENNDS; however, no specific allergen has yet been identified in e-liquids *(49)*. In 2019, a further case of hypersensitivity pneumonitis was reported in the United Kingdom *(50)*. This case was not related to e-cigarette- and vaping-associated lung injury (EVALI) (see HSP2) but indicates that inhalation of e-cigarette aerosol can induce allergic responses, although rarely. The etiology of these cases should therefore be elucidated, with monitoring of future developments. The increasing variety of e-liquids should be regarded as a potential hazard, as many of the constituents can generate thermal degradation products and undergo chemical conversions. Indeterminate chemical reactions are facilitated by coil temperatures of ≥ 250 °C *(51)*.

Carbonyls and thermal degradation products. Carbonyl compounds, including formaldehyde, acetaldehyde and acrolein, are considered the most relevant toxic emissions from commercial ENDS. They originate from degradation of glycerol and propylene glycol, depending on the device temperature. In low-powered devices, carbonyls occur mainly under dry-puff conditions due to overheating of the wire in the absence of liquid *(12,52)*. In high-wattage devices, carbonyls are formed according to the applied power. Talih et al. *(45)* demonstrated enhanced carbonyl formation in high-power single-coil conventional ENDS and sub-ohm devices that varied with the coil surface and the amount of liquid consumed. The study showed total aldehyde emissions of ≤ 400 μg per 15 puffs. The carbonyl content of the aerosol can approach that of conventional cigarettes but is highly variable. For example, it may be strongly enhanced by flavours *(53)*, and puffing topography can account for variations in formaldehyde levels from 20 to 255 ng/puff *(45)*. In general, carbonyls in ENDS and ENNDS occur at levels from hardly detectable to several micrograms per puff. Acetaldehyde not only has carcinogenic and other hazardous properties but also increases the addictiveness of nicotine *(54)* by inhibiting monoamine oxidase. Moderate levels of carbonyls might therefore increase consumer satisfaction by making products more attractive.

Adult smokers are less likely than non-smoking adolescents to use JUUL as an alternative to cigarettes, although this device delivers high levels of nicotine *()*. The trend to increasing the wattage to variable, higher levels may be partly to increase user satisfaction, as higher levels of carbonyls can enhance the effects of nicotine and possibly its addictiveness. Toxicological assessments should be conducted of various groups of ENDS and ENNDS products. A special category might be considered for ENDS that operate at ≥ 15 W, as these devices produce dense aerosols and markedly higher levels of toxicants than low-power devices. It has been reported that high-power ENDS and ENNDS increase the risk for lung injuries, including transient inflammation, and disturb gas exchange *(56)*. Regulators should be aware that it would be difficult to impose upper limits on the levels of ingredients in high-capacity open-system ENDS and ENNDS, as they could be compensated by switching to a higher power *(57)*. For example, the intake of nicotine per puff can be increased when a higher volume of liquid is aerosolized, increasing the risk to health.

Metals. Metals that occur in the aerosols of ENDS and ENNDS usually originate from the devices themselves *(58)*. The levels remain below the limits of toxicological concern *(59,60)* but can be increased at high-power settings *(61)*. As redox-active metals might increase the levels of reactive oxygen species, cardiovascular risk might be increased, especially in the presence of a high level of nicotine. Haddad et al. *(62)* showed that high-wattage ENDS can generate levels of reactive oxygen species that are comparable to those from combustible cigarettes.

In summary, hazard analysis has confirmed that nicotine, carbonyls and metals are relevant risk factors, implying that e-cigarette use could enhance the risks for cardiovascular diseases. The high nicotine content, application of high wattages and low-quality standards also affect the levels of metals in emissions. As aerosols may contain irritants, the risks of respiratory diseases should also be assessed. ENDS generate levels of carcinogens that are lower than those typical for tobacco smoke, except those of carbonyls *(63,64)*. This finding was recently confirmed for sub-ohm and high-wattage devices in an industry-sponsored study *(65)* but should be confirmed in independent research.

8.2.2 Potential health effects of ENDS and ENNDS

Smoking can increase the risks for arteriosclerosis, myocardial infarction, stroke and other cardiovascular diseases. The risk factors include a number of chemical species, such as carbon monoxide, nitrogen oxides, metals and particulate matter. Pathogenesis is usually linked to inflammation, and some nicotine-related effects, such as increased blood pressure and decreased perfusion of coronary vessels, can exacerbate cardiovascular injuries *(30,31)*. There is still, however, some debate about whether e-cigarette use also increases cardiovascular risk. A systematic review of studies on the use of e-cigarettes and cardiovascular disease indicated potentially increased risks for thrombosis and atherosclerosis *(66)*.

Short-term exposure to e-cigarette emissions can trigger vascular oxidative stress and dysfunction *(67)*. e-Cigarettes can have an effect even after a single use and even without nicotine, with a transient effect on endothelial function *(68)*. In a meta-analysis, Skotsimara et al. *(69)* found associations between e-cigarette use and endothelial damage, arterial stiffness and a long-term risk for coronary events; however, these findings were made in single studies and were not confirmed in others. Short-term effects on vascular function do not necessarily progress to clinically relevant disease. Analysis of data collected in the Behavioral Risk Factor Surveillance System in 2016 showed substantially increased risks for stroke, myocardial infarction and coronary artery disease among 70 000 respondents who reported use of e-cigarettes *(70,71)*. Further, analysis of National Health Interview Surveys (2014–2016) confirmed a higher risk for myocardial infarction after adjustment for cigarette smoking and other risk factors *(72)*. Analysis of pooled Behavioral Risk Factor Surveillance System data collected in 2016 and 2017 did not confirm an increased cardiovascular risk for e-cigarette users who never smoked cigarettes *(73)*. No specific data are yet available on devices that might enhance such risks. A comprehensive review of preclinical findings and epidemiological evidence on the effects of e-cigarettes on cardiovascular and general health was inconclusive, and more data are needed. The authors concluded that, while it is reasonable to consider e-cigarettes less hazardous than combustible tobacco products, no smoke is better than electronic smoke *(70)*. One report showed that smokers who switched to e-cigarettes had significantly improved endothelial function and vascular health *(74)*.

Other health effects include irritation of the respiratory tract, mainly the upper airways *(75)*. Effects on lung function have also been reported, with decreased lung function capacity among e-cigarette users than non-users and possibly increased lung resistance *(76,77)*. Furthermore, e-cigarette users may have a reduced response to infections *(78)*. An analysis of the data in the Behavioral Risk Factor Surveillance System for 2016–2017 indicated that e-cigarettes enhance pulmonary toxicity. Increased risks of respiratory diseases have also been confirmed by others *(79)*. Another study suggested that the health of patients with chronic obstructive pulmonary disease who smoked tobacco improved after they switched completely to e-cigarettes, including better outcomes and reversal of some of the harm caused by smoking *(80)*. This illustrates the different perspectives of smokers and non-smokers on the health risks of e-cigarettes. A link between e-cigarette use and asthma has been reported; however, the high prevalence of e-cigarette use by adults with asthma could be related to quit attempts in this group *(73)*.

In the summer of 2019, a series of cases of serious lung injury associated with use of vaping or e-cigarette product use was reported in the USA. (See also section 12.) The number of incident cases related to e-cigarette use increased, and fatalities with no history of lung disease were reported *(81)*. The disease, named "E-cigarette- and Vaping-Associated Lung Injury" (EVALI), was restricted mainly to the USA. By 18 February 2020, 2807 hospitalized cases had been reported, including 68 EVALI-associated deaths *(82,83)*. The patients presented similar clinical characteristics, such as dyspnoea and cough, and were hypoxaemic, with bilateral airspace opacities on chest imaging *(84)*. As these symptoms are similar to those of other respiratory diseases, these cases were not immediately linked to use of specific types of e-cigarettes *(85)*. Many, but not all, patients had used THC-containing e-liquids, and further research indicated that vitamin E acetate, which is added to THC-containing e-liquids as a thickener, was the probable cause of the respiratory injuries *(86–88)*. The patients with respiratory disease were found to have used e-liquids containing THC more often than non-patients and, perhaps more importantly, reported more frequently obtaining products from informal sources *(89)*. Cases also occurred, however, in patients who did not use THC-containing liquids. Patients showed respiratory effects within a few days to several weeks of inhaling vitamin E acetate, which allowed identification of the cause of the disease; however, an association is much more difficult to establish with long-term health effects. In one study that confirmed the association *(88)*, exceptionally high levels of vitamin E acetate were found in products collected from EVALI patients. The reported concentrations were 31–88%, while the THC content was often lower than that advertised *(90)*. It should be noted that these products have little in common with typical e-cigarettes. Both risk assessment and public education are necessary to respond to the increasing proliferation of unconventional, often illicit uses of e-cigarettes to deliver drugs other than nicotine. Risk assessment would benefit from a clear distinction between these products and regular ENDS, which would be difficult to achieve in practice.

8.2.3 Risks for bystanders

Bystanders may be exposed to emissions of ENDS, ENNDS or HTP exhaled by users. Their actual exposure will depend largely on the size and ventilation of the room. Bystanders could experience irritation of the respiratory tract as a result of exposure to propylene glycol and glycerol. Systemic effects of nicotine might also be expected if nicotine-containing e-liquid is used, including heart palpitations and raised systolic blood pressure. Because tobacco-specific nitrosamines are present in some e-liquids, bystanders might be at increased risk of tumours in a worst-case scenario *(91)*. Another study concluded that the health of bystanders was unlikely to be at risk due to e-cigarette use *(92)*. Studies of bystander exposure to e-cigarettes and to tobacco cigarettes have shown that the health risk associated with second-hand e-cigarette aerosol was lower than that associated with second-hand smoke *(93,94)*. Other studies have shown that the concentration of nicotine in the air is much lower when e-cigarettes are used than with tobacco cigarettes *(95)*, although bystanders can take up nicotine from second-hand smoke *(96)*. Use of HTPs is considered to result in higher exposure of bystanders than use of e-cigarettes *(97)*. Whether exposure of bystanders to second-hand HTP or e-cigarette emissions could have adverse effects is largely unknown. One study suggested an association between second-hand exposure to e-cigarettes and reported asthma attacks in the past 12 months *(98)*. This indicates that some population groups, such as patients with airway disease and young children, are more vulnerable to adverse effects on exposure to second-hand emissions.

8.3 Methods for quantifying risk

Tobacco products differ not only in type but also in emissions and use, resulting in different health risks. Even products within a class, like ENDS and ENNDS, have different effects on health. Smokers of tobacco cigarettes who wish to switch to a potentially less harmful product require information on the relative risks of such products. A product that is potentially less harmful for a smoker is more harmful for a non-smoker. Quantitative hazard characterization, which includes a dose– or concentration–response relation, can be used to determine the potential health effects at population level when information is available on the number of users and their use pattern. Generalization of the health risk of a tobacco product is, however, not scientifically appropriate in view of the differences among products in a class and among users *(99)*. Risk assessment for ENDS and ENNDS should therefore be conducted separately for groups of devices or even for individual devices and liquids. A wide range of e-cigarette use parameters must also be considered in estimating human exposure.

Quantification of the risks of chemical mixtures is inherently difficult, and the appearance of novel nicotine and tobacco products adds to the challenges of the wide variety of products, compositions and diversity of use. Differences among products make quantification of both exposure and hazard uncertain. To measure exposure, information is necessary on the topography of e-cigarette use, the emissions inhaled and, depending on the method used, particle size and particle size distribution, which determines deposition in the respiratory tract and therefore local exposure and effects *(100,101)*. Information on the chemical composition of the emissions is also necessary to identify the compounds to which users are exposed. The emissions from tobacco products contain many different compounds, which depend on the topography of e-cigarette use and device settings such as temperature and power *(62,102)*. The compounds in emissions must be characterized and quantified in order to assess the risk of these products *(9)*. Unfortunately, information on ingredients (contents) alone is insufficient, as they may degrade or burn during aerosolization or may originate from the device.

Information on hazard can be derived from toxicological studies of compounds, preferably administered by inhalation, the most relevant route of exposure for ENDS, ENNDS and tobacco products. Information on hazard is not available for all the compounds in each of these products. e-Liquids commonly contain flavours that have been tested for toxicity by oral administration for use in foods, and no information on toxicity resulting from inhalation of these compounds is available. Such information is necessary, as, in toxicity studies, some compounds, like diacetyl and cinnamaldehyde, can have local effects on the respiratory tract when inhaled *(103,104)*. Furthermore, users of ENDS, ENNDS and HTP products are exposed to mixtures of compounds that may or may not interact biologically.

The risk associated with mixtures can be quantified by combining data on the hazard of individual compounds with the quantities present in the emissions of products. Alternatively, the risks of ENDS, ENNDS or HTP could be quantified case by case in toxicity studies; however, studies in experimental animals may not provide meaningful results for assessing the risks that tobacco products pose to humans. In addition, exposure in experimental studies is generally for 6 h/day, 5 days/week, which is not comparable to use of ENDS, ENNDS or HTP, which result in irregular peak exposure for 7 days/week. The development and use of alternative models, such as cell models, are increasing rapidly, but this has not yet allowed hazard characterization of mixtures.

Some work has been conducted on the toxicological effects of mixtures *(105)*. In a project by the European Union *(106)*, EuroMix (https://www.euromix project.eu/), which ended in 2019, compounds were classified according to their target organ and their mechanism of action. Compounds with similar and dissimilar modes of action were assessed in assays specific for a target organ to determine whether the effect of the mixture was different from those of the individual compounds. The project resulted in a toolbox for exposure scenarios and testing strategies outlined in a handbook of practical guidance. As in the EuroMix project, most studies of the toxicity of mixtures have addressed binary combinations; however, no reliable method is available even for these relatively simple mixtures to predict quantitatively the effects of any combination of the two compounds in the mixture without the input of at least some experimental data. It is difficult to predict whether the compounds in a mixture will interact, and this may be obscured by variations in the biological response. Binary mixtures often have an additive effect, such that the potency-adjusted doses of the individual compounds can be summed to predict their combined effect. Mathematical models are available to determine whether the toxicity of a mixture is due to interaction between the compounds. The two commonly used mathematical models used are the dose–concentration addition model and the independent action–response addition model *(107–109)*. The dose–concentration addition model is based on the assumption that all compounds have a similar mode of action but may have different potency to induce an effect. Once the potency of one compound is determined relative to that of another, the concentrations of both compounds can be expressed relative to that of one of the compounds as a reference. When the concentrations are summed, they can be used to determine the effect of the binary mixture on the dose–response curve of the reference compound *(110)*. In the concentration addition model, a similar mode of action is assumed, whereas the independent action model can be used to determine the effects of compounds that are due to different mechanisms or modes of action *(111)*. Mathematical models can be used to determine whether compounds interact, which is the case in synergism or antagonism. These models cannot be used to predict the effects of complex mixtures such as tobacco smoke.

Risk assessment of the complex emissions of tobacco products, ENDS and ENNDS is even more complicated and is similar to the assessment of other complex mixtures, such as petroleum-derived and cement-like products. Generally, the approach to assessing the risks of such products is to assess the toxicity of a representative sample of the mixture as a whole, primarily in experimental animals. A similar approach may be used to assess the hazard of ENDS and ENNDS, while differentiating classes of products. Although studies in experimental animals are ethically debatable, various (tobacco) products have been tested; however, translating the results into human effects and assessing different products within a class remains a challenge.

8.3.1 Threshold of toxicological concern

One approach to assessing exposure to complex mixtures is the threshold of toxicological concern (TTC) *(112)*. In this approach, originally developed for assessing carcinogenicity, the compounds of greatest toxicological concern in a mixture are identified from structure–activity relations and read-across. TTC values (in μg/person or μg/kg body weight per day) have been defined for three classes (Cramer classes I–III) according to structural elements, but only after oral exposure. Cramer class III indicates the highest carcinogenic risk and consequently the lowest TTC value *(113)*. The risk of a mixture is then assessed by comparing exposure to these compounds, either alone or summed, with the appropriate TTC value. This approach has been applied to complex mixtures such as botanical extracts *(114)*, flavour complexes *(115)* and inhaled toxicants *(116,117)*. This method does not indicate a risk to health but indicates that further testing is required if a compound exceeds the TTC threshold; otherwise, the probability of a health risk is low. This method might be used when no hazard data are available. As it is already known that tobacco products have adverse health effects, the TTC approach for risk assessment cannot be used to quantify health effects. For ENDS and ENNDS, however, it might be used to identify compounds that pose a health risk and to prioritize them for further testing.

8.3.2 Risk assessment based on individual compounds

Information on exposure to and the hazard of individual components could be used to assess the risk of a product as a whole. For cigarettes, compounds could be selected for their hazardous potential *(6,118,119)*. For ENDS, the number of compounds in emissions is limited, and they do not necessarily overlap with known tobacco toxicants. Compounds that are generated by thermal degradation, such as aldehydes, should also be considered for ENDS and ENNDS. This method has been applied to assess the toxicity of e-cigarettes for users and bystanders *(91)*. The data on hazards used in this approach are derived from studies for setting a safe level of exposure. In emissions from tobacco products, and in some cases also from ENDS and ENNDS, the concentrations are often above the safe level of exposure, so that the information can be used as an indicator of potential concern but not for quantifying risk. The effect seen in experimental animals may not directly reflect a similar effect at a similar exposure level in humans, as the exposure regime in experimental studies differs from human exposure patterns. Addition of the risks of individual compounds in order to estimate the risk of a complex mixture probably results in underestimates of health risks, as interactive effects among compounds in the mixture are ignored. Risk assessment based on individual compounds does not allow comparison of the risks of different (tobacco) products because of the design of many experimental studies. To compare the severity of effects, detailed information is necessary on the relations between exposure and health effects and how they can be extrapolated to effects in humans.

8.3.3 Relative risk approaches

Studies have also been conducted to estimate the carcinogenic potency of a tobacco product as a whole and relative to a (reference) tobacco cigarette *(3,24)*. In this approach, data from carcinogenicity studies in rodents are used to determine the carcinogenic potency of a compound, by using a modelled linear relation between exposure level and the number of tumours induced. For example, benchmark dose methods can be used to determine a lower confidence bound (usually 95%) of the effective dose that results in a tumour incidence of 10%. Quantitative data on carcinogenic potential can be used to calculate the relative carcinogenic potency of each compound according to technical guidance documents, such as that published by the Office of Environmental Health Hazard Assessment in California, USA *(120)*. The total relative carcinogenic potency of mixtures or aerosols can then be calculated by adding the values for individual compounds. In this approach, it is assumed that relative potency factors are equal over the entire dose range and the mechanisms are comparable. Although the mechanisms of action differ among carcinogens, the outcomes may give an indication of carcinogenic risk and allow comparison of carcinogenic risk among products. The validity of the assumptions is assessed during validation of the method.

This relative risk approach depends on the availability of data on both emissions and carcinogenicity. Data on the emissions of as many compounds as possible is required, as the inclusion of more compounds improves the calculation of relative risk. The methods used to analyse emissions should be similar for both products, and information on variations in the emissions must be available to determine the uncertainty in the relative risk. Information on carcinogenic potency should be derived from studies in rodents; alternatively, data from other sources can be used that are indicators of carcinogenic potency, if data on dose and response are available. This approach currently appears to be the best for quantifying the (relative) risk of products; however, for systematic application, more data should be available on emissions and on hazard level. A similar method might be applicable to other health effects, such as cardiovascular disease or chronic obstructive pulmonary disease.

8.3.4 Margin-of-exposure approach

The margin-of-exposure approach is based on the ratio of the exposure level and the dose at which no effects occur or the dose at which a predefined adverse effect occurs (e.g. a benchmark dose level). The larger the margin of exposure of a compound, the higher the risk may be. This approach has been applied to compounds in tobacco products *(121,122)* and can be used to prioritize compounds that should be reduced in tobacco smoke emissions and to assess individual compounds in the emissions of ENDS and ENNDS. A margin of exposure is calculated for each compound from data on inhaled emissions and information on hazard and depends on the quality of the data. The approach does not result in a quantitative risk characterization. Its main goal is to determine whether a specific exposure is of concern. The magnitude of the margin of exposure is not a measure of risk.

8.3.5 Non-carcinogenic effects

Many methods address risk evaluation of carcinogenic effects, which are easier to model than non-carcinogenic effects, as the end-point (cancer) is easier to compare and the dose–response curves for carcinogens are comparable. Inherently, non-carcinogenic effects are diverse. Local effects should be differentiated from systemic effects; different compounds have effects on different organs and, even if compounds have an effect on the same organ, their mechanism and result may be different. Quantification of non-carcinogenic effects might have to be conducted at the level of the mechanism of action *(106)*; however, this approach requires dose–response data on the mechanism of action of compounds, which is limited. Non-cancer risk indices have, however, been generated for the cardiovascular and respiratory effects of cigarette smoke *(4)*.

Alternatively, toxicological assays of emissions may be used to characterize the risk of a product. As mentioned above, bioassays in experimental animals may not be preferable, and the results of cellular assays are difficult to translate into effects in humans. In addition, not only should the effects (read-out parameters) be extrapolated to human effects, but the exposure should resemble human exposure. This includes smoking topography and, in the case of a lung model, deposition in the airways. This field is evolving, but in vitro assays for characterizing non-carcinogenic effects of ENDS, ENNDS and HTPs remain in the future.

8.3.6 Evaluation frameworks

Assessment of the health effects of ENDS and ENNDS could be based on an appropriate evaluation framework. In this approach, expert judgement is used to score aspects of a product in order to identify the most important risks of, for example, drugs *(123)*. The Netherlands National Institute for Public Health and the Environment has developed an evaluation framework for tobacco products that summarizes all the factors that may influence the attractiveness, addictiveness and toxicity of a product and can be used to identify knowledge gaps or prioritize research on a specific product *(124)*. These models allow evaluation of a product even when limited data are available. Evaluation improves with increasing knowledge about a product, but this approach is limited to qualitative results based on expert judgement.

The methods described above can be used to compare the hazards of different products. Table 8.1 summarizes the methods, their data requirements and their application. The feasibility of applying the methods is determined by the information available and not by the product itself. The information necessary for full risk characterization is currently not available. In some methods, a weight-of-evidence approach can be used for data of different quality. Many methods for risk quantification also require data on emissions, an indicator of human exposure. Tobacco smoke and emissions from ENDS and ENNDS are dynamic. Emissions cool as they pass from a heating element to the exit of the device, resulting in condensation of volatile compounds and agglomeration of particles. In addition, the inhaled air, which includes the emission, is humidified in the upper airways. These processes occur simultaneously and determine local deposition in the airways, which can result in high doses at specific locations in the airways, which could have site-specific adverse effects. Modelling of airway deposition of tobacco smoke and ENDS or ENNDS emissions is under way *(100,125)*. The outcome of a qualitative risk assessment depends on the quality of dosimetry, which is limited.

Table 8.2 summarizes the limitations and advantages of the methods described in this paper. It should be noted that all the methods are intended for assessment of risk to users. Similar methods could be used to assess the risk of bystanders, provided that information is available on their exposure.

Table 8.1. Summary of methods for quantitative risk assessment of ENDS and ENNDS according to individual compounds or their mixture (1), the data required (2 and 3) and their application (4–7)

Method	1. Individual compounds or mixture?	2. Dependent on emission characterization?	3. Dependent on available hazard data?	4. Allows quantification of risk of single compounds?	5. Allows quantification of risk of the product as a whole?	6. Allows quantification of risk of the product as a whole for vulnerable groups?	7. Data required for a modified product
Threshold of toxicological concern	Individual	Yes	No	No, allows identification of potential hazardous compounds	No	No	Emission data
Risk assessment based on individual compounds	Individual	Yes	Yes	Yes	Yes, but only if data for all compounds are available	Yes, but only if data for all compounds and specific hazard data on vulnerable groups are available	Emission data
Relative risk approaches	Mixture	Yes	Yes	Yes	Yes, but only if data for all compounds are available	Yes, but only if data for all compounds and specific hazard data on vulnerable groups are available	Emission data

Method	1. Individual compounds or mixture?	2. Dependent on emission character-ization?	3. Dependent on available hazard data?	4. Allows quantifica-tion of risk of single compounds?	5. Allows quantifica-tion of risk of the product as a whole?	6. Allows quantifica-tion of risk of the product as a whole for vulner-able groups?	7. Data required for a modified product
Margin-of-exposure approach	Individual	Yes	Yes	No	No	No	Emission data
Bioassays for non-carcinogenic effects	Mixture	No	No	No	Yes	Yes, but only if specific data on hazards for vulnerable groups are available	Toxicity bioassays
Evaluation frameworks	Mixture	No	Yes	No	No, only non-quantitative risk	No, only non-quantitative risk if data are available on vulnerable groups	Re-evaluation of e.g. emission data

Column 1: Applicability of the output of the method to individual compounds or to the emission as a mixture
Column 2: Requirement for quantitative data on (ideally) all compounds in the emission
Column 3: Requirement for data on dose–response related hazard
Column 4: Applicability of the method to quantify the risk of exposure to one of the compounds in the emission
Column 5: Applicability of the method to quantify the risk of exposure to the product
Column 6: Applicability of the method to quantify the risk of exposure to the product pf vulnerable groups, such as infants
Column 7: Data required to quantify the risk of a slightly changed product, such as a new flavour in an e-liquid or technical adaptation of the device

Table 8.2. Limitations and advantages of each method for quantifying the health risk of ENDS and ENNDS

Method	Main limitations	Main advantages	Potential application for ENDS, ENNDS and HTPs
Threshold of toxicological concern	Cannot assess risk of complete product No quantification of risk	Information on possible risk from exposure	Prioritization of compounds for further testing
Risk assessment of individual compounds	Cannot assess risk of complete product	Identification of compounds with highest health risk	Health risk assessment based on available data
Relative risk approaches	Currently only for carcinogens	Allows comparison of risks between products	Health risk assessment based on available data and allows comparison of products
Margin-of-exposure approach	Cannot assess risk of complete product	Information on exposure in relation to health concern	Prioritization of compounds for further testing
Bioassays for non-carcinogenic effects	Extensive testing required and extrapolation of exposure and results to humans	Does not require data on emissions or hazard	Health risk assessment based on available bioassays
Evaluation frameworks	Most subjective method	Requires limited data; more data will improve outcomes	Non-quantitative health risk assessment, can be used for setting priorities

8.4 Heated tobacco products

Methods used to assess the risks of ENDS and ENNDS can in principle be applied to HTPs. HTPs currently vary less than the e-systems, and some reliable, independent data are available on the composition of the aerosol *(9)*. Industry data must be verified independently before it can be used for risk assessment. Previous investigations addressed the toxicants that typically occur in cigarette smoke, and further independent research is required for a comprehensive analysis of emissions and health risks. The methods used to analyse cigarette smoke and standardization of inhalation topography must still be adapted to obtain user-representative measurements. Further, some toxicants might not be relevant in cigarette smoke but could occur preferentially in the aerosol of HTPs. This question has been addressed with untargeted screening methods *(126–128)*, which have also been used by industry scientists *(129)*. Some components of concern, including glycidol (classified in 2A by a working group at the International Agency for Research on Cancer) and furfuryl alcohol (classified in 2B) have been identified in HTP aerosol, and data for specified HTPs would allow risk modelling and comparative or quantitative assessment. Stephens *(24)* modelled the carcinogenic potency of aerosols from cigarettes, ENDS and an HTP device, and comparative modelling approaches have since been refined *(3)* to determine the relative cancer potency of individual compounds and product emissions, with confidence intervals. The ratio or change in cumulative exposure can then be calculated with a probabilistic approach for two products. For HTPs, the change in cumulative exposure to selected compounds was 10–25 times lower than from smoking cigarettes. With relevant information on human dose responses, the change in cumulative exposure can be translated into an associated health impact for each device. This approach was initially used for eight carcinogens that occur in the aerosol of HTP and in cigarette smoke but should be extended to compounds that are found at higher levels in HTP aerosols than in cigarette smoke.

These calculations illustrate the differences in the composition of HTP aerosol and that of tobacco smoke, which may affect the health risks. As HTPs continue to evolve, standards of performance and upper limits for key toxicants should be considered the most useful, feasible options for regulation. The methods for quantitative risk assessment of ENDS and ENNDS described above are based on data on emissions and hazards and can therefore be applied to HTPs as well, if the data are available.

8.5 Implications for regulation

Current methods for quantifying the health risk of ENDS, ENNDS and tobacco products are not yet adequate for use in regulation. Some approaches can be used, however, to obtain an indication of the absolute health risk of a product. These methods are based on the risk associated with specific or unique compounds in emissions, which depends on the availability of data. Compounds in emissions should be identified and quantified, ideally in user-representative settings. In addition, (human) toxicological data on these compounds is required to quantify health risk; however, data are lacking for both these parameters. The risk assessment approaches described in this paper could be considered for use in regulation; however, currently, because of lack of data, this stage has not been reached, as the model is only as good as the quality of the data.

Quantified health risks could be used in models for estimating health risks at a population level. Although this has not yet been done, the feasibility of modelling population health effects has been explored *(130)*. The health impact of ENDS, ENNDS or HTPs in smokers, non-smokers and former smokers can be estimated when monitoring their popularity and use and used as a reason for legislative measures to limit use of the product or as a basis for public education. The outcomes depend strongly on the input data, which, in this case, will also include epidemiological data. The observation that epidemiological data on product use and switching between products is inconsistent should be considered when applying population modelling.

Modelling has indicated that switching from cigarettes to HTPs can affect human health; however, the effect depends on the compliance of devices with substantially lower levels of previously documented toxicants in the emissions. Regulators should consider setting mandatory upper limits for carbonyls, carbon monoxide and other key toxicants to ensure that devices meet technical and performance standards.

Effective regulation of ENDS and ENNDS and their characteristics could also contribute to limiting their health risks. Priority should be given to closed systems, like pods or sealed cartridges that cannot be easily manipulated by consumers. Problematic ingredients and constituents (diacetyl, sucralose, essential oils and all carcinogenic, mutagenic and teratogenic compounds) should be prohibited, and upper limits for nicotine in the emissions of closed-system ENDS might also be considered. Although prohibitions of hazardous compounds should apply to liquids in general, the emissions of flexible, usually high-powered open systems are difficult to regulate. The focus might therefore be shifted to technical features, such as overheating controls, maximum wattage or temperature. Further regulation would require detailed assessment of ENDS and ENNDS subcategories, especially of high-powered products, including aerosol chemistry, toxicology and the design and performance of devices.

The ambiguous terminology and definition of ENNDS are also matters of concern. For example, nicotine-free devices can be used to inhale compounds such as cannabidiol, for which some health benefits have been claimed. The impression that certain ENNDS might be beneficial for health, however, might distract from and confuse current assessments and should be avoided. Use of

ENDS to combine nicotine with other pharmacologically or physiologically active compounds should be prohibited, as this could increase the attractiveness of nicotine consumption. The terminology and definition of ENNDS should be specified, to identify them as products that are the same as ENDS products but without nicotine. This should be applied to both devices and consumables. The term "ENNDS" would still cover conventional e-liquids offered as nicotine-free versions. The regulatory framework for ENDS and ENNDS might still include other electronic inhalation systems as related products, thus defining all devices equally. A clear separation of nicotine and tobacco substitutes from other electronic inhalation products would, however, be beneficial, for several reasons. First, specific rules could be adopted for ENDS, ENNDS and other electronic inhalation systems used to inhale other substances or materials. Second, the clarification would require specific risk communication, thus preventing a misleading generalization, as observed early in the EVALI episode. Third, regulators might gain more flexibility for dealing with any novel inhalation systems developed in the future.

It should be noted that quantification of health risks is not a static outcome but remains an estimation based on the available knowledge. Information on ENDS and ENNDS is increasing, as is, probably even more important for ENDS and ENNDS, the wide variety of devices, user settings and e-liquids, which will influence health risks. Providing public information on the health effects of ENDS and ENNDS is a challenge. It is difficult to convey to the general public the changing (relative) risks over time due to differences in devices or information. Furthermore, information on health risks could be used by the industry to promote alternative products inappropriately.

Several of the methods described in this paper are promising for assessing the risks of ENDS, ENNDS and HTPs, although probably more than one model will be required for a full assessment. At this time, not enough scientific data are available to make definitive assessments. As many of the methods require a substantial amount of data on hazards and exposure, we could prepare for the future by collecting those data and conducting standardized assays so that the results are suitable for feeding into a database for future use. Non-targeted screening can be used to identify product-specific compounds, and their hazard could be derived from the database. Information on the relation between actual human exposure and the occurrence of adverse effects is necessary for risk characterization. Development of risk assessment models should continue, and, at some point, they should be validated with human data. Models of airway deposition should also be developed for application in risk assessment, as this is a crucial step between emission quantification and hazard characterization. Ultimately, this would require only chemical analysis of a novel product, which, combined with models of deposition and risk assessment, would allow determination of the health effects.

8.6 Discussion

A causal relation between ENDS or ENNDS use and acute effects (short-term health risk) is generally easier to identify, as the time between exposure and effect is short. In many cases, when users stop using the product the adverse effects are reversed. Assessment of the health risk of ENDS and ENNDS would benefit from data on health effects in long-time e-cigarette users; unfortunately, such data are not yet available, as e-cigarettes have not been available for the time necessary to develop chronic health effects such as cancer. In addition, current ENDS and ENNDS users are often former smokers. Thus, if an ENDS or ENNDS user develops disease, it may be a delayed effect of smoking and not necessarily related to ENDS and ENNDS use. The most robust data for assessing health risk would be for ENDS or ENNDS users who are not former smokers. The wide variety of devices and e-liquids, which continue to evolve rapidly, and the lack of information on the products in current use obviate conclusions on the risk posed by the group of ENDS and ENNDS as a whole. The lack of long-term data and of information on non-smokers may change over time as the products remain on the market for longer and if more non-smokers start using ENDS and ENNDS. The variation in products is not expected to stop; on the contrary, more and more products are entering the market. On the Dutch market, as in other countries, more than 20 000 different e-liquids are already available. A pragmatic approach would be to identify compounds in e-liquids or device settings that adversely affect health, as mentioned previously. The health risks of individual compounds could indicate the health risk of the product as a whole, and such information could inform policy on permissible constituents, device design features and the levels of certain constituents.

Risk modelling, epidemiological studies and assessments of individual compounds have resulted in consensus among some experts that unadulterated ENDS are less harmful than conventional cigarettes *(131)*. Nevertheless, a number of health risks remain, as acknowledged and summarized by WHO *(10)*. Although quantitative assessments remain difficult *(70)*, ENDS use results in a substantial decrease in carcinogenic risk, as confirmed by modelling *(24)*. ENDS are, however, associated with a high risk of nicotine addiction and increased risks for respiratory diseases and other adverse health effects that are especially relevant for children, adolescents and people who have never smoked. The rapid increase in use of pod systems in the USA is a concern *(16)*; however, the potential benefits for established smokers who use ENDS as a substitute for cigarettes should also be considered. For example, patients with chronic obstructive pulmonary disease who switched to ENDS showed improvement *(80)*. There is some evidence that ENDS are useful for smoking cessation *(132)*; however, a general conclusion is not scientifically justified, as ENDS and ENNDS are a heterogeneous, changing group of products *(99)*. Communication of health risks should avoid imprecise generalizations, especially because of the diversity of the products and their possible effects.

ENDS and ENNDS are often still considered as a single group of products in terms of risk assessment and legislation. Regulators and ENDS users should, however, be aware of the highly variable risks, which might range from very low to levels comparable to those for HTPs with devices that reach ≥ 250 °C. The health risks even of individual compounds are highly variable, depending on the device and performance settings. Major subgroups of ENDS (i.e. pod, sub-ohm) should therefore be categorized separately to ensure more specific terms of reference for both risk assessment and regulation.

Hazardous ingredients or device settings that lead to formation of (more) hazardous compounds could also be regulated. Although this would be beneficial from a health perspective, toxicological information on exposure to e-liquid ingredients by inhalation is very limited. Regulation or prohibition of specific ingredients on the basis of toxicological information might be a useful option. As noted above, flexible strategies will be required to cover different product groups and open and closed systems. Importantly, restrictions only on liquids should not imply that other ingredients are safe, as their risks especially in the context of inhalation are not yet known.

Furthermore, ENDS and ENNDS should be distinguished from inhalation devices used to deliver cannabidiol and other substances for their pharmacological or physiological effects and responses. Health risks and injuries caused, for example, by unconventional liquid constituents and modified devices that can aerosolize oils and waxes or are used to inhale illicit drugs must not be considered to be adverse outcomes related to vaping or e-cigarette use. Combination of nicotine with other physiologically or pharmacologically active compounds in ENDS should be prohibited, as this could increase the attractiveness of nicotine consumption. Tobacco control should include definition and specification of the framework for assessing ENDS and ENNDS according to these new challenges.

8.7 Recommendations

Risk assessment of mixtures is of great interest, not only for ENDS and ENNDS but also for tobacco products. Several models are available that can be used to assess the risk of mixtures, although most address carcinogenic effects. Implementation is limited by lack of data, which also implies that the models cannot yet be validated with data on human use. A few recommendations based on the findings of this review are listed below.

- Data should be collected on the emissions, toxicity, use and effects of ENDS, ENNDS and HTPs on exposed populations for application of quantitative methods of risk assessment.
- Characterization of toxicants should include non-targeted screening approaches to identify product-specific compounds that are not usually measured in tobacco smoke.

- Appropriate studies in experimental animals and epidemiological studies should be conducted on long-term adverse effects and of product switching.
- The models used to justify any claim of a positive health impact of HTPs and in their marketing should be verified.

Each product and change in product may result in a change in risk. In addition, users can flexibly adapt the systems to their requirements, which may alter their exposure from that used in risk assessments and change their health risk. Other recommendations for consideration by regulators are listed below.

- Limit product variations.
- Regulators should define subcategories or classes and terminology for ENDS and ENNDS as a basis for differentiated risk assessments. Different product groups, such as "single-use" products, open, refillable and highly powered systems, might warrant specified technical standards or regulatory approaches to minimize health risks and addictiveness.
- Neither ENDS nor ENNDS should contain any compound that might mediate drug-related effects or potentially lead to health effects, except for nicotine in ENDS.
- The term "ENNDS" should not be applied to electronic "vaping" products that contain pharmacologically active ingredients such as cannabidiol or hemp oils.

8.8 Conclusions

ENDS and ENNDS are highly variable categories of products, although they share many features. The variation in e-liquids and devices makes it impossible to assess the health risks of this group of products. Risks should therefore be assessed for each individual device, liquid and use. Alternatively, indicators of risk could be highlighted, such as specific ingredients or specific settings of a device, that could be applied to many products.

Several approaches have been used to quantify the health risk of tobacco products, either the absolute risk or that relative to a tobacco cigarette. Currently, the most promising approaches are those based on the relative potency of compounds in the emissions. Their applicability depends, however, on the availability of data, which are often limited. None of the methods is ready to be used in regulation, although this may be possible in due time, and care should be used in communicating health risks to the general public to ensure that the message is clear.

8.9 References

1. Tobacco. Geneva: World Health Organization; 2019 (https://www.who.int/news-room/fact-sheets/detail/tobacco, accessed 10 January 2021).
2. Sinha DN, Palipudi KM, Gupta PC, Singhal S, Ramasundarahettige C, Jha P et al. Smokeless tobacco use: a meta-analysis of risk and attributable mortality estimates for India. Indian J Cancer. 2014;51(Suppl 1):S73–7.
3. Slob W, Soeteman-Hernández LG, Bil W, Staal YCM, Stephens WE, Talhout R. A method for comparing the impact on carcinogenicity of tobacco products: a case study on heated tobacco versus cigarettes. Risk Anal. 2020;40(7):1355–66.
4. Fowles J, Dybing E. Application of toxicological risk assessment principles to the chemical constituents of cigarette smoke. Tob Control. 2003;12(4):424–30.
5. The scientific basis of tobacco product regulation. Second report of a WHO study group (WHO Technical Report Series, No. 951). Geneva: World Health Organization; 2008:277 (https://www.who.int/tobacco/global_interaction/tobreg/publications/9789241209519.pdf?ua=1, accessed 10 January 2021).
6. Burns DM, Dybing E, Gray N, Hecht S, Anderson C, Sanner T et al. Mandated lowering of toxicants in cigarette smoke: a description of the World Health Organization TobReg proposal. Tob Control. 2008;17(2):132–41.
7. Work in progress in relation to Articles 9 and 10 of the WHO FCTC. Geneva: World Health Organization; 2014:21 (https://apps.who.int/gb/fctc/PDF/cop4/FCTC_COP4_ID2-en.pdf, accessed 10 January 2021).
8. Hoffmann D, Hoffmann I, El-Bayoumy K. The less harmful cigarette: a controversial issue. a tribute to Ernst L. Wynder. Chem Res Toxicol. 2001;14(7):767–90.
9. Mallock N, Pieper E, Hutzler C, Henkler-Stephani F, Luch A. Heated tobacco products: A review of current knowledge and initial assessments. Front Public Health. 2019;7:287.
10. Electronic nicotine delivery systems and electronic non-nicotine delivery systems (ENDS and ENNDS). Geneva: World Health Organization; 2016.
11. Allen JG, Flanigan SS, LeBlanc M, Vallarino J, MacNaughton P, Stewart JH et al. Flavoring chemicals in e-cigarettes: Diacetyl, 2,3-pentanedione, and acetoin in a sample of 51 products, including fruit-, candy-, and cocktail-flavored e-cigarettes. Environ Health Perspect. 2016;124(6):733–9.
12. Hutzler C, Paschke M, Kruschinski S, Henkler F, Hahn J, Luch A. Chemical hazards present in liquids and vapors of electronic cigarettes. Arch Toxicol. 2014;88(7):1295–308.
13. Noel JC, Rainer D, Gstir R, Rainer M, Bonn G. Quantification of selected aroma compounds in e-cigarette products and toxicity evaluation in HUVEC/Tert2 cells. Biomed Chromatogr. 2020;34(3):e4761.
14. Williams M, Talbot P. Design features in multiple generations of electronic cigarette atomizers. Int J Environ Res Public Health. 2019;16(16):e2904.
15. Yingst J, Foulds J, Zurlo J, Steinberg MB, Eissenberg T, Du P. Acceptability of electronic nicotine delivery systems (ENDS) among HIV positive smokers. AIDS Care. 2020;32(10):1224–8.
16. Fadus MC, Smith TT, Squeglia LM. The rise of e-cigarettes, pod mod devices, and JUUL among youth: Factors influencing use, health implications, and downstream effects. Drug Alcohol Depend. 2019;201:85–93.
17. Walley SC, Wilson KM, Winickoff JP, Groner J. A public health crisis: Electronic cigarettes, vape, and JUUL. Pediatrics. 2019;143(6):e20182741.
18. Goniewicz ML, Boykan R, Messina CR, Eliscu A, Tolentino J. High exposure to nicotine among adolescents who use Juul and other vape pod systems ("pods"). Tob Control. 2019;28(6):676–7.
19. Giroud C, de Cesare M, Berthet A, Varlet V, Concha-Lozano N, Favrat B. e-Cigarettes: A review of new trends in cannabis use. Int J Environ Res Public Health. 2015;12(8):9988–10008.

20. Pisanti S, Malfitano AM, Ciaglia E, Lamberti A, Ranieri R, Cuomo G et al. Cannabidiol: State of the art and new challenges for therapeutic applications. Pharmacol Ther. 2017;175:133–50.
21. Poklis JL, Mulder HA, Peace MR. The unexpected identification of the cannabimimetic, 5F-ADB, and dextromethorphan in commercially available cannabidiol e-liquids. Forensic Sci Int. 2019;294:e25–7.
22. Owusu D, Huang J, Weaver SR, Pasch KE, Perry CL. Patterns and trends of dual use of e-cigarettes and cigarettes among US adults, 2015–2018. Prev Med Rep. 2019;16:101009.
23. Godtfredsen NS, Prescott E, Osler M. Effect of smoking reduction on lung cancer risk. JAMA. 2005;294(12):1505–10.
24. Stephens WE. Comparing the cancer potencies of emissions from vapourised nicotine products including e-cigarettes with those of tobacco smoke. Tob Control. 2018;27(1):10–7.
25. Mayer B. How much nicotine kills a human? Tracing back the generally accepted lethal dose to dubious self-experiments in the nineteenth century. Arch Toxicol. 2014;88(1):5–7.
26. Seo AD, Kim DC, Yu HJ, Paulus MC, van Heel DAM, Bomers BHA et al. Accidental ingestion of e-cigarette liquid nicotine in a 15-month-old child: an infant mortality case of nicotine intoxication. Korean J Pediatr. 2016;59(12):490–3.
27. Wang B, Liu S, Persoskie A. Poisoning exposure cases involving e-cigarettes and e-liquid in the United States, 2010–2018. Clin Toxicol. 2020;58(6):488–94.
28. Morley S, Slaughter J, Smith PR. Death from ingestion of e-liquid. J Emerg Med. 2017;53(6):862–4.
29. Viswam D, Trotter S, Burge PS, Walters GI. Respiratory failure caused by lipoid pneumonia from vaping e-cigarettes. BMJ Case Rep. 2018;2018:bcr2018224350.
30. Benowitz NL, Burbank AD. Cardiovascular toxicity of nicotine: Implications for electronic cigarette use. Trends Cardiovasc Med. 2016;26(6):515–23.
31. Messner B, Bernhard D. Smoking and cardiovascular disease: mechanisms of endothelial dysfunction and early atherogenesis. Arterioscler Thromb Vasc Biol. 2014;34(3):509–15.
32. Hansson J, Galanti MR, Hergens MP, Fredlund P, Ahlbom A, Alfredsson L et al. Use of snus and acute myocardial infarction: pooled analysis of eight prospective observational studies. Eur J Epidemiol. 2012;27(10):771–9.
33. MacDonald A, Middlekauff HR. Electronic cigarettes and cardiovascular health: What do we know so far? Vasc Health Risk Manag. 2019;15:159–74.
34. Boffetta P, Straif K. Use of smokeless tobacco and risk of myocardial infarction and stroke: Systematic review with meta-analysis. BMJ. 2009;339:b3060.
35. Teo KK, Ounpuu S, Hawken S, Pandey MR, Valentin V, Hunt D et al. Tobacco use and risk of myocardial infarction in 52 countries in the INTERHEART study: a case–control study. Lancet. 2006;368:647–58.
36. Babic M, Schuchardt M, Tölle M, van der Giet M. In times of tobacco-free nicotine consumption: The influence of nicotine on vascular calcification. Eur J Clin Invest. 2019;49(4):e13077.
37. The 2004 United States Surgeon General's report: The health consequences of smoking. New S Wales Public Health Bull. 2004;15(5–6):107.
38. Dwyer JB, McQuown SC, Leslie FM. The dynamic effects of nicotine on the developing brain. Pharmacol Ther. 2009;122(2):125–39.
39. Joschko MA, Dreosti IE, Tulsi RS. The teratogenic effects of nicotine in vitro in rats: A light and electron microscope study. Neurotoxicol Teratol. 1991;13(3):307–16.
40. Bruin JE, Gerstein HC, Holloway AC. Long-term consequences of fetal and neonatal nicotine exposure: a critical review. Toxicol Sci. 2010;116(2):364–74.
41. Bergman BC, Perreault L, Hunerdosse D, Kerege A, Playdon M, Samek AM et al. Novel and reversible mechanisms of smoking-induced insulin resistance in humans. Diabetes. 2012;61(12):3156–66.
42. Hess R, Bartels MJ, Pottenger LH. Ethylene glycol: an estimate of tolerable levels of exposure

based on a review of animal and human data. Arch Toxicol. 2004;78(12):671–80.
43. Chung S, Baumlin N, Dennis JS, Moore R, Salathe SF, Whitney PL et al. Electronic cigarette vapor with nicotine causes airway mucociliary dysfunction preferentially via TRPA1 receptors. Am J Respir Crit Care Med. 2019;200(9):1134–45.
44. Pisinger C, Døssing M. A systematic review of health effects of electronic cigarettes. Prev Med. 2014;69:248–60.
45. Wieslander G, Norbäck D, Lindgren T. Experimental exposure to propylene glycol mist in aviation emergency training: Acute ocular and respiratory effects. Occup Environ Med. 2001;58(10):649–55.
46. Talih S, Salman R, Karaoghlanian N, El-Hellani A, Saliba N, Eissenberg T et al. "Juice monsters": sub-ohm vaping and toxic volatile aldehyde emissions. Chem Res Toxicol. 2017;30(10):1791–3.
47. Duell AK, McWhirter KJ, Korzun T, Strongin RM, Peyton DH. Sucralose-enhanced degradation of electronic cigarette liquids during vaping. Chem Res Toxicol. 2019;32(6):1241–9.
48. Matura M, Goossens A, Bordalo O, Garcia-Bravo B, Magnusson K, Wrangsjö A et al. Oxidized citrus oil (R-limonene): a frequent skin sensitizer in Europe. J Am Acad Dermatol. 2002;47(5):709–14.
49. Sommerfeld CG, Weiner DJ, Nowalk A, Larkin A. Hypersensitivity pneumonitis and acute respiratory distress syndrome from e-cigarette use. Pediatrics. 2018;141(6):e20163927.
50. Gallagher J. Vaping nearly killed me, says British teenager. BBC News, 12 November 2019 (https://www.bbc.com/news/health-50377256).
51. Chen W, Wang P, Ito K, Fowles J, Shusterman D, Jaques PA et al. Measurement of heating coil temperature for e-cigarettes with a "top-coil" clearomizer. PLoS One. 2018;13(4):e0195925.
52. Farsalinos KE, Voudris V, Spyrou A, Poulas K. e-Cigarettes emit very high formaldehyde levels only in conditions that are aversive to users: A replication study under verified realistic use conditions. Food Chem Toxicol. 2017;109(1):90–4.
53. Gillman IG, Pennington ASC, Humphries KE, Oldham MJ. Determining the impact of flavored e-liquids on aldehyde production during vaping. Regul Toxicol Pharmacol. 2020;112:104588.
54. Talhout R, Opperhuizen A, van Amsterdam JG. Role of acetaldehyde in tobacco smoke addiction. Eur Neuropsychopharmacol. 2007;17(10):627–36.
55. Patel M, Cuccia A, Willett J, Zhou Y, Kierstead EC, Czaplicki L et al. JUUL use and reasons for initiation among adult tobacco users. Tob Control. 2019;28(6):681–4.
56. Chaumont M, van de Borne P, Bernard A, Van Muylem A, Deprez G, Ullmo J et al. Fourth generation e-cigarette vaping induces transient lung inflammation and gas exchange disturbances: results from two randomized clinical trials. Am J Physiol Lung Cell Mol Physiol. 2019;316(5):L705–19.
57. Gotts JE. High-power vaping injures the human lung. Am J Physiol Lung Cell Mol Physiol. 2019;316(5):L703–4.
58. Gray N, Halstead M, Gonzalez-Jimenez N, Valentin-Blasisni L, Watson C, Pappas RS. Analysis of toxic metals in liquid from electronic cigarettes. Int J Environ Res Public Health. 2019;16(22):4450.
59. Farsalinos KE, Rodu B. Metal emissions from e-cigarettes: A risk assessment analysis of a recently-published study. Inhal Toxicol. 2018;30(7–8):321–6.
60. Olmedo P, Goessler W, Tanda S, Grau-Perez M, Jarmul S, Aherrera A et al. Metal concentrations in e-cigarette liquid and aerosol samples: The contribution of metallic coils. Environ Health Perspect. 2018;126(2):027010.
61. Zhao D, Navas-Acien A, Ilievski V, Slavkovich V, Olmedo P, Adria-Mora B et al. Metal concentrations in electronic cigarette aerosol: Effect of open-system and closed-system devices and power settings. Environ Res. 2019;174:125–34.
62. Haddad C, Salman R, El-Hellani A, Talih S, Shihadeh A, Saliba NA. Reactive oxygen species emissions from supra- and sub-ohm electronic cigarettes. J Anal Toxicol. 2019;43(1):45–50.

63. Goniewicz ML, Knysak J, Gawron M, Kosmider L, Sobczak A, Kurek J et al. Levels of selected carcinogens and toxicants in vapour from electronic cigarettes. Tob Control. 2014;23(2):133–9.
64. Goniewicz ML, Smith DM, Edwards KC, Blount BC, Caldwell KL, Feng J et al. Comparison of nicotine and toxicant exposure in users of electronic cigarettes and combustible cigarettes. JAMA Netw Open. 2018;1(8):e185937.
65. Belushkin M, Tafin Djoko D, Esposito M, Korneliou A, Jeannet C, Lazzerini M et al. Selected harmful and potentially harmful constituents levels in commercial e-cigarettes. Chem Res Toxicol. 2020: doi:10.1021/acs.chemrestox.9b00470.
66. Kennedy CD, van Schalkwyk MCI, McKee M, Pisinger C. The cardiovascular effects of electronic cigarettes: A systematic review of experimental studies. Prev Med. 2019;127:105770.
67. Kuntic M, Oelze M, Steven S, Kröller-Schön S, Stamm P, Kalinovic S et al. Short-term e-cigarette vapour exposure causes vascular oxidative stress and dysfunction: Evidence for a close connection to brain damage and a key role of the phagocytic NADPH oxidase (NOX-2). Eur Heart J. 2019;41(26):2472–83.
68. Caporale A, Langham MC, Guo W, Johncola A, Chatterjee S, Wehrli SW. Acute effects of electronic cigarette aerosol inhalation on vascular function detected at quantitative MRI. Radiology. 2019;293(1):97–106.
69. Skotsimara G, Antonopoulos AS, Oikonomou E, Siasos G, Ioakeimidis N, Tsalamandris S et al. Cardiovascular effects of electronic cigarettes: A systematic review and meta-analysis. Eur J Prev Cardiol. 2019;26(11):1219–28.
70. D'Amario D, Migliaro S, Borovac JA, Vergallo R, Galli M, Restivo A et al. Electronic cigarettes and cardiovascular risk: Caution waiting for evidence. Eur Cardiol. 2019;14(3):151–8.
71. Ndunda PM, Muutu TM. Electronic cigarette use is associated with a higher risk of stroke. Stroke. 2019;50(Suppl 1):A9.
72. Alzahrani T, Pena I, Temesgen N, Glantz SA. Association between electronic cigarette use and myocardial infarction. Am J Prev Med. 2018;55(4):455–61.
73. Osei AD, Mirbolouk M, Orimoloye OA, Dzaye O, Iftekhar Uddin SM, Dardari ZA et al. The association between e-cigarette use and asthma among never combustible cigarette smokers: Behavioral risk factor surveillance system (BRFSS) 2016 & 2017. BMC Pulmon Med. 2019;19(1):180.
74. George J, Hussain M, Vadiveloo T, Ireland S, Hopkinson P, Struthers AD et al. Cardiovascular effects of switching from tobacco cigarettes to electronic cigarettes. J Am Coll Cardiol. 2019;74(25):3112–20.
75. Jordt SE, Jabba S, Ghoreshi K, Smith GJ, Morris JB. Propylene glycol and glycerin in e-cigarettes elicit respiratory irritation responses and modulate human sensory irritant receptor function. Am J Respir Crit Care Med. 2019;201:A4169.
76. Meo SA, Ansary MA, Barayan FR, Almusallam AS, Almehaid AM, Alarifi NS et al. Electronic cigarettes: Impact on lung function and fractional exhaled nitric oxide among healthy adults. Am J Mens Health. 2019;13(1):1557988318806073.
77. Stockley J, Sapey E, Gompertz S, Edgar R, Cooper B. Pilot data of the short-term effects of e-cigarette vaping on lung function. Eur Respir J. 2018;52(Suppl 62):PA2420.
78. Rebuli ME, Glista-Baker E, Speen AM, Hoffmann JR, Duffney PF, Pawlak E et al. Nasal mucosal immune response to infection with live-attenuated influenza virus (LAIV) is altered with exposure to e-cigarettes and cigarettes. Am J Respir Crit Care Med. 2019;199(9):A4170.
79. Wills TA, Pagano I, Williams RJ, Tam EK. e-Cigarette use and respiratory disorder in an adult sample. Drug Alcohol Depend. 2019;194:363–70.
80. Polosa R, Morjaria JB, Prosperini U, Russo C, Pennisi A, Puleo R et al. Health effects in COPD smokers who switch to electronic cigarettes: A retrospective-prospective 3-year follow-up. Int J Chron Obstruct Pulmon Dis. 2018;13:2533–42.
81. Layden JE, Ghinai I, Pray I, Kimball A, Layer M, Tenforde MW et al. Pulmonary illness related to

e-cigarette use in Illinois and Wisconsin – Preliminary report. N Engl J Med. 2020;382(10):903–16.
82. Moritz ED, Zapata LB, Lekiachvili A, Glidden E, Annor FB, Werner AK et al. Update: Characteristics of patients in a national outbreak of e-cigarette, or vaping, product use-associated lung injuries – United States, October 2019. Morb Mortal Wkly Rep. 2019;68(43):985–9.
83. Outbreak of lung injury associated with e-cigarette use, or vaping. Atlanta (GA): Centers for Disease Control and Prevention; 2020 (https://www.cdc.gov/tobacco/basic_information/e-cigarettes/severe-lung-disease.html).
84. Kalininskiy A, Bach CT, Nacca NE, Ginsberg G, Marraffa J, Navarette KA et al. e-Cigarette, or vaping, product use associated lung injury (EVALI): case series and diagnostic approach. Lancet Respir Med. 2019;7(12):1017–26.
85. Blagev DP, Harris D, Dunn AC, Guidry DW, Grissom CK, Lanspa MJ. Clinical presentation, treatment, and short-term outcomes of lung injury associated with e-cigarettes or vaping: A prospective observational cohort study. Lancet. 2019;394(10214):2073–83.
86. Blount BC, Karwowski MP, Morel-Espinosa M, Rees J, Sosnoff C, Cowan E et al. Evaluation of bronchoalveolar lavage fluid from patients in an outbreak of e-cigarette, or vaping, product use-associated lung injury – 10 states, August–October 2019. Morb Mortal Wkly Rep. 2019;68(45):1040–1.
87. Boudi FB, Patel S, Boudi A, Chan C. Vitamin E acetate as a plausible cause of acute vaping-related illness. Cureus. 2019;11(12):e6350.
88. Blount BC, Karwowski MP, Shields PG, Morel-Espinosa M, Valentin-Blasini L, Gardner M et al. Vitamin E acetate in bronchoalveolar-lavage fluid associated with EVALI. N Engl J Med. 2020;382(8):697–705.
89. Navon L, Jones CM, Ghinai I, King BA, Briss PA, Hacker KA et al. Risk factors for e-cigarette, or vaping, product use-associated lung injury (EVALI) among adults who use e-cigarette, or vaping, products – Illinois, July–October 2019. Morb Mortal Wkly Rep. 2019;68(45):1034–9.
90. Lewis N, McCaffrey K, Sage K, Cheng CJ, Green J, Goldstein L et al. e-Cigarette use, or vaping, practices and characteristics among persons with associated lung injury – Utah, April–October 2019. Morb Mortal Wkly Rep. 2019;68(42):953–6.
91. Visser WF, Klerx WN, Cremers HWJM, Ramlal R, Schwillens PL, Talhout R. The health risks of electronic cigarette use to bystanders. Int J Environ Res Public Health. 2019;16(9):1525.
92. McAuley TR, Hopke PK, Zhao J, Babaian S. Comparison of the effects of e-cigarette vapor and cigarette smoke on indoor air quality. Inhal Toxicol. 2012;24(12):850–7.
93. Avino P, Scungio M, Stabile L, Cortelessa G, Buonnano G, Manigrasso M. Second-hand aerosol from tobacco and electronic cigarettes: Evaluation of the smoker emission rates and doses and lung cancer risk of passive smokers and vapers. Sci Total Environ. 2018;642:137–47.
94. Hess IM, Lachireddy K, Capon A. A systematic review of the health risks from passive exposure to electronic cigarette vapour. Public Health Res Pract. 2016;26(2):e2621617.
95. Czogala J, Goniewicz ML, Fidelus B, Zielinska-Danch W, Travers MJ, Sobczak A. Secondhand exposure to vapors from electronic cigarettes. Nicotine Tob Res. 2014;16(6):655–62.
96. Ballbe M, Martínez-Sánchez JM, Sureda X, Fu M, Pérez-Ortuña R, Pascual JA et al. Cigarettes vs. e-cigarettes: Passive exposure at home measured by means of airborne marker and biomarkers. Environ Res. 2014;135:76–80.
97. Protano C, Manigrasso M, Avino P, Vitali M. Second-hand smoke generated by combustion and electronic smoking devices used in real scenarios: Ultrafine particle pollution and age-related dose assessment. Environ Int. 2017;107:190–5.
98. Bayly JE, Bernat D, Porter L, Choi K. Secondhand exposure to aerosols from electronic nicotine delivery systems and asthma exacerbations among youth with asthma. Chest. 2019;155(1):88–93.
99. Smoking cessation: A report of the Surgeon General. Rockville (MD): Department of Health and Human Services; 2020 (https://www.hhs.gov/sites/default/files/2020-cessation-sgr-full-re-

port.pdf).
100. Kane DB, Asgharian B, Price OT, Rostami A, Oldham M. Effect of smoking parameters on the particle size distribution and predicted airway deposition of mainstream cigarette smoke. Inhal Toxicol. 2010;22(3):199–209.
101. Son Y, Mainelis G, Delnevo C, Wackowski OA, Schwander S, Meng Q. Investigating e-cigarette particle emissions and human airway depositions under various e-cigarette-use conditions. Chem Res Toxicol. 2020;33(2):343–52.
102. Beauval N, Verriele M, Garat A, Fronval I, Dusautoir R, Anthérieu S et al. Influence of puffing conditions on the carbonyl composition of e-cigarette aerosols. Int J Hyg Environ Health. 2019;222(1):136–46.
103. Hubbs AF, Kreiss K, Cummings KJ, Fluharty KL, O'Connell R, Cole A et al. Flavorings-related lung disease: A brief review and new mechanistic data. Toxicol Pathol. 2019;47(8):1012–6.
104. Clapp PW, Lavrich KS, van Heusden CA, Lazarowski ER, Carson JL, Jaspers I. Cinnamaldehyde in flavored e-cigarette liquids temporarily suppresses bronchial epithelial cell ciliary motility by dysregulation of mitochondrial function. Am J Physiol Lung Cell Mol Physiol. 2019;316(3):L470–86.
105. Bopp SK, Barouki R, Brack W, Dalla Costa S, Dorne JCM, Drakvik PE et al. Current EU research activities on combined exposure to multiple chemicals. Environ Int. 2018;120:544–62.
106. EFSA Scientific Committee, More SJ, Bampidis V, Benford D, Bennekou SH, Bragard C et al. Guidance on harmonised methodologies for human health, animal health and ecotoxicological risk assessment of combined exposure to multiple chemicals. EFSA J. 2019;17(3):5634.
107. Ukić S, Sigurnjak M, Cvetnić M, Markić M, Novak Stankov M et al. Toxicity of pharmaceuticals in binary mixtures: Assessment by additive and non-additive toxicity models. Ecotoxicol Environ Saf. 2019;185:109696.
108. Gao Y, Feng J, Kang L, Xu X, Zhu L. Concentration addition and independent action model: Which is better in predicting the toxicity for metal mixtures on zebrafish larvae. Sci Total Environ. 2018;610-611:442–50.
109. Jonker MJ, Svendsen C, Bedaux JJ, Bongers M, Kammenga JE. Significance testing of synergistic/antagonistic, dose level-dependent, or dose ratio-dependent effects in mixture dose-response analysis. Environ Toxicol Chem. 2005;24(10):2701–13.
110. Altenburger R, Backhaus T, Boedeker W, Faust M, Scholze M, Grimme LH. Predictability of the toxicity of multiple chemical mixtures to Vibrio fischeri: Mixtures composed of similarly acting chemicals. Environ Toxicol Chem. 2000;19(9):2341–7.
111. Baas J, Van Houte BP, Van Gestel CA, Kooijman SALM. Modeling the effects of binary mixtures on survival in time. Environ Toxicol Chem. 2007;26(6):1320–7.
112. Leeman WR, Krul L, Houben GF. Complex mixtures: Relevance of combined exposure to substances at low dose levels. Food Chem Toxicol. 2013;58:141–8.
113. EFSA Scientific Committee. Scientific opinion on exploring options for providing advice about possible human health risks based on the concept of threshold of toxicological concern (TTC). EFSA J. 2012;10(7).
114. Kawamoto T, Fuchs A, Fautz R, Morita O. Threshold of toxicological concern (TTC) for botanical extracts (botanical-TTC) derived from a meta-analysis of repeated-dose toxicity studies. Toxicol Lett. 2019;316:1–9.
115. Rietjens I, Cohen SM, Eisenbrand G, Fukushima S, Gooderham NJ, Guengerich FP et al. FEMA GRAS assessment of natural flavor complexes: Cinnamomum and Myroxylon-derived flavoring ingredients. Food Chem Toxicol. 2020;135:110949.
116. Tluczkiewicz I, Kühne R, Ebert RU, Batke M, Schüürmann G, Mangelsdorf I et al. Inhalation TTC values: A new integrative grouping approach considering structural, toxicological and mechanistic features. Regul Toxicol Pharmacol. 2016;78:8–23.
117. Schüürmann G, Ebert RU, Tluczkiewicz I, Escher SE, Kühne R. Inhalation threshold of toxico-

logical concern (TTC) – Structural alerts discriminate high from low repeated-dose inhalation toxicity. Environ Int. 2016;88:123–32.
118. The scientific basis of tobacco product regulation. Report of a WHO Study Group (WHO Technical Report Series, No. 945). Geneva: World Health Organization; 2007 (https://www.who.int/tobacco/global_interaction/tobreg/who_tsr.pdf, accessed 10 January 2021).
119. Wagner KA, Flora JW, Melvin MS, Avery KC, Ballentine RM, Brown AP et al. An evaluation of electronic cigarette formulations and aerosols for harmful and potentially harmful constituents (HPHCs) typically derived from combustion. Regul Toxicol Pharmacol. 2018;95:153–60.
120. Air Toxics Hot Spots Program risk assessment guidelines. Part II. Technical support document for cancer potency factors: Methodologies for derivation, listing of available values, and adjustments to allow for early life stage exposures. Sacramento (CA): California Environmental Protection Agency, Office of Environmental Health Hazard Assessment, Air Toxicology and Epidemiology Branch; 2008 (https://oehha.ca.gov/media/downloads/crnr/tsd062008.pdf).
121. Cunningham FH, Fiebelkorn S, Johnson M, Meredith C. A novel application of the margin of exposure approach: segregation of tobacco smoke toxicants. Food Chem Toxicol. 2011;49(11):2921–33.
122. Soeteman-Hernández LG, Bos PM, Talhout R. Tobacco smoke-related health effects induced by 1,3-butadiene and strategies for risk reduction. Toxicol Sci. 2013;136(2):566–80.
123. Risicobeoordeling lachgas [Risk assessment on nitrous oxide]. Bilthoven: Coördinatiepunt Assessment en Monitoring nieuwe drugs; 2019 (https://www.vvgn.nl/wp-content/uploads/2019/12/risicobeoordelingsrapport-lachgas-20191209-beveiligd.pdf).
124. Staal YCM, Havermans A, van Nierop L et al. Evaluation framework to qualitatively assess the health effects of new tobacco and related products [submitted 2020].
125. Sosnowski TR, Kramek-Romanowska K. Predicted deposition of e-cigarette aerosol in the human lungs. J Aerosol Med Pulm Drug Deliv. 2016;29(3):299–309.
126. Savareear B, Escobar-Arnanz J, Brokl M, Saxton MJ, Wright C, Liu C et al. Non-targeted analysis of the particulate phase of heated tobacco product aerosol and cigarette mainstream tobacco smoke by thermal desorption comprehensive two-dimensional gas chromatography with dual flame ionisation and mass spectrometric detection. J Chromatogr A. 2019;1603:327–37.
127. Salman R, Talih S, El-Hage R, Haddad C, Karaoghlanian N, El-Hellani A et al. Free-base and total nicotine, reactive oxygen species, and carbonyl emissions from IQOS, a heated tobacco product. Nicotine Tob Res. 2019;21(9):1285–8.
128. Cancelada L, Sleiman M, Tang X, Russell ML, Montesinos VN, Litter MI et al. Heated tobacco products: Volatile emissions and their predicted impact on indoor air quality. Environ Sci Technol. 2019;53(13):7866–76.
129. Bentley MC, Almstetter M, Arndt D, Knorr A, Martin E, Pospisil P et al. Comprehensive chemical characterization of the aerosol generated by a heated tobacco product by untargeted screening. Anal Bioanal Chem. 2020;412(11):2675–85.
130. Apelberg BJ, Feirman SP, Salazar E, Corey CG, Ambrose BK, Paredes A et al. Potential public health effects of reducing nicotine levels in cigarettes in the United States. N Engl J Med. 2018;378(18):1725–33.
131. Chen J, Bullen C, Dirks K. A comparative health risk assessment of electronic cigarettes and conventional cigarettes. Int J Environ Res Public Health. 2017;14(4):382.
132. Hajek P, Phillips-Waller A, Przulj D, Pesola F, Myers Smith K, Bisal N et al. A randomized trial of e-cigarettes versus nicotine-replacement therapy. N Engl J Med. 2019;380(7):629–37.

9. Flavours in novel and emerging nicotine and tobacco products

Ahmad El-Hellani, Department of Chemistry, Faculty of Arts and Sciences, American University of Beirut, Beirut, Lebanon; Center for the Study of Tobacco Products, Virginia Commonwealth University, Richmond (VA), USA

Danielle Davis, Department of Psychiatry, Yale School of Medicine, Yale University, New Haven (CT), USA

Suchitra Krishnan-Sarin, Department of Psychiatry, Yale School of Medicine, Yale University, New Haven (CT), USA

Contents

Abstract
9.1 Background and introduction
9.2 Epidemiology (frequency, patterns and reasons for use by sociodemographic variables) of flavoured products
    9.2.1 Electronic nicotine and non-nicotine delivery systems
    9.2.2 Traditional smoked and smokeless tobacco products
    9.2.3 Heated tobacco products
9.3 Effects of flavours on appeal, experimentation, uptake and sustained use
    9.3.1 Adolescents and young adults
    9.3.2 Adults
9.4 Common flavours, properties, health effects and implications for public health
    9.4.1 Common flavours in electronic nicotine delivery systems (ENDS) and tobacco products
    9.4.2 Chemical and physical properties of common flavours in flavoured products
    9.4.3 Toxicity of flavours
9.5 Regulation of flavoured products
    9.5.1 Global regulation of flavoured ENDS
    9.5.2 Global regulation of flavoured tobacco products
    9.5.3 Pros and cons of common approaches
    9.5.4 Impact of regulation of specific flavoured products among all nicotine products
    9.5.5 Future regulation of flavours
9.6 Discussion
9.7 Conclusions
9.8 Research gaps, priorities and questions
9.9 Policy recommendations
9.10 References

Abstract

Nicotine and tobacco products contain characterizing flavours that mask their harshness, ease their use and increase their acceptability. Recent estimates indicate that flavour use is common, and thousands of flavours are now available for electronic nicotine delivery systems. Non-traditional flavours, like fruit and confectionery, are particularly appealing to young people, and use of these flavours is also increasing among adults, especially among adult smokers who are trying to quit smoking. Flavours in all nicotine and tobacco products have been shown not only to increase the appeal and first use of the products but may also contribute to extent of use, progression from experimental to regular use and dependence. Another concern is that chemicals in flavours may contribute to the toxicity of these products. Flavours are not regulated uniformly in nicotine and tobacco products or among countries. Some countries ban all flavours in all products, others have bans on only some flavours (e.g. excluding menthol), while others include only certain products (e.g. traditional products, such as cigarettes and smokeless tobacco) in regulations. This report on flavours in nicotine and tobacco products calls for adoption of common terminology for flavours in nicotine and tobacco products and consideration of policy for reducing the availability of flavoured nicotine and tobacco products on the market to those for which there is clear evidence of benefit in assisting smokers in quitting use of traditional smoked tobacco products.

9.1 Background and introduction

Addition of flavours to tobacco products dates back to the nineteenth century, when fruity flavours were added to smokeless tobacco products *(1)*. Now, almost all nicotine and tobacco products, including traditional smoked tobacco products such as cigarettes and cigars, newer products such as snuff and snus, heated tobacco products (HTPs) and electronic nicotine delivery systems (ENDS) are available with a variety of flavours, which contributes to their increased appeal, the prevalence of use and the perception of reduced harm *(2,3)*. Flavourings mask the harshness of nicotine and tobacco, ease their use and also reduce second-hand exposure to harsh odours, thus increasing their acceptability and potentially leading to progression and maintenance of dependence on these products. Flavourings also enhance the appeal of nicotine and tobacco products to novice users and vulnerable populations, thus increasing initiation and progression in the use of these products *(4)*. The dual role that flavours play in enticing novice users, especially young people, to initiate nicotine and tobacco product use and in prolonging use by current users should be addressed by health authorities and public health communities to ensure the best regulations to address the inclusion of flavours in nicotine and tobacco products *(5–8)*.

Flavoured nicotine and tobacco products are used all over the world. There is limited systematic evidence about their availability and use globally, and preferences for such products are often specific to countries and regions. For example, traditional cigarettes that contain cloves and oils (*kretek*) are highly popular in Indonesia, and spiced smokeless tobacco containing tobacco mixed with spices, oils, flavouring, betel nut and other ingredients (*pan masala, gutka*) is widely used in India. Hookah smoking, which involves heating heavily flavoured, sweetened tobacco, originated in India and the Middle East but has become increasingly popular among young people in Europe and North America.

Restrictions on the use of flavours in nicotine and tobacco products differ by country. Some countries, such as Brazil, Chile, Ethiopia, the Republic of Moldova, and some Canadian provinces restrict all flavours in nicotine and tobacco products, including menthol, although flavours that impart a port, wine, rum or whisky flavour are allowed in Canada *(9)*. In May 2020, the European Union, with 28 Member States, implemented a ban on menthol cigarettes and roll-your-own tobacco *(10)*. Brazil, Canada, Chile and Ethiopia include non-cigarette smoked products, such as little cigars and cigarillos, in flavour restrictions, while others, including those in the European Union, the Republic of Moldova and Turkey, do not extend the restrictions to products other than cigarettes and roll-your-own tobacco *(11)*. There is also variation in whether menthol is included in restrictions; for example, while Brazil has banned all characterizing flavours (defined as flavours with a taste or aroma, apart from tobacco, distinguishable before, during or after tobacco consumption *(12)*), including menthol, in all tobacco products, in the USA, characterizing flavours except menthol are banned only in cigarettes and roll-your-own tobacco in most jurisdictions *(13)*, although some localities (e.g. San Francisco, California) have banned all characterizing flavours, including menthol *(14)* in tobacco products. In the European Union, as in the Republic of Moldova, Turkey and the USA, this ruling does not extend to other tobacco products, such as cigars, cigarillos, little cigars and smokeless tobacco products *(15)*.

ENDS may be regulated differently from tobacco products, as they do not contain tobacco. While WHO does not consider them to be tobacco products *(16)*, some countries include these products under existing tobacco product laws, while others consider these products separately from tobacco products.

The sale of these products is banned in 30 countries, and several countries have regulations to restrict the availability of flavours *(17)* and limit the maximal nicotine concentration. Restrictions on the addition of flavours to pod-based ENDS were instituted in the USA in late 2019, but many states have local restrictions on sales of flavoured nicotine and tobacco products *(18)*. The sale of nicotine-containing e-cigarettes is banned in some countries, including Australia and Japan *(19)*. e-Cigarettes that do not contain nicotine are referred to as electronic non-nicotine delivery systems (ENNDS) and, depending on regulations, are still available with flavours. In countries in which ENDS are available, the rate of ENNDS use is reported to be low *(20)*; as a result, they are often included with ENDS in evaluations of flavoured electronic systems. In this review, we have distinguished the two where possible in descriptions of the patterns of ENDS use.

Since the introduction of ENDS onto the global market in 2003, there has been renewed interest in the role of flavours. The popularity of ENDS has grown globally, but they are arguably most popular in Canada, Europe and the USA *(21)*. ENDS are available in an ever-growing range of customizable e-liquid flavours. A study of the online ENDS marketplace in 2013–2014 found 7764 unique e-liquid flavours *(22)*, and a follow-up in 2018 showed that the figure had doubled, to 15 586 flavours *(23)*. The availability of ENDS has led to debate in the public health field about use of these devices to reduce harm by helping smokers quit use of more harmful traditional smoked tobacco products. The debate has carried over into one on the risk versus benefit of flavours in ENDS. Some researchers and advocates argue that addition of flavours to ENDS products is beneficial, as it may potentially help smokers to quit smoking *(24–26)*, while others argue that the presence of flavours only enhances the appeal of these devices to young people and leads to increased use and dependence *(26–29)*.

ENDS are not the only relatively new product on the market. Heated tobacco products (HTPs), which have unique characteristics but are tobacco products, emerged in their current iteration in 2013. These products do not have the variety of ENDS flavours but are marketed similarly, in that they are advertised as potential alternatives to traditional tobacco products.

Given the relatively recent appearance of HTPs on the market and the fact that they may be marketed as a lower-risk product, it is important to understand how flavours contribute to their appeal and use, especially among non-users and young people *(30)*.

This report is an update to that in the seventh report of the WHO Study Group on Tobacco Product Regulation *(31)*. We describe the epidemiology of flavoured products, the impact of flavours on appeal, experimentation and continual use of nicotine and tobacco products, the most commonly used flavours, their health effects and current regulation of flavours in these products.

9.2 Epidemiology of flavoured products (frequency, patterns and reasons for use by sociodemographic variables)

Although flavoured nicotine and tobacco products are used globally, there is limited systematic information on their use. Most of the information is for the USA, where surveys indicate high rates of use. The National Adult Tobacco Survey in 2013–2014 indicates that an estimated 10.2 million ENDS users (68.2%), 6.1 million hookah users (82.3%), 4.1 million cigar smokers (36.2%) and 4.0 million smokeless tobacco users (50.6%) had used flavoured products in the past 30 days *(32)*. In the same survey, among cigarette smokers, the use of cigarettes flavoured with menthol (the only characterizing flavour in cigarettes in the USA at the time of the survey) was relatively high, comprising 39% of cigarettes used *(33)*. Similar results were reported from a population-based survey in 2014–2015, which showed that 41.4% of nicotine and tobacco users reported use of flavoured products, ranging by product from 28.3% of cigar smokers to 87.2% of hookah users *(34)*.

Limited data exist on global use of flavoured nicotine and tobacco products and are sorely needed to evaluate the scope of flavour use worldwide. Information should be generated on use of flavoured nicotine and tobacco products elsewhere in the world. Although data for the USA provide a reasonable estimate, the types of flavoured nicotine and tobacco products used and regulation of these products differ by country, which may affect the use of different flavours. The patterns of use of flavoured tobacco products may differ according to their availability and popularity. Nevertheless, there is a higher prevalence of use among adolescents and young adults than among older adults. Evidence on the epidemiology of flavour use in different tobacco products is described below.

9.2.1 Electronic nicotine and non-nicotine delivery systems

Flavours are increasingly available for use in ENDS, because of both the dynamic growth in the number of flavours available *(23)* and their popularity among adolescents and young adults *(8)*. The Population Assessment of Tobacco and Health (PATH), a longitudinal national survey conducted in the USA, indicated that use of flavoured ENDS by current vapers was most prevalent among adolescents aged 12–17 (97%), followed by young adults aged 18–24 (97%) and adults aged ≥ 25 years (81%). A similar pattern was observed in initiation of ENDS products, 93% of adolescents, 84% of young adults and 55% of adults reporting initiation of ENDS with a flavoured product *(35)*. These patterns of use of flavoured ENDS are confirmed by other studies *(36–38)* and recent reviews *(39,40)*, which suggest that use of flavoured ENDS use is more prevalent among adolescents and young adults than among older adults. A report from the National Youth Tobacco Survey in the USA on tobacco product use by middle- and high-school students in 2019 indicated that use of flavours by current ENDS users was most prevalent among non-Hispanic white adolescents (77%) and similar among males (71%) and females (69%) *(41)*.

Use of non-traditional flavours (i.e. other than tobacco or menthol flavours) is more prevalent among adolescents and young adults; however, a shift in flavour preferences towards non-traditional flavours appears to be occurring in all age groups. Wave 2 of the PATH study suggested that fruit flavours were the category most commonly reported by adolescents, and menthol or mint, a flavour traditionally found in tobacco products, was most commonly reported by adults *(42)*. A follow-up study of wave 3 showed that fruit flavours were those most commonly used by both young and adult users and that dessert or confectionery was also highly popular among young and adult users *(29)*. These findings are consistent with other recent work; specifically, evidence that fruit flavours are the most highly endorsed flavour category across age groups, including youth, young adults and adults *(36,43,44)*. A study of long-term use of flavours in ENDS in the USA found that a preference for tobacco and menthol or mint flavours decreased over time, preference for fruit flavours remained stable, and preference for dessert and sweet flavours increased *(45)*.

Research has also been conducted on the flavour preferences of ENDS users who are current or former users of other tobacco products (i.e. cigarettes). This is important because, particularly in Europe and the USA, there is debate about whether and how flavours reduce the appeal of ENDS for young people or enhance their appeal for users of tobacco products, such as cigarette smokers, who wish to switch to ENDS. Fruit flavours appear to be popular regardless of the type of tobacco use. A global Internet survey of former and current adult cigarette smokers found that, while fruit and sweet flavours were the most popular (69% and 61%, respectively), current cigarette smokers were more likely to report use of tobacco flavours and less likely to report use of fruit and sweet flavours than former cigarette smokers *(24)*. In a study in New Zealand, fruit flavour was preferred by exclusive ENDS users, former cigarette smokers and current cigarette smokers *(46)*. Even as flavour preferences appear to be shifting to non-traditional flavours in all age groups, it is important to note that the prevalence of use of non-traditional flavours in ENDS products appears still to be the highest among young people *(47)*.

Many users of flavoured ENDS report that they use several flavours (e.g. fruit, dessert or confectionery, menthol or mint). A study of ENDS users in the USA suggested that multiple flavour use is more prevalent among adolescents (46%) than adults (32%) *(29)*. In India, adult ENDS users reported a relatively high rate (65%) of use of several flavours *(48)*. The global Internet survey of current and former adult cigarette smokers who used ENDS *(24)* indicated that switching between flavours was common; 68% reported switching at least daily, 16% weekly and 10% less than weekly.

Although there is limited evidence on use of flavoured ENNDS, a study among adolescents and young adults (18–29 years) in Japan, where these products are banned, indicated use by 4.3% of the group. Although flavour use was not directly studied, the main reason reported for using the product use was fruit flavours, suggesting that flavours are a reason for using ENNDS *(49)*.

9.2.2 Traditional smoked and smokeless tobacco products

Although ENDS offer by far the most flavours of all nicotine and tobacco products, flavours are also used in other nicotine and tobacco products, both smokeless and smoked. As for ENDS, use of flavoured tobacco products is more prevalent among adolescents and young adults than adults. In a nationally representative sample of Canadian young people who used tobacco products (cigarettes, pipes, cigars, cigarillos, *bidis*, smokeless tobacco, hookah, blunts, roll-your-own cigarettes), 52% reported using flavoured products *(50)*. Similarly, a national survey in Poland showed that younger smokers were more likely to use flavoured cigarettes *(51)*. In the PATH study in the USA, about half of users of all ages who reported current use of cigarillos and filtered cigars said that they used a flavoured product, while relatively low proportions of current cigar smokers reported use of flavoured products, from 24% of adults, 27% of adolescents to 36% of young adults. In the same study, the proportions of those who reported that their first tobacco product had been flavoured were 68% of adolescents, 63% of young adults and 42% of adults for cigarillos; and 56% of adolescents, 54% of young adults and 40% of adults for filtered cigars. Among traditional cigar users, the overall prevalence was lower, but the graded effect of age was still evident, with 39% of adolescents, 35% of young adults and 22% adults reporting that their first product had been flavoured *(35)*.

The types of flavours used in more traditional tobacco products were shown in 2013–2014 in the National Adult Tobacco Survey in the USA to be menthol or mint in smokeless tobacco (77%), fruit in hookah tobacco (74%); fruit (52%), confectionery, chocolate and other sweet flavours (22%) and alcohol (14%) in cigars, cigarillos and filtered little cigars; and fruit (57%), confectionery, chocolate and other sweet flavours (26%) and menthol or mint (25%) in pipe tobacco *(32)*.

Capsules have been inserted in cigarettes to incorporate flavours other than traditional tobacco and menthol, and capsule cigarettes appear to be capturing a growing portion of the global market *(52)*. The flavours range from menthol to green tea and whisky and others. Capsule cigarettes are popular among adolescents and young adult smokers. Over half of adolescent cigarette smokers in Australia reported capsule use *(53)*, and young adults in the United Kingdom and the USA expressed a preference for these products *(53,54)*. This may change, as flavour capsules have been banned in Brazil and the European Union since spring 2020 *(55)*.

9.2.3 Heated tobacco products

HTPs have been developed since the 1960s in the USA and globally since the 1980s; however, the early products were unsuccessful. A new generation of HTPs has been marketed since 2013 *(30)*, and they are currently available in about 50 countries, including Canada, Israel, Italy, Japan, the Republic of Korea and, most recently, the USA *(56,57)*. HTPs contain flavours, which may increase their appeal and use *(58)*. These devices produce aerosols containing nicotine and toxic chemicals when tobacco is heated or when a device containing tobacco is activated, and users inhale the aerosol by sucking or smoking. The products have been marketed as a safer alternative to combustible cigarettes *(59)*. They have fewer flavour options than ENDS, the main choices being either tobacco or menthol; other flavour options include "mentholated-fruity" and coffee.

Japan has been a recent test markets for these products. Data from wave 1 of the International Tobacco Control Japan Survey, collected in 2018, indicated that the prevalence of HTP use was only 2.7%. The popularity of flavours appeared to be similar among exclusive HTP users and HTP users who used other products, menthol flavour being the most popular, followed by tobacco and mentholated-fruity flavour *(60)*. Given the evidence that flavours increase the appeal of other nicotine and tobacco products, particularly for adolescents and young adults who do not use tobacco, the alarm has been sounded that these products might appeal as a function of their flavours *(58)*. Comprehensive research should be conducted on the use of flavours in these products as the market for HTPs is extended to other countries.

9.3 Effects of flavours on appeal, experimentation, uptake and sustained use (See also section 3.)

9.3.1 Adolescents and young adults

Appeal

As adolescent and young adult tobacco product users appear to use flavoured products and ENDS at greater rates than adults *(36,51)*, it is important to understand the appeal of these products to adolescents and how it may differ from that to adult users. Comparisons of smoked product use by age show that flavoured cigars, cigarettes and hookahs are more appealing to young than to adult users *(61,62)*. Young people have reported that flavours are the main reason for both initiation and continued use of ENDS *(63,64)*, and sweet and fruit-flavoured e-liquid solutions were more appealing to adolescents and young adults than non-sweet (e.g. tobacco-flavoured) e-liquids *(65,66)*. Similarly, young people reported that flavours were the main reason for cigarillo use *(67)*.

Notably, appeal for flavoured tobacco products may begin even before using the flavour. ENDS. Flavours in both smoked tobacco products and ENDS may appeal not only because of a positive experience associated with flavours *(68–70)* but also because they reduce the perceived risk of the harm of these products *(71–75)*. ENDS are not only provided in non-traditional tobacco flavours but are also advertised with colourful images and appealing descriptions of the flavours. Functional magnetic resonance imaging showed that, in young adults susceptible to using ENDS, just viewing advertisements showing fruit, mint and sweet flavours for ENDS products increased activity in the nucleus acumens to a greater extent than advertisements for tobacco flavours. Heightened activity was also seen when participants viewed advertisements for non-ENDS fruit, mint and sweet flavours, indicating that the appeal of non-traditional flavours may begin before an ENDS is sampled and the advertisements may lead to initiation *(76)*.

Experimentation, uptake and sustained use

Flavours may play an important role in initiation of ENDS use and progression by adolescents and young adults. Thus, initial use of flavoured ENDS was associated not only with continued use but also with more days of use, suggesting heavier use with time *(77)*. In another study, preference for specific flavours and the total number of flavours used were associated with more days of ENDS use by young people and not by adult users *(28)*, indicating that flavour preferences play a different role in adolescent use from that of adults. Sweet non-traditional flavours, in particular, appeal to young people and may contribute to the uptake and use of ENDS *(28,37,63,78,79)*. In a study in young people, use of flavoured ENDS was associated with non-traditional flavours (i.e. not tobacco, mint or menthol) and continuation but not with the number of days of ENDS use over time, suggesting that use of non-traditional flavours may sustain use *(37)*. This may be due in part to perceived sensory effects of non-traditional flavours. In a laboratory study of ENDS with six commercially available flavours, fruit flavours were considered the sweetest and tobacco flavour the most bitter. When flavours were rated for coolness and harshness, sweetness and coolness were positively correlated and harshness and bitterness negatively correlated with liking *(80)*.

Flavours in nicotine and tobacco products may increase their addictive potential, which will sustain their use *(81)*. It has been reported that green apple tobacco flavour in ENDS alters smoking behaviour, which may be associated with upregulation of nicotinic acetylcholine receptors *(82)*, as suggested by a study of the biological mechanisms by which menthol alters tobacco smoking behaviour, including reinforcing sensory cues associated with nicotine, upregulating nicotinic acetylcholine receptors and altering nicotine metabolism to increase its bioavailability *(83)*.

Initiation with flavoured ENDS results in continued, heavier use by young people, and this pattern may also occur with other tobacco products and in older age groups. Longitudinal results from the PATH study indicated that first use of a flavoured (i.e. menthol) cigarette was associated with continued cigarette use in adolescents, young adults and adult smokers, and, in young adults and adults, this pattern extended to other tobacco products. Thus, first use of a flavoured ENDS, cigar, cigarillo, filtered cigar, hookah or any smokeless tobacco was associated with continued use of the product *(38)*.

9.3.2 Adults
Appeal

In adults, flavours also appear to contribute to the appeal of tobacco products. As in young people, flavoured ENDS products in particular appear to be highly appealing. In studies of ENDS users in the Netherlands, New Zealand and the USA, ENDS were rated as appealing as a function of the availability of flavours

(46,84). A study of university students (from undergraduates to doctoral candidates) in the Asian–Pacific rim found that 34% of e-cigarettes users used the products because of the flavours offered *(85)*. Flavours may contribute to the appeal in adults by raising positive expectancy about the product *(86)* and an overall positive perception among both users and non-users *(87)*.

Experimentation, uptake and sustained use

Flavours play a role in uptake not only among adolescents and young adults but also among older adults. In a sample of adults (≥ 25 years), regular current use of flavoured cigarettes, cigars, cigarillos, filtered cigars, hookah, smokeless tobacco and e-cigarettes was associated with flavoured products but not with first use of a non-flavoured product, as among adolescents and younger adults *(38)*. Smokeless tobacco users had a similar pattern of flavour use, whereby those who started with an unflavoured product were likely to switch to a flavoured product, while those who started with a flavoured product were likely to continue using it *(88)*.

There is evidence in adults that the use of flavoured tobacco products leads to greater dependence, which contributes to sustained use. Two markers of dependence on tobacco products, daily use and time to first use in the morning, were associated with use of flavoured products in a survey in the USA in 2014–2015. The same study indicated that use of flavoured ENDS was more likely to be daily than use of unflavoured products. More users of flavoured cigars (large cigars, cigarillos and little cigars) than of unflavoured cigars reported first use in the morning within 30 min of waking *(89)*.

Flavoured products may sustain use because they influence the reward and reduce the aversiveness of nicotine, the dependence-producing constituent in these products *(90)*. There is extensive literature on the interaction of nicotine and menthol in combustible cigarettes, which suggests that menthol serves as a cue for the sensory effects of nicotine *(91)* and enhances both the reward from nicotine *(92)* and the withdrawal symptoms *(93)*. This finding may indicate why menthol cigarette use is growing among smokers of combustible cigarettes even as overall use decreases *(94)* and why smokers of menthol cigarettes have more difficulty in quitting smoking *(95)*. Fruit flavours in ENDS have been shown to enhance the reward and reinforcing effects of nicotine *(78,96)* and to suppress its aversive effects *(97)*. Menthol flavours improve the taste of e-liquids in ENDS and make higher nicotine concentrations less aversive and more rewarding *(96–98)*. Similar patterns have been observed for combustible products. Tobacco industry documents suggest that flavoured cigar products increase their appeal to naïve users by reducing throat irritation and making emissions easier to inhale *(99)*.

Another mechanism that may sustain use is the perception that, in general, flavoured tobacco products, including ENDS, are less harmful than unflavoured tobacco products. Flavours in ENDS give a false perception of safety not only to users but also to bystanders *(100)*. This attitude is, however, changing, as more voices rise for the inclusion of use of ENDS in designated smoke-free indoor areas, as second-hand aerosols from flavoured ENDS can leave pungent odours *(101)*. Chemical and toxicological assessment of second-hand aerosols from flavoured ENDS is lagging behind their excessive use indoors and in public spaces *(102,103)*.

Questions remain about the role of flavours in switching from conventional smoked products to other products, such as ENDS or HTPs. No studies have yet been reported on the relation between HTPs, flavours and switching behaviour. The flavours in ENDS may appeal to users of smoked products, and use of flavoured ENDS may be associated with a greater likelihood of short and longer attempts to quit use of smoked tobacco products *(104,105)*. There have been no experimental investigations of the specific role ENDS flavours play in switching from traditional smoked products, although many describe the abuse liability of ENDS in adolescents and young adults and in users of non-combustible products. Proof is required to determine the role flavours in ENDS play in product switching.

9.4 Common flavours, properties, health effects and implications for public health

9.4.1 Common flavours in electronic nicotine delivery systems and tobacco products

The National Institute for Public Health and the Environment in The Netherlands has published a "flavour library" of flavours added to tobacco cigarettes and roll-your-own products *(106)*. The flavours are listed in eight main categories: fruit, spice, herb, alcohol, menthol, sweet, floral and miscellaneous. A year later, a similar report was issued in which the authors classified flavours in ENDS into 13 categories: tobacco, menthol, fruit, dessert, alcohol, nut, spices, confectionery, coffee/tea, beverages, sweet-like flavours, unspecified and unflavoured *(107)*. As noted above, the number of unique ENDS flavours on the market has increased dramatically in recent years, from 7764 in 2013–2014 to 15 586 in 2016–2017 *(22,23)*. Unique to ENDS products, users commonly mix and match flavours in refillable ENDS *(108)*, and "do-it-yourself" is common practice, sometimes with the addition of illicit substances *(109)*, and researchers should also consider the impact of such additives on ENDS use.

9.4.2 Chemical and physical properties of common flavours in flavoured products

Flavour ingredients in tobacco cigarettes

The above-mentioned flavour library *(106)* includes chemical analyses of flavours in tobacco cigarettes and roll-your-own products. This complements analyses of the total tobacco matrix in cigarette filler. Identification of flavour ingredients in tobacco cigarettes and studies to assess their fate after pyrolysis have been reported *(110)*.

Flavour additives in ENDS

Several studies have reported chemical profiling of flavoured ENDS liquids *(111,112)*, indicating the complexity that flavours add to the ENDS matrix *(113)*. Although no comprehensive study is available of the thousands of possible flavours, analysis of commercial flavours showed that some chemicals are commonly used in more than one flavour *(114)*. A meta-analysis of the reported literature *(115–117)* was recently published by one of the authors of this report of the recurrence of certain chemicals, which indicates the frequency of chemicals such as ethyl maltol (47%), vanillin (37%), menthol (29%), ethyl vanillin (23%), linalool (23%), benzaldehyde (22%), benzyl alcohol (21%), maltol (20%), cinnamaldehyde (20%), ethyl butanoate (19%) and hydroxyacetone (16%). The work also describes the possible contributions of these flavouring additives to the toxicity of the aerosols generated by ENDS activation *(115)*. Flavours are either distilled intact into the aerosol (their contribution to toxicity depends on their properties and emission levels) *(118)*, react with ENDS carriers (propylene glycol and glycerol) to form acetal compounds with unique toxicological properties *(119)* or undergo thermal degradation to toxicants such as carbonyls *(120)*, reactive oxygen species *(121)* and volatile organic compounds *(122)*. The gas–particle-phase partitioning coefficients of several flavour ingredients have been determined; they are relevant to assessment of toxicity as they may determine the site of absorption of these chemicals into the body *(123)*.

Flavours added to ENDS not only impart a specific flavour but also increase sweetness (e.g. sucrose) *(124)*, as observed in other products, including smokeless products *(125)* and cigars *(126)*. Sweeteners, including artificial ones (e.g. sucralose), can increase the appeal of nicotine and tobacco products *(65)*, although there is limited evidence for direct effects of sweeteners *(127)*.

Flavours in other tobacco products

Flavour additives in smokeless tobacco products have been addressed in two studies *(128,129)*. Flavour compounds have also been identified and quantified in *bidis (130)*, clove cigarettes *(131)* and flavoured waterpipe tobacco *(132)*.

9.4.3 Toxicity of flavours

Flavours in ENDS can strongly increase the general toxicity of the aerosols *(133)*. Several targeted analyses of ENDS liquids included quantification of diacetyl in nutty-flavoured ENDS liquids *(134)* and found that emission of this chemical and its inhalation by ENDS users increased their risk of bronchiolitis obliterans or "popcorn lung" disease *(135)*. Some cherry-flavoured ENDS liquids expose users to benzaldehyde, albeit at low levels *(136)*. To assess disease risk, the levels of these toxicants in ENDS aerosols are usually compared with workspace exposure limits. Toxicological assessment of flavours in ENDS liquids and aerosols, in cell lines and in animals, showed that flavours increase the toxicity of ENDS aerosols in various ways *(121,137)*. One report showed that adducts of flavours with ENDS carriers are more cytotoxic than their parent flavour compounds *(119)*.

Flavours may also increase the toxicity of ENDS aerosols by adding to toxicant emissions. For example, flavours in liquids increase emissions of carbonyl compounds and other compounds that are known or possible human carcinogens *(120)*. Flavours also increase toxicity by disturbing the oxidative balance in the body, as they increase the presence of radicals and reactive oxygen species in ENDS aerosols over that produced by plain liquids composed only of carriers *(138)*. This type of contribution to the toxicity of aerosols depends on the device operating parameters, such as power input and liquid composition *(139)*, as higher power increases the temperature of the heating coil, resulting in greater degradation of flavour compounds.

At present, the net effect of flavours on the toxicity of emissions from tobacco and nicotine products cannot be determined for products other than ENDS.

9.5 Regulation of flavoured products

9.5.1 Global regulation of flavoured ENDS

The contentious issue of the impact of flavours on the acceptability of ENDS and satisfaction among smokers who are seeking cessation and on the appeal of ENDS for experimentation and continued use by young people is highly polarized and intense *(81,140,141)*. This controversy is reflected in the different approaches used to tackle the ENDS use epidemic by public health authorities in different countries: flavours are banned, restricted or allowed (Table 9.1), depending on the authorities' assessment of the available information and view of the arguments on both sides of the debate *(142,143)*. Currently, about 100 countries regulate use of ENDS with new regulations or, mainly, by adapting regulations for other tobacco products *(144)*. About 30 countries ban the marketing and sale of ENDS *(145,146)*. A search in June 2020 on the website of the Institute for Global Tobacco Control at the Johns Hopkins Bloomberg School of Public Health in Baltimore (USA) indicated that the policies of 35 countries on ENDS referred to ingredients or flavours *(147)*. Most focus on labelling and ensuring that high-quality ingredients are used, and only five, Canada, Finland, Luxembourg, Saudi Arabia and the USA, had specific regulations on flavours in ENDS *(142)*. Finland bans characterizing flavours (e.g. fruity, confectionery) in all ENDS; Luxembourg's policy prohibits additives that influence the perceptions of ENDS users with regard to health; Saudi Arabia allows fruit flavours and menthol but prohibits other characterizing flavours (e.g. cocoa, vanilla, coffee, tea, spices, confectionery, chewing-gum, cola and alcohol) *(149)*, while the US Food and Drug Administration (FDA) has issued a policy banning all flavoured cartridge ENDS (except tobacco- and menthol-flavoured products) on the basis of evidence that flavours strongly influence young people's use of ENDS, especially the extremely popular cartridge products (e.g. JUUL) *(150)*. It also exempts the flavoured liquids used in open-system ENDS *(150,151)*. Canada restricts marketing flavours that may appeal to young people, and, recently, the province of Nova Scotia banned flavoured ENDS; other provinces are considering doing the same *(152,153)*.

Table 9.1. Country regulatory approaches to flavoured e-cigarettes (as of June 2020)

Country	Regulation
Argentina, Australia, Azerbaijan, Bahrain, Barbados, Brazil, Brunei Darussalam, Cambodia, Chile, Colombia, Costa Rica, Ecuador, Egypt, El Salvador, Fiji, Gambia, Georgia, Honduras, Hungary, Iceland, India, Indonesia, Iran (Islamic Republic of), Israel, Jamaica, Japan, Jordan, Kuwait, Lao People's Democratic Republic, Lebanon, Malaysia, Maldives, Mauritius, Mexico, Nepal, New Zealand, Nicaragua, Norway, Oman, Palau, Panama, Paraguay, Philippines, Qatar, Republic of Korea, Republic of Moldova, Seychelles, Singapore, South Africa, Sri Lanka, Suriname, Switzerland, Syrian Arab Republic, Tajikistan, Thailand, Timor-Leste, Togo, Turkey, Turkmenistan, Uganda, Ukraine, United Arab Emirates, Uruguay, Venezuela (Bolivarian Republic of), Viet Nam	No specific regulation
Austria, Belgium, Bulgaria, Croatia, Cyprus, Czechia, Denmark, Estonia, France, Germany, Greece, Ireland, Italy, Latvia, Lithuania, Malta, Netherlands, Poland, Portugal, Romania, Slovakia, Slovenia, Spain, Sweden, United Kingdom (England, Northern Ireland, Wales)	e-Liquid should not contain certain additives (not specified). High-quality ingredients should be used in e-cigarette manufacture. Only ingredients that do not pose a risk to human health in heated or unheated form can be used.
Canada	Marketing and sale of e-cigarettes that contain certain additives is prohibited (not specified). Restrictions on the marketing of flavours that may appeal to young people (including flavour suggestions, confectionery, dessert, cannabis, soft drink and energy drink).
Finland	e-Liquid should not contain certain additives and characterizing flavours (such as confectionery or fruit flavours). High-quality ingredients should be used in e-cigarette manufacture. Only ingredients that do not pose a risk to human health in heated or unheated form can be used.
Saudi Arabia	Flavours in e-cigarette liquids are partially prohibited. Fruit flavours and menthol are allowed, but cocoa, vanilla, coffee, tea, spices, confectionery, chewing-gum, cola and alcohol flavours are banned.
USA	All flavoured cartridge-based ENDS except tobacco- and menthol-flavoured products are banned. Flavoured liquids used in open-system ENDS are exempted.

Country	Regulation
Luxembourg	Additives that may create the impression that an e-cigarette product has a health benefit or presents a reduced health risk are prohibited (e.g. vitamins). Caffeine, taurine and other stimulants associated with energy or vitality are prohibited. Any additives that add colour, alter the properties of emissions or facilitate inhalation or nicotine uptake are prohibited. Only ingredients that do not pose a risk to human health in heated or unheated form can be used. Additives that have carcinogenic, mutagenic or reproductive toxic properties in their unburnt form are also prohibited. High-quality ingredients should be used in e-cigarette manufacture.

9.5.2 Global regulation of flavoured tobacco products

As for ENDS, regulation of other flavoured tobacco products also differs among countries. Table 9.2 lists countries and territories with different regulatory approaches from an analysis of regulations by the Campaign for Tobacco-Free Kids *(154)*. These can be summarized as no regulations, partial bans on specific categories of flavours (with or without menthol) and full bans on all characterizing flavours (the definition may differ by jurisdiction), which may include menthol.

Table 9.2. Countries and territories approaches to regulation of the contents of tobacco products, including flavours (As of June 2020)

Countries and Territories	Regulation
Afghanistan, Argentina, Azerbaijan, Bangladesh, Belarus, Benin, Bhutan, Botswana, Burkina Faso, Burundi, Cambodia, Cameroon, Cabo Verde, Chad, China, Colombia, Comoros, Congo, Côte d'Ivoire, Djibouti, Egypt, El Salvador, Eritrea, Eswatini, Fiji, France, Gabon, Germany, Ghana, Guinea, Iceland, India, Iran (Islamic Republic of), Israel, Jamaica, Japan, Jordan, Kazakhstan, Lao People's Democratic Republic, Latvia, Lebanon, Liberia, Madagascar, Malawi, Malaysia, Maldives, Mali, Mauritius, Mexico, Myanmar, Namibia, Nepal, New Zealand, Norway, Oman, Pakistan, Panama, Peru, Philippines, Poland, Qatar, Saudi Arabia, Seychelles, Singapore, Solomon Islands, South Africa, Spain, Suriname, Sweden, Syrian Arab Republic, Timor-Leste, Togo, Turkmenistan, United Arab Emirates, Venezuela (Bolivarian Republic of), Viet Nam, occupied Palestinian territory, including east Jerusalem	The law does not grant authority to regulate the contents of cigarettes.
Algeria, Brunei Darussalam, Chile, Costa Rica, Democratic Republic of the Congo, Ecuador, Gambia, Georgia, Guatemala, Guyana, Honduras, Indonesia, Iraq, Kenya, Rwanda, Thailand, Ukraine, United Republic of Tanzania, Uruguay	The law grants authority to regulate the contents of cigarettes; however, no regulations have been issued.
Australia	The contents and ingredients of cigarettes are not regulated at national level; however, fruit- and confectionery-flavoured cigarettes are banned in all states and territories. Mint is banned in at least one state.
Armenia, Russian Federation	The law regulates specified contents of cigarettes, including banning mint and some herbs, and other, unspecified flavourings.

Country	Regulation
Brazil, Canada,[a] Ireland,[a] Italy, Mauritania, Republic of Moldova, Niger,[b] Romania,[a] Senegal, Slovenia,[a] Sri Lanka, Turkey, Uganda, United Kingdom (England, Northern Ireland, Wales)[a]	The law regulates specified contents of cigarettes, including banning of sugars and sweeteners, characterizing flavours, menthol, mint and spearmint, spices and herbs, ingredients that facilitate nicotine uptake, ingredients that create the impression of health benefits, other flavourings, ingredients associated with energy and vitality and colouring agents.
Ethiopia, Nigeria	The law regulates specified contents of cigarettes, including banning characterizing flavours, ingredients that create the impression of health benefits and ingredients associated with energy and vitality.

[a] Menthol as a characterizing flavour was banned as of 20 May 2020.
[b] Menthol is not prohibited.

In 2009, the FDA banned all characterizing flavours in cigarettes except for menthol *(13)*. In 2014, the European Union followed suit, with a series of policies banning flavours other than menthol in cigarettes and roll-your-own tobacco *(28,155)*, and the Tobacco Products Directive banned menthol in cigarettes and roll-your-own tobacco in May 2020 *(156,157)*, although the regulations do not apply to other tobacco products *(28)*. Other countries have banned flavours in tobacco products *(158)*, including cigarettes; however, the regulations are either still at legislative level, as in Brazil *(159)* and Uganda *(160)*, or were recently implemented, as in Turkey *(161)*. In 2014, Singapore prohibited fruit flavours in waterpipe tobacco *(63)*. In 2010, Canada prohibited the sale of all flavoured cigarettes and little cigars but exempted menthol cigarettes, and banned all flavours in other tobacco products, including waterpipe, smokeless tobacco and *bidis*. In 2017, the provinces of Alberta and Ontario banned the sale of menthol cigarettes *(161)*.

In a study of the response to regulations, quitting behaviour was observed in Canadians who reported daily, some or "never" (i.e. users of non-menthol cigarettes) use of menthol cigarettes after the menthol product ban and found that more daily and occasional users than non-menthol product users had attempted to or had quit cigarette use *(162)*. In a study of residents of San Francisco, USA, menthol product and e-cigarette use decreased among young (18–24 years) and older adults (25–34 years) after a ban on menthol flavour, but cigarette smoking increased among young adults, and 65% of participants did not believe that the ban was enforced uniformly across the city *(14)*. Ethiopia provides an ideal example of a strict tobacco control strategy, with a ban on the sale and distribution of all flavoured tobacco products, including those with menthol, and all ENDS are banned in a comprehensive approach to protect public health *(163)*. Use of tobacco products in the country is lower than in other countries with a low human development index *(164)*.

9.5.3 Pros and cons of common approaches

Flavours in tobacco products increase their appeal and the perceptions of users and bystanders of their safety *(100,165–166)*. National regulatory offices have therefore attempted to reduce the effect of use of these products on public health. As noted above, only a few countries ban characterizing flavours in nicotine and tobacco products, either to respond to their obligations under the WHO Framework Convention on Tobacco Control or to protect young people and public health *(158)*.

The different approaches to regulating flavours globally present a mosaic of policies that could be weighed according to their estimated benefit to public health. Some do not mention flavours, tobacco ingredients or content in general. Such lack of specificity could leave loopholes through which the tobacco industry could address young people. Partial banning of some flavours or banning of flavours in some nicotine and tobacco products also leaves a wide margin for the tobacco industry to advertise other flavours or alternative flavoured nicotine and tobacco products and thus challenge the work of regulators. A full ban on all flavours in all nicotine and tobacco products would appear to be a strong approach to curbing young people's use of tobacco products, although regulators should consider the potential argument that flavours might be tools to accommodate switching from use of traditional smoked products to other products which could be substitutes. Regulators should make sure that customizable products that can be used to deliver nicotine in products such as open-system ENDS are removed from the market; otherwise, users will add unorthodox and illicit additives to their products *(167)*. There have also been several calls for the removal of flavours from ENDS, as the associated risks outweigh their potential public health benefit *(168,169)*. Both the supply and demand should be addressed in all regulations through widespread advocacy and awareness campaigns to seek support from the public to enforce implementation *(170–172)*.

9.5.4 Impact of regulation of specific flavoured nicotine products

As most regulations on flavoured tobacco products were introduced recently, limited information about their impact is available *(173–175)*. Lessons can nevertheless be learnt of the effects of policies to restrict flavours in tobacco products on their consumption *(176–178)*. For example, assessment of the effect of restrictions of the sale of flavoured cigars (< 1.4 g) in Canada on the sale of other cigars showed an overall decrease. Furthermore, after the ban on mentholated combustible products, more menthol cigarette smokers quit and made quit attempts than smokers of non-mentholated products *(162)*, although the authors noted that the exemption of certain flavours and product types might have reduced the effectiveness of the policy *(179)*. Similarly, evaluation of the effect of the ban on flavoured tobacco products (e.g. cigars, little cigars, roll-your-own) in New York City, USA, showed a significant decrease (28%) in the odds of adolescents using any tobacco product *(180)*, although evaluation of the effect of the ban on menthol in San Francisco, USA, indicated a decrease in e-cigarette use but not cigarette use among younger people *(14)*. Another report suggested that making flavoured tobacco products less accessible and less affordable could help reduce the use of all tobacco products *(181)*.

As there are no or only relatively new regulations on flavouring in ENDS, there are no longitudinal data on their use in ENDS or other tobacco products *(182)*. Impact analysis and modelling have been used to estimate the possible effect of a regulation on use of flavours in ENDS *(143,183,184)*, including the net effect on use of all tobacco products *(185)*. One report showed that restrictive regulations on ENDS flavours could increase the intention of young adults to use cigarettes and both ENDS and combustible cigarettes *(183)*, although re-evaluation of the data showed that the net effect of regulation of both products is favourable to public health *(185)*. Estimation of the impact of a ban on all flavours in ENDS, menthol in cigarettes or all flavours in cigarettes showed that the measure that would reduce both smoking and vaping rates would be a ban on all flavours in both products, although use of cigarettes would still be 2.7% higher than the status quo *(169)*. As HTPs are considered to be tobacco products and are included with ENDS in some policies, there are no specific regulations on flavours in these products *(186)*. We have demonstrated that flavours in ENDS increase the abuse liability of these products, specifically among adolescents and young adults, and we have no evidence that specific flavours in ENDS would help cigarette smokers to quit. In considering regulatory measures, it might be important to consider whether unflavoured ENDS have an effect similar to flavoured ENDS in supporting attempts to quit combustible cigarette use and reducing the abuse liability of these products.

9.5.5 Future regulation of flavours

Flavours increase the appeal, continued use, extent of use, dependence and toxicity of nicotine and tobacco products and increase the risk of new generations of nicotine and tobacco addicts. Addiction to nicotine and exposure to the other toxicants emitted place a significant burden on public health *(187)*. The only way in which flavours could benefit public health would be in tobacco products proven to be less toxic, less risky and that support reduced use of combustible tobacco *(81)*. Even so, users of such alternative flavoured tobacco products should be encouraged to stop using them in order to withdraw from nicotine addiction and to avoid any lapse or relapse to use of tobacco or nicotine products.

The section in the seventh report of the WHO Study Group on Tobacco Product Regulation *(31)* on flavours in tobacco and nicotine products noted that the research priorities were systematic monitoring of the global epidemiology of flavoured conventional, traditional, new and emerging tobacco products, identification of how flavours contribute to the appeal of these products and identification of flavour chemicals, their toxicity and their health effects. This report confirms that flavours are still widely prevalent in all nicotine and tobacco products, that the popularity of ENDS products has increased, that, while there are regulations on the availability of flavours in nicotine and tobacco products, they vary widely by country and that the global epidemiology of flavoured nicotine and tobacco product use should be monitored systematically.

This report also raises concern about the appeal of flavoured products to adolescents and young adults. ENDS are the most commonly used nicotine and tobacco products in these groups, and they have the highest use of flavours of all age groups. Flavours in ENDS may therefore be uniquely appealing to adolescents and young adults. One explanation may be the wide availability of non-traditional flavours *(23)*, which are more popular in these groups than among older adults *(47)*. It has been shown that fruit flavours in ENDS enhance the reward and reinforcing effects of nicotine delivered to the user *(24,91)*.

To better understand the impact of ENDS flavours on use, they have been grouped into categories, similar to those for combustible tobacco *(107)*. For example, flavours such as blueberry and green apple are considered fruit flavours, while muffin and cupcake flavours are categorized as "dessert". This grouping is useful for interpreting results for products with a wide, growing variety of flavour options.

Use of flavoured products is associated with a greater likelihood of use of other tobacco products *(188)*, especially among young people *(189)*. Policies to reduce all tobacco use must therefore be based on actual use patterns *(190)*. The effect of flavours on the appeal and use of ENDS is controversial from a regulatory perspective, as there is hot debate about whether the availability of many flavours assists in switching from combustible products to ENDS or increases the uptake of ENDS by naïve young people. It has been shown that the availability of flavours in ENDS is an important consideration for acceptance of these products by cigarette smokers *(24,191,192)*.

Flavours in nicotine and tobacco products, especially ENDS, are marketed by vivid descriptions of the taste and sensory experiences associated with them *(193)*. Perhaps not surprisingly, the overwhelming majority of ENDS users – adolescents, young adults and adults – endorsed "Come in flavours I like" as a reason for using ENDS *(35)*, and ENDS users ranked a choice of flavours and unique flavours as two of the most important factors in choosing between competing vape shops *(157)*. The availability of many flavours may make it more likely that users will find a product that appeals to them and may explain why flavours are used to a greater extent in ENDS products than in other tobacco and nicotine products. Flavours are still prevalent in other products with demonstrated appeal and sustained use, such as smokeless and combustible products and newer tobacco products such as HTPs. As tobacco regulations globally increasingly focus on ENDS, regulatory agencies should continue to monitor use of other nicotine and tobacco products in order to reduce their use.

Regulation of ENDS began some time after the major tobacco companies began their production *(194)*. Some advocacy groups have criticized the FDA for not acting fast enough to prevent young people from using ENDS, which may have contributed to the high rates of use by young people in the USA, and health organizations won a lawsuit that obliged the FDA to bring forward the deadline for submission of studies on the safety of ingredients in premarketing applications for ENDS products from 2022 to 12 May 2020 *(196–198)*. Regulation of flavours in other tobacco products, including cigarettes, also lagged behind their spread in the population *(37)*. Flavours were introduced in smokeless tobacco in the nineteenth century, while flavouring of cigarettes flourished only a few decades ago *(1,4,199)*. Flavours are used extensively in other tobacco products such as waterpipes and *bidis (2,32)*, and the wide choice of flavours contributes to their popularity, especially among young people *(27,61)*. Flavours not only increase the addictiveness of tobacco products but also increase their toxic emissions exponentially. Regulation of flavours is therefore at the intersection of harm reduction and a precautionary approach in tobacco regulation *(200)*. The best regulatory approach to stop the tobacco epidemic is to develop complementary policies on all flavoured tobacco products that will eventually bring this epidemic to an end *(201,202)*.

9.6   Discussion

The use of flavours in nicotine and tobacco products is controversial, as they have been clearly shown to contribute to the use and appeal of these products, particularly among young people. ENDS products continue to be a major concern, as their popularity is growing. A major feature of their appeal is the wide variety of flavours, which promote experimentation and prolonged use. Additionally, emerging evidence suggests that flavours may contribute to the toxicity of newer products such as ENDS in unique ways. Increased use of tobacco and nicotine due to flavours increases the burden on public health; however, flavours might be used to reduce the burden, as some adult smokers have reported that the flavours in products like ENDS contribute to their efforts to stop or reduce cigarette use. Policy-makers should consider this aspect when regulating flavours in tobacco products. Regulation of flavouring in tobacco products should be a priority in all regulatory approaches to limit the spread and progression of nicotine and tobacco use and to reduce use of combustible tobacco products.

9.7 Research gaps, priorities and questions

Research is necessary on the following aspects of flavoured nicotine and tobacco products, especially ENDS:

- surveillance studies on the global epidemiology of use of flavoured nicotine and tobacco products;
- longitudinal studies of ENDS use characteristics, reasons for uptake, flavour use over time and continued use;
- scientific classification to provide a means to categorize flavoured tobacco products and their chemical constituents;
- consensus on a current definition of a "characterizing flavour";
- impact analyses of regulations, restrictions and bans on flavours in new and emerging nicotine and tobacco products, especially ENDS, including modelling of responses to flavour-related policies in hypothetical scenarios and tasks;
- biomedical and behavioural studies on the impact of flavour on the experience of reward with use of nicotine and tobacco products, by age group and tobacco use status; and
- the toxicity of individual flavour ingredients and of chemicals in nicotine and tobacco products and of newly formed combined moieties.

Global priorities are to:

- build evidence of the impact of flavours on use of nicotine and tobacco products by age group and on use of different products in different countries;
- determine the impact of flavours in nicotine and tobacco products on the decades-long effort to reduce nicotine and tobacco use in the population; and
- exchange experience among countries in the regulation of flavours in tobacco products.

Questions to policy-makers and international organizations such as WHO include:

- How can concerns about the use of flavours in nicotine and tobacco products be addressed rapidly to prevent a new generation of users from becoming dependent on nicotine and tobacco?
- What can be done to outpace industry manoeuvres to use flavours to enhance the appeal and use of their products?
- Can a robust, self-developing, sustainable regulatory model be built to address similar concerns for any new or modified risk tobacco product?

9.8 Policy recommendations

A piecemeal approach to regulating flavoured nicotine and tobacco products will not turn the tide of the tobacco epidemic. A multi-pronged combination of various policy tools with a panoramic view of all nicotine and tobacco product use will help health agencies to address the issue of flavoured products *(203,204)*. ENDS could be used as an opportunity to increase the regulation of all tobacco products *(205)* to achieve the ultimate objective of nicotine- and tobacco-free future generations *(206)*. Policies on flavours in novel and emerging nicotine and tobacco products should include the following.

- Where flavours are not banned, their regulation in nicotine and tobacco products should be consistent globally; i.e. the availability of flavours should be regulated similarly for all nicotine and tobacco products rather than for each product.
- Research should be conducted on the possible role of characterizing flavours in products like ENDS or HTPs in helping smokers to quit.

9.9 References

1. Kostygina G, Ling PM. Tobacco industry use of flavourings to promote smokeless tobacco products. Tob Control. 2016;25(Suppl 2):ii40–9.
2. Ben Taleb Z, Breland A, Bahelah R, Kalan ME, Vargas-Rivera M, Jaber R et al. Flavored versus nonflavored waterpipe tobacco: A comparison of toxicant exposure, puff topography, subjective experiences, and harm perceptions. Nicotine Tob Res. 2018;21(9):1213–9.
3. Meernik C, Baker HM, Kowitt SD, Ranney LM, Goldstein AO. Impact of non-menthol flavours in e-cigarettes on perceptions and use: an updated systematic review. BMJ Open. 2019;9(10):e031598.
4. Carpenter CM, Wayne GF, Pauly JL, Koh HK, Connolly GN. New cigarette brands with flavors that appeal to youth: tobacco marketing strategies. Health Aff (Millwood). 2005;24(6):1601–10.
5. Gendall P, Hoek J. Role of flavours in vaping uptake and cessation among New Zealand smokers and non-smokers: a cross-sectional study. Tob Control. 2020: 055469.
6. Yao T, Jiang N, Grana R, Ling PM, Glantz SA. A content analysis of electronic cigarette manufacturer websites in China. Tob Control. 2016;25(2):188–94.
7. He G, Lin X, Ju G, Chen Y. Mapping public concerns of electronic cigarettes in China. J Psychoactive Drugs. 2019:52(1):1–7.
8. Cullen KA, Gentzke AS, Sawdey MD, Chang JT, Anic GM, Wang TW et al. e-Cigarette use among youth in the United States, 2019. JAMA. 2019;322(21):2095–2103.
9. Prohibited additives. In: Tobacco and Vaping Products Act (S.C. 1997, c. 13). Ottawa: Department of Justice; 1997 (https://laws.justice.gc.ca/eng/acts/T-11.5/page-12.html).
10. Hiscock R, Silver K, Zatoński M, Gilmore AB. Tobacco industry tactics to circumvent and undermine the menthol cigarette ban in the UK. Tob Control. 2020; 055769.
11. Farrelly MC, Loomis BR, Kuiper N, Han B, Gfroerer J, Caraballo RS et al. Are tobacco control policies effective in reducing young adult smoking? J Adolesc Health. 2014;54(4):481–6.
12. Talhout R, van de Nobelen S, Kienhuis AS. An inventory of methods suitable to assess additi-

ve-induced characterising flavours of tobacco products. Drug Alcohol Depend. 2016;161:9–14.
13. Food and Drug Administration. Enforcement of general tobacco standard special rule for cigarettes. Fed Reg. 2009;E9:23144 (https://www.federalregister.gov/documents/2009/09/25/E9-23144/enforcement-of-general-tobacco-standard-special-rule-for-cigarettes, accessed 18 February 2020).
14. Yang Y, Lindblom EN, Salloum RG, Ward KD. The impact of a comprehensive tobacco product flavor ban in San Francisco among young adults. Addict Behav Rep. 2020;11:100273.
15. Family Smoking Prevention and Tobacco Control. Division A. Public Law. 22 June 2009;111-31 (https://www.govinfo.gov/content/pkg/PLAW-111publ31/pdf/PLAW-111publ31.pdf, accessed 30 November 2020).
16. Tobacco. Geneva: World Health Organization; 2020 (https://www.who.int/news-room/fact-sheets/detail/tobacco, accessed 18 August 2020).
17. Kennedy RD, Awopegba A, De León E, Cohen JE. Global approaches to regulating electronic cigarettes. Tob Control. 2017;26(4):440–5.
18. FDA finalizes enforcement policy on unauthorized flavored cartridge-based e-cigarettes that appeal to children, including fruit and mint. Silver Spring (MD): Food and Drug Administration; 2020 (https://www.fda.gov/news-events/press-announcements/fda-finalizes-enforcement-policy-unauthorized-flavored-cartridge-based-e-cigarettes-appeal-children, accessed 23 June 2020).
19. Drope J, Schulger NW, editors. The Tobacco Atlas, sixth edition. Atlanta (GA): American Cancer Society; 2018.
20. Dawkins L, Turner J, Roberts A, Soar K. "Vaping" profiles and preferences: an online survey of electronic cigarette users. Addiction. 2013;108(6):1115–25.
21. WHO global report on trends in prevalence of tobacco smoking 2015. Geneva: World Health Organization; 2015.
22. Zhu SH, Sun JY, Bonnevie E, Cummins SE, Gamst A, Yin L et al. Four hundred and sixty brands of e-cigarettes and counting: implications for product regulation. Tob Control. 2014;23(suppl 3):iii3–9.
23. Hsu G, Sun JY, Zhu SH. Evolution of electronic cigarette brands from 2013–2014 to 2016–2017: Analysis of brand websites. J Med Internet Res. 2018;20(3):e80.
24. Farsalinos KE, Romagna G, Tsiapras D, Kyrzopoulos S, Spyrou A, Voudris V. Impact of flavour variability on electronic cigarette use experience: an internet survey. Int J Environ Res Public Health. 2013;10(12):7272–82.
25. Jones DM, Ashley DL, Weaver SR, Eriksen MP. Flavored ENDS use among adults who have used cigarettes and ENDS, 2016–2017. Tob Regul Sci. 2019;5(6):518–31.
26. Bold KW, Krishnan-Sarin S. e-Cigarettes: Tobacco policy and regulation. Curr Addict Rep. 2019;6(2):75–85.
27. Ambrose BK, Day HR, Rostron B, Conway KP, Borek N, Hyland A et al. Flavored tobacco product use among US youth aged 12–17 years, 2013–2014. JAMA. 2015;314(17):1871–3.
28. Morean ME, Butler ER, Bold KW, Kong G, Camenga DR, Simon P et al. Preferring more e-cigarette flavors is associated with e-cigarette use frequency among adolescents but not adults. PLoS One. 2018;13(1):e0189015.
29. Schneller LM, Bansal-Travers M, Goniewicz ML, McIntosh S, Ossip D, O'Connor RJ. Use of flavored e-cigarettes and the type of e-cigarette devices used among adults and youth in the US – Results from wave 3 of the Population Assessment of Tobacco and Health Study (2015–2016). Int J Environ Res Public Health. 2019;16(16):2991.
30. Heated tobacco products (HTPs) market monitoring information sheet 2018. Geneva: World Health Organization; 2018 (https://apps.who.int/iris/bitstream/handle/10665/273459/WHO-NMH-PND-18.7-eng.pdf?ua=1, accessed 18 August 2020).

31. Krishnan-Sarin S, Green BG, Jordt SE, O'Malley SS. The science of flavour in tobacco products. In: WHO Study Group on Tobacco Product Regulation, Seventh report (WHO Technical Report Series No. 1015). Geneva: World Health Organization; 2019.
32. Bonhomme MG, Holder-Hayes E, Ambrose BK, Tworek C, Feirman SP, King BA et al. Flavoured non-cigarette tobacco product use among US adults: 2013–2014. Tob Control. 2016;25(Suppl 2):ii4–13.
33. Villanti AC, Mowery PD, Delnevo CD, Niaura RS, Abrams DB, Giovino GA. Changes in the prevalence and correlates of menthol cigarette use in the USA, 2004-2014. Tob Control. 2016;25(Suppl 2):ii14–20.
34. Odani S, Armour B, Agaku IT. Flavored tobacco product use and its association with indicators of tobacco dependence among US adults, 2014–2015. Nicotine Tob Res. 2020;22(6):1004–15.
35. Rostron BL, Cheng YC, Gardner LD, Ambrose BK. Prevalence and reasons for use of flavored cigars and ENDS among US youth and adults: Estimates from wave 4 of the PATH study, 2016–2017. Am J Health Behav. 2020;44(1):76–81.
36. Harrell MB, Weaver SR, Loukas A, Creamer M, Marti CN, Jackson CD et al. Flavored e-cigarette use: Characterizing youth, young adult, and adult users. Prev Med Rep. 2017;5:33–40.
37. Leventhal AM, Goldenson NI, Cho J, Kirkpatrick MG, McConnell RS, Stone MD et al. Flavored e-cigarette use and progression of vaping in adolescents. Pediatrics. 2019;144(5):e20190789.
38. Villanti AC, Johnson AL, Glasser AM, Rose SW, Ambrose BK, Conway KP et al. Association of flavored tobacco use with tobacco initiation and subsequent use among US youth and adults, 2013–2015. JAMA Netw Open. 2019;2(10):e1913804.
39. Goldenson NI, Leventhal AM, Simpson KA, Barrington-Trimis JL. A review of the use and appeal of flavored electronic cigarettes. Curr Addict Rep. 2019;6(2):98–113.
40. Zare S, Nemati M, Zheng Y. A systematic review of consumer preference for e-cigarette attributes: Flavor, nicotine strength, and type. PLoS One. 2018;13(3):e0194145.
41. Wang TW, Gentzke AS, Creamer MR, Cullen KA, Holder-Hayes E, Sawdey MD et al. Tobacco product use and associated factors among middle and high school students – United States, 2019. Morbid Mortal Wkly Rep. 2019;68(12):1–22.
42. Schneller LM, Bansal-Travers M, Goniewicz ML, McIntosh S, Ossip D, O'Connor RJ. Use of flavored electronic cigarette refill liquids among adults and youth in the US – Results from wave 2 of the Population Assessment of Tobacco and Health Study (2014–2015). PLoS One. 2018;13(8):e0202744.
43. Russell C, McKeganey N, Dickson T, Nides M. Changing patterns of first e-cigarette flavor used and current flavors used by 20,836 adult frequent e-cigarette users in the USA. Harm Reduct J. 2018;15(1):33.
44. Berg CJ. Preferred flavors and reasons for e-cigarette use and discontinued use among never, current, and former smokers. Int J Public Health. 2016;61(2):225–36.
45. Du P, Bascom R, Fan T, Sinharoy A, Yingst J, Mondal P et al. Changes in flavor preference in a cohort of long-term electronic cigarette users. Ann Am Thorac Soc. 2020;17(5):573–81.
46. Gendall P, Hoek J. Role of flavours in vaping uptake and cessation among New Zealand smokers and non-smokers: a cross-sectional study. Tob Control. 2020; doi.org/10.1136/tobaccocontrol-2019-055469.
47. Soneji SS, Knutzen KE, Villanti AC. Use of flavored e-cigarettes among adolescents, young adults, and older adults: Findings from the Population Assessment for Tobacco and Health Study. Public Health Rep. 2019;134(3):282–92.
48. Sharan RN, Chanu TM, Chakrabarty TK, Farsalinos K. Patterns of tobacco and e-cigarette use status in India: a cross-sectional survey of 3000 vapers in eight Indian cities. Harm Reduct J. 2020;17(1):21.
49. Okawa S, Tabuchi T, Miyashiro I. Who uses e-cigarettes and why? e-Cigarette use among older

adolescents and young adults in Japan: JASTIS study. J Psychoactive Drugs. 2020;52(1):37–45.
50. Minaker LM, Ahmed R, Hammond D, Manske S. Flavored tobacco use among Canadian students in grades 9 through 12: prevalence and patterns from the 2010–2011 youth smoking survey. Prev Chronic Dis. 2014;11:E102.
51. Kaleta D, Usidame B, Szosland-Faltyn A, Makowiec-Dabrowska T. Use of flavoured cigarettes in Poland: data from the global adult tobacco survey (2009–2010). BMC Public Health. 2014;14:127.
52. Moodie C, Thrasher JF, Cho YJ, Barnoya J, Chaloupka FJ. Flavour capsule cigarettes continue to experience strong global growth. Tob Control. 2019;28(5):595–6.
53. White V, Williams T. Australian secondary school students' use of tobacco in 2014. Victoria: Centre for Behavioural Research in Cancer, Cancer Council Victoria; 2014.
54. Emond JA, Soneji S, Brunette MF, Sargent JD. Flavour capsule cigarette use among US adult cigarette smokers. Tob Control. 2018;27(6):650–5.
55. Directive 2014/40/EU of the European Parliament and of the Council of 3 April 2014 on the approximation of the laws, regulations and administrative provisions of the Member States concerning the manufacture, presentation and sale of tobacco and related products and repealing Directive 2001/37/EC. Offic J Eur Union. 2014; L127:1–38.
56. Kim M. Philip Morris International introduces new heat-not-burn product, IQOS, in South Korea. Tob Control. 2018;27(e1):e76–8.
57. Valinsky J. A new non-vaping, non-smoking way to get nicotine has come to America. CNN Business, 4 October 2019.
58. McKelvey K, Popova L, Kim M, Chaffee BW, Vijayaraghaven M, Ling P et al. Heated tobacco products likely appeal to adolescents and young adults. Tob Control. 2018;27(Suppl 1):s41–7.
59. Sustainability. Lausanne: Philip Morris International; 2017 (https://www.pmi.com/sustainability, accessed 1 November 2017).
60. Sutanto E, Miller C, Smith DM, O'Connor RJ, Quah ACK, Cummings KM et al. Prevalence, use behaviors, and preferences among users of heated tobacco products: Findings from the 2018 ITC Japan survey. Int J Environ Res Public Health. 2019;16(23):10.3390.
61. Klein SM, Giovino GA, Barker DC, Tworek C, Cummings KM, O'Connor RJ. Use of flavored cigarettes among older adolescent and adult smokers: United States, 2004–2005. Nicotine Tob Res. 2008;10(7):1209–14.
62. King BA, Tynan MA, Dube SR, Arrazola R. Flavored-little-cigar and flavored-cigarette use among US middle and high school students. J Adolesc Health. 2014;54(1):40–6.
63. Kong G, Morean ME, Cavallo DA, Camenga DR, Krishnan-Sarin S. Reasons for electronic cigarette experimentation and discontinuation among adolescents and young adults. Nicotine Tob Res. 2015;17(7):847–54.
64. Kong G, Bold KW, Morean ME, Bhatti H, Camenga DR, Jackson A et al. Appeal of JUUL among adolescents. Drug Alcohol Depend. 2019;205:107691.
65. Goldenson NI, Kirkpatrick MG, Barrington-Trimis JL, Pang RD, McBeth JF, Pentz MA et al. Effects of sweet flavorings and nicotine on the appeal and sensory properties of e-cigarettes among young adult vapers: Application of a novel methodology. Drug Alcohol Depend. 2016;168:176–80.
66. Jackson A, Green B, Erythropel HC, Kong G, Cavallo DA, Eid T et al. Influence of menthol and green apple e-liquids containing different nicotine concentrations among youth e-cigarette users. Exp Clin Psychopharmacol. 2020:10.1037.
67. Kong G, Bold KW, Simon P, Camenga DR, Cavallo DA, Krishnan-Sarin S. Reasons for cigarillo initiation and cigarillo manipulation methods among adolescents. Tob Regul Sci. 2017;3(2 Suppl 1):S48–58.
68. Sharma E, Clark PI, Sharp KE. Understanding psychosocial aspects of waterpipe smoking

among college students. Am J Health Behav. 2014;38(3):440–7.
69. Choi K, Fabian L, Mottey N, Corbett A, Forster J. Young adults' favorable perceptions of snus, dissolvable tobacco products, and electronic cigarettes: findings from a focus group study. Am J Public Health. 2012;102(11):2088–93.
70. Moodie C, Ford A, Mackintosh A, Purves R. Are all cigarettes just the same? Female's perceptions of slim, coloured, aromatized and capsule cigarettes. Health Educ Res. 2015;30(1):1–12.
71. Griffiths MA, Harmon TR, Gilly MC. Hubble bubble trouble: The need for education about and regulation of hookah smoking. J Public Policy Marketing. 2011;30(1):119–32.
72. Roskin J, Aveyard P. Canadian and English students' beliefs about waterpipe smoking: A qualitative study. BMC Public Health. 2009;9:10.
73. Hammal F, Wild TC, Nykiforuk C, Abdullahi K, Mussie D, Finegan BA. Waterpipe (hookah) smoking among youth and women in Canada is new, not traditional. Nicotine Tob Res. 2016;18(5):757–62.
74. Kowitt SD, Meernik C, Baker HM, Osman A, Huang LL, Goldstein AO. Perceptions and experiences with flavored non-menthol tobacco products: A systematic review of qualitative studies. Int J Environ Res Public Health. 2017;14(4):338.
75. Sterling KL, Fryer CS, Fagan P. The most natural tobacco used: A qualitative investigation of young adult smokers' risk perceptions of flavored little cigars and cigarillos. Nicotine Tob Res. 2016;18(5):827–33.
76. Garrison KA, O'Malley SS, Gueorguieva R, Krishnan-Sarin S. A fMRI study on the impact of advertising for flavored e-cigarettes on susceptible young adults. Drug Alcohol Depend. 2018;186:233–41.
77. Audrain-McGovern J, Rodriguez D, Pianin S, Alexander E. Initial e-cigarette flavoring and nicotine exposure and e-cigarette uptake among adolescents. Drug Alcohol Depend. 2019;202:149–55.
78. Audrain-McGovern J, Strasser AA, Wileyto EP. The impact of flavoring on the rewarding and reinforcing value of e-cigarettes with nicotine among young adult smokers. Drug Alcohol Depend. 2016;166:263–7.
79. Krishnan-Sarin S, Morean ME, Camenga DR, Cavallo DA, Kong G. e-Cigarette use among high school and middle-school adolescents in Connecticut. Nicotine Tob Res. 2015;17(7):810–8.
80. Kim H, Lim J, Buehler SS, Brinkman MC, Johnson NM, Wilson L et al. Role of sweet and other flavours in liking and disliking of electronic cigarettes. Tob Control. 2016;25(Suppl 2):ii55–61.
81. Landry RL, Groom AL, Vu THT, Stokes AC, Berry KM, Kesh A et al. The role of flavors in vaping initiation and satisfaction among US adults. Addict Behav. 2019;99:106077.
82. Avelar AJ, Akers AT, Baumgard ZJ, Cooper SY, Casinelli GP, Henderson BJ. Why flavored vape products may be attractive: Green apple tobacco flavor elicits reward-related behavior, upregulates nAChRs on VTA dopamine neurons, and alters midbrain dopamine and GABA neuron function. Neuropharmacology. 2019;158:107729.
83. Wickham RJ. How menthol alters tobacco-smoking behavior: A biological perspective. Yale J Biol Med. 2015;88(3):279–87.
84. Romijnders KA, Krüsemann EJ, Boesveldt S, Graaf K, Vries H, Talhout R. e-Liquid flavor preferences and individual factors related to vaping: A survey among Dutch never-users, smokers, dual users, and exclusive vapers. Int J Environ Res Public Health. 2019;16(23):4661.
85. Wipfli H, Bhuiyan MR, Qin X, Gainullina Y, Palaganas E, Jimba M et al. Tobacco use and e-cigarette regulation: Perspectives of university students in the Asia-Pacific. Addict Behav. 2020;107:106420.
86. Ashare RL, Hawk LW Jr, Cummings KM, O'Connor RJ, Fix BV, Schmidt WC. Smoking expectancies for flavored and non-flavored cigarettes among college students. Addict Behav. 2007;32(6):1252–61.

87. Feirman SP, Lock D, Cohen JE, Holtgrave DR, Li T. Flavored tobacco products in the United States: A systematic review assessing use and attitudes. Nicotine Tob Res. 2016;18(5):739–49.
88. Oliver AJ, Jensen JA, Vogel RI, Anderson AJ, Hatsukami DK. Flavored and nonflavored smokeless tobacco products: rate, pattern of use, and effects. Nicotine Tob Res. 2013;15(1):88–92.
89. Odani S, Armour B, Agaku IT. Flavored tobacco product use and its association with indicators of tobacco dependence among US adults, 2014–2015. Nicotine Tob Res. 2020;22(6):1004–15.
90. Lynn WR, Davis RM, Novotny TE, editors. The health consequences of smoking: Nicotine addiction: A report of the Surgeon General. Atlanta (GA): Center for Health Promotion and Education, Office on Smoking and Health; 1988.
91. DeVito EE, Valentine GW, Herman AI, Jensen KP, Sofuoglu M. Effect of menthol-preferring status on response to intravenous nicotine. Tob Regul Sci. 2016;2(4):317–28.
92. Henderson BJ, Wall TR, Henley BM, Kim CH, McKinney S, Lester HA. Menthol enhances nicotine reward-related behavior by potentiating nicotine-induced changes in nAChR function, nAChR upregulation, and DA neuron excitability. Neuropsychopharmacology. 2017;42(12):2285–91.
93. Alsharari SD, King JR, Nordman JC, Muldoon PP, Jackson A, Zhu AZX et al. Effects of menthol on nicotine pharmacokinetic, pharmacology and dependence in mice. PLoS One. 2015;10(9):e0137070.
94. Villanti AC, Johnson AL, Ambrose BK, Cummings KM, Stanton CA, Rose SW et al. Flavored tobacco product use in youth and adults: Findings from the first wave of the PATH study (2013–2014). Am J Prev Med. 2017;53(2):139–51.
95. Ahijevych K, Garrett BE. The role of menthol in cigarettes as a reinforcer of smoking behavior. Nicotine Tob Res. 2010;12(Suppl 2):S110–6.
96. DeVito EE, Jensen KP, O'Malley SS, Gueorguieva R, Krishnan-Sarin S, Valentine G et al. Modulation of "protective" nicotine perception and use profile by flavorants: Preliminary findings in e-cigarettes. Nicotine Tob Res. 2020;22(5):771–81.
97. Leventhal AM, Goldenson NI, Barrington-Trimis JL, Pang RD, Kirkpatrick MG. Effects of non-tobacco flavors and nicotine on e-cigarette product appeal among young adult never, former, and current smokers. Drug Alcohol Depend. 2019;203:99–106.
98. Krishnan-Sarin S, Green BG, Kong G, Cavallo DA, Jatlow P, Gueorguieva R et al. Studying the interactive effects of menthol and nicotine among youth: An examination using e-cigarettes. Drug Alcohol Depend. 2017;180:193–9.
99. Kostygina G, Glantz SA, Ling PM. Tobacco industry use of flavours to recruit new users of little cigars and cigarillos. Tob Control. 2016;25(1):66–74.
100. Romijnders KAGJ, van Osch L, de Vries H, Talhout R. Perceptions and reasons regarding e-cigarette use among users and non-users: A narrative literature review. Int J Environ Res Public Health. 2018;15(6):1190.
101. Wilson N, Hoek J, Thomson G, Edwards R. Should e-cigarette use be included in indoor smoking bans? Bull World Health Organ. 2017;95(7):540–1.
102. Shearston J, Lee L, Eazor J, Meherally S, Park SH, Vilcassim MR et al. Effects of exposure to direct and secondhand hookah and e-cigarette aerosols on ambient air quality and cardiopulmonary health in adults and children: protocol for a panel study. BMJ Open. 2019;9(6):e029490.
103. Visser WF, Klerx WN, Cremers HWJM, Ramlal R, Schwillens PL, Talhout R. The health risks of electronic cgarette use to bystanders. Int J Environ Res Public Health. 2019;16(9):1525.
104. Hajek P, Phillips-Waller A, Przulj D, Pesola F, Myers Smith K, Bisal N et al. A randomized trial of e-cigarettes versus nicotine-replacement therapy. N Engl J Med. 2019;380(7):629–37.
105. Glasser A, Vojjala M, Cantrell J, Levy DT, Giovenco DP, Abrams D et al. Patterns of e-cigarette use and subsequent cigarette smoking cessation over two years (2013/2014 to 2015/2016) in the Population Assessment of Tobacco and Health (PATH) study. Nicotine Tob Res. 2020: doi: 10.1093/ntr/ntaa182.

106. Krüsemann EJ, Visser WF, Cremers JW, Pennings JL, Talhout R. Identification of flavour additives in tobacco products to develop a flavour library. Tob Control. 2018;27(1):105–11.
107. Krüsemann EJZ, Boesveldt S, de Graaf K, Talhout R. An e-liquid flavor wheel: A shared vocabulary based on systematically reviewing e-liquid flavor classifications in literature. Nicotine Tob Res. 2019;21(10):1310–9.
108. Cooper M, Harrell MB, Perry CL. A qualitative approach to understanding real-world electronic cigarette use: Implications for measurement and regulation. Prev Chronic Dis. 2016;13:E07.
109. Cox S, Leigh NJ, Vanderbush TS, Choo E, Goniewicz ML, Dawkins L. An exploration into "do-it-yourself" (DIY) e-liquid mixing: Users' motivations, practices and product laboratory analysis. Addict Behav Rep. 2019;9:100151.
110. Baker RR, Bishop LJ. The pyrolysis of tobacco ingredients. J Anal Appl Pyrolysis. 2004;71(1):223–311.
111. Aszyk J, Kubica P, Kot-Wasik A, Namieśnik J, Wasik A. Comprehensive determination of flavouring additives and nicotine in e-cigarette refill solutions. Part I: Liquid chromatography–tandem mass spectrometry analysis. J Chromatogr A. 2017;1519:45–54.
112. Aszyk J, Woźniak MK, Kubica P, Kot-Wasik A, Namieśnik J, Wasik A. Comprehensive determination of flavouring additives and nicotine in e-cigarette refill solutions. Part II: Gas-chromatography–mass spectrometry analysis. J Chromatogr A. 2017;1517:156–64.
113. Aszyk J, Kubica P, Woźniak MK, Namieśnik J, Wasik A, Kot-Wasik A. Evaluation of flavour profiles in e-cigarette refill solutions using gas chromatography–tandem mass spectrometry. J Chromatogr A. 2018;1547:86–98.
114. Krüsemann EJZ, Havermans A, Pennings JLA, de Graaf K, Boesveldt S, Talhout R. Comprehensive overview of common e-liquid ingredients and how they can be used to predict an e-liquid's flavour category. Tob Control. 2020:055447.
115. Hua M, Omaiye EE, Luo W, McWhirter KJ, Pankow JF, Talbot P. Identification of cytotoxic flavor chemicals in top-selling electronic cigarette refill fluids. Sci Rep. 2019;9(1):2782.
116. Omaiye EE, McWhirter KJ, Luo W, Tierney PA, Pankow JF, Talbot P. High concentrations of flavor chemicals are present in electronic cigarette refill fluids. Sci Rep. 2019;9(1):2468.
117. Salam S, Saliba NA, Shihadeh A, Eissenberg T, El-Hellani A. Flavor–toxicant correlation in e-cigarette: A meta-analysis. Chem Res Toxicol. 2020.
118. El-Hage R, El-Hellani A, Salman R, Talih S, Shihadeh A, Saliba NA. Fate of pyrazines in the flavored liquids of e-cigarettes. Aerosol Sci Technol. 2018;52(4):377–84.
119. Erythropel HC, Jabba SV, DeWinter TM, Mendizabal M, Anastas PT, Jordt SE et al. Formation of flavorant–propylene glycol adducts with novel toxicological properties in chemically unstable e-cigarette liquids. Nicotine Tob Res. 2019;21(9):1248–58.
120. Khlystov A, Samburova V. Flavoring compounds dominate toxic aldehyde production during e-cigarette vaping. Environ Sci Technol. 2016;50(23):13080–5.
121. Lerner CA, Sundar IK, Yao H, Gerloff J, Ossip DJ, McIntosh S et al. Vapors produced by electronic cigarettes and e-juices with flavorings induce toxicity, oxidative stress, and inflammatory response in lung epithelial cells and in mouse lung. PLoS One. 2015;10(2):e0116732.
122. Pankow JF, Kim K, McWhirter KJ, Luo W, Escobedo JO, Strongin RM et al. Benzene formation in electronic cigarettes. PLoS One. 2017;12(3):e0173055.
123. Pankow JF, Kim K, Luo W, McWhirter KJ. Gas/particle partitioning constants of nicotine, selected toxicants, and flavor chemicals in solutions of 50/50 propylene glycol/glycerol as used in electronic cigarettes. Chem Res Toxicol. 2018;31(9):985–90.
124. Kubica P, Wasik A, Kot-Wasik A, Namieśnik J. An evaluation of sucrose as a possible contaminant in e-liquids for electronic cigarettes by hydrophilic interaction liquid chromatography–tandem mass spectrometry. Anal Bioanal Chem. 2014;406(13):3013–8.
125. Miao S, Beach ES, Sommer TJ, Zimmerman JB, Jordt SE. High-intensity sweeteners in alterna-

126. Erythropel HC, Kong G, deWinter TM, O'Malley SS, Jordt SE, Anastas PE et al. Presence of high-intensity sweeteners in popular cigarillos of varying flavor profiles. JAMA. 2018;320(13):1380–3.
127. DeVito EE, Jensen KP, O'Malley SS, Gueorguieva R, Krishnan-Sarin S, Valentine G et al. Modulation of "protective" nicotine perception and use profile by flavorants: Preliminary findings in e-cigarettes. Nicotine Tob Res. 2020;22(5):771–81.
128. Lisko JG, Stanfill SB, Watson CH. Quantitation of ten flavor compounds in unburned tobacco products. Anal Meth. 2014;6(13):4698–704.
129. Chen C, Isabelle LM, Pickworth WB, Pankow JF. Levels of mint and wintergreen flavorants: Smokeless tobacco products vs. confectionery products. Food Chem Toxicol. 2010;48(2):755–63.
130. Stanfill SB, Brown CR, Yan X, Watson CH, Ashley DL. Quantification of flavor-related compounds in the unburned contents of bidi and clove cigarettes. J Agric Food Chem. 2006;54(22):8580–8.
131. Polzin GM, Stanfill SB, Brown CR, Ashley DL, Watson CH. Determination of eugenol, anethole, and coumarin in the mainstream cigarette smoke of Indonesian clove cigarettes. Food Chem Toxicol. 2007;45(10):1948–53.
132. Schubert J, Luch A, Schulz TG. Waterpipe smoking: Analysis of the aroma profile of flavored waterpipe tobaccos. Talanta. 2013;115:665–74.
133. Behar RZ, Luo W, McWhirter KJ, Pankow JF, Talbot P. Analytical and toxicological evaluation of flavor chemicals in electronic cigarette refill fluids. Sci Rep. 2018;8(1):8288.
134. Allen JG, Flanigan SS, LeBlanc M, Vallarino J, MacNaughton P, Stewart JH et al. Flavoring chemicals in e-cigarettes: Diacetyl, 2,3-pentanedione, and acetoin in a sample of 51 products, including fruit-, candy-, and cocktail-flavored e-cigarettes. Environ Health Perspect. 2016;124(6):733–9.
135. Kreiss K, Gomaa A, Kullman G, Fedan K, Simoes EJ, Enright PL. Clinical bronchiolitis obliterans in workers at a microwave-popcorn plant. NEJM. 2002;347(5):330–8.
136. Kosmider L, Sobczak A, Prokopowicz A, Kurek J, Zaciera M, Knysak J et al. Cherry-flavoured electronic cigarettes expose users to the inhalation irritant, benzaldehyde. Thorax. 2016;71(4):376–7.
137. El-Hage R, El-Hellani A, Haddad C, Salman R, Talih S, Shihadeh A et al. Toxic emissions resulting from sucralose added to electronic cigarette liquids. Aerosol Sci Technol. 2019;53(10):1197–203.
138. Muthumalage T, Prinz M, Ansah KO, Gerloff J, Sundar IK, Rahman I. Inflammatory and oxidative responses induced by exposure to commonly used e-cigarette flavoring chemicals and flavored e-liquids without nicotine. Front Physiol. 2018;8:1130.
139. Sleiman M, Logue JM, Montesinos VN, Russell ML, Litter MI, Gundel LA et al. Emissions from electronic cigarettes: Key parameters affecting the release of harmful chemicals. Environ Sci Technol. 2016;50(17):9644–51.
140. Wagener TL, Meier E, Tackett AP, Matheny JD, Pechacek TF. A proposed collaboration against Big Tobacco: Common ground between the vaping and public health community in the United States. Nicotine Tob Res. 2015;18(5):730–6.
141. Gartner C. How can we protect youth from putative vaping gateway effects without denying smokers a less harmful option? Addiction. 2018;113(10):1784–5.
142. Kennedy RD, Awopegba A, De León E, Cohen JE. Global approaches to regulating electronic cigarettes. Tob Control. 2017;26(4):440–5.
143. Doan TTT, Tan KW, Dickens BSL, Lean YA, Yang Q, Cook AR. Evaluating smoking control policies in the e-cigarette era: a modelling study. Tob Control. 2019:054951.
144. Country laws regulation e-cigarettes: A policy scan. Baltimore (MD): Johns Hopkins Bloomberg School of Public Health, Institute for Global Control; 2020 (https://www.globaltobaccocontrol.org/e-cigarette/country-laws/view?field_policy_domains_tid%5B%5D=119, accessed 17 February 2020).

145. Villanti AC, Byron MJ, Mercincavage M, Pacek LR. Misperceptions of nicotine and nicotine reduction: The importance of public education to maximize the benefits of a nicotine reduction standard. Nicotine Tob Res. 2019;21(Suppl 1):S88–90.
146. Country laws regulation e-cigarettes: Policy domains: Sale. Baltimore (MD): Johns Hopkins Bloomberg School of Public Health, Institute for Global Tobacco Control; 2020 (https://www.globaltobaccocontrol.org/e-cigarette/sale, accessed 15 June 2020).
147. Ingredients/flavors. Baltimore (MD): Johns Hopkins Bloomberg School of Public Health, Institute for Global Tobacco Control; 2020 (https://www.globaltobaccocontrol.org/category/policy-domains/ingredientsflavors, accessed 15 June 2020).
148. Ollila E. See you in court: Obstacles to enforcing the ban on electronic cigarette flavours and marketing in Finland. Tob Control. 2019:055260.
149. Country laws regulating e-cigarettes: Saudi Arabia. Baltimore (MD): Johns Hopkins Bloomberg School of Public Health, Institute for Global Tobacco Control; 2020 (https://www.globaltobaccocontrol.org/e-cigarette/saudi-arabia, accessed 15 June 2020).
150. Enforcement priorities for electronic nicotine delivery systems (ENDS) and other deemed products on the market without premarket authorization. Silver Spring (MD): Food and Drug Administration; 2020 (https://www.fda.gov/media/133880/download, accessed 17 February 2020).
151. Tanne JH. FDA bans most flavoured e-cigarettes as lung injury epidemic slows. BMJ. 2020;368:m12.
152. Doucette K. Nova Scotia first province to ban flavoured e-cigarettes and juices. CTV News Atlantic, 5 December 2019 (https://atlantic.ctvnews.ca/nova-scotia-first-province-to-ban-flavoured-e-cigarettes-and-juices-1.4716489, accessed 17 February 2020).
153. Ontario considering ban on flavoured vape products: health minister. The Canadian Press, 5 December 2019 (https://www.cbc.ca/news/canada/toronto/ontario-considering-ban-on-flavoured-vape-products-health-minister-1.5386305, accessed 17 February 2020).
154. Tobacco control laws. Legislation. Washington (DC): Campaign for Tobacco-Free Kids; 2020 (https://www.tobaccocontrollaws.org/legislation, accessed 17 June 2020).
155. Vardavas CI, Bécuwe N, Demjén T, Fernández E, McNeill A, Mons U et al. Study protocol of European Regulatory Science on Tobacco (EUREST-PLUS): Policy implementation to reduce lung disease. Tob Induced Dis. 2018;16(2):2.
156. Zatoński M, Herbeć A, Zatoński W, Przewozniak K, Janik-Koncewicz K, Mons U et al. Characterising smokers of menthol and flavoured cigarettes, their attitudes towards tobacco regulation, and the anticipated impact of the Tobacco Products Directive on their smoking and quitting behaviours: The EUREST-PLUS ITC Europe Surveys. Tob Induced Dis. 2018;16:A4.
157. Sussman S, Garcia R, Cruz TB, Baezconde-Garbanati L, Pentz MA, Unger JB. Consumers' perceptions of vape shops in Southern California: an analysis of online Yelp reviews. Tob Induced Dis. 2014;12(1):22.
158. Erinoso O, Clegg Smith K, Iacobelli M, Saraf S, Welding K, Cohen JE. Global review of tobacco product flavour policies. Tob Control. 2020:055404.
159. Oliveira Da Silva AL, Bialous SA, Albertassi PGD, Arquete DADR, Fernandes AMMS, Moreira JC. The taste of smoke: tobacco industry strategies to prevent the prohibition of additives in tobacco products in Brazil. Tob Control. 2019;28(e2):e92–101.
160. Agaku IT, Odani S, Armour BS, King BA. Adults' favorability toward prohibiting flavors in all tobacco products in the United States. Prev Med. 2019;129:105762.
161. Soule EK, Lopez AA, Guy MC, Cobb CO. Reasons for using flavored liquids among electronic cigarette users: A concept mapping study. Drug Alcohol Depend. 2016;166:168–76.
162. Chaiton MO, Nicolau I, Schwartz R, Cohen JE, Soule E, Zhang B et al. Ban on menthol-flavoured tobacco products predicts cigarette cessation at 1 year: a population cohort study. Tob Con-

trol. 2020;29(3):341–7.
163. Tobacco Control Directive. Addis Ababa: Ethiopian Food, Medicine and Healthcare Administration and Control Authority. 2015 (https://www.tobaccocontrollaws.org/files/live/Ethiopia/Ethiopia%20-%20Tobacco%20Ctrl.%20Dir.%20No.%2028_2015%20-%20national.pdf, accessed 18 February 2020).
164. Palandri F, Breccia M, Bonifacio M, Polverelli N, Elli EM, Benevolo G et al. Life after ruxolitinib: Reasons for discontinuation, impact of disease phase, and outcomes in 218 patients with myelofibrosis. Cancer. 2020;126(6):1243–52.
165. Gorukanti A, Delucchi K, Ling P, Fisher-Travis R, Halpern-Felsher B. Adolescents' attitudes towards e-cigarette ingredients, safety, addictive properties, social norms, and regulation. Prev Med. 2017;94:65–71.
166. Huang LL, Baker HM, Meernik C, Ranney LM, Richardson A, Goldstein AO. Impact of non-menthol flavours in tobacco products on perceptions and use among youth, young adults and adults: a systematic review. Tob Control. 2017;26:709–19.
167. Eissenberg T, Soule E, Shihadeh A. "Open-system" electronic cigarettes cannot be regulated effectively. Tob Control. 2020:055499.
168. Drazen JM, Morrissey S, Campion EW. The dangerous flavors of e-cigarettes. N Engl J Med. 2019;380:679–80.
169. Buckell J, Marti J, Sindelar JL. Should flavours be banned in cigarettes and e-cigarettes? Evidence on adult smokers and recent quitters from a discrete choice experiment. Tob Control. 2018:054165.
170. Keklik S, Gultekin-Karakas D. Anti-tobacco control industry strategies in Turkey. BMC Public Health. 2018;18(1):282.
171. McKee M. Evidence, policy, and e-cigarettes. NEJM. 2016;375(5):e6.
172. Jongenelis MI, Kameron C, Rudaizky D, Pettigrew S. Support for e-cigarette regulations among Australian young adults. BMC Public Health. 2019;19(1):67.
173. Courtemanche CJ, Palmer MK, Pesko MF. Influence of the flavored cigarette ban on adolescent tobacco use. Am J Prev Med. 2017;52(5):e139–46.
174. Yang Y, Lindblom EN, Salloum RG, Ward KD. The impact of a comprehensive tobacco product flavor ban in San Francisco among young adults. Addict Behav Rep. 2020;11:100273.
175. Chaiton M, Schwartz R, Cohen JE, Soule E, Eissenberg T. Association of Ontario's ban on menthol cigarettes with smoking behavior 1 month after implementation. JAMA Intern Med. 2018;178(5):710–1.
176. Lidón-Moyano C, Martín-Sánchez JC, Saliba P, Graffelman J, Martínez-Sánchez JM. Correlation between tobacco control policies, consumption of rolled tobacco and e-cigarettes, and intention to quit conventional tobacco, in Europe. Tob Control. 2017;26(2):149–52.
177. Harrell MB, Loukas A, Jackson CD, Marti CN, Perry CL. Flavored tobacco product use among youth and young adults: What if flavors didn't exist? Tob Regul Sci. 2017;3(2):168–73.
178. Chaiton MO, Nicolau I, Schwartz R, Cohen JE, Soule E, Zhang B et al. Ban on menthol-flavoured tobacco products predicts cigarette cessation at 1 year: a population cohort study. Tob Control. 2019:054841.
179. Chaiton MO, Schwartz R, Tremblay G, Nugent R. Association of flavoured cigar regulations with wholesale tobacco volumes in Canada: an interrupted time series analysis. Tob Control. 2019;28(4):457–61.
180. Farley SM, Johns M. New York City flavoured tobacco product sales ban evaluation. Tob Control. 2017;26(1):78–84.
181. Agaku IT, Odani S, Armour B, Mahoney M, Garrett BE, Loomis BR et al. Differences in price of flavoured and non-flavoured tobacco products sold in the USA, 2011–2016. Tob Control. 2019;29:537–47.

182. Glantz SA, Bareham DW. e-Cigarettes: Use, effects on smoking, risks, and policy implications. Annu Rev Public Health. 2018;39:215–35.
183. Pacek LR, Rass O, Sweitzer MM, Oliver JA, McClernon FJ. Young adult dual combusted cigarette and e-cigarette users' anticipated responses to hypothetical e-cigarette market restrictions. Substance Use Misuse. 2019;54(12):2033–42.
184. Brock B, Carlson SC, Leizinger A, D'Silva J, Matter CM, Schillo BA. A tale of two cities: exploring the retail impact of flavoured tobacco restrictions in the twin cities of Minneapolis and Saint Paul, Minnesota. Tob Control. 2019;28(2):176–80.
185. Glantz SA. Net effect of young adult dual combusted cigarette and e-cigarette users' anticipated responses to hypothetical e-cigarette marketing restrictions. Substance Use Misuse. 2020;55(6):1028–30.
186. Krishnan-Sarin S, Jackson A, Morean M, Kong G, Bold KW, Camenga DR et al. e-Cigarette devices used by high-school youth. Drug Alcohol Depend. 2019;194:395–400.
187. Jankowski M, Krzystanek M, Zejda JE, Majek P, Lubanski J, Lawson JA et al. e-Cigarettes are more addictive than traditional cigarettes – A study in highly educated young people. Int J Environ Res Public Health. 2019;16(13):2279.
188. Pacek LR, Villanti AC, McClernon FJ. Not quite the rule, but no longer the exception: Multiple tobacco product use and implications for treatment, research, and regulation. Nicotine Tob Res. 2020;22(11):2114–7.
189. Harrell PT, Naqvi SMH, Plunk AD, Ji M, Martins SS. Patterns of youth tobacco and polytobacco usage: The shift to alternative tobacco products. Am J Drug Alcohol Abuse. 2017;43(6):694–702.
190. Horn K, Pearson JL, Villanti AC. Polytobacco use and the "customization generation"– New perspectives for tobacco control. J Drug Educ. 2016;46(3-4):51–63.
191. Shiffman S, Sembower MA, Pillitteri JL, Gerlach KK, Gitchell JG. The impact of flavor descriptors on nonsmoking teens' and adult smokers' interest in electronic cigarettes. Nicotine Tob Res. 2015;17(10):1255–62.
192. Tackett AP, Lechner WV, Meier E, Grant DM, Driskill LM, Tahirkheli NN et al. Biochemically verified smoking cessation and vaping beliefs among vape store customers. Addiction. 2015;110(5):868–74.
193. Soule EK, Sakuma K-LK, Palafox S, Pokhrel P, Herzog TA, Thompson N et al. Content analysis of internet marketing strategies used to promote flavored electronic cigarettes. Addict Behav. 2019;91:128–35.
194. Cox E, Barry RA, Glantz S. e-Cigarette policymaking by local and state governments: 2009–2014. Milbank Q. 2016;94(3):520–96.
195. Printz C. US Food and Drug Administration considers comments on proposed nicotine product regulations. Cancer. 2018;124(20):3959–60.
196. Printz C. FDA launches public education effort to prevent youth e-cigarette use. Cancer. 2018;124(7):1313–4.
197. Jenssen BP, Walley SC. e-Cigarettes and similar devices. Pediatrics. 2019;143(2):e20183652.
198. Jaffe S. Will Trump snuff out e-cigarettes? Lancet. 2019;394(10213):1977–8.
199. Carpenter CM, Wayne GF, Pauly JL, Koh HK, Connolly GN. New cigarette brands with flavors that appeal to youth: Tobacco marketing strategies. Health Affairs. 2005;24(6):1601–10.
200. Green SH, Bayer R, Fairchild AL. Evidence, policy, and e-cigarettes – Will England reframe the debate? NEJM. 2016;374(14):1301–3.
201. McDaniel PA, Smith EA, Malone RE. The tobacco endgame: a qualitative review and synthesis. Tob Control. 2016;25(5):594–604.
202. Malone RE. Tobacco endgames: what they are and are not, issues for tobacco control strategic planning and a possible US scenario. Tob Control. 2013;22(Suppl 1):i42–4.
203. Balaji S. Electronic cigarettes and its ban in India. Indian J Dental Res. 2019;30:651.

204. Guliani H, Gamtessa S, Çule M. Factors affecting tobacco smoking in Ethiopia: evidence from the demographic and health surveys. BMC Public Health. 2019;19(1):938.
205. Hall W, Gartner C, Forlini C. Nuances in the ethical regulation of electronic nicotine delivery systems. Addiction. 2015;110(7):1074–5.
206. Beaglehole R, Bonita R, Yach D, Mackay J, Reddy KS. A tobacco-free world: a call to action to phase out the sale of tobacco products by 2040. Lancet. 2015;385(9972):1011–8.

10. Global marketing and promotion of novel and emerging nicotine and tobacco products and their impacts

Meagan Robichaud, Caleb Clawson and Ryan David Kennedy, Institute for Global Tobacco Control, Department of Health, Behavior and Society, Johns Hopkins Bloomberg School of Public Health, Baltimore (MD), USA

Contents

    Abstract
    10.1 Background
    10.2 Electronic nicotine delivery systems and electronic non-nicotine delivery systems
        10.2.1 Introduction
        10.2.2 Markets, products and strategies used in marketing ENDS and ENNDS
        10.2.3 Global use of ENDS and prevalence of use
        10.2.4 Trends in advertising, promotion and sponsorship of ENDS products
        10.2.5 Measures to control advertising, promotion and sponsorship of ENDS products
        10.2.6 Recommendations
    10.3 Heated tobacco products
        10.3.1 Introduction
        10.3.2 Market players, products and strategies
        10.3.3 Global use and prevalence of use of HTPs
        10.3.4 Trends in advertising, promotion and sponsorship of HTPs
        10.3.5 Measures to control advertising, promotion and sponsorship of HTPs
        10.3.6 Recommendations for monitoring trends in marketing, advertising, promoting and sponsorship of HTPs
        10.3.7 Recommendations for regulators
        10.3.8 Gaps in research on HTPs
    10.4 Summary
    10.5 References

Abstract

Global tobacco use has decreased in the past few decades due, in large part, to successful work by the public health community to discourage use through evidence-based tobacco control strategies. Recently, nicotine and tobacco manufacturers have developed novel products, including electronic nicotine delivery systems (ENDS), electronic non-nicotine delivery systems (ENNDS) and heated tobacco products (HTPs). The introduction of these products is complicating global progress in tobacco control. In many markets, these devices are particularly popular among adolescents and young adults. Many users and non-users perceive these devices to be harmless, despite evidence of the potential harm of tobacco and nicotine use. The marketing of these products leads to experimentation, including by adolescents and young adults who have never used tobacco; robust, worldwide surveillance of product advertising, marketing, promotion and use is therefore essential. Continuous surveillance of ENDS, ENNDS and HTP marketing, including advertising in traditional media, direct-to-consumer marketing, point-of-sale marketing, online marketing (including social media), cross-border marketing and sponsorship, may prove to be a valuable comprehensive strategy to prevent the use of novel and emerging nicotine and tobacco products from undermining work to reduce the global public health burden of tobacco.

10.1 Background

Progress has been made in reducing the public health burden of tobacco use in many countries, due in large part to decreasing use. The global prevalence of tobacco use among people aged ≥ 15 years fell from 33.3% in 2000 to 24.9% in 2015 and is projected to fall to 20.9% by 2025 *(1)*. Surveillance of tobacco use indicates a significant turning-point in 2018, the first year in which a decrease in tobacco use was observed among males, who accounted for approximately 81% of global tobacco users in 2015 *(1)*. An estimated 1.05 billion men used tobacco products in 2000, and this number increased by 22 million between 2000 and 2005, 13 million between 2005 and 2010 and 7 million between 2010 and 2015. The number of male tobacco users in 2018 was 1.093 billion, and this number is expected to drop by 2 million by 2020 and by another 4 million by 2025 *(1)*.

In 2018, 23.6% of the global population aged ≥ 15 years used tobacco, 18.9% used combusted tobacco, and 16.1% used cigarettes *(1)*. Thus, approximately 80% of global tobacco users in 2018 used combusted products. Globally, more than 5.3 trillion cigarettes were sold in 2018 *(2)*. While global cigarette sales are expected to fall by approximately 7% by 2023 *(2)*, the emergence of novel nicotine and tobacco products, such as ENDS, ENNDS and HTPs, has raised concern. It is estimated that 1.2% of adults globally were current ENDS users in 2018, with significant variation in ENDS use by country, WHO region and demographic group *(3)*. The use of these products by children and young adults, including some who had never used tobacco previously, is a particular concern. There is evidence that ENDS use is associated with later use of combusted tobacco, raising concern that both ENDS and HTPs may contribute to re-"normalizing" smoking after decades of work to discourage tobacco use *(4,5)*.

In the USA, increasing use of ENDS by children and young adults contributed to the first increase in overall tobacco use measured in recent decades *(6)* (Fig. 10.1). Results from the 2019 National Youth Tobacco Survey show that the rate of ENDS use among young people has continued to increase dramatically *(7)*. The percentage of high-school students in the USA who currently use e-cigarettes increased from 20.8% in 2018 to 27.5% in 2019, and the rate among middle-school students increased from 4.9% in 2018 to 10.5% in 2019 *(6,7)*. These increases in ENDS use among the young have motivated federal and state policy-makers to look more closely at strategies to reduce the appeal of these products *(8)*. The results of the 2020 survey had not been published at the time of writing.

Fig. 10.1. Proportions of middle- and high-school students in the USA who currently use e-cigarettes and any tobacco product, 2011–2018

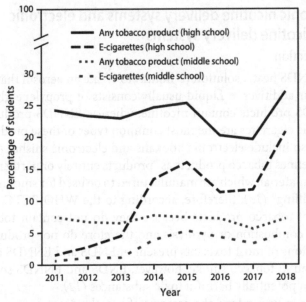

Source: reference 6.

Policy-makers around the world are considering and implementing various policies to include ENDS and HTPs in existing tobacco prevention frameworks or to regulate or ban these products specifically. The evolving market for such products and the popularity of ENDS among children and young adults lends urgency to sharing information and evidence on the effects of such policies. A substantial body of evidence links marketing of tobacco and nicotine products to greater susceptibility to use of these products and increased rates of product use among both young people and adults. An increased focus on the role of marketing in promoting nicotine and tobacco use is therefore vital for ensuring effective public health measures to reduce tobacco use.

This report extends a background paper prepared for the third meeting of the Global Tobacco Regulators Forum,[1] at which ENDS use and marketing

---

1   Kennedy RD, Clawson C. Global landscape of electronic nicotine delivery system (ENDS) marketing and promotion. Paper prepared for the Third Meeting of the Global Tobacco Regulators Forum, Geneva, 11–12 September 2019 (unpublished).

was discussed, to include marketing and promotion of ENNDS and HTPs. This report also makes a distinction between ENDS/ENNDS and HTPs with regard to marketing strategies. We address the decisions of the seventh and eighth sessions of the Conference of the Parties to the WHO Framework Convention on Tobacco Control (FCTC/COP7 and FCTC/COP8) to continue monitoring and reporting on market developments, including advertising and promotion, for ENDS and ENNDS as well as HTPs *(9,10)*.

10.2 Electronic nicotine delivery systems and electronic non-nicotine delivery systems

10.2.1 Introduction

ENDS and ENNDS heat a solution (e-liquid) to create an aerosol that frequently contains flavour additives. e-Liquid usually consists of propylene glycol and/or glycerine. ENDS products contain nicotine, whereas ENNDS products do not. While electronic cigarettes are the most common types of these products, ENDS and ENNDS also include electronic hookahs and electronic shishas *(11,12)*. The WHO FCTC defines tobacco products as "products entirely or partly made of leaf tobacco as raw material, which are manufactured to be used for smoking, sucking, chewing or snuffing" *(13)*; therefore, according to the WHO FCTC, ENDS and ENNDS are not tobacco products because they do not contain tobacco. They do not involve combustion or pyrolysis and therefore do not produce "smoke". The concentrations of most toxicants present in ENDS and ENNDS aerosols are much lower than in tobacco smoke; however, ENDS and ENNDS solutions and aerosols contain potentially harmful toxic substances *(14)*.

In order to understand the patterns of use, the diverse array of products on the market and the nonstandard nomenclature of ENDS and ENNDS devices must be identified *(4)*. Many devices resemble traditional tobacco products, such as cigarettes, pipes, hookahs and cigars, while others resemble non-tobacco products, including pens and USB flash drives. Many terms are used to refer to ENDS and ENNDS, including "e-cigarettes", "e-cigs", "cigalikes", "e-hookahs", "mods", "vape pens", "vapes", "shisha pens" and "tank systems". In this report, the terms "ENDS" and "ENNDS" are used to refer to a heterogeneous class of products in a rapidly evolving market.

ENDS and ENNDS have been widely marketed and sold in recent years by the major transnational tobacco companies, with soaring uptake by adolescents in Europe and North America, to levels high enough to alarm public health experts, parents and elected officials *(15)*. According to Euromonitor market research published in 2017, the consumption of "e-vapour" products grew by 818% between 2011 and 2016 *(16)*. Between 2011 and 2014, expenditure for marketing ENDS products increased by nearly 10 times in some markets (e.g.

from US$12 million to > US$ 125 million in the USA for e-cigarettes), stimulating a sharp rise in ENDS use in many countries *(17,18)*. In this section, we focus on major ENDS markets, products and strategies, the prevalence of ENDS use globally, regulation of marketing, monitoring and surveillance of marketing and measures to control advertising, promotion and sponsorship of ENDS.

10.2.2 Markets, products and strategies used in marketing ENDS and ENNDS

During the past two decades, the global market for combusted cigarettes has seen consolidation of manufacturers into powerful transnational tobacco companies. In 2001, the market share of combusted cigarette sales of the five largest transnational tobacco companies, Philip Morris, British American Tobacco (BAT), Japan Tobacco International (JTI), Reemsta and Altadis, was 43% *(19)*. By 2017, the market share of the five largest companies, China National Tobacco Corporation, Philip Morris International (PMI), BAT, Japan Tobacco, Inc. (parent company of JTI) and Imperial Tobacco Group had grown to 81% *(2)*. PMI, BAT, Japan Tobacco and Imperial Tobacco held four of the top six market positions for combusted cigarette retail volume in 2019 *(2)*, in addition to offering a variety of ENDS products. New entrants to the ENDS market, such as JUUL Labs, have received significant investment from the transnational tobacco companies in order to maintain their influence and reach into the global market *(20)*. Significant investment in ENDS innovation by these companies may further complicate efforts to discourage tobacco and nicotine use and to achieve public health goals.

Market players, products and market share

JUUL Labs Inc. (35% owned by Altria). JUUL currently occupies the largest (26.2%) share of the world market for ENDS with its popular nicotine salt variant and the small, ergonomic design of their devices *(16)*. In 2019, JUUL announced plans to launch its product line in India and the Philippines *(21)*; however, the Indian Government banned the production, manufacture, import, export, sale and distribution of ENDS because of concerns about trends in use among young people *(22)*. JUUL is currently testing an app that would allow users to track their nicotine consumption and allow the company to track second-hand sales of newly manufactured products *(23)*. In October 2018, JUUL Labs bought V2 and the parent company VMR Products LLC for US$ 75 million *(24)*, and V2 Cigs closed permanently on 1 November 2018. JUUL Labs currently markets its products in 20 countries *(25)*.

Altria. In December 2018, Altria, the parent company of Philip Morris USA, announced its decision to refocus its work on innovative products, including discontinuation of the production and distribution of all Nu Mark ENDS products, such as MarkTen and Green Smoke products, which had a significant market share in Canadian and US markets in recent years *(26)*. Additionally, on 20 December 2018, Altria purchased shares of non-voting convertible common stock of JUUL Labs for US$ 12.8 billion through a wholly owned subsidiary, representing a 35% economic interest in JUUL. Altria has since generally agreed not to compete with JUUL in ENDS marketing for at least 6 years *(20)*.

PMI. The PMI website for its "smoke-free" product line notes that the company is "exploring new e-vapour products" and states that

> We are also developing products inspired by technology that we acquired in 2011…. Our scientists continue to develop this technology to replicate the feel and ritual of smoking without tobacco and without burning. One of these products under development is called STEEM…[which] unlike an e-cigarette… generates a nicotine-containing vapor in the form of a nicotine salt *(27)*.

Euromonitor reports sales of other PMI ENDS brands, including Solaris in Spain and Nicocig, Vivid Vapour and MESH in the United Kingdom *(28)*.

JTI. JTI's annual report for 2018 states that "RRP [Reduced Risk Products] is one of the key pillars of our growth strategy in the tobacco business, and we will prioritize allocation of resources into the category" *(29)*. Logic is the company's flagship e-cigarette brand, with products available in 26 countries, including the United Kingdom and the USA *(30)*. On 17 September 2018, JTI launched Logic Compact, a pocket-sized device, in the United Kingdom. Its design bears a striking resemblance to that of JUUL, an ENDS device that has taken a dramatic share of the ENDS market in recent years. Logic Compact has since become available in 25 countries *(29)*. JTI also produces "E-lites", another ENDS product, which is available in Bulgaria and Germany.

BAT. BAT claims that it "is at the forefront of the development and sale of a whole range of potentially reduced-risk products that provide much of the enjoyment of smoking without burning tobacco" *(32)*. Its growing portfolio of what they claim as "potentially reduced-risk products" includes a range of ENDS products. BAT launched their flagship brand Vype in 2013 *(33)*. In 2017, BAT acquired Reynolds Vapor Co., which had launched Vuse in 2013. While BAT has since acquired a number of ENDS brands, including Ten Motives (United Kingdom), Chic (Poland) and VIP (United Kingdom), the company announced on 28 November 2019 that it was migrating its ENDS brands, when possible, to Vuse during 2020 in order to simplify its "new category product portfolio", the other new category products being Velo for "modern oral products" and glo for HTPs *(34)*. By the end of December 2019, BAT's ENDS products were available in 27 markets *(33)*.

Reynolds American Inc. (now owned by BAT). Vuse was the number one e-cigarette product sold in convenience shops in the USA in 2016 *(35)*. The company noted that "The future success of Vuse and other RJR Vapor e-cigarette offerings, including Vuse Vibe, will depend on the ability to innovate in an evolving category of alternative tobacco products". Vuse products are sold in the USA.

Imperial Brands. The company stated in its annual report in 2018 that, "Through our growing portfolio of Next Generation Products we are providing adult smokers with a range of less harmful alternatives to cigarettes, with a particular focus on the vapour category". The company has prioritized "building a presence in the specialist vape channel and online" *(36)*. Its flagship e-cigarette brand is blu, and it launched myblu and myblu Intense in 2018 *(36)* as the brand's closed-system e-cigarettes, with prefilled pods. myblu Intense is a nicotine salt variant, which the company claims "more closely replicates the experience and satisfaction of smoking a cigarette" *(37)*. Both myblu and myblu Intense have nicotine-free variants *(36,38)*. The blu brand also includes an open-system product, blu pro, and Imperial Brands launched another open-system product, blu ACE, in 2018, which has since been discontinued *(36,39,40)*. At the end of 2019, blu products were available in 16 markets *(38)*.

Marketing strategies to promote sales of ENDS and ENNDS

Nicotine and tobacco companies use a wide range of strategies to market ENDS and ENNDS. These marketing strategies have demonstrated a trend in aggressive marketing to youth, with teenagers increasingly exposed to ENDS advertising from a variety of sources *(41–43)*. Additionally, e-liquids containing nicotine are marketed in thousands of flavours, including confectionery and fruit flavours that appeal to young people. Unless marketing of ENDS and ENNDS is regulated, their use could re-normalize nicotine and tobacco use *(44)*. The following general marketing strategies are used for ENDS and ENNDS.

- Advertisement *(45)*
    - online, including social media (e.g. Facebook, Instagram, Twitter) *(46)* and use of social media influencers
    - television, cinemas *(42)*
    - radio *(42)*
    - print media (e.g. newspapers, magazines) *(47)*
    - billboards and posters *(47)*
    - displays and advertisements at points of sale *(48)*
- Sponsorship
    - sports, cultural and artistic events *(49,50)*
    - events, including school programmes *(51)*
- Youth-oriented marketing tactics

- use of cartoon characters *(52)*
- flavours, especially confectionery, fruit and other sweet flavours *(53)*
- marketing near schools *(54)*
- targeting schools and youth camps *(55)*
- marketing with popular online or mobile games (e.g. Pokémon Go) *(56)*
- Glamourizing product use
  - endorsement by celebrities *(45)*
  - promotion at "glamorous" events, e.g. free handouts at New York Fashion Week *(57)*
- Pricing strategies
  - coupons, discounts, discount codes, rebates *(58)*
  - "multi-buy", e.g. buy one, get one free *(59)*
  - free samples *(60)*
- Product innovation *(61)*
- Product design
  - ease of concealment (especially for young people) *(62)*
  - customization with colours and patterns *(63)*
  - sleek, modern design *(64)*
- Sexualization of product use *(49,65)*
- Claims of health or harm reduction *(66–68)*
- ENDS-branded merchandise *(69)*
- Funding of front groups, including *(70)*:
  - think tanks (e.g. European Policy Information Center)
  - public relations firms (e.g. Blue Star Strategies)
- Lobbying and hiring others to lobby on behalf of the industry *(71)*
- Corporate social responsibility and philanthropy to boost the image of the industry *(72)*

Common strategies

Of the six major market players that do not produce only ENDS and ENNDS products, five intend to extend innovation and/or production of their product ventures. The abrupt, substantial surge in ENDS use in some jurisdictions has newly motivated development of novel nicotine and tobacco products, and many recently introduced products that offer higher nicotine concentrations than previous generations of ENDS. For example, JUUL and NJOY offer products containing 5% and 6% nicotine, respectively*(73)*; and, in 2018, JUUL's manufacturer claimed that one 5% nicotine JUUL pod contained approximately the same amount of nicotine as 20 cigarettes *(74)*. In ENDS products available in 2013–2015, the highest concentration of nicotine was 4.9% *(73)*. Furthermore, as the market moves towards nicotine salt variants, popularized by JUUL, many can deliver high concentrations of nicotine more effectively than previous generations of these products *(75)*. These include PMI's STEEM, JTI's Logic Compact, Imperial Brand's Myblu and less influential brands such as Shenzhen IVPS Technology's Smok Nord *(64)*. The products being developed by the major market players are exclusively closed systems, meaning that users are not intended to refill their devices with e-liquid but must instead purchase refills in the form of pods or capsules. Former smokers and those attempting to quit smoking combusted cigarettes have shown a preference for open systems *(76)*, although this may change with the rising popularity of closed systems with nicotine salts *(77)*. The tobacco industry is purchasing stock and/or majority shares in competitive companies, as seen in the cases of Altria, JUUL Labs and VMR. Expansion into new markets has been a priority, and Imperial Brands, JTI and JUUL have explicitly announced geographical development.

e-Cigarette companies also use indirect marketing tactics to reach consumers, including the young. This is often achieved through front groups, which are defined as organizations that claim to be independent but in reality "serve [an]other party or interest whose sponsorship is hidden or rarely mentioned" *(70)*. Notable front groups include the Foundation for a Smoke-Free World, which has been funded solely by PMI since 2019, and the Freedom Organisation for the Right to Enjoy Smoking Tobacco (Forest), which has fought against revision of the European Union Tobacco Products Directive that would require licensing of e-cigarettes containing nicotine above a certain level *(78,79)*. Front groups also include think tanks, public relations firms and lobbying groups *(70)*.

e-Cigarette manufacturers also use corporate social responsibility strategies to boost their public image and to promote their brands. Corporate social responsibility "refers to voluntary corporate action that claims to act in the public interest by prioritising social goals" *(72)*. The companies frequently use philanthropy as a partial demonstration of their corporate social responsibility, including charitable donations to youth-oriented organizations and to causes relevant to other groups that are disproportionately impacted by tobacco use, including LGBTQ+ communities and racial and ethnic minorities *(72,80)*. For example, in 2018, Altria made charitable contributions to the National Museum of African American History and Culture and to the Boys and Girls Clubs of America *(80)*.

10.2.3 Global use of ENDS and prevalence of use

Sales of ENDS worldwide are increasing rapidly. The global market reached US$ 2.76 billion in sales in 2014 *(81)*, US$ 9.39 billion in 2017 and is expected to reach up to US$ 58.32 billion by 2026 *(82)*.

In 2018, approximately one third of the world's men (32.4%) and 5.5% of women were smokers *(1)*. It is estimated that, in 2018, 1.2% of adults worldwide used ENDS, comprising 1.7% of men and 0.7% of women *(3)*. Use of combusted tobacco and ENDS varies by WHO region. The prevalence of smoking is highest in the European Region (26.2%), and ENDS use is highest in the Western Pacific Region (2.4%) (Table 10.1).

Table 10.1. Prevalence of smoking of combusted tobacco and of ENDS use by WHO region, 2018

Region and country	Smoking (%)			ENDS use (%)		
	All	Males	Females	All	Males	Females
African Region (3 of 46 countries represented)						
Algeria	18.4	33.5	3.2	3.8	7.0	0.6
Nigeria	10.8	16.1	5.4	0.0	0.0	0.0
South Africa	18.9	31.1	7.3	0.4	0.6	0.2
Americas Region (9 of 35 countries represented)						
Canada	14.8	17.1	12.6	3.5	3.7	3.3
Chile	31.8	36.7	27.2	0.5	0.5	0.5
Colombia	11.6	17.0	6.6	0.1	0.1	0.0
Costa Rica	12.5	17.0	8.0	0.2	0.3	0.2
Dominican Republic	9.3	10.4	8.3	0.1	0.1	0.1
Ecuador	15.2	21.8	8.7	0.1	0.1	0.1
Guatemala	6.8	11.4	2.5	0.2	0.3	0.2
Peru	11.4	19.3	3.8	0.4	0.4	0.3
USA	13.7	15.5	11.9	3.8	4.1	3.5
Eastern Mediterranean Region (5 of 22 countries represented)						
Egypt	30.6	54.9	5.3	0.7	1.3	0.1
Morocco	20.3	37.6	3.7	0.7	1.4	0.0
Pakistan	21.0	34.6	6.9	0.0	0.0	0.0
Saudi Arabia	29.8	39.2	15.5	0.3	0.4	0.3
Tunisia	32.1	54.4	11.0	0.6	1.1	0.1
European Region (38 of 53 countries represented)						
Austria	26.2	28.1	24.4	1.2	1.3	1.1
Azerbaijan	26.0	33.1	19.3	0.3	0.5	0.1
Belarus	24.8	46.2	7.0	2.1	3.3	1.1
Belgium	22.0	23.0	21.0	4.4	4.8	4.0

Region and country	Smoking (%)			ENDS use (%)		
	All	Males	Females	All	Males	Females
Bosnia and Herzegovina	38.2	44.5	32.3	1.3	1.3	1.3
Bulgaria	32.2	36.6	28.1	1.1	1.9	0.4
Croatia	27.5	35.5	20.3	1.2	1.6	0.8
Czechia	33.4	37.0	30.0	5.7	6.9	4.5
Denmark	21.1	21.0	21.1	4.8	4.8	4.8
Estonia	23.4	31.0	17.0	1.6	2.2	1.1
Finland	12.2	12.8	11.6	2.0	2.8	1.3
France	26.2	30.5	22.3	4.3	5.0	3.6
Georgia	28.5	54.1	6.2	1.4	2.9	0.1
Germany	21.4	24.0	19.0	5.3	6.6	4.1
Greece	42.1	52.5	32.6	2.4	3.2	1.7
Hungary	28.6	34.5	23.4	2.1	2.9	1.3
Ireland	19.5	19.8	19.3	5.4	5.4	5.3
Israel	23.5	31.3	16.0	0.4	0.5	0.3
Italy	21.1	25.5	17.0	1.5	1.8	1.3
Kazakhstan	30.3	42.0	20.0	4.0	8.2	0.3
Latvia	27.0	42.5	14.4	1.1	2.0	0.4
Lithuania	26.3	37.4	17.1	1.6	1.9	1.3
Netherlands	22.5	25.7	19.3	3.6	4.0	3.3
North Macedonia	31.0	33.0	29.0	1.0	1.1	0.9
Norway	11.0	11.5	10.5	1.5	1.5	1.5
Poland	33.1	34.1	32.2	5.4	7.7	3.3
Portugal	19.1	26.4	12.8	2.2	3.2	1.3
Romania	30.2	39.8	21.2	3.3	5.0	1.8
Russian Federation	33.3	44.4	24.2	1.5	2.2	1.0
Serbia	32.3	35.3	29.5	0.5	0.6	0.5
Slovakia	31.4	45.5	18.3	1.9	2.2	1.6
Slovenia	23.4	26.1	20.9	0.9	1.5	0.2
Spain	25.5	28.7	22.4	1.7	1.8	1.7
Sweden	10.1	11.0	9.2	1.1	1.2	1.0
Switzerland	25.0	28.5	21.6	1.8	2.5	1.2
Ukraine	28.8	36.3	22.7	2.0	2.5	1.6
United Kingdom	14.7	16.5	13.0	6.1	7.7	4.6
Uzbekistan	11.3	19.9	3.1	0.5	0.2	0.7
South-East Asia Region (2 of 11 countries represented)						
India	3.8	6.9	0.6	0.1	0.1	0.0
Indonesia	36.4	67.9	5.0	0.5	1.1	0.0

Region and country	Smoking (%)			ENDS use (%)		
	All	Males	Females	All	Males	Females
Western Pacific Region (7 of 27 countries represented)						
Australia	13.7	15.5	11.9	0.8	1.0	0.6
China	27.8	51.9	2.8	0.2	0.5	0.0
Japan	18.2	29.0	8.1	0.2	0.3	0.1
Malaysia	21.5	40.1	1.5	2.7	4.3	1.0
New Zealand	14.9	16.2	13.6	4.1	2.0	6.1
Philippines	23.3	42.0	4.8	0.3	0.6	0.1
Republic of Korea	22.0	36.7	7.5	4.1	7.1	1.1

Source: reference 3.

The total prevalence of ENDS use ranged from 0.0% to 6.1% for both sexes, 0.0% to 8.2% for men and 0.0% to 6.1% for women. National data available for 2018 showed that ENDS use tended to be higher in countries in the European Region, in Canada and the USA in the Americas, in New Zealand and the Republic of Korea in the Western Pacific and in Algeria in the African Region. These numbers are consistent with earlier published data (83–85). Nigeria and Pakistan reported no use of ENDS. Consistent with trends in combusted tobacco use, ENDS use rates were typically higher among men than women. ENDS use by WHO region is listed below.

- *African Region.* Information on the prevalence of ENDS use was available for only three of the 46 countries in the Region: Algeria, Nigeria and South Africa. ENDS use among males in Algeria was notably high, at 7.0%.
- *Region of the Americas.* ENDS data were available for nine countries. The highest reported prevalence of ENDS use was in the USA (3.8%), followed by Canada (3.5%). The rates in the other countries in the Region were < 1.0%. The rates were similar by gender, except in Canada and the USA.
- *Eastern Mediterranean Region.* The prevalence of ENDS use was available for Egypt, Morocco, Pakistan, Saudi Arabia and Tunisia. Although smoking rates remain high in several countries in the Region, ENDS have yet to penetrate the market in any significant way, remaining at < 1% in all countries for which data were available.
- *European Region.* The prevalence of ENDS use was available for 38 countries. The highest total prevalence was in the United Kingdom, at 6.1%. The rates were ≥ 5.0% in Czechia, Ireland and Poland and relatively high among males in Kazakhstan (8.2%), France (5.0%), Germany (6.6%) and Romania (5.0%). Consistent with the demographics of smoking, the prevalence of ENDS use in the Nordic states

was similar for men and women. Interestingly, the prevalence in Uzbekistan was higher among women (0.7%) than among men (0.2%)
- *South-East Asia Region.* Data on ENDS use were available only for India and Indonesia. While both countries report high smoking rates, particularly among males, ENDS use is almost non-existent.
- *Western Pacific Region.* ENDS data were available for seven countries. Use is particularly high among men in the Republic of Korea (7.1%). In New Zealand, men are more likely to smoke than women, but women are more likely to use ENDS (6.1%) than men (2.0%).

While these data show the prevalence of ENDS use among adults, in several countries, more young adults aged 18–24 years than adults aged ≥ 25 years have ever or currently use ENDS, and the rate has been increasing steadily in recent years *(84–86)*. The prevalence of ever use of ENDS was 23.5% among adults aged 18–24 years in the USA in 2016 *(875)*, 28% among adults in this age group in the European Union and 29% among adults aged 20–24 years in Canada in 2017 *(84–86)*. Data for 2017–2018 suggest that the introduction into the Canadian and US markets of new-generation products with refillable or disposable pods (pod mods) containing nicotine salt has contributed to recent, more dramatic increases in use of ENDS in the previous 30 days among high-school students (28% in 2019) *(8,88)*. Canada and the USA also reported greater increases in the prevalence of ENDS use by young people between 2017 and 2019 than in the United Kingdom (England), where more comprehensive policies regulate access to and distribution of ENDS *(75)*.

10.2.4 Trends in advertising, promotion and sponsorship of ENDS products
How ENDS manufacturers advertise in markets through social media

Social media platforms represent an important channel for advertising ENDS, with sites that can promote these products worldwide. A scan of ENDS advertisements and promotions was conducted on Instagram and Twitter, the search terms consisting of hashtags with either the brand names of specific ENDS (e.g. #JUUL) or generic terms associated with ENDS use (e.g. #vape), accompanied by the name of a WHO Member State (e.g. #Belarus). As Instagram's search function allows only one search term, the ENDS brand name or search term was combined with the name of a WHO Member State (e.g. #VapeCanada), whereas Twitter's search function accommodated multiple search terms and Boolean operators (e.g. #JUUL AND #Canada).

The survey showed that ENDS are marketed on the social media platforms Twitter and Instagram in at least 149 (77%) WHO Member States. A content analysis of up to six advertisements (three Twitter, three Instagram) per country revealed several common advertising strategies used in the six WHO regions (Table 10.2).

Table 10.2. Common ENDS advertising strategies by WHO region, 2019

WHO region	No. of countries	Data available No. (%)	No. of advertisements	Health warning No. (%)	Nicotine lexical No. (%)	Flavour lexical No. (%)	Design feature lexical No. (%)	Image device No. (%)	Image of e-liquid No. (%)
Africa	46	26 (57)	84	1 (1)	24 (29)	63 (75)	15 (18)	22 (26)	61 (73)
Americas	35	26 (74)	135	3 (2)	14 (10)	55 (41)	72 (53)	86 (64)	59 (44)
South-East Asia	11	9 (82)	42	0	3 (7)	19 (45)	21 (50)	25 (60)	19 (45)
European	53	53 (100)	267	5 (2)	44 (16)	141 (53)	117 (44)	142 (53)	148 (55)
Eastern Mediterranean	22	21 (95)	102	1 (1)	17 (16)	50 (49)	47 (46)	51 (50)	50 (49)
Western Pacific	27	14 (52)	70	3 (4)	11 (16)	34 (49)	36 (51)	42 (60)	35 (50)
All	194	149 (77)	700	13 (2)	113 (16)	362 (52)	308 (44)	368 (53)	372 (53)

Very few of the advertisements identified in this search included a health warning. The advertisements commonly had images of devices or e-liquids and lexical content describing flavours or other design features.

Impact of COVID-19 on ENDS marketing strategies

During the COVID-19 pandemic, several ENDS product manufacturers, retailers and users have aligned their marketing and promotion strategies on social media with messages relevant to the pandemic and containment strategies. Several themes are found in ENDS marketing on Instagram and Twitter.

Coping with boredom and isolation. For example, one Twitter ad by an e-liquid manufacturer stated "Covid lockdowns got you feeling blue? We've got the Antidote. Blue Raspberry and Mango ice in perfect sync – for your dipper, tank or favorite pod system."

Online shopping. For example, one Instagram post from Vuse Middle East states, "Order online with Instashop! We're all hangin' around these days, so while you're staying in, your Vuse order will come right to you."

Working from home. Retailers and manufacturers have used this theme to encourage users to purchase their products online when they are busy "working from home". ENDS users have posted photos of their ENDS products in their home offices.

Obtaining protective equipment and supplies. Some manufacturers have posted advertisements offering protective equipment and supplies, such as masks and hand sanitizer. For example, one Instagram advertisement for the e-cigarette brand MOTI America states, "Compared with cigarettes, #vapes are 95% less harmful to the #lungs. During #Covid_19 pandemic, we recommend using

MOTI to alternate cigarettes for your #health. [What can you get?] 2 pieces of Disposable #SurgicalMasks" *(66)*.

Staying healthy (especially promoting lung health). The above example from MOTI America also shows that brands have also capitalized on a harm reduction theme during the COVID-19 pandemic *(66)*.

Supporting businesses affected by the pandemic. Manufacturers and retailers have posted messages of support for businesses affected by the pandemic in their advertisements. For example, INNOPHASE, a manufacturer and exporter of ENDS products, stated, "Currently, we have a great promotion on the VPOD, to help our partners in these difficult COVID-19 times" in an advertisement on Twitter.

10.2.5 Measures to control advertising, promotion and sponsorship of ENDS products

Many countries currently restrict advertising, promotion or sponsorship of ENDS and ENNDS; however, the regulatory strategies vary significantly. For example, eight countries (Costa Rica, Ecuador, Georgia, Japan, Mexico, New Zealand, Palau and Republic of Moldova) regulate marketing of ENDS products but not ENNDS products, in that the advertising restrictions apply only to "e-cigarettes that contain nicotine or that are regulated as medicines" *(89)*. In European Union Member States, bans on distinctive branding elements are intended to reduce advertising potential, and some have further reduced that potential by requiring out-of-sight retail sales and reduced branding opportunities on packaging. In the USA, the Food and Drug Administration passed several regulations in 2016 on the marketing and promotion of e-cigarettes, including prohibiting free sampling of e-liquid solutions inside shops *(90)*. The Food and Drug Administration also passed measures to prohibit false or misleading advertising (e.g. use of descriptors such as "light", "mild" or "low") and require manufacturers to submit applications for authorization as "modified risk tobacco products", with a full scientific review of the impact of marketing of the product on population health before it can be marketed as modified risk *(90,91)*.

Several countries have focused specifically on marketing to young people. As increasing rates of use among the young are of particular concern, the United States Food and Drug Administration issued a policy in January 2020 that prioritizes enforcement of regulations on flavoured "cartridge-based e-cigarettes" (excluding menthol and tobacco flavours) in an attempt to limit the access of young people to certain flavoured ENDS products *(92)*. Canada has banned all marketing, packaging elements that indicate flavour and design attributes that would appeal to young people *(93,94)*.

Regulation of marketing towards the young, and indeed marketing of any kind, is important in light of evidence linking exposure to tobacco and nicotine product marketing, advertising and promotion with susceptibility to use of ENDS and ENNDS. It has been noted that social media influencers have been used to promote ENDS. A minimum age is required to open an account on most social media platforms, and several platforms do not accept or run advertisements for tobacco products; however, it is unclear if these policies limit social media influencers from promoting ENDS products. Discussions in the United Kingdom have identified possible solutions for regulating social media influencers, including a minimum number of followers in order to be defined as an "influencer" and requiring online influencers to disclose payment for endorsing any product.

Increased exposure to tobacco advertising and access to price promotions has been associated with increased susceptibility to use of both ENDS and combusted products among adults (95), and increased exposure to ENDS advertisements specifically is associated with greater susceptibility to use and an increased likelihood of current use of ENDS among both adults and young people (96,97). Receptivity to ENDS advertising has also been shown to increase with exposure (96).

Certain aspects of ENDS advertising appear to be associated with an increased likelihood of reporting interest in using ENDS and later ENDS use by young people. ENDS advertising with a social rather than a health message and advertising seen on social media platforms were associated with increased interest in using ENDS and increased ENDS use, respectively (98,99). Endorsements by inspirational figures or celebrities in advertisements are also associated with an increased likelihood of use (100). The newer generation of high-tech ENDS devices associated with some of these advertising strategies and marketing of confectionery- and fruit-flavoured products have quickly become popular among adolescents and young adults (43,52,101).

10.2.6 Recommendations

Recommendations for monitoring trends in marketing, advertising, promotion and sponsorship of ENDS and ENNDS

- **Better surveillance of ENDS and ENNDS marketing, with attention to social media, marketing at points of sale and sponsorship**

    Social media. In order that policy-makers fully understand the marketing tactics of the industry, it is important to monitor traditional and social media advertising channels, paying attention to how the practices are changing over time. Monitoring can be performed independently within government ministries or by using media and Internet monitoring services, industry reports and population-level surveys. Reports should emphasize the extent to which young people are exposed to marketing.

Point-of-sale marketing. The tobacco industry has long used marketing at points of sale as an opportunity for promotion campaigns. This strategy can be monitored through surveillance on the ground.

Sponsorship. Sponsorship continues to be integral to the marketing campaigns of nicotine and tobacco industries. All events should be monitored for sponsorship and the use of testimonials in advertising to understand how the industry uses sponsorship as a promotional tactic and the populations who are exposed to this type of marketing.

- **Collaboration in monitoring marketing trends among governments**

   Cross-border advertising, including through media into bordering jurisdictions, will require collaboration among governments.

- **Monitoring of the access of young people to direct marketing**

   Direct-to-consumer marketing (through the post and e-mail) is a key strategy of the tobacco industry. Policies could be implemented to ensure that material from marketing campaigns is received only by adults and only those who consent to receive such material. Population-level surveys could aid regulators in monitoring this type of advertising.

- **Monitoring of policies for regulating ENDS and ENNDS globally**

   Evidence is lacking on how differences in policies and in the marketing of different products and product characteristics affect perceptions of risk or harm associated with use of ENDS and ENNDS and differences in use of tobacco and nicotine products. Some countries regulate ENNDS differently from ENDS, adding complexity and ambiguity to the regulation of new and emerging products *(89)*. More robust reporting of policy developments will result in more timely, more effective strategies for all Parties to the WHO FCTC.

- **Monitoring of disparities in ENDS and ENNDS marketing**

   Given the tobacco industry's history of targeted advertising to specific demographic groups, such as low-income communities, racial and ethnic minorities and sexual and gender minorities *(102,103)*, differences in both the volume and content of ENDS and ENNDS advertisements in communities and in print and digital media should be monitored.

Recommendations for regulators

- **Consider supporting state, provincial and local regulation of ENDS and ENNDS products.**

    State, provincial and local health departments and local coalitions play important roles in advancing tobacco control and decreasing the burden of tobacco use *(104–106)*.key informants (e.g. local public health center directors Local coalitions have shifted social norms on tobacco use, built support for tobacco control policies and enforced tobacco control measures *(105)*, and such local actors can be used to regulate marketing of ENDS and ENNDS.

- **Consider strategies and policies to protect tobacco control from industry interference.**

    e-Cigarette companies have used various strategies to undermine the regulation of ENDS and ENNDS and efforts to prevent young people from using these products, including sponsoring prevention programmes in schools *(51)*, lobbying against policies to regulate ENDS and ENNDS and corporate social responsibility and philanthropic activities *(107,108)*. Regulators can take measures to implement Article 5.3 of the WHO FCTC, which requires that

    > in setting and implementing their public health policies with respect to tobacco control, Parties shall act to protect these policies from commercial and other vested interests of the tobacco industry in accordance with national law *(109)*.

    These steps include avoiding entering into partnerships with companies that produce ENDS and ENNDS and initiatives funded by those companies (e.g. Foundation for a Smoke-Free World, which received initial funding from PMI) *(110)*, refusing industry contributions (financial or otherwise) and prohibiting industry sponsorship of events, particularly for youth-oriented events *(107,108)*.

- **Remain focused on evidence-based smoking prevention strategies.**

    Governments and health organizations should maintain use of evidence-based measures to reduce smoking (as outlined in the WHO FCTC) and should not be distracted from action in these areas by the promotion and marketing of novel products such as ENDS and ENNDS.

- **Consider cost-effective counter-marketing strategies.**

  Given the global reach of promotion of ENDS and ENNDS on social media, both governments and social media platforms have difficulty in effectively regulating such content, particularly when generated by users. Therefore, counter-marketing may be the most feasible option. Counter-marketing strategies can take various forms, including social media campaigns, which may be more cost-effective than traditional media campaigns. Counter-marketing may also include educating the public about industry activities, in addition to discouraging nicotine and tobacco use.

- **Consider banning all tobacco advertising, promotion and sponsorship, where possible.**

  Throughout its history, the tobacco industry has circumvented nearly all restrictions on advertising, promotion and sponsorship to reach consumers, including young people. Therefore, a complete ban on tobacco marketing might be necessary to minimize exposure of young people to marketing for nicotine and tobacco products. Such a ban would ensure that most information about these products came from national and local governments and public health agencies.

- **Foster collaboration among governments and government sectors in considering, implementing and enforcing marketing regulations for ENDS and ENNDS.**

  Cross-border advertising, including through the media, will require collaboration among governments.

- **Maintain awareness of industry strategies to market ENDS and ENNDS, particularly to young people.**

  In order that policies on ENDS and ENNDS marketing are effective, regulators must learn to recognize the strategies used by the industry to market these products, including targeted advertising, sponsorship and price promotions.

10.2.7 Research gaps for ENDS and ENNDS

- Additional research on ENDS and ENNDS marketing, especially on social media, is necessary to inform regulators about the marketing and promotion strategies used by companies and retailers.

As advertising on social media platforms has been associated with increased interest in using ENDS, researchers should continue to build evidence on the marketing tactics used by ENDS and ENNDS manufacturers and retailers on social media *(98,99)*.

- **Additional research should be conducted specifically on ENNDS marketing and its impact on perceptions of risk.**

As ENNDS are regulated differently from ENDS in some jurisdictions, they may be marketed differently. Many people incorrectly believe that nicotine is the main carcinogen in cigarettes, and, in one study, many participants believed that a cigarette with very low nicotine content was less carcinogenic than currently available cigarettes *(111)*. Studies should therefore be conducted on consumers' perceptions of risk associated with ENNDS and with ENDS and how the marketing and promotion of ENNDS shapes those perceptions.

- **Additional research should be conducted on social media content on ENDS and ENNDS generated by users and its potential effects on risk perceptions, product use and the effectiveness of regulations on marketing.**

Social media user content has increased the presence of these products. For example, content related to JUUL continued to be posted widely on social media among peers, even after JUUL stopped posting its own content *(112)*. Given that user-generated content has been used as an important marketing tactic for ENDS and ENNDS companies, it should be monitored in order to understand its impact on risk perception, product use and the effectiveness of marketing regulations.

10.3 Heated tobacco products

10.3.1 Introduction

HTPs "produce aerosols containing nicotine and toxic chemicals when tobacco is heated or when a device containing tobacco is activated" *(113)*. The distinction between HTPs and ENDS is that ENDS deliver nicotine derived from tobacco, whereas HTPs heat tobacco to deliver nicotine to the user *(5,114)*. They also contain non-tobacco additives and are often flavoured. HTPs mimic conventional cigarette smoking behaviour, and some are specially designed as cigarettes that contain tobacco for heating. Although HTP technology has existed since the 1980s, the early products were unsuccessful (Fig. 10.2) *(115)*.

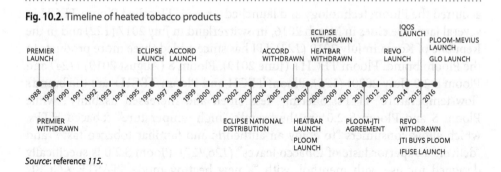

Fig. 10.2. Timeline of heated tobacco products

Source: reference 115.

10.3.2 Market players, products and strategies

On the basis of data and trends in tobacco sales in 2016, Euromonitor International predicted that the proportion of total tobacco sales represented by combusted cigarettes will continue to decrease but will be offset by market gains from novel and emerging nicotine and tobacco products, such as HTPs *(116)*. Sales of HTPs are expected to grow rapidly, to a market value of US$ 22 billion globally by 2024, from US$ 6.3 billion in 2018 *(117)*. The global market for ENDS was valued at US$ 9.39 billion in 2017 and is expected to reach US$ 58.32 billion by 2026 *(82)*. While continued rapid growth is projected for both ENDS and HTPs, the global market for combusted cigarettes still dwarfs both, as it was valued at US$ 888 billion in 2018 and is expected to reach US$ 1124 billion by 2024 *(118)*.

Market players, products and market share

The HTP market is currently dominated by three leading manufacturers: PMI, JTI and BAT *(115)*. As mentioned above, these three transnational tobacco companies also ranked among the top six manufacturers in terms of combusted cigarette retail sales volume in 2019 *(119)* and have invested significantly in ENDS production globally to augment their product portfolios. Diversification of their production to include ENDS and HTP allows significant consolidation of market power and complicates efforts to combat tobacco use.

PMI. PMI launched IQOS in Japan at the end of 2014. As of 30 June 2020, IQOS was available in 57 markets *(120)*. PMI's report for the second quarter of 2020 and its website note that it is investing in not only the next generation of IQOS products but also new HTPs such as TEEPS, which has a carbon source to heat tobacco *(120,121)*. The total estimated number of IQOS users reached 15.4 million in the second quarter of 2020 *(120)*. Profit margins for PMI's IQOS are 30–50% higher than those for conventional cigarettes *(115)*, and IQOS now makes up more than 10% of PMI's sales volume *(120)*.

JTI. JTI presented the first new-generation HTP in 2013, with the launch of Ploom, which was developed in a joint venture with a company of that name in the USA, which is now called Pax Labs. After dissolution of the partnership, JTI acquired the Ploom technology and launched a new HTP called Ploom TECH in several Japanese cities in March 2016, in Switzerland in July 2017 *(122)* and in the Republic of Korea in July 2019 *(123)*. JTI has since added three more products to the Ploom brand: Ploom TECH+ (June 2019), Ploom S (August 2019) *(124)* and Ploom S 2.0 (July 2020) *(125)*. Ploom TECH and Ploom TECH+ are the brand's "low-temperature" HTPs, offering "less smell and increased usability", while Ploom S and Ploom S 2.0 are the brand's "high-temperature" tobacco HTPs, which allow consumers "to enjoy an authentic and familiar tobacco taste" and "delivers a superior taste of tobacco leaves" *(126,127)*. Ploom S 2.0 is specifically designed for use with menthol, with "a new heating mode, 'TASTE ACCEL', which lengthens the duration of the peak heating temperature, compared to that of the current Ploom S" *(127)*.

BAT. BAT was the third entrant into the new-generation HTP market, with the introduction of iFuse in Romania in 2015 *(115)*. In 2016, BAT developed and launched glo in Japan and has since launched additional products under the glo brand *(128)*. In 2019, the company launched glo pro which has induction heating instead of the "two-zone heating chamber" of previous glo products in order to improve "consumer satisfaction and their sensorial experience" *(129,130)*. BAT also launched glo nano, a slimmer device, and glo sens, a hybrid product which "combines vaping technology with real tobacco" *(131)*. In 2020, the company launched glo Hyper in Germany, Italy, Japan, Romania and the Russian Federation. glo Hyper is designed to work with the company's "Neo demi-slim range" products, which "contain 30% more tobacco than the existing Neo sticks" *(132,133)*. BAT's tobacco heating products were available in 17 markets at the end of 2019 *(130)*. Its growing portfolio of what it claims to be "potentially reduced-risk products" includes a range of HTPs under the names of five subsidiaries *(32)*.

Korea Tobacco and Ginseng Corporation (KT&G). KT&G entered the HTP market with the launch of lil in the fourth quarter of 2017 in the Republic of Korea. KT&G is the country's leading cigarette producer, in a market that has witnessed the rapid conversion of cigarette users to HTPs; lil was intended to create a domestic presence *(115)*. In 2018, the company launched three products with HTP technology under the lil brand: lil plus, lil mini and lil hybrid *(134)*. Lil plus and lil mini are exclusively HTPs, while lil hybrid has both HTP and ENDS technology *(135)*. In January 2020, KT&G and PMI reached an agreement that will allow PMI to distribute KT&G's smoke-free products, including the HTPs under the lil brand and lil's ENDS product, lil Vapour *(135)*.

Marketing strategies to promote sales of HTPs

Tobacco companies have used a wide range of marketing strategies to promote HTPs, often targeting adolescents and young adults *(136,137)*. The strategies for HTPs include those listed below.

- Advertisements
    - Online, including social media (e.g. Facebook, Instagram, Twitter) *(138)*
    - Television *(138)*
    - Radio *(138)*
    - Newspapers and magazines *(138)*
    - Billboards and posters *(138)*
    - Displays and advertisements at points of sale *(139)*
    - Dedicated retail stores for HTPs *(115)*
    - Bars and pubs *(138)*
- Emphasis on similarities to cigarettes *(115)*
- Acknowledgment of the harms of cigarettes, while presenting HTPs as "cleaner alternatives" *(140)*
    - In the USA, capitalizing on this potential has led manufacturers to try to circumvent stringent regulations on advertising language by applying for designation of HTPs as "modified risk tobacco products" *(92)* For example, in July 2020, PMI successfully obtained an "exposure modification" order from the US Food and Drug Administration for the IQOS system and three of its Marlboro Heatsticks, which "permits the marketing of the products with certain claims".
- Use of brand "ambassadors" (in person and on social media) and demonstrations *(115,141)*
- Product design
    - Sleek, high-tech appearance *(138,142)*
    - Rapid charging *(115)*
    - Less odour *(142)*
    - Less emission of second-hand smoke *(141)*
    - Customization with colours and limited-edition designs *(115)*
- Sponsorship *(141)*
    - Sporting events
    - Art shows
    - Concerts
    - Food and wine festivals
- Pricing strategies
    - "Bait and hook" pricing: discounted prices for devices and recurrent cost for specially designed refills or inserts *(115)*

- Free samples *(141)*
- Customer service
  - Call centre support *(115)*
  - Dedicated brand retail stores and websites *(115)*
  - Apps to help customers locate nearby stores and to troubleshoot their device *(141,143)*
- Marketing to young people
  - Placement of HTPs near youth-oriented merchandize at points of sale *(139)*
  - Sponsorship of youth-oriented events (e.g. Tel Aviv's TLV Student Day) *(141)*
- Funding front groups (e.g. Foundation for a Smoke-Free World) *(70)*
- Lobbying *(144)*
- Corporate social responsibility to boost industry image *(72)*

Common strategies

The latest generation of HTPs are not only targeted at a specific sub-segment of tobacco users but are marketed and distributed non-traditionally. In expectation of increased sales, tobacco firms are investing heavily in increasing their HTP portfolios. For example, BAT is creating additional features for its glo HTPs, which includes the next generation of devices, additional flavours and blending technologies. PMI's website notes that it is following a similar strategy, with investments in not only the next generation of IQOS products but also new HTPs like TEEPS, which has an alternative source to heat tobacco *(121)*.

HTP manufacturers also use indirect marketing tactics to reach consumers, including young people. This often involves the use of front groups, including the Foundation for a Smoke-Free World, and lobbying *(70)*. HTP companies also use corporate social responsibility strategies to boost their public image and to promote their brands *(72)*, citing philanthropy as evidence of their corporate social responsibility, often making charitable donations to environmental causes, child labour prevention organizations and organizations that extend access to education *(145)*.

10.3.3 Global use and prevalence of use of HTPs

While there are currently no robust, publicly available data from global surveillance of trends in the prevalence of HTP use, national and regional trends have been reported. The Asia–Pacific region currently reports the largest share of revenue from HTP sales and use concentrated in the age group 18–39 years.

Japan accounted for the largest share of revenue at 85% of the global HTP market in 2018 *(146)*, and the fastest rate of growth in HTP revenue was in the Republic of Korea *(147)*. In Japan in 2018, 2.7% of adults had used in HTPs in the previous 30 days and 1.7% had used them daily; nearly all the HTP users surveyed were also current or former smokers of combusted cigarettes *(148)*.

In the Republic of Korea, ever and current use of HTPs among young adults aged 19–24 years grew rapidly after IQOS was introduced in 2017; 5.7% of those surveyed reported ever use after it had been on the market for only 3 months, and 3.5% reported current use *(149)*. HTP inserts accounted for 2.2% of cigarette sales in 2017 and for 9.6% by 2018. One year after HTPs were introduced onto market, 2.8% of Korean adolescents aged 12–18 years reported ever use of HTPs *(150)*.

Market projections for 2019–2025 show significant investment by manufacturers and expectations of robust sales in Europe due to growth in Croatia, Germany, Italy, Poland and the Russian Federation *(151)*. In 2019, Euromonitor reported that Italy represented the largest HTP market outside the Asia–Pacific region; other countries with rapid growth in the market between 2018 and 2019 were Czechia, Germany, Romania, Russian Federation and Ukraine *(152)*. In 2017, only 0.7% of adults in the USA reported ever use of HTPs (2.7% of current smokers of combusted cigarettes); however, that rate increased significantly in only 1 year to 2.4% of adults in 2018 (6.7% of current cigarette smokers) *(124)*. Euromonitor data have also been used to predict the market value of HTPs by 2021; Germany, Japan, the Republic of Korea, Turkey and the USA were those predicted to be highest *(153)*.

While many countries are beginning to survey and report on trends in use of HTPs in adults, evidence is lacking on use of these products by young people and on the preferences and use patterns of adult users. Both are critical areas for future research.

10.3.4 Trends in advertising, promotion and sponsorship of HTPs

How HTPs are advertised in markets through social media

Social media platforms are an emerging channel for advertising HTPs, as sites such as Twitter and Instagram are used by both adult and young populations worldwide. A scan of HTP advertisements and promotions was conducted on Instagram and Twitter with search terms consisting of hashtags with HTP brand names (e.g. #IQOS) and the name of a WHO Member State (e.g. #Belarus). As Instagram's search function allows only one search term, the HTP brand name or search term was combined with the WHO Member State name (e.g. #IQOSCanada), whereas Twitter's search function accommodates several search terms and Boolean operators (e.g. #IQOS AND #Canada). The search showed that HTPs are marketed on Twitter and Instagram in at least 95 WHO Member

States (Table 10.3). Analysis of the content of up to six advertisements (three Twitter, three Instagram) per country indicated use of similar advertisement strategies in the six WHO regions.

Table 10.3. Common HTP advertising strategies on social media, by WHO region, 2020

WHO region	No. of countries	Data available No. (%)	No. of advertisements	Health warning No. (%)	Nicotine lexical No. (%)	Flavour lexical No. (%)	Design feature lexical No. (%)	Image device No. (%)	Image of e-liquid No. (%)
African	46	3 (7)	9	1 (11)	3 (33)	0	0	7 (78)	4 (44)
Americas	35	12 (34)	42	17 (40)	3 (7)	12 (29)	9 (21)	38 (90)	12 (29)
South-East Asia	11	6 (56)	16	11 (69)	3 (19)	3 (19)	4 (25)	14 (89)	4 (25)
European	53	48 (91)	186	86 (46)	21 (11)	41 (22)	22 (12)	146 (78)	30 (16)
Eastern Mediterranean	22	15 (68)	62	12 (19)	17 (27)	17 (27)	4 (6)	55 (89)	18 (29)
Western Pacific	27	11 (41)	39	13 (33)	2 (5)	8 (21)	13 (33)	38 (97)	8 (21)
All	194	95 (49)	354	140 (40)	49 (14)	81 (23)	52 (15)	298 (84)	76 (21)

HTP advertisements were present in approximately half of WHO Member States, and most included an image of the device. Less than half of the social media advertisements for HTPs included a health warning, and almost one quarter mentioned a flavour.

Impact of COVID-19 on marketing strategies for HTPs

During the COVID-19 pandemic, several HTP manufacturers, retailers and users have posted messages relevant to COVID-19 and related containment strategies in marketing and promotion strategies on social media. Themes related to the pandemic and containment that emerged in HTP marketing on Instagram and Twitter included the following.

Coping with boredom and isolation: Manufacturers have marketed their products as a means for users to enjoy themselves safely at home during lockdown measures. For example, an Instagram post from glo's worldwide account stated "Left brain says stay in. Right brain wants to go out. We're loving these new tools that let you party from home with no compromise. #BreakBinary #NetflixParty #Discoverglo #Myglo" *(66).*

Stocking up on essential supplies: Users have implied that people should "stock up" on their favourite HTPs in preparation for lockdown or quarantine, equating these products to "essential" supplies for the duration of the lockdown. For example, one user posted a photo of several HTP devices (primarily IQOS) and several packages of HEETS (heated tobacco units inserted into IQOS devices), with the statement "I'm ready for quarantine!" Another user posted a photo on both Instagram and Twitter of an IQOS device and three packages of HEETS, with the caption "#lockdown #essentials and/or #quarantine #addictions. Thanks for the gift @iqos_it".

Stay at home campaigns. Some HTP brands are using "stay at home" campaigns and messages on social media to promote their products. For example, glo Greece held a "stay home challenge", in which participants could win prizes *(66)*. This post capitalized on the #menoumespiti message on Instagram, a popular hashtag meaning "we stay home" in Greek.

Obtaining protective equipment and supplies. Some manufacturers are offering branded face masks and hand sanitizer with the purchase of their products. For example, one Instagram advertisement by glo Kazakhstan shows a woman wearing a glo-branded face mask, with the caption "...Can't find a mask anywhere? We'll give it to you [winking face emoji]. Until Thursday 26 March, you can get a mask for free in the glo space on Nazarbayev 100G. Well, after that you can get the same mask by placing a purchase of the device on the website or in our outlet..." *(66)*.

10.3.5 Measures to control advertising, promotion and sponsorship of HTPs

In the countries that regulate advertising, promotion or sponsorship of HTPs, many have done so by including HTPs in existing regulations, some as novel tobacco products and some in other categories, whereas others have specific regulations for these products. Furthermore, some countries consider HTPs to be tobacco products rather than giving them a separate designation, so that they are subjected to all current national tobacco product regulations *(113,154)*. Continued monitoring of such trends and their effects on use will provide valuable insight for policy-makers across the world who are observing trends in HTP use and for the design of a comprehensive regulatory framework or for updating existing regulations to reduce the harm associated with tobacco use.

While no peer-reviewed evidence is yet available on an association between HTP advertising and promotion and use, researchers have pointed to youth-friendly HTP advertising that has also been used to market ENDS, such as high-tech, novel design features, claims that HTPs are less harmful than combusted products and messages that HTPs may be more socially acceptable than combusted products *(136,150)*. Given the evidence of successful use of these strategies to advertise ENDS products, HTP advertising, promotional messaging and media should be closely monitored, with trends in the susceptibility of children and adolescents to using HTPs.

10.3.6 Recommendations for monitoring trends in marketing, advertising, promoting and sponsorship of HTPs

- **Better surveillance of trends in HTP use and sales**

  To increase the impact of the recommendations below, more robust data should be collected in various countries on trends in HTP use, including demographic data and product preferences. While the new generation of HTPs are relatively new on the tobacco product market, the rapid rise in the popularity of ENDS indicates that more rapid surveillance of HTP use and sales and reporting on national policies will be critical to including HTPs in existing control frameworks.

- **Better surveillance of HTP marketing, with particular attention to social media**

  The robust presence of marketing for alternative tobacco products on youth-friendly social media platforms suggests that increased surveillance of marketing trends is necessary to better understand whether exposure to such marketing is associated with attitudes about and use of HTPs. This should include surveillance of content generated by HTP companies, retailers and users.

- **Scanning of how governments globally are regulating HTPs, including their advertisement, promotion and sponsorship**

  Additional information should be collected on the regulation, marketing and promotion of HTPs in order to understand the global picture. Better reporting of key policy developments across the world will assist in the development of timely, effective strategies for all Parties to the WHO FCTC.

10.3.7 Recommendations for regulators

- **Remain focused on evidence-based smoking prevention strategies.**
  Governments and health organizations should maintain a focus on evidence-based measures to reduce smoking (as outlined in the WHO FCTC) and should not be distracted from action by the promotion and marketing of novel products such as HTPs.

- **Consider cost-effective counter-marketing strategies.**
  Given the global reach of HTP advertising on social media, neither

governments nor social media platforms can effectively regulate the content, particularly when generated by users. Counter-marketing may be the most feasible option. Various forms could be used, including social media campaigns, which may be more cost-effective than traditional media campaigns. Counter-marketing can include education about industry activity, in addition to discouraging nicotine and tobacco use.

- **Apply relevant lessons learnt from regulation of ENDS and ENNDS**
  As HTP companies have adopted many of the same marketing strategies that they used to promote ENDS and ENNDS, regulators can apply lessons learnt from ENDS and ENNDS marketing to regulating HTP marketing and promotion.

- **Foster collaboration among governments and government sectors for implementing and enforcing marketing regulations for HTPs.**
  Cross-border advertising, including transmission of media into neighbouring countries, should be addressed collaboratively at government level.

- **Consider banning all tobacco advertising, promotion and sponsorship, where possible.**
  The tobacco industry has circumvented nearly all restrictions on advertising, promotion and sponsorship to reach consumers, including young people. Therefore, a complete ban on tobacco marketing may be necessary to minimize exposure to marketing for nicotine and tobacco products. Such a ban would ensure that most communication about these products came from national and local governments and public health agencies.

- **Keep informed of industry strategies to market HTPs, particularly to young people.**
  In order for policies to control HTP marketing to be effective, regulators must learn to recognize the strategies used by the industry, including use of health-related claims, sponsorship and price promotions.

10.3.8 Gaps in research on HTPs

- Additional evidence is required to understand the relations among HTP regulation, perception of the risk of these products and product use.

    Much of the evidence on novel and emerging nicotine and tobacco products addresses the association between policies and trends in use; however, evidence specific to HTPs is insufficient. Further evaluation is necessary of how different policies and product characteristics change perceptions of risk and/or product use (particularly among young people).

- Additional research is necessary on HTP marketing, with particular attention to social media, to inform regulators of the marketing and promotion strategies used by HTP companies and retailers.

    Social media platforms are an important channel for advertising and promoting HTPs worldwide. As advertising seen on social media platforms has been associated with increased interest in using tobacco and nicotine products and also their actual use, research should be conducted on the impact of HTP marketing and promotion on these platforms *(98,99)*.

- Additional research should be conducted on user-generated HTP content on social media and its effect on risk perception, product use and the effectiveness of marketing regulations.

    User-generated content has been used as an important marketing tactic for ENDS and ENNDS companies. Therefore, trends in user-generated content related to HTPs should be monitored, with studies of its impact on risk perceptions and product use and on the effectiveness of marketing regulations.

10.4 Summary

Globally, ENDS, ENNDS and HTPs are marketed through both traditional and emerging channels, such as social media. The evidence gathered for this report indicates that ENDS and HTPs are heavily marketed on Twitter and Instagram. The strategies used to regulate the marketing and promotion of these products differ widely among countries, some banning certain products, some regulating only products containing nicotine and others imposing restrictions on flavours, packaging and advertisements. Global systems are necessary to monitor marketing of ENDS, ENNDS and HTP to understand how these products are advertised; additional evidence is necessary to understand how marketing of these products influences product perception and product use, particularly among adolescents and young adults. All levels of government should regulate the advertising, promotion and sponsorship of ENDS, ENNDS and HTPs when such regulations do not exist.

10.5 References

1. WHO global report on trends in prevalence of tobacco use 2000–2025. Third edition. Geneva: World Health Organization; 2019.
2. The global cigarette industry. Washington (DC): Campaign for Tobacco-Free Kids; 2019.
3. Smoking population – number of adult smokers. London: Euromonitor International Ltd; 2019 (http://www.portal.euromonitor.com.proxy1.library.jhu.edu/portal/statisticsevolution/index).
4. e-Cigarette use among youth and young adults: A report of the Surgeon General. Washington (DC): Department of Health and Human Services; 2016.
5. 6 important things to know about IQOS, the new heated cigarette product. Washington (DC): Truth Initiative; 2019 (https://truthinitiative.org/research-resources/emerging-tobacco-products/6-important-things-know-about-iqos-new-heated, accessed 18 December 2020).
6. Cullen KA, Ambrose BK, Gentzke A, Apelberg BJ, Jamal A, King BA. Notes from the field: Use of electronic cigarettes and any tobacco product among middle and high school students – United States, 2011–2018. Morb Mortal Wkly Rep. 2018;67:629–33.
7. Youth tobacco use: Results from the National Youth Tobacco Survey. Silver Spring (MD): Food and Drug Administration; 2020 (https://www.fda.gov/tobacco-products/youth-and-tobacco/youth-tobacco-use-results-national-youth-tobacco-survey#1, accessed 18 August 2020).
8. Trump administration combating epidemic of youth e-cigarette use with plan to clear market of unauthorized, non-tobacco-flavored e-cigarette products. Silver Spring (MD): Food and Drug Administration; 2019 (https://www.fda.gov/news-events/press-announcements/trump-administration-combating-epidemic-youth-e-cigarette-use-plan-clear-market-unauthorized-non#:~:text=%E2%80%9CThe%20Trump%20Administration%20is%20making,Human%20Services%20Secretary%20Alex%20Azar, accessed 10 January 2021).
9. Conference of the Parties to the WHO Framework Convention on Tobacco Control. Decision FCTC/COP7(9) Electronic nicotine delivery systems and electronic non-nicotine delivery systems. Geneva: World Health Organization; 2016 (https://www.who.int/fctc/cop/cop7/FCTC_COP7_9_EN.pdf?ua=1, accessed 10 January 2021).
10. Conference of the Parties to the WHO Framework Convention on Tobacco Control. Decision FCTC/COP8(22) Novel and emerging tobacco products. Geneva: World Health Organization; 2018 (https://www.who.int/fctc/cop/sessions/cop8/FCTC_COP8(22).pdf?ua=1, accessed 13 August 2020).
11. Public health consequences of e-cigarettes. Washington (DC): National Academies of Sciences Engineering and Medicine; 2018. doi:10.17226/24952.
12. Williams M, Bozhilov K, Ghai S, Talbot P. Elements including metals in the atomizer and aerosol of disposable electronic cigarettes and electronic hookahs. PLoS One. 2017;12(4):e0175430.
13. World Health Assembly resolution 56.1. Geneva: World Health Organization; 2003 (https://www.who.int/tobacco/framework/final_text/en/). https://www.who.int/tobacco/framework/final_text/en/index3.html, accessed August 16, 2020.
14. Conference of the Parties to the WHO Framework Convention on Tobacco Control. Electronic

nicotine delivery systems and electronic non-nicotine delivery systems (ENDS/ENNDS): Report by WHO. Geneva: World Health Organization; 2016 (https://www.who.int/fctc/cop/cop7/FCTC_COP_7_11_EN.pdf, accessed 13 August 2020).
15. Stone E, Marshall H. Tobacco and electronic nicotine delivery systems regulation. Transl Lung Cancer Res. 2019;8(S1):S67–76.
16. MacGuill S. Growth in vapour products. London: Euromonitor International; 2017 (https://blog.euromonitor.com/growth-vapour-products/, accessed 13 December 2020).
17. Kim AE, Arnold KY, Makarenko O. e-Cigarette advertising expenditures in the US, 2011–2012. Am J Prev Med. 2014;46(4):409–12.
18. Electronic cigarettes: An overview of key issues. Washington (DC): Campaign for Tobacco-Free Kids; 2020 (https://www.tobaccofreekids.org/assets/factsheets/0379.pdf).
19. Tobacco companies. In: Mackay J, Eriksen M, editors. The tobacco atlas. Geneva: World Health Organization; 2002:51 (https://www.who.int/tobacco/en/atlas18.pdf?ua=1, accessed 13 August 2020).
20. Exhibit 99.1. Altria makes $12.8 billion minority investment in Juul to accelerate harm reduction and drive growth. Washington (DC): United States Securities Exchange Commission; 2019 (https://www.sec.gov/Archives/edgar/data/764180/000119312518353970/d660871dex991.htm, accessed 13 August 2020).
21. Kalra A, Kirkham C. Exclusive: Juul plans India e-cigarette entry with new hires, subsidiary. Reuters, 30 January 2019 (https://www.reuters.com/article/us-juul-india-exclusive-idUSKCN1PO0VV, accessed 11 November 2020).
22. Gupta S. India is banning all e-cigarettes over fears about youth vaping. CNN, 18 September 2019 (https://edition.cnn.com/2019/09/18/health/india-e-cigarette-ban-intl/index.html#:~:text=India%20is%20banning%20all%20e%2Dcigarettes%20over%20fears%20about%20youth%20vaping&text=Electronic%20cigarette%20devices%20on%20display,risk%2C%20especially%20to%20young%20people).
23. Kastrenakes J, Garun N. Juul launches a Bluetooth e-cigarette that tracks how much you vape. The Verge, 6 August 2019 (https://www.theverge.com/2019/8/6/20754655/juul-c1-bluetooth-e-cigarette-vape-monitor-consumption-age-restriction, accessed 13 August 2020).
24. Turning Point Brands, Inc. announces third quarter 2018 results. Business Wire, 7 November 2018 (https://www.businesswire.com/news/home/20181107005281/en/Turning-Point-Brands-Announces-Quarter-2018-Results, accessed 13 August 2020).
25. Find your country's online store. San Francisco (CA): JUUL Labs (https://www.juul.com/global, accessed 16 August 2020).
26. 2018. Altria Group, Inc. Annual Report. Richmond (VA): Altria Group, Inc.; 2018 (https://www.annualreports.com/HostedData/AnnualReportArchive/a/NYSE_MO_2018.pdf, accessed 10 January 2021).
27. STEEM: Exploring new e-vapor products. Lausanne: Philip Morris International (https://www.pmi.com/smoke-free-products/steem-creating-a-vapor-with-nicotine-salt, accessed 13 August 2020).
28. Brand shares: E-Vapour products. London: Euromonitor International Inc.; 2018 (https://www-portal-euromonitor-com.proxy1.library.jhu.edu/portal/statisticsevolution/index, accessed 13 August 2020).
29. Japan Tobacco Inc. Annual Report 2018. Tokyo: Japan Tobacco Inc.; 2019 (https://www.jt.com/investors/results/annual_report/pdf/2018/annual.fy2018_E2.pdf, accessed 13 August 2020).
30. 2019 earnings report. Geneva: Japan Tobacco International; 2020 (https://www.jti.com/sites/default/files/press-releases/documents/2020/jt-group-2019-financial-results-and-2020-forecast.pdf, accessed 13 August 2020).
31. JTI launches Logic Compact in the UK – the first market to launch this new premium e-ciga-

rette: The moment vaping "just clicks" has arrived. Geneva: Japan Tobacco International; 2018 (https://www.jti.com/jti-launches-logic-compact-uk-first-market-launch-new-premium-e-cigarette, accessed 13 August 2020).
32. British American Tobacco plc. Transforming tobacco: Annual report and Form 20-F 2018. London; 2018 (https://www.bat.com/group/sites/UK__9D9KCY.nsf/vwPagesWebLive/DOAWWGJT/$file/Annual_Report_and_Form_20-F_2018.pdf, accessed 16 December 2020).
33. Vapour products. London: British American Tobacco plc; 2020 (https://www.bat.com/ecigarettes, accessed 13 August 2020).
34. British American Tobacco to focus on three global new category brands to further accelerate their growth. London: British American Tobacco plc; 2019 (https://www.bat.com/group/sites/UK__9D9KCY.nsf/vwPagesWebLive/DOBJBMSE, accessed 16 August 2020).
35. Form 10-K. Reynolds American Inc. Securities registered pursuant to Section 12(b) of the Securities Exchange Act of 1934; Washington (DC): United States Securities Exchange Commission; 2017 (https://www.sec.gov/Archives/edgar/data/1275283/000156459017001245/rai-10k_20161231.htm, accessed 13 August 2020).
36. Something better: Annual report and accounts 2018. London; Imperial Brands plc; 2018 (https://www.imperialbrandsplc.com/content/dam/imperial-brands/corporate/investors/annual-report-and-accounts/2018/annual-report-and-accounts-2018.pdf, accessed 13 August 2020).
37. Imperial Brands CAGNY presentation 21 02 2019. London: Imperial Brands plc; 2019 (https://www.imperialbrandsplc.com/content/dam/imperial-brands/corporate/investors/presentations/conferences/2019/CAGNY Script.pdf.downloadasset.pdf, accessed 13 August 2020).
38. Imperial Brands plc. Annual report and accounts 2019. London: Imperial Brands plc; 2019 (https://www.imperialbrandsplc.com/content/dam/imperial-brands/corporate/investors/annual-report-and-accounts/2019/Annual Report 2019.pdf, accessed 13 August 2020).
39. What happened to the blu ACE device? Amsterdam: Fontem Ventures BV; 2020 (https://support-uk.blu.com/hc/en-gb/articles/360015460600-What-happened-to-the-blu-ACE-device-, accessed 13 August 2020).
40. Imperial Brands plc interim results for the six months ended 31 March 2018. London: Imperial Brands plc; 2018 (https://www.imperialbrandsplc.com/content/dam/imperial-brands/corporate/investors/results-centre/2018/2018-05-09 Interims 18 script.pdf, accessed 13 August 2020).
41. Padon AA, Lochbuehler K, Maloney EK, Cappella JN. A randomized trial of the effect of youth appealing e-cigarette advertising on susceptibility to use e-cigarettes among youth. Nicotine Tob Res. 2018;20(8):954–61.
42. Mantey DS, Cooper MR, Clendennen SL, Pasch KE, Perry CL. e-Cigarette marketing exposure is associated with e-cigarette use among US youth. J Adolesc Health. 2016;58(6):686–90.
43. Jackler RK, Chau C, Getachew BD, Whitcomb MM, Lee-Heidenreich J, Bhatt AM et al. JUUL advertising over its first three years on the market. Berkeley (CA): Stanford Research into the Impact of Tobacco Advertising; 2019 (http://tobacco.stanford.edu/tobacco_main/publications/JUUL_Marketing_Stanford.pdf).
44. Sebastian M. e-Cig marketing budgets growing by more than 100% year over year. AdAge, 14 April 2014 (https://adage.com/article/media/e-cig-companies-spent-60-million-ads-year/292641/).
45. Duke JC, Lee YO, Kim AE, Watson KA, Arnold KY, Nonnemaker JM et al. Exposure to electronic cigarette television advertisements among youth and young adults. Pediatrics. 2014;134(1):e29–36.
46. Gregory A. Tobacco companies "pushing e-cigs on youngsters via Facebook and Twitter". The Mirror, 27 November 2013 (https://www.mirror.co.uk/lifestyle/health/tobacco-compani-

es-pushing-e-cigs-youngsters-2854699, accessed 13 August 2020).
47. Chen-Sankey JC, Unger JB, Bansal-Travers M, Niederdeppe J, Bernat E, Choi K. e-Cigarette marketing exposure and subsequent experimentation among youth and young adults. Pediatrics. 2019;144(5):e20191119.
48. Carpenter CM, Wayne GF, Pauly JL, Koh HK, Connolly GN. New cigarette brands with flavors that appeal to youth: Tobacco marketing strategies. Health Aff. 2005;24(6):1601–10.
49. 7 ways e-cigarette companies are copying big tobacco's playbook. Washington (DC): Campaign for Tobacco-Free Kids; 2013 (https://www.tobaccofreekids.org/blog/2013_10_02_ecigarettes, accessed August 16, 2020).
50. 4 marketing tactics e-cigarette companies use to target youth. Washington (DC): Truth Initiative; 2018 (https://truthinitiative.org/research-resources/tobacco-industry-marketing/4-marketing-tactics-e-cigarette-companies-use-target, accessed 16 August 2020).
51. Evidence brief: Tobacco industry sponsored youth prevention programs in schools. Atlanta (GA): Centers for Disease Control and Prevention; 2019 (https://www.cdc.gov/tobacco/basic_information/youth/evidence-brief/index.htm, accessed 16 August 2020).
52. Jackler RK, Ramamurthi D. Unicorns cartoons: marketing sweet and creamy e-juice to youth. Tob Control. 2017;26(4):471–5.
53. Ashley DL, Spears CA, Weaver SR, Huang J, Eriksen MP. e-Cigarettes: How can they help smokers quit without addicting a new generation? Prev Med. 2020:106145.
54. Giovenco DP, Casseus M, Duncan DT, Coups EJ, Lewis MJ, Delnevo CD. Association between electronic cigarette marketing near schools and e-cigarette use among youth. J Adolesc Health. 2016;59(6):627–34.
55. Kaplan S. Juul targeted schools and youth camps, House Panel on Vaping Claims. The New York Times, 25 July 2019 (https://www.nytimes.com/2019/07/25/health/juul-teens-vaping.html, accessed 13 August 2020).
56. Kirkpatrick MG, Cruz TB, Goldenson NI, Allem JP, Chu KH, Pentz MA et al. Electronic cigarette retailers use Pokémon Go to market products. Tob Control. 2017;26(E2):e147.
57. Cardellino C. Electronic cigarettes available for free at Fashion Week. Cosmopolitan. 5 September 2013 (https://www.cosmopolitan.com/style-beauty/fashion/advice/a4736/nyfw-2013-cigarettes/, accessed 13 August 2020).
58. Ali FRM, Xu X, Tynan MA, King BA. Use of price promotions among US adults who use electronic vapor products. Am J Prev Med. 2018;55(2):240–3. 0
59. D'Angelo H, Rose SW, Golden SD, Queen T, Ribisl KM. e-Cigarette availability, price promotions and marketing at the point-of sale in the contiguous United States (2014–2015): National estimates and multilevel correlates. Prev Med Rep. 2020;19:101152.
60. Wadsworth E, McNeill A, Li L, Hammond D, Thrasher JF, Yong HH et al. Reported exposure to e-cigarette advertising and promotion in different regulatory environments: Findings from the International Tobacco Control Four Country (ITC-4C) survey. Prev Med. 2018;112:130–7.
61. De Andrade M, Hastings G, Angus K. Promotion of electronic cigarettes: tobacco marketing reinvented? BMJ. 2013;347:f7473.
62. Lee SJ, Rees VW, Yossefy N, Emmons KM, Tan ASL. Youth and young adult use of pod-based electronic cigarettes from 2015 to 2019: A systematic review. JAMA Pediatr. 2020;174(7):714–20.
63. Bach L. JUUL and youth: Rising e-cigarette popularity. Washongton DC: TobaccoFree Kids; 2019 (https://www.tobaccofreekids.org/assets/factsheets/0394.pdf, accessed 16 August 2020).
64. Hammond D, Wackowski OA, Reid JL, O'Connor RJ. Use of JUUL e-cigarettes among youth in the United States. Nicotine Tob Res. 2018;22(5):827–32.
65. Jivanda T. "Put it in my mouth": Viewers outraged by apparent reference to oral sex in VIP e-cig advert. Independent, 4 December 2013 (https://www.independent.co.uk/news/media/advertising/put-it-my-mouth-viewers-outraged-apparent-reference-oral-sex-vip-e-cig-ad-

vert-8982918.html, accessed 10 January 2021).
66. Hickman A. "Big tobacco" using COVID-19 messaging and influencers to market products. PR Week, 15 May 2020 (https://www.prweek.com/article/1683314/big-tobacco-using-covid-19-messaging-influencers-market-products, accessed 13 August 2020).
67. Yao T, Jiang N, Grana R, Ling PM, Glantz SA. A content analysis of electronic cigarette manufacturer websites in China. Tob Control. 2016;25(2):188–94.
68. Klein EG, Berman M, Hemmerich N, Carlson C, Htut S, Slater M. Online e-cigarette marketing claims: A systematic content and legal analysis. Tob Regul Sci. 2016;2(3):252–62.
69. Hrywna M, Bover Manderski MT, Delnevo CD. Prevalence of electronic cigarette use among adolescents in New Jersey and association with social factors. JAMA Netw Open. 2020;3(2):e1920561.
70. Tobacco tactics. Front groups. Bath: University of Bath; 2019 (https://tobaccotactics.org/wiki/front-groups, accessed 14 October 2020).
71. Tobacco tactics. Lobby groups. Bath: University of Bath; 2019 (https://tobaccotactics.org/topics/lobby-groups/, accessed 14 October 2020).
72. Tobacco tactics. CSR strategy. Bath: University of Bath; 2020 (https://tobaccotactics.org/wiki/csr-strategy/, accessed 15 October 2020).
73. Romberg AR, Miller Lo EJ, Cuccia AF, Willett GJ, Xiao H, Hair EC et al. Patterns of nicotine concentrations in electronic cigarettes sold in the United States, 2013–2018. Drug Alcohol Depend. 2019;203:1–7.
74. How much nicotine is in JUUL? Washington (DC): Truth Initiative; 2019 (https://truthinitiative.org/research-resources/emerging-tobacco-products/how-much-nicotine-juul, accessed December 2020).
75. Hammond D, Rynard VL, Reid JL. Changes in prevalence of vaping among youths in the United States, Canada, and England from 2017 to 2019. JAMA Pediatr. 2020;174(8):797–800.
76. Chen C, Zhuang YL, Zhu SH. e-Cigarette design preference and smoking cessation. Am J Prev Med. 2016;51(3):356–63.
77. e-Cigarettes: Facts, stats and regulations. Washington (DC): Truth Initiative; 2019 (https://truthinitiative.org/research-resources/emerging-tobacco-products/e-cigarettes-facts-stats-and-regulations#:~:text=Between%202012%20and%202013%2C%202.4,among%20adults%20aged%2045%2D54).
78. Tobacco tactics. Forest. Bath: University of Bath; 2020 (https://tobaccotactics.org/wiki/forest/, accessed 15 October 2020).
79. Tobacco tactics. Foundation for a Smoke-Free World. Bath: University of Bath (https://tobaccotactics.org/wiki/foundation-for-a-smoke-free-world/, accessed 10 January 2021).
80. 2018 recipients of charitable contributions from the Altria family of companies. Richmond (VA): Altria; 2018 (https://web.archive.org/web/20200320145936/https://www.altria.com/-/media/Project/Altria/Altria/responsibility/investing-in-communities/2018Grantees.pdf, accessed 15 October 2020).
81. Statistics and market data on consumer goods & FMCG: Tobacco. New York City (NY): Statista; 2020.
82. Electronic cigarette – global market outlook (2017–2026). Dublin: Research and Markets; 2019 (https://www.researchandmarkets.com/reports/4827644/electronic-cigarette-global-market-outlook, accessed 10 January 2021).
83. Bao W, Xu G, Lu J, Snetselaar LG, Wallace RB. Changes in electronic cigarette use among adults in the United States, 2014–2016. JAMA. 2018;319(19):2039.
84. Wang T, Asman K, Gentzke A, Cullen KA, Holder-Hayes E, Reyes-Guzman C et al. Tobacco product use among adults — United States, 2017. Morb Mortal Wkly Rep. 2017;67(44):1225–32.
85. Canadian Tobacco, Alcohols and Drugs Survey (CTADS): summary of results for 2017. Ottawa:

86. Special Eurobarometer 458: Attitudes of Europeans toward tobacco and electronic cigarettes. Brussels: European Commission; 2017.
87. Schoenborn CA, Clarke TC. QuickStats: Percentage of adults who ever used an e-cigarette and percentage who currently use e-cigarettes, by age group – National Health Interview Survey, United States, 2016. Morb Mortal Wkly Rep. 2017;66(33):892.
88. Hammond D, Reid JL, Rynard VL, Fong GT, Cummings KM, McNeill A et al. Prevalence of vaping and smoking among adolescents in Canada, England, and the United States: repeat national cross sectional surveys. BMJ. 2019;365:l2219.
89. Country laws regulating e-cigarettes. Advertising, promotion and sponsorship. Baltimore (MD): Institute for Global Tobacco Control; 2020 (https://www.globaltobaccocontrol.org/e-cigarette/advertising-promotion-and-sponsorship, accessed 16 August 2020).
90. e-Cigarettes at the point of sale. CounterTobacco.Org; 2020 (https://countertobacco.org/resources-tools/evidence-summaries/e-cigarettes-at-the-point-of-sale/, accessed 16 August 2020).
91. Modified risk tobacco product applications. Silver Spring (MD): Food and Drug Administration, Center for Tobacco Products; 2020.
92. FDA finalizes enforcement policy on unauthorized flavored cartridge-based e-cigarettes that appeal to children, including fruit and mint. Silver Spring (MD): Food and Drug Administration; 2020 (https://www.fda.gov/news-events/press-announcements/fda-finalizes-enforcement-policy-unauthorized-flavored-cartridge-based-e-cigarettes-appeal-children, accessed 17 August 2020).
93. Notice of intent – Potential measures to reduce the impact of vaping product advertising on youth and non-users of tobacco products. Ottawa: Health Canada; 2019 (https://www.canada.ca/en/health-canada/programs/consultation-measures-reduce-impact-vaping-products-advertising-youth-non-users-tobacco-products/notice-document.html).
94. Canada Gazette, Part I, Volume 153, Number 25: Vaping Products Labelling and Packaging Regulations. Ottawa: Health Canada; 2019 (http://gazette.gc.ca/rp-pr/p1/2019/2019-06-22/html/reg4-eng.html).
95. Nicksic NE, Snell LM, Rudy AK, Cobb CO, Barnes AJ. Tobacco marketing, e-cigarette susceptibility, and perceptions among adults. Am J Health Behav. 2017;41(5):579–90.
96. Nicksic NE, Snell LM, Barnes AJ. Does exposure and receptivity to e-cigarette advertisements relate to e-cigarette and conventional cigarette use behaviors among youth? Results from wave 1 of the Population Assessment of Tobacco and Health Study. J Appl Res Child. 2017;8(2):3.
97. Cho YJ, Thrasher JF, Reid JL, Hitchman S, Hammond D. Youth self-reported exposure to and perceptions of vaping advertisements: Findings from the 2017 International Tobacco Control Youth Tobacco and Vaping Survey. Prev Med. 2019;126:105775.
98. Pokhrel P, Fagan P, Herzog TA, Chen Q, Muranaka N, Kehl L et al. e-Cigarette advertising exposure and implicit attitudes among young adult non-smokers. Drug Alcohol Depend. 2016;163:134–40.
99. Camenga D, Gutierrez KM, Kong G, Cavallo D, Simon P, Krishnan-Sarin S. e-Cigarette advertising exposure in e-cigarette naïve adolescents and subsequent e-cigarette use: A longitudinal cohort study. Addict Behav. 2018;81:78–83.
100. Phua J, Jin SV, Hahm JM. Celebrity-endorsed e-cigarette brand Instagram advertisements: Effects on young adults' attitudes towards e-cigarettes and smoking intentions. J Health Psychol. 2018;23(4):550–60.
101. Behind the explosive growth of JUUL. Washington (DC): Truth Initiative; 2019.
102. Marketing tobacco to LGBT communities. Washington (DC): National Institutes of Health; 2020

(https://smokefree.gov/marketing-tobacco-lgbt-communities, accessed 17 August 2020).
103. Tobacco industry marketing. Atlanta (GA): Centers for Disease Control and Prevention; 2020 (https://www.cdc.gov/tobacco/data_statistics/fact_sheets/tobacco_industry/marketing/index.htm, accessed 17 August 2020).
104. Berg CJ, Dekanosidze A, Torosyan A, Grigoryan L, Sargsyan Z, Hayrumyan V et al. Examining smoke-free coalitions in Armenia and Georgia: Baseline community capacity. Health Educ Res. 2019;34(5):495–504.
105. Berg CJ. Local coalitions as an underutilized and understudied approach for promoting tobacco control in low- and middle-income countries. J Glob Health. 2019;9(1):10301.
106. Restricting tobacco advertising. Counter Tobacco; 2020 (https://countertobacco.org/policy/restricting-tobacco-advertising-and-promotions/, accessed 17 August 2020).
107. Public health groups and leaders worldwide urge rejection of Philip Morris International's new foundation. Washington (DC): Campaign for Tobacco-Free Kids; 2020 (https://www.tobaccofreekids.org/what-we-do/industry-watch/pmi-foundation/compilation, accessed 17 August 2020).
108. Spinning a new tobacco industry: How Big Tobacco is trying to sell a do-gooder image and what Americans think about it. Washington (DC): Truth Initiative; 2019 (https://truthinitiative.org/research-resources/tobacco-industry-marketing/spinning-new-tobacco-industry-how-big-tobacco-trying, accessed 17 August 2020).
109. Guidelines for implementation of Article 5.3 of the WHO Framework Convention on Tobacco Control. Geneva: World Health Organization; 2011 (https://www.who.int/fctc/guidelines/article_5_3.pdf?ua=1, accessed 17 August 2020).
110. Van der Eijk Y, Bero LA, Malone RE. Philip Morris International-funded "Foundation for a Smoke-Free World": analysing its claims of independence. Tob Control. 2019;28:712–8.
111. Byron MJ, Jeong M, Abrams DB, Brewer NT. Public misperception that very low nicotine cigarettes are less carcinogenic. Tob Control. 2018;27(6):712–4.
112. Most JUUL-related Instagram posts appeal to youth culture and lifestyles, study finds. Washington (DC): Truth Initiative; 2019 (https://truthinitiative.org/research-resources/emerging-tobacco-products/most-juul-related-instagram-posts-appeal-youth-culture, accessed 15 October 2020).
113. Heated tobacco products. Information sheet. 2nd edition. Geneva: World Health Organization; 2020 (https://www.who.int/publications/i/item/WHO-HEP-HPR-2020.2).
114. Heated tobacco products. Atlanta (GA): Centers for Disease Control and Prevention; 2020 (https://www.cdc.gov/tobacco/basic_information/heated-tobacco-products/index.html, accessed 10 January 2021).
115. Heated tobacco products (HTPs) market monitoring information sheet. Geneva: World Health Organization; 2018 (https://apps.who.int/iris/bitstream/handle/10665/273459/WHO-NMH-PND-18.7-eng.pdf?ua=1).
116. Cigarettes to record US$7.7 billion loss by 2021 as heated tobacco grows 691 percent. Euromonitor International, 24 June 2017 (https://blog.euromonitor.com/cigarettes-record-loss-heated-tobacco-grows-691-percent/).
117. Heat not burn tobacco products market – forecasts from 2019 to 2024. Dublin: Research and Markets; 2019 (https://www.researchandmarkets.com/reports/4849645/heat-not-burn-tobacco-products-market-forecasts).
118. Cigarette market: Global industry trends, share, size, growth, opportunity and forecast 2019–2024. Dublin: Research and Markets; 2019 (https://www.researchandmarkets.com/reports/4752264/cigarette-market-global-industry-trends-share#:~:text=The%20global%20cigarette%20market%20was,4%25%20during%202019%2D2024).
119. The global cigarette industry. Washington (DC): Campaign for Tobacco-Free Kids; 2019 (htt-

ps://www.tobaccofreekids.org/assets/global/pdfs/en/Global_Cigarette_Industry_pdf.pdf, accessed 13 August 2020).

120. 2020 second-quarter & year-to-date highlights. Lausanne: Philip Morris International; 2020 (https://philipmorrisinternational.gcs-web.com/static-files/b88146c3-a4e9-421e-b2aa-b470e74fcdf2, accessed 13 August 2020).

121. Carbon heated tobacco product: TEEPS. Lausanne: Philip Morris International; 2020.

122. Cigarettes in Japan: Country report. London: Euromonitor International; 2020 (https://www.euromonitor.com/cigarettes-in-japan/report, accessed 13 December 2020).

123. Korea – Ploom Tech. Geneva: Japan Tobacco International; 2020 (https://www.jti.com/node/5242, accessed 13 August 2020).

124. Japan Tobacco Inc. Integrated report 2019. Tokyo: Japan Tobacco Inc.; 2020 (https://www.jti.com/sites/default/files/global-files/documents/jti-annual-reports/integrated-report-2019v.pdf, accessed 13 August 2020).

125. Japan Tobacco International. 2020 second quarter results. Lausanne; 2020 (https://www.jt.com/investors/results/forecast/pdf/2020/Second_Quarter/20200731_07.pdf, accessed 13 August 2020).

126. JT launches two new tobacco vapor products under its Ploom brand. Tokyo: Japan Tobacco Inc.; 2019 (https://www.jti.com/sites/default/files/press-releases/documents/2019/JT-launches-two-new-tobacco-vapor-products-under-its-ploom-brand_0.pdf, accessed 13 August 2020).

127. JT launches Ploom S 2.0, an evolved heated tobacco device, in Japan on July 2nd and introduces two new Camel menthol tobacco stick products. Tokyo: Japan Tobacco Inc., 2020 (https://www.jt.com/media/news/2020/0601_01.html, accessed 13 August 2020).

128. Tobacco heating products: Innovating to lead the transformation of tobacco. London: British American Tobacco plc; 2020 (https://www.bat.com/group/sites/UK__9D9KCY.nsf/vwPages-WebLive/DOAWUGNJ, accessed 17 August 2020).

129. BAT Science: Introduction to tobacco heating products. London: British American Tobacco plc; 2020 (https://www.bat-science.com/groupms/sites/BAT_B9JBW3.nsf/vwPagesWebLive/DOBB3CEX, accessed 17 August 2020).

130. Delivering for today & investing in the future: Annual report and form 20-F 2019. London: British American Tobacco plc; 2020 (https://www.bat.com/ar/2019/pdf/BAT_Annual_Report_and_Form_20-F_2019.pdf, accessed 14 August 2020).

131. News release: British American Tobacco expands its new category portfolio with the launch of three innovative new tobacco heating products. London: British American Tobacco plc; 2019 (https://www.bat.com/group/sites/UK__9D9KCY.nsf/vwPagesWebLive/DOBFPCWP, accessed 14 August 2020.

132. British American Tobacco plc half-year report to 30 June 2020. London: British American Tobacco plc; 2020 (https://www.bat.com/group/sites/uk__9d9kcy.nsf/vwPagesWebLive/DO72TJQU/$FILE/medMDBRZQL4.pdf?openelement, accessed 17 August 2020).

133. Introduction to tobacco heating products. London: British American Tobacco plc; 2020 (https://www.bat-science.com/groupms/sites/BAT_B9JBW3.nsf/vwPagesWebLive/DOBB3CEX, accessed 14 August 2020).

134. lil: a little is a lot. Seoul: Korea Tobacco & Ginseng Corp., 2019 (http://en.its-lil.com/brand/main_brand. Published 2019, accessed 14 August 2020).

135. Philip Morris International Inc. announces agreement with KT&G to accelerate the achievement of a smoke-free future. Lausanne: Philip Morris International; 2020 (https://www.pmi.com/media-center/press-releases/press-release-details/?newsId=21786, accessed 14 August 2020).

136. McKelvey K, Popova L, Kim M, Chaffee BW, Vijayaraghavan M, Ling P et al. Heated tobacco products likely appeal to adolescents and young adults. Tob Control. 2018;27(Suppl 1):s41–7.

137. Heated tobacco products. Washington (DC): Campaign for Tobacco-Free Kids; 2020 (https://www.tobaccofreekids.org/what-we-do/global/heated-tobacco-products, accessed 13 August 2020).
138. Gravely S, Fong GT, Sutanto E, Loewen R, Ouimet J, Xu SS et al. Perceptions of harmfulness of heated tobacco products compared to combustible cigarettes among adult smokers in Japan: Findings from the 2018 ITC Japan survey. Int J Environ Res Public Health. 2020;17(7):2394.
139. Bar-Zeev Y, Levine H, Rubinstein G, Khateb I, Berg CJ. IQOS point-of-sale marketing strategies in Israel: A pilot study. Isr J Health Policy Res. 2019;8(1):11.
140. Bialous SA, Glantz SA. Heated tobacco products: Another tobacco industry global strategy to slow progress in tobacco control. Tob Control. 2018;27(Suppl 1):s111–7.
141. Jackler RK, Ramamurthi D, Axelrod AK, Jung JK, Louis-Ferdinand NG, Reidel JE et al. Global marketing of IQOS: The Philip Morris campaign to popularize "heat not burn". Berkeley (CA): Stanford Research into the Impact of Tobacco Advertising; 2020 (http://tobacco.stanford.edu/tobacco_main/publications/IQOS_Paper_2-21-2020F.pdf, accessed 17 August 2020).
142. Hair EC, Bennett M, Sheen E, Cantrell J, Briggs J, Fenn Z et al. Examining perceptions about IQOS heated tobacco product: Consumer studies in Japan and Switzerland. Tob Control. 2018;27(Suppl 1):s70–3.
143. IQOS App. Lausanne: Philip Morris International Management SA; 2020 (https://play.google.com/store/apps/details?id=com.pmi.store.PMIAPPM06278, accessed 14 August 2020).
144. IQOS in the US. Washington (DC): Truth Initiative; 2020 (https://truthinitiative.org/research-resources/emerging-tobacco-products/iqos-us, accessed 15 October 2020).
145. 2018 charitable contributions at a glance. Lausanne: Philip Morris International; 2018 (https://www.pmi.com/resources/docs/default-source/our_company/transparency/charitable-2018.pdf?sfvrsn=d97d91b5_2, accessed 15 October 2020).
146. Uranaka T, Ando R. Philip Morris aims to revive Japan sales with cheaper heat-not-burn tobacco. Reuters, 23 October 2018.
147. Asia Pacific heat-not-burn tobacco product market (2019–2025): Market forecast by product type, by demography, by sales channels, by countries, and competitive landscape. Dublin: Research and Markets; 2019 (https://www.researchandmarkets.com/reports/4911922/asia-pacific-heat-not-burn-tobacco-product-market).
148. Sutanto E, Miller C, Smith DM, O'Connor RJ, Quah ACK, Cummings KM et al. Prevalence, use behaviors, and preferences among users of heated tobacco products: Findings from the 2018 ITC Japan survey. Int J Environ Res Public Health. 2019;16(23):4730.
149. Kim J, Yu H, Lee S, Paek YJ. Awareness, experience and prevalence of heated tobacco product, IQOS, among young Korean adults. Tob Control. 2018;27(Suppl 1):s74–7.
150. Kang H, Cho S. Heated tobacco product use among Korean adolescents. Tob Control. 2020;29:466–8.
151. Heated tobacco products market size, share & trends analysis report by product (stick, leaf), by distribution channel (online, offline), by region, and segment forecasts, 2019–2025. San Francisco (CA): Grand View Research; 2019 (https://www.grandviewresearch.com/industry-analysis/heated-tobacco-products-htps-market).
152. Tobacco tactics. Heated tobacco products. Bath: University of Bath; 2020 (https://tobaccotactics.org/topics/heated-tobacco-products/, accessed 10 January 2021).
153. MacGuill S, Genov I. Emerging and next generation nicotine products (webinar). London: Euromonitor International; 2017 (http://go.euromonitor.com/rs/805-KOK-719/images/Emerging-and-Next-Generation-Nicotine-Products-Euromonitor.pdf, accessed 6 December 2020).
154. Countries that regulate heated tobacco. Baltimore (MD): Institute for Global Tobacco Control; 2020.

Supplementary sections

11. Forms of nicotine in tobacco plants, chemical modifications and implications for electronic nicotine delivery systems products

Nuan Ping Cheah, Pharmaceutical, Cosmetics and Cigarette Testing Laboratory, Health Sciences Authority, Singapore

Najat Saliba, Department of Chemistry, Faculty of Arts and Sciences, American University of Beirut, Beirut, Lebanon; Center for the Study of Tobacco Products, Virginia Commonwealth University, Richmond (VA), USA

Ahmad El-Hellani, Department of Chemistry, Faculty of Arts and Sciences, American University of Beirut, Beirut, Lebanon; Center for the Study of Tobacco Products, Virginia Commonwealth University, Richmond (VA), USA

Contents
Abstract
11.1 Background
11.2 Chemical modification of nicotine and influence on nicotine delivery
 11.2.1 Brief summary of the effect of curing on nicotine
 11.2.2 Modification with alkali
 11.2.3 Modification with acid
11.3 Implications for ENDS products and diversity
 11.3.1 Free-base nicotine vs nicotine salt in ENDS
 11.3.2 Feasible concentrations and abuse liability
 11.3.3 Potential masking of the harshness of products
 11.3.4 Health implications and potential regulations
11.4 Discussion
11.5 Research gaps, priorities and questions to members regarding further work or a full paper
11.6 Recommendation
11.7 Considerations
11.8 References

Abstract

The impact of the form of nicotine, i.e. free base or salt, on its delivery from tobacco products and electronic nicotine delivery systems (ENDS) has been debated by scientists and regulators. In this paper, we briefly discussed the various ways of modifying the ratio of nicotine forms in tobacco products and ENDS. We focus on partitioning of nicotine forms in ENDS liquids, especially in the recently introduced pod-based ENDS. We discuss the influence of various parameters on nicotine delivery from ENDS, such as the form of nicotine, counter-anions in nicotine salts, the power output of devices and user puffing topography. Recommendations are made on means to avoid capping nicotine concentrations in ENDS liquids as the sole measure for regulating nicotine delivery from ENDS and on adoption of "nicotine flux" as a regulatory tool that accounts for all the parameters that affect nicotine delivery from ENDS. We highlight the importance of including the form of nicotine in constructing a nicotine flux model and of minimizing possible customization of ENDS by users. While research is still necessary on methods for testing nicotine forms in ENDS liquids and aerosols and on absorption of the different forms by the body in the presence or absence of flavours, it is recommended that WHO urge countries to include nicotine flux and form in ENDS regulation to better inform users about nicotine delivery from their devices.

11.1 Background

This paper served as a "horizon" paper for the 10th meeting of the WHO Study Group on Tobacco Product Regulation and a platform for discussion and consideration of nicotine in ENDS products.

Nicotine can exist either as a free base or in combination with organic acids as various salts. This report provides a brief review of relevant published scientific information on nicotine, its presence in tobacco products, its modifications before manufacture and its forms in finished tobacco products. We also discuss the implications of the form of nicotine in ENDS for their appeal, addictive potential and health impact. The effect of the form of nicotine on tobacco control, regulation and research is also presented. In addition, we briefly discuss the relevance of nicotine flux for regulation of nicotine delivery from ENDS, arguing that the form of nicotine could be incorporated into the flux model.

Nicotine is the primary alkaloid in tobacco (1) and is the most abundant pyridine alkaloid in the leaves of 33 species; nornicotine is the most abundant in 24 species, anabasine in two species (*N. glauca* and *N. debneyi*) and anatabine in one species (*N. otophora*). In the roots, nicotine predominates in 51 species, nornicotine in two species (*N. alata* and *N. africana*) and anabasine in seven species (*N. glauca*, *N. solanifolia*, *N. benavidesii*, *N. cordifolia*, *N. debneyi*, *N. maritima* and *N. hesperis*) (2). Nicotine is synthesized in the roots of tobacco (3) by an enzymatic pathway, with condensation of nicotinic acid (pyridine ring) and N-methyl-Δ^1-pyrrolinium cation (pyrrolidine ring) (4). The amounts of nicotine and three other major pyridine alkaloids in selected *Nicotiana* species are shown in Table 11.1.

Table 11.1. Nicotine and major alkaloids contents in selected Nicotiana species

Subgenus, section	Species	Content (mg/g dry weight)		Percentage of total							
				Nicotine		Nornicotine		Anabasine		Anatabine	
		Leaves	Root	Leaves	Root	Leaves	Root	Leaves	Root	Leaves	Root
Rustica, Paniculatae	N. glauca	8 872	5 246	12.5	35.5	1.5	2.8	85.1	51.3	0.9	10.4
	N. solanifolia Walpers	848	9 326	3.2	27.7	81.4	10.0	15.4	60.3	Trace	2.0
	N. benavidesii Goodspeed*	2 166	14 666	82.7	44.9	1.3	0.8	14.8	48.1	1.2	6.2
	N. cordifolia Philippi	789	13 435	58.4	26.4	6.1	2.5	29.0	64.4	6.5	6.7
Rustica, Rusticae	N. rustica L.	7 752	8 439	96.4	81.6	0.9	1.7	1.1	6.6	1.6	10.1
Tabacum, Tomentosae	N. otophora Grisebach*	377	7 924	6.9	61.3	32.9	27.0	Trace	0.6	60.2	11.1
Tabacum, Genuinae	N. tabacum L.	11 462	2 176	94.8	81.3	3.0	6.0	0.3	1.7	1.9	11.0
Petunioides, Alatae	N. alata Link & Otto	26	1 998	100	37.7	Trace	46.4	–	Trace	–	15.8
Petunioides, Suaveolentes	N. debneyi Domin	2 457	3 038	31.1	34.7	15.8	1.4	46.0	53.2	7.1	10.7
	N. maritima Wheeler	608	14 030	7.2	20.8	70.4	30.0	15.8	44.5	6.6	4.6
	N. hesperis Burbridge	4 108	1 930	52.1	22.1	0.4	1.2	44.3	74.9	3.2	1.8
	N. africana Merxmuller & Buttler*	6 776	7 698	4.7	45.0	92.4	45.1	0.3	1.0	2.6	8.9

Source: reproduced from reference 2.
* Species did not bloom.

The amount of biosynthesized nicotine in cultivated tobacco (*N. tabacum* L. and *N. rustica* L.) has been through gene modifications *(5)* and targeted gene manipulation for industrial applications *(6–8)*. *N. tabacum* L. and *N. rustica* L. are the major species used in the manufacturing of tobacco products due to the abundant level of nicotine present in these species *(9,10)*. Typical nicotine concentrations range from 15 to 35 mg per g tobacco *(3)*, with total alkaloid concentrations reaching up to 79 mg per g tobacco *(11)*.

The structure of nicotine comprises a pyrrolidine ring connected to a pyridine ring. The form of nicotine, free base, monoprotonated or diprotonated nicotine, depends on protonation of two nitrogen centres by naturally occurring acids in the leaves (Fig. 11.1). Protonated forms, also known as nicotine salts, predominate in unprocessed tobacco leaves; however, tobacco products have different ratios of free base to salts. At pH 7, 8 and 9, free base is present in the solution at 9%, 49% and 90.5%, respectively; correspondingly, the nicotine salt is available at 91%, 51% and 9.5%.

Fig. 11.1. Free-base nicotine and monoprotonated and diprotonated nicotine salts

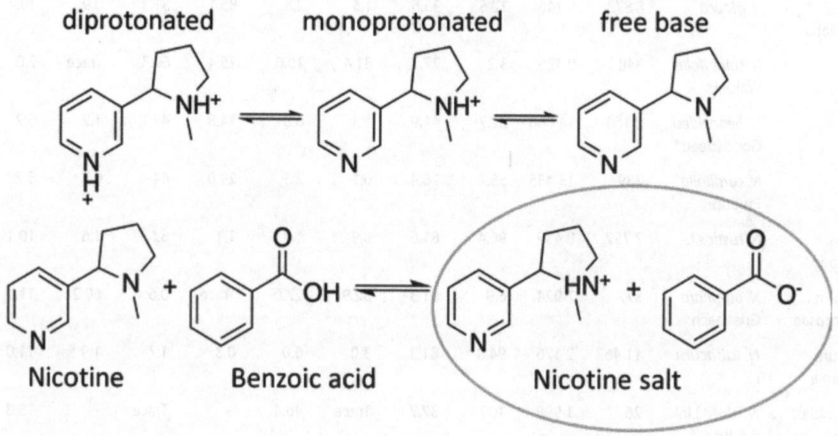

Source: references 12 and 13.
The lower part of the figure shows protonation of nicotine with benzoic acid, which may occur naturally in tobacco leaves or during manufacture of tobacco products.

The stimulatory and addictive effects of nicotine are attributed to the action of the pyridine alkaloid on neuronal nicotinic acetylcholine receptors in the brain. Currently, cigarette smoking is the most effective form of nicotine delivery. Nicotine in mainstream smoke from combusted tobacco is rapidly absorbed into the lungs and can reach the brain in as little as 7 s *(14)*. Although tobacco smoking has been the most prevalent form of nicotine intake for decades, alternative tobacco products recently introduced onto the market (e.g. ENDS, heated tobacco products) are becoming popular around the globe, leading to an overall increase in total use of nicotine and tobacco products, especially among vulnerable populations such as young people. Thus, despite significant progress in tobacco control and prevention, nicotine and tobacco product use continues to grow. This report addresses nicotine abuse liability with one of the most widely marketed and most popular nicotine delivery products: ENDS. ENDS products, which are marketed in myriad combinations, allow users more customization than other nicotine and tobacco products *(14)*, hence the challenge of implementing one standard set of regulations. The impact of the form of nicotine (free base vs salt) on delivery from these devices is therefore being investigated *(15)*.

11.2 Chemical modification of nicotine and influence on nicotine delivery

11.2.1 Brief summary of the effect of curing on nicotine

The main determinant of the dependence potential of a nicotine and tobacco product is its ability to deliver pharmacologically active levels of nicotine rapidly *(16,17)*. Nicotine dosage is carefully controlled by manufacturers to ensure that it is sufficient to produce the desired effects, such as relaxation and mental acuity, while minimizing the risk of undesirable effects, such as nausea and intoxication *(18)*. Nicotine constitutes 2–8% of the dry weight of cured tobacco leaf, with wider ranges for some *Nicotiana* species *(19)*. The three main types of tobacco leaf used in commercial cigarettes are flue-cured, Burley and oriental *(20, Table 11.2)*. Conventional cigarettes are made up of blends of these tobaccos. The primary blend differs among countries, but flue-cured and Burley tobaccos are used in the highest volume in commercial cigarettes. Most cigarettes contain primarily either flue-cured tobacco (e.g. in Canada) or a mixture of mainly flue-cured and Burley, with a minor amount of oriental tobacco (American blend) added.

Table 11.2. Types of tobacco, curing processes and nicotine content in tobacco leaf from the upper stalk position reported in literature

Tobacco type	Curing process	Tobacco products	Nicotine content (mg/g)
Virginia (or Bright)	Flue-cured by hanging the tobacco leaves in an enclosed area with heated air for 1 week	Blended cigarettes Virginia cigarettes	6.52–60.4
Burley	Air-cured in an air-ventilated area for 4–8 weeks	Blended cigarettes Kretek cigarettes (clove-flavoured)	35.6–47.73
Oriental	Cured by hanging tobacco leaves in the sun for 2 weeks	Blended cigarettes	1.80–12.6

Source: reference 21.

Air-curing may decrease the final level of nicotine in tobacco leaves *(22–24)* due to oxidation of nicotine to cotinine or other oxidation products and conversion of nicotine to nornicotine by demethylation *(23,25,26)*. Flue-curing retains higher levels of sugars in the leaves; as these are precursors of organic acids in tobacco smoke, there is a smaller fraction of free-base nicotine in the smoke *(27)*.

11.2.2 Modification with alkali

Tobacco and "smoke pH" can be raised by using ammonia compounds (e.g. diammonium phosphate) and other substances (e.g. calcium carbonate) in tobacco processing. Calcium and sodium carbonates are added to cigarette filters to increase "smoke pH", possibly eliminating the addition of bases to tobacco filler *(28,29)*. Ammonia has been used in the manufacture of tobacco since the accidental finding in the 1960s that elevated pH facilitates nicotine absorption, increasing free-base nicotine in cigarette smoke and tobacco products *(30–32)*, despite industry denial *(33,34)*. Ammonia also reacts with natural organic hydroxy compounds from tobacco and improves the quality of smoke, giving a smoother, "chocolate-like", less acidic taste *(35,36)*.

Other commonly used alkaline substances that increase smoke pH and improve smoke flavour include urea, diammonium phosphate, ethanolamines and carbonates *(37,38)*. In an alkaline or high-pH environment, nicotine in its un-ionized (free base) form is rapidly absorbed across mucous membranes; however, this rapid flux is irritating to the user.

11.2.3 Modification with acid

When cigarette smoke is perceived as too harsh, smokers inhale less deeply *(39)*. Additives such as levulinic acid make smoke appear smoother to the upper respiratory tract by lowering the fraction of free-base nicotine. As a result, the smoke is easier to inhale into the lungs *(40)*. Levulinic acid has also been reported to increase nicotine yield *(41)*. Addition of inorganic salts such as magnesium nitrate was found to lower the transfer of nicotine to tobacco smoke *(42)*. In acidic conditions, nicotine is ionized (protonated) and therefore crosses biological membranes much more slowly and is less irritating *(43)*. Table 11.3 summarizes the results of studies of the addition of acids to tobacco products.

Table 11.3. Results of adding acids to tobacco products

Type of acid	Purpose of addition
Lactic acid	Decreased harshness and bitterness, resulting in a sweeter flavour and smoothness *(44)*
Citric acid	Reduced harshness, modified flavour, lowered smoke pH, "neutralized" the impact of nicotine, enhanced sheet formation in reconstituted tobacco *(45,46)*
Tartaric acid	Similar to lactic acid, reduced the pH of smoke *(46,47)*
Malic acid	Did not promote migration in typical construction and storage procedures *(48,49)*
Formic acid	Increased nicotine delivery, but had a distinct sour taste and failed to improve subjective performance *(50,51)*
Levulinic acid	Reduced harshness without decreasing nicotine delivery in smoke and with no unpleasant taste. Nicotine salt of levulinic acid also increased smoke nicotine delivery *(41,52)*.
Benzoic and sorbic acid	Reduced harshness, increased nicotine delivery. Form salts with nicotine.
Pyruvic and lauric acid	Typically added to form nicotine salt *(53,54)*

11.3 Implications for ENDS products and diversity

11.3.1 Free-base nicotine vs nicotine salt in ENDS

The previous section showed that the form and dosage of nicotine in combustible cigarettes can be controlled by the manufacturer during tobacco curing or processing. We reported above that the distribution of different forms of nicotine in tobacco smoke affects the inhalability of nicotine. Free-base nicotine is readily absorbed in the upper respiratory tract, while nicotine salts are delivered to the bronchioalveolar region (55). Although the sites of absorption of the different nicotine forms are known, there is still controversy about how the site affects the rate of nicotine delivery to the brain (56).

ENDS allow users greater customization of their experience with nicotine in terms of dose and form. The ratios of nicotine forms in ENDS liquids results in pH in the entire range (5.3–9.3) (57), although the aerosol of the most popular ENDS (e.g. pod-based ENDS) has a low pH and high levels of nicotine salts. Added acids, such as levulinic acid, form monoprotonated and diprotonated nicotine forms, making inhalation of aerosols from ENDS smoother on the throat and upper airways. Other common acids used to form nicotine salts are lactic, benzoic, sorbic, pyruvic, salicylic, malic, lauric and tartaric acids (54). One report stated that flavour additives such as phenols, vanillin and ethyl vanillin can act as protonating agents in e-liquids (58). A study based on a randomized controlled trial (59) indicated that the presence of nicotine salts in ENDS reduces craving to the same extent as conventional cigarettes. Pharmacokinetics and subjective data demonstrate that nicotine lactate delivers nicotine via the pulmonary route for rapid absorption, albeit with a maximum nicotine level that does not exceed that in conventional cigarettes, and also showed acceptable subjective satisfaction and relief of a desire to smoke (59).

Nicotine in its three forms is considered the main addictive chemical in tobacco products. By the 1990s, it was increasingly accepted that tobacco products without nicotine would not sustain addiction (16,60). It is therefore important to focus on the interaction between flavours and nicotine in user perceptions of ENDS aerosols. Flavours could reduce the upper respiratory tract irritation of high nicotine levels in ENDS aerosols or contribute to the sensory impact of aerosols with low nicotine levels, as was shown with menthol (61). Moreover, published data show that some flavours, like apple, may increase the reinforcing effects of nicotine in ENDS aerosols (62), as was shown to be the case with menthol in cigarette smoke (63).

Nicotine salts such as nicotine benzoate are monoprotonated salts. These are reported to produce a high degree of satisfaction in users, as evidenced by the popular JUUL product (64), a patented formulation with benzoate salts (54) and similar ENDS (65). During aerosolization, nicotine salt dissociates during evaporation to give free-base nicotine and acid molecules that recombine upon contact with ambient air to condense into inhalable aerosols (54). Chromatography identifies the counter-anions derived from added acids, such as salicylate, tartrate, levulinate and malate in e-liquid or aerosols (66,67).

11.3.2 Feasible concentrations and abuse liability

Several investigators have adapted or developed analytical methods for determining the nicotine content of ENDS liquids and aerosols *(68–70)*. The concentrations in ENDS liquids and prefilled cartridges range widely, from 0 in nicotine-free cartridges and liquids to about 130 mg/mL in some "do-it-yourself" liquids *(71–73)*. In the early years of the ENDS epidemic, cartridge-based ENDS (closed systems) had lower nicotine concentrations than open systems *(71)*; however, the recently introduced pod-based ENDS contain very high levels of nicotine of > 60 mg/mL *(69,74)*. The appeal of ENDS is related not only to the nicotine concentration but also to the form of nicotine, free base or salt *(75–77)*. The few reports to date of analyses of the form of nicotine in ENDS liquids showed a wide range of pH values and ratios of nicotine forms *(68,78,79)*. A variety of approaches was used in these studies, including pH measurement and then estimation of the ratio of nicotine forms with the Henderson–Hasselbalch equation or an organic solvent extraction of nicotine from ENDS liquid dissolved in water to measure the different forms of nicotine by gas chromatography *(68)* or determining the ratios of different forms of nicotine by proton nuclear magnetic resonance spectroscopy *(80)*. Moreover, nicotine present in the liquid is transferred efficiently to the aerosol and subsequently influences the resulting subjective effects *(67)*. Recent work by the authors of this report showed that the form of nicotine does not affect the total yield of nicotine delivered in aerosols *(81)*.

Several factors contribute to the appeal and continued use of ENDS, including flavours, high "customizability" and nicotine delivery *(82,83)*. In the USA, use of ENDS, unlike any other non-cigarette tobacco product in the past decade, has surpassed cigarette smoking among children and young adults *(84)*. The possibility of unlimited combinations of operating parameters (i.e. power, liquid composition, puff topography) in some ENDS allows delivery of nicotine at doses ranging from trace amounts to orders of magnitude higher than those delivered by a combustible cigarette *(71,85,86)*. This wide range of nicotine delivery may increase the risk of abuse liability and nicotine dependence for users *(87)*. Addiction to nicotine, like other drugs, is a function of the dose delivered and the speed of delivery *(88)*. Smoking a combustible tobacco cigarette is an efficient, rapid means of nicotine delivery, hence its addictive character *(89)*. The same applies to ENDS: recent brain imaging studies showed that ENDS can deliver nicotine to the brain at a rate similar to that of a combustible cigarette *(90,91)*.

It is important to note that ENDS (and heated tobacco products) are often used as complements to cigarette smoking and not as substitutes, especially in smoke-free environments *(92)*. Thus, dual use of ENDS and cigarettes is a common practice that sustains nicotine dependence *(93)*. Longitudinal studies have also shown that ENDS users concurrently smoke combustible cigarettes, perhaps due to greater nicotine dependence *(94,95)*. There is thus a growing trend of dual use of ENDS with combustible cigarettes *(96,97)*.

11.3.3 Potential masking of the harshness of products

Nicotine delivery is one of the main reasons for using ENDS *(98)*. Another factor that contributes to the popularity of ENDS is the wide availability of unique flavours *(99,100)*, the number of which has grown dramatically in recent years *(101)*. The most common flavours are tobacco, menthol or mint, fruit, candy or dessert and beverages *(102–104)*. Flavours can mask the harshness associated with inhaled free-base nicotine, especially if they induce a cool sensation, such as mint and menthol *(105)*. An interesting area of research would therefore be the correlation between flavour choice and nicotine form used by novice and by experienced users.

11.3.4 Health implications and potential regulations

ENDS are commonly perceived as safer and less addictive than cigarette smoking, which may contribute to their rising popularity *(106,107)*. Global use of ENDS has increased rapidly in the past decade, especially among young people, and there is evidence of nicotine dependence in this population *(81,108–113)*. Early ENDS devices were deemed inefficient in delivering nicotine to the user *(114)*; however, as the devices evolved and users became more experienced in their use, the efficiency of nicotine delivery greatly increased *(115–117)*.

ENDS products are notably heterogeneous, with differences in materials, configurations, electrical power output, solvents and composition *(118)*. Nicotine yield therefore depends on a combination of variables, such as device power, liquid composition and user puffing behaviour *(84,119)*. As noted above, ENDS provide users unprecedented opportunities for customizing the nicotine concentration and sometimes form, and thus their nicotine dose, by preparing their own liquids and modifying operating parameters.

Nicotine delivery from ENDS is a subject of much debate among scientists and policy-makers. Some argue that, if ENDS products deliver nicotine at a dose and rate comparable to those of a combustible cigarette, ENDS may help smokers to reduce or quit smoking and subsequently reduce their exposure to the associated harm *(120)*. Comparably efficient nicotine delivery may, however, make ENDS users, including previously nicotine-naïve individuals, more addicted to nicotine *(110,121,122)*. A recent study showed that imposing a limit on the nicotine concentration of ENDS liquid is not sufficient to ensure nicotine yields lower than those of a combustible cigarette *(123)*, as users can increase the power of the device to obtain the levels of nicotine in combustible cigarettes. Moreover, users who switch to higher-power devices inhale more aerosol and thus more toxicants, with unintended health consequences *(86)*.

Most countries have not revised their legislation to include regulations on ENDS. Researchers and regulators, mainly in Europe, have considered that a first regulatory measure to mitigate risk could be to cap the nicotine content of ENDS liquids to control nicotine delivery. Such a policy has been effective in the European Union since 2014, with a limit of 20 mg/mL on the nicotine concentration in ENDS liquids. Similar approaches could be considered in other jurisdictions. This policy does not, however, account for variations in ENDS product characteristics, such as power and puff topography. For example, users in a jurisdiction in which the nicotine concentration is limited may circumvent the aim of the regulation by choosing devices with higher power to obtain a nicotine yield that exceeds that of a combustible cigarette. Shihadeh and Eissenberg (124) proposed measurement of "nicotine flux", which is the amount of nicotine delivered from a mouthpiece in a unit time (mg/s), as a suitable regulatory tool that encapsulates all the relevant operating parameters of ENDS that affect the rate and dose of nicotine delivered to the user. Current work from this group focuses on incorporating the form of nicotine into the nicotine flux construct. Clinical studies of the addictiveness and abuse liability of ENDS will determine the impact of the form of nicotine on the speed of delivery. The form of nicotine may also affect toxicity, as demonstrated by Pankow et al. (125), who showed that use of benzoic acid in the preparation of nicotine salt may lead to formation of benzene in ENDS emissions.

11.4 Discussion

Industry has used many approaches to enhance the efficiency of nicotine delivery from cigarettes, including manipulating the ratio of nicotine forms in cigarette filler. Nonetheless, a balance between nicotine delivery and harshness in the generated smoke dictated the manufacture of conventional cigarettes. A similar approach has recently been taken in the design of ENDS (126). The liquids used in ENDS that have the largest market share have a lower pH, due to the addition of organic acids, which masks the harshness of the large quantity of nicotine delivered during aerosolization. Nicotine salts, however, may contribute to the toxicity of electronic cigarette aerosols due to the degradation of counter-anions (125).

11.5 Research gaps, priorities and questions to members regarding further work or a full paper

The evidence indicates that various forms of nicotine, i.e. free-base nicotine and monoprotonated and diprotonated nicotine salts, are already available on the market of nicotine and tobacco products and are spreading quickly to other products. Recently, the industry has begun to manufacture synthetic nicotine, which is becoming cheaper to produce than previous technologies. It has already been used in e-liquids (127). This may be a challenge for regulation in certain jurisdictions, as the nicotine is not of tobacco origin.

Research and development are required to:

- develop and/or validate standard methods for measuring free-base nicotine and determining the ratio of free base to protonated and diprotonated nicotine in e-liquids and aerosols of ENDS;
- quantify total and different forms of nicotine and organic acids in ENDS liquids and aerosols *(68,81)*;
- determine the impact of the form of nicotine on nicotine delivery to ENDS users and the dependence potential, including maintenance of addiction in a study conducted in the presence and absence of confounders;
- investigate the health implications of the use of organic acids to change nicotine pharmacokinetics *(128)* and the impact on toxicity; and
- validate nicotine flux, with nicotine form, as a tool for regulating nicotine delivery from ENDS.
- If these gaps are addressed, the global priorities could be to:
- ensure that the ratio of nicotine forms in ENDS helps smokers of combustible cigarettes to quit and does not lead novice users to become nicotine addicts; and
- on the basis of rigorous evidence, restrict manipulation of nicotine concentration and form by manufacturers.

In view of the importance of nicotine delivery from ENDS and the possible combined effect of nicotine form and concentration in ENDS liquids on its appeal, attractiveness and addictiveness, consideration should be given to requesting a full paper in the future.

11.6   Recommendation

Consideration should be given to preparation of a full paper on nicotine forms for a future meeting, if the topic is considered a priority and sufficient information is available.

11.7   Considerations

When countries strengthen their tobacco regulatory framework, a primary goal should be to reduce exposure to nicotine, the most important addictive substance in tobacco. Many countries have the authority to regulate products made of or derived from tobacco, as covered in Article 1 of the WHO Framework Convention on Tobacco Control. So far, the only regulation on nicotine delivery from ENDS is that of the European Union, which limits the nicotine content of

ENDS liquids. We argue above that this may be ineffective in addressing all the capabilities of this new category of nicotine delivery product, as it does not reflect the fact that the nicotine yield from ENDS is a result of many factors, such as nicotine concentration, power output and user puffing regime. Nicotine flux may be an option for regulatory consideration.

Moreover, the impact of the form of nicotine on its delivery, pharmacokinetics and pharmacodynamics is not well studied, and more research is necessary. Regulators should also consider minimizing the extent of possible customization of ENDS by users in terms of nicotine load and form and other features such as flavours.

Countries might have to review their legislation to regulate nicotine-containing products comprehensively, regardless of the origin of the nicotine, in the interests of public health. If nicotine delivery products are not regulated, the work of countries and WHO in reducing tobacco use and nicotine abuse liability may be compromised. Finally, WHO should consider discussion of the inclusion of ENDS in the WHO Framework Convention on Tobacco Control or a provision within the Convention to address regulatory issues specific to these products.

11.8 References

1. Wernsman E. Time and site of nicotine conversion in tobacco. Tob Sci. 1968;12:226–8.
2. Saitoh F, Noma M, Kawashima N. The alkaloid contents of sixty Nicotiana species. Phytochemistry. 1985;24(3):477–80.
3. Collins W, Hawks S. Principles of flue-cured tobacco production. Raleigh (NC): HarperCollins Publishers; 1993.
4. Sun B, Tian YX, Zhang F, Chen Q, Zhang Y, Luo Y et al. Variations of alkaloid accumulation and gene transcription in Nicotiana tabacum. Biomolecules. 2018;8(4):114.
5. Liu H, Kotova TI, Timko MP. Increased leaf nicotine content by targeting transcription factor gene expression in commercial flue-cured tobacco (Nicotiana tabacum L.). Genes. 2019;10(11):10.3390/genes10110930.
6. Saunders JW, Bush LP. Nicotine biosynthetic enzyme activities in Nicotiana tabacum L. genotypes with different alkaloid levels. Plant Physiol. 1979;64(2):236–40.
7. Moghbel N, Ryu B, Ratsch A, Steadman KJ. Nicotine alkaloid levels, and nicotine to nornicotine conversion, in Australian Nicotiana species used as chewing tobacco. Heliyon. 2017;3(11):e00469.
8. Griffith R, Valleau W, Stokes G. Determination and inheritance of nicotine to nornicotine conversion in tobacco. Science. 1955;121(3140):343–4.
9. Chaplin JF, Burk L. Agronomic, chemical, and smoke characteristics of flue-cured tobacco lines with different levels of total alkaloids 1. Agron J. 1984;76(1):133–6.
10. Chaplin JF, Burk L. Genetic approaches to varying chemical constituents in tobacco and smoke. Beitr Tabakforsch. 1977;9(2):102–6.
11. Tso TC. Production, physiology, and biochemistry of tobacco plant. Beltsville (MD): Ideals; 1990.
12. Baxendale IR, Brusotti G, Matsuoka M, Ley SV. Synthesis of nornicotine, nicotine and other functionalised derivatives using solid-supported reagents and scavengers. J Chem Soc Perkin Trans. 2002;2002(2):143–54.
13. Domino EF, Hornbach E, Demana T. The nicotine content of common vegetables. N Engl J

Med. 1993;329(6):437.
14. Maisto SA, Galizio M, Connors GJ, editors. Drug use and abuse. Seventh edition. Belmont (CA): Cengage Learning; 2014.
15. Gholap VV, Kosmider L, Golshahi L, Halquist MS. Nicotine forms: Why and how do they matter in nicotine delivery from electronic cigarettes? Expert Opin Drug Delivery. 2020;1–10.
16. Benowitz NL, Henningfield JE. Establishing a nicotine threshold for addiction – the implications for tobacco regulation. New Engl J Med. 1994;331(2):123–5.
17. Henningfield J, Benowitz N, Slade J, Houston T, Davis R, Deitchman S. Reducing the addictiveness of cigarettes. Tob Control. 1998;7(3):281.
18. Carpenter CM, Wayne GF, Connolly GN. Designing cigarettes for women: new findings from the tobacco industry documents. Addiction. 2005;100(6):837–51.
19. Gorrod JW, Jacob P III, editors. Analytical determination of nicotine and related compounds and their metabolites. Amsterdam: Elsevier Science; 1999.
20. Paschke T, Scherer G, Heller WD. Effects of ingredients on cigarette smoke composition and biological activity: a literature overview. Beitr Tabakforsg. 2002;20(3):107–244.
21. Djordjevic MV, Gay SL, Bush LP, Chaplin JF. Tobacco-specific nitrosamine accumulation and distribution in flue-cured tobacco alkaloid isolines. J Agric Food Chem. 1989;37(3):752–6.
22. Burton H, Bush L, Hamilton J. Effect of curing on the chemical composition of burley tobacco. Rec Adv Tob Sci. 1983;9:91–153.
23. Wahlberg I, Karlsson K, Austin DJ, Junker N, Roeraade J, Enzell C et al. Effects of flue-curing and ageing on the volatile, neutral and acidic constituents of Virginia tobacco. Phytochemistry. 1977;16(8):1217–31.
24. Burton HR, Kasperbauer M. Changes in chemical composition of tobacco lamina during senescence and curing. 1. Plastid pigments. J Agric Food Chem. 1985;33(5):879–83.
25. Bush L. Physiology and biochemistry of tobacco alkaloids. Recent Adv Tob Sci. 1981;7:75–106.
26. Frankenburg WG, Gottscho AM, Mayaud EW, Tso TC. The chemistry of tobacco fermentation. I. Conversion of the alkaloids. A. The formation of 3-pyridyl methyl ketone and of 2,3´-dipyridyl. J Am Chem Soc. 1952;74(17):4309–14.
27. Elson L, Betts T, Passey R. The sugar content and the pH of the smoke of cigarette, cigar and pipe tobaccos in relation to lung cancer. Int J Cancer. 1972;9(3):666–75.
28. Leffingwell JC. Tobacco, production, chemistry and technology. Oxford: Blackwell;. 1999.
29. Douglas C. Tobacco manufacturers manipulate nicotine content of cigarettes to cause and sustain addiction. Tobacco: the growing epidemic. Berlin: Springer; 2000:209–11.
30. Brambles Australia Ltd (Brambles) v British American Tobacco Australia Services Ltd (BATAS). 2007. Bates No.: proctorr20071203b (https://www.industrydocuments.ucsf.edu/docs/tmyg0225, accessed 18 March 2020).
31. Fant RV, Henningfield JE, Nelson RA, Pickworth WB. Pharmacokinetics and pharmacodynamics of moist snuff in humans. Tob Control. 1999;8(4):387–92.
32. Henningfield JE, Radzius A, Cone EJ. Estimation of available nicotine content of six smokeless tobacco products. Tob Control. 1995;4(1):57.
33. Stevenson T, Proctor RN. The secret and soul of Marlboro. Am J Public Health. 2008;98(7):1184–94.
34. Hind JD, Seligman RB. Tobacco sheet material. US Patent. 3,353,541. 1967. Filed 16 June 1966 and issued 21 November 1967.
35. Teague CE Jr. Modification of tobacco stem materials by treatment with ammonia and other substances. Winston-Salem (NC): RJ Reynolds; 1954. Bates no. 504175083-5084 (http://legacy.library.ucsf.edu/tid/gpt58d00, accessed 9 November 2020).
36. Pankow JF, Mader BT, Isabelle LM, Luo W, Pavlick A, Liang C. Conversion of nicotine in tobacco smoke to its volatile and available free-base form through the action of gaseous ammonia. Environ Sci Technol. 1997;31(8):2428–33.

37. Seligman R. The use of alkalis to improve smoke flavor. Philip Morris. 1965. Bates no. 2026351158-2026351163 (https://www.industrydocuments.ucsf.edu/docs/tfpv0109, accessed 4 December 2020).
38. Armitage AK, Dixon M, Frost BE, Mariner DC, Sinclair NM. The effect of tobacco blend additives on the retention of nicotine and solanesol in the human respiratory tract and on subsequent plasma nicotine concentrations during cigarette smoking. Chem Res Toxicol. 2004;17(4):537–44.
39. Henningfield JE, Fant RV, Radzius A, Frost S. Nicotine concentration, smoke pH and whole tobacco aqueous pH of some cigar brands and types popular in the United States. Nicotine Tob Res. 1999;1(2):163–8.
40. Scientific Committee on Emerging and Newly Identified Health Risks. Addictiveness and attractiveness of tobacco additives. Brussels: European Commission, Directorate-General for Health and Consumers; 2010 (https://ec.europa.eu/health/scientific_committees/emerging/docs/scenihr_o_029.pdf, accessed 9 November 2020).
41. Keithly L, Ferris Wayne G, Cullen DM, Connolly GN. Industry research on the use and effects of levulinic acid: a case study in cigarette additives. Nicotine Tob Res. 2005;7(5):761–71.
42. Kobashi Y, Sakaguchi S. The influence of inorganic salts on the combustion of cigarette and on the transfer of nicotine on to smoke. Agric Biol Chem. 1961;25(3):200–5.
43. van Amsterdam J, Sleijffers A, van Spiegel P, Blom R, Witte M, van de Kassteele J et al. Effect of ammonia in cigarette tobacco on nicotine absorption in human smokers. Food Chem Toxicol. 2011;49(12):3025–30.
44. Meyer L. Lactic acid sprayed cellulose acetate filters. 1963, Bates No.: 1003105149-1003105150 (https://www.industrydocuments.ucsf.edu/docs/qjwd0107, accessed 5 April 2020).
45. Morris P. Citric acid. 1989. Bates No.: 2028670128-2028670129 (https://www.industrydocuments.ucsf.edu/tobacco/docs/#id=nzfm0111, accessed 5 April 2020).
46. Winterson WD, Cochran TD, Holland TC, Torrence KM, Rinehart S, Scott GR. Method of making pouched tobacco product. US Patent no. US 7,980,251 B2; 2011.
47. JD Backhurst. The effects of ameliorants on the smoke from Burley tobacco. 1968. Bates No.: 302075334-302075350 (https://www.industrydocuments.ucsf.edu/tobacco/docs/#id=jnvj0189, accessed 5 April 2020).
48. Sudholt MA. Increased impact of cellulose–malic acid filter containing migrated nicotine. 1985. Bates No.: 81122186-81122188 (https://www.industrydocuments.ucsf.edu/tobacco/docs/#id=jnfk0037, accessed 5 April 2020).
49. Sudholt MA. Continuation of study of malic acid treated cellulose filters. 1985. Bates No.: 80551009-80551010 (https://www.industrydocuments.ucsf.edu/tobacco/docs/#id=kmfk0037, accessed 5 April 2020).
50. Cipriano JJ, Kounnas CN, Spielberg HL. Manipulation of nicotine delivery by addition of acids to filler. 1979. Bates No.: 3990037328-3990037339 https://www.industrydocuments.ucsf.edu/tobacco/docs/#id=kpxp0180, accessed 5 April 2020).
51. Cipriano JJ, Kounnas CN, Spielberg HL. Manipulation of nicotine delivery by addition of acids to filter. 1975. Bates No.: 3990093180-3990093188 (https://www.industrydocuments.ucsf.edu/tobacco/docs/#id=zpgp0180, accessed 5 April 2020).
52. Levulinic acid. RJ Reynolds Collection. 1991. Bates No.: 512203269-512203295 (https://www.industrydocuments.ucsf.edu/tobacco/docs/#id=rymk0089, accessed 5 April 2020).
53. Bowen A, Xing C. Nicotine salt formulations for aerosol devices and methods thereof. US Patent no. USS 9,215,895 B2; 2015.
54. Bowen A, Xing C. Inventors; JUUL Labs, Inc., assignee (2016). Nicotine salt formulations for aerosol devices and methods thereof. US patent 20,160,044,967. 28 October 2015.
55. Armitage A, Dollery C, Houseman T, Kohner E, Lewis P, Turner D. Absorption of nicotine by

56. man during cigar smoking. Br J Pharmacol. 1977;59(3):493P.
56. Pankow JF. A consideration of the role of gas/particle partitioning in the deposition of nicotine and other tobacco smoke compounds in the respiratory tract. Chem Res Toxicol. 2001;14(11):1465–81.
57. El-Hellani A, Salman R, El-Hage R, Talih S, Malek N, Baalbaki R et al. Nicotine and carbonyl emissions from popular electronic cigarette products: correlation to liquid composition and design characteristics. Nicotine Tob Res. 2018;20(2):215–23.
58. Pankow JF, Duell AK, Peyton DH. Free-base nicotine fraction αfb in non-aqueous versus aqueous solutions: electronic cigarette fluids without versus with dilution with water. Chem Res Toxicol. 2020;33(7):1729–35.
59. O'Connell G, Pritchard JD, Prue C, Thompson J, Verron T, Graff D et al. A randomised, open-label, cross-over clinical study to evaluate the pharmacokinetic profiles of cigarettes and e-cigarettes with nicotine salt formulations in US adult smokers. Intern Emerg Med. 2019;14(6):853–61.
60. Henningfield J, Pankow J, Garrett B. Ammonia and other chemical base tobacco additives and cigarette nicotine delivery: issues and research needs. Nicotine Tob Res. 2004;6(2):199–205.
61. Rosbrook K, Green BG. Sensory effects of menthol and nicotine in an e-cigarette. Nicotine Tob Res. 2016;18(7):1588–95.
62. Avelar AJ, Akers AT, Baumgard ZJ, Cooper SY, Casinelli GP, Henderson BJ. Why flavored vape products may be attractive: Green apple tobacco flavor elicits reward-related behavior, upregulates nAChRs on VTA dopamine neurons, and alters midbrain dopamine and GABA neuron function. Neuropharmacology. 2019;158:107729.
63. Ahijevych K, Garrett BE. The role of menthol in cigarettes as a reinforcer of smoking behavior. Nicotine Tob Res. 2010;12(Suppl 2):S110–6.
64. Fadus MC, Smith TT, Squeglia LM. The rise of e-cigarettes, pod mod devices, and Juul among youth: Factors influencing use, health implications, and downstream effects. Drug Alcohol Depend. 2019;201:85–93.
65. Barrington-Trimis JL, Leventhal AM. Adolescents' use of "pod mod" e-cigarettes – urgent concerns. New Engl J Med. 2018;379(12):1099–102.
66. Tracy M, Liu X. Analysis of nicotine salts by HPLC on an anion-exchange/cation-exchange/reversed-phase tri-mode column. Sunnyvale (CA): Dionex Corp.; 2011 (https://www.chromatographyonline.com/view/analysis-nicotine-salts-hplc-anion-exchangecation-exchangereversed-phase-tri-mode-column, accessed 7 March 2020).
67. Sádecká J, Polonský J. Determination of organic acids in tobacco by capillary isotachophoresis. J Chromatogr A. 2003;988(1):161–5.
68. El-Hellani A, El-Hage R, Baalbaki R, Salman R, Talih S, Shihadeh A et al. Free-base and protonated nicotine in electronic cigarette liquids and aerosols. Chem Res Toxicol. 2015;28(8):1532–7.
69. Omaiye EE, McWhirter KJ, Luo W, Pankow JF, Talbot P. High-nicotine electronic cigarette products: toxicity of JUUL fluids and aerosols correlates strongly with nicotine and some flavor chemical concentrations. Chem Res Toxicol. 2019;32(6):1058–69.
70. Duell AK, Pankow JF, Peyton DH. Free-base nicotine determination in electronic cigarette liquids by 1H NMR spectroscopy. Chem Res Toxicol. 2018;31(6):431–4.
71. El-Hellani A, Salman R, El-Hage R, Talih S, Malek N, Baalbaki R et al. Nicotine and carbonyl emissions from popular electronic cigarette products: correlation to liquid composition and design characteristics. Nicotine Tob Res. 2018;20(2):215–23.
72. Grana RA, Ling PM. "Smoking revolution": a content analysis of electronic cigarette retail websites. Am J Prev Med. 2014;46(4):395–403.
73. Davis B, Dang M, Kim J, Talbot P. Nicotine concentrations in electronic cigarette refill and do-it-yourself fluids. Nicotine Tob Res. 2015;17(2):134–41.
74. Talih S, Salman R, El-Hage R, Karam E, Karaoghlanian N, El-Hellani E et al. Characteristics and

75. St Helen G, Ross KC, Dempsey DA, Havel CM, Jacob P, Benowitz NL. Nicotine delivery and vaping behavior during ad libitum e-cigarette access. Tob Regul Sci. 2016;2(4):363–76.
76. Kong G, Bold KW, Morean ME, Bhatti H, Camenga DR, Jackson A et al. Appeal of JUUL among adolescents. Drug Alcohol Depend. 2019;205:107691.
77. Seeman JI, Fournier JA, Paine JB, Waymack BE. The form of nicotine in tobacco. Thermal transfer of nicotine and nicotine acid salts to nicotine in the gas phase. J Agric Food Chem. 1999;47(12):5133–45.
78. Stepanov I, Fujioka N. Bringing attention to e-cigarette pH as an important element for research and regulation. Tob Control. 2015;24(4):413–4.
79. Lisko JG, Tran H, Stanfill SB, Blount BC, Watson CH. Chemical composition and evaluation of nicotine, tobacco alkaloids, pH, and selected flavors in e-cigarette cartridges and refill solutions. Nicotine Tob Res. 2015;17(10):1270–8.
80. Gholap VV, Heyder RS, Kosmider L, Halquist MS. An analytical perspective on determination of free base nicotine in e-liquids. J Anal Meth Chem. 2020;2020:6178570.
81. Talih S, Salman R, El-Hage R, Karaoghlanian N, El-Hellani A, Saliba N et al. Effect of free-base and protonated nicotine on nicotine yield from electronic cigarettes with varying power and liquid vehicle. Sci Rep.2020;10:16263.
82. Cobb CO, Lopez AA, Soule EK, Yen MS, Rumsey H, Lester Scholtes R et al. Influence of electronic cigarette liquid flavors and nicotine concentration on subjective measures of abuse liability in young adult cigarette smokers. Drug Alcohol Depend. 2019;203:27–34.
83. Camenga DR, Morean M, Kong G, Krishnan-Sarin S, Simon P, Bold K. Appeal and use of customizable e-cigarette product features in adolescents. Tob Regul Sci. 2018;4(2):51–60.
84. Youth and tobacco use. Atlanta (GA): Centers for Disease Control and Prevention; 2019 (https://www.cdc.gov/tobacco/data_statistics/fact_sheets/youth_data/tobacco_use/index.htm, accessed 21 January 2020).
85. Talih S, Balhas Z, Salman R, El-Hage R, Karaoghlanian N, El-Hellani A et al. Transport phenomena governing nicotine emissions from electronic cigarettes: Model formulation and experimental investigation. Aerosol Sci Technol. 2017;51(1):1–11.
86. Voos N, Goniewicz ML, Eissenberg T. What is the nicotine delivery profile of electronic cigarettes? Expert Opin Drug Delivery. 2019;16(11):1193–203.
87. Dawkins L, Cox S, Goniewicz M, McRobbie H, Kimber C, Doig M et al. "Real-world" compensatory behaviour with low nicotine concentration e-liquid: subjective effects and nicotine, acrolein and formaldehyde exposure. Addiction. 2018;113(10):1874–82.
88. Allain F, Minogianis EA, Roberts DCS, Samaha AN. How fast and how often: The pharmacokinetics of drug use are decisive in addiction. Neurosci Biobehav Rev. 2015;56:166–79.
89. Berridge MS, Apana SM, Nagano KK, Berridge CE, Leisure GP, Boswell MV. Smoking produces rapid rise of [11C]nicotine in human brain. Psychopharmacology. 2010;209(4):383–94.
90. Baldassarri SR, Hillmer AT, Anderson JM, Jatlow P, Nabulsi N, Labaree D et al. Use of electronic cigarettes leads to significant beta2-nicotinic acetylcholine receptor occupancy: Evidence from a PET imaging study. Nicotine Tob Res. 2017;20(4):425–33.
91. Solingapuram Sai KK, Zuo Y, Rose JE, Garg PK, Garg S, Nazih R et al. Rapid brain nicotine uptake from electronic cigarettes. J Nuclear Med. 2019;10.2967.
92. Shi Y, Cummins SE, Zhu SH. Use of electronic cigarettes in smoke-free environments. Tob Control. 2017;26(e1):e19–22.
93. Owusu D, Huang J, Weaver SR, Pechacek TF, Ashley DL, Nayak P et al. Patterns and trends of dual use of e-cigarettes and cigarettes among US adults, 2015–2018. Prev Med Rep. 2019;16:101009.
94. Stanton CA, Sharma E, Edwards KC, Halenar MJ, Taylor KA, Kasza KA et al. Longitudinal trans-

itions of exclusive and polytobacco electronic nicotine delivery systems (ENDS) use among youth, young adults and adults in the USA: findings from the PATH Study waves 1–3 (2013–2016). Tob Control. 2020;29(Suppl 3):s147–54.
95. Aleyan S, Hitchman SC, Ferro MA, Leatherdale ST. Trends and predictors of exclusive e-cigarette use, exclusive smoking and dual use among youth in Canada. Addict Behav. 2020:106481.
96. Piper ME, Baker TB, Benowitz NL, Jorenby DE. Changes in use patterns over 1 year among smokers and dual users of combustible and electronic cigarettes. Nicotine Tob Res. 2020;22(5):672–80.
97. Roh EJ, Chen-Sankey JC, Wang MQ. Electronic nicotine delivery system (ENDS) use patterns and its associations with cigarette smoking and nicotine addiction among Asian Americans: Findings from the national adult tobacco survey (NATS) 2013–2014. J Ethnicity Subst Abuse. 2020:1–19.
98. Kong G, Morean ME, Cavallo DA, Camenga DR, Krishnan-Sarin S. Reasons for electronic cigarette experimentation and discontinuation among adolescents and young adults. Nicotine Tob Res. 2014;17(7):847–54.
99. Soule EK, Lopez AA, Guy MC, Cobb CO. Reasons for using flavored liquids among electronic cigarette users: A concept mapping study. Drug Alcohol Depend. 2016;166:168–76.
100. Romijnders KAGJ, van Osch L, de Vries H, Talhout R. Perceptions and reasons regarding e-cigarette use among users and non-users: a narrative literature review. Int J Environ Res Public Health. 2018;15(6):1190.
101. Soule EK, Sakuma KLK, Palafox S, Pokhrel P, Herzog TA, Thompson N et al. Content analysis of internet marketing strategies used to promote flavored electronic cigarettes. Addict Behav. 2019;91:128–35.
102. Krüsemann EJZ, Boesveldt S, de Graaf K, Talhout R. An e-liquid flavor wheel: a shared vocabulary based on systematically reviewing e-liquid flavor classifications in literature. Nicotine Tob Res. 2019;21(10):1310–9.
103. Krüsemann EJZ, Havermans A, Pennings JLA, de Graaf K, Boesveldt S, Talhout R. Comprehensive overview of common e-liquid ingredients and how they can be used to predict an e-liquid's flavour category. Tob Control. 2020: 055447.
104. E-Cigarette market – growth, trends and forecast (2019–2024). Hyderabad: MordorIntelligence; 2018.
105. Wickham RJ. How menthol alters tobacco-smoking behavior: a biological perspective. Yale J Biol Med. 2015;88(3):279–87.
106. Huang J, Feng B, Weaver SR, Pechacek TF, Slovic P, Eriksen MP. Changing perceptions of harm of e-cigarette vs cigarette use among adults in 2 US national surveys from 2012 to 2017. JAMA Network Open. 2019;2(3):e191047.
107. Amrock SM, Lee L, Weitzman M. Perceptions of e-cigarettes and noncigarette tobacco products among US youth. Pediatrics. 2016;138(5):e20154306.
108. Cullen KA, Gentzke AS, Sawdey MD, Chang JT, Anic GM, Wang TW et al. e-Cigarette use among youth in the United States, 2019. JAMA. 2019;322(21):2095–103.
109. Cullen KA, Ambrose BK, Gentzke AS, Apelberg BJ, Jamal A, King BA. Notes from the field: use of electronic cigarettes and any tobacco product among middle and high school students – United States, 2011–2018. Morb Mortal Wkly Rep. 2018;67:1276–7.
110. Soule EK, Lee JGL, Egan KL, Bode KM, Desrosiers AC, Guy MC et al. "I cannot live without my vape": Electronic cigarette user-identified indicators of vaping dependence. Drug Alcohol Depend. 2020:107886.
111. Gentzke AS, Creamer M, Cullen KA, Ambrose BK, Willis G, Jamal A et al. Vital signs: tobacco product use among middle and high school students – United States, 2011–2018. Morbid Mortal Wkly Rep. 2019;68(6):157–64.

112. Dai H, Leventhal AM. Prevalence of e-cigarette use among adults in the United States, 2014–2018. JAMA. 2019;322(18):1824–7.
113. WHO global report on trends in prevalence of tobacco smoking. Geneva: World Health Organization; 2015 (https://apps.who.int/iris/bitstream/handle/10665/156262/9789241564922_eng.pdf?sequence=1&isAllowed=y, accessed 9 November 2020).
114. Eissenberg T. Electronic nicotine delivery devices: ineffective nicotine delivery and craving suppression after acute administration. Tob Control. 2010;19(1):87–8.
115. Vansickel AR, Eissenberg T. Electronic cigarettes: effective nicotine delivery after acute administration. Nicotine Tob Res. 2013;15(1):267–70.
116. Dawkins L, Corcoran O. Acute electronic cigarette use: nicotine delivery and subjective effects in regular users. Psychopharmacology. 2014;231(2):401–7.
117. Dawkins L, Kimber C, Puwanesarasa Y, Soar K. First- versus second-generation electronic cigarettes: predictors of choice and effects on urge to smoke and withdrawal symptoms. Addiction. 2015;110(4):669–77.
118. Breland A, Soule E, Lopez A, Ramôa C, El-Hellani A, Eissenberg T. Electronic cigarettes: what are they and what do they do? Ann NY Acad Sci. 2017;1394(1):5–30.
119. Talih S, Balhas Z, Eissenberg T, Salman R, Karaoghlanian N, El-Hellani A et al. Effects of user puff topography, device voltage, and liquid nicotine concentration on electronic cigarette nicotine yield: measurements and model predictions. Nicotine Tob Res. 2015;17(2):150–7.
120. Hajek P, Phillips-Waller A, Przulj D, Pesola F, Myers Smith K, Bisal N et al. A randomized trial of e-cigarettes versus nicotine-replacement therapy. New Engl J Med. 2019;380(7):629–37.
121. Jankowski M, Krzystanek M, Zejda JE, Majek P, Lubanski J, Lawson JA et al. E-cigarettes are more addictive than traditional cigarettes – A study in highly educated young people. Int J Environ Res Public Health. 2019;16(13):2279.
122. Vogel EA, Prochaska JJ, Ramo DE, Andres J, Rubinstein ML. Adolescents' e-cigarette use: increases in frequency, dependence, and nicotine exposure over 12 months. J Adolesc Health. 2019;64(6):770–5.
123. Talih S, El-Hage R, Karaoghlanian A, Saliba NA, Salman R, Karam EE et al. Might limiting liquid nicotine concentration result in more toxic electronic cigarette aerosols? Tob Control. 2020: 10.1136/tobaccocontrol-2019-055523.
124. Shihadeh A, Eissenberg T. Electronic cigarette effectiveness and abuse liability: predicting and regulating nicotine flux. Nicotine Tob Res. 2015;17(2):158–62.
125. Pankow JF, Kim K, McWhirter KJ, Luo W, Escobedo JO, Strongin RM et al. Benzene formation in electronic cigarettes. PLoS One. 2017;12(3):e0173055.
126. Duell AK, Pankow JF, Peyton DH. Nicotine in tobacco product aerosols: "It's déjà vu all over again". Tobacco Control. 2019;29(6):656–62.
127. Process for preparing racemic nicotine. United States Patent US 8,884,021 B2; 2014 (https://patentimages.storage.googleapis.com/d1/cd/24/4141350fb210fc/US8884021.pdf).
128. El-Hellani A, El-Hage R, Salman R, Talih S, Shihadeh A, Saliba NA. Carboxylate counteranions in electronic cigarette liquids: Influence on nicotine emissions. Chem Res Toxicol. 2017;30(8):1577–81.

12. EVALI: e-cigarette or vaping product use-associated lung injury

Farrah Kheradmand, Professor of Medicine, Immunology and Pathology at Baylor College of Medicine and Staff Physician at Michael E DeBakey VA, Houston (TX), USA

Laura E. Crotty Alexander, Associate Professor of Medicine, University of California at San Diego and Section Chief of Pulmonary and Critical Care at Veterans Administration San Diego Healthcare System, San Diego (CA), USA

Contents
Abstract
12.1 Background
    12.1.1 Respiratory effects associated with e-cigarettes and vaping
    12.1.2 E-cigarette or vaping product use-associated lung injury (EVALI)
    12.1.3 Products and chemicals implicated in EVALI
12.2 EVALI
    12.2.1 Detailed description and history
    12.2.2 Symptoms
    12.2.3 Clinical presentation
    12.2.4 Reported cases
12.3 Identification of EVALI
12.4 Surveillance for EVALI
    12.4.1 National surveillance mechanisms
    12.4.2 Regional surveillance mechanisms
    12.4.3 International surveillance mechanisms and validation
12.5 Discussion
12.6 Considerations
12.7 Recommendations
    12.7.1 Key recommendations
    12.7.2 Other recommendations
12.8 References

Abstract

The beneficial or detrimental effects of electronic nicotine delivery systems (ENDS, e-cigarettes) on lung health have been heavily debated over the past decade, both in academic circles and by the press. In the summer of 2019, the debates took a new direction after reports of several clusters of e-cigarette users who presented with acute respiratory failure, resulting in hospitalization and, in some cases, death. We describe the outbreak of e-cigarette or vaping product use-associated lung injury (EVALI), the latest information on its clinical features and lung pathology and investigations of the causal chemicals identified in many commercial and/or illicit products used by affected individuals. Although news reports highlighted several clusters of EVALI in the USA, isolated case reports of vaping-associated respiratory failure have also been reported in other jurisdictions, including European countries. Many unanswered questions remain about the acute and long-term effects of exposure to the chemicals in the aerosols of e-cigarettes and other vaping products. A comprehensive approach, guided by epidemiological, translational and basic research, is necessary to assess the risks associated with inhalation of the emissions of these products, which are used in many countries around the world.

12.1 Background

12.1.1 Respiratory effects associated with e-cigarettes and vaping

Use of electronic nicotine delivery system (ENDS), commonly known as e-cigarettes or vape devices, is relatively new but is rapidly evolving among people of all ages in many countries and especially among young people in some countries, such as Canada and the USA. Vaping devices deliver nicotine to the lungs by aerosolizing liquid carriers that contain hydrophilic solvents, propylene glycol and vegetable glycerine. As the taste of the combinations of these heated chemicals is not appealing, > 99% of e-liquids contain chemical flavourings. While the long-term respiratory and/or systemic effects of these devices remain unknown, their use has been associated with acute and subacute effects on the lungs, including eosinophilic pneumonia, hypersensitivity pneumonitis, lipoid pneumonia, acute respiratory distress syndrome and diffuse alveolar haemorrhage *(1,2)*. "Vaping", an informal term used to refer to the use of these products, has also been reported to exacerbate pre-existing lung disease, particularly airway hyperreactivity and cough in asthma *(3)*. This report provides information on the outbreak of lung injury associated with vaping products in 2019.

N.B. The term "vaping" may have positive connotations because of the association with "water vapour", which may imply that the products are risk-free; however, e-cigarettes are not harmless.

12.1.2 E-cigarette or vaping product use-associated lung injury (EVALI)

In the summer of 2019, several clusters of lung injuries caused by vaping or e-cigarette use were recognized in the USA. The term EVALI was coined by the US Centers for Disease Control and Prevention (US CDC) on 11 October 2019. Epidemiological analyses in the USA have confirmed that the specific disease entity EVALI did not exist before 2019. Although lung diseases induced by e-cigarette or vaping were reported before 2019, EVALI is believed to have been caused by different chemical exposures, via disparate pathological mechanisms *(4)*. EVALI reached epidemic levels in September 2019; however, although the number of emergency department visits associated with EVALI has decreased, cases are still occurring across the country. As of February 2020, more than 2800 cases requiring hospitalization had been confirmed, with 68 deaths. Various types of lung pathology have been identified, including diffuse alveolar damage (5).

12.1.3 Products and chemicals implicated in EVALI

Survivors of EVALI reported using various e-cigarettes, vaping devices, e-liquids and flavours, and no specific brand or device was common to all cases. Over 80% of affected individuals reported that marijuana and other cannabinoids (Δ^9-tetrahydrocannabinol; THC) were present in their e-liquids, and half vaped both THC and nicotine. All THC-containing e-liquids have been identified as potentially dangerous. e-Liquids containing vitamin E acetate have also been incriminated in this illness, as this substance was detected in the majority of e-liquids used by the affected patients as well as in their bronchoalveolar lavage fluid (6). Although many brands of e-liquid have been associated with this disease, dealers often fill empty cartridges with "in-house blends" of e-liquids. Therefore, no e-devices or e-liquids can be considered safe.

As over 80% of patients with EVALI reported using THC, and most of the e-liquids they used tested positive for THC, THC-containing e-cigarette or vaping products are believed to have played a major role in the outbreak. Although 14% of the people with EVALI vaped only nicotine, experts believe that they had other forms of vaping-induced lung injury and not EVALI. This conclusion is supported by the broad, nonspecific definition of EVALI and the fact that most of the patients in this nicotine-only cohort were older women. The carriers used for these products, particularly those from in-person or online dealers, have been strongly implicated. Of the 152 different THC-containing product brands identified, Dank Vapes (cartridges containing THC liquids of unknown source sold at many sites) are the most common in north-east and southern USA; TKO and Smart Cart brands have been reported in the west, and Rove has been found in mid-west states. These findings suggest that EVALI is associated with THC-containing products and is probably not due to a single brand. Both public health agencies and the US Food and Drug Administration have identified vitamin E acetate as the chemical most strongly associated with EVALI, as vitamin E acetate was detected in 48 of 51 bronchoalveolar lavage fluid samples obtained from EVALI patients (6).

12.2 EVALI

12.2.1 Detailed description and history

The first outbreaks of EVALI in the United States were identified in Illinois and Wisconsin in July 2019 (7,8). The features of EVALI cases were identified as:

vaping within 90 days of symptom onset, bilateral lung infiltrates on either chest X-ray or chest computed topography and the absence or unlikely evidence of an infectious cause. The US CDC then categorized EVALI cases as either probable, when microbial studies were positive but were unlikely to have caused the clinical presentation, or confirmed, if clinicians could rule out a respiratory infection *(9)*. For example, if *Staphylococcus aureus* grew from a sputum culture, the case could be considered probably EVALI, with *S. aureus* as merely a colonizer and not the cause of the symptoms. If the most common tests for respiratory microbes were negative (e.g. for influenza, other respiratory viruses, including SARS-CoV2, sputum Gram stain and culture, *Streptococcus pneumonia* urine antigen, and *Legionella pneumonia* urine antigen), the case would meet the criteria for confirmed EVALI.

After categorization by the US CDC, several hundred additional cases were confirmed across the USA *(10,11)*. The peak of the epidemic occurred in September 2019, since when the numbers of reported and confirmed cases has dropped. Possible explanations for the subsequent decrease include the following.

- The intense media interest in EVALI may have led vapers to quit e-cigarettes, buy them from reliable sources or quit vaping THC-containing products in particular.
- Fewer health care professionals were reporting cases, as the diagnosis is not unique or novel.
- Makers of e-cigarette or vaping liquid may have stopped adding chemicals associated with EVALI to e-liquids.

12.2.2 Symptoms

The concurrent presence of gastrointestinal and respiratory symptoms is the most specific sign of EVALI. The most frequent findings in EVALI cases include abdominal pain, nausea, vomiting or diarrhoea with shortness of breath, cough, dyspnoea on exertion or chest pain. Less specific symptoms include fever, malaise, fatigue and weight loss *(12,13)*.

12.2.3 Clinical presentation

Some patients sought clinical care within hours of the appearance of their first symptoms, but others had symptoms for weeks to months before their initial presentation, complicating understanding of variations in disease onset. Half of hospitalized EVALI patients have hypoxia that requires admission to an intensive care unit, and approximately half require mechanical ventilation and or extracorporeal membrane oxygenation. Moderate cases in which patients require 2–6 L of oxygen are quite common. Mild cases are increasingly recognized in which patients do not require supplemental oxygen to maintain oxygen saturation > 94% but have symptoms and findings similar to those in moderate and severe cases. Pneumothorax and pneumomediastinum have been reported commonly, raising the spectre of damage to the lung parenchyma leading to bronchopulmonary fistulas *(12,13)*. Recently, several EVALI patients have been reported to have died 2–3 days after discharge from hospital, perhaps due to sudden pneumothorax.

12.2.4 Reported cases

The US CDC reported only the number of hospitalized patients from January 2020, reflecting a bias to reporting moderate-to-severe cases, and stopped reporting the numbers of EVALI cases completely on 25 February 2020. Canada, Japan, Mexico and the United Kingdom are among other countries that have reported cases of vaping-associated lung injury. Some of these reports predate the description of EVALI. Most patients required hospitalization, and the radiographic and clinical descriptions of their illness mirrored those of EVALI. As vaping was known to cause a wide variety of lung diseases before the emergence of EVALI and inhalation-induced lung diseases have similar presentations, it is likely that the cases were not related to THC or vitamin E acetate and were not EVALI. No systematic review of cases of vaping-induced lung disease, trends, prevalence or clusters of case reports has been reported from other countries.

12.3 Identification of EVALI

A universal definition of EVALI could be established if it has one causal agent, as for the global consensus among international pulmonary experts on the definition of acute respiratory distress syndrome. Any definition will have to be more specific and detailed than the current one *(9)*, which includes everyone who has vaped within 90 days, has bilateral lung infiltration and no clear infectious cause. Ideally, the definition of EVALI will exclude other vaping-related lung diseases as well as non-vaping-related diseases. The addition of specific testing to exclude lung diseases that are idiopathic or vaping-related but not EVALI would be helpful. The pathological picture of EVALI is broad and overlaps with those of acute interstitial pneumonia, hypersensitivity pneumonitis, diffuse alveolar haemorrhage, lipoid pneumonia and adult respiratory disease syndrome.

12.4 Surveillance for EVALI

12.4.1 National surveillance mechanisms

It has been difficult to collect all the cases in the USA, as the US CDC and the Food and Drug Administration rely on local and regional public health departments, which are robust and reliable but decide to share their data according to local priorities and state health privacy laws. Another difficulty has been in educating health care providers, as most have not asked specifically about the use of e-cigarettes, vaping devices or THC-specific devices. It has been proposed that an open portal for reporting be established for both patients and health care providers in order to identify more cases *(14)*. Privacy issues are the main deterrent; however, a similar method has been used for other diseases, such as lead poisoning and infectious diseases.

12.4.2   Regional surveillance mechanisms

A regional surveillance mechanism was described recently *(11)*. Regional surveillance could help each health care system to accurately detect and track all EVALI cases, and the data could then be shared with regional public health departments and with the US CDC and the Food and Drug Administration in the USA. The United Kingdom has a system known as the "yellow card scheme", which is available on the website of the Medicines and Healthcare Products Regulatory Agency *(15)*. The scheme allows reporting of information on adverse or suspected adverse events related to medicines and makes provision for reporting of any side-effects of the use of e-cigarettes or safety concerns related to these products or their e-liquids. Members of the public and health care personnel in each country of the United Kingdom can file reports through the system.

12.4.3   International surveillance mechanisms and validation

One challenge in surveillance is identifying cases in both medical centres and rural areas in similar investigations and with similar confirmation methods to control and respond in each case. WHO sponsors a global network of over 250 institutions dedicated to responding and raising awareness about acute public health events, the Global Outbreak Alert and Response Network *(16)*.

Another difficulty is in defining EVALI, as the current definition is broad, making identification difficult. To improve surveillance, standards (e.g. case definitions) and training in recognition of EVALI are required at national and international levels. The US CDC has reported that nearly 3% of EVALI patients have required rehospitalization, and nearly one in seven deaths from EVALI has occurred after hospital discharge, particularly among people with one or more chronic diseases *(17)*.

12.5   Discussion

It is estimated that 35–40 million adults and children globally vape, indicating a large number of people who are vulnerable to EVALI and other vaping-associated health outcomes. e-Cigarette use or vaping itself carries health risks and may harm health beyond the lungs. Further, the risks are increased by lacing of products with drugs and other substances, and the products should be properly regulated.

The outbreak of EVALI in the USA highlights the importance of broadening the definition of e-cigarette toxicity beyond that of smoking, as vaping results in disease risks that are different from those associated with smoking *(18)*. A major concern during the coronavirus disease 2019 (COVID-19) pandemic is the delay in recognizing and reporting EVALI in patients admitted with respiratory failure. For example, in April 2020, eight patients admitted to hospital for respiratory failure met the US CDC case definition of EVALI, but physicians first considered EVALI in differential diagnoses only 1–8 days (median day 3) after hospitalization *(19)*. This report highlights some of the difficulties in recognizing respiratory failure due to EVALI, which may therefore be underdiagnosed during COVID-19 pandemic.

A priority at national and international level is to establish registries of EVALI patients that can be accessed by researchers and clinicians in order to assess the long-term effects and clinical outcomes. Further, acquisition and analysis of specimens from humans (e.g. whole blood, tracheal aspirates, bronchoalveolar lavage fluid, lung biopsy specimens, urine and autopsy specimens) could provide insight into the pathophysiology of EVALI. More mechanistic studies should be conducted to understand the toxic effect of vaping products in the lungs. Specifically, it is unclear whether and how vaping of propylene glycol or vegetable glycerine before exposure to vitamin E acetate causes lung injury. Further, vaping temperature, especially in high-powered devices, plays a role in lung injury *(20)*. Animal models of EVALI would be useful for studying the potential causes of toxicity to the lungs related to vaping and use of THC products *(14)*.

12.6 Considerations

A full report on the many aspects of the EVALI disease spectrum and its potential contribution to lung injury in other countries is recommended. Several key questions should be addressed to improve current understanding of care for patients who develop EVALI and respiratory failure.

- A significant challenge is ensuring that physicians recognize an EVALI case, particularly during the COVID-19 pandemic. Physicians should be made aware of the risk of EVALI and ask about nicotine or THC use in electronic products as part of a routine history for patients who present with respiratory failure.
- EVALI must be better defined to guide primary care and paediatric physicians in the correct diagnosis.
- Studies should address whether e-cigarette use also increases vulnerability to SARS-CoV-2 infection in adults and children. Preclinical models of e-cigarette product use have provided strong evidence that chronic use alters lung defence immunity against influenza, a common viral pathogen *(18)*.

12.7 Recommendations

12.7.1 Key recommendations

- National and international registries of EVALI patients and people with other vaping-associated lung diseases should be established to improve monitoring of long-term clinical outcome in survivors.
- A strong international campaign should be organized to alert parents, children and young adults to the hazards of inhaling the chemicals contained in the aerosols of electronic vaping products.
- Consideration should be given to writing a full paper on EVALI when more information becomes available.

12.7.2 Other recommendations

- Specimens such as blood, tracheal aspirates, bronchoalveolar lavage fluid, lung tissue and urine should be obtained and analysed in order to better understand the pathophysiology of EVALI.
- Animal models of EVALI should be developed to gain insight into the cellular and molecular mechanisms underlying the toxicity caused in the lungs and systemically by vaping both nicotine and THC and also the health effects of the vehicle and flavour chemicals contained in these products.
- Research should be pursued to define the mechanisms by which vaping products harm the lungs.

12.8 References

1. Arter ZL, Wiggins A, Hudspath C, Kisling A, Hostler DC, Hostler JM. Acute eosinophilic pneumonia following electronic cigarette use. Respir Med Case Rep. 2019;27:100825.
2. Sommerfeld CG, Weiner DJ, Nowalk A, Larkin A. Hypersensitivity pneumonitis and acute respiratory distress syndrome from e-cigarette use. Pediatrics. 2018;141(6):20163927.
3. Lappas AS, Tzortzi AS, Konstantinidi EM, Teloniatis SI, Tzavara CK, Gennimata SA et al. Short-term respiratory effects of e-cigarettes in healthy individuals and smokers with asthma. Respirology. 2018;23:291–7.
4. McCauley L, Markin C, Hosmer D. An unexpected consequence of electronic cigarette use. Chest. 2012;141:1110–3.
5. Butt YM, Smith ML, Tazelaar HD, Vaszar LT, Swanson KL, Cecchini MJ et al. Pathology of vaping-associated lung injury. N Engl J Med. 2019;381(18):1780–1.
6. Blount BC, Karwowski MP, Shields PG, Morel-Espinosa M, Valentin-Blasini L, Gardner M et al. Vitamin E acetate in bronchoalveolar-lavage fluid associated with EVALI. N Engl J Med. 2020;382:697–705.
7. Layden JE, Ghinai I, Pray I, Kimball A, Layer M, Tenforde M et al. Pulmonary illness related to

8. e-cigarette use in Illinois and Wisconsin – preliminary report. N Engl J Med. 2020;382:903–16.
9. Ghinai I, Pray IW, Navon L, O'Laughlin K, Saathoff-Huber L, Hoots B et al. e-Cigarette product use, or vaping, among persons with associated lung injury – Illinois and Wisconsin, April–September 2019. Morb Mortal Wkly Rep. 2019;68:865–9.
10. 2019 lung injury surveillance. Primary case definitions. September 18, 2019. Atlanta (GA): Centers for Disease Control and prevention; 2019 (https://www.cdc.gov/tobacco/basic_information/e-cigarettes/assets/2019-Lung-Injury-Surveillance-Case-Definition-508.pdf, accessed 10 January 2021).
11. Blagev DP, Harris D, Dunn AC, Guidry DW, Grissom CK, Lanspa MJ. Clinical presentation, treatment, and short-term outcomes of lung injury associated with e-cigarettes or vaping: a prospective observational cohort study. Lancet. 2019;394(10214):2073–83.
12. Alexander LEC, Perez MF. Identifying, tracking, and treating lung injury associated with e-cigarettes or vaping. Lancet. 2019;394(10214):2041–3.
13. Zou RH, Tiberio PJ, Triantafyllou GA, Lamberty PE, Lynch MJ, Kreit JW et al. Clinical characterization of e-cigarette, or vaping, product use associated lung injury in 36 patients in Pittsburgh, PA. Am J Respir Crit Care Med. 2020;201(10):1303–6.
14. Chatham-Stephens K, Roguski K, Jang Y, Cho P, Jatlaoui TC, Kabbani S et al. Characteristics of hospitalized and nonhospitalized patients in a nationwide outbreak of e-cigarette, or vaping, product use-associated lung injury – United States, November 2019. Morb Mortal Wkly Rep. 2019;68:1076–80.
15. Crotty Alexander LE, Ware LB, Calfee CS, Callahan SJ, Eissenberg T, Farver C et al. NIH workshop report: E-cigarette or vaping product use associated lung injury: Developing a research agenda. Am J Respir Crit Care Med. 2020;202(6):795–802.
16. Yellow card. London: Medicines and Healthcare Products Regulatory Agency; 2020 (https://yellowcard.mhra.gov.uk/the-yellow-card-scheme/#:~:text=The%20Yellow%20Card%20scheme%20is,by%20health%20professionals%20and%20patients, accessed 10 January 2021).
17. Global Outbreak Alert and Response Network. Geneva: World Health Organization; 2020 (https://extranet.who.int/goarn/, accessed 10 January 2021).
18. Evans ME, Twentyman E, Click ES, Goodman AB, Weissman DN, Kiernan E et al. Update: Interim guidance for health care professionals evaluating and caring for patients with suspected E-cigarette, or vaping, product use-associated lung injury and for reducing the risk for rehospitalization and death following hospital discharge – United States, December 2019. Morb Mortal Wkly Rep.2020;68(5152):1189–94.
19. Madison MC, Landers CT, Gu BH, Chang CY, Tung HY, You R et al. Electronic cigarettes disrupt lung lipid homeostasis and innate immunity independent of nicotine. J Clin Invest. 2019;129:4290–304.
20. Armatas C, Heinzerling A, Wilken JA. Notes from the field: e-cigarette, or vaping, product use-associated lung injury cases during the COVID-19 response – California, 2020. Morb Mortal Wkly Rep. 2020;69:801–2.
21. Chaumont M, Bernard A, Pochet S, Mélot C, El Khattabi C, Reye F et al. High-wattage e-cigarettes induce tissue hypoxia and lower airway injury: A randomized clinical trial. Am J Crit Care Med. 2018;198(1):123–6.

13. Overall recommendations

The WHO Study Group on Tobacco Product Regulation publishes reports to provide a scientific basis for tobacco product regulation. In line with Articles 9 and 10 of the WHO Framework Convention on Tobacco Control (FCTC), the reports identify evidence-based approaches to the regulation of the contents, emissions and design features of tobacco products.

The 10th meeting of the Study Group, the deliberations, outcomes and recommendations of which are included in this report, specifically addressed novel and emerging nicotine and tobacco products, including electronic nicotine delivery systems (ENDS) and electronic non-nicotine delivery systems (ENNDS) and heated tobacco products (HTPs). Despite this focus, which is partly informed by the decision of the Eighth Conference of the Parties (COP8) to the WHO FCTC on novel and emerging tobacco products (decision FCTC/COP8(22) *(1)*, all tobacco products fall under the remit of the Study Group. This allows a comprehensive approach to synthesizing and making available evidence to countries on both conventional and newer products to address challenges in tobacco control, which remains a global priority.

Regulators are reminded that tobacco kills more than 8 million people a year *(2,3)*, with more than 7 million of those deaths attributed to direct tobacco use and about 1.3 million to exposure of non-smokers to second-hand smoke *(4,5)*. Tobacco also eventually kills up to half of its users and therefore remains a global health emergency *(2)*. The introduction of novel and emerging nicotine and tobacco products, which are aggressively marketed and promoted by manufacturers, including to children and adolescents, in some jurisdictions further complicates tobacco control and has been a distraction for regulators. Thus, the recommendations of this report are intended to be taken in the context of wider tobacco control and to complement the recommendations of the Study Group in other reports on tobacco product regulation *(6–12)*, which addressed cigarettes, smokeless tobacco and waterpipe tobacco.

The aggressive marketing and promotion of novel and emerging nicotine and tobacco products poses a serious threat to tobacco control. The Study Group, having considered the requests by countries for technical support on regulating these products, concluded that a focus must be maintained on wider tobacco control and that regulators should not allow themselves to be distracted by tobacco and related industry tactics and aggressive promotion of these products. The report highlights the importance of the following:

- good science and verification of industry research;
- full disclosure of product information to regulators;
- clarification of the source of research funding to identify undue influence;

- independent research;
- application of tobacco control laws to all tobacco products, without exception;
- monitoring the activities of tobacco and related industries; and
- protecting policies from the influence of nicotine and tobacco industries, especially in the context of Article 5.3 of the WHO FCTC and its guidelines *(13,14)*.

Sections 2–12 of the report provide scientific information, policy recommendations and guidance to bridge regulatory gaps in tobacco control. The report also identifies areas for further work and research, with a focus on the regulatory needs of countries while accounting for regional differences, thus providing a strategy for continued, targeted technical support to all countries, especially WHO Member States. The main recommendations of the Study Group are outlined below.

13.1 Main recommendations

The main recommendations to policy-makers and all other interested parties are the following:

- to ensure continued focus on evidence-based measures to reduce tobacco use as outlined in the WHO FCTC and seek to avoid being distracted from tobacco industry actions to promote novel and emerging tobacco products, such as heated tobacco products;
- to use existing regulations for tobacco products to regulate heated tobacco products (including the device) and consider broadening the scope of existing regulations in which regulatory loopholes may be exploited by the tobacco industry, including in countries in which heated tobacco products are currently not legally available;
- to apply the most restrictive tobacco control regulations to heated tobacco products (including the device), as appropriate within national laws, taking into account a high level of protection for human health;
- to prohibit all manufacturers and associated groups from making claims about reduced harm of heated tobacco products, as compared with other products, or portraying heated tobacco products as an appropriate approach for cessation of use of any tobacco product and to ban their use in public spaces unless robust independent evidence emerges to support a change in policy;
- to ensure that the public is well informed about the risks associated with use of heated tobacco products, including the risks of dual use

with conventional cigarettes and other smoked tobacco products, and also their use during pregnancy; to correct false perceptions, counter misinformation and stress that reduced exposure does not necessarily mean reduced harm;

- to rely on independent data and to support continuing independent research on the public health impact of use of heated tobacco products, with critical analysis and interpretation of tobacco industry-funded data, including data on the emissions and toxicity of heated tobacco products and associated exposures and effects in users and non-users;

- to require tobacco manufacturers to disclose all product information – including product design, chemical profile, total nicotine content, nicotine forms, toxicity, other findings of product testing and testing methods – to appropriate regulatory agencies at least once a year; any modifications to products should require updating of the report;

- to ban all commercial marketing of electronic nicotine delivery systems, electronic non-nicotine delivery systems and heated tobacco products, including in social media and through organizations funded by and associated with the tobacco industry;

- to prohibit the sale of electronic nicotine delivery systems and electronic non-nicotine delivery systems in which the user can control device features and liquid ingredients (that is, open systems);

- to prohibit the sale of electronic nicotine delivery systems with a higher abuse liability than conventional cigarettes, for example by restricting the emission rate or flux of nicotine; and

- to prohibit the addition of pharmacologically active substances such as cannabis and tetrahydrocannabinol (in jurisdictions where they are legal), other than nicotine in electronic nicotine delivery systems, to electronic nicotine delivery systems and electronic non-nicotine delivery systems.

Countries are urged to implement the above recommendations, as there is enough information about nicotine and tobacco products for countries to act to protect the health of their populations, especially the younger generation. While the report acknowledges that still more is to be learnt about these products and emphasizes that continued independent research is necessary to build further intelligence on the products, including their marketing, features, prevalence of use and availability, and on the promotional strategies of tobacco and related industries, there are more than a billion tobacco users, and millions of people use the newer products. Therefore, the public health community should answer the call for continued acceleration of evidence-based policies and recommendations, such as those in the WHO FCTC, WHO MPOWER measures and the relevant COP reports. Countries should thus implement proven policy measures and, in addition, consider implementing the recommendations in this report. Specific recommendations on each of the topics considered can be found in sections 2.8, 3.5.4, 4.9, 5.11, 6.7, 7.11, 8.5, 8.7, 9.8, 10.2.6, 10.3,7, 11.6 and 12.7.

13.2 Significance for public health policies

The Study Group's report provides helpful guidance to the science, research and evidence on all tobacco products, including cigarettes, smokeless tobacco and waterpipe tobacco. Recently, the Study Group has extended its work by providing much needed information to regulators on the contents, emissions, variation in and features of novel and emerging nicotine and tobacco products, in particular, ENDS, ENNDS and HTPs. The report highlights the public health impact of these products and their features on users and non-users, including: their addictive potential; perception and use of the products; their attractiveness; their potential role in initiating and stopping tobacco use; marketing, including promotional strategies and impacts; claims of reduced harm; variation in products; quantification of risk to the health of individuals and populations; regulatory mapping and the experience of selected countries; impact on tobacco control; and research gaps. The Study Group's recommendations, outlined above, directly address some of the unique regulatory challenges faced by certain Member States because of the penetration of these products into their markets. Further, the report will help Member States to update their knowledge on novel and emerging nicotine and tobacco products and aid in the formulation of effective regulatory strategies for nicotine and tobacco products.

The Study Group, because of its unique composition of regulatory, technical and scientific experts, navigates and distils complex data and research and synthesizes them into policy recommendations, which inform policy development at national, regional and global levels. This authoritative report by a multidisciplinary team of experts goes to the heart of the challenges faced by governments on novel and emerging products. The nature of the report means that regulators, governments and interested parties can rely on the science and evidence presented to counter the arguments of tobacco and related industries, as appropriate. The identification of gaps in policy and research on nicotine and tobacco products indicates areas in which there is insufficient information. Countries, in formulating their research agendas, could focus on areas pertinent to their policy goals, objectives and national context. This is a critical role of the Study Group, especially for governments with inadequate resources or capacity to navigate technical information on tobacco product regulation.

The recommendations made in the report promote international coordination of regulatory efforts and the adoption of best practices in product regulation, strengthen capacity in product regulation in all WHO regions, provide a ready resource for Member States that is based on sound science and support implementation of the WHO FCTC by its States Parties. Tobacco product regulation complements other provisions of the WHO FCTC on demand reduction. The recommendations of the Study Group, if effectively implemented, would contribute to reducing tobacco use, thus reducing tobacco use prevalence and promoting good health.

13.3 Implications for the Organization's programmes

The report fulfils the mandate of the WHO Study Group on Tobacco Product Regulation to provide the Director-General with scientifically sound, evidence-based recommendations for Member States about tobacco product regulation,[1] which is a highly technical area of tobacco control in which Member States face complex regulatory challenges. The outcomes of the Study Group's deliberations and main recommendations will improve Member States' understanding of ENDS, ENNDS and HTPs and the implications of the proliferation of these products on markets in many countries in the broader context of tobacco control.

The contribution of the report to the body of knowledge on product regulation will inform the work of the tobacco programme in WHO's Department of Health Promotion, especially in providing technical support to Member States. It will also contribute to updating Member States and regulators through meetings of the WHO Global Tobacco Regulators' Forum and information-sharing via the Forum's EZCollab network. States Parties to the WHO FCTC will be updated through a comprehensive report on research and evidence on novel and emerging tobacco products, which was requested by the Conference of the Parties at its eighth session.[2] The comprehensive report will include the key messages and recommendations of the eighth report of the Study Group. All of these will contribute to meeting target 3.a of the Sustainable Development Goals (that is, strengthening implementation of the WHO Framework Convention on Tobacco Control) and the triple billion targets of WHO's Thirteenth Global Programme of Work.

The report, which is a WHO global public health good (i.e. an initiative developed or undertaken by WHO that is of benefit either globally or to many countries in many regions *(15)*), is available to all countries to help drive impact at country level and globally, towards reducing tobacco use and improving overall public health.

---

1 In November 2003, the Director-General formalized the status of the former Scientific Advisory Committee on Tobacco Product Regulation from a scientific advisory committee to a study group.
2 See decision FCTC/COP8(22), paragraph 2(a).

13.4 References

1. Novel and emerging tobacco products. Decision FCTC/COP8(22) of the Conference of the Parties to the WHO Framework Convention on Tobacco Control at its Eighth session. Geneva: World Health Organization; 2018 (https://www.who.int/fctc/cop/sessions/cop8/FCTC__COP8(22).pdf?ua=1, accessed 19 December 2020).
2. WHO report on the global tobacco epidemic, 2019. Geneva: World Health Organization; 2019 (http://www.who.int/tobacco/global_report/en, accessed 19 December 2020).
3. GBD 2019 Risk Factors Collaborators. Global burden of 87 risk factors in 204 countries and territories, 1990–2019: a systematic analysis for the Global Burden of Disease Study 2019. Lancet 2020;396:1223–49.
4. Tobacco. Fact sheet. Geneva: World Health Organization; 2020 (https://www.who.int/news-room/fact-sheets/detail/tobacco, accessed 23 December 2020).
5. Findings from the Global Burden of Disease Study 2017: GBD Compare. Seattle (WA): Institute for Health Metrics and Evaluation; 2018 (http://vizhub.healthdata.org/gbd-compare, accessed 19 December 2020).
6. The scientific basis of tobacco product regulation. Report of a WHO study group (WHO Technical Report Series, No. 945). Geneva: World Health Organization; 2007 (https://www.who.int/tobacco/global_interaction/tobreg/who_tsr.pdf, accessed 10 January 2021).
7. The scientific basis of tobacco product regulation. Second report of a WHO study group (WHO Technical Report Series, No. 951). Geneva: World Health Organization; 2008 (https://apps.who.int/iris/bitstream/handle/10665/43997/TRS951_eng.pdf?sequence=1, accessed 10 January 2021).
8. WHO Study Group on Tobacco Product Regulation. Report on the scientific basis of tobacco product regulation: Third report of a WHO study group (WHO Technical Report Series, No. 955). Geneva: World Health Organization; 2009 (https://apps.who.int/iris/bitstream/handle/10665/44213/9789241209557_eng.pdf?sequence=1, accessed 10 January 2021).
9. WHO Study Group on Tobacco Product Regulation. Report on the scientific basis of tobacco product regulation: Fourth report of a WHO study group (WHO Technical Report Series, No. 967). Geneva: World Health Organization; 2012 (https://apps.who.int/iris/bitstream/handle/10665/44800/9789241209670_eng.pdf?sequence=1, accessed 10 January 2021).
10. WHO Study Group on Tobacco Product Regulation. Report on the scientific basis of tobacco product regulation: Fifth report of a WHO study group (WHO Technical Report Series, No. 989). Geneva: World Health Organization; 2015 (https://apps.who.int/iris/bitstream/handle/10665/161512/9789241209892.pdf?sequence=1, accessed 10 January 2021).
11. WHO Study Group on Tobacco Product Regulation. Report on the scientific basis of tobacco product regulation: Sixth report of a WHO study group (WHO Technical Report Series, No. 1001). Geneva: World Health Organization; 2017 (https://apps.who.int/iris/bitstream/handle/10665/260245/9789241210010-eng.pdf?sequence=1, accessed 10 January 2021).
12. WHO Study Group on Tobacco Product Regulation. Report on the scientific basis of tobacco product regulation: Seventh report of a WHO study group (WHO Technical Report Series, No. 1015). Geneva: World Health Organization; 2019 (https://apps.who.int/iris/bitstream/handle/10665/329445/9789241210249-eng.pdf, accessed 10 January 2021).
13. Article 5.3. WHO Framework Convention on Tobacco Control. Geneva: World Health Organization; 2005 (https://apps.who.int/iris/bitstream/handle/10665/42811/9241591013.pdf;jsessionid=1DE94B49AE5482D746B237CAC38D5658?sequence=1, accessed 19 December 2020).
14. Guidelines for implementation of Article 5.3 of the WHO Framework Convention on Tobacco Control on the protection of public health policies with respect to tobacco control from commercial and other vested interests of the tobacco industry. Geneva: World Health Organization;

2008 (https://www.who.int/fctc/guidelines/article_5_3.pdf, accessed 19 December 2020).

15. Thirteenth General Programme of Work 2019–2023. Geneva: World Health Organization; 2019 (https://apps.who.int/iris/bitstream/handle/10665/324775/WHO-PRP-18.1-eng.pdf, accessed 29 December 2020).